Probabilistic Structural Mechanics: Advances in Structural Reliability Methods

International Union of Theoretical and Applied Mechanics

P. D. Spanos, Y.-T. Wu (Eds.)

Probabilistic Structural Mechanics: Advances in Structural Reliability Methods

IUTAM Symposium, San Antonio, Texas, USA
June 7 – 10, 1993

Springer-Verlag
Berlin Heidelberg New York
London Paris Tokyo
Hong Kong Barcelona Budapest

Prof. Pol D. Spanos

Rice University
L. B. Ryon Chair in Engineering
P. O. Box 1892
Houston, TX 77251
USA

Dr. Y.-T. Wu

Southwest Research Institute
6220 Culebra Road
San Antonio, TX 78228
USA

ISBN-13 : 978-3-642-85094-3 **e-ISBN-13 : 978-3-642-85092-9**
DOI : 10.1007 / 978-3-642-85092-9

Typesetting: Camera ready by authors
SPIN: 10133839 61/3020 – 5 4 3 2 1 0 – Printed on acid-free paper.

Preface

This symposium is the seventh of a series of IUTAM sponsored symposia which focus on probabilistic methods in mechanics. It is the sequel to the series of meetings in Coventry, UK (1972), Southhampton, UK (1976), Frankfurt/Oder, Germany (1982), Stockholm, Sweden (1984), Innsbruck/Igls, Austria (1987), and Turin, Italy (1991). The symposium focused on advances in the area of probabilistic mechanics with direct application to structural reliability issues. The contributed papers address collectively the four components of a structural reliability problem. They are: characterization of stochastic loads, description of material properties in terms of fatigue and fracture, response determination, and quantitative assessment of the reliability of the structural system. Four Keynote Lectures by V. Bolotin (Russia), O. Ditlevsen (Denmark), R. Heller (USA), and F. Ziegler (Austria) were delivered; the remaining contributed papers were organized in ten technical sessions.

A reception was hosted by Dr. Y. Wu the first day of the symposium; the second day of the symposium a banquet was hosted by Dr. P. Spanos, with Dr. N. Abramson serving as the banquet speaker. Closing remarks were provided by the IUTAM Secretary General, Dr. F. Ziegler.

On behalf of the Scientific and Organizing Committees of the symposium, we would like to thank the IUTAM bureau for promoting and sponsoring this symposium. Thanks are also expressed to the sponsors of this symposium who have assisted in defraying its cost. Of course, the contribution of the authors and participants in the symposium is most appreciated. The assistance in processing the manuscipts provided by Mrs. L. Richardson and Messrs. R. Eberle, S. Miller, and J. D. Rowatt of Rice University is acknowledged.

From a purely editorial perspective, a concerted effort has been made through an iterative process between the editors and the authors to normalize and homogenize the papers which are included in this volume. However, beyond a certain stage of iteration, it was decided to respect the preferences and even idiosyncratic style of some of the authors.

P. D. Spanos, Editor
Y.-T. Wu, Co-Editor

Closing Remarks by F. Ziegler, IUTAM-Secretry General

IUTAM is proud of developments in Stochastic Mechanics, discussed in documents such as the Proceedings of the IUTAM *Symposium on Nonlinear Stochastic Mechanics*, (Torino, Italy, 1-5 July 1991), edited by N. Bellomo and F. Casciati, and published by Springer-Verlag, Berlin, 1992. Also here in San Antonio, the Scientific Committee has put together a fine and selective program with a proper mixture of contributions by younger and/or distinguished speakers. Participants from 12 countries and from all five continents are registered. Emphasis is on Reliability, but the Mechanics background is more or less visible. Personally, I think that the Mechanics community must break through the "classical" boundaries. However, theoretical and applied Mechanics must remain the source of our considerations. Lively discussions and devoted speakers made this symposium a success also in the sense of a IUTAM-Symposium. The Session-Chairmen had an easy task due to the 30 minutes allocated to each speaker and due to their discipline; the very good personal relations within that truly international group guaranteed a stimulating atmosphere. The thanks of IUTAM go to the chairman, Dr. Y.-T. Wu, to the Scientific Committee, and last but not least, to the Speakers and active Participants who did their best to bring up the latest achievements in the field. Pol Spanos will seek a publisher of the Proceedings of our Symposium which has come to an end.

In closing, I would like to call your attention to the next IUTAM-Symposium on "Advances in Nonlinear Stochastic Mechanics", to be held 3-7 July 1995 in Trondheim, Norway under the co-chairmanship of S. Krenk and A. Naess. If we would like to keep the subject within IUTAM and to have a Symposium in the year 1997, a proposal should be sent to the IUTAM-Secretary General by the end of 1993. The decision would be made by the General Assembly in Amsterdam, on 28 August 1994 after considering the suggestions made by the Symposia Panel. Another earlier option would be to organize a Satellite Session, a new "invention", at the 19th ICTAM, Kyoto, 25-30 August 1996. The Congress President is Professor T. Tatsumi, President of the Kyoto Institute of Technology.

F. Ziegler

Contents

Scientific Committee

A. H-S. Ang, USA

T. A. Cruse, USA

A. Der Kiureghian, USA

O. Ditlevsen, Denmark

R. A. Heller, USA

H. Ishikawa, Japan

R. E. Melchers, Australia

R. Rackwitz, Germany

K. Sobczyk, Poland

P. D. Spanos, USA

Y.-T. Wu, USA

F. Ziegler, Austria

Local Organizing Committee

Y.-T. Wu

U. S. Lindholm

O. H. Burnside

P. D. Spanos

Sponsors

International Union of Theoretical and Applied Mechanics

Southwest Research Institute, USA

National Science Foundation, USA

Department of Energy, USA

Deriving Probability Models for Stress Analysis

Suneung Ahn, Stephen Chick†and Max B. Mendel‡

Department of Industrial Engineering and Operations Research, University of California, Berkeley, CA 94720, USA

†Supported by Department of Defense NDSEG fellowship.
‡Research partially supported by the Army Research Office (DAAL03-91-G-0046) grant to the University of California at Berkeley.

Abstract. This paper presents an approach to derive probability models for use in structural reliability studies. Two main points are made. First, that it is possible to translate engineering and physics knowledge into a requirement on the form of a probability model. And second, that making assumptions about a probability model for structural failure implies either explicit or hidden assumptions about material and structural properties. The work is foundational in nature, but is developed with explicit examples taken from planar and general stress problems, the von Mises failure criterion, and a modified Weibull distribution.

Keywords. Probability models, static stress, material reliability

1 Introduction

An important design parameter for structural analysis is the probability that a structure or component will fail. This number is generally calculated from material constants, the probable stresses to be applied to the component, a failure criterion (such as the von Mises or Tresca criteria), experimentation, and the selection of a probability model (such as the normal, exponential or Weibull distributions). Although there has been much discussion on the relative merits of the various failure criteria, and much experimental work in describing material constants such as Young's modulus and the Poisson ratio, relatively little work has been done to direct the selection of an appropriate probability model, or to connect model parameters to physically measurable values.

Probability models are often selected by choosing one of the standard analytical models, such as the normal, exponential or Weibull models, and fitting the parameters to experimental data. This paper discusses an alternate approach to selecting a probability model. It is a method of *deriving* models that are consistent with physical laws and established engineering judgement. Similarly, the approach illustrates a technique for determining the physical assumptions that are implied by assuming a particular probability model. The work is foundational in nature, but applications to failure analysis for isotropic materials are given. Because of this, the results are relevant to a wide variety of structural engineering problems.

2 Problem definition

Frequently, a critical point (CP) of failure is identified in a component, and the probability of failure p(fail) for the component is taken to be the probability of failure at the CP. Assume for the time being, that the CP has been identified, and that information regarding the probable states of stress S is available. The probability that the material will fail can then be found by applying the law of total probability

$$\mathrm{p(fail)} = \int \mathrm{p(fail} \mid S)\, f(S)\, dS,$$

where $\mathrm{p}(S) \overset{\text{def}}{=} f(S)dS$ is the probability of being in a given stress state S, with probability density $f(S)$. The problem is to understand how the components of this generic probability model relate to the underlying question of material failure. The present work restricts its attention to yielding under static loading. We intend to extend the techniques, however, to dynamic loading conditions in future work.

First consider the probability model $\mathrm{p}(S)$ for the stresses S which are applied to the CP. One factor that is important in determining a successful model is historical or design data. Two other factors which are equally important are one, the physical design and symmetries of the material near the CP, and two, the choice of coordinates for expressing the state of stress. For instance, many failure criteria use information about S in terms of the principal stresses $(\sigma_1, \sigma_2, \sigma_3)$.* On the other hand, sensors may report information regarding the observed strains or stresses $(\sigma_x, \tau_{xy}, \ldots)$. In addition, there may be information regarding the orientation of the principal stresses, which may be expressed relative to the Euler angles (ϕ, θ, ψ) of the principal stress directions. A successful probability model for the state of stress $\mathrm{p}(S)$ will take into consideration each of these data requirements. There are difficulties in constructing such a model, however.

A general state of stress for an isotropic material is a 6 dimensional quantity. Therefore, the probability model should have six degrees of freedom. But more than 6 coordinates have been identified as relevant to the problem of stress, including the principal stresses, measurements, and principal

*The maximum stress criteria and von Mises criteria are most simply expressed in terms of 3 coordinates-the principal stresses, even though more complicated forms which are expressed in terms of 6 coordinates-$(\sigma_x, \tau_{xy}, \ldots)$.

stress directions. Section 3 describes how to encapsulate the information available about each of these parameters into a single coherent probability model by discussing stress states of linear elastic isotropic materials in 2 and 3 dimensions.

Consider next the probability of material failure under a particular state of stress, p(fail | S). This probability is often taken to be 0 or 1, depending on whether the selected failure criterion predicts failure. For instance, with the von Mises criterion, failure is predicted with probability 1 if the distortional strain energy exceeds a given parameter Θ. If Θ is not known, however, p(fail | S) may be any value in $[0, 1]$. Lindquist[1] derives a Weibull-type model for Θ, and in Section 4 we discuss and extend the result by indicating what assumptions on the material are implied by assuming different parameter values in the Weibull model.

3 Probability Models for an Observed Stress

In this section, we examine by way of two examples, the translation of engineering statements about material stresses into probability models. In the first example, planar stresses are analyzed, and the second example treats stresses in 3-dimensional materials. Both examples make a similar assumption, namely, it is assumed that the principal stresses are equally likely to come from any direction. In the plane, this means that the principal stress coordinate frame is rotated uniformly from 0 to π radians. A similar statement holds for 3-dimensional materials.

In both cases, the goal is to find a probability model which is parametrized in terms of some observations (such as σ_x, τ_{xz}) and the principal stresses. The model must satisfy the additional assumption regarding the randomness of the principal stress directions. Such a model can then be used to update the probability of failure, given that certain measurements have been made. For instance, it may be of value to calculate

$$\text{p(fail} \mid \sigma_x) = \int \text{p(fail} \mid S, \sigma_x)\, f(S \mid \sigma_x)\, dS.$$

Since σ_x is information which is contained in the stress tensor S, the formula can be slightly simplified as shown in the equation

$$\text{p(fail} \mid \sigma_x) = \int \text{p(fail} \mid S)\, f(S \mid \sigma_x)\, dS.$$

3.1 Planar stress states

A planar stress tensor has three degrees of freedom, $(\sigma_x, \sigma_y, \tau_{xy})$. Equivalently, these degrees of freedom may be specified by the principal stresses and a rotation angle specifying their direction $(\sigma_1, \sigma_2, \phi)$. This notation is summarized on a sketch of the familiar Mohr circle (Fig. 3.1).

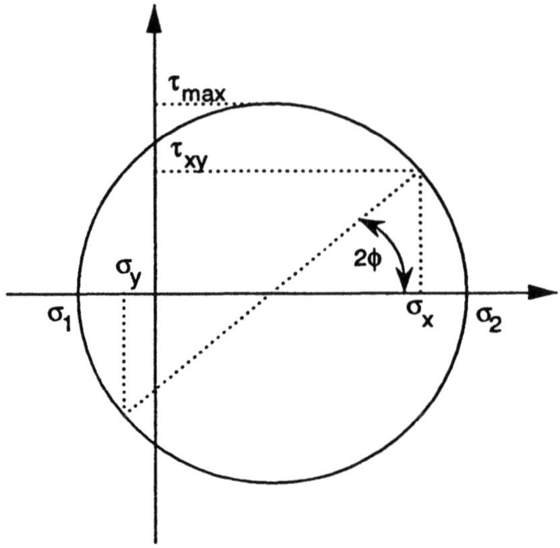

Fig. 3.1 A Mohr circle for an arbitrary planar stress.

The statement that ϕ is randomly selected is the same as saying that ϕ is uniformly distributed on $(0, \pi)$. It also implies that the rotation angle ϕ is independent of the principal stresses. If the joint probability of the principal stresses is given by $f(\sigma_1, \sigma_2)d\sigma_1 d\sigma_2$, the probability of observing a given state of stress is then

$$\begin{aligned} \mathrm{p}(S) &= \mathrm{p}(\sigma_1, \sigma_2, \phi) \\ &= \mathrm{p}(\sigma_1, \sigma_2 \mid \phi)\,\mathrm{p}(\phi) \\ &= \mathrm{p}(\sigma_1, \sigma_2)\,\mathrm{p}(\phi) \\ &= \frac{f(\sigma_1, \sigma_2)}{\pi} d\sigma_1 d\sigma_2 d\phi. \end{aligned}$$

Measurements of a planar state of stress are typically made in terms of τ_{xy}, σ_x, or σ_y rather than the rotation angle. Assume the observation τ_{xy} has been made. To make predictions about the principal stresses given this information, rewrite $\mathrm{p}(S)$ in terms of the coordinates $(\sigma_1, \sigma_2, \tau_{xy})$. This can be done by the Jacobian rule for probability densities and the formulas

$$\begin{aligned} \tau_{xy} &= (\sigma_2 - \sigma_1)\cos\phi\,\sin\phi, \\ d\sigma_1 d\sigma_2 d\tau_{xy} &= (\sigma_2 - \sigma_1)\cos(2\phi) d\sigma_1 d\sigma_2 d\phi. \end{aligned}$$

Since $\sin(2\phi) = \tau_{xy} / \mid \sigma_1 - \sigma_2 \mid /2$, the probability model can be written

$$\mathrm{p}(S) = \frac{f(\sigma_1, \sigma_2)}{(\sigma_2 - \sigma_1)\pi} \left(1 - \frac{\tau_{xy}^2}{(\frac{\sigma_2 - \sigma_1}{2})^2}\right)^{-1/2} d\sigma_1 d\sigma_2 d\tau_{xy}, \tag{1}$$

where the stress tensor S is expressed in terms of $(\sigma_1, \sigma_2, \tau_{xy})$.

Although there is a great deal of freedom in modeling the strength of the principal stresses with an arbitrary $f(\sigma_1, \sigma_2) d\sigma_1 d\sigma_2$, the rest of the probability model is determined by the assumption that the principal stress directions are completely random. Any model which can be written in the form of Equation (1) is consistent with this assumption, and any model which can not be expressed in this form is *not* consistent with the assumption. In this sense, we have derived Equation (1) based on the physical assumptions of the problem. This probability model does not normally appear in introductory probability texts, but it has been described before. It is a special case of the l_2-isotropic probability model discussed in Mendel[2].

The same process produces a similar result if σ_x is the given data, rather than τ_{xy}. In Equation (1), replace τ_{xy} with $\sigma_x - \mid \sigma_1 + \sigma_2 \mid /2$, and $d\tau_{xy}$ with $d\sigma_x$. This result is easily seen relative to the Mohr circle, since τ_{xy} and $\sigma_x - \mid \sigma_1 + \sigma_2 \mid /2$ are coordinates for the 2-d state of stress, whose distance is a constant from the center of the Mohr circle.

The above models break down when the principal stresses are the same, $\sigma_1 = \sigma_2$. This special case, however, represents a pure hydrostatic stress, and is handled in a straightforward manner by most criteria for failure.

3.2 General stress states

A more complicated problem is that of general stress in a solid, isotropic component. Here, there are six degrees of freedom, often specified by the observable stress tensor components $(\sigma_x, \sigma_y, \sigma_z, \tau_{xy}, \tau_{xz}, \tau_{yz})$. Alternately, the stress tensor can be specified by the principal stresses $(\sigma_1, \sigma_2, \sigma_3)$ and their directions. Since these directions are orthogonal, they can be specified by the Euler angles (ϕ, θ, ψ) in 3 dimensions. In this section, we derive a probability model which is consistent with the notion that the 3 principal stress directions may be oriented randomly. This is the same as saying that the set of all stress tensors with the same principal stresses should somehow be 'uniformly' distributed. The model will be expressed in terms of the principal stresses and three observable stress components to facilitate inference on the probability of failure once the value of some, but not all, stress components are known.

We analyze first what is meant by a random orientation. The Euler angle triple, (ϕ, θ, ψ), represents a rotation of $\phi \in (-\pi, \pi)$ radians about the z-axis, followed by a rotation of $\theta \in (0, \pi)$ about the new x-axis, and then a rotation of $\psi \in (-\pi, \pi)$ about the new z-axis. Each Euler angle rotation can be thought of as an orientation for the principal stress directions, and it can be shown (Miles[3]) that a random orientation is given by

$$\mathrm{p}(\phi, \theta, \psi) = \frac{1}{8\pi^2} \sin\theta \, d\phi d\theta d\psi,$$

where the constant normalizes the probability to 1.

Therefore, a complete probability model is given by

$$\mathrm{p}(S) = \frac{f(\sigma_1, \sigma_2, \sigma_3)}{8\pi^2} \sin\theta \, d\sigma_1 d\sigma_2 d\sigma_3 d\phi d\theta d\psi,$$

where $f(\sigma_1, \sigma_2, \sigma_3)$ is any probability density for the principal stresses. Again, proceed using the Jacobian rule to replace the angle measurements with observables from the stress matrix. This gives the following conditional probability model for $\sigma_z, \tau_{xz}, \tau_{yz}$ conditioned on knowing the values of the principal stresses:

$$\mathrm{p}(\sigma_z, \tau_{xz}, \tau_{yz} | \sigma_1, \sigma_2, \sigma_3) = \frac{d\sigma_z d\tau_{xz} d\tau_{yz}}{8\pi^2(\sigma_1 - \sigma_2)(\sigma_1 - \sigma_3)(\sigma_2 - \sigma_3) \sin 2\phi \sin^2\theta}.$$

The complete model is therefore

$$\begin{aligned}\mathrm{p}(S) &= \frac{f(\sigma_1, \sigma_2, \sigma_3)\, d\sigma_1 d\sigma_2 d\sigma_3 d\sigma_z d\tau_{xz} d\tau_{yz}}{8\pi^2(\sigma_1 - \sigma_2)(\sigma_1 - \sigma_3)(\sigma_2 - \sigma_3) \sin 2\phi \sin^2\theta} \\ &= \frac{f(\sigma_1, \sigma_2, \sigma_3)\, d\sigma_1 d\sigma_2 d\sigma_3 d\sigma_z d\tau_{xz} d\tau_{yz}}{16\pi^2(-\alpha_{12}\alpha_{23}\alpha_{31})^{1/2}},\end{aligned}$$

where the stress tensor S is parametrized in terms of the principal stresses and three observables, $\sigma_z, \tau_{xz}, \tau_{yz}$, and the α_{ij} are given by

$$\alpha_{ij} = (\sigma_i - \sigma_z)(\sigma_j - \sigma_z) + \tau_{xz}^2 + \tau_{yz}^2.$$

A derivation of this is due to Tsai[4].

Once again, this model gives the entire class of probability distributions which are consistent with the assumption that the principal stresses may come from any direction whatsoever with equal probability. It also enables the inference on the probability of structural failure by giving a conditional probability model for observables in terms of the principal stresses. Bayes rule can be applied to calculate the probability of failure given the observables. This is particularly helpful in a dynamic environment where preventative action may be taken if an undesirable situation arises.

Notice that the model presented here is not valid when some of the principal stresses are the same, because a division by 0 would be required. This is handled by two cases. First, suppose that all three principal stresses were the same. This represents the case of pure hydrostatic stress, and no special probability model is required. If exactly two of the three principal stresses are the same, there are only 4 degrees of freedom to the stress tensor. Two of the degrees of freedom are $\sigma_1 = \sigma_2$ and σ_3, the other two being the principal stress direction for σ_3, which can be given by the first two Euler angles (ϕ, θ). The ψ is neglected because rotations in the σ_1, σ_2 stress plane result in the same stress tensor. The general model which satisfies the given assumptions is therefore

$$\mathrm{p}(S \mid \sigma_1 = \sigma_2) = \frac{f(\sigma_1, \sigma_3)}{4\pi} \sin\theta \, d\sigma_1 d\sigma_3 d\phi d\theta.$$

4 Probability Models for Failure Given the Stress

Consider the problem of describing the probability of failure given that a certain state of stress is applied to it. Timoshenko[5] presents a number of criteria as a basis, including the maximum stress, maximum strain, maximum shearing stress (Tresca), and maximum distortion energy (von Mises-Hencky) criteria. At face value, these criteria predict failure if their parameter exceeds a certain critical level. For instance, with the von Mises criterion, material failure is predicted if the total distortion energy U_{dist} exceeds the mean distortion energy capacity θ of the material. This section discusses two different approaches to determine what θ should be, and describes the connection between the parameters of these models and hidden assumptions regarding the distribution of distortion energy in a material.

For the sake of preciseness, assume the problem is that of studying the failure of a ductile, linear, isotropic material with the von Mises criterion. † In this case, the model predicts material failure if the total distortion energy exceeds the mean distortion energy capacity θ of a material.

If the mean distortion energy capacity θ of a material were known precisely, the von Mises model for failure p(fail | S) could be modeled as

$$\mathrm{p}(\mathrm{fail} \mid S) = \begin{cases} 1 \text{ if } U_{dist} \geq \theta \\ 0 \text{ if } U_{dist} < \theta \end{cases} .$$

However, since variability in material strength exists due to the variability of material properties or potential flaws and cracks, θ cannot be specified by a fixed value. Further, there may be restrictions on the amount of uniaxial stress test data which is available for estimating θ.

A probability model can be used to describe this inherent uncertainty regarding θ. With such a model for θ, the probability of failure can be rewritten as

$$\mathrm{p}(\mathrm{fail} \mid S) = \mathrm{p}\left(\theta \geq U_{dist}\right).$$

Weibull[6] proposed the following model which is applied to a wide variety of problems in engineering, including the determination of a probability density for yielding stress, and in turn for θ. The model gives the probability for component failure under a uniaxial stress x, and has 3 parameters as seen in the equation

$$\mathrm{p}(x \mid x_u, m, x_0) = \frac{m}{x_0}(x - x_u)^{m-1} \exp\left(-\frac{(x - x_u)^m}{x_0}\right) dx. \qquad (2)$$

†Note that although the von Mises criterion is discussed in particular, a similar analysis for the other failure models can be made.

Here, x_u is the zero strength, measured in units of stress, m is a shape parameter for the model, and x_0 is correlated with θ.‡

Although Equation (2) is largely motivated by engineering judgement, Weibull admits that the function $(x - x_u)^m/x_0$ was chosen in an ad hoc fashion. His criteria was to choose a function that disappears at the zero strength x_u, is easy to write, and is positive and non-decreasing. He did not see a way to derive, from engineering judgements, any of these 3 parameters.

Recently, Lindquist[1] proposed that the function in the exponent of Equation (2) can be derived from assumptions about how distortion energy is distributed in a material. In particular, he shows that the function $(x-x_u)^m/x_0$ is more reasonably written as $(x^2 - x_u^2)/(6G\theta)$, where G is the shearing modulus. This differs in two ways from Weibull's equation. First, the exponent applies to the individual terms rather than their difference, and second, the parameters are identified with physically meaningful quantities—x_0 is shown to be a function of the shearing modulus and the mean distortion energy capacity, and m is 2 because distortion energy is proportional to the square of the uniaxial stress ($u = \sigma^2/6G$). Here, $x_u^2/(6G)$ is interpreted as the minimum distortion energy capacity of a material.

It is useful to tersely summarize Linquist's derivation before making additional comments. First, assume that a batch of material is sufficiently large to make N stress test rods. If rod i breaks with distortion energy u_i due to the uniaxial stress σ_i, then the average distortion energy is

$$\theta = \frac{1}{N}\sum_{i=1}^{N} u_i = \frac{1}{N}\sum_{i=1}^{N} \frac{\sigma_i^2}{6G}.$$

From Mendel[2], the following likelihood is derived for the first n test rods, conditioned on the average density. Note that we maintain the assumption that a rod never fails with a distortion energy less than $e_u = \sigma_u^2/6G$.

$$\mathrm{p}(u_1,\ldots,u_n \mid \theta) = c_N \left(1 - \frac{\sum_{i=1}^{n}(u_i - e_u)}{N\theta}\right)^{N-1} du_1 \cdots du_n.$$

The proportionality factor c_N normalizes the probability to 1. For very large batch sizes, and with a change of variables, this model can be approximated by the exponential function given in the equation

$$\mathrm{p}(\sigma_1,\ldots,\sigma_n \mid \theta) = c\prod_{i=1}^{n} \frac{\sigma_i}{6G\theta - \sigma_u^2} \exp\left(-\frac{\sigma_i^2 - \sigma_u^2}{6G\theta - \sigma_u^2}\right) d\sigma_1 \cdots d\sigma_n. \quad (3)$$

This derivation explicitly points out that a slightly modified form of Weibull's distribution is required, given certain physical considerations and the assumption that average distortion energy alone is relevant to predicting failure.

‡The parameters are traditionally found by statistical parameter estimation or Bayesian inference based on Equation (2) and a set of uniaxial stress test data taken from experiments with sample rods.

Although many strong assumptions were made in this section, there are several interesting points which arise as results of the derivation of Equation (3). First, we reiterate that Weibull chose the function $(x - x_u)^m/x_0$ somewhat arbitrarily, although it posesses many qualitative features which seem to match the problem of failure prediction. Next, Lindquist has shown how a slightly modified form of Weibull's function can be derived from the physical notion of average distortion energy. In this framework, parameters assume an operational meaning. In addition, we see that if the new model's analog of m is assumed to be something other than 2, the notion that average distortion energy in the samples is relevant to the failure criterion is violated. This may indicate that another quantity, such as the average applied stress when $m = 1$, is relevant, or that there are additional factors involved with the failure process. This seems to be the case for non-ductile, nonlinear, or non-isotropic materials, where standard statistical fitting techniques have given values for m over a wide range (say, from 2 to 26). Additional parameters, therefore, may be required to model the probability of failure in these cases. In any event, other probability models certainly make physical assumptions about the failure process, although they may be more difficult to mathematically describe.

5 Conclusions

Although the models developed in this paper made strong assumptions which do not apply to all structural reliability problems, several widely applicable conclusions may be drawn. One, it is possible to convert physical properties and engineering judgements into classes of probability models. It is not necessary to arbitrarily select them. In the case here, a statement about principal stress directions was used to derive a class of models. A key tool is the Jacobian rule for probability densities, which permits engineering statements to be preserved regardless of the coordinate system chosen to express them. Two, parameters in a probability or statistical model have meaning relative to the engineering assumptions that apply to a system. In particular, we saw how the Weibull parameters for a material strength model relate to the way strength is assumed to be distributed through a collection of components made from the same material. Much work remains to be done, however, in deriving other model types and determining the physical assumptions inherent in other probability models. Three, it is unnecessary to base models on parameters which have an abstract and non-measurable definition. All of the parameters derived here have meaning relative to the problem of structural reliability. To be sure, for the sake of computational efficiency, it may be simpler to use such an abstract parameter. However, the use of such a parameter *does* make hidden assumptions about the physics of the problem.

References

1. Lindquist, E., Strength of Materials and the Weibull Distribution. Technical Report ESRC 92-32, UC Berkeley, Berkeley, CA, 1992. Engineering Research Systems Center.
2. Mendel, M.B., *Development of Bayesian Parametric Theory with Applications to Control.* PhD thesis, MIT, Cambridge, MA., June 1989. Dept. of Mechanical Engineering, unpublished.
3. Miles, R.E., On Random Rotations in $\Re^3$. *Biometrika*, 52:636–639, 1965.
4. Tsai, P.S., Predicting the State of Stress under Static Loading. Technical Report ESRC 93-17, UC Berkeley, Berkeley, CA, 1993. Engineering Research Systems Center.
5. Timoshenko, S.P., *Strength of Materials*, volume 2. Rober E. Krieger Publishing Co., Inc., New York, 3rd edition, 1956.
6. Weibull, W., A Statistical Distribution Function of Wide Applicability. *Journal of Applied Mechanics*, 18:293–297, 1951.

Fatigue Life Prediction of Structures

V.V.Bolotin

Institute of Mechanical Engineering, Russian Academy of Sciences, Griboyedova str. 4, 101830 Moscow, Russia

Abstract

Fatigue life prediction problems are discussed from the viewpoint of probabilistic mechanics of solids and structures. A survey of random factors in damage and fracture is presented with a special emphasis on the statistical scatter of mechanical properties of materials and/or structural components. A model of fatigue crack initiation is discussed based on the concepts of continuum damage mechanics as well as the assumption that fatigue crack nuclei form Poissonian sets distributed in a body. To describe macroscopic fatigue crack propagation, the theory proposed by author that takes into account the interaction between the microdamage accumulation along the crack trajectories and the global balance of forces, energy, etc. in the cracked body under loading is used. Various approaches to the modeling of material property fields are considered, and the influence of random factors from different sources on fatigue life is studied by means of numerical simulation. Discussion of alternative probabilistic approaches to modeling of fatigue damage is presented.

Keywords. Fatigue, fracture, fatigue crack initiation, fatigue crack growth, microdamage accumulation.

1. Survey of random factors in fatigue life prediction

Fatigue damage of mechanical components occurs under the influence of a large number of factors that interact in a complex and not easily predicted way. A part of these factors may be considered as random and they may be described in the framework of probabilistic models. The second part of factors forms the group usually called "uncertainties". Not going into scholastic discussion about differences between randomnesses and uncertainties, recall that most of factors given with uncertainties may be treated quite satisfactorily using probabilistic models. Moreover, even pure probabilistic models contain uncertainties since the parameters of probabilistic distributions, spectral characteristics, etc., are not known with complete assurance. Using confidence intervals, verisimilitude techniques, and other tools of mathematical statistics, we, as a matter of fact, try to manage with a large amount of uncertainties. Later on we mention random factors only and discuss them in terms of probabilistic models.

The influence of random factors on fatigue life can be divided in three groups. The first group is composed of random mechanical properties of materials, the second one of random defects and imperfections of structural components, and the third one of random loads, actions and environmental conditions. The borders between the groups are rather vague. For example, it is a point of disagreement where cracks, according to their size, are to be attributed as microcracks and to be related to the first group, and where cracks are to be considered as macroscopic cracks, i.e. related to the second group. Similarly, it is an open question as how to classify residual stress and strain fields born as a result of the former life of a structure - to treat them as defects of the structure or to consider them as a result of the former loads and actions. Nevertheless, although intersections can exist between the three groups of random factors, the classification is useful. It is seems to be reflected even in the distribution of interests of professional people. Those who are engaged into the analysis of loading processes (e.g., in spectral presentation and cycle counting in fatigue) are rarely active in the study of mechanical properties, and vice versa. In practical design, all the three groups of factors enter, at least in principle, jointly in the fatigue life prediction [1]. Such approaches are used especially for structures of high responsibility: in aerospace engineering, nuclear power engineering, etc. It is possible to gather in a single pile all randomnesses and uncertainties on the academic stage of analysis. But there is nothing to expect in this from the viewpoint of new concepts, methods, and phenomena. In the author's opinion, shared by most of the researcher community, it is more expedient to perform analysis of particular classes of random factors. Among them, maybe, the most interesting and least studied is the randomness of material properties that characterize the material resistance against mechanical damage and fracture. Later on, we discuss this group of random properties with special emphasis on the resistance against fatigue and related phenomena (fracture under sustained loading, creep fatigue, corrosion fatigue, etc.).

2. Randomness of material properties

It is expedient to distinguish three kinds of material randomnesses. They are called later within-specimen, specimen-to-specimen, and batch-to-batch randomnesses. The within-specimen randomness is inherent to the microstructure of materials, i.e. to nonhomogeneities, flaws and imperfections on the level of grains, fibers, etc. This kind of randomness is inevitable due to the nature of materials and methods of their manufacturing. The scale of this randomness varies within a wide range. The scale of crystal lattice, molecular structure of polymers, and dislocations is of the order of $10^{-10}...10^{-6}$ m. On that level the randomness is a subject matter of solid-state physics. Referring to microstructure, we mean the scales beginning from 10^{-6} m and larger up to larger scales that are nevertheless small compared with characteristic sizes of structural components. They may be of the order of 10^{-3} m, and even higher, e.g. of the scale of particles in large concrete structures. Essentially, all of these elements of microstructure may be described within the framework of micromechanics understood as a branch of continuum mechanics.

Not only properties distributed in the volume of a body, but also properties of the surface, e.g. its small-scale roughness must be included into this group of random factors. Random fields with various properties are used to describe the randomness of microstructure on this level. The fields are continuous for amorphous polymers, piece-continuous for multi-phase materials or materials with multiple microcracking, etc.

The specimen-to-specimen randomness of mechanical properties is well known to experimentors. They observe a significant scatter of mechanical properties even if all specimens are made from the same sheet and carefully controlled before testing. Since the characteristic size of microstructure is small compared with the size of specimens, and most of the specimen properties expose themselves in an averaged form, only a part of the inherent material randomness is responsible for specimen-to-specimen scatter. Another part is born from instabilities and imperfections of manufacturing processes. This links together the specimen-to-specimen and batch-to-batch randomness. In fact, mechanical properties of commercial materials vary in a rather large range even when they are strongly specified and controlled. Moreover, on the design stage of a structure, an engineer cannot be sure that the specified properties will be realized in the future. This uncertainty can be described with probabilistic models, and parameters of the models, at least in principle, can be estimated from experimental data.

The two latter types of randomness have a lot in common as far as the corresponding probabilistic models are concerned. Opposite to the within-specimen randomness described with random fields, the specimen-to-specimen and batch-to-batch randomness can be described with a set of random variables. Among them are those characterizing the random fields of local material properties. For example, the fracture toughness of material may be a random function of coordinates within a specimen. Meanwhile the parameters of this function (mean fracture toughness, its variance, space correlation constant, etc.) may be random values characterizing the specimen-to-specimen and/or batch-to-batch scatter of material properties.

From the point of probabilistic models, the most important distinction between the within-specimen randomness and the two other types is that the former is to be discribed with random fields - functions of coordinates (and, depending on the loading conditions, on time or cycle number). Meanwhile, the specimen-to-specimen and batch-to-batch scatters are more natural to describe using random values. The common account of all three types leads to conditional random fields, i.e. to functions of coordinates, time or cycle number which depend on a set of random variables with given distributions.

As it usually happens, this classification includes inevitable intersections and other ambiguous points. For example, analysing fatigue crack propagation, we must know not only the material properties distribution but the parameters characterizing initial conditions (the position and the size of an initial crack, the level of damage at its tip, etc.). These data, being treated as random, are to be attributed to the specimen-to-specimen scatter, meanwhile the properties ahead of the crack are discribed as a random field. Another example is presented with the distribution of properties among the specimens produced of a single but large metal sheet. It is known that, due to peculiarities of the manufacturing process, properties of the metal may vary within a sheet significantly. It means that these properties form within the sheet a random field. Cutting comparatively small specimens, we in fact produce a sample of random variables generated from the random field. The number of analogous examples could be extended.

There is another criterion for the classification of randomness when resistance against mechanical damage and fracture is considered. This classification, discussed in detail in the book [2] is based on the type of interaction between the elements of microstructure, i.e. between grains, fibers, etc. Two extreme types of behaviour in damage and fracture are to be distinguished: the brittle-type and plastic-type behavior.

We say that the behaviour is brittle (ideal, perfectly brittle) if the weakest or/and the most stressed element of the microstructure is responsible for the body as a whole. The simplest and typical example is the fracture of a body of an ideally brittle material when the fracture of a single elements results in a total failure. Such idealized models are described, in a natural way, with asymptotic distributions of extremes, and the Weibull distribution is the most widely used among them. Another example of this type of model is fatigue crack nucleation in a macroscopically flawless and comparatively smooth specimen. The time and place of the fatigue crack initiation is impossible to predict beforehand: it is a typical random event. There are a number of microscopic defects in the specimen including surface microcracks that most frequently play the part of the nuclei of macroscopic fatigue cracks. Only a single or at least a few of many microcracks develop into macroscopic cracks, and it is a typical weakest-link behaviour. One cannot wonder that the Weibull distribution and its generalizations appear to be the most consistent tool to describe fatigue crack nucleation [3].

The opposite extreme type of response is the plastic-type one. While the brittle-type behaviour corresponds to a series (from the system reliability viewpoint) distribution of elements in the system, the plastic-type one corresponds to a parallel composition of elements. It means a high level of redundancy and load redistribution in the case of some elements becoming ruptured. A typical example is the plastic deformation of a specimen of a mild steel in tension. Generally, in each case when averaged, bulk material properties are responsible for a body as a whole, we might say the material exhibits plastic-type behaviour.

The terms "brittle-type" and "plastic-type" are rather conditional and, maybe, not the most successful. In particular, elastic deformations are to be related to the second type since elastic moduli characterize averaged properties but not local ones. On the other hand, a crack propagating through the cross section of a specimen or a structural component meets on its way a large number of grains, fibers, and other elements of microstructure. In the process of propagation, local properties are averaged. The resulting process is usually almost continuous and is characterized with averaged material properties. The crack front acts here as a kind an averager resulting in crack behaviour which might be labelled as "plastic-type".

From the viewpoint of probabilistic models, the plastic--type behaviour is governed mainly with the averaged characteristics of the material. Since most commercial materials are nonhomogeneous, and the scales of nonhomogeneities are small compared with the cross section scales, the averaging has significant effects on decreasing the scatter of damage and fracture parameters such as fatigue life, residual strength, etc. Denote with n the number of elements participating in an event simultaneously. Then the standard deviation of the resulting parameters will decrease as $n^{1/2}$. This means that for very large n the scatter might become insignificant. We stress that this discussion concerns the within-specimen randomness only and not, say, the batch-to-batch one. For example, Young's modulus may be assumed deterministic within a batch of metal, but, generally, it is a random value when we consider metal produced by different plants. Since (and if) the variance of parameters corresponding to the plastic-type behaviour is small, the normal and related distributions are valid for this kind of material response.

In reality most of the phenomena relate not to the extreme, but to intermediate types of behaviour. In particular, while fatigue crack nucleation relates to the first type, and fatigue crack growth through the cross section to the second type, the intermediate stage is certainly of a mixed type.

A special attention is to be paid to so-called "short" cracks [4]. They are cracks which depth is of the order of magnitude of several sizes of grains or other elements of structural nonhomogeneity. Recently, the term "small" crack is used more widely than "short" crack. There are sufficient reasons to change the terminology. To say "short" is more appropriate to cracks that are shallow but not necessarily short looking from the body's surface (see Fig.2.1,a where a is depth of the crack and 2b is surface

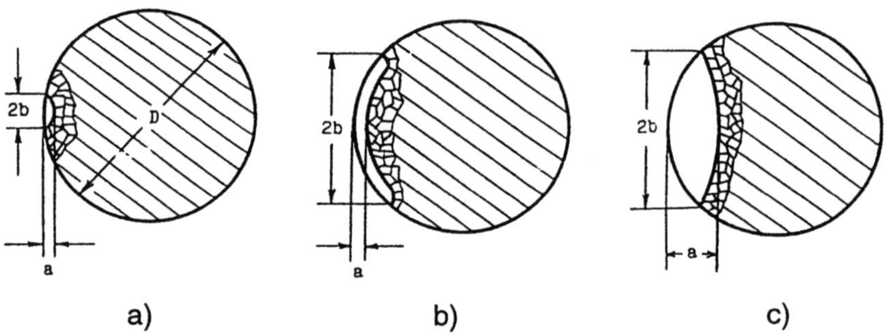

Fig.2.1. Various types of cracks: a - small crack, b - shallow but long at the surface crack, c - large crack.

length). If a crack is long measured on the surface but nevertheless shallow (Fig.2.1,b), its propagation mechanism differs from that of the crack shown in Fig.2.1,a. In fact, only a small number of grains are involved in the growth of a "small" crack. Opposite, a distinct averaging mechanism is present in the propagation of the crack that is long at the surface. From that point, there is no essential difference between the cracks shown in Fig. 2.1,b and c although the latter is much deeper than the former one. As a result, we distinguish "small" cracks where growth is closer to the brittle-type behaviour, and "large" cracks where propagation is accompanied with significant averaging of local mechanical nonhomogeneities. Here and later we use the terms "small" and "large" cracks without quotes commas.

The mechanism of small fatigue crack growth is illustrated in Fig.2.2. The nucleation of small cracks is usually ex-

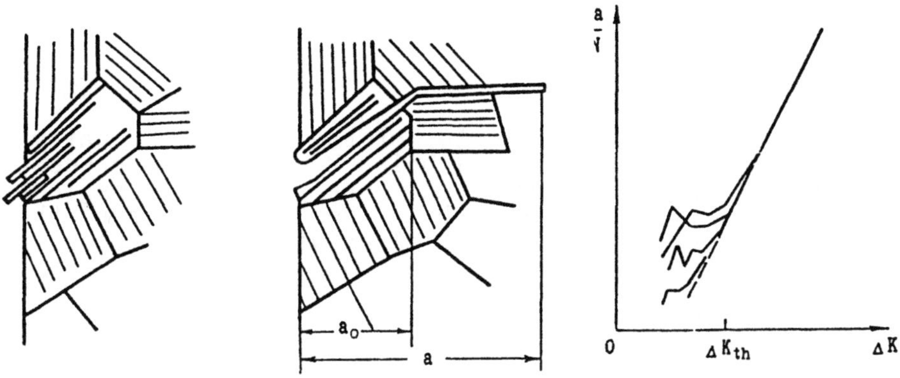

a) b) c)

Fig.2.2. Fatigue crack nucleation and initial phase of propagation: a - extrusion/intrusion mechanism; b - initial mixed mode stage resulting into the large crack propagation; c - crack growth rate diagramme with the initial stage of growth.

plained [4] in terms of an extrusion/intrusion mechanism (Fig.2.2,a). The initial mode of small cracks is, as a rule, close to the shear or related mixed mode. In particular, if a specimen is subjected to cyclic tension, a small crack begins to grow at angles near to 45^0 (Fig.2.2,b). After crossing a several number of grains, the crack begins to propagate in the opening mode. It is very probable that nuclei of cracks are located near the weakest points. They might be grain aggregates that are comparatively weak and/or unfavourably oriented with respect to the macroscopic stress-strain field. They might also be grain aggregates located near the local microstress concentrators. Since grains are vulnerable with respect to intra- or intergranular shear, a crack begins to grow in the shear mode. After a large number of grains become involved, the factor of averaging begins to dominate. It results into the turn of the crack to the regular direction as it is shown in Fig.2.2,b.

There are some pecularities in the part of the fatigue crack growth rate diagram corresponding to small cracks (Fig.2.2, c). Among them is a larger scatter of experimental data in the area of small stress intensity factor range ΔK. Another feature is that, at least in some experiments, the crack growth rate for small cracks is higher than it is natural to expect from the extrapolation of results concerning larger cracks. These features are known. They are usually explained with the influence of the regular component of the stress [4]. In fact, the stress concentration factor at the tip of a crack with depth a and tip curvature radius ρ is

$$\kappa = 1 + Z(a/\rho)^{1/2} \tag{2.1}$$

where Z is nondimensional factor of the order of unity. For example, for an elliptical hole $Z = 2$. To present experimental data in terms of ΔK, one needs to neglect unity in Eq.(2.1) compared with the second term. Then the applied (remote) stress σ multiplied with the factor κ produces the complex $\sigma\,(a)^{1/2}$, i.e. the core of the stress intensity factor. Keeping unity in Eq.(2.1), we come to a more complicated model.

The opinion is spread widely that the pecularities of small fatigue crack propagation can be explained with the contribution of regular terms of the stress field. In the author's opinion (and experience, too), this explanation appears to be insufficient. It is more probable that small fatigue cracks show irregular behaviour because they are in a major degree weakest-link formations (see Fig.2.1,a). As they lose this property, they propagate in a direct, regular way.

The above statements concern fatigue cracks in polycrystalline or granular materials. The qualitative picture, say, in unidirectional fiber composites is, of course, quite different although small and large cracks are to be distinguished there, too. While the dispersed damage and macroscopic crack formation proceed in the brittle-type manner, the growth of developed cracks is governed mainly with the averaged properties of fibers and fiber-matrix interfaces. A probabilistic model of crack initiation and growth in unidirectional fiber composite under tension along fiber was suggested in [5,6]. Due to the comparatively low resistance of matrix and fiber-matrix interfaces against splitting of fibers, a transverse crack rapidly develops into a brush-like domain spread along fibers (Fig.2.3,a). The transfer (ineffective) length λ plays a dominant part in formation of this domain. Significant fiber strength scale effect also influences the elongation of the damaged region. But, opposite to small cracks where only a small number of ruptured fibers are involved, the growth of cracks (Fig.2.3,b) is governed mainly by the averaged properties of fibers.

3. Fatigue crack nucleation

A number of probabilistic models of fatigue crack initiation are proposed beginning from the probabilistic generalization of the Miner summation rule up to more sophisticated models with account of a materials's microstructure, the damage accumulation in elements of microstructure, and the interaction of damaged elements. Later on we discuss

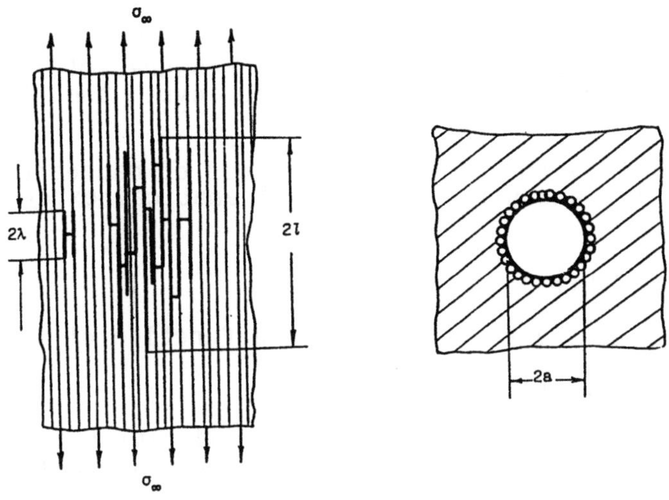

Fig 2.3. Transverse crack in unidirectional fiber composite under tension along fibers: a - brush-like damage; b - schematized front of the crack in the cross section.

one of the approaches based on continuum damage mechanics [2,6]. The main idea is to connect the microdamage field $\omega(x,t)$ with the probability of formation of even a single nucleus of the macroscopic fatigue crack in the considered domain of a body.

Introduce a scalar variable $\|\omega(x,t)\|$ characterizing the local microdamage measure $\omega(x,t)$. In the simplest case $\omega(x,t)$ is a scalar field with magnitudes from [0,1]. Then $\|\omega(x,t)\| = \omega(x,t)$. In other cases it may be a norm of a vector, an invariant of a tensor or a function of such invariants, etc. The limitations are that $\|\omega\| = 0$ for nondamaged material, $\|\omega\| = 1$ for completely damaged material, and $\|\omega(x,t)\|$ grows in time during the microdamage accumulation. We call this variable the local norm of the field $\omega(x,t)$, although the field $\omega(x,t)$ and $\|\omega(x,t)\|$, generally, do not satisfy all the requirements put on normalized spaces.

In the beginning consider macroscopically homogeneous deterministic stress, strain and microdamage fields in a certain domain of the body M_0, say, in the volume or on the surface area of a standard specimen. Experimenters usually consider a specimen as damaged when they detect the first macroscopic crack. Therefore, if $\|\omega\| = 1$ in the whole domain with the measure M_o, we may expect that in average one macroscopic crack appears in this domain. Moreover, we assume that the expectad number of macrocracks initiated in the domain M_0 on the time segment $[0,t]$ is a function of the norm $\|\omega(t)\|$ of the measure $\omega(t)$, whose magnitudes, generally, are extended over the interval $(1, \ )$.

Notate with $k(t)$ the point-wise process of macrocrack initiation in M_0, and with $\mu(t)$ its mathematical expectation, i.e.

$$\mu(t) = E[k(t)]. \tag{3.1}$$

Postulate that the process $\mu(t)$ is a function of $\|\omega(t)\|$:

$$\mu(t) = f\ \|\omega(t)\|. \tag{3.2}$$

The function $f(\ )$ is differentiable and satisfies conditions $f'(\ ) > 0$, $f(0) = 0$, $f(1) = 1$. A typical example is

$$f(\omega) = \omega^{\beta} \tag{3.3}$$

where ω is a scalar measure, and $\beta > 0$.

Eq.(3.2) corresponds to a homogeneous distribution of microdamage in the domain with the measure M_0. The natural generalization of Eq (3.2) upon nonhomogeneous fields is

$$\mu(t) = \int_M f\|\omega(x,t)\| \frac{dM}{M_0}. \tag{3.4}$$

If the domain M consists of subdomains M_1, M_2, ... with, generally, different dimensions, we obtain that

$$\mu(t) = \sum_i \int_{M_i} f_i \|\omega_i(x,t)\| \frac{dM_i}{M_{oi}}. \tag{3.5}$$

As a rule, in the further discussion we will refer to Eq. (3.2) with the scalar measure ω assuming $\|\omega\|\ \ \omega$. Generalizations based on Eqs.(3.4), (3.5) and tensorial measures ω are easy to imagine.

To link the probability of formation of macrocracks with the expected value of their number, a Poisson model was used [7]. In fact, the formation of nuclei of macroscopic cracks is a rare event, their characteristic size a_* is small compared with the sizes of the body, and the interaction between neighbouring nuclei may be considered as negligible. Then the probability of formation of k macrocracks in the domain M_0 is determined as

$$Q_k = \frac{\mu^k}{k!} \exp(-\mu) \quad (k = 0,1,\ldots). \tag{3.6}$$

Substitution Eq.(3.2) in Eq.(3.6) at $\|\omega(x,t)\|$ $\omega(t)$ yields

$$Q(t) = \frac{f^k[\omega(t)]}{k!} \exp\left\{-f[\omega(t)]\right\} \quad (k = 0,1,\ldots). \tag{3.7}$$

The probability of the event that at the time instant t at least one macrocrack initates is

$$Q(t) = 1 - \exp\left\{-f[\omega(t)]\right\}. \tag{3.8}$$

The right-hand side of Eq.(3.8) coincides with the distribution function $F_*(t_*)$ of times t_* up to the appearance of the first macrocrack. A similar equation is valid with respect to the cycle number N_* until the macrocrack nucleation.

The discussed model is illustrated in Fig.3.1,a where the damage measure $\omega(t)$ and the expected number of macrocracks $\mu(t)$ are shown varying in time. A realization of the step-wise process k(t) is presented there also. It is assumed that in Eq.(3.2) $\mu = \omega^2$. Fig.3.1,b shows the growth of the sizes of macroscopic cracks from the initial value a_* up to the critical value a_{**}.

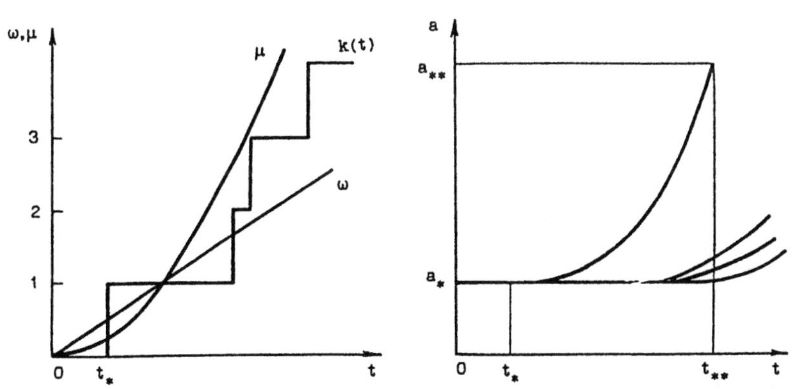

a) b)

Fig 3.1. Fatigue crack nucleation model: a - microdamage accumulation; b - macrocracks growth.

In a more general case when the damage field $\omega(x,t)$ is nonhomogeneous in the domain M but varies there sufficiently slowly, we obtain with the use of Eq.(3.4) that

$$F_*(t_*) = 1 - \exp\left\{ - \int_M f[\omega(t)] \frac{dM}{M_0} \right\}. \tag{3.9}$$

Here the notation of the left-hand side is changed putting $Q(t_*) \quad F_*(t_*)$.

The distributions given with Eqs.(3.8) and (3.9) are related to the Weibull distribution, and in special cases coincide with the latter, say, with the distribution from the book [3]. Actually, the equation of damage accumulation has the form

$$\frac{\partial \omega}{\partial t} = \begin{cases} 0, \; s \le s_0 \text{ or } t \le t_0 \\ \dfrac{1}{t_c} \left[\dfrac{s - s_0}{s_c} \right]^m, \; s > s_0, \; t > t_0 \end{cases} \tag{3.10}$$

where $t_c > 0$ is a time constant (e.g. the duration of a single cycle), $t_0 \ge 0$ is a certain threshold time necessary to initiate the damage accumulation process, and $m > 0$. At $s = \text{const} > s_0$, $\omega(0) = 0$, Eq.(3.10) results into

$$\omega(t) = \left[\frac{s - s_0}{s_c} \right]^m \left[\frac{t - t_0}{t_c} \right]. \tag{3.11}$$

Substitution of Eqs.(3.3) and (3.11) into Eq.(3.9) gives

$$Q(t) = 1 - \exp\left[- \int_M \left[\frac{s - s_0}{s_c} \right]^\alpha \left[\frac{t - t_0}{t_c} \right]^\beta \frac{dM}{M_0} \right] \tag{3.12}$$

where the notation $\alpha = m\beta$ is used. If instead of time t we use the cycle number N and the associated material constants N_c and ΔN_{th}, Eq.(3.12) transfers into equation presented in the book [3].

Eq.(3.12) is not a unique consistent relationship between the number of macroscopic cracks nucleated due to microdamage and the microdamage measure. Moreover, if the microdamage $\omega(x,t)$ is a random function of space and/or of time (for example, if loading is stochastic), Eq.(3.2) becomes invalid since $\mu(t)$ is by definition a deterministic function. At least three ways to overcome this obstacle may be listed under which the basic relationship (3.2) remains valid for deterministic functions $\omega(t)$. First, $\omega(t)$ can be replaced with its expected value. Secondly, the averaging procedure can be applied to the right-hand side of Eq.(3.2). In the third, most consistent way, one can interpret the probabilities $Q_k(t)$, $Q(t)$ and $F_*(t)$ in Eqs. (3.6) - (3.9) as conditional. For example, in the latter approach instead of Eq.(9) we obtain

$$Q(t) = 1 - \int_0^\infty \exp[-f(\omega)]dF_\omega(\omega;t) \tag{3.13}$$

where $F_\omega(\omega;t)$ is the cumulative distribution function of the measure ω at the instant t.

A more general equation for the probability $Q(t)$ is as follows:

$$Q(t) = P\{ \sup \|\omega(x,t)\| \geq \omega_*;\ x \in M\}. \tag{3.14}$$

Both randomness of the microdamage field, and its nonhomoheneity in the domain M are taken into account. Note that special forms of the governing equations such as Eqs. (3.3), (3.10), etc. were given here as examples only to demonstrate possibilities to obtain final equations similar to those commonly used in the statistical treatment of fatigue data.

4. Fatigue crack growth

Fatigue crack growth is governed with a number of random factors from all the three groups discussed above: random material properties, random defects and imperfections, random loads and actions. Only a small part of them is illustrated schematically in Fig.4.1 in terms of fatigue crack growth under regular (deterministic) loading. The cracked specimen under loading is shown in Fig.4.1,a. In Fig.4.1,b the relationship is shown between the crack size a, stress intensity factor K and a certain material parameter K_f. The latter can be interpreted either as a resistance property against fatigue (fatigue toughness) or that against brittle fracture (fracture toughness). The initial crack size a_0 is a random value with the probability density function $p_0(a_0)$ presented on Fig.4.1,b. During the crack growth the maximum (within the cycle) stress intensity factor $K = Y(a/b)\sigma^{max}(\pi a)^{1/2}$ increases as $a^{1/2}$ on earlier stages and more rapidly, due to the increase of the form-factor $Y(a/b)$, at later stages. A current probability density function $p(a)$ is shown in Fig.4.1,b, too,

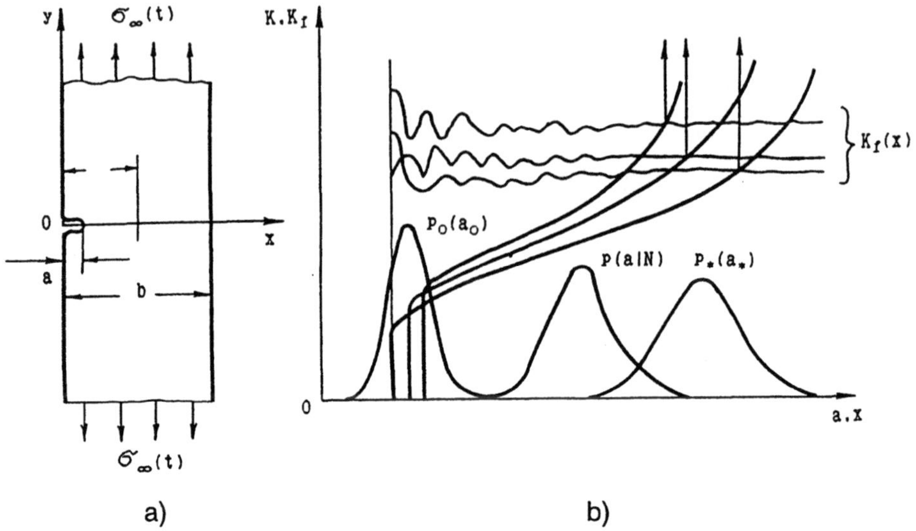

Fig. 4.1. Fatigue crack growth with account of random within-specimen and specimen-to-specimen properties: a - scheme of loading; b - loading and material parameters as functions of crack size a and coordinate x along the crack trajectory. Other explanations in text.

as well as the analogous function $p_*(a_*)$ of critical crack sizes corresponding to the first crossings of processes K(a) and $K_f(a)$. Note that the curves $K_f(x)$ drawn in Fig. 4.1,b correspond generally to mechanical properties homogeneously distributed in a specimen. The fluctuations of $K_f(x)$ which are significant in the beginning, become smaller when the crack size increases. This is a result of transition from small cracks to large ones. Note also that both within-specimen and specimen-to-specimen randomness are presented in Fig. 4.1,b.

The picture becomes more complicated if characteristic cycle stresses, say, σ^{max} vary in time, moreover, randomly [2,3,8] Even in the case of stationary deterministic cyclic loading the problem of fatigue life assessment is in fact a first-crossing problem for two random processes, K(a) and $K_f(x)$ at x = a. But the first process is smooth and monotonic in Fig. 4.1,b, and fluctuations of the second process are not so intensive. The situation changes when loading is a random process such as wave action on ship and offshore structures, turbulence action on aircraft structures, and seismic actions on buildings and structures. Another source of complexity is ageing and other kinds of deterioration of materials resulting in the time dependence of $K_f(x)$. Some methods of simplification were shown in [2]. They include, in particular, the wide use of conditional distributions, the application (when it is allowable) of the central limit theorem, etc. Below we pay attention to effects of material property randomness on the fatigue crack growth rate.

Two kinds of probabilistic models are used to include the randomness of material properties into the quantitative fatigue life analysis. The first way is to introduce randomness with a direct randomization of corresponding deterministic equations. The account of the batch-to-batch and specimen-to-specimen scatter of material properties requires a slight modification of deterministic equations by means of replacement of some material parameters with their stochastic equivalents - random variables. But to take into consideration the random variability of properties within a specimen or a structural component, these parameters are to be replaced with random functions of coordinates measured along the cracks trajectories, and, maybe, randomly varying in time, too.

The second way is to use available mathematical models that seem to be suitable to describe irreversible processes of any physical nature. There are many such models beginning with elementary Markov chains and including more sophisticated stochastic diffusion processes. A survey of both kinds of models can be found in [9]. The book by Bogdanoff and Kozin [10], however, is dedicated completely to the models of the second kind.

A typical equation for a single-parameter fatigue crack with the size a has the form

$$\frac{da}{dN} = f(\sigma_{max}, \sigma_{min}, a). \tag{4.1}$$

Here σ_{max} and σ_{min} are extremal magnitudes of the characteristic stress within each cycle usually replaced with the stress range $\Delta\sigma = \sigma_{max} - \sigma_{min}$ and the stress ratio $R = \sigma_{min} / \sigma_{max}$. Moreover, the right-hand side of Eq.(4.1) is usu- ally replaced with a function of the stress intensity factor $K = Y\sigma (\pi a)^{1/2}$, more exactly, its range $\Delta K = Y\Delta\sigma (\pi a)^{1/2}$. Here Y is a calibration factor, usually of the order of unity. The most popular is the Paris-Erdogan equation

$$\frac{da}{dN} = \text{const} \left[\Delta K\right]^m \tag{4.2}$$

with power index m taking, as a rule, the magnitude from two to four.

Many proposals to randomize Eq.(4.2) and related equations

include replacement ot the constant in Eq.(4.2) either with a random value or a random function of a, N or both. Such equations are discussed in [8,11,12]. Some simple approaches to randomization lead to models of the second kind. For example, assuming that the right-hand side of Eq.(4.2) contains an additional multiplier which is a normal stationary white noise, we arrive at a stochastic differential equation corresponding to the simplest continuous diffusion process. Obviously, this model is not sound from the physical viewpoint, but it shows the direction of how to develop more sophisticated models.

5. Mechanics of fatigue fracture

In this paper we develop probabilistic models using the theory of fatigue fracture suggested in [13], presented in detail in the book [2] and survey paper [14]. The theory is based on the concepts of analytical (rational) mechanics. Any system cracked body - loading or loading device is in fact a mechanical system of a special kind. Firstly, two groups of generalized coordinates are needed for such systems. The first group describes the displacement field under the condition that all of the crack parameters are fixed. The second group consists of independent parameters characterizing sizes, shapes and positions of cracks. These parameters we call, in honor of Griffith, Griffith's generalized coordinates (later on, for brevity, G-coordinates). Since cracks in common materials are irreversible, G-coordinates can be chosen in such a way that their variations are non-negative. Variations of Lagrangian generalized coordinates consist of two parts: one part are common Lagrangian variations, and another part is generated due to variations of G-coordinates. It is essential that constraints put on G-coordinates are unilateral and, generally, nonholonomic. We may assume these constraints ideal, and relate the resistance against crack propagation to additional, explicitly given forces.

Consider a cracked body in equilibrium under quasistatic loading. According to the principle of virtual work for systems with unilateral constraints, a system under quasistatic loading is in equilibrium if the work produced on all small admissible displacements is non-positive:

$$\delta A \le 0. \tag{5.1}$$

Since by definition the considered, unperturbed state and neighboring, perturbed states are equilibriums (in the common sense), the virtual work produced on Lagrangian variations is identically equal to zero. Eq.(5.1) takes the form

$$\delta_G A \equiv \delta_G A_e + \delta_G A_i + \delta_G A_f \le 0 \tag{5.2}$$

where index G shows that all virtual displacements are born from variations of G-coordinates, $\delta_G A_e$ and $\delta_G A_i$ are components of virtual work produced by external and internal forces, respectively. The term $\delta_G A_f$ takes into account the fracture work within processing zones located at crack tips.

Let the number of G-coordinates be finite, and denote them $a_1, \ldots, a_m$. Re-write the terms of Eq (5.2) as follows:

$$\delta_G A_e + \delta_G A_i \equiv \sum_{j=1}^{m} G_j \delta a_j, \qquad \delta_G A_f = \sum_{j=1}^{m} R_j \delta a_j. \tag{5.3}$$

Eqs (20) contain two kinds of generalized forces: the active, driving forces $G_1, \ldots, G_m$, and the corresponding passive, resistance forces $R_1, \ldots, R_m$. In the simplest case of the standard Griffith - Irwin problem, the force G is in fact the energy release rate, and the force R is the corresponding critical value of that rate, i.e. the fracture toughness characteristics notated commonly G_c.

States of cracked bodies under loading were classified [13] in the two levels: with respect to equilibrium and with respect to stability. Depending on the sign of $\delta_G A$, sub-equilibrium, equilibrium and non-equilibrium states are to be distinguished. In terms of generalized forces the classification is performed depending on the sign in the relationships

$$G_j \lesseqgtr R_j \quad (j = 1, \ldots, m) \tag{5.4}$$

with account of additional conditions put on derivatives of generalized forces.

To include fatigue crack growth into the analytical fracture mechanics, it is sufficient to take into account microdamage accumulation near the crack tips (and, maybe, in the far field, too). It means that all generalized forces and especially resistance forces are to be calculated with account of microdamage. Then fatigue crack growth can be represented as an alternating transition of the system from one sub-equilibrium state to the next one going each time through an equilibrium state. If jumps are small enough, the process may be replaced with a continuous one. The only difference, and an essential one, is that, generally, all the forces, G_j and R_j depend on certain measures that characterize microdamage near crack tips. The step-wise propagation of fatigue cracks is described with alternating conditions given with Eq.(5.4).

Additional variables are to be added to extend the fracture mechanics concepts to the theory of fatigue crack growth. First, special variables are needed to describe the microdamage field. They may be presented as scalar, vector or tensor fields [15]. Denoting the microdamage field measure with $\omega(x,N)$ where x is a space reference vector, and N is loading cycle number, the equation of microdamage accumulation may be written as

$$\omega(x,N) = \underset{n=0}{\overset{n=N}{\Omega}} \left\{ s(n),\ a(n) \right\}. \tag{5.5}$$

Here s(n) is a set of variables that characterize the loading process (loading vector), and the vector process a(n) is composed of G-coordinates $a_1(n), \ldots, a_m(n)$. The right--hand side of Eq.(5.5) is a hereditary functional of the loading and crack growth history.

Microdamage at the crack tips is of crucial significance in the fatigue theory. Introduce a special value microdamage vector $\psi(N)$ characterizing the level of microdamage at the tips. This vector depends on the magnitudes of $\omega(x,N)$ at $x = x[a_j(N)]$, $j = 1, \ldots, m$. If $\omega(x,N)$ is a scalar measure, and $m = 1$,

$$\psi(N) = \omega \left\{ x[a(n)],\ N \right\}. \tag{5.6}$$

In any case, discussing microdamage at the crack tips, we must abstain from the presentation of cracks as mathematical cuts. Instead, we assume the curvature of the tips to be finite. Denote with $\rho_1, \ldots, \rho_m$ curvature radii corresponding to G-coordinates $a_1, \ldots, a_m$. Vector process $\rho(N)$ composed of these radii at the cycle number N is a hereditary functional similar to that given with Eq.(5.5)

$$\rho(N) = \underset{n=0}{\overset{n=N}{P}} \{ a(n),\ \psi(n) \} \tag{5.7}$$

with an implicit dependence on the loading process s(n). Note that the latter assumption does not mean that crack faces are presented as smooth surfaces. The quantities ρ_j, or more precisely, a_j/ρ_j are just measures of the stress concentration near the tips. Introduction of additional parameters with the dimension of length is inherent to any consistent theory of fatigue crack growth. Even those authors who consider the damage accumulation around a crack treated as a mathematical cut, usually choose a point located not on the tip but at a certain characterisric distance from the tip. That distance is in fact an equivalent of the finite tip radius. Of course, no such difficulties will be awaited if a material is elasto-plastic, e.g. in the framework of Dugdale model. But, generally, any consistent theory of fatigue is to be applicable for all kinds of solids including linear elastic and linear visco-elastic media. This implies considering cracks with finite curvature on their tips. We do not mention corrosion fatigue and stress corrosion cracking where tip sharpening and blunting are an essential part of crack propagation [16].

Eqs.(5.5) - (5.7) together with the relatioships

$$G_j[\, s(N),\ a(N),\ \psi(N),\ \rho(N)] \lessgtr R_j\, [\, s(N),\ a(N),\ \psi(N),\ \rho(N)] \qquad (\, j = 1,\ldots, m) \qquad (5.8)$$

and the initial conditions put on a(N), ω(N) and ρ(N) form a set of constitutive functional equations describing the fatigue crack growth process beginning from the crack initiation stage up to the final failure due to the unstable crack propagation. Under some additional assumptions [13], the problem can be reduced approximately to differential equations with respect to $a_1,\ldots,a_m$, i.e. to equivalents of the well-known equations of fatigue crack growth, say, Paris-Erdogan equation (4.2).

Later on, exlusively for simplification, consider a single-parameter crack omitting indices at a_j, G_j and R_j. Then Eq. (5.8) takes the form

$$G[s(N),\ a(N),\ \psi(N),\ \rho(N)] \lessgtr R[s(N),\ a(N),\ \psi(N),\ \rho(N)] \qquad (5.9)$$

where, for simplification also, variables are considered as onedimensional.

6. Numerical simulation of fatigue crack growth

For numerical simulation, assume that the body's material is linear isotropic elastic with the only difference that it is subjected to microdamage effecting the resistance against fatigue. Let the processes a(N), ω(N), etc. be sufficiently slow to treat them as continuous functions of the cycle number N. Assume the equation of microdamage accumulation (22) in the form

$$\frac{\partial \omega}{\partial N} = \begin{cases} 0, & \Delta\sigma < \Delta\sigma_{th}, \\ \left[\dfrac{\Delta\sigma - \Delta\sigma_{th}}{\sigma_f}\right]^m, & \Delta\sigma \geq \Delta\sigma_{th}. \end{cases} \qquad (6.1)$$

Here $\Delta\sigma$ is the range of the opening stress $\sigma(x,0,t)$ during a cycle at the prolongation of the crack $x \geq a$, $y = 0$; σ_f is resistance stress against microdamage; $\Delta\sigma_{th}$ is threshold resistance stress; m is a positive number relative to the power exponent entering into S-N curves or into the Paris-Erdogan equation. Generally, these material parameters depend on temperature and stress ratio $R = \sigma_{min}/\sigma_{max}$.

Neglecting the effect of microdamage on the bulk material properties, the energy release rate is given by the Irwin formula, i.e.

$$G = K^2(1 - \nu^2)/E \tag{6.2}$$

where E is Young's modulus and ν is Poisson's ratio. This value takes in this problem the part of the active generalized force. Let the specific fracture work per unit crack area (without doubling of the upper and lower faces of the crack) be $\gamma = \gamma_0(1 - \psi^\alpha)$. Here γ_0 is the specific fracture work for undamaged material, and $\alpha > 0$. Hence the resistance generalized force

$$R = \gamma_0(1 - \psi^\alpha). \tag{6.3}$$

$\rho(N)$ is governed with equation

$$\frac{d\rho}{dN} = \frac{\rho_s - \rho}{\lambda_\rho}\frac{da}{dN} + (\rho_b - \rho)\frac{d\psi}{dN} \tag{6.4}$$

where ρ_s is the radius of a "sharp" crack, and ρ_b is that of a "blunt" crack. The first term in the right-hand of Eq.(6.3) describes sharpening of the tip due to crack growth acceleration. The second term describes blunting due to the microdamage accumulation and, implicitly, crack growth retardation. The length parameter λ_ρ entering on Eq.(6.4) may be of the order of the size of the processing zone at the tip. Eq. (6.4) is related to similar equations used in the treatment of corrosion fatigue and stress corrosion cracking problems [16].

For numerical simulation, it is expedient to chose an appropriate crack size increment Δa, and to evaluate the cycle number increment necessary to attain the new stable equilibrium state. Equalizing G and R at $a + \Delta a$, we obtain from Eqs. (6.2) and (6.3) that the microdamage corresponding to the equilibrium state is

$$\psi = [1 - K^2(1-\nu^2)/(E\gamma_0)]^{1/\alpha}. \tag{6.5}$$

It means that the increment $\Delta\psi$ at the moving tip is known, and Eq.(6.4) presents, in the first approximation, the increment (decrement) of the tip radius:

$$\Delta\rho \approx \frac{\rho_s - \rho}{\lambda_\rho} \Delta a + (\rho_b - \rho)\, \Delta\psi \tag{6.6}$$

Now we can estimate the stress concentration at the tip and, generally, the stress distribution along the crack trajectory. Using Eq.(6.1), we find the first approximation for the cycle number increment ΔN:

$$\Delta N = \left[\Delta\psi - \frac{\partial\omega}{\partial x}\Big|_a \Delta a\right]\left[\frac{\Delta\sigma - \Delta\sigma_{th}}{\sigma_f}\right]^{-m}. \tag{6.7}$$

Here $\Delta\sigma_t$ is the stress at the tip, and the term with $\partial\omega/\partial x$ accounts for the change of microdamage when the tip advances into a less damaged part of the body.

Evidently, in the coarse approximation calculations might be terminated on that stage since all of the values, including the crack growth rate da/aN or $\Delta a/\Delta N$, the microdamage measure $\psi + \Delta\psi$ and the tip radius $\rho + \Delta\rho$ are estimated. But in the fatigue life prediction, it could result into a systematic error accumulation. The more reliable way is to solve jointly Eqs.(6.1) and (6.4) at the known stress distribution $\sigma = f(x/a, \rho/a)$, and to estimate ΔN from the equation $G = R$. Including into the computional procedure an iterative cycle with respect to ρ, $\omega(x)$, ψ and $\sigma_f(x)$, we can make the computational error as small as needed. However, the same result might be obtained with the use of sufficiently small crack increment Δa, say $\Delta a = \min\{\rho_s, \lambda_0\}\ 10^{-2}$ where λ_0 is the characteristic scale of material nonhomogeneity.

7. Randomization of material resistance against fatigue

There is a large liberty in choice of probabilistic models for distribution of material properties along the crack trajectory. Generally, almost all of the parameters entering into Eqs.(6.1) - (6.4) ought to be treated as random. To avoid, on this stage of study, unnecessary complexities, assume as the leading material parameter the resistance stress σ_f. Other parameters will be connected with σ_f, say, $\Delta\sigma_{th} = 0.05\sigma_f$, etc., or will be assumed deterministic.

There are two sources of randomness of σ_f. One is the variability of σ_f along the crack trajectory, and therefore σ_f is a random function of the coordinate x measured along the crack path, i.e. of a. The second source is the batch-to-batch and/or specimen-to-specimen scatter. Hence, if we introduce a random function $\sigma_f(x)$ of the running coordinate $x = a(N)$, it must be described with conditional distributions. For example, suppose that $\sigma_f(x)$ is a homogeneous normal function. That is an acceptable assumption if fluctuations of $\sigma_f(x)$ are sufficiently small. Then the mean and the variance of this function ought to be considered as random. That concerns also, in the general case, parameters entering into the power spectral density. If a Markov model is used for $\sigma_f(x)$, transition probabilities and initial probability distribution should be taken depending on random parameters that describe the scatter of properties between specimens or structural components. Both sources of randomness are covered with the conditional random function $\sigma_f(x|I_0, I_1, \ldots)$ where I_0, $I_1, \ldots$ are random parameters with the joint probability density function $p(I_0, I_1, \ldots)$. There area lot of ways to describe these functions fitting experimental data. For instance, $\sigma_f(x)$ may be either a continuous or a discrete random function. Although the microstructure of most structural materials is discontinuous (from the viewpoint of continuum mechanics), a crack, as a rule, propagates rather slowly. The crack front acts here as a kind of averager, smoothing irregularities of the microstructure. On the other hand, the computerized version of Eqs.(4.1) or (4.2) is a discrete one. For example, if we assume a continuous Markov model for $\sigma_f(x)$, after discretization it becomes a kind of Markov or semi-Markov discrete (chain) model. The well-known situations in fatigue crack growth appear on the stage when elasto-plastic behaviour is to be taken into account. The jump-like propagation of cracks is also covered with the mechanics of fatigue fracture. To simulate this process, we must return to alternating relationships (5.8) between the generalized forces. Transition from one state of equilibrium to another is governed with these relationships under variation of the sign between the left- and right-hand sides. It is easily to simulate numerically this discrete process, but it is not at all a Markov process. As an example assume the resistance stress $\sigma_f(x)$ in the form

$$\sigma_f(x) = \sigma_f^0 + (\sigma_f^+ - \sigma_f^-)v + \sigma_{f1}u(x) \tag{7.1}$$

Here σ_f^0, σ_f^+, σ_f^-, and σ_{f1} are deterministic values, v is a random variable, and u(x) is a normalized specified random function of the coordinate x measured along the crack path. Values σ_f^+, σ_f^- and v as well as values entering into the power spectral density $S_u(k)$ of the function u(x) characterize the specimen-to-specimen scatter, meanwhile the within-specimen randomness is accounted for u(x). If v and u(x) are non-negative, then σ_f^0 means the lower bound of the resistance stress.

Let u(x) be a nonlinear transform of a normalized stationary ergodic Gaussian process with the power spectral density

$$S_u(k) = \frac{2k_1k_0}{\pi} \frac{1}{(k^2 - k_0^2)^2 + 4k_1^2k^2}. \tag{7.2}$$

Here k_0 is an analogue of frequency - the wave number connected with the characteristic length λ_0 of nonhomogeneities as $k_0 = 2\pi/\lambda_0$, and k_1 describes the space correlation of nonhomogeneities. We consider for simplification k_0 and k_1 fixed. The beta-distribution is natural to assume for v:

$$p(v) = \frac{\Gamma(\mu + \nu) v^{\mu-1}(1 - v)^{\nu-1}}{\Gamma(\mu)\Gamma(\nu)}. \tag{7.3}$$

Here $\mu > 0$, $\nu > 0$, $\Gamma(\bullet)$ is gamma-function.

For numerical simulation of the resistance stress distribution along the crack path assume the following set of data: $\sigma_f^0 = 9$ GPa, $\sigma_f^+ - \sigma_f^- = 2$ GPa, $\sigma_{f1} = 1$ GPa, $\Delta\sigma_{th} = 0.025\,\sigma_f^0$, $\lambda_0 = \lambda_1 = 10^{-3}$m, $\mu = \nu = 4$. The function u(x) entering in Eq. (7.1) is taken in the form $u(x) = [n_1^2(x) + n_2^2(x)]^{1/2}$ where $n_1(x)$ and $n_2(x)$ are two independent Gaussian processes with zero mean and unity variance. Hence, u(x) is a Rayleigh process. A sample function $\sigma_f(x)$ computed according to Eq. (7.1) is shown in Fig.7.1.

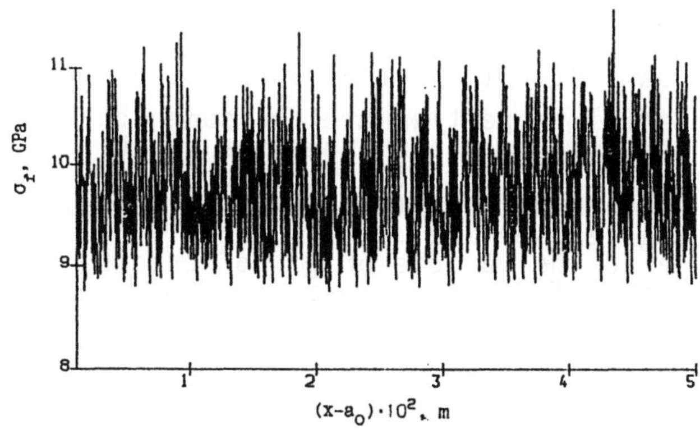

Fig.7.1. Sample function of the material parameter σ along the crack trajectory.

As to the conditions at the crack tip, assume the "sharp" tip radius $\rho_s = 10^{-5}$m, and the "blunt" one $\rho_b = 10^{-4}$ with the scale from Eq.(6.4) $\lambda_\rho = 10^{-2}$m. Attributing the material toughness against quasibrittle fracture to the minimal tip radius, estimate the fracture toughness as K_{IC} $(Y/2)\sigma_f^0 (\pi\rho_s)^{1/2}$ where Y is form-factor in the equation for stress intensity factors. Substituting Y = 1, σ_f^0 = 9 GPa, $\rho_s = 10^{-5}$m, we obtain K_{IC} 28 MPa $m^{1/2}$. The corresponding specific fracture work of undamaged material (at E = = 200 GPa) is of the order γ_0 K_{IC}^2/E 3.9 kJm^{-2} . Taking instead of ρ_s the "blunt" radius ρ_b, we obtain K_{IC} 89 MPa $m^{1/2}$ and γ_0 39 kJm^{-2} , respectively. These estimates show that the assumed numerical data are quite consistent.

8. Discussion

Computations were performed for a central crack in a stripe subjected to self-centered tension with remote tensile stress σ . The width of the stripe b = 500 mm, and the stress range $\Delta\sigma$ was assumed constant during each considered fatigue life. The load level $\Delta\sigma$, intial half-length of the crack a_0, and initial tip radius ρ_0 were varied in a wide range. Parameters characterizing mechanical properties were kept the same in all the computations. Special questions arise concerning initial conditions. For simplification, the initial crack size a_0, and the microdamage at the tip, $\psi \equiv \omega(a_0, 0)$ are supposed to be deterministic. In particular, $\psi(0) = 0$. Initial distribution of the resistance stress $\sigma_f(x)$ is also to be known, or, in probabilistic terms, the probability density function of $\sigma_f(a_0)$ is to be given. For example, one can assume that $\sigma_f(a_0)$ follows the same distribution as $\sigma_f(x)$ at $|x| > a_0$. It means that material properties are the same both near the tip of the initial crack and along the crack path, in the bulk of the material. In fact, it is not true even for laboratory specimens with initial cracks specially implanted before testing. In service, it is much more complicated situation due to the various origin of cracks, environmental actions, etc. The randomness of initial conditions that does not coincide with the randomness of material properties along the crack tip may be interpreted as the specimen-to-specimen scatter and treated correspondingly. Generally, there are no obstacles to assume arbitrary initial values for $\sigma_f(a_0)$, but it is out of the frame of this paper.

The plan of computations includes an analysis of the influence of the stress level $\Delta\sigma$, initial radius ρ_0 and initial crack length a_0 on the total fatigue life as well as on the duration of different stages of this life. Special attention was paid to discriminate the influence of the within-specimen and specimen-to-specimen scatters of material properties. Like the results of physical experimentation with a broad program, the results of numerical experimentation are difficult to discuss in a compact form. Some results are presented in the recent paper [17] where, opposite to this study, a stochastic differential equation was used to simulate fatigue crack growth in a media with random mechanical properties. The latter was quantified with the random parameters that describe the specimen-to-specimen scatter.

The main observation from the computed results is as follows. Fluctuations of $\sigma_f(x)$ along the path of the crack result in variability of the crack growth rate, and, therefore, in variability of the crack size plotted against the cycle number N. The scatter of fatigue growth curves a(N) is significant even if only within-specimen randomness of material properties is taken into account. Behavior of samples corresponding to one family is of interest. It appears that only a pair of intersecting samples happens to be observed among the 15 - 20 trials. It is essential that the divergence of sample functions begins on the earlier stage of the crack growth. It signifies that the conditions near the initial crack tip play an important part in the within-specimen variability of fatigue crack growth. Account of the specimen-to-specimen scatter does not change this feature considerably.

The computational experimention [17] allows for reconsidering the relationship between experimental data and numerical simulation of fatigue crack growth. The most popular are data by Virkler et al. [18]. They were obtained for central crack tension specimens of aluminium alloy cut from the same sheet and tested with the same initial crack size. Looking at the experimental diagrammes where the crack size a is plotted against the cycle number N, one cannot miss a striking point: the curves corresponding to various specimens intersect very rarely. It means that the scatter of crack growth is born not only from the point-to-point randomness of mechanical properties, but also, and not in a lesser degree, from the specimen-to-specimen scatter. Another explanation is that such a behaviour of curves is caused with the scatter of initial conditions, i.e. with that of the resistance against crack growth at the beginning of loading. But in fact it is the same specimen-to-specimen scattering.

The features of experimental fatigue crack growth curves demonstrated with numerical simulation seem to be obvious.

It is strange that, to the author's knowledge, it has not been emphasized when the comparison was performed between experimental data and mathematical models. For example, in the book [10] where a Markov chain model was proposed for damage accumulation processes, both Virkler's and the corresponding numerically simulation curves were presented. There is an evident difference in the behaviour of sample functions in experiment and in simulation: opposite to experimental samples, simulated ones intersect violently. It means that the model suggested in the mentioned book describes a random process quite different from the experimental one, despite a satisfactory agreement that has been achieved on the level of means and mean squares.

The computational results presented below are obtained using a model of fatigue damage and crack growth rather different from that in the paper [17]. This model is based directly on the relationship governing stability/instability of cracks with the microdamage accumulation both near the tip and in the far field as well as with account of sharpening/blunting of the tip in the process of crack growth. But the approach to randomization of material properties is almost the same. The difference is, which parameter, σ_f or K_f is to be chosen as a random function of x varying randomly from one specimen to another.

Typical samples of fatigue crack growth curves a(N) are presented in Fig.8.1. They are obtained for $\Delta\sigma$ = 100 MPa and $\rho_0 = 5 \cdot 10^{-5}$ m, and other numerical data were mentioned earlier. Three sample curves from a family of 20 curves were chosen: the "worst", corresponding to a shortest fatigue life, the "best", corresponding to a longest fatigue life, and an "average". It is noteworthy that the deviation of curves begins on the earlier stages of the fatigue life, and almost no intersections of curves were observed on the latter stages. Additional information is shown in Figs.8.2 and 8.3. In Fig. 8.2 three samples of microdamage ψ at the moving crack tip are plotted versus the cycle number N. Opposite to Fig.8.1, three stages of fatigue life are quite distinctive in Fig. 8.2. It is the initiation stage, up to the $N_* = 3 \cdot 10^4 \ldots 6 \cdot 10^4$ when damage is accumulating at the fixed tip $a = a_0$. The crack growth begins when the damage level attains the level very close to the critical magnitude $\psi_* = 1$. This stage is considerably long taking up to 10^6 cycles. It is to be noted that the crack growth at this stage could be calculated with a simplified procedure just assuming that a crack grows under the condition that $\psi_* = 1$. A similar simplified approach is used widely in numerical simulation of creep fracture [19]. The last stage, looking shortly in Fig.8.2 due to the log scale, corresponds to the accelerating fracture process resulting in final failure. The total fatigue life varies

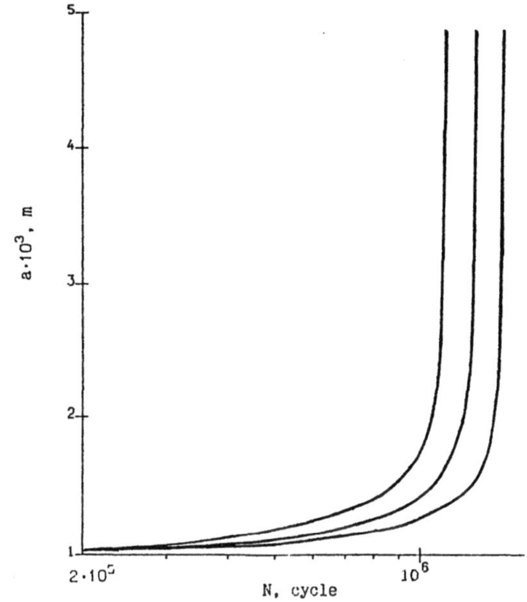

Fig.8.1. Three samples of fatigue crack growth: the "worst", the "best", and an "average" chosen from the array of 20 samples.

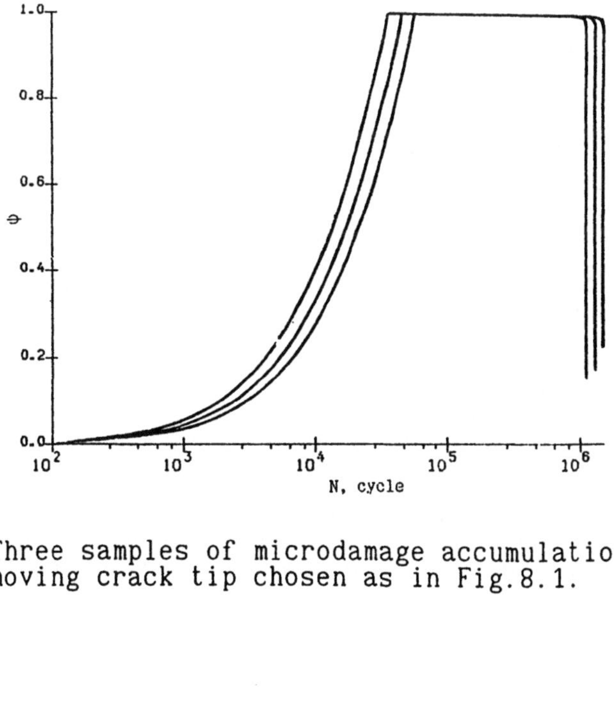

Fig.8.2. Three samples of microdamage accumulation at the moving crack tip chosen as in Fig.8.1.

between 10^6 and $1.2 \cdot 10^6$ cycles within the family of specimens. It is essential that despite randomly mixed properties ahead of the tip, the initial scatter remains and even increases.

The variation of the tip curvature radius ρ during the fatigue life is illustrated in Fig.8.3. This picture is ob-

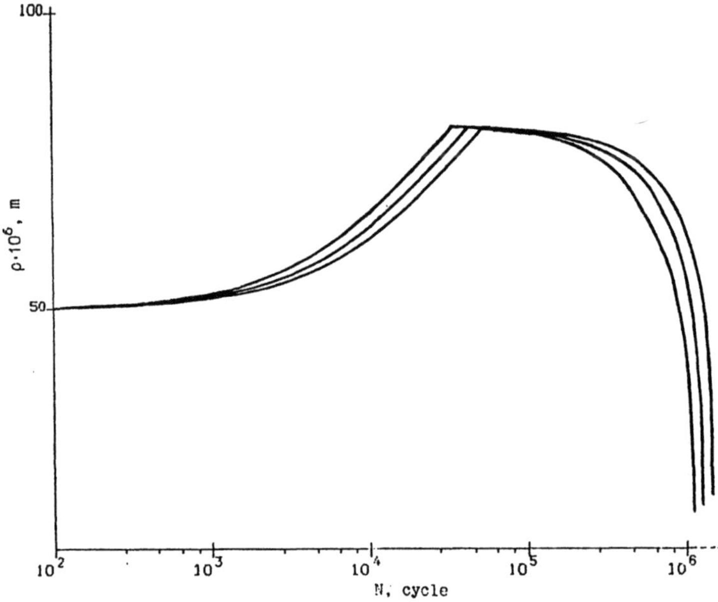

Fig.8.3. Three samples of crack tip curvature radius chosen as in Fig.8.2.

tained using Eq.(6.4) that is, of course, only one of many suitable models. Comparing Figs.8.2 and 8.3, we see their correspondance. At the initiation stage the crack tip is blunting due to the microdamage accumulation. From the start of crack growth, a gradual sharpening of the tip begins. Close to the final stage of the fatigue life this sharpening turns to be intensive: a crack advances into almost nondamaged material.

For evaluation of fatigue life, the crack growth rate diagrams are useful relating the rate da/dN to the stress intensity factor range ΔK (or to a related loading parameter). A set of samples of such diagrams obtained with account of both within-specimen and specimen-to-specimen scatter of material properties is shown in Fig.8.4. The scatter of rate da/dN is moderate near the fatigue threshold and attains a half of the order of magnitude at the middle stage. The scatter becomes huge close to the final failure. The source of such a scatter is not only the randomness of mechanical properties but the high sensivity of specimens to fluctuations of fracture toughness when a specimen is cracked significantly. However, it ought to be noted that Fig.8.4 is plotted using very small crack in-

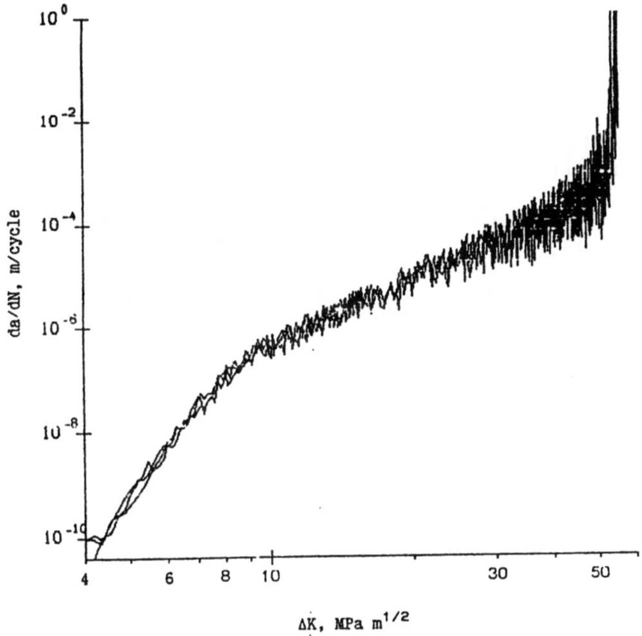

Fig.8.4. A set of samples of fatigue crack growth diagrams obtained by numerical simulation

crements without any smoothing. A set of samples of fatigue crack growth diagrams obtained by numerical simulation.

A large number of questions and problems remain to be open for further research. Among them are the account of "natural" averaging of material properties, the choice of models for microdamage accumulation, and crack tip conditions with the assessment of involved parameters, etc. But even in the performed scope, the performed analysis shows that the statistical scatter of the fatigue crack growth rate and that of the cycle number up to the final failure mostly originate on the initial stage of crack propagation and, therefore, depend significantly on conditions near the initial crack tip. Therefore, more attention must be paid to initial conditions and specimen-to-specimen scatter of mechanical properties to obtain more reliable prediction of fatigue life of structural components.

Acknowledgement

This work was supported by the Russian Foundation for Basic Research (grant 93 - 013 - 16486). Computations were performed by B.V.Minakov, and the camera-ready form of the manuscript by A.S.Ryabtsev and A.S.Shubin. Author thanks both for financial support and technical assistance.

References

1. Heller, R.A. and G.H. Stevens: Bayesian estimation of crack initiation times from service data, J. of Aircraft, 11 (1978), 794-798.

2. Bolotin, V.V.: Prediction of Service Life for Machines and Structures, ASME Press, New York 1989.

3. Bolotin, V.V.: Statistical Methods in Structural Mechanics, Holden Day, San Francisco 1968.

4. Miller, K.J. and E.R. de los Rios: The Behaviour of Short Cracks, Mech. Eng. Publ., London 1986.

5. Bolotin, V.V.: Some mathematical and experimental fracture models, Strength Problems, 2 (1971), 13-20 (in Russian).

6. Bolotin, V.V.: A unified model of fracture of composite materials under sustained loads, Mechanics of Composite Materials, 3 (1981), 405-420 (in Russian).

7. Bolotin, V.V.: On the theory of delayed fracture, Soviet Mech. Solids (MTT), 1 (1981), 137-146 (in Russian).

8. Ditlevsen, O.: Random fatigue crack growth. A first passage problem, Eng. Fracture Mech., 23 (1986), 467-477.

9. Sobczyk, K. and B.F. Spencer: Random Fatigue: from Data to Theory, Academic Press, New York 1992.

10. Bogdanoff, J.L. and F. Kozin: Probabilistic Models of Cumulative Damage, John Wiley, New York 1985.

11. Lin, Y.K. and J.N.Yang: A stochastic theory of fatigue crack propagation, AIAA J., 1 (1985), 117-129.

12. Dolinski, K.: Stochastic loading and material inhomogeneity in fatigue crack propagation, Eng. Fracture Mech., 25 (1986), 809-818.

13. Bolotin, V.V.: Equations of fatigue crack growth, Soviet Mech. Solids (MTT), 4 (1983), 153-160 (in Russian).

14. Bolotin, V.V.: Mechanics of fatigue fracture, Nonlinear Fracture Mechanics,in: CISM course 314 (Ed. M. Wnuk), Springer, Berlin 1989, 1-69.

15. Krajcinovic, D.: Continuum damage mechanics, AMR, 1 (1984), 1-6.

16. Bolotin, V.V., Ryabtsev, A.S., and A.S. Shubin: Prediction of corrosion fatigue crack growth, Mashinovedeniye, 2 (1989), 16-23, (in Russian).

17. Bolotin, V.V.: Fatigue crack growth as a random process Russian Mech. Solids (MTT), 4 (1993) (in Russian, in print).

18. Virkler, D.A., Hillberry B.M. and P.K. Guel The statistical nature of fatigue crack propagation, Trans. ASME, J. Eng. Math. Techn., 1 (1979), 148-153.

19. Riedel, H.: Fracture at High Temperatures, Springer, Berlin 1987.

Stochastic Boundary Element Methods: Computational Methodology and Applications

Tadeusz Burczyński

Department of Engineering Mechanics, Silesian Technical University, Konarskiego 18a, 44-100 GLIWICE, Poland

Abstract. Basic concepts of computational methodology of the stochastic boundary element method are presented. Applications of stochastic boundary elements to dynamic problems, stochastic media and stochastic shape sensitivity analysis are presented.

Keywords. Stochastic boundary elements, random vibration, stochastic media, sensitivity analysis

1 Introduction

The functionality of many modern engineering systems depends to a large extent on their ability to perform adequately and with a high level of reliability under not absolutely controllable conditions. In response to these problems, computer methods have been developed to deal with the statistical nature of loads, material properties and the shape of a structure. Stochastic boundary element method (SBEM), as an alternative numerical technique to the stochastic finite element method,[22,24] belongs to these methods.

Many different interpretations are possible for terminology of the SBEM. This term is used here to refer to the boundary element method which accounts for uncertainties in boundary conditions or material properties of a structure as well as the shape of a boundary. Such uncertainties are usually distributed on a boundary (Γ) or within a domain (Ω) of the structure and should be modelled as spatial $\mathbf{r}(\mathbf{x},\gamma)$ or spatially-temporal $\mathbf{r}(\mathbf{x},t,\gamma)$ stochastic fields, where $\mathbf{x}=(x_k)$ denotes spatial position in Ω or Γ, $t \in T$ is time, and γ is an elementary event ($\gamma \in \mathcal{G}$). Such stochastic fields are defined on the probability space ($\mathcal{G},\mathcal{F},\mathcal{P}$), where $\mathcal{G}$ is the space of elementary events, $\mathcal{F}$ is a σ-algebra of subsets of $\mathcal{G}$ and $\mathcal{P}$ is a probability.

Applications of the SBEM appear to have been initiated in the early 1980's. The earliest application used the boundary integral equation method to solve stochastic boundary value problems of elastostatics[1]. Next the SBEM was applied to stochastic potential problems[2], stochastic heat conduction problems[19], groundwater flow[17] and dynamical problems[3,6,13,14,15,30]. SBEM was also extended to problems with random media[4,20,21,25] and to problems with uncertain boundaries[5,7,9,27,28].

A concise presentation of the SBEM for problems with stochastic boundary conditions, stochastic material properties, stochastic shape sensitivity and identification problems was given in the form of chapters by Burczyński[8,10,11,12].

2 Computational Methodology of SBEM

Using SBEM to solve boundary value problems with random boundary conditions in the form of prescribed stochastic fields of displacements $\mathbf{u}(\mathbf{x},\gamma)=\overset{o}{\mathbf{u}}(\mathbf{x},\gamma)$, $\mathbf{x}\in\Gamma_1$, and tractions $\mathbf{p}(\mathbf{x},\gamma)=\overset{o}{\mathbf{p}}(\mathbf{x},\gamma)$, $\mathbf{x}\in\Gamma_2$, respectively, where $\Gamma=\Gamma_1\cup\Gamma_2$ and $\Gamma_1\cap\Gamma_2=\emptyset$, a vector stochastic boundary integral equation is obtained:

$$\mathbf{c}(\mathbf{x})\mathbf{u}(\mathbf{x},\gamma)=\int_{\Gamma}\mathbf{U}(\mathbf{x},\mathbf{y})\mathbf{p}(\mathbf{y},\gamma)d\Gamma(\mathbf{y})-\int_{\Gamma}\mathbf{P}(\mathbf{x},\mathbf{y})\mathbf{u}(\mathbf{y},\gamma)d\Gamma(\mathbf{y}), \tag{1}$$

where $\mathbf{U}(\mathbf{x},\mathbf{y})$ and $\mathbf{P}(\mathbf{x},\mathbf{y})$ are deterministic fundamental solutions for elastostatics.

Practically, stochastic solutions of these equations are not sought, but their moments (i.e. expectations, covariances) are. In order to solve this problem, it is possible to distinguish two general approaches, namely continuous and discrete approaches.

In the continuous approach, deterministic boundary integral equations for moments are formulated. Boundary integral equations for expectations are identical to those for deterministic systems. Boundary integral equations for covariances differ from the traditional ones in that they involve double, instead of single, boundary integrations[1,2,17,23]. This approach can be useful for problems with the Dirichlet (essential) or Neumann (natural) type of stochastic boundary conditions.

In the case of mixed stochastic boundary conditions it is convenient to use the discrete approach in which stochastic boundary integral equations are discretized into a system of random algebraic equations:

$$[\mathbf{A}]\ \{\mathbf{X}(\gamma)\} = [\mathbf{B}]\ \{\mathbf{Y}(\gamma)\}, \tag{2}$$

where $[\mathbf{A}]$ and $[\mathbf{B}]$ are deterministic matrices dependent on boundary integrals of fundamental solutions $\mathbf{P}$ and $\mathbf{U}$, $\{\mathbf{X}(\gamma)\}$ is a column matrix of unknown random nodal values of boundary displacements and tractions, and $\{\mathbf{Y}(\gamma)\}$ is a column matrix of random nodal values of boundary displacements and tractions prescribed by boundary conditions.

For the boundary element discretization, it is necessary to approximate stochastic fields of boundary displacements and tractions by random vector representations $\{\mathbf{X}(\gamma)\}$ and $\{\mathbf{Y}(\gamma)\}$. Two methods of discretization can be proposed.

The midpoint method

The field value of $\mathbf{r}(\mathbf{x},\gamma)$, $\mathbf{r}=\mathbf{u},\mathbf{p}$, over a boundary element Γ^e is represented by its value at the midpoint $\mathbf{x}^e$ of the element Γ^e:

$$\mathbf{r}^e(\gamma) = \mathbf{r}(\mathbf{x}^e,\gamma), \quad \mathbf{x}^e \in \Gamma^e \quad . \tag{3}$$

The boundary averaging method

The value for a boundary element Γ^e is represented by the spatial average of the random field over the element Γ^e:

$$\mathbf{r}^e(\gamma) = \frac{1}{|\Gamma^e|} \int_{\Gamma^e} \mathbf{r}(\mathbf{x},\gamma) d\Gamma(\mathbf{x}), \tag{4}$$

where $|\Gamma^e|$ is the area of Γ^e.

Discretization of random fields of displacements $\mathbf{u}(\mathbf{x},\gamma)$ and tractions $\mathbf{p}(\mathbf{x},\gamma)$ implies that the column matrices $\{\mathbf{X}(\gamma)\}$ and $\{\mathbf{Y}(\gamma)\}$ consist of variables whose covariance depends on the boundary element mesh. Consequently, the stochastic character of the boundary element values may be expected to depend significantly on the chosen mesh geometry. Thus, the important problem in using the discrete approach is the selection of the mesh size. Numerical calculations show[10,11,12] that the midpoint discretization method tends to over-represent the variability within the boundary element, whereas the spatial averaging method tends to under-represent the same variability. The two methods coincide when the random field mesh is sufficiently fine in relation to the correlation length.

The system of random algebraic equations (2) is the base for evaluating moments. The mean value is given by the equation

$$\{\mathbf{m}_X\} = [\mathbf{A}]^{-1}[\mathbf{B}]\ \{\mathbf{m}_Y\}, \tag{5}$$

while the covariance matrix can be evaluated by

$$[\mathbf{K}_X] = [\mathbf{A}]^{-1}[\mathbf{B}]\ [\mathbf{K}_Y]\ [\mathbf{B}]^T[\mathbf{A}]^{-T}. \tag{6}$$

The cross-covariance matrices are evaluated as follows:

$$[\mathbf{K}_{XY}] = [\mathbf{A}]^{-1}[\mathbf{B}]\ [\mathbf{K}_Y], \qquad [\mathbf{K}_{YX}] = [\mathbf{K}_Y]\ [\mathbf{B}]^T[\mathbf{A}]^{-T}. \tag{7}$$

Having obtained probabilistic characteristics of unknown random values of displacements and tractions, together with the specified values of tractions and displacements, the interior values of displacements and stresses can be calculated[6,10,11].

3 Application of SBEM to Dynamic Problems

Two different boundary element approaches to dynamic analysis can be distinguished, namely the time domain approach and the integral transform (Laplace or Fourier) domain approach. Both can be used for stochastic dynamic analysis, but it seems that the application of the Fourier integral transform domain approach offers a special convenience consisting in the possibility of employing spectral densities in the description of spatially-temporal stochastic boundary fields of displacements $\mathbf{u}(\mathbf{x},t,\gamma)$ and tractions $\mathbf{p}(\mathbf{x},t,\gamma)$. This spectral approach was used in boundary element analyses of stochastic vibration of elastic and visco-elastic systems by Burczyński[3,6], Burczyński and John[13,14,15]. An alternative time domain approach to stochastic vibration of elastic systems was also proposed by Burczyński and John[14].

The stochastic boundary integral equation for dynamic problems in the Fourier transform domain has the form:

$$\mathbf{c}(\mathbf{x})\bar{\mathbf{u}}(\mathbf{x},\omega,\gamma)=\int_{\Gamma}\bar{\mathbf{U}}(\mathbf{x},\mathbf{y},\omega)\bar{\mathbf{p}}(\mathbf{y},\omega,\gamma)d\Gamma(\mathbf{y})-\int_{\Gamma}\bar{\mathbf{P}}(\mathbf{x},\mathbf{y},\omega)\bar{\mathbf{u}}(\mathbf{y},\omega,\gamma)d\Gamma(\mathbf{y}), \tag{8}$$

where $\bar{\mathbf{u}}(\mathbf{x},\omega,\gamma)$ and $\bar{\mathbf{p}}(\mathbf{x},\omega,\gamma)$ are the Fourier transforms of stochastic spatially-temporal fields of displacements and tractions, respectively:

$$\bar{\mathbf{r}}(\mathbf{x},\omega,\gamma) = \int_{-\infty}^{\infty}\mathbf{r}(\mathbf{x},t,\gamma)\exp(-i\omega t)dt, \qquad \mathbf{r}=\mathbf{u},\mathbf{p} \tag{9}$$

and $\bar{\mathbf{U}}(\mathbf{x},\mathbf{y},\omega)$ and $\bar{\mathbf{P}}(\mathbf{x},\mathbf{y},\omega)$ are the Fourier transforms of deterministic fundamental solutions $\mathbf{U}(\mathbf{x},\mathbf{y},t)$ and $\mathbf{P}(\mathbf{x},\mathbf{y},t)$, respectively.

The discrete version of equation (8) for $\omega=(\omega^l)$, $l=1,2,..L$, takes the form:

$$[\bar{\mathbf{A}}(\omega)]\ \{\bar{\mathbf{X}}(\omega,\gamma)\} = [\bar{\mathbf{B}}(\omega)]\ \{\bar{\mathbf{Y}}(\omega,\gamma)\}, \tag{10}$$

where $[\bar{\mathbf{A}}(\omega)]$ and $[\bar{\mathbf{B}}(\omega)]$ are deterministic complex square matrices dependent on boundary integrals of Fourier transforms of fundamental solutions $\bar{\mathbf{P}}$ and $\bar{\mathbf{U}}$, $\{\bar{\mathbf{X}}(\omega,\gamma)\}$ is a column matrix of unknown Fourier transforms of random nodal values of boundary displacements and tractions, and $\{\bar{\mathbf{Y}}(\omega,\gamma)\}$ is a column matrix of Fourier transforms of random boundary displacements and tractions prescribed by boundary conditions.

If $\det|\bar{\mathbf{A}}(\omega_k)|\neq 0$ for each $\omega_k=(\omega_k^l)$, $k=1,2$; $l=1,2,..L$, then the spectral density of the solution to equation (10) has the form:

$$[\mathbf{S}_X(\omega_1,\omega_2)] = [\bar{\mathbf{A}}(\omega_1)]^{-1}[\bar{\mathbf{B}}(\omega_1)][\mathbf{S}_Y(\omega_1,\omega_2)][\bar{\mathbf{B}}(\omega_2)]^{H}[\bar{\mathbf{A}}(\omega_2)]^{-H} . \tag{11}$$

Cross-spectral densities are expressed as follows:

$$[\mathbf{S}_{XY}(\omega_1,\omega_2)] = [\bar{\mathbf{A}}(\omega_1)]^{-1}[\bar{\mathbf{B}}(\omega_1)][\mathbf{S}_Y(\omega_1,\omega_2)], \tag{12}$$

$$[\mathbf{S}_{YX}(\omega_1,\omega_2)] = [\mathbf{S}_Y(\omega_1,\omega_2)][\bar{\mathbf{B}}(\omega_2)]^H[\bar{\mathbf{A}}(\omega_2)]^{-H}, \tag{13}$$

for $\omega_k=(\omega_k^l)$, $k=1,2$, $l=1,2,..L$. The $[\cdot]^H$ denotes conjugation and transposition of the matrix $[\cdot]$.

4 Application of SBEM to Random Media

The stochastic nature of a material results mainly from the nonhomogeneity and indeterminacy of the media structure. The complexity and irregularity of the properties of real media leads to a stochastic description of these media[30].

It is difficult to obtain fundamental solutions for stochastic media because of the complexity of their mathematical formulations. To solve this problem, perturbation techniques can be used[4,20,21,25]. However, this approach requires the assumption that random fluctuations of stochastic properties of a medium are small.

It is possible to apply another approach[10] which does not require such an assumption. This approach consists in using the isotropic deterministic fundamental solutions corresponding to a reference elastic model $\mathbf{C}^o$, whose properties may be found by averaging the stochastic medium:

$$\mathbf{C}(\mathbf{x},\gamma) = \mathbf{C}^o + \bar{\mathbf{C}}(\mathbf{x},\gamma), \qquad \mathbf{x}\in\Omega, \tag{14}$$

where $\mathbf{C}^o=\mathrm{E}[\mathbf{C}(\mathbf{x},\gamma)]$ is the mean value of the elastic moduli tensor, and $\bar{\mathbf{C}}(\mathbf{x},\gamma)$ is a random field characterizing the medium fluctuations with mean value $\mathrm{E}[\bar{\mathbf{C}}(\mathbf{x},\gamma)]=0$ and the correlation moment $\mathbf{K}_{\bar{C}}(\mathbf{x}_1,\mathbf{x}_2)=[\mathrm{K}_{kjmnprst}(\mathbf{x}_1,\mathbf{x}_2)]$. $\mathrm{E}[\cdot]$ indicates expectation. In the general case for three-dimensional problems, $\mathbf{C}$ is a 6x6 matrix and the number of different elastic constants in $\bar{\mathbf{C}}(\mathbf{x},\gamma)$ is 21.

As a result the following vector integral equation is obtained:

$$\mathbf{c}(\mathbf{x})\mathbf{u}(\mathbf{x},\gamma)= \int_\Gamma \mathbf{U}(\mathbf{x},\mathbf{y})\mathbf{p}(\mathbf{y})d\Gamma-\int_\Gamma \mathbf{P}(\mathbf{x},\mathbf{y})\mathbf{u}(\mathbf{y})d\Gamma-\int_\Omega \mathbf{R}(\mathbf{x},\mathbf{y})\bar{\sigma}(\mathbf{y},\gamma)d\Omega(\mathbf{y}), \tag{15}$$

where $\mathbf{U}$, $\mathbf{P}$ and $\mathbf{R}$ are deterministic fundamental solutions for the medium with material constants $\mathbf{C}^o$, and

$$\bar{\sigma}(\mathbf{y},\gamma) = \bar{\mathbf{C}}(\mathbf{y},\gamma)\varepsilon(\mathbf{y}), \qquad \mathbf{y}\in\Omega. \tag{16}$$

Equation (15) is similar to deterministic problems with only the addition of a new stochastic domain term which depends on the elastic moduli tensor $\bar{\mathbf{C}}(\mathbf{x},\gamma)=\bar{\mathrm{C}}_{kjmn}(\mathbf{x},\gamma)$ characterizing the random fluctuations of the medium. This term can be integrated over the domain using internal cells.

The problem is solved iteratively by finding the first solution with $\bar{\sigma}_{kj}(y,\gamma)=0$ and then resolving equation (15) as many times as required.

5 Application of SBEM to Shape Design Sensitivity Analysis

One knows that boundaries bounding real bodies are very complicated as far as their geometrical shape is concerned. Usually they are uneven, and the irregularities do not easily lead to a unique deterministic description. Therefore, boundaries of such bodies can be defined stochastically. SBEM is a very useful and natural technique for modelling such problems. In order to solve these problems it is possible to use an approach based on the idea of *stochastic shape design sensitivity analysis*[5,7,9,10,11,12]. The application of this approach to examining stresses, strains, displacements, natural frequencies and, in the general case, an arbitrary functional with respect to the stochastic shape of the boundary is presented. For the simplicity of further considerations, boundary conditions have been assumed deterministic.

One assumes that the stochastic shape of the boundary $\overset{*}{\Gamma}$ may be defined by prescribing a stochastic vector field $\delta\mathbf{g}(\mathbf{x},\gamma)=(\delta g_k(\mathbf{x},\gamma))$, so that:

$$\overset{*}{\mathbf{x}}(\gamma) = \mathbf{x} + \delta\mathbf{g}(\mathbf{x},\gamma), \quad \mathbf{E}[\delta\mathbf{g}]=0, \tag{17}$$

where the deterministic variable $\mathbf{x}$ is related to the baseline of the boundary Γ.

The stochastic transformation field $\mathbf{g}(\mathbf{x},\mathbf{a}(\gamma))$ modifies the external boundary Γ, where $\mathbf{a}(\gamma)=(a_r(\gamma))$, $r=1,2,..R$, is a set of random shape parameters which specifies the actual stochastic tolerance range of the structure.

The variation of the transformation field $\delta\mathbf{g}$ is expressed as

$$\delta g_k = \overset{r}{v}_k \delta a_r(\gamma), \tag{18}$$

where $\overset{r}{v}_k=\partial g_k/\partial a_r$ can be considered as a velocity transformation field which is associated with a shape design parameter $a_r(\gamma)$. One assumes that random shape parameters can be expressed as follows:

$$\mathbf{a}(\gamma) = \mathbf{a}_o + \delta\mathbf{a}(\gamma), \quad \mathbf{E}[\delta\mathbf{a}(\gamma)]=0, \tag{19}$$

where variation $\delta\mathbf{a}(\gamma)$ represents the fluctuation of random parameters, and $\mathbf{a}_o=\mathbf{E}[\mathbf{a}(\gamma)]$ is the mean value of shape parameters $\mathbf{a}(\gamma)$.

If $\mathbf{a}(\gamma)$ has the Gaussian distribution, then it is completely described by the mean value $\mathbf{a}_o$ and the covariance matrix $[\mathbf{K}]=\mathbf{E}[\delta\mathbf{a}\delta\mathbf{a}^T]$.

Due to small random variations of the boundary, the resulting stochastic fields of stresses $\sigma(\mathbf{x},\gamma)$, strains $\varepsilon(\mathbf{x},\gamma)$, displacements $\mathbf{u}(\mathbf{x},\gamma)$ and natural frequency $\omega(\gamma)$ can be expressed as follows:

$$\sigma(\mathbf{x},\gamma) = \sigma_0(\mathbf{x}) + \delta\sigma(\mathbf{x},\gamma), \quad E[\delta\sigma(\mathbf{x},\gamma)]=0, \quad \mathbf{x}\in\Omega, \tag{20}$$

$$\varepsilon(\mathbf{x},\gamma) = \varepsilon_0(\mathbf{x}) + \delta\varepsilon(\mathbf{x},\gamma), \quad E[\delta\varepsilon(\mathbf{x},\gamma)]=0, \quad \mathbf{x}\in\Omega, \tag{21}$$

$$\mathbf{u}(\mathbf{x},\gamma) = \mathbf{u}_0(\mathbf{x}) + \delta\mathbf{u}(\mathbf{x},\gamma), \quad E[\delta\mathbf{u}(\mathbf{x},\gamma)]=0, \quad \mathbf{x}\in\Omega \text{ or } \mathbf{x}\in\Gamma, \tag{22}$$

$$\omega(\gamma) = \omega_0 + \delta\omega(\gamma), \quad E[\delta\omega(\gamma)]=0, \tag{23}$$

where $\mathbf{q}_0=E[\mathbf{q}(\mathbf{x},\gamma)]$, $\mathbf{q}=\sigma,\varepsilon,\mathbf{u}$ and $\omega_0=E[\omega(\gamma)]$ are identified with the mean value of state fields calculated for the untransformed boundary Γ with the deterministic base shape parameters $\mathbf{a}_0=(a_{0r})$, $r=1,2,..R$.

In the general case it is possible to consider an arbitrary functional

$$J= \int_{\Omega(\mathbf{a})}\Psi(\sigma,\varepsilon,\mathbf{u})d\Omega + \int_{\Gamma(\mathbf{a})}\varphi(\mathbf{u},\mathbf{p})d\Gamma, \tag{24}$$

where $\Psi(\sigma,\varepsilon,\mathbf{u})$ is an arbitrary function of stresses σ, strains ε and displacements $\mathbf{u}$ within the domain $\Omega^*=\Omega(\mathbf{a})$, and $\varphi(\mathbf{u},\mathbf{p})$ is an arbitrary function of displacements $\mathbf{u}$ and tractions $\mathbf{p}$ on the boundary $\Gamma^*=\Gamma(\mathbf{a})$.

The functional J can express global mechanical characteristics (e.g. potential or complementary energy) as well as local stresses, strains or displacements[9].

Due to stochastic shape variation the functional J can be expressed as follows:

$$J(\mathbf{a}_0 + \delta\mathbf{a}(\gamma)) = J(\mathbf{a}_0) + \delta J(\gamma), \tag{25}$$

where the first variation of the functional $\delta J(\gamma)$ can be expressed analytically utilizing the adjoint approach.

Independently of the primary system, a concept of an *adjoint system* with the adjoint solution $\mathbf{u}^a$, ε^a and σ^a is introduced.
The adjoint system is an elastic body with identical configuration and physical properties as the primary system but with other boundary conditions and body forces. On the boundary Γ there are prescribed boundary conditions in the form:

$$\mathbf{u}^{ao} = -\frac{\partial\phi(\mathbf{u},\mathbf{p})}{\partial\mathbf{p}} \quad \text{on } \Gamma_1, \qquad \mathbf{p}^{ao} = \frac{\partial\phi(\mathbf{u},\mathbf{p})}{\partial\mathbf{u}} \quad \text{on } \Gamma_2, \tag{26}$$

and within the domain Ω initial strain ε^{ai}, and stress σ^{ai} fields and body forces $\mathbf{b}^a$ are specified:

$$\overset{ai}{\varepsilon} = \frac{\partial \Psi\ (\sigma,\varepsilon,u)}{\partial \sigma}, \quad \overset{ai}{\sigma} = \frac{\partial \Psi\ (\sigma,\varepsilon,u)}{\partial \varepsilon}, \quad \overset{a}{b} = \frac{\partial \Psi\ (\sigma,\varepsilon,u)}{\partial u} \tag{27}$$

The constitutive law for the adjoint system has the form:

$$\overset{a}{\sigma} = C(\overset{a}{\varepsilon} - \overset{ai}{\varepsilon}) - \overset{ai}{\sigma}, \tag{28}$$

where $C\overset{a}{\varepsilon}=c_{ijkl}\varepsilon_{kl}$, $c_{ijkl}=\lambda\delta_{ij}\delta_{kl}+ \mu(\delta_{ik}\delta_{jl}+ \delta_{il}\delta_{jk})$, λ and μ are the Lame constants.

The first variation of the functional J can be expressed as:

$$\delta J(\gamma) = \{S\}^T\{\delta a(\gamma)\}, \tag{29}$$

where

$$\{\delta a(\gamma)\} = [\delta a_1(\gamma),\delta a_2(\gamma),..,\delta a_r(\gamma),..\delta a_R(\gamma)]^T \tag{30}$$

is the matrix of variation of random shape parameters, and

$$\{S\} = \left[S_1,S_2,..,S_r,..S_R\right]^T \tag{31}$$

is the sensitivity matrix whose elements S_r, $r=1,2,..R$, are expressed by the total material derivative of the functional J with respect to the shape parameters, i.e. $S_r=DJ/Da_r$.

The total material derivatives of J have the form[26,9,11,12]:

$$S_r = \int_\Gamma \left[\Psi - \sigma\cdot\overset{a}{\varepsilon} + b\cdot\overset{a}{u} + (\phi + p\cdot\overset{a}{u})_{,n} - 2(\phi + p\cdot\overset{a}{u})\mathcal{K}\right] n_k \overset{r}{v}_k d\Gamma$$

$$+\int_{\Gamma_1}\left(\frac{\partial\phi}{\partial u} - \overset{a}{p}\right)\left(\frac{D\overset{o}{u}}{Da_r} - \overset{o}{u}_{,k}\overset{r}{v}_k\right)d\Gamma_1 + \int_{\Gamma_2}\left(\frac{\partial\phi}{\partial p} + \overset{a}{u}\right)\left(\frac{D\overset{o}{p}}{Da_r} - \overset{o}{p}_{,k}\overset{r}{v}_k\right)d\Gamma_2 \tag{32}$$

$$+ \int_L [\![\phi + p\cdot\overset{a}{u}]\!]\overset{r}{v}_\upsilon dL,$$

where integrand $[\![\phi+p\cdot\overset{a}{u}]\!]=\left(\phi+p\cdot\overset{a}{u}\right)^+ - \left(\phi+p\cdot\overset{a}{u}\right)^-$ represents the discontinuity of $(\phi+p\cdot\overset{a}{u})$ along the curve L, which separates two parts of the boundary Γ_1 and Γ_2, $n=(n_k)$ is the unit normal vector, $\mathcal{K}$ is the mean curvature of the boundary.

It is seen from (32) that sensitivities of J depend only on boundary state variables of the primary system and the adjoint system. This fact gives significant advantages in numerical calculations by means of the boundary element method.

Consider now a special kind of shape transformation in the form of a translation, a rotation and a scale change (expansion or contractions). This case is especially interesting when a body contains inhomogeneity in the form of internal defects such as a crack, a cavity or an inclusion. Then the dependence of the functional J on the stochastic shape and location of such internal defects can be examined.

One assumes that the stochastic shape transformation field $g(x,a(\gamma))$ modifies a given initial shape of the internal defect by

(1) a stochastic translation (T) by prescribing $\delta g_k(x,\gamma)=\delta b_k(\gamma)$, $k=1,2,3$, where $b_k(\gamma)$ are random translation parameters,

(2) a stochastic rotation (R) by prescribing $\delta g_k(x,\gamma)=x_l e_{kpl}\delta\omega_p(\gamma)$, $p=1,2,3$, where $\omega_p(\gamma)$ are random rotation parameters, e_{kpl} denotes the permutation tensor, and

(3) a stochastic scale change (expansion or contraction) (S) of the crack by prescribing $\delta g_k(x,\gamma)=x_k\delta\eta(\gamma)$, where $\eta(\gamma)$ is a random scale change parameter.

Now the total material derivatives DJ/Da_r, where $a(\gamma)=(a_r(\gamma))\equiv(b_1,b_2,b_3,\omega_1,\omega_2,\omega_3\eta)$, take the form of the path-independent integrals:

$$S_r \equiv \frac{DJ}{Da_r} = \int_{\Gamma_*} Z_L^r(\sigma,\varepsilon,u,\overset{a}{\sigma},\overset{a}{\varepsilon},\overset{a}{u})\,d\Gamma_*, \tag{33}$$

$$r=1,2,3,4,5,6,7; \quad L=T,R,S$$

where

$$Z_T^k = (\Psi\delta_{kj} + \sigma_{ij}\overset{a}{u}_{i,k} + \overset{a}{\sigma}_{ij}u_{i,k} - \sigma_{il}\overset{a}{\varepsilon}_{il}\delta_{kj})n_j, \tag{34}$$

for translation ($k=1,2,3$),

$$Z_R^{p+3} = e_{kpl}(\Psi x_l\delta_{kj} - \sigma_{iq}\overset{a}{\varepsilon}_{iq}x_l\delta_{kj} + \sigma_{ij}\overset{a}{u}_{i,k}x_l + \sigma_{lj}\overset{a}{u}_k + \overset{a}{\sigma}_{lj}u_k + \overset{a}{\sigma}_{ij}u_{i,k}x_l)n_j, \tag{35}$$

for rotation ($p=1,2,3$), and

$$Z_S^7 = \left[\frac{2}{\alpha}\,\overset{a}{\sigma}_{ij}u_i - \left(\frac{2}{\alpha} - \beta\right)\sigma_{ij}\overset{a}{u}_i + x_k\overset{a}{\sigma}_{ij}u_{i,k} - x_k\sigma_{ij}\overset{a}{u}_{i,j}\delta_{jk} + x_k\sigma_{ij}\overset{a}{u}_{i,k}\right]n_j, \tag{36}$$

for expansion or contraction.

In the last case, the form of Z_S^7 is valid when $\Psi=0$ and $\varphi=\varphi(u)$ is a homogeneous function of order α and $\beta=1$ for 3-D and $\beta=0$ for 2-D. Γ_* is an arbitrary closed surface (or contour for 2-D) enclosing the defect. Primary $(u_i,\varepsilon_{ij},\sigma_{ij})$ and adjoint $(\overset{a}{u}_i,\overset{a}{\varepsilon}_{ij},\overset{a}{\sigma}_{ij})$ solutions along Γ_* can be obtained using boundary element procedures. Enclosing the defect by a surface Γ_*, derivatives of J can be determined by calculating the respective path-independent integrals along any surface Γ_*. In particular, the surface Γ_* can be identified with the external boundary surface Γ.

Internal defects can introduce local gradients of displacements or stresses which reach great or infinite values as compared to respective values within the structure domain. Calculations of derivatives of J by means of path-independent integrals along fixed contours far from such singularities (e.g. cracks) ensure good accuracy of these derivatives[16].

The variation of natural frequency can be expressed by:

$$\delta\omega(\gamma) = \{S^{\omega}\}^{T}\{\delta \mathbf{a}(\gamma)\}, \tag{37}$$

where elements of the sensitivity matrix $\{S^{\omega}\}$ are evaluated by the total material derivative of natural frequencies with respect to shape parameters[5,7,9,10,11,12]:

$$S_r^{\omega} = \frac{D\omega}{D a_r} = \frac{1}{2\omega_o}\int_{\Gamma} [\sigma(\mathbf{u})\cdot\varepsilon(\mathbf{u}) - \omega_o^2\,\rho\;\mathbf{u}\cdot\mathbf{u}]n_k v_k^r d\Gamma, \tag{38}$$

where ρ is the mass density and $\mathbf{u}(\mathbf{x})$ is the displacement amplitude.

It is interesting to notice that $\delta\omega(\gamma)$ is expressed by the boundary integral and depends on the natural frequency ω_o, the mode $\mathbf{u}(\mathbf{x})$ and the stochastic fluctuation of shape parameters $\delta\mathbf{a}(\gamma)$. This is very important in numerical calculations using boundary elements.

The mean value ω_o can be evaluated using the dual reciprocity approach for the free vibration problem. After discretization of the boundary Γ by means of boundary elements the following equation is obtained:

$$[\mathbf{H}]\;\{\mathbf{u}\} = \omega_o^2\;[\mathbf{M}]\{\mathbf{u}\}, \tag{39}$$

where $\{\mathbf{u}\}$ is the column matrix of nodal values of displacement amplitude, $[\mathbf{M}]$ is the mass matrix.

The covariance of stochastic stresses, strains and displacements for the two fixed points $\mathbf{x}_1$ and $\mathbf{x}_2$ is calculated from:

$$K_q(\mathbf{x}_1,\mathbf{x}_2) = \mathbf{E}\Big\{[q(\mathbf{x}_1,\gamma) - q_o(\mathbf{x}_1)][q(\mathbf{x}_2,\gamma) - q_o(\mathbf{x}_2)]\Big\} =$$

$$= \mathbf{E}[\delta q(\mathbf{x}_1,\gamma)\delta q(\mathbf{x}_2,\gamma)] = \{\mathbf{S}(\mathbf{x}_1)\}\;[\mathbf{K}]\;\{\mathbf{S}(\mathbf{x}_2)\}^{T} \quad . \tag{40}$$

$$q = \sigma_{qs},\ \varepsilon_{qs},\ u_q$$

The variance of the random natural frequency can be calculated using the equation:

$$\mathrm{Var}(\omega) = \mathbf{E}[\omega(\gamma) - \omega_o]^2 = \mathbf{E}[\delta\omega(\gamma)]^2 = \{\mathbf{S}^{\omega}\}\;[\mathbf{K}]\;\{\mathbf{S}^{\omega}\}^{T}. \tag{41}$$

In order to calculate the sensitivity matrices $\{\mathbf{S}\}$ or $\{\mathbf{S}^{\omega}\}$ one should determine the transformation velocity field v_k^r which is associated with the shape parameter $a_r(\gamma)$. The selection of shape design parameters is the key element in the shape sensitivity analysis and optimal design.

6 Numerical Examples

Example 1

The chain link (Fig.1a) made up of the visco-elastic material is loaded by the slowly changing, spatially-temporal, stationary stochastic field of tractions $\mathbf{p}(\mathbf{x},t,\gamma)$ whose spectral density is given by

$$S_p(\mathbf{x}_1,\mathbf{x}_2,\omega) = 2\,\delta(\mathbf{x}_1-\mathbf{x}_2)\,\sigma_p^2\,\alpha(\beta^2+\alpha^2)/[\pi(\omega^2-\beta^2-\alpha^2)^2+4\alpha^2\omega^2]$$

where $\sigma_p^2 = 289[N^2]$, $\alpha=0.15[s^{-1}]$, $\beta=7[s^{-1}]$.

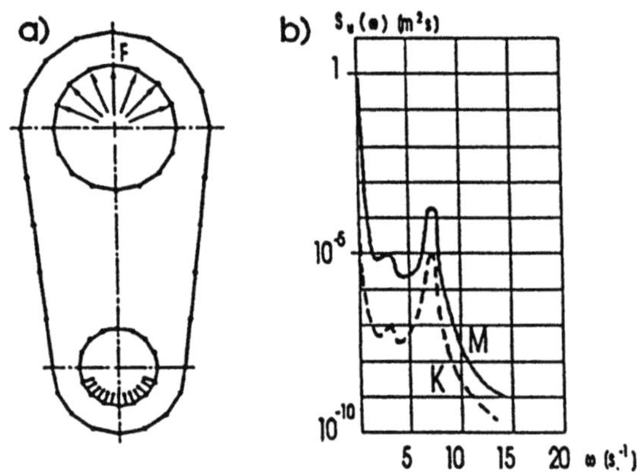

Fig.1

Numerical results of spectral densities of vertical displacements of the point F for the Maxwell (M) and Kelvin (K) models are presented in Fig.1b.

Example 2

The problem of stochastic shape sensitivity analysis for a vibrating steel arch (Fig.2a) is considered. The boundary element model consists of 38 linear elements. It is assumed that the boundary undergoes small random variation measured along the radius with the covariance $K(\mathbf{x},\mathbf{y})=\sigma^2\exp[(-|x_1-x_2|-|y_1-y_2|)/a]$, where $\sigma^2=15.10^{-6}[m^2]$. The mean values of the first, second and third natural frequencies are $\omega_{01}=3034[s^{-1}]$, $\omega_{02}=4788[s^{-1}]$ and $\omega_{03}=7743[s^{-1}]$, respectively.

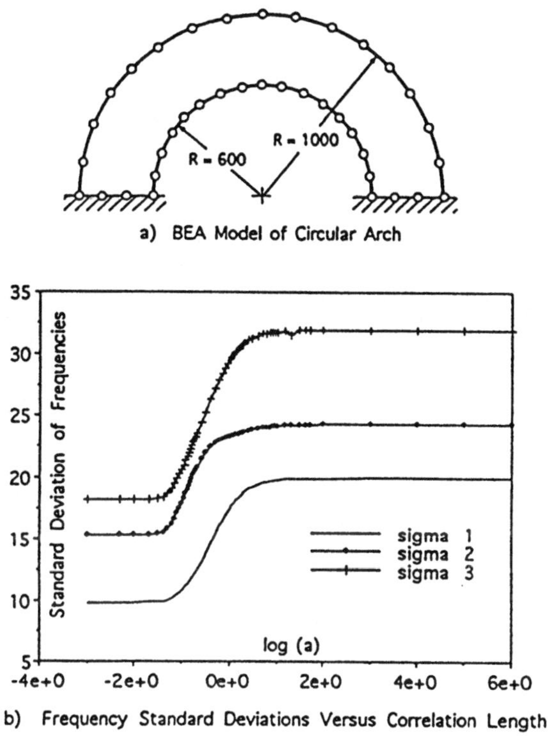

Fig.2

Fig.2b shows the dependence of standard deviations of natural frequencies as a function of the radius of correlation a [m].

7 Conclusions

This paper was concerned with the computational methodology and applications of the stochastic boundary element method to continuous static and dynamic systems with boundary conditions, random media, and stochastic shape design sensitivity analysis.

Stochastic problems are more computationally time-consuming and burdensome than deterministic ones. The SBEM formulation, which reduces the size of the problem by one and requires only solutions for stochastic boundary variables, appears promising in various probabilistic structural mechanics problems and belongs to the rapidly advancing fields of computational mechanics.

Acknowledgements

The author wishes to acknowledge the support of the Polish Committee of Scientific Research KBN under grant No PB 1057/P4/92/02.

References

1. Burczyński,T., Stochastic boundary value problems of elastostatics in terms of the boundary integral equation method. *Papers of Inst. Civil Engng.*, Wrocław Technical University, Vol. 1, No.28, Wrocław, 1981, pp.61-68, (in Polish).
2. Burczyński,T., The boundary element method for stochastic potential problems. *Applied Math. Modelling*, **9** (1985) 189-194.
3. Burczyński,T., Stochastic formulation of wheelset dynamics by means of the boundary element method. *Proc. 8th International Wheelset Congress*, Madrid, Vol.1, 1985, pp.3/1-3/15.
4. Burczyński,T., Stochastic boundary element method for wave propagation through continuous random media. *BETECH'86-MIT*, ed. J.J.Connor & C.A.Brebbia, CMP, Southampton, 1986, pp.733-741.
5. Burczyński,T., The boundary element procedure for dependence of eigenvalues with respect to stochastic shape of elastic systems. *Proc. 25th Symp. on Modelling in Mechanics - PTMTS*, Vol.2, Gliwice-Kudowa, 1986, 235-238.
6. Burczyński,T., The boundary element method in stochastic problems of elasticity. *Mechanics & Computer*, **7** (1988) 125-147 (in Polish)
7. Burczyński,T., Boundary element method for deterministic and stochastic shape design sensitivity analysis. *Advanced Boundary Element Methods*, ed. T.A.Cruse, Springer-Verlag, Berlin, 1988, pp.73-80.
8. Burczyński,T., The Boundary Element Method for Stochastic Analysis Problems. Chapter in *The Boundary Element Method for Selected Analysis and Optimization Problems of Deformable Bodies*, T.Burczyński, Silesian Tech. Univ. Publications, Gliwice, 1989, pp.94-124 (in Polish).
9. Burczyński,T., Shape sensitivity analysis of uncertain static and vibrating systems using stochastic boundary elements. *Boundary Element Methods - Fundamentals and Applications*, ed. S.Kobayashi & N.Nishimura, Springer-Verlag, Berlin, 1992, pp.49-58.
10. Burczyński,T., Stochastic Boundary Element Methods. Chapter in *Advances in Boundary Element Techniques*, ed. J.H.Kane, G.Maier, N.Tosaka & S.N.Atluri, Springer-Verlag, Berlin, 1993, pp.30-54.
11. Burczyński,T., Stochastic Boundary Elements to Analysis, Shape Design Sensitivity and Identification Problems. Chapter in *Computational Stochastic Mechanics*, ed. A.H-D.Cheng & C.Y.Yang, CMP, Southampton, Elsevier Applied Science, London, 1993, pp.569-593.
12. Burczyński,T., Stochastic Boundary Element Approach to Shape Design Sensitivity Analysis. Chapter in *Design Sensitivity Analysis*, ed. T.Hisada & M.Kleiber, Atlanta Technology Publications, Atlanta, 1993, pp.68-92.
13. Burczyński,T. & John,A., Stochastic dynamic analysis of elastic and viscoelastic systems by means of the boundary element method. *Boundary Elements VII*, ed. C.A.Brebbia & G.Maier, Springer-Verlag, Berlin, Vol.I, 1985, pp.6/53-6/61.
14. Burczyński,T. & John,A., The stochastic boundary integral equation method for random vibrations of continuous media. *Opole Technical University Publications*, Opole, No.109, 1985, pp.61-74.

15. Burczyński,T. & John,A., Application of the stochastic boundary element method to analysis of elastic and viscoelastic structures. *Proc. X Polish Conf. Computer Methods in Mechanics*, Świnoujście, 1991, pp.47-54.
16. Burczyński,T. & Polch,E.Z., Sensitivity analysis of cracks using boundary and path-independent integrals. *Proc. International Symposium on Boundary Element Methods*, Technical University of Braunschweig, Germany, 16-19 August, 1993.
17. Cheng,A.H-D. & Lafe,O.E., Boundary element solution for stochastic groundwater flow: random boundary condition and recharge. *Water Resources*, **27** (1991) 231-242.
18. Dems,K. & Mróz,Z., On a class of conservation rules associated with sensitivity analysis in linear elasticity. *Int. J. Solids Structures*, **22** (1986) 737-758.
19. Drewniak,J., Boundary elements for random heat conduction problems. *Engineering Analysis*, **2** (1985) 168-170.
20. Ettouney,M., Benaroya,H. & Wright,J., Probabilistic Boundary Element Methods. Chapter in *Computational Mechanics of Probabilistic and Reliability Analysis*, ed. W.K.Liu & T.Belytschko, Elmepress Int., Lausanne, 1989, pp.142-165.
21. Ettouney,M.M., Daddazio,R.P. & Abboud,N.N., Probabilistic Boundary Element Method in Soil Dynamics. Chapter in *Computational Stochastic Mechanics*, ed. A.H-D.Cheng & C.Y.Yang, CMP, Southampton, Elsevier Applied Science, London, 1993, pp.623-641.
22. Ghanem,R.G. & Spanos,P.D., *Stochastic Finite Elements: A Spectral Approach*. Springer-Verlag, Berlin, 1991.
23. Lafe,O. & Cheng,A.H-D., Stochastic Indirect Boundary Element Method. Chapter in *Computational Stochastic Mechanics*, ed. A.H-D.Cheng & C.Y.Yang, CMP, Southampton, Elsevier Applied Science, London, 1993, pp.301-319.
24. Kleiber,M. & Hien,T.D., *The Stochastic Finite Element Method*. Wiley, Chichester, 1992.
25. Manolis,G.D., Acoustic and Seismic Wave Propagation in Random Media. Chapter in *Computational Stochastic Mechanics*, ed. A.H-D.Cheng & C.Y.Yang, CMP, Southampton, Elsevier Applied Science, London, 1993, pp.585-622.
26. Mróz,Z., Variational Approach to Shape Sensitivity Analysis and Optimal Design. Chapter in *The Optimum Shape*, ed. J.SA.Bennet & M.E.Botkin, Plenum Press, New York, 1986, pp.79-110.
27. Nakagiri,S., Suzuki,K. & Hisada,T., Stochastic boundary element method applied to stress analysis. *Boundary Elements V*, ed. C.A.Brebbia, T.Futugami & M.Tanaka, Springer-Verlag, Berlin, 1983, pp.439-448.
28. Nakagiri,S. & Suzuki,K., Indeterminate shape modification for stress reduction based on boundary element sensitivity analysis. *Computer Aided Optimum Design of Structures: Recent Advances*, ed. C.A.Brebbia & S.Hernandez, CMP, Southampton, Springer-Verlag, Berlin, 1989, pp.85-94.
29. Sobczyk,K., *Stochastic Wave Propagation*. PWN, Warsaw, Elsevier, Amsterdam, 1984.
30. Spanos,P.D. & Ghanem,R., Boundary element formulation for random vibration problems. *J. Engng. Mech. ASCE*, **117** (1991) 409-423.

Hysteretic Systems Subjected to Delta Correlated Input

S. Caddemi and M. Di Paola

Dipartimento di Ingegneria Strutturale e Geotecnica
Università degli Studi di Palermo
Viale delle Scienze, I–90128 Palermo, Italy

Abstract. The paper deals with the evaluation of the probabilistic response of a single degree of freedom elastic–perfectly plastic system subjected to a delta correlated input process. The probabilistic characterisation of the response is here obtained by considering the accumulated plastic deformations as a compound homogeneous Poisson process independent of the external input. In this case the former can be considered as an external noise acting on the linear system. A closed form solution is also obtained and the analytic expression is compared with the customary Monte–Carlo method.

Keywords. Stochastic dynamics, Elastic–perfectly plastic system, delta correlated process.

1 INTRODUCTION

The analysis of structural systems showing an hysteretic behaviour subjected to stochastic excitation has received considerable attention in view of several engineering applications. Several constitutive models have been studied in order to describe the elastic-plastic behaviour exhibited by real materials and formulated in a rate form able to be used for stochastic analysis[1–4]. The most diffused elastic-plastic constitutive model is the bilinear hysteretic model showing a sharp corner of the constitutive law between the elastic and plastic phases. Systems with a residual stiffness in the plastic phase, though subjected to accumulation of plastic deformations, are still able to develop increments of elastic deformations and, hence, show a hardening behaviour. On the other hand if the system possesses a zero post-yielding stiffness, it is called *elastic–perfectly plastic*, and beyond its elastic limit it is subjected only to accumulation of plastic deformations without any increment of the restoring force. Whatever elastic plastic constitutive model is adopted, the hysteretic behaviour requires the adoption of a hysteretic variable which retains the main features of the physical nonlinearity of the system and whose evolution is ruled by a first-order nonlinear differential equation.

Single degree of freedom hysteretic systems subjected to white noise or filtered stochastic excitations have been analysed by means of approximate methods due to their highly nonlinear behaviour. Several approaches have been developed so far in order to evaluate the statistics of the response of such systems but to the authors' knowledge all of them (apart from those based on the solution of the Fokker–Plank differential equation) can be framed within three main branches: the classical equivalent linearization approach, the non-Gaussian closure technique and the so-called yield increment approach.

Although the above classification may not be satisfactory, an attempt to interpret the main features of the large variety of methods proposed, on the base of the above-mentioned general approaches, will be provided in the following.

A solution procedure to solve nonlinear systems subjected to stochastic processes is the equivalent linearization method proposed by Caughey[5] and adopted by many researchers to treat different hysteretic constitutive models[4,6–8]. The equivalent linearization technique tends to underestimate the mean square displacement yet produces reasonably accurate results for velocity responses. It is furthermore expected that the equivalent linearization technique provides minor errors for light nonlinearity but produces totally inaccurate results for nearly elastic–perfectly plastic systems. In this case, in fact, the variance of the displacement increases linearly under a stationary excitation while the equivalent linearization technique leads to a stationary solution[8,9].

Another class of solution procedures[1,10] aimed at providing higher order statistics of the response is based on the formulation of a set of differential equations ruling the evolution of the stochastic moments of the response (non–Gaussian closure techniques). These procedures lead to a set of differential equations in terms of moments which is not closed, hence it requires a cumulant neglect closure technique to be solved.

An approach which is totally different from those previously mentioned is the so-called *yield increment approach* proposed by Karnopp and Scharton[11], and further developed by Paparizos and Iwan[9] and Grossmayer[12], to study elastic–perfectly plastic oscillators subjected to a stationary white noise. Karnopp and Scharton[11] thought of the response of an elastic–perfectly plastic oscillator as composed by two phases: the first represented by a process which is almost always linear obtained by removing the yielding time intervals, hence not showing accumulation of plastic deformations; the second composed by a set of all the plastic time intervals in which the energy loss and the accumulation of plastic deformations occur. On this base, by means of some additional hypothesis on the response process, the probability density function of the accumulated plastic deformations has been provided. Within the context of the yield increment approach, Paparizos and Iwan[9] studied the properties of the yield increment process by means of the theory of discrete Markov process models and stated the Brownian motion nature of the accumulated plastic deformations. They proved the yield increments to be exponentially distributed and proposed an analytical expression for the mean square value of the accumulated plastic deformations able to reproduce the linear divergent behaviour shown by digital simulations. By means of numerical simulations, Vanmarcke et al.[13] noted that the accumulated plastic deformation process can be considered as a homogeneous compound renewal process (marked renewal point process). If the exceedances are sufficiently rare, the homogeneous compound renewal process may even be taken as a homogeneous compound Poisson process. By means of such an assumption, Ditlevsen[14] focussed his attention in evaluating the distribution of the single jump of plastic movements.

The methods previously presented, based on the yield increment approach, aim at capturing the principal features of the accumulated plastic deformation process, but they seem quite limited by allowing only a second order stochastic analysis. For this reason, for an elastic perfectly plastic oscillator, an approach capable of evaluating the higher order statistics of the response process is proposed. The linear system, according to the pseudo–force method, is loaded by two processes. The first one is the external input process here considered as a normal, delta correlated noise, while the second is an internal load that is a compound homogeneous Poisson process which accounts for the nonlinearity of the system. At last, by means of the Itô stochastic calculus, the statistics of the response are easily evaluated and compared with those obtained by digital simulation.

2 ELASTIC PERFECTLY PLASTIC OSCILLATOR UNDER WHITE NOISE EXCITATION

The motion equation of a single degree of freedom oscillator subjected to a zero mean stationary white noise $W^0(t)$ can be written as follows

$$\ddot{X} + 2\xi\omega\dot{X} + F = W^0(t) \quad ; \quad X(0) = X_0 \quad , \quad \dot{X}(0) = \dot{X}_0 \quad , \tag{1}$$

where ω is the natural frequency, ξ is the damping coefficient, $X(t)$ is the displacement of the system, the upper dot means time derivative and F is the restoring force. In order to provide a mathematically well–defined expression of the restoring force for an oscillator showing an elastic–perfectly plastic behaviour, an hysteretic component $Y(t)$ is introduced as follows

$$F = \omega^2\, Y(t) \quad , \tag{2}$$

where the evolution of the hysteretic component $Y(t)$ is ruled by the following nonlinear, non–analytical differential equation

$$\dot{Y} = k(\dot{X}, Y)\, \dot{X} \quad , \tag{3}$$

where

$$k(\dot{X}, Y) = 1 - \left[1 - H(-\dot{X})\right]\, H(Y - Y_0) - \left[1 - H(\dot{X})\right]\, H(-Y - Y_0) \tag{4}$$

is given as a combination of Heaviside unit step functions $H(\cdot)$ ($H(X)=1$ for $X \geq 0$, $H(X)=0$ for $X<0$). Y_0 appearing in Eq.(4) is the elastic limit displacement.

The system represented by Eqs.(1)–(4) can be thought of as a rigid unit mass with a Jenkin's element and a viscous element also included, as depicted in Fig.1a. The Jenkin's element consists of a linear spring with stiffness ω^2 in series with a Coulomb slip element which has a maximum allowable force $F^*=\omega^2 Y_0$, and shows, hence, an elastic–perfectly plastic behaviour. The total displacement of the mass is $X(t)$ while the hysteretic component $Y(t)$ is the elongation of the linear spring of the Jenkin's element. When the elastic restoring force reaches the value F^*, the Coulomb slip element is active and causes the development of plastic deformations, while the hysteretic component, whose evolution is depicted in Fig.1b, cannot exceed the value Y_0. The difference between the total displacement $X(t)$ and the hysteretic component $Y(t)$ represents the process of ***accumulated plastic deformations*** (Drift) $C(t)=X(t)-Y(t)$ occurring at the Coulomb slip element.

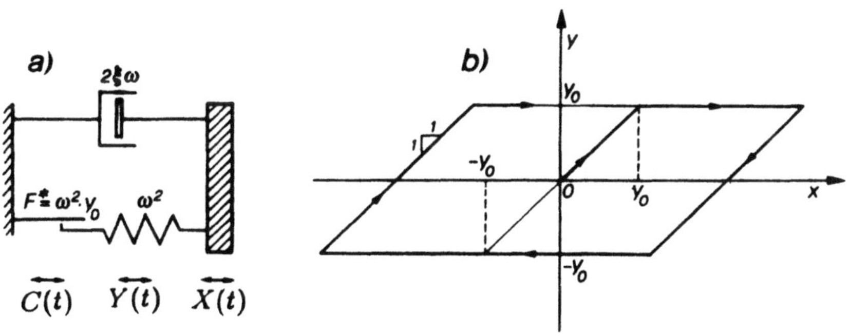

Fig.1. *Hysteretic model*

3 CHARACTERISATION OF THE ACCUMULATED PLASTIC DEFORMATION (DRIFT) PROCESS

Vanmarcke et al.[13], by means of numerical simulations, stated that the process $C(t)$ can be modelled as a homogeneous compound renewal process (marked renewal point process). However, if the exceedances of the elasticity limit are rare, then the drift process can be also considered as a homogeneous compound Poisson process independent of the white noise input process $W^0(t)$. In the latter case the derivetive of the plastic deformation (Drift) process $\dot{C}(t)=W(t)$ is a Poisson delta correlated one, that is, the $W(t)$ process can be written as follows

$$W(t) = \sum_{k=1}^{N(t)} R_k \delta(t - t_k) \quad , \tag{5}$$

where $N(t)$ is a counting process giving the number of impulses in the time interval $[0, t]$ with deterministic initial condition $N(0)=0$. The plastic deformation (yield) increments R_k are assumed to be a family of zero mean random variables mutually independent and independent of the random instants t_k. It follows that the Poisson process $W(t)$ is completely characterised by the distribution of R_k and by the expected rate $\lambda(t)$ of events (impulse occurrences), that is, the mean number of the random impulses in the infinitesimal interval $[t, t+dt]$ that will be denoted by $E[N(dt)]$ and given as

$$E\left[N(dt)\right] = \lambda(t)dt + O(dt) \quad . \tag{6}$$

The correlation function of $W(t)$ is given by dirac's delta product in the form

$$k_j\left[W(t_1)W(t_2)\ldots W(t_j)\right] = \tilde{q}_j\delta(t_2-t_1)\delta(t_3-t_1)\cdots\delta(t_j-t_1) \quad , \tag{7}$$

where the strenght $\tilde{q}_j$ is given as

$$\tilde{q}_j = \lambda(t)E\left[R^j\right] \quad . \tag{8}$$

From Eq.(8) we recognise that the only quantities we need for describing the process $W(t)$ are λ and the probability density function of R. Paparizos and Iwan[9] and Ditlevsen[14] dealt with the evaluation of this function stating that, with fairly good approximation, it can be approximated as an exponential function and can be written as

$$p_R(r) = \frac{\alpha}{2}\exp\left(-\alpha\,|r|\right) \quad , \tag{9}$$

where $\alpha>0$ is an unknown coefficient.

In view of Eq.(9) we obtain

$$E\left[R^{2j}\right] = \frac{2j}{\alpha^{2j}} \quad , \quad E\left[R^{2j+1}\right] = 0 \quad , \quad j = 1, 2, 3, \cdots \tag{10}$$

and from Eq.(10) the parameter α is given as

$$\alpha = \frac{\sqrt{2}}{\sigma_R} \quad , \tag{11}$$

where σ_R is the standard deviation of the random increments of the plastic deformations.

From Eqs.(8) and (11) we recognize that the only quantities of interest for the complete description of the drift process are σ_R and $\lambda(t)$. The determination of these quantities will be discussed later.

4 ANALYTICAL TREATMENT OF ELASTIC–PERFECTLY PLASTIC SYSTEMS

It has been pointed out in the previous section that the equation of motion of an elastic perfectly plastic oscillator subjected to a normal white noise process $\beta W^0(t)$ can be written in the form

$$\ddot{X} + 2\xi\omega\dot{X} + \omega^2 X = \beta W^0(t) + \omega^2 C(t) \quad , \tag{12}$$

where $\beta W^0(t)$ is the input process and $\omega^2 C(t)$ is the pseudo–force process that is the effect of the accumulated plastic deformations; the latter can be conveniently assumed as a compound homogeneous Poisson process. In order to evaluate the cumulants of the response, in view of the linearity of Eq.(1), we can split Eq.(1) into a couple of differential equations. $X_1(t)$ is the response of the linear oscillator subjected to βW^0 and $X_2(t)$ is the response to the compound homogeneous Poisson process $\omega^2 C(t)$

$$\ddot{X}_1 + 2\xi\omega\dot{X}_1 + \omega^2 X_1 = \beta W^0(t) \quad . \quad \ddot{X}_2 + 2\xi\omega\dot{X}_2 + \omega^2 X_2 = \omega^2 C(t) \quad . \tag{13}$$

The total response $X(t)$ will be the sum of $X_1(t)$ and $X_2(t)$, that is,

$$X(t) = X_1(t) + X_2(t) \quad . \tag{14}$$

The differential equation (12) can be transformed into a new set of first order differential equations having complex coefficients in the form

$$\dot{Z} = \nu Z + \rho\left(\beta W^0 + \omega^2 C(t)\right) \quad , \tag{15a}$$

$$\dot{Z}^* = \nu^* Z^* + \rho^* \left(\beta W^0 + \omega^2 C(t)\right) \quad , \tag{15b}$$

where the star means complex conjugate, and ν,ρ are given as

$$\nu = -\xi\omega + i\bar{\omega} \quad ; \quad \rho = \frac{i}{2\bar{\omega}} \quad , \tag{16}$$

where i is the imaginary unit ($i = \sqrt{-1}$) and $\bar{\omega}$ is the damped circular radian frequency ($\bar{\omega} = \omega\sqrt{1-\xi^2}$). Once $Z(t)$ is evaluated, the response in terms of displacements and velocities can be adequately obtained as

$$X(t) = Z(t) + Z^*(t) = 2\Re[Z(t)] \quad , \tag{17a}$$

$$\dot{X}(t) = \nu Z(t) + \nu^* Z^*(t) = 2\Re[\nu Z(t)] \quad , \tag{17b}$$

where $\Re[\cdot]$ means real part of $[\cdot]$. According to Eq.(13), $Z(t)$ can be split into two parts Z_1 and Z_2 and its complex conjugate, in the form

$$\dot{Z}_1 = \nu Z_1 + \rho\beta W^0(t) \quad ; \quad \dot{Z}_1^* = \nu^* Z_1^* + \rho^*\beta W^0(t) \quad , \tag{18a}$$

$$\dot{Z}_2 = \nu Z_2 + \rho\omega^2 C(t) \quad ; \quad \dot{Z}_2^* = \nu^* Z_2^* + \rho^*\omega^2 C(t) \quad . \tag{18b}$$

Now, the differential equation (18a) can be converted into an Itô type stochastic differential equation

$$dZ_1 = \nu Z_1 dt + \rho\beta dB \quad , \tag{19}$$

where B is the unit Wiener process. In view of the linearity of the differential equation (19), only the second cumulant of Z_1 is different from zero and can be evaluated by means of the differential equation[15]

$$\dot{k}_2\left[Z_1(t)\right] = 2\nu k_2\left[Z_1(t)\right] + \rho^2\beta^2 \quad . \tag{20}$$

Moreover,

$$\dot{k}_2\left[Z_1(t)Z_1^*(t)\right] = 2\Re[\nu]k_2\left[Z_1(t)Z_1^*(t)\right] + \rho\rho^*\beta^2 \quad . \tag{21}$$

If β is constant and for zero start condition, the transient response is easily obtained in the form

$$k_2\left[Z_1(t)\right] = -\frac{\beta^2\rho^2}{2\nu}\left[1 - \exp(2\nu t)\right] \quad , \tag{22a}$$

$$k_2\left[Z_1(t)Z_1^*(t)\right] = -\frac{\beta^2\rho\rho^*}{2\Re[\nu]}\left[1 - \exp\left(2\Re[\nu]t\right)\right] \quad . \tag{22b}$$

At last, the second order cumulant of X_1 is written as

$$k_2\left[X_1\right] = k_2\left[Z_1\right] + k_2\left[Z_1^*\right] + 2k_2\left[Z_1Z_1^*\right] \quad , \tag{23}$$

where $k_2\left[Z_1^*\right] = k_2^*\left[Z_1\right]$.

In order to characterise the process Z_2 from a probabilistic point of view, we differentiate Eq.(18b) to obtain

$$\dot{Q}(t) = \nu Q(t) + \rho\omega^2 W(t) \quad , \tag{24}$$

where $W(t)$ is a delta correlated process and $Q(t)=\dot{Z}(t)$. Equation (24) can be converted into an Itô type stochastic differential equation

$$dQ = \nu Q dt + \rho\omega^2 dC \quad . \tag{25}$$

The equation ruling the evolution of the 2-nd order cumulant is simply written in the form

$$\dot{k}_2\left[Q\right] = 2\nu k_2\left[Q\right] + \rho^2\omega^4\tilde{q}_2 \quad . \tag{26}$$

In the case under examination, $Q(t)$ is a zero mean stochastic process and $\tilde{q}_2=\lambda E[R^2]$, the second strength of the delta-correlated process, is a constant. Then,

$$k_2\left[Q\right] = -\frac{\rho^2\omega^4\tilde{q}_2}{2\nu}\left[1 - \exp\left(2\nu t\right)\right] \quad . \tag{27}$$

Moreover, the differential equation of the second cumulant $k_2[Q(t)Q^*(t)]$ is given in the form

$$\dot{k}_2[Q(t)Q^*(t)] = (\nu+\nu^*)\, k_2[Q(t)Q^*(t)] + \rho\rho^*\omega^4\tilde{q}_2 \quad . \tag{28}$$

It follows that the transient response is simply given as

$$k_2[Q(t)Q^*(t)] = -\frac{\rho\rho^*\omega^4}{(\nu+\nu^*)}\{1-\exp[(\nu+\nu^*)t]\} \quad . \tag{29}$$

Once the cumulant $k_2[Q(t)Q^*(t)]$ is evaluated, we can easily evaluate the second cumulant of Z_2; in fact,

$$k_2[Z_2(t)] = \int_0^t\int_0^t k_2[Q(t_1)Q(t_2)]\, dt_1\, dt_2 \quad . \tag{30}$$

The kernel of Eq.(30) can be evaluated[16] by means of the following relationship

$$\begin{aligned} k_2[Q(t_1)Q(t_2)] &= e^{\nu(t_1-t_m)}\, e^{\nu(t_2-t_m)}\, k_2[Q(t_m)] \quad , \\ t_m &= \min(t_1,t_2) \quad . \end{aligned} \tag{31}$$

At last, the cumulant $k_2[Z_2(t)Z_2^*(t)]$ can be obtained in the form

$$k_2[Z_2(t)Z_2^*(t)] = \int_0^t\int_0^t k_2[Q(t_1)Q^*(t_2)]\, dt_1\, dt_2 \quad , \tag{32}$$

and

$$\begin{aligned} k_2[Q(t_1)Q^*(t_2)] &= e^{\nu(t_1-t_m)}\, e^{\nu^*(t_2-t_m)}\, k_2[Q(t_m)Q^*(t_m)] \quad , \\ t_m &= \min(t_1,t_2) \quad . \end{aligned} \tag{33}$$

The transient second cumulant of $Z_2(t)$ is then given in the form

$$k_2[Z_2] = -\frac{\omega^4\tilde{q}_2}{\nu^2}\left[-\frac{2}{\nu}+2\frac{e^{\nu t}}{\nu}-t-te^{\nu t}\right] \quad , \tag{34}$$

$$k_2[Z_2Z_2^*] = |\rho|^2\omega^4\left[\frac{t}{|\nu|^2}+\frac{3\Re^2[\nu]+\Im^2[\nu]}{\Re[\nu]\,|\nu|^4}-\frac{\Re[\nu e^{\nu^* t}]}{\Re[\nu]\,|\nu|^4}+\frac{e^{2\Re[\nu]}}{|\nu|^2}\right] \quad , \tag{35}$$

where $\Im[\cdot]$ means imaginary part of $[\cdot]$. At last, the second cumulant of the response X in virtue of the independence between W^0 and W is given as

$$k_2[X] = k_2[Z_1]+k_2[Z_1^*]+2k_2[Z_1Z_1^*]+k_2[Z_2]+k_2[Z_2^*]+2k_2[Z_2Z_2^*] \quad . \tag{36}$$

It has to be emphasized that Eq.(36) accounts for the linear divergent behaviour of the response already noted by several authors[8,9].

5 NUMERICAL RESULTS AND COMPARISON WITH DIGITAL SIMULATION

It has been pointed out in the previous section, that the hypothesis of rare plastic occurrences allows us to evaluate in an analytical form the second cumulant of the response of an elastic–perfectly plastic oscillator subjected to a white noise input. In order to test the above mentioned analytical expression, a digital simulation, by directly integrating Eq.(1), has been performed first. It has been considered an elastic–perfectly plastic oscillator with ω=10, ξ=0.3, Y_0=0.05, subjected to a zero mean stationary white noise with unit intensity. The digital simulation provided a standard deviation of the plastic deformation increments σ_R=0.02527, while the probability density function p_R of the plastic deformation increments $R(t)$ is depicted in Fig.2 together with the suitable exponential approximation obtained by means of Eqs.(9) and (11). In Fig.3 the first two cumulants $k_1[N(t)]$ and $k_2[N(t)]$ of $N(t)$ are presented. They show a linear divergent behaviour typical of a counting Poisson process, with a common slope giving the expected rate of impulses λ=0.65 .

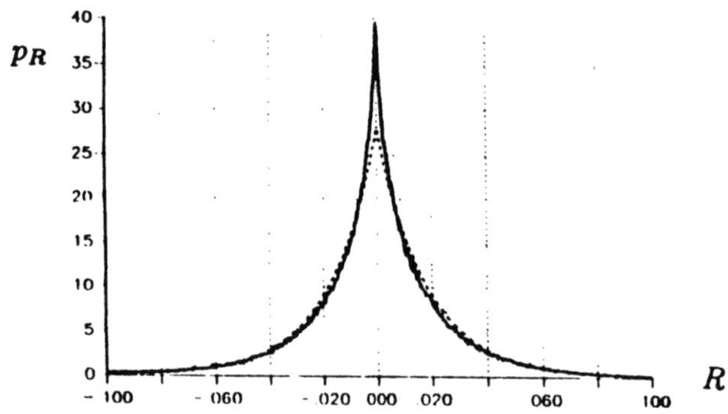

Fig.2. ***Probability density function of the yield increments: digital simulation with* 100000 *samples (solid line); exponential approximation (broken line)***

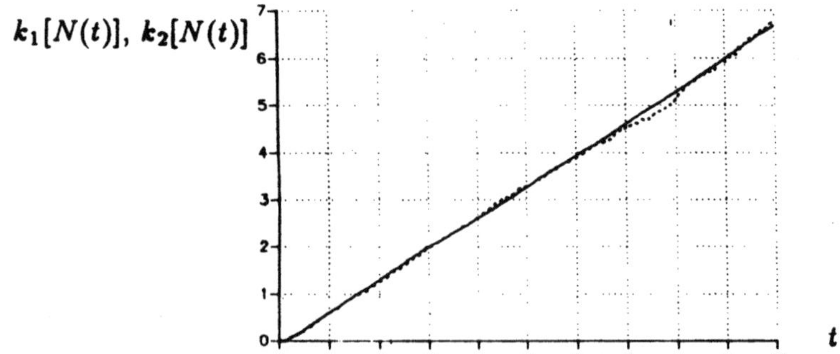

Fig.3. ***Cumulants of the yield increment counting process obtained by means of a digital simulation with* 1000 *samples: $k_1[N(t)]$ (solid line); $k_2[N(t)]$ (broken line)***

Once the parameters σ_R and λ have been evaluated by means of the digital simulation, the second cumulant of the total displacement $X(t)$ can also be evaluated by means of the closed form expression presented in the previous section. The evolution of the second cumulant $k_2[N(t)]$ of the total displacement $X(t)$ obtained by means of both the analytical treatment presented and the digital simulation is depicted in Fig.4. The good agreement of the results shown by Fig.4 proves the capability of the analytical treatment to provide correct results (under the hypothesis of rare plastic occurrences).

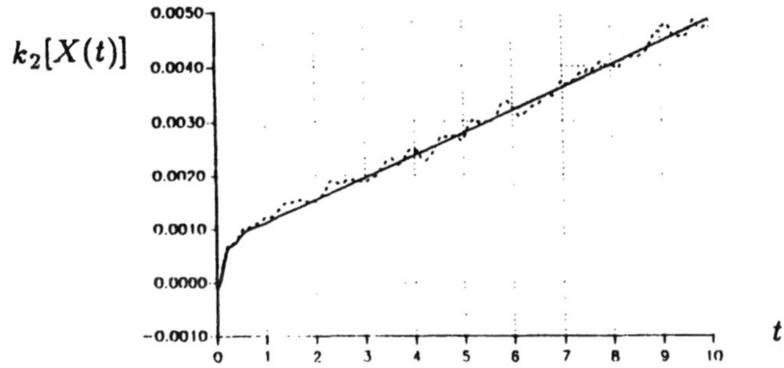

Fig.4. ***Second cumulant of displacement: response of the hysteretic system under the normal white noise (digital simulation with 1000 samples) (broken line); response of the linear system excited by the normal white noise and the compound homogeneous Poisson process (analytical treatment) (solid line)***

However, in order to avoid the evaluation of the parameters σ_R and λ by means of an expensive digital simulation, a digital simulation involving very few samples can be employed. In fact, by making use of a digital simulation involving only 10 samples, the standard deviation σ_R takes the value 0.02499, while the expected rate of impulses can be evaluated by interpolating the evolution of the second cumulant $k_2[N(t)]$ of $N(t)$ as shown in Fig.5 and takes the value of λ=0.65.

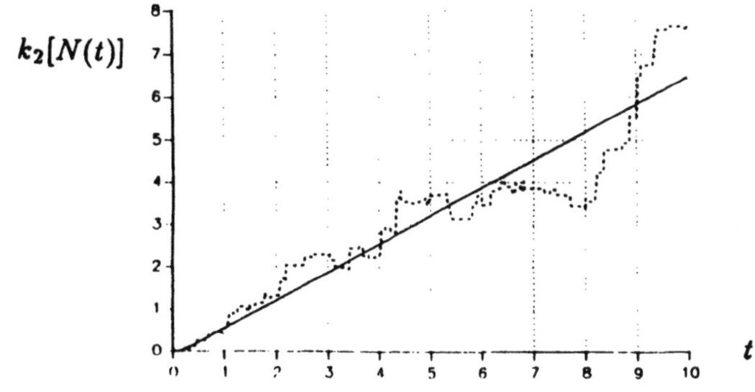

Fig.5. ***Second cumulant of the yield increment counting process obtained by means of a digital simulation with 10 samples (broken line), linear interpolation (solid line)***

In summary, a possible procedure for evaluating the statistics of the response of a hysteretic oscillator can be pursued as follows: first, with very few sample functions, evaluate the mean slope of $k_2[N(t)]$ and the mean number of jumps of the drift response; then, by means of the analytic procedure described, evaluate the response statistics. It has to be pointed out that an efficient test aiming to check the validity of rare plastic occurrences can be easily performed by verifying that the slope of $k_2[N(t)]$ is fairly equal to the mean slope of $k_1[N(t)]$.

6 CONCLUSIONS

In the framework of the yield increment approach, by considering that the events of plastic movements are rare and independent of the input process, the nonlinear elastic–perfectly plastic system has been considered as a linear system driven by two distinct and independent processes: the external one and a compound homogeneous Poisson process accounting for the internal pseudo–forces due to yielding excursion. Extensive studies by means of digital simulation these statements. By assuming the plastic deformation process as a compound homogeneous Poisson process, the mathematical treatment for the evaluation of the cumulants of every order of the response has been also presented and tested against the digital simulation procedure.

REFERENCES

1. Suzuki, Y., and Minai, R., Application of stochastic differential equations to seismic reliability analysis of hysteretic structures, *Prob. Eng. Mech.*, **3** (1988) pp. 43–52.
2. Iwan, W.D., A distributed element model for hysteresis and its steady–state dynamic response, ASME *J. App. Mech.*, (1966) pp. 893–900.
3. Wen, Y.K., Method for random vibration of hysteretic systems, ASCE *J. Eng. Mech.*, **102** (1976) pp. 249–263.
4. Casciati, F., and Faravelli, L., *Fragility analysis of complex structural systems*, J.B.Roberts Ed., Research Studies Press Ltd., England, 1991.
5. Caughey, T.K., Equivalent linearization techniques, *J. Acoustical Society of America*, **35** (1963) pp. 1706–1711.
6. Spanos, P.T.D., Hysteretic structural vibrations under random load, *J. Acoustical Society of America*, **65** (1979) pp. 404–410.
7. Wen, Y.K., Equivalent linearization for hysteretic systems under random excitation, ASME *J. App. Mech.*, **47** (1980) pp. 150–154.
8. Park, Y.J., Equivalent linearization for seismic responses. I: formulation and error analysis, ASCE *J. Eng. Mech.*, **118** (1992) pp. 2207–2226.
9. Paparizos, L.G., and Iwan, W.D., Some observations on the random response of an elasto–plastic system, ASME *J. App. Mech.*, **55** (1988) pp. 911–917.

10. Iwankiewicz, R., and Nielsen, S.R.K., Dynamic response of hysteretic systems to Poisson-distributed pulse trains, *Prob. Eng. Mech.*, **7** (1992) pp. 135–148.

11. Karnopp, D., and Scharton, T.D., Plastic deformation in random vibration, *J. Acoustical Society of America*, **39** (1966) pp. 1154–1161.

12. Grossmayer, R.L., Elastic-plastic oscillators under random excitation, *J. Sound and Vibration*, **65** (1979) pp. 353–379.

13. Vanmarcke, E.H., Yanev, P.I., De Estrada, M.B., Response of simple hysteretic systems to random excitation, MIT, Dept. of Civil Engineering, Research Report R70–66, Cambridge, 1970.

14. Ditlevsen, O., Elasto-plastic oscillator with Gaussian excitation, ASCE *J. Eng. Mech.*, **112** (1986) pp. 386–406.

15. Lutes, L.D., and Chen, D.C.K., Stochastic response moments for linear systems, *Prob. Eng. Mech.*, **7** (1992) pp. 165–173.

16. Falsone, G., Cumulants and correlations for linear systems under non-stationary delta-correlated processes, *Prob. Eng. Mech.*, (1993) to appear.

A New Second-Order Approximate Reliability Method Based on Hyper-Parabolic Failure Surfaces

G. Q. Cai[1] and I. Elishakoff[2]

[1] Center for Applied Stochastics Research, Florida Atlantic University, Boca Raton, FL 33431-0991, USA
[2] Center for Applied Stochastics Research and Department of Mechanical Engineering, Florida Atlantic University, Boca Raton, FL 33431-0991, USA

Abstract. The first-order and second-order methods for structural reliability analysis are reviewed with attendant new interpretation for the first-order method and a new approximate result for the second-order method. Exact and approximate solutions for the example case - reliability of a shaft subject to random bending moments and a random torque are obtained. The comparison of the approximate results with exact ones shows that the first-order approximation is only applicable to the case where the failure surface is "far" from the origin, while the suggested second-order approximation yields quite accurate results even when the failure surface is "close" to the origin.

Keywords. structural reliability, exact solution, first and second-order approximations

1 Introduction

In structural design and reassessment of the safety of existing structures, reliability analysis is usually performed when a structure is subject to random loads, variable geometric and material properties and other uncertainties. Let each uncertain component be denoted by a random variable X_i, and the random vector $X = \{X_1, X_2, ..., X_n\}$ describe the uncertainties of the structure. If the probability density of the random vector X, $p(x)$, can be obtained by measurements or according to past experience, the reliability, defined as the probability of the structure being in a safe state, is evaluated as follows

$$R = \int_S p(x)dx = 1 - \int_F p(x)dx \tag{1}$$

where S is the domain in which the structure operates safely, whereas F is the domain in which the structure fails. The surface separating the safe domain S and the failure domain F is called *failure surface* or limit state surface.

Generally, the safe domain, failure surface and failure domain can be constructed in the following manner

$$g(x) = g(x_1, x_2, ..., x_n) \begin{cases} > 0, & x \in \text{safe domain } S \\ = 0, & x \in \text{ failure surface} \\ < 0, & x \in \text{failure domain } F. \end{cases} \tag{2}$$

The function $g(x)$ is called the *performance function*. For convenience, it is assumed that the origin is within the safe domain S. In equations (1) and (2), the lower case symbol x represents the possible values of the upper case random vector X.

The reliability index β is defined by[1,2]

$$\beta = \Phi^{-1}(R) \tag{3}$$

where Φ is the cumulative normal distribution function, namely,

$$\Phi(u) = \frac{1}{\sqrt{2\pi}} \int_{-\infty}^{u} \exp(-\frac{v^2}{2})dv. \tag{4}$$

By suitable transformations[3,4], the random variables X_1 , X_2 , ..., X_n can be transformed into a set of independent, standard normally distributed random variables. Accordingly, the limit state surface is also mapped to one in the new state space. Therefore, in what follows, we assume, without loss of generality, that the random variables X_1 , X_2 , ..., X_n are independent, standard, normally distributed random variables.

The multi-dimensional integration in (1) can be carried out analytically only for a very limited number of cases[2,5-8]. For most of practical cases, the integration domain S is such that the integration is impossible to be carried out analytically. Numerical methods can generally be performed to evaluate the integration, but it often turns out to be rather tedious and costly. Of course, Monte Carlo simulation is a universal method to deal with stochastic problems, including reliability analysis. It can be used either to evaluate the integration in (1) or to obtain the reliability directly from the original structure model. However, the method is known to be computer time consuming. Thus, relatively simple and accurate approximate techniques to evaluate the reliability are needed.

The most widely used approximate method is the so-called first-order method[1,5,9], in which the performance function $g(x)$ is approximated as a linear function of x, namely, the failure surface is replaced by a hyper-plane. The reliability for the replacing linear performance function can be found exactly by carrying out the integration in (1) and then considered as an approximate one for the original structure. The first-order approximation yields accurate results if the failure surface is nearly a hyper-plane or if it is "far away" from the origin. The effective numerical techniques and their critical comparisons were given by

Schuëller and his associates[10-12]. The accuracy of the approximate method has been investigated by Elishakoff[6], and Elishakoff and Hasofer[7,8] by comparing *exact* and *approximate* results for some structures. However, if the failure surface is not far from the origin and has large curvatures, the reliability estimated according the first-order method may considerably deviate from the true value, and more accurate approximate methods have to be applied. Fiessler *et al.*[13], Breitung[14,15], Tvedt[16], and Madsen *et al.*[2] have generalized the approximation to second order by using a quadratic failure surface to replace the original failure surface, and developed efficient algorithms to calculate the approximate reliability. However, as shown later, even the best approximate methods developed up to now, namely those derived by Breitung[14,15], Tvedt[16], and Madsen *et al.*[2] are not applicable for some often encountered cases.

In the present investigation, a new derivation for the first-order method, which appears to be theoretically more rigorous than the previous versions, are presented. For the second-order method, a new analytical approximate result presumably applicable to all cases is devised. The reliability of a shaft subject to random bending moments in two directions and a random toque is evaluated both in exact and approximate terms. It is shown that, for this problem, the first-order approximate results are accurate only when the failure surface are far from the origin, while the second-order approximate results are quite accurate for all the cases.

2 First-Order Analysis

Consider a point $\boldsymbol{x}_0$ on the failure surface, namely,

$$g(\boldsymbol{x}_0) = g(x_{10}, x_{20}, \dots, x_{n0}) = 0 . \tag{5}$$

Linearize $g(\boldsymbol{x})$ around $\boldsymbol{x}_0$ and denote the linearized function as $g_1(\boldsymbol{x})$

$$g_1(\boldsymbol{x}) = \sum_{j=1}^{n} \alpha_j (x_j - x_{j0}) \tag{6}$$

where

$$\alpha_j = \left(\frac{\partial g}{\partial x_j} \right)_{\boldsymbol{x}=\boldsymbol{x}_0} . \tag{7}$$

Since the X_i are assumed to be independent, standard, normally distributed random variables, the random variable

$$G_1 = g_1(\boldsymbol{X}) = \sum_{j=1}^{n} \alpha_j (X_j - x_{j0}) \tag{8}$$

is also a normally distributed random variable with the probability density

$$p(g_1) = \frac{1}{\sqrt{2\pi}\,\sigma} \exp\left[-\frac{(g_1 - \mu)^2}{2\sigma^2} \right] \tag{9}$$

where the mean μ and standard deviation σ are, respectively,

$$\mu = \mu(x_0) = -\sum_{j=1}^{n} \alpha_j x_{j0} = -\sum_{j=1}^{n} x_{j0} \left(\frac{\partial g}{\partial x_j} \right)_{x=x_0} \tag{10}$$

$$\sigma = \sigma(x_0) = \left(\sum_{j=1}^{n} \alpha_j^2 \right)^{1/2} = \left[\sum_{j=1}^{n} \left(\frac{\partial g}{\partial x_j} \right)^2_{x=x_0} \right]^{1/2}. \tag{11}$$

For the hyper-plane $g_1(x) = 0$, which is tangent to the failure surface at point x_0,

$$\int_{g_1>0} p(g_1)dg_1 = \int_0^{\infty} \frac{1}{\sqrt{2\pi}\,\sigma} \exp\left[-\frac{(g_1 - \mu)^2}{2\sigma^2} \right] dg_1 = \Phi\left(\frac{\mu}{\sigma}\right) \tag{12}$$

where the function Φ is given in (4). As is known from analytical geometry, the quantity μ/σ represents the distance from the origin to the hyper-plane $g_1(x) = 0$. Figs. 2.1(a) and (b) illustrate the failure surface $g(x) = 0$ and its tangent hyper-plane in the two-dimensional case. In Fig. 2.1(a), the failure surface is concave toward the origin, while in Fig. 2.1(b), it is convex toward the origin. An example for concave failure surface is that the safe region is defined by $g(x_1, x_2) = x_1^2 + x_2^2 < a^2$. On the other hand, if the safe region is given by $g(x_1, x_2) = x_1 x_2 < b^2$, the failure surface is convex toward the origin, as shown in Fig. 1(b). It can be seen from these figures that

$$\Phi\left(\frac{\mu}{\sigma}\right) \begin{cases} > R\,, & g(x) = 0 \text{ is concave} \\ = R\,, & g(x) = 0 \text{ is a hyperplane} \\ < R\,, & g(x) = 0 \text{ is convex}\,. \end{cases} \tag{13}$$

Therefore, in the case of a concave failure surface, the approximate reliability should be the minimum value of $\Phi(\mu/\sigma)$ for all points on the failure surface $g(x) = 0$, whereas the opposite is valid if the failure surface is convex, namely,

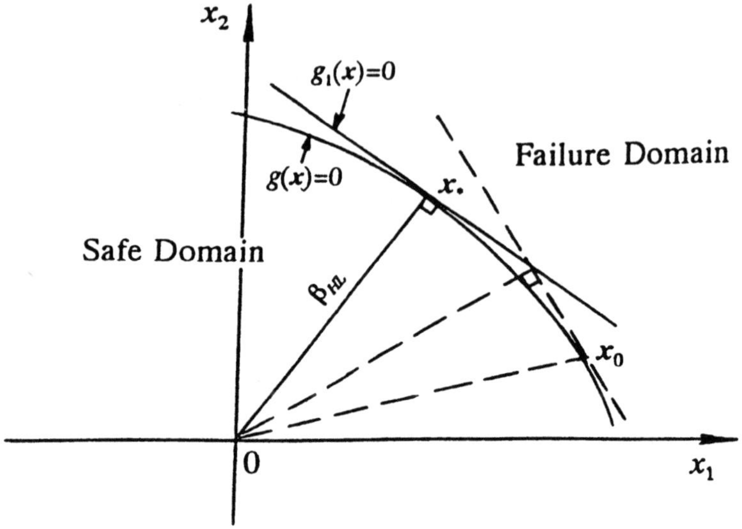

(a) Concave failure surface

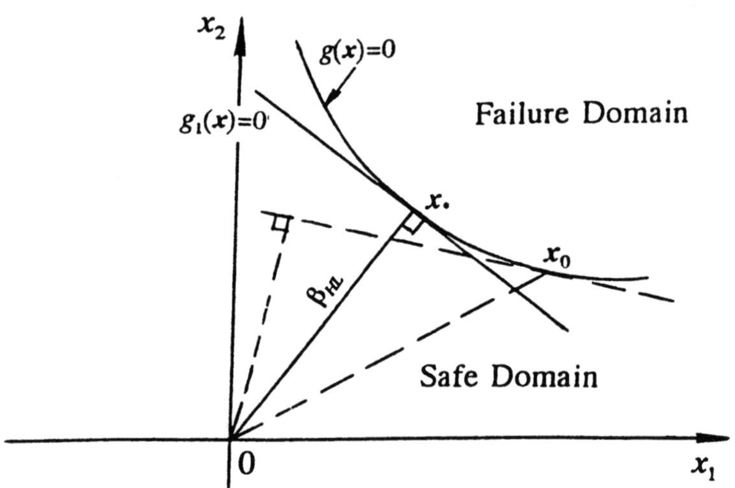

(b) Convex failure surface

Fig. 2.1. Linear approximation of failure surface

$$R \approx \begin{cases} \min\limits_{g(x_0)=0}\left[\Phi\left(\frac{\mu}{\sigma}\right)\right] = \Phi\left[\min\limits_{g(x_0)=0}\left(\frac{\mu}{\sigma}\right)\right], & g(x)=0 \text{ is concave} \\ \max\limits_{g(x_0)=0}\left[\Phi\left(\frac{\mu}{\sigma}\right)\right] = \Phi\left[\max\limits_{g(x_0)=0}\left(\frac{\mu}{\sigma}\right)\right], & g(x)=0 \text{ is convex} \end{cases} \tag{14}$$

where use has been made of the fact that the cumulative distribution function Φ is a monotonic function.

Assume point x_* is such a point on the failure surface that the distance from the origin to x_* is the shortest one of those from the origin to all points on the failure surface. Then, it is known from geometry that vector x_* is normal to the failure surface, that the distance from the origin to point x_* is equal to that from the origin to the hyper-plane tangent to the failure surface at point x_*, and that

$$\left(\sum_{j=1}^{n} x_{j*}^2\right)^{1/2} = \frac{\mu(x_*)}{\sigma(x_*)} = \begin{cases} \min\limits_{g(x)=0}\left(\frac{\mu}{\sigma}\right), & g(x)=0 \text{ is concave} \\ \max\limits_{g(x)=0}\left(\frac{\mu}{\sigma}\right), & g(x)=0 \text{ is convex}\ . \end{cases} \tag{15}$$

From equations (14) and (15),

$$R \approx \Phi(\beta_{HL}), \quad \text{or} \quad \beta \approx \beta_{HL} \tag{16}$$

where

$$\beta_{HL} = \min_{g(x)=0}\left(\sum_{j=1}^{n} x_j^2\right)^{1/2} = \left(\sum_{j=1}^{n} x_{j*}^2\right)^{1/2} = \frac{\mu(x_*)}{\sigma(x_*)} \tag{17}$$

β_{HL} is called the *Hasofer-Lind reliability index*[5], and point x_* is called the design point. Interestingly, analogous ideas to shortest-distance reliability index have been expressed by Freudenthal[17]. Murzewski[18] is attributing this method to Levi[19].

In the present case, namely the origin is located in the safe domain, the Hasofer-Lind reliability index β_{HL} and the reliability index β have the following relationship

$$
\beta_{HL} \begin{cases} > \beta\,, & g(x) = 0 \text{ is concave} \\ = \beta\,, & g(x) = 0 \text{ is a hyperplane} \\ < \beta\,, & g(x) = 0 \text{ is convex}\,. \end{cases} \tag{18}
$$

It is obvious that the closer the failure surface to a hyper-plane, the more accurate result the method yields. For a failure surface with large curvatures, the first-order approximation is reasonable only if the failure surface is sufficiently far from the origin. In this case, the replacement of a curved surface by a hyper-plane tangent at the design point would not lead to a significant difference in the integration (1) since values of the probability density $p(x)$ at points far from the origin are small and no longer important.

Equation (17) shows that the Hasofer-Lind reliability index is the solution of an optimization problem with a constraint, namely,

$$
\begin{aligned} &\text{Minimize:} && d(x) = \sum_{i=1}^{n} x_i^2 \\ &\text{Subject to:} && g(x) = 0\,. \end{aligned} \tag{19}
$$

In (19) we choose the target function to be the square of the distance for convenience. Numerical algorithms have been devised for solving such an optimization problem, for example, by Ditlevsen[3], Hohenbichler and Rackwitz[4], Rackwitz and Fiessler[20], Madsen *et al.*[2] and Shinozuka[9]. If there is only a single local minimum distance between the origin and the points on the failure surface, then this minimum distance is clearly the Hasofer-Lind reliability index. When a failure surface is partly concave and partly convex, or more complicated, then several local minimum distances may exist. In this case, the Hasofer-Lind reliability index is the smallest one of these local minimum distances, and the corresponding point is the design point.

It is noted that the value of $\mu(x_*)$ is always positive for the present case where the origin is contained in the safe domain. Thus, β_{HL} is also positive and truly represents the minimum distance from the origin to the failure surface. On the other hand, if the origin is located in the failure domain, $\mu(x_*)$ and β_{HL} are both negative, and the absolute value of β_{HL} is the minimum distance from the origin to the failure surface. However, equation (16) can still be applied to find an approximate value for the reliability.

3 Second-Order Analysis

To improve the accuracy of the first-order analysis, expand the performance function $g(x)$ into the Taylor series around the design point x_* up to the second order, and denote the expanded series as $g_2(x)$,

$$g_2(x) = \left(\frac{\partial g}{\partial x_j}\right)_{x=x_*}(x_j - x_{j*}) + \frac{1}{2}\left(\frac{\partial^2 g}{\partial x_j \partial x_k}\right)_{x=x_*}(x_j - x_{j*})(x_k - x_{k*})$$

$$= a^T(x - x_*) + \frac{1}{2}(x - x_*)^T b (x - x_*) \tag{20}$$

where the superscript T denote a transposition, and elements of vector a and matrix b are given by

$$a_j = \left(\frac{\partial g}{\partial x_j}\right)_{x=x_*}, \quad b_{jk} = \left(\frac{\partial^2 g}{\partial x_j \partial x_k}\right)_{x=x_*}. \tag{21}$$

Denote e as the unit vector in the same direction as vector x_*, namely,

$$x_* = \beta_{HL} e = \frac{\mu_*}{\sigma_*} e \tag{22}$$

where $\mu_* = \mu(x_*)$ and $\sigma_* = \sigma(x_*)$. As is known, vector a, which is normal to the failure surface at the design point x_*, is in the direction opposite to x_* or e; thus,

$$a = -\sigma_* e . \tag{23}$$

By substituting (22) and (23) into (20), the approximate failure surface is given by

$$g_2(x) = -\sigma_*(e^T x - \beta_{HL}) + \frac{1}{2}(x - \beta_{HL} e)^T b (x - \beta_{HL} e) = 0 \tag{24}$$

which represents a general second-order surface. However, even if we use such an approximate failure surface, the integration in (1) is still impossible to be carried out analytically. Therefore, further approximation is in order.

Performing an orthogonal transformation

$$y = H x \tag{25}$$

where matrix H is a standard orthogonal matrix with the last (nth) row equal to the transposition of vector e, and noticing that

$$e^T x = e^T H^T y = y_n \tag{26}$$

$$H(x - \beta_{HL} e) = (y_1, y_2, ..., y_{n-1}, y_n - \beta_{HL})^T \tag{27}$$

equation (24) becomes

$$y_n = \beta_{HL} + \frac{1}{2\sigma_*}(\bar{y}^T \quad y_n - \beta_{HL})\, H\, b\, H^T \begin{pmatrix} \bar{y} \\ y_n - \beta_{HL} \end{pmatrix} \tag{28}$$

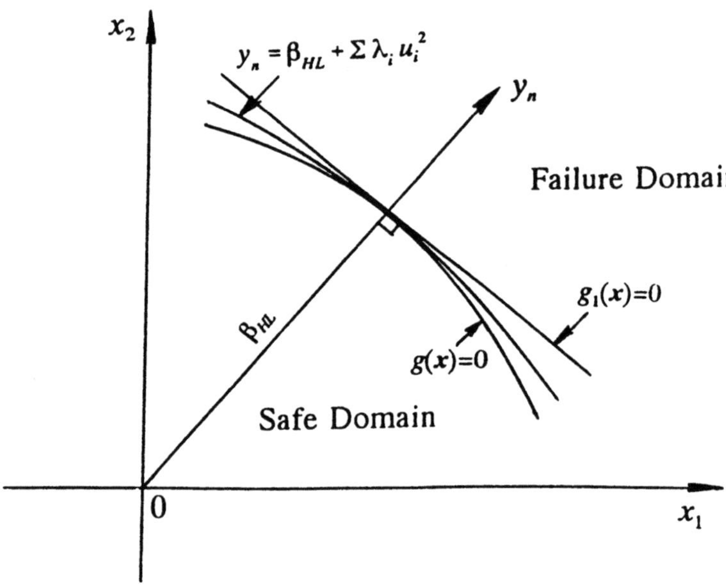

Fig. 3.1. Quadratic approximation of failure surface

where $\bar{y} = (y_1, y_2, ..., y_{n-1})^T$ is a (n-1)-dimensional vector. Physically, the transformation is a coordinate rotation making the y_n axis coincident with vector e, as shown in Fig. 3.1. Since expansion (20) is obtained near point x_*, y_1, y_2, ..., y_{n-1} and $y_n - \beta_{HL}$ are all small quantities. Solving for y_n from equation (28) and only keeping up to second-order terms in y_j, j = 1, 2, ..., n-1, we obtain

$$y_n = \beta_{HL} + \bar{y}^T c\, \bar{y} \tag{29}$$

where elements of the $(n\text{-}1)\times(n\text{-}1)$ matrix c are given by

$$c_{jk} = \frac{1}{2\sigma_*}(HbH^T)_{jk}\,, \qquad j,k = 1, 2, \dots, n-1\,. \tag{30}$$

By a further orthogonal transformation H_1,

$$u = H_1\,\bar{y} \tag{31}$$

equation (29) can be reduced to

$$y_n = \beta_{HL} + \sum_{j=1}^{n-1} \lambda_j\, u_j^2 \tag{32}$$

where the λ_j are eigenvalues of matrix c. Equation (32) represents a hyper-parabolic surface, which is now considered as an approximate failure surface.

The approximate reliability is now calculated from

$$R \approx \int_{-\infty}^{\infty} \dots\dots \int_{-\infty}^{\infty} \left[\int_{-\infty}^{\beta_{HL}+\sum_{j=1}^{n-1}\lambda_j u_j^2} p(u, y_n)\, dy_n \right] du_1 du_2 \dots du_{n-1}\,. \tag{33}$$

Since X_1, X_2, ..., X_n are independent random variables with standard normal distribution, and H and H_1 are both orthogonal transformations, it can be shown that random variables Y_1, Y_2, ..., Y_n and U_1, U_2, ..., U_{n-1}, Y_n are two sets of independent and standard normally distributed random variables. Thus, equation (33) can be rewritten as

$$R = \int_{-\infty}^{\infty} \dots\dots \int_{-\infty}^{\infty} \phi(u_1)\,\phi(u_2)\dots\phi(u_{n-1})\, du_1 du_2 \dots du_{n-1} \int_{-\infty}^{\beta_{HL}+\sum_{j=1}^{n-1}\lambda_j u_j^2} \phi(y_n) dy_n \tag{34}$$

where ϕ is the standard normal probability density function

$$\phi(v) = \frac{1}{\sqrt{2\pi}} \exp(-\frac{v^2}{2})\,. \tag{35}$$

An approximation to (34) was obtained by Breitung[14,15] as follows

$$R = 1 - \Phi(-\beta_{HL}) \prod_{j=1}^{n-1} (1 - \beta_{HL}\kappa_j)^{-1/2} \tag{36}$$

where

$$\kappa_j = -2\lambda_j \tag{37}$$

are the main curvatures of the failure surface at the design point x_*. The condition for the approximate result (36) to be valid is

$$\beta_{HL}\kappa_j < 1\,, \quad j = 1, 2, \ldots, n-1\,. \tag{38}$$

Equation (36) was also deduced by Tvedt[16] using different notations.

Unfortunately, condition (38) is not always valid. One example is a shaft under random bending moments and a torque (see the next section). Also, if some $1 - \beta_{HL}\kappa_j$ is very small, (36) may lead to an enormous error, even leading to a negative R. In the following, we will derive an approximate result for (34), which is applicable for all cases.

The inner integral in (34) can be divided into two parts:

$$\int_{-\infty}^{\beta_{HL}+\sum\lambda_j u_j^2} \phi(y_n)dy_n = \Phi(\beta_{HL}) + \int_{\beta_{HL}}^{\beta_{HL}+\sum\lambda_j u_j^2} \frac{1}{\sqrt{2\pi}}\exp(-\frac{y_n^2}{2})\,dy_n$$

$$= \Phi(\beta_{HL}) + \frac{1}{\sqrt{2\pi}}\exp(-\frac{\beta_{HL}^2}{2})\int_0^{\sum\lambda_j u_j^2}\exp(-\beta_{HL}v - v^2/2)\,dv \tag{39}$$

where $v = y_n - \beta_{HL}$. In equation (39) and in what follows, we omit the indices for the summations for simplicity. Expanding $\exp(-\beta_{HL}v - v^2/2)$ into the Taylor series at point $v = 0$ and carrying out the last integration in (39), we obtain

$$\int_0^{\sum\lambda_j u_j^2} e^{-\beta_{HL}v - v^2/2}dv = \sum\lambda_j u_j^2 - \frac{1}{2}\beta_{HL}\left(\sum\lambda_j u_j^2\right)^2 + \frac{1}{6}(\beta_{HL}^2 - 1)\left(\sum\lambda_j u_j^2\right)^3 - \ldots\,. \tag{40}$$

Successive substitutions of (40) into (39) and (39) into (34) lead to

$$R = \Phi(\beta_{HL}) + \frac{1}{\sqrt{2\pi}}\exp(-\frac{\beta_{HL}^2}{2})(D_1 + D_2 + D_3 + \ldots) \tag{41}$$

where

$$D_1 = \sum\lambda_j \tag{42}$$

$$D_2 = -\frac{1}{2}\beta_{HL}\left(3\sum\lambda_j^2 + \sum_{j\neq k}\lambda_j\lambda_k\right) \tag{43}$$

$$D_3 = \frac{1}{6}(\beta_{HL}^2 - 1)\left(15\sum \lambda_j^3 + 9\sum_{j\neq k} \lambda_j^2\lambda_k + \sum_{j\neq k\neq l} \lambda_j\lambda_k\lambda_l\right) \qquad (44)$$

....

Equation (41) tends to (34) asymptotically by taking more terms. For practical purposes, three term approximation is usually accurate enough.

4 Reliability of a Shaft under Random Bending Moments and Torque

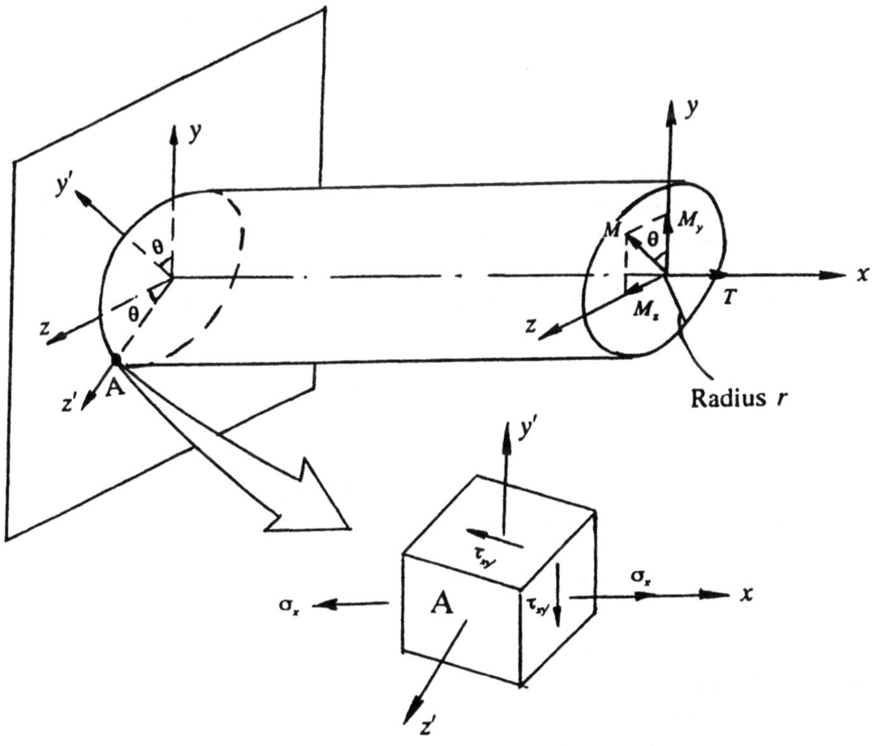

Fig. 4.1. A circular shaft subjected to external bending moments M_y, M_z and torque T.

Consider a circular shaft with one end clamped and one end free, and subject to two bending moments M_y and M_z and a torque T, as shown in Fig. 4.1. The moments M_y, M_z and torque T are assumed to be independent random variables with normal distributions: $M_y \sim N(\mu_1, \sigma_1^2)$, $M_z \sim N(\mu_2, \sigma_2^2)$, $T \sim N(\mu_3, \sigma_3^2)$. It is known that the bending moments M_y and M_z are equivalent to their resultant M, and both the tensile stress and shear stress reach their maximum values at point A on the clamped end, as shown in Fig. 4.1. Therefore, the reliability of the shaft depends on the stresses at point A. In the coordinate system x-y'-z' shown in Fig. 3, the stress at point A are found to be

$$\sigma_x = 4\frac{\sqrt{M_y^2 + M_z^2}}{\pi r^3}, \quad \sigma_{y'} = \sigma_{z'} = 0$$
$$\tau_{xy'} = \frac{2T}{\pi r^3}, \quad \tau_{xz'} = \tau_{y'z'} = 0\,. \tag{45}$$

Thus, the failure criterion may be expressed for several failure modes as follows

$$M_{eq}^2 = M_y^2 + M_z^2 + \delta T^2 \le M_{yield}^2 = \left(\frac{\pi r^3}{4}\sigma_{yield}\right)^2 \tag{46}$$

where M_{eq} is an equivalent bending moment, replacing the three moments acting on the shaft, δ is a constant depending on the failure mode. For example, $\delta = 1$ for the maximum shear stress criterion, and $\delta = 3/4$ for the von Mises criterion (maximum distortion energy criterion). A simpler case of a single bending moment and a torque was studied by Bolotin[21]. The reliability of the shaft in calculated from

$$R = \frac{1}{(2\pi)^{3/2}\sigma_1\sigma_2\sigma_3}\iiint_S \exp\left[-\sum_{i=1}^{3}\frac{(m_i-\mu_i)^2}{2\sigma_i^2}\right]dm_1 dm_2 dm_3 \tag{47}$$

where m_1, m_2 and m_3 are state variables of M_y, M_z and T, respectively, and the safe domain S is an ellipsoid described by

$$m_1^2 + m_2^2 + \delta\, m_3^2 \le M_{yield}^2\,. \tag{48}$$

The reliability can be then obtained accurately by performing integration in equation (47) at least numerically.

Fig. 4.2 shows the reliability for the shaft with $\mu_1 = \mu_2 = \mu_3 = \mu$, $\sigma_1 = \sigma_2 = \sigma_3 = \sigma$, calculated by directly integrating equation (47). Both the maximum shear stress and von Mises criteria were employed. As expected, the von Mises criterion yields greater reliability than the maximum shear stress criterion. However, when the ratio μ/σ is not large, both criteria result in almost the same reliability.

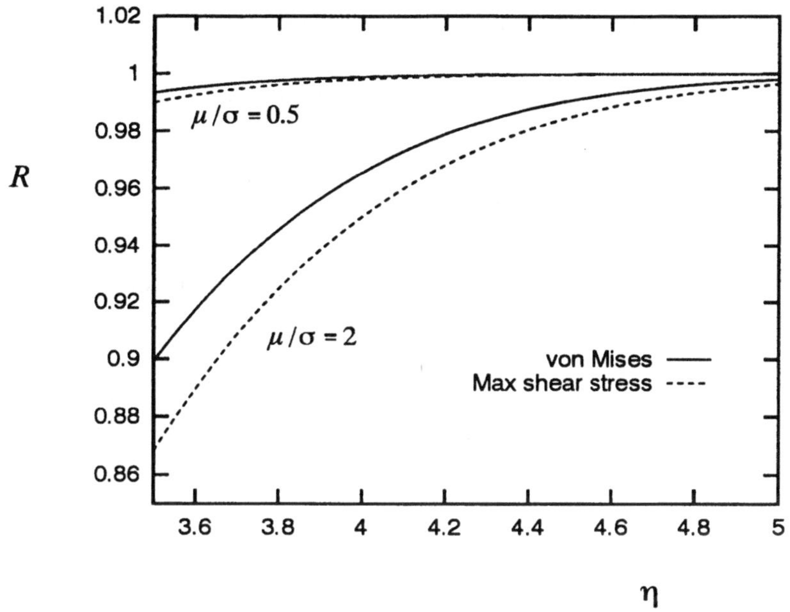

Fig. 4.2. Reliability of a shaft under random bending moments and torque with $\sigma_1 = \sigma_2 = \sigma_3 = \sigma$ and $\mu_1 = \mu_2 = \mu_3 = \mu$

The first-order approximation is described by the following optimization problem:

$$\text{Minimize:} \quad d(x) = x_1^2 + x_2^2 + x_3^2$$
$$\text{Subject to:} \quad (\sigma_1 x_1 + \mu_1)^2 + (\sigma_2 + \mu_1)^2 + \delta\,(\sigma_3 + \mu_3)^2 = M_{yield}^2 \tag{49}$$

where x_1, x_2 and x_3 are defined by

$$x_i = \frac{m_i - \mu_i}{\sigma_i}, \qquad i = 1, 2, 3 \ . \tag{50}$$

The solution of problem (49) can be easily obtained by using, for example, the IMSL program NCONG.

To apply the results of the second-order approximation, an orthogonal transformation H in (25) has to be determined. For the present three-dimensional case, a general rotational transformation can be expressed as

$$H=\begin{bmatrix} (\cos\psi\cos\varphi-\cos\theta\sin\psi\sin\varphi) & (\sin\psi\cos\varphi+\cos\theta\cos\psi\sin\varphi) & \sin\theta\sin\varphi \\ (-\cos\psi\sin\varphi-\cos\theta\sin\psi\cos\varphi) & (-\sin\psi\sin\varphi+\cos\theta\cos\psi\cos\varphi) & \sin\theta\cos\varphi \\ \sin\theta\sin\psi & -\sin\theta\cos\psi & \cos\theta \end{bmatrix}. \tag{51}$$

Letting $\varphi = 0$ and the three elements in the last row be e_1, e_2 and e_3 respectively, we obtain

$$H = \frac{1}{\sqrt{1-e_3^2}}\begin{bmatrix} -e_2 & e_1 & 0 \\ -e_1 e_3 & -e_2 e_3 & 1-e_3^2 \\ e_1\sqrt{1-e_3^2} & e_2\sqrt{1-e_3^2} & e_3\sqrt{1-e_3^2} \end{bmatrix}. \tag{52}$$

Then, the second-order approximate reliability can be calculated following the procedures: (1) solve problem (49) for design point $\boldsymbol{x}_*$ and β_{HL}, and then compute μ_*, σ_* and $\boldsymbol{e} = \boldsymbol{x}_* / \beta_{HL}$; (2) find $\boldsymbol{H}$ according to (52), $\boldsymbol{b}$ from (21), and $\boldsymbol{c}$ from (30); (3) find the eigenvalues λ_1 and λ_2 for matrix $\boldsymbol{c}$ and calculate D_1, D_2 and D_3 from (42)-(44), and R from (41).

The following numerical investigation is based on the von Mises criterion. The exact and approximate values of the reliability are calculated for the shaft with $\mu_1 = \mu_2 = \mu_3 = \mu$, $\sigma_1 = \sigma_2 = \sigma_3 = \sigma$. The results are plotted against $\eta = M_{yield}/\sigma$ in Fig. 4.3(a)-(c) for three parameter values of μ/σ. For all three cases, the first-order (Hasofer-Lind) approximate results are accurate when η exceeds four or five, namely the failure surfaces are far from the origin. Breitung's second-order approximation, equation (36), yields good results when γ is not small, as shown in Figs. 4.3(b) and 4.3(c). However, it is seen from Fig. 4.3(a) that Breitung's method produces large errors when γ is small. In the case of $\gamma = 0$, namely all moments have zero means, Breitung's method is not applicable. With a different extent of accuracy, the present second-order approximation, equation (41), is applicable for almost all γ values even with $\eta \sim 2.5$ and low values of R. In computations, three correction terms were retained in equation (41). It is found that adding more correction terms in equation (43) did not improve the accuracy of the results, indicating that the error arises from the replacement of the original failure surface by a second-order failure surface, and the further replacement of the second-order failure surface by a hyper-parabolic failure surface.

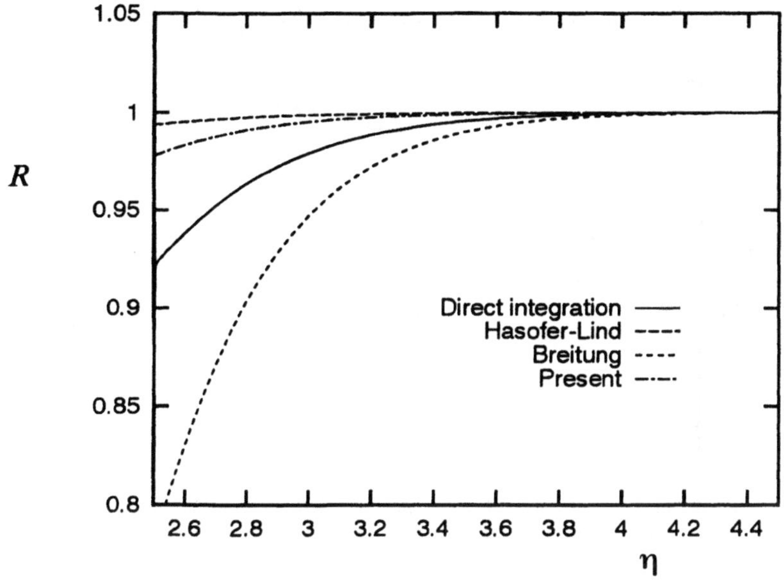

(a) $\mu/\sigma = 0.01$

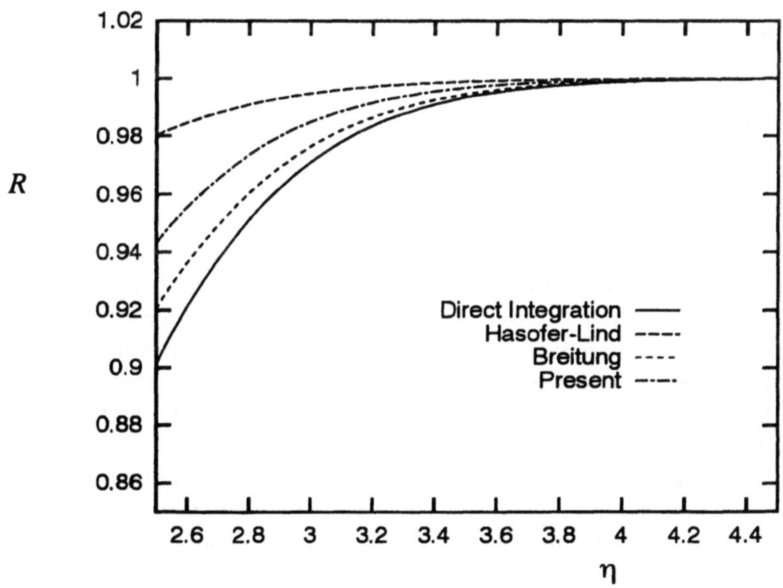

(b) $\mu/\sigma = 0.5$

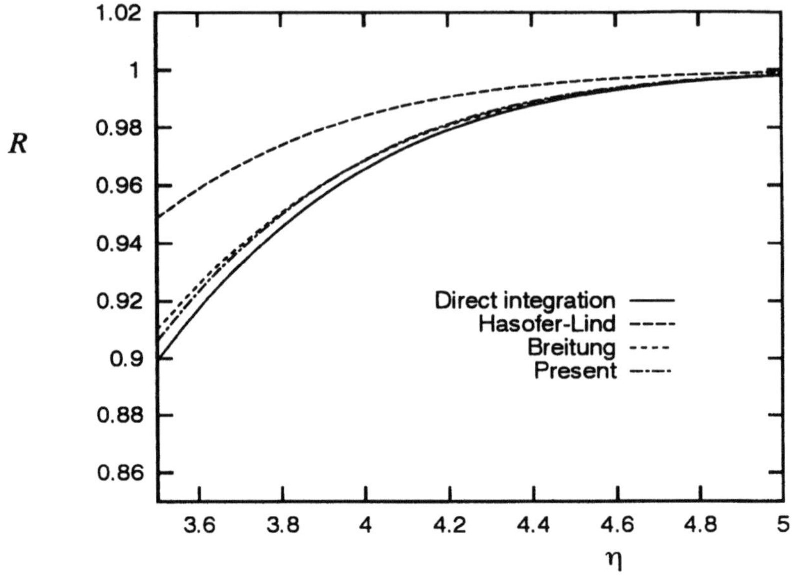

(c) $\mu/\sigma = 2$

Fig. 4.3. Reliability of a shaft under random bending moments and torque with $\sigma_1 = \sigma_2 = \sigma_3 = \sigma$ and $\mu_1 = \mu_2 = \mu_3 = \mu$, using von Mises failure criterion

5 Conclusion

A new derivation of the first-order approximation for reliability analysis is given by considering two types of failure surfaces - concave and convex failure surfaces. If a failure surface is concave toward the origin, the approximate reliability index is the minimum distance from the origin to all hyper-planes tangent to the failure surface. On the other hand, for the case of a convex failure surface, the approximate reliability index is the maximum distance between the origin and the tangent hyper-plane. In both cases, the approximate reliability index is the minimum distance from the origin to the points on the failure surface, namely, the so-called Hasofer-Lind reliability index.

The second-order approximation is necessary if the failure surface is far from the origin and with large curvatures. To obtain an analytical result, a hyper-parabolic surface is usually used to replace the original failure surface. In this paper, a new expression is developed to calculate the second-order approximate reliability.

Exact and approximate values of the reliability for a circular shaft under two random bending moments and a random torque are evaluated. It is shown that the second-order approximation yields accurate results even when the failure surface is not far from the origin and with large curvatures.

6 Acknowledgment

This study has been supported by National Science Foundation under award No. MSS-9215698 (Dr. Ken Chong - Program Director). Any opinions, findings, and conclusions or recommendations expressed by this publication are those of the authors and do not necessarily reflect the views of the National Science Foundation. Discussion with Dr. Karl Breitung is appreciated.

7 References

1. Ditlevsen, O., Generalized second moment reliability index. *Journal of Structural Mechanics*, **7** (1979) 435-451.
2. Madsen, H. O., Krenk, S. and Lind, N. C., *Methods of Structural Safety*, Prentile-Hall Inc., Englewood Cliffs, NJ., 1986.
3. Ditlevsen, O., Principle of normal tail approximation. *Journal of Engineering Mechanics Division*, ASCE, **107** (1980) 1191-1206.
4. Hohenbichler, M. and Rackwitz, R., Non-normal dependent vectors in structural safety. *Journal of Engineering Mechanics Division*, ASCE, **107** (1980) 1227-1238.
5. Hasofer, A. M. and Lind, N. C., Exact and invariant second-moment code format. *Journal of Engineering Mechanics Division*, ASCE, **100** (1974) 111-121.
6. Elishakoff, I., *Probabilistic Methods in the Theory of Structures*, Wiley-Interscience, New York, 1983.
7. Elishakoff, I. and Hasofer, A. M., On the accuracy of Hasofer-Lind reliability index. *Proceedings of International Conference on Structural Safety and Reliability*, Volume I, Kobe, Japan, 1985, pp. 229-239.
8. Elishakoff, I. and Hasofer, A. M., Exact versus approximate determination of structural reliability. *Applied Scientific Research*, **44** (1987) 303-312.
9. Shinozuka, M., Basic analysis of structural safety. *Journal of Structural Engineering*, ASCE, **109** (1980) 721-740.
10. Schuëller, G. I. and Strix, R., A critical appraisal of methods to determine failure probabilities. *Structural Safety*, **4** (1987) 293-309.
11. Schuëller, G. I., Bucher, C. G., Bourgund, U., and Ouypornprasert, W., On efficient computational schemes to calculate structural failure reliabilities. *probabilistic Engineering Mechanics*, **4** (1989) 10-18.

12. Schuëller, G. I. and Bayer, V., Computational procedures in structural reliability. *Uncertainty Modeling and Analysis*, ed. B. Ayyub, IEEE Computer Society Press, Los Alamitos, CA, 1993, pp. 552-559.
13. Fiessler, B., Neumann, H-J. and Rackwitz, R., Quadratic limit states in structural reliability. *Journal of Engineering Mechanics Division*, ASCE, **105** (1979) 661-676.
14. Breitung, K., An asymptotic formula for the failure probability. *Reliability Theory of Structural Engineering Systems*, ed. G. Mohr. Danish Engineering Academy, Civil Department, Lyngby, Denmark, 1982, pp. 19-45.
15. Breitung, K., Asymptotic approximation for multinormal integrals. *Journal of Engineering Mechanics Division*, ASCE, **110** (1984) 357-366.
16. Tvedt, L., Two second-order approximations to the failure probability, Veritas Report R74-33, Det norse Veritas, Oslo, Norway, 1983.
17. Freudenthal, A. M., Safety and probability of structural failure. *Transactions ASCE*, **121** (1965) 1337-1375.
18. Murzewski, J., *Niezawodnosc Konstrukcji Inzyniernich*, Arkady Publishing House, Warszawa, Poland, 1989 (in Polish).
19. Levi, R., Calculs probabilistes de la security des constructions. *Ann. Ponts et chaussees*, **119** (1949) 493-539 (in French).
20. Rackwitz, R. and Fiessler, B., Structural reliability under combined random load sequences. *Computers and Structures*, **9** (1978) 489-494.
21. Bolotin, V. V., *Application of the Methods of the Theory of Probability and the Theory of Reliability to Analysis of Structures*, State Publishing House for Buildings, Moscow, 1971. (English translation: FTD-MT-24-771-73, Foreign Technology Division, Wright-Patterson AFB, Ohio, 1974).

Fatigue Lifetime Prediction for Uncertain Systems

Fabio Casciati and Pierluigi Colombi

Department of Structural Mechanics, University of Pavia,
via Abbiategrasso 211, I27100, Pavia (Italy)

Abstract. Fatigue life prediction deals with factors which can be modeled as random variables, random processes or stochastic fields. Given a probabilistic definition of these factors, the life distribution results from a reliability analysis. The advanced mean value first order (AMVFO) method was suggested to minimize the number of functions to be evaluated in the computation of the probability of failure, but it works only if all the uncertainties can be regarded as random variables. A more powerful method must be used when random processes and random fields are present; the potentialities of an extended response surface method are illustrated in this paper. The cumulative distribution function (CDF) of the number of cycles to failure is evaluated by both methods and the results are compared and discussed.

Keywords. Fatigue, lifetime prediction, response surface, structural reliability

1 INTRODUCTION

Experimental results show a large scatter for the fatigue strength of any given material. This variability is due to material inhomogeneities and to flaws introduced by the production process. A source of uncertainty is also associated with the nature of the external loads. This is why stochastic models are widely recognized to be unreplaceable in fatigue life prediction.

A stochastic model of fatigue crack growth should be based on fracture mechanics. The simplest model regards as random variables all the parameters involved in the crack growth process [9]. It is unable however to describe the material inhomogeneities and the external loads randomness which must be described by random processes and/or random fields.

A more accurate model regards the crack growth law as a stochastic differential equation by introducing temporal fluctuations into the fatigue parameters. The results are easily obtained, in terms of the cumulative distribution function (CDF) of the number of cycles to failure, when the crack length is approximated by a Markov process. Some fluctuation models, although convenient from a probabilistic point of view, show mathematical inconsistencies, e.g. negative crack growth rates [8,10,12] ! Moreover even if these inconveniences are avoided [4,13], system uncertainties cannot be incorporated into a simple stochastic fluctuation. A reliability assessment approach must be considered for this purpose. The advanced mean value first order (AMVFO) method [17] is appropriate but it is unable to deal with random processes and fields. Conversely, if an extended version of the response surface methodology [3,7] is used, any kind of fatigue crack growth equation can be adopted and the parameters can be described by random variables, random processes or random fields [5]. The potentialities of this approach are discussed in this paper, thus providing also a better understanding of the original AMVFO.

2 CRACK PROPAGATION MODEL

The crack propagation law adopted in this paper is the Paris law [2]

$$\frac{da}{dN} = C \cdot (\Delta K)^m = C \cdot [Y(a) \cdot S \cdot \sqrt{\pi a}]^m \quad , \tag{1}$$

but any of the several available alternative models could be used. In Eq. (1) ΔK is the stress intensity factor range, $Y(a)$ is the geometry factor, S the stress range and C and m are material constants.

Given a structural component, it will generally experience variable amplitude stress. Such a situation can be approached by two different techniques: the equivalent stress range method [11] and the cycle-by-cycle counting method [2].

1. Assume that any sequence effect is negligible; then the concept of equivalent stress range can be introduced. It can be defined in the ergodic case as

$$S_{eq} = (E[\underline{S}^m])^{1/m} \quad , \tag{2}$$

where

$$E[\underline{S}^m] = \int_0^\infty S^m p_{\underline{S}}(S) dS \quad . \tag{3}$$

In Eqs. (2) and (3) $\underline{S}$ is a random variable modeling the stress range, and $p_{\underline{S}}(S)$ is its probability density function. The term S_{eq} permits one to cast the fracture mechanics equations in the characteristic S-N fatigue format [2].

2. When large overloads are present in the load history, sequence effects may become significant. The concepts of crack closure and a cycle-by-cycle counting must therefore be used [1].

This paper is restricted within the first approach. Counting technique schemes are presently under investigation [6].

Accuracy requires that a threshold level ΔK_{th} is introduced for ΔK [14]. Below this level the crack will not propagate. The consequent threshold stress range level is (see Eq. (1))

$$S_L(a) = \frac{\Delta K_{th}}{Y(a) \cdot \sqrt{\pi a}} \quad . \tag{4}$$

This means that the crack growth only occurs for stress cycles of intensity $S > S_L(a)$, where, of course, S_L, being a function of a, changes as the crack grows. One can then introduce the random variable $\underline{S}_d$, denoting the stress range above S_L, and, hence, the equivalent stress range for cycles which cause damage

$$S_{d_{eq}}(a) = E[\underline{S}_d{}^m|a]^{1/m} \quad . \tag{5}$$

In Eq. (5) one has

$$E[\underline{S}_d{}^m|a] = \int_{S_L(a)}^{\infty} S^m p_{\underline{S}_d}(S|a)dS \quad , \tag{6}$$

where $p_{\underline{S}_d}(S|a)$ is the probability density function of the damaging stress range. The Paris equation becomes

$$\frac{da}{dN_d} = C \cdot [Y(a) \cdot S_{d_{eq}}(a) \cdot \sqrt{\pi a}]^m \quad , \tag{7}$$

where N_d is the number of damaging cycles only. Now it is possible to rearrange the fatigue crack equation in order to express it in terms of the total applied stresses and cycles [14]. Let

$$S_0{}^m(a) = \int_{S_L(a)}^{\infty} S^m p_{\underline{S}}(S)dS \tag{8}$$

and assume that crack growth is slow. Then the fatigue crack growth rate is approximated by

$$\frac{da}{dN} = C \cdot [Y(a) \cdot S_0(a) \cdot \sqrt{\pi a}]^m \quad . \tag{9}$$

This equation can be integrated to provide an expression for the number of duty cycles to failure. Integration from a_0, the initial crack length, to a_F, the failure crack length, leads to the number of cycles to failure N_F

$$N_F = \frac{1}{C} \int_{a_0}^{a_F} \frac{da}{(S_0(a))^m \cdot (Y(a))^m \cdot (\pi a)^{m/2}} \quad . \tag{10}$$

Note that Eq. (10) is formally equivalent to what is done in the standard S-N approach.

3 ADVANCED MEAN VALUE FIRST ORDER METHOD

Consider the expression of the fatigue life given by Eq. (10). The goal is to evaluate the cumulative distribution function (CDF) of N_F. Let $\mathbf{X}$ be the vector of random variables in Eq. (10). Following the advanced mean value concept [16,17,15], the CDF of $N_F(\mathbf{X})$, N_F being an implicit function of $\mathbf{X}$, is evaluated by introducing a first-order Taylor's series expansion of N_F about the mean value $\mu_{\mathbf{X}}$ of $\mathbf{X}$. This requires solutions at $\mu_{\mathbf{X}}$ and at small perturbations about $\mu_{\mathbf{X}}$ in order to evaluate the parameters of the linear function. The function $N_F(\mathbf{X})$ is then replaced by a linear function $N'_F(\mathbf{X})$

$$N'_F(\mathbf{X}) = c_0 + \sum_{i=1}^{k} c_i(X_i - \mu_{X_i}) \quad , \tag{11}$$

where

$$c_i = [\frac{\partial N_F}{\partial X_i}]_{\mu_{\mathbf{X}}} \quad , \tag{12}$$

The corresponding estimates for the mean value μ_{N_F} and the standard deviation σ_{N_F} of N_F are

$$\mu_N = N(\mu_{\mathbf{X}}) \quad , \tag{13}$$

$$\sigma_N = \left(\sum_{i=1}^{n} [\frac{\partial N}{\partial x_i}^2]_{\mu_{\mathbf{X}}} \cdot \sigma_{x_i}{}^2\right)^{1/2} \quad . \tag{14}$$

The partial derivatives are computed using the perturbation data. The CDF is approximated at discrete points N_{F_i}. The number of points is arbitrary, but it was shown that three or four points on both side of the mean are adequate to define the CDF when the probability of failure is expected to be a smooth function. A first estimate of the CDF is made from the linear form. The limit state function at each N_{F_i} is

$$g_i(\mathbf{X}) = N'_F(\mathbf{X}) - N_{F_i} \quad . \tag{15}$$

The probability of failure is then computed by

$$P_F(N_{F_i}) = \Phi(-\beta_i) \quad , \tag{16}$$

where β_i is the reliability index associated with each g_i, $P_F(N_{F_i})$ is the probability of failure after N_{F_i} duty cycles,and $\Phi(\cdot)$ the standard normal CDF. This is the mean value first order (MVFO) method; it leads one to estimate the design point $\mathbf{X}^*_i$ for the linear approximation of Eq. (11). To improve the estimate of the CDF, the actual lifetime $N_F(\mathbf{X}^*_i)$ is evaluated at each design point of the sample space identified. These values are associated with the probability of failure $P_F(N_{F_i})$ obtained from Eq. (16). The corresponding plot is regarded as an improved form of the searched CDF. This improvement gives rise to the advanced mean value first order (AMVFO) method. The total number of function evaluations is given by

$$J = n_v + n_p + 1 \quad , \tag{17}$$

where n_v is the total number of random variables,and n_p is the number of points used to define the CDF. The estimate of $P_F(N_F)$ could be improved by introducing a new linear approximation for N_F at each design point $\mathbf{X}^*_i$. The reliability analysis is then performed for each of the linear functions; as a result the probability estimate at each value of N_{F_i} is more accurate. Still another improvement could be made by evaluating the real function at each design point. This process can be repeated, but experience suggests that the CDF obtained by the first function evaluation (AMVFO) is generally very close to the exact solution [17].

4 RESPONSE SURFACE METHOD

An alternative method for the evaluation of the CDF of the number of cycles to failure N_F is based on the evaluation of a response surface (response surface method). This surface describes N_F by a polynomial form of the input variables $\mathbf{X}$. This result is pursued by regression analysis on the output of some numerical experiments, which must be appropriately planned. The number of experiments to be performed could be rather high if the problem depends on several input random variables. The extended response surface methodology adopted in this paper is able to work with several random variables and to incorporate stochastic processes and random fields. The reader is referred to [3] for the details of the procedure which is just summarized in this section.

The input random variables are grouped in two classes as the result of a preliminary sensitivity analysis: primary random variables and secondary random variables. The primary random variables are the random variables whose randomness deeply influences the randomness of the structural response. The secondary random variables are the variables whose randomness is of marginal importance for the randomness of the structural response.

Let $\mathbf{X}'$ the vector of the primary random variables. Suitable transformations $\bar{N_F}$ of N_F and $\bar{\mathbf{X}}$ of $\mathbf{X}'$ are introduced in order to make the model more flexible. The relation between $\bar{N_F}$ and $\bar{\mathbf{X}}$ is described using a second order polynomial. In matrix notation one has:

$$\bar{N_F} = \alpha_0 + \bar{\mathbf{X}}^T \boldsymbol{\alpha}_1 + \bar{\mathbf{X}}^T \boldsymbol{\alpha}_2 \bar{\mathbf{X}} + \epsilon \quad . \tag{18}$$

The coefficients $\boldsymbol{\alpha}_0$, $\boldsymbol{\alpha}_1$ and $\boldsymbol{\alpha}_2$ are evaluated by a regression analysis over the results of the numerical experiments in a central composite plan [3]. The term ϵ on the right-hand side takes into account the model error and the effect of the secondary random variables $\mathbf{X}''$ as well as the presence of stochastic processes and random fields. The basic differences from the previously discussed AMVFO method can be envisaged:

1. Eq. (11) is linear while Eq. (18) is quadratic;
2. Eq. (11) is written in the original space of the variable $\mathbf{X}$ while Eq. (18) is written in a transformed space (generally the standardized one);
3. A deterministic error term η was not written explicitly in Eq. (11); however for a given vector $\mathbf{X}^*{}_i$, the probability $P_F(N_{Fi})$ is associated with $N_F(\mathbf{X}^*{}_i)$ instead of $N'_F(\mathbf{X}^*{}_i)$, where

$$N_F(\mathbf{X}^*{}_i) = N'_F(\mathbf{X}^*{}_i) + \eta \quad . \tag{19}$$

Therefore, Eq. (11) assumes that η can be computed, while the error term ϵ in Eq. (18) is a random variable, the variance of which is estimated through statistical analysis. This makes possible to incorporate in it the effect of all the uncertainties which affect N_F.

5 A NUMERICAL EXAMPLE

A simple numerical example is developed to illustrate the different potentialities of the two fatigue reliability methods illustrated in the previous sections, namely the AMVFO and the response surface approach. The cumulative distribution function (CDF) of the number of cycles to failure is computed by the two methods, and the results are compared. Assume that the geometry factor $Y(a)$ in Eq. (10) has the form [14]

$$Y(a) = (\psi - 1) \cdot e^{-\gamma \cdot a^{\nu}} + 1 \quad , \tag{20}$$

where ψ is the theoretical stress concentration factor, and γ and ν are constants. Eq. (20) provides a reasonable fit to $Y(a)$ for cracks which grow in a region of stress concentration. Moreover assume that the long term stress range $\underline{S}$ is Weibull distributed

$$p_{\underline{S}}(S) = \frac{\xi}{\delta} \cdot \left(\frac{S}{\delta}\right)^{\xi-1} \cdot e^{-(\frac{S}{\delta})^{\xi}} \quad , \qquad S \geq 0 \quad , \tag{21}$$

where ξ is the Weibull shape parameters, and δ is given by

$$\delta = \bar{S_0} \cdot (lnN_T)^{-1/\xi} \quad . \tag{22}$$

In Eq. (22) N_T is the total number of cycles, and $\bar{S_0}$ is the design stress range defined by $P(S > \bar{S_0}) = \frac{1}{N_T}$.

The numerical data of the terms in Eq. (10) are given in Table I.

5.1 AMVFO Solution

Consider first the solution by the AMVFO method. Here $\bar{S_0}$, ΔK_{th}, C and a_0 are identified as the random design factors. Eq. (8) can be solved analytically to produce

$$S_0{}^m(a) = \delta^m \cdot Q(1 + \frac{m}{\xi}; S_L(a)) \quad , \tag{23}$$

where $Q(\cdot)$ is the incomplete gamma function. The number of cycles to failure N_F is computed by solving Eq. (10) numerically.

N_F is computed at the mean values of the random variables and then at perturbated values (arbitrally chosen as the median plus 0.10 standard deviations); its linear expansion is about the mean values. The resulting first order approximation to N_F is

$$N'_F = 3.09 \cdot 10^9 - 1.08 \cdot 10^8 \cdot \bar{S_0} + 3.18 \cdot 10^8 \cdot \Delta K_{th} - 2.65 \cdot 10^{18} \cdot C - 3.94 \cdot 10^{10} \cdot a_0 \quad . \tag{24}$$

First the MVFO method is used to compute $P(\underline{N}_F \leq N_i)$ at preselected values of N_i. This solution is shown in Fig. 1 as a dashed line. The subsequent evaluation at the design points of the actual values of N_F, AMVFO, provides the improved solution; it is shown by solid line in Fig. 1. The Monte Carlo solution is eventually superimposed in order to check the accuracy of the computed CDF.

5.2 Response Surface Solution

The CDF of the number of cycles to failure $\underline{N}_F$ is now evaluated by using the response surface technique of section 4. The primary random variables are $\underline{\bar{S_0}}$, $\Delta \underline{K}_{th}$, $\underline{C}$ and $\underline{a}_0$ (see Table I). Moreover, the final crack length $\underline{a}_F$ is regarded as a secondary random variable. Its probabilistic definition is given in Table I. The numerical experiment plan requires that 31 experiments be

Table I. Numerical data for the deterministic and random parameters in the fatigue crack growth equation (see Eq. 10).

Variable	Distribution	Mean Value (Mpa units)	Standard deviation (Mpa units)
$\underline{S}$	Weibull	see Eqs. (20) and (21)	
$\underline{\bar{S}}_0$	Type I extreme value	172.05	13.764
$\Delta \underline{K}_{th}$	Normal	121.394	12.1394
$\underline{C}$	Weibull	1.2723E-13	1.2723E-14
$\underline{a}_0$	Lognormal	25.4E-02	1.27E-02
ξ	Constant	1.	–
N_T	Constant	1.0E+8	–
m	Constant	3.	–
$\underline{a}_F$	Normal	25.4	2.54E-01
ψ	Constant	3.3	–
γ	Constant	4.42	–
ν	Constant	0.44	–

carried out. The central point of the planned experiments corresponds to a number of cycles to failure equal to 5.92E+08; this value is evaluated by solving numerically Eq. (10), where $S_0(a)$ is evaluated analytically by Eq. (23). A second order polynomial is then fitted on the results of the numerical experiments. The probability of failure for a given value N_i of the number of cycles is then computed using a first order reliability method. Fig. 2 shows the resulting CDF of N_F to be compared with the Monte Carlo solution. In Fig. 3 the result obtained by the AMVFO is compared with the response surface estimate.

Since the low tail of the CDF is of practical interest, a second response surface [7] is evaluated by shifting the numerical experiment plan toward the design point for which the number of cycles to failure is equal to 1.82E+08 with a probability of failure 7.E-03.

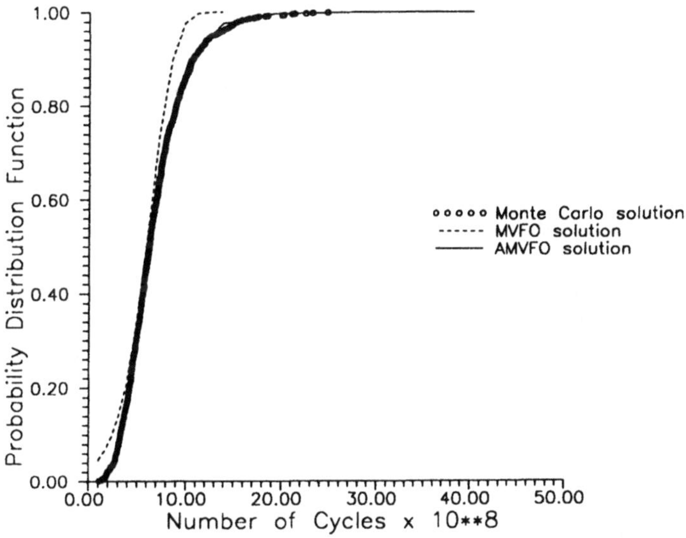

Fig. 1. Comparison between the AMVFO solution, the MVFO solution and the Monte Carlo Solution.

The corresponding CDF is given in Fig. 4, where it is compared with the original response surface solution and the Monte Carlo plot. The match of this second solution with the simulation result is excellent.

Introduce now a stochastic characterization of the process $C(a)$, i.e. a random dependence of the material parameter C on the crack length a. In particular, consider

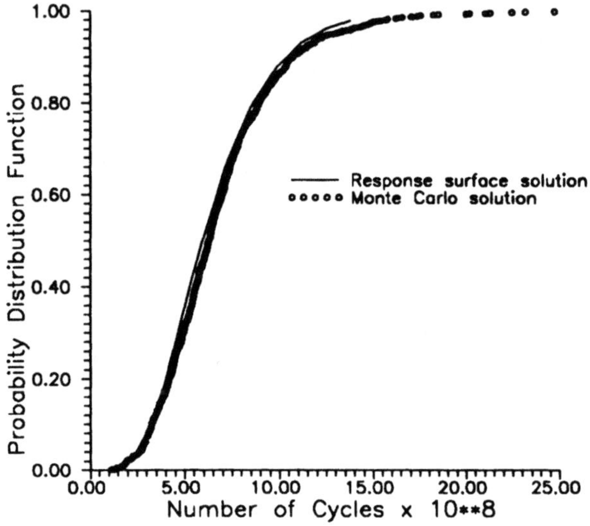

Fig. 2. Comparison between the response surface solution and the Monte Carlo solution.

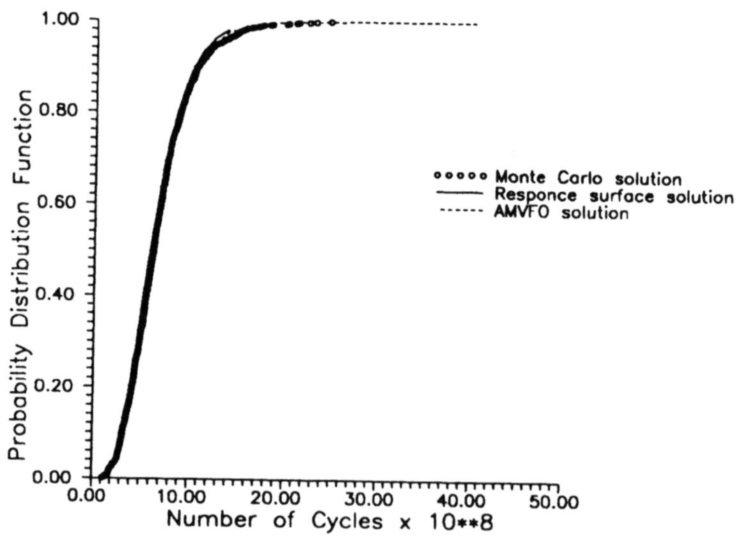

Fig. 3. Comparison between the response surface solution, the AMVFO solution and the Monte Carlo solution.

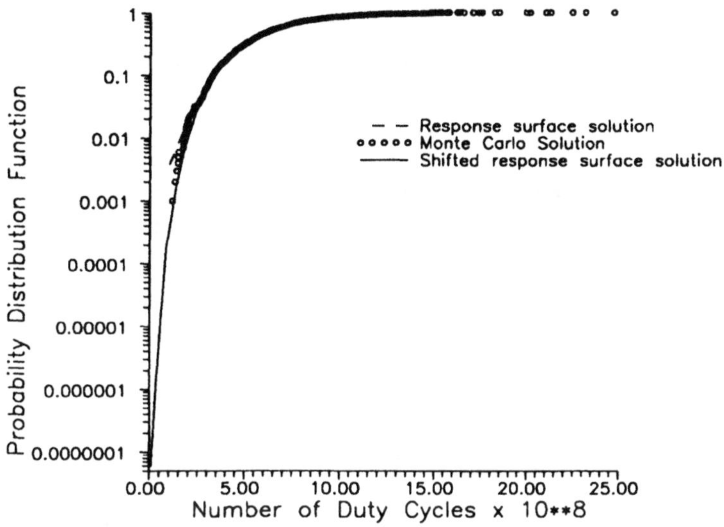

Fig. 4. Comparison in logarithmic scale between the shifted response surface solution, the original response surface solution and the Monte Carlo solution.

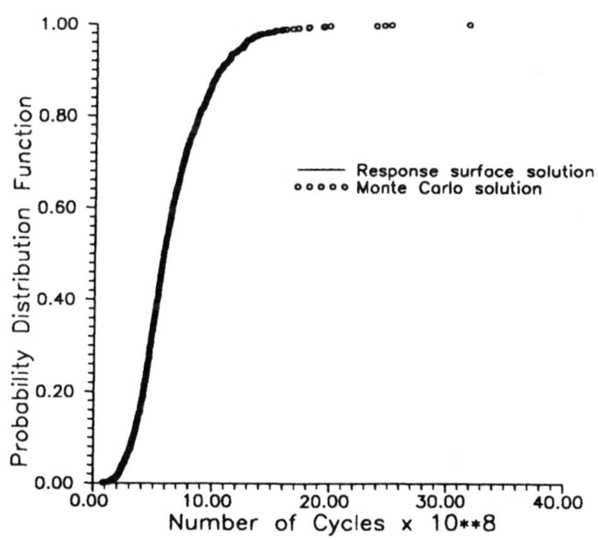

Fig. 5. Comparison between the response surface solution and the Monte Carlo solution when C(a) is modeled as a stochastic process.

$$C(a) = C \cdot \gamma(a) \quad , \tag{25}$$

where C is the random variable defined in Table I,and $\gamma(a)$ is a filtered Gaussian white noise of mean value 1. The parameters of the force-displacement filter were selected in such a way that one cuts the high frequencies and makes physical the white noise. Its finite standard deviation is 0.05,and the process intensity is selected of consequence. The AMVFO method is unable to provide any solution, whereas the response surface scheme gives the result of Fig. 5. Once more the agreement between calculated and simulated distributions is very satisfactory.

6 CONCLUSIONS

Two different techniques of fatigue reliability analysis are studied, namely the AMVFO method [17] and the extended response surface approach [3]. The AMVFO method can only model the randomness present in fatigue crack propagation law by random variables. Conversely, random processes can be easily introduced to describe the spatial variability of the fatigue resistance when the CDF of the number of cycles to failure $\underline{N}_F$ is evaluated by the response surface scheme.

Furthermore, the AMVFO method is shown to be a special form of the response surface scheme in which the error term is regarded as a deterministic function and, hence, estimable on the basis of specific numerical experiments. All the methods make use of a fatigue crack propagation law where the concept of equivalent stress range is applied. When large overloads are present, a more refined crack propagation law should be used. Ongoing research is attempting to introduce the so-called retardation effect. An extension of the response surface scheme to incorporate it does not seem to offer serious difficulties.

7 ACKNOWLEDGEMENTS

This research was supported from funds by both the Italian Ministry of University and Scientific and Technological Research (MURST) and the Italian Spatial Agency (ASI).

8 REFERENCES

1. Bignonnet, A. & Olagnon, M., Fatigue Life Prediction for Variable Amplitude Loading, *X Offshore Mechanics and Artic Engineering Conference*, ed. Salama M.M., Ree H.C., Kam J.P.C., Booth G.S., Williams J.G., ASME, New York, 1991, pp. 395-401.
2. Broek D., *Elementary Engineering Fracture* Mechanics, Martinus Nijhoff Publishers, New York, 1982.
3. Casciati F. & Faravelli L. , *Fragility Analysis of Complex Structural System*, Research Studies Press Ltd., Taunton, 1991.
4. Casciati F., Colombi P. & Faravelli L., Filter Technique for Stochastic Crack Growth, *Fatigue Fract. Engng. Mater. Struct.*, **15 (5)**, (1992), 463-475.
5. Casciati F., Colombi P. & Faravelli L., Lifetime Prediction of Fatigue Sensitive Structural Elements, *Structural Safety*, **12**, (1993), 105-111.
6. Casciati F. & Colombi P., Load Combination and Related Fatigue Reliability Problems, *Structural Safety*, in press.
7. Faravelli L., Structural Reliability via Response Surface, *Nonlinear Stochastic Mechanics*, ed. Bellomo N. & Casciati F., Springer-Verlag, Berlin, 1991, pp. 213-223.
8. Ishikawa H., Tsurui A., Tanaka H. & Ishikawa H., Reliability Assesment of Structures Based Upon Probabilistic Fracture Mechanics, *Probabilistic Engineering Mechanics*, **8**, (1993), 43-56.
9. Kozin F. & Bogdanoff J.L., A Critical Analysis of Some Probabilistic Models of Fatigue Crack Growth, *Eng. Frac. Mech.*, **14**, (1981), 59-89.
10. Lin Y.K. & Yang J.N., On Statistical Moments of Fatigue Crack Propagation, *Eng. Fract. Mech.*, **18**, (1983), 243-256.
11. Schijve J., Observation on the Prediction of Fatigue Crack Growth Propagation under Variable Amplitude Loading., Fatigue Crack Growth Under Spectrum Loads, *ASTM SPT 595*, 1976.
12. Solomos G.P., First-Passage Solution in Fatigue Crack Propagation, *Prob. Eng. Mech.*, **4**, (1989), 32-39.
13. Tang J. & Spencer B.F. Jr., Reliability Solution for the Stocahstic Fatigue Crack Growth Problem, *Eng. Fract. Mech.*, **34**, (1989), 419-433.
14. Wirshing P.H., Ortiz K. & Chen Y.N., Fracture Mechanics Fatigue Model in a Reliability Format, *VI Int. Symp. Offshore Mech. and Artic Eng.*, ASME, New York, 1987, pp. 331-337.
15. Wirshing P.H., Torng T.Y. & Martin W.S., Advanced Fatigue Reliability Analysis, *Int. J. Fatigue*, **13 (5)**, (1991), 389-394.
16. Wu Y.T., Burnside O.H. & Dominquez J., Efficient Probabilistic Fracture Mechanics Analysis, *Fourth Int. Conf. Numerical Methods in Fracture Mechanics*, Pineridge Press, Swansea, 1987, pp. 85-100.
17. Wu Y.T., Millwater H.R. & Cruse T.A., An Advanced Probability Structural Analysis Method for Implicit Performance Functions, *30th Structures, Structural Dynamics and Materials Conference, Mobile, Alabama*, 1989.

Probabilistic Assessment of Composite and Smart Composite Structures

Christos C. Chamis[1] and Michael C. Shiao[2]

[1] NASA Lewis Research Center, Cleveland, Ohio 44135, U.S.A.
[2] Sverdrup Technology, Inc., Brook Park, Ohio 44142, U.S.A.

Abstract. A general purpose methodology for integrated probabilistic assessment of composite structures and smart composite structures is discussed and demonstrated using aircraft fuselage and wing type structures with rectangular cutouts. The computational simulation was performed for the probabilistic assessment of the structural behavior including global displacements and local stresses. The scatter in the structural response is simulated based on the inherent uncertainties in the primitive (independent random) variables at the fiber/matrix constituent, ply, laminate and structural scales that describe the composite structures. The effect of uncertainties due to fabrication process variables such as fiber volume ratio, void volume ratio, ply misalignment, and ply thickness as well as those for sensors and controls is also included. Results show that in a composite structure with rectangular cutouts the uncertainties in the longitudinal ply stress is mainly due to the uncertainty in the laminate thickness. Results also show that a smart composite wing can be controlled to minimize distortions and to have specified stress levels in the presence of cutouts.

Keywords. Composite mechanics, computer programs, fuselage components, wing sections, sensitivity factors, cutouts

1. Introduction

The technology of advanced fiber composites has matured to the point where these composites are prime contenders for various aeronautic/aerospace related applications. Their outstanding mechanical properties are very attractive to the aerospace industry especially the high strength/stiffness to density ratio. They also possess excellent fatigue strength and ability to resist corrosion and impact. However, the material properties of composites depend on a wide variety of primitive (random) variables. The variables are, for example, fiber and matrix material properties at the constituent scale and fabrication variables at the ply scale such as fiber volume ratio, void volume ratio, ply orientation, and ply thickness. Additionally, the service environments that these structures must endure are also statistical in nature. Since the uncertainties of the material properties of a typical composite can not be quantified deterministically, the behavior of composite structures exhibits considerable scatter from the mean or deterministic value.[1] The probabilistic composite mechanics and the probabilistic composite structural analysis methodologies have been integrated into a single code IPACS (Integrated Probabilistic Assessment of Composite Structures)[2]. Also recent developments in the application of smart structures concepts by the use of sensor/control material, show the potential to enhance the structural performance as well as durability/reliability[3,4]. These can be readily integrated into a smart composite structure by using combinations of intraply and interply hybrid composites[5]. Thus making it possible to ascertain that smart composite structures will operate in the design-specified range. The intraply hybrid mechanics for composites are embedded in the computer code ICAN[6] for integrated composite analysis.

The objective of this paper is to (1) describe the methodology and the capabilities of the IPACS code, (2) demonstrate its effectiveness for an aircraft fuselage type (stiffened composite cylindrical shell) structure with the rectangular cutouts, (3) describe and demonstrate the adaptation of the intraply hybrid composite concept to smart composite structures. The description of the IPACS computer code is in Section 2 while that of the smart composite structure is in Sections 3 and 4.

2. Integrated Probabilistic Assessment of Composite Structures, IPACS

The methodology embedded into IPACS is shown in Figure 2.1.

IPACS integrates several computer codes as shown in Figure 2.2 COBSTRAN (COmposite Blade STructural ANalyzer)[7] is a dedicated finite element model generator for composite structures. PICAN (Probabilistic Integrated Composite Analyzer)[8] enables the computation of the perturbed and probabilistic composite material properties/structural responses at the constituent, ply and laminate scales. NESSUS (Numerical Evaluation of Stochastic Structures Under Stress)[9] simulates the perturbed and probabilistic structural response at global (structural) scale. PICAN and NESSUS share the FPI (Fast Probability Integrator)[10] module for the application of fast probability integration algorithm to obtain the probability density functions of the materials behavior and structural response.

The probabilistic assessment of composite structures using IPACS starts with the identification of primitive variables at the micro and macro composite scales including the fabrication process. These variables are selectively perturbed several times in order to create a database for the determination of the relationships between the desired materials behavior/structural response and the primitive variables. The composite micromechanics are used to carry over the scatter in primitive variables to the ply and laminate scales (steps A and B in Figure 2.1). Laminate theory is then used to determine the scatter in the material behavior at the laminate scale (step C). This leads to the perturbed resultant force/moment -strain/curvature relationships used in the structural analysis. Next, the finite element analysis is performed to determine the perturbed structural responses corresponding to the selectively perturbed primitive variables (step D). This completes the description of the hierarchical composite material/structure synthesis shown on the left side of Figure 2.1. The multi-scale progressive decomposition of the structural response to the laminate, ply, and to fiber/matrix constituent scales is shown on the right side of Figure 2.1 via steps E through G. After the decomposition, the perturbed fiber, matrix and ply stresses can be determined. An important feature of IPACS, depicted at the bottom of Figure 2.1, is the nonlinear multi-factor interaction model for computing fiber/matrix constituent material properties including the effects of prevailing service environments.

The fast probability integrator (FPI)[10] is used to determine the functional relationship between the response and the primitive variables. The cumulative distribution function of the response is then calculated with the numerically determined functional relationship and the known (or assumed) probability density function of the primitive variables. The sensitivity factors of the primitive variables to each response's cumulative probability are also determined.

IPACS is used to probabilistically assess a stiffened composite cylindrical shell with several rectangular cutouts. The composite is made from graphite fibers and epoxy matrix. It consists of eight vertical stringers, five circumferential frames and a cylindrical shell with 12 cutouts as shown in Figure 2.3. The diameter of the cylindrical shell is 20 feet and its length is 50 feet. The shell is supported at one end by a set of translational and torsional spring and is unconstrained at the other end where random loads are applied as shown in Figure 2.4. The modeling of supports using translational and torsional springs allows for uncertainties in the boundary conditions enabling simulation of real-life conditions. The composite configurations of the shell skin, stringers and frames are $[\pm45/0_2/\pm45/0_2/\pm45/0/90]_s$, $[0_{24}]$ and $[0_{24}]$, respectively. The structure finite element model has 1256 nodes (6 degrees of freedom per node) and 1196 four-noded shell elements. The probabilistic assessment is carried out by identifying the primitive variable at the constituent, ply and structure scales. Table 2.1 shows the probability information of the primitive variables related to the fiber/matrix constituent materials. Table 2.2 provides the same information for the primitive variable related to the fabrication process.

The probability density functions (PDF) of the displacements at locations A and B along the shell circumference at the free end are shown in Figure 2.5. The mean displacements at location A and B are shown in Table 2.3. It is readily seen that mean displacements at tip are increased by 11 to 19 percent with rectangular cutouts. The displacement corresponding to the 0.999 cumulative probability at location B is determined to be 2.57 inches. This type of information can be used to assess the reliability with respect to the design allowable for the displacement tolerance. For example, if the design reliability is set at 0.999 with the allowable displacement to be 3 inches, than the corresponding failure probability is less than 0.001. However, if the allowable displacement is set to be 2 inches, the failure probability will be greater than 0.001. This latter reliability requirement is therefore violated. Sensitivity factors, ranking of the importance of random variables to the probability distribution of the displacement, are shown in Figure 2.6. The result indicates that at 0.999 probability level, the random force in Y direction has the most probability contribution (sensitivity factor = 0.93), followed by laminate thickness, fiber and matrix moduli (sensitivity factor = 0.26, 0.15, 0.15 respectively).

The PDFs of the stresses in 45°, -45° and 0° plies near the root between the cuts (location C in Figure 2.5), normalized with respect to the mean stress in the 0° ply are plotted in Figure 2.7. The maximum longitudinal stress occurs in the 0° ply. The mean value of this maximum stress is increased by 7 percent with rectangular cutouts as shown in Table 2.3. The stresses in the fiber direction in the +45° and -45° plies are 40% and 20% of that of the 0° ply, respectively. It indicates that 0° plies carry most external loads. The sensitivity factors for the longitudinal stress in 0° ply are shown in Figure 2.8. At 0.001 probability, the skin laminate thickness contributes the most probability to the ply stress (sensitivity factor = 0.84). The next highest contribution is that due to the random forces in axial and lateral directions (0.37 and 0.34 in sensitivity factors).

3. Probabilistic Smart Composite Structures

Figure 3.1 depicts a conceptual diagram of a smart composite wing system. The essential parts of a smart structure include (1) a structure (2) strategically located sensors, (3) signal processors which process the signals generated by the sensors, (4) dedicated computers with suitable firmware and software which continuously check the structural response magnitude and compares it to predetermined (red line) values and which provide desired corrections to the controller, (5) controller which signals the actuator to implement the desired corrections and (6) actuators. In the concept described herein the sensors are also the actuators.

The adaption of the intraply hybrid composite concept[5] to smart composite structures is depicted schematically in Figure 3.2. In Figure 3.2(a), the intraply hybrid configuration is shown while in Figure 3.2(b), its adaptation to smart composite structures is shown. It is observed in Figure 3.2 that the intraply hybrid composite consists of plies which have strips, of a regular (host) composite material and strips of material for sensor/actuator devices. Actuators, made of control materials such as piezoelectric ceramic, are used to control the behavior of the composite structure by expanding (positive induced strain) or contracting (negative induced strain) the sensor/actuator strips to achieve the requisite design and operational goals. However, the strains induced by the actuator are affected by uncertainties in several factors that can only be quantified probabilistically. These include: (1) inaccurate measurements made by the sensors, (2) application of incorrect electric field, (3) uncertain induced strain-electric field strength relationship, (4) uncertain material properties for the sensor/actuator materials, (5) uncertain electric field strength, and (6) improper location of the sensor/actuator materials. Because of these factors, the use of sensor/actuator devices increases the uncertainty in the already uncertain composite structural behavior. To properly quantify the benefits of the induced strain, a comprehensive probabilistic assessment is needed. This type of assessment is readily performed by the use of the IPACS computer code where the uncertainties of the sensors and controls are also incorporated as described below.

The probabilistic assessment of the smart composite structure concept as described previously is demonstrated by using it to configure a smart composite wing. Comparable assessment for a composite plate is described in[11]. The geometry of the composite wing's internal structure is shown in Figure 3.3. The wing is loaded with nonuniform pressure and temperature which vary parabolically from root to tip with 5 psi at the root and 4.5 psi at the tip. The composite wing is assumed to be made from AS graphite-fiber/epoxy-matrix composite which is the same as in Section 2 and have the same constituent materials properties and fabrication scatter shown in Tables 2.1 and 2.2. Those for the sensors/actuators are summarized in Table 3.1. One interesting point to observe is that the scatter range for all the participating variables is from 0 (zero) to 5 percent. This scatter is realistic since measuring most of these properties requires rather complex testing procedures and relatively small specimens. See[11] for a more detailed description of the wing.

The uncertainties in the angle of attack were evaluated as the range of the probable scatter from a 0 (zero) reference position. Results for the probable scatter in the angle of attack at midspan are shown in Figure 3.4(a) for a wing with and without actuators. Corresponding results at the tip are shown in Figure 3.4(b). There are two important observations to be seen: (1) actuators can be configured to change the angle of attack substantially - 0.5 percent versus 1.0 percent induced strain; and (2) the spread in the scatter increases as the magnitude of the control induced action increases.

The probabilistic scatter in the vertical displacement with two magnitudes of control induced strain and without are shown in Figure 3.5(a) at midspan leading edge and in Figure 3.5(b) at midspan trailing edge. It can be seen that substantial changes as much as three times in vertical displacement can be induced. Comparable results for the vertical displacement at the tip are shown in Figure 3.6. The collective results in Figures 3.4, 3.5 and 3.6 demonstrate that the intraply hybrid composite concept is an effective means to implement smart composite structures to controlled displacements type structural responses.

4. Probabilistic Assessment - Composite Wing with Defects

In the previous sections, smart composite wings with controlled displacements were described. In this section the concept is extended to composite wings with rectangular openings.

The composite wing is the same as that described previously. The location of the rectangular openings is near the root where the stresses are high (Fig. 4.1).

The probabilistic scatter of normalized ply longitudinal stresses are shown in Figure 4.2 for all the plies. The control ply for this case is the 0° ply. Note that for this case the stresses without the opening labelled without hole are also shown for comparisons. As can be seen, the stress magnitudes in the control ply (0° ply) can be modified to approach those of the wing without defects.

All three ply stresses for the control (0°) -ply are shown in Figure 4.3. The important observation in this figure is that the magnitudes of the transverse and intralaminar shear stresses are in the order of 5 percent or less of the longitudinal stress, and more importantly, the transverse stresses are compressive. The important conclusion is that the matrix-dominated ply stresses are relatively insignificant in the control plies of smart composite structure with defects.

Probabilistic sensitivity factors are an important feature of probabilistic assessment of smart composite structures. These factors provide quantifiable information on which "design parameters" in the smart composite structural concept are most sensitive. Subsequently, these design parameters can be modified to obtain the "best" benefit with minimum alteration. The sensitivity factors for the control and controlled ply are discussed below to illustrate their effectiveness.

The sensitivity factors for the longitudinal stress in the (+45°) ply are shown in Figure 4.4 for the four different cases investigated. It is readily observed that: (1) the composite design parameters (longitudinal modulus, the fiber volume ratio and the ply misorientation) dominate this stress without the controls in the cases of with and without hole, (2) the control design parameters (electric field and the sensor/actuator volume ratio) dominate in the cases with the actuators and (3) the dominance of control design parameters becomes more significant as the actuation induced strain increases. The sensitivity factors for the longitudinal stress in the control (0°) ply are shown in Figure 4.5. It is interesting to note that for this case the design parameters for the skin laminate and the wing inner structure dominate up to an induced strain of 0.5 percent.

The control design parameters become relatively significant when the actuator induced strain approaches 1 percent. Two important conclusions follow from the discussion of the sensitive factors, (1) the design variables of a composite wing internal structure can be modified to reduce skin laminate stress concentration; and (2) the relative significance of actuator design variables becomes comparable to composite structural variables as the induced strain approaches 1 percent.

5. Concluding Remarks

A formal procedure is described for the probabilistic assessment of composite and smart composite structures. This procedure integrates composite mechanics, finite element methods and probability theory for a comprehensive probabilistic analysis considering the uncertainties in all composite scales (constituent, ply, laminate, structural). A stiffened composite cylindrical shell with cutouts is used to demonstrate the effectiveness of the methodology. The displacements at tip and stress near the root increase due to the cutouts. The scatter range of the composite structural responses is adequately quantified. It is found that the uncertainty in the longitudinal stress is mainly due to the uncertainty in the laminate thickness.

The procedure is also applied to smart composite structures and demonstrated by its use to a smart composite wing. Important results from that demonstration include: (1)The scatter range in the distortions (angle of attack and vertical displacements) are probabilistically quantified. This scatter is sensitive to actuator induced strain which can be adjusted as needed. (2) Small defects (near the root about 20 percent chord) have negligible effect on the scatter range of wing distortions. (3) The uncertainties in the range of the ply longitudinal stress in the control (+45°) ply increase as the control induced strain increases. (4) The uncertainties in the longitudinal (0°) ply stress decrease with increasing actuator induced strain in stress concentration regions. Stresses in other plies must be evaluated for specific cases. (5) Probabilistic sensitivity factors indicate that the actuator/sensor electric field design variables dominate the stresses in the control ply and become more dominant as the actuator induced strain increases. (6) Probabilistic sensitivity factors for the control ply indicate that composite laminate and configuration, and internal wing construction dominate the ply stresses in that ply.

References

1. C. C. Chamis, "Probabilistic Structural Analysis Methods for Space Propulsion System Components." NASA TM 88861, June 1986.
2. C. C. Chamis and M. C. Shiao, "IPACS--Integrated Probabilistic Assessment of Composite Structures: Code Development and Applications," Third NASA Advanced Composites Technology Conference (1992).
3. B. K. Wade, "Adaptive Structures, 30th Structures, Structural Dynamics and Materials Conference," Part 1, AIAA, 1989, pp. 1-11.
4. O. Song, et. al., "Application of Adaptive Technology to Static Aeroelastic Control of Wing Structures," AIAA Journal, Vol. 30, No. 12, December 1992, pp. 282-289.
5. C. C. Chamis and J. H. Sinclair, "INHYD: Computer Code for Intraply Hybrid Composite Design," NASA TP 2239, December 1983.

6. P. L. N. Murthy and C. C. Chamis, "Interlaminar Fracture Toughness: Three-Dimensional Finite-Element Modeling for End-Notch and Mixed-Mode Flexure." NASA TM 87138, March 1985.
7. R. A. Aiello, "COmposite Blade STRuctural Analyzer (COBSTRAN) User's Manual," NASA TM-101461 (1989).
8. D. G. Liaw, M. C. Shiao, S. N. Singhal and C. C. Chamis, "Probabilistic Simulation of Multi-Scale Composite Behavior," 6th U.S.-Japan Conference on Composite Material (1992).
9. "Probabilistic Structural Analysis for Select Space Propulsion System Structural Components." Annual Report by Southwest Research Institute, Rocketdyne, University of Arizona, Marc Analysis Corporation and Columbia University for NASA Lewis Research Center, Contract NAS3-24389, (1986).
10. Y. T. Wu, "Demonstration of a New, Fast Probability Integration Method for Reliability Analysis," in Advances in Aerospace Structural Analysis, O.H. Burnside and C. H. Parr, ed., ASME, New York, pp. 63-73, (1985).
11. M. C. Shiao, et. al. , "Probabilistic, Assessment of Uncertain Adaptive Hybrid Composites," NASA TM in publication.
12. M. C. Shiao, and C. C. Chamis, "IPACS - Integrated Probabilistic Assessment of Composite Structures: Code Development and Applications," NASA TM in publication.
13. M. C. Shiao, P. L. N. Murthy, "Design for Cyclic Loading Endurance of Composites," The 24th International SAMPE Technical Conference,(1992).
14. M. C. Shiao, S. N. Singhal, "Probabilistic Assessment of Composite Structural Design," The 24th International SAMPE Technical Conference(1992).

Table 2.1. Material Properties at Constituent Level for both Skin and Frames

	Unit	Distribution Type	Mean	Coefficient Of Variation
E_{f11}	Msi	Normal	31.0	0.05
E_{f22}	Msi	Normal	2.0	0.05
G_{f12}	Msi	Normal	2.0	0.05
G_{f23}	Msi	Normal	1.0	0.05
ν_{f12}	---	Normal	0.2	0.05
ν_{f23}	---	Normal	0.25	0.05
E_m	Msi	Normal	0.5	0.05
G_m	Msi	Normal	0.185	0.05
ν_m	---	Normal	0.35	0.05

Table 2.2. Fabrication variables

	unit	Distribution type	mean	Coefficient Of Variation
fvr	---	Normal	0.60	0.05
vvr	---	Normal	0.02	0.05
θ_p	degree	Normal	0.00	1.00 (stdv)
t_{psk}	in	Normal	0.005	0.05
t_{pst}	in	Normal	0.020	0.05

Table 2.3. Mean Displacements and Stress of the Stiffened Composite Shell with/without Cutouts (Normalized by the Respective Value without Cutouts) (Location A, B, and C are Shown in Figure 5)

	Displacement at Location A	Displacement at Location B	Longitudinal Stress in 0° Ply at Location C
Without Cutouts	1	1	1
With Cutouts	1.11	1.19	1.07

Table 3.1. Uncertainties in the Sensor/Control Device

variable	Distribution type	mean	Coefficient of variation
secondary composite system volume ratio	Normal	0.50	0.05
sensor/control volume ratio	Normal	0.60	0.05
sensor/control modulus (Msi)	Normal	12.4	0.05
sensor/control strain coefficient (in/Volt)	Normal	2.0×10^{-8}	0.05
electric field strength (Volt/in)	Normal	1.0×10^{6}	0.05

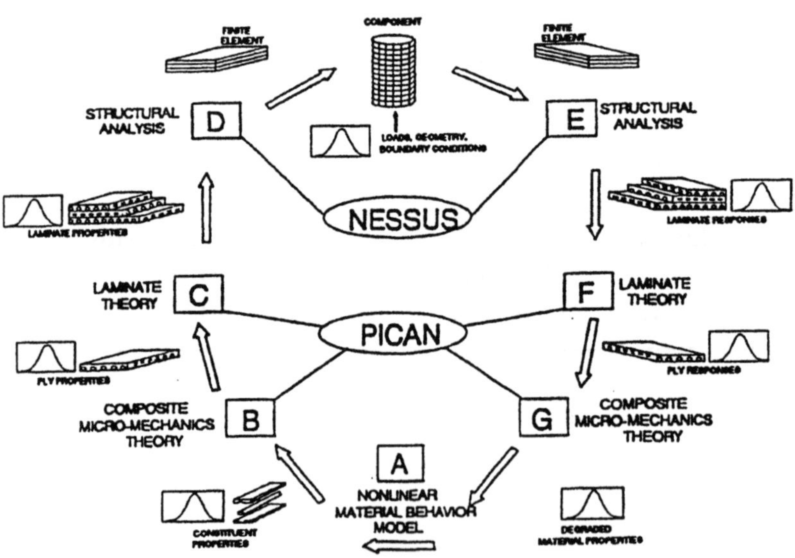

Fig. 2.1. Concept of Probabilistic Assessment of Composite Structures

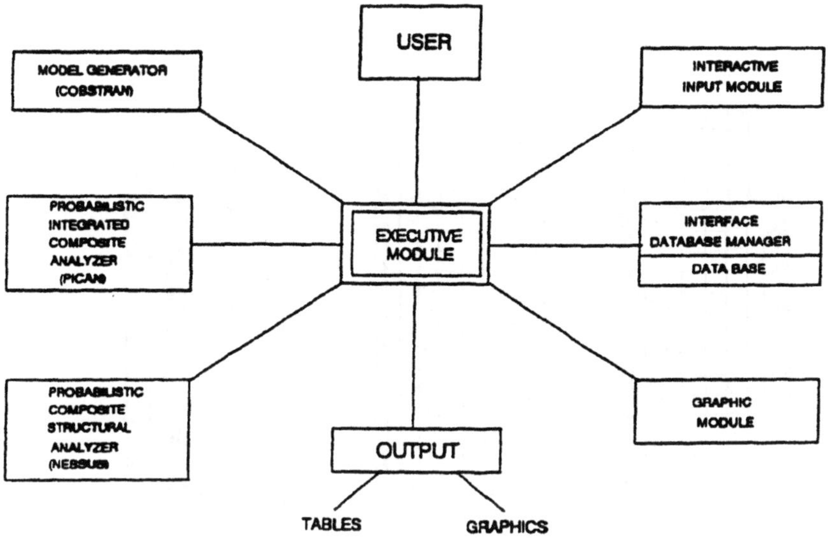

Fig. 2.2. Architecture of IPACS Software System
(IPACS: Intergrated Probabilistic Assessment of Composite Structures)

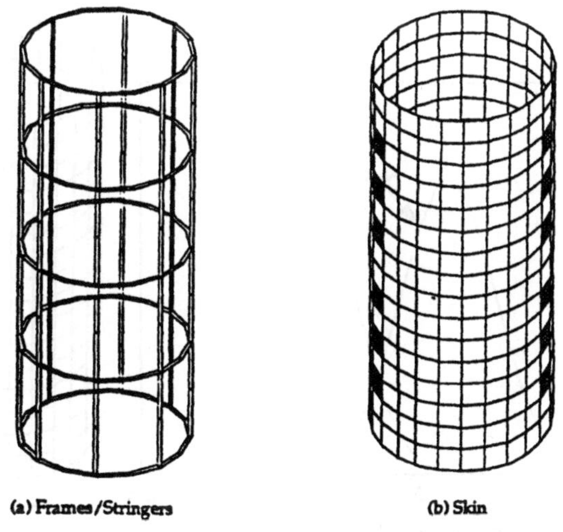

Fig. 2.3. Finite Element Model of a Composite Cylindrical Shell

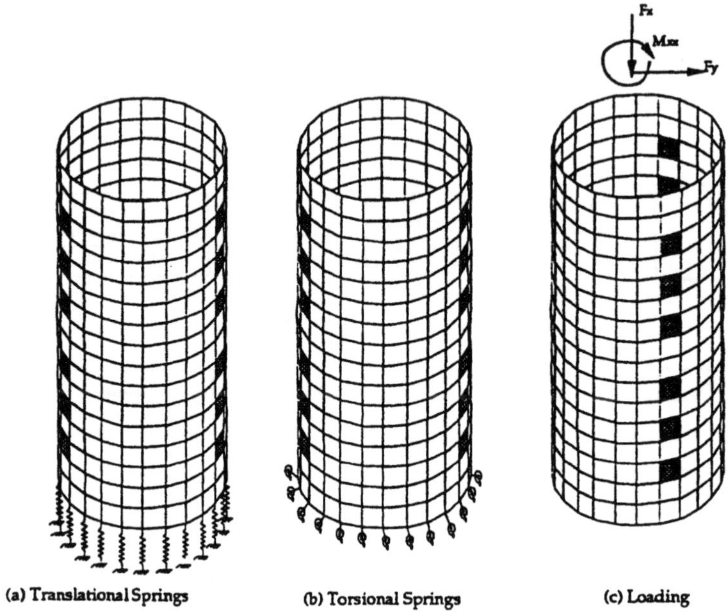

Fig. 2.4. Boundary and Loading Conditions of a Composite Cylindrical Shell

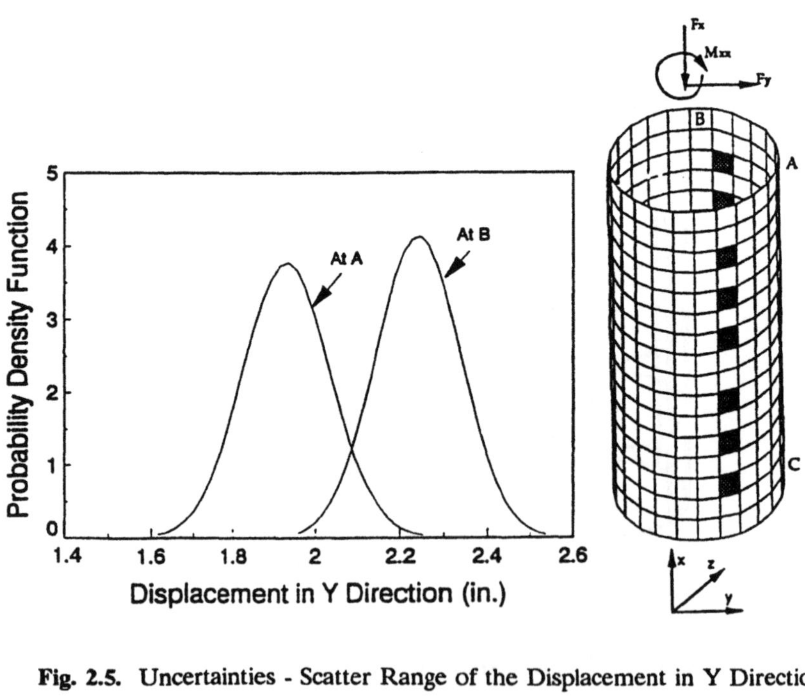

Fig. 2.5. Uncertainties - Scatter Range of the Displacement in Y Direction

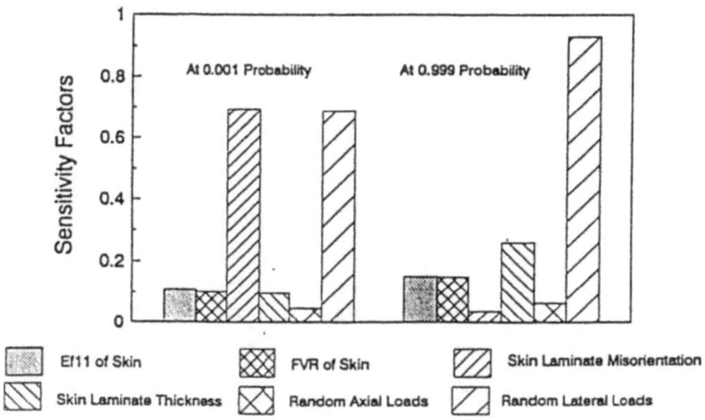

Fig. 2.6. Sensitivities of the Probabilistic Displacement in Y Direction (At Location B in Fig. 2.5) to Significant Uncertain Variables

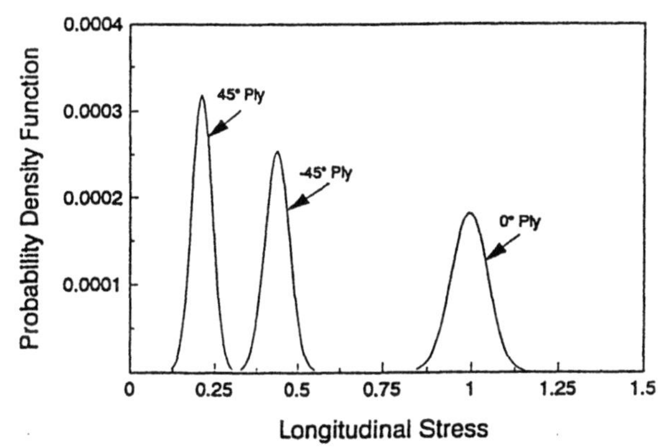

Fig. 2.7. Uncertainties - Scatter Range of Normalized Longitudinal Stresses (At Location C in Fig. 2.5 Normalized by Mean Stress in 0° Ply)

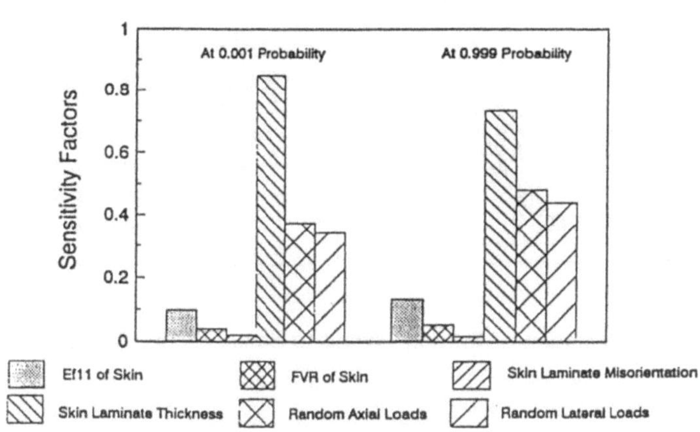

Fig. 2.8. Sensitivities of the Probabilistic 0° Ply Longitudinal Stress to Significant Uncertain Variables

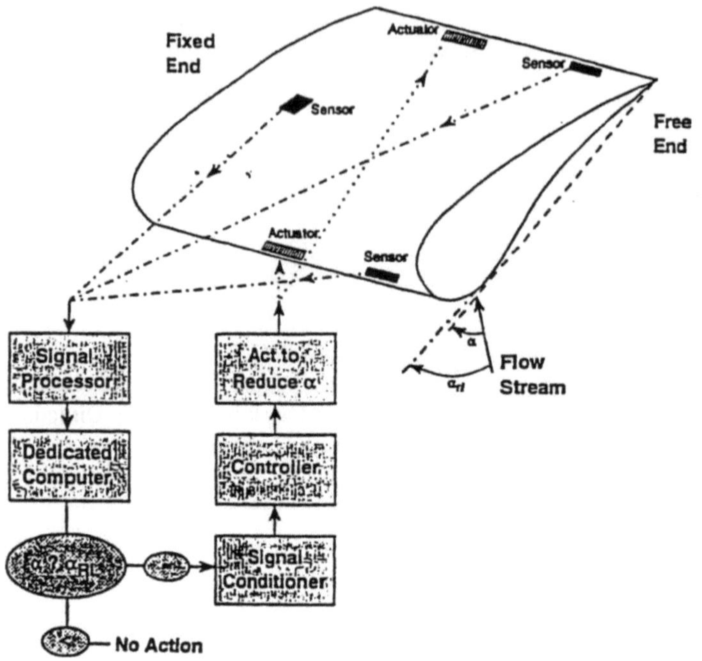

Fig. 3.1. Conceptional Diagram of Smart Composite Aircraft Wing System

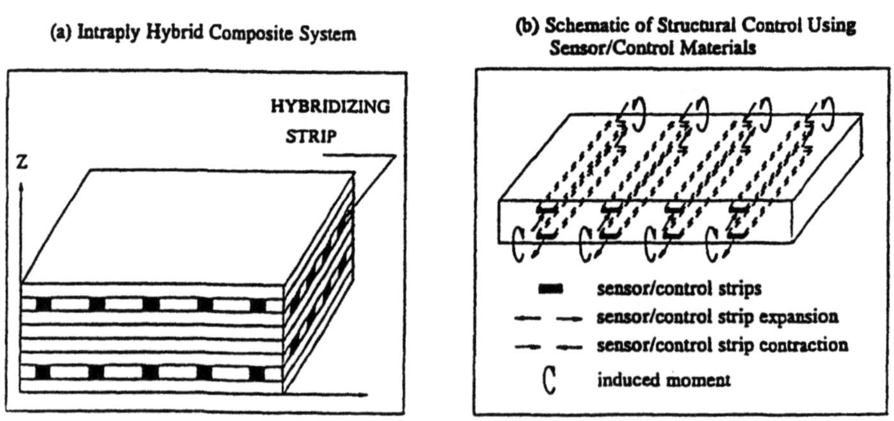

Fig. 3.2 Adaptation of Intraply Hybrid to Smart Composite System

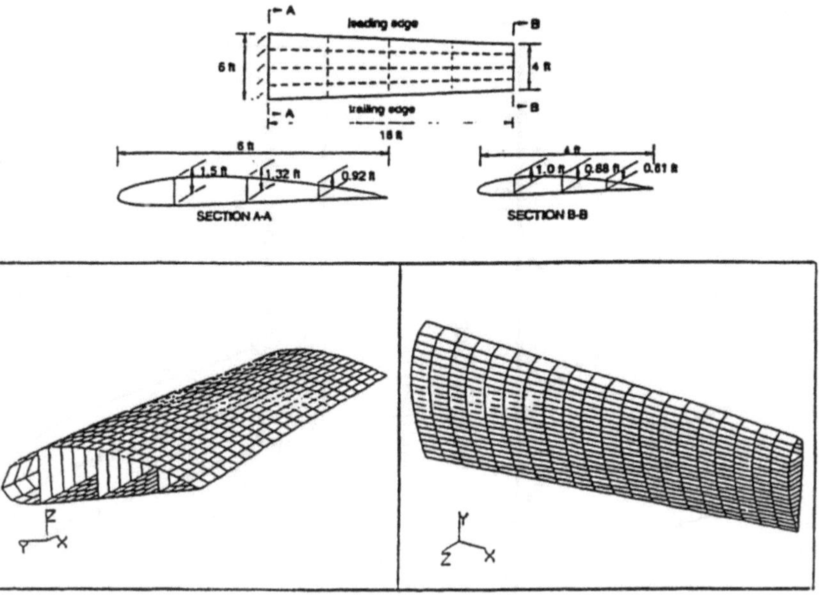

Fig. 3.3. Geometry and Finite Element Model for a Composite Wing

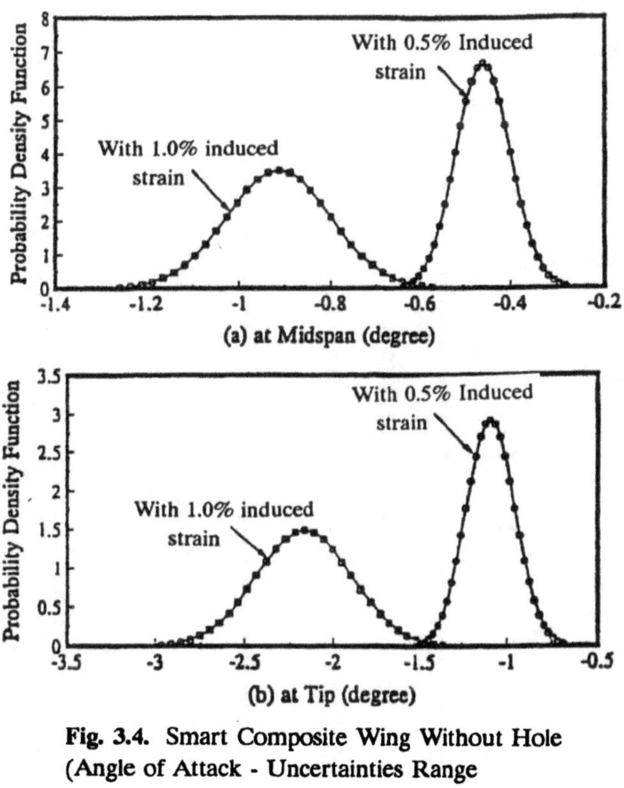

Fig. 3.4. Smart Composite Wing Without Hole (Angle of Attack - Uncertainties Range

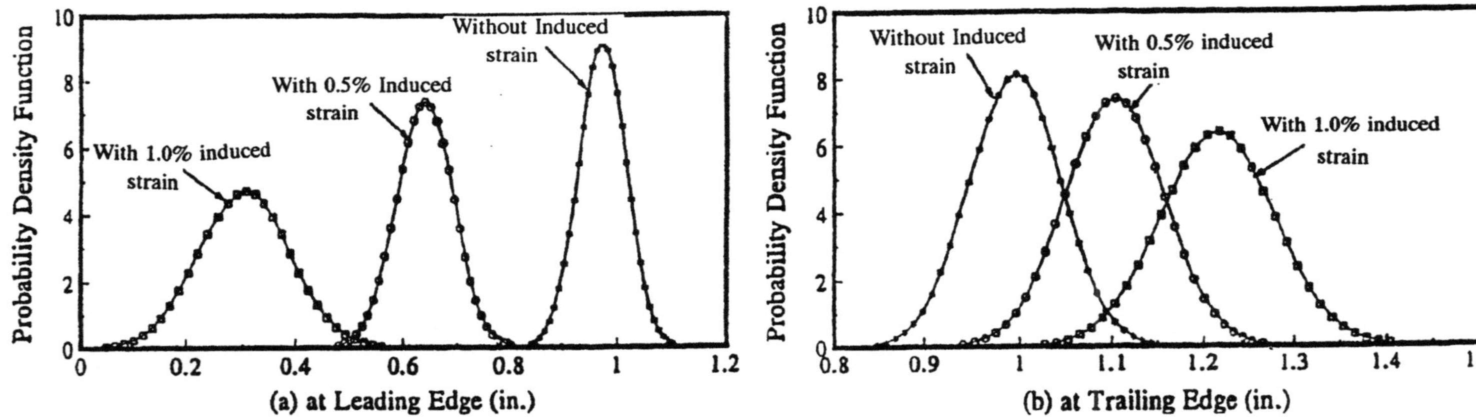

Fig. 3.5. Smart Composite Wing Without Hole (Vertical Displacements - Uncertainties Range at Midspan)

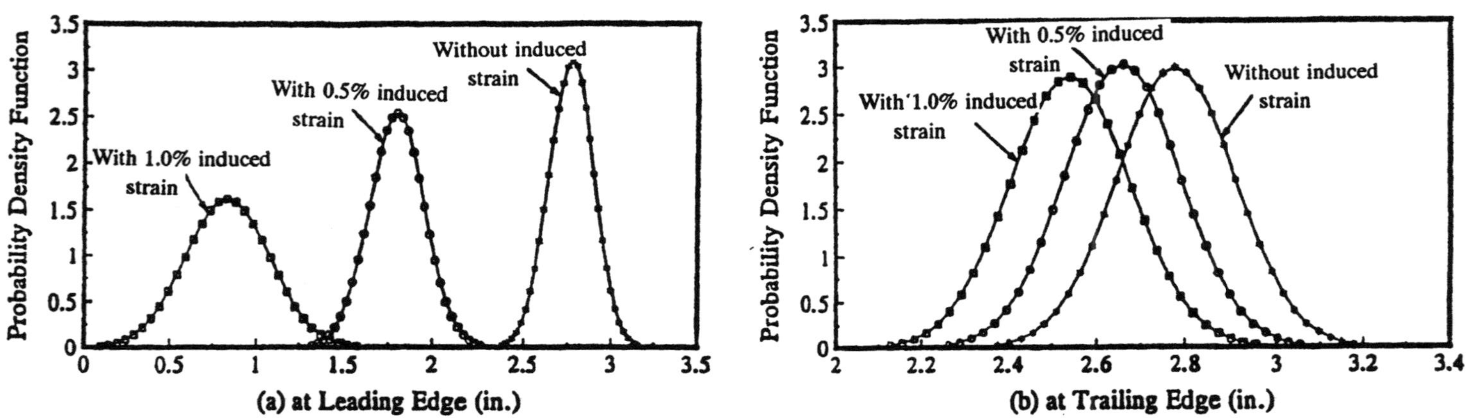

Fig. 3.6. Smart Composite Wing Without Hole (Vertical Displacements - Uncertainties Range at Tip)

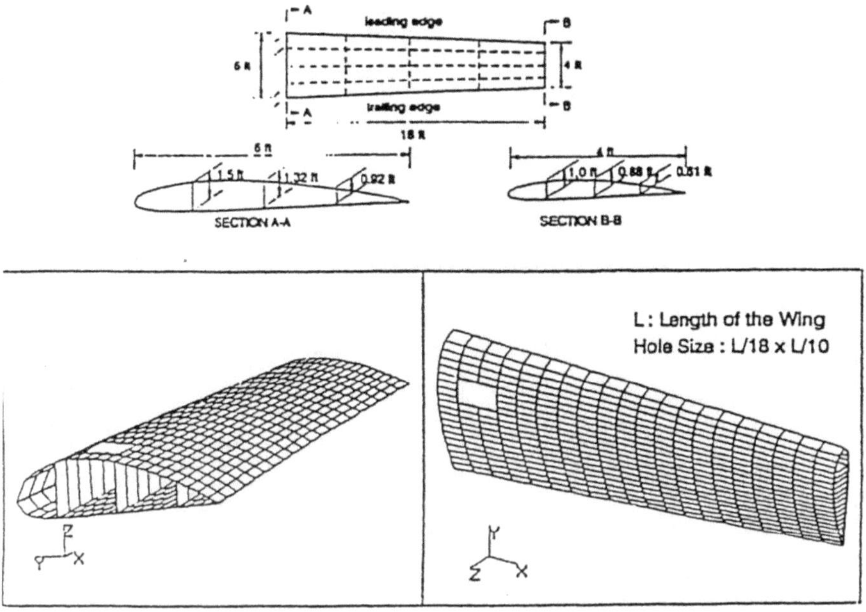

Fig. 4.1. Geometry and Finite Element Model for a Composite Wing

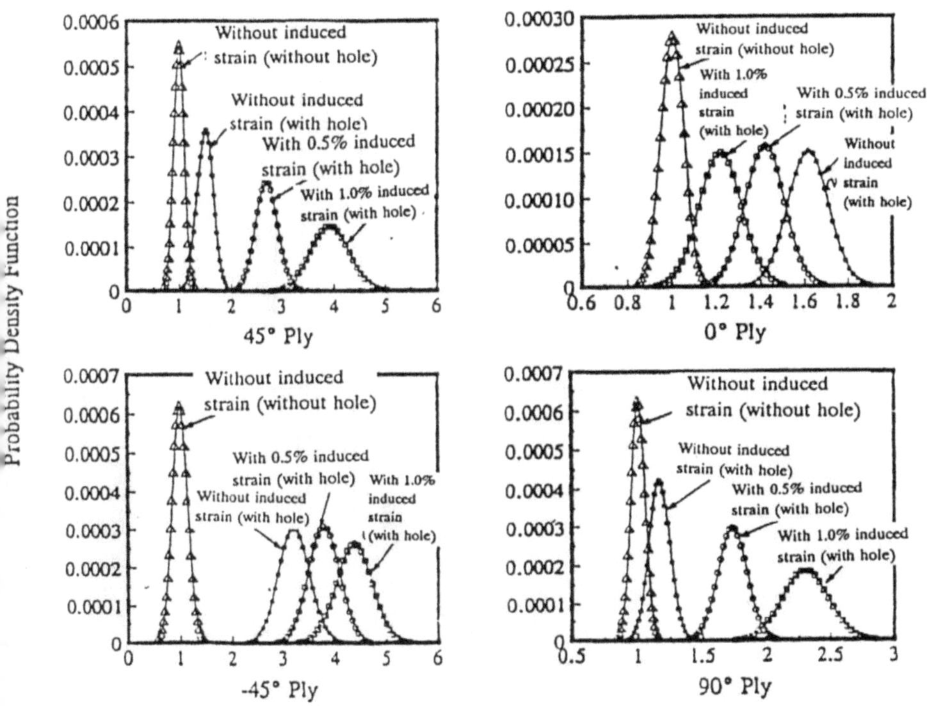

Fig. 4.2. Smart Composite Wing With/Without Rectangular Hole (Normalized Longitudinal Stress - Uncertainties Range With/Without Control)

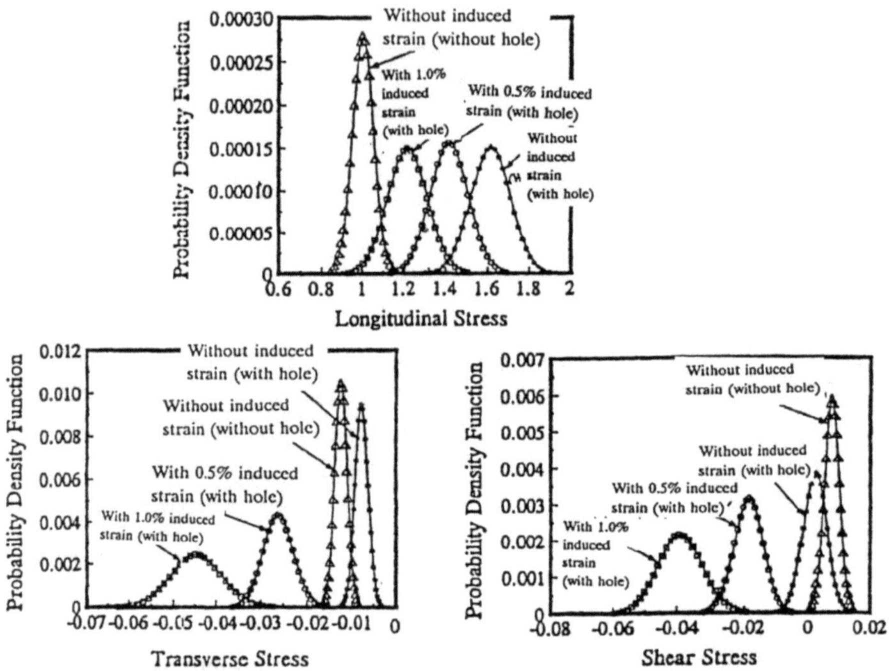

Fig. 4.3. Normalized Ply Stresses in 0° Ply
(A Smart Composite Wing With Rectangular Cutout)

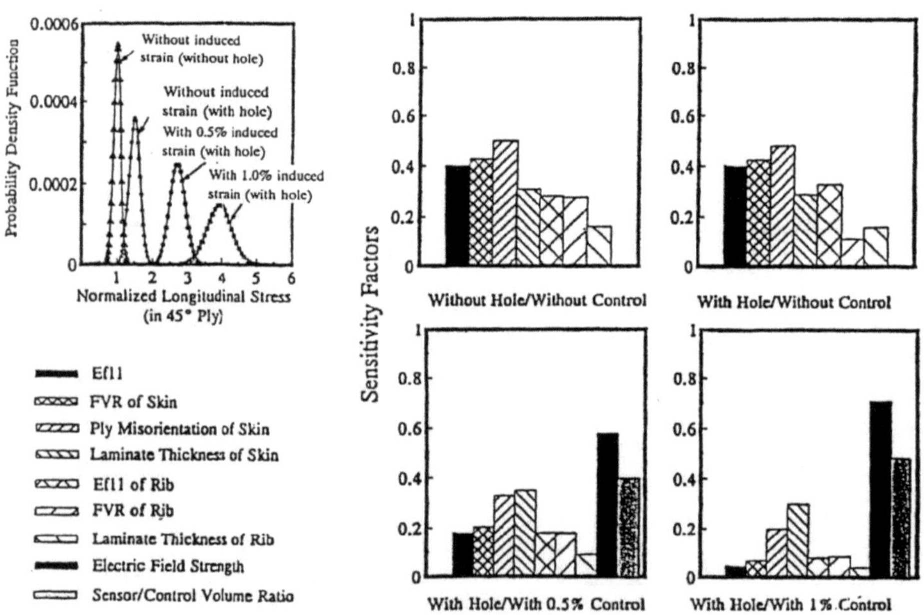

Fig. 4.4. Sensitivities of the Probabilistic 45° Ply Longitudinal Stress to Significant Uncertain Variables
(A Smart Composite Wing With Rectangular Cutout)

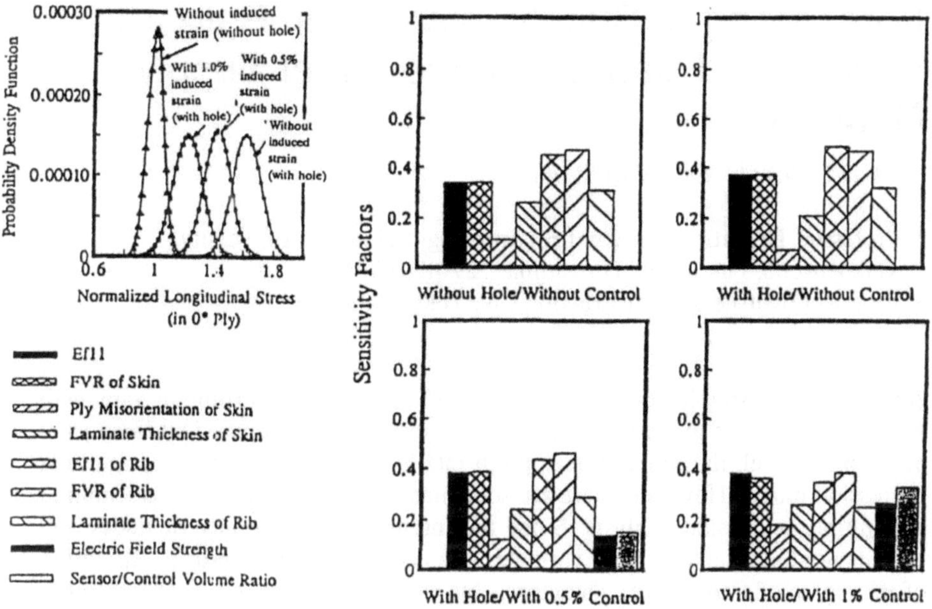

Fig. 4.5. Sensitivities of the Probabilistic 0° Ply Longitudinal Stress to Significant Uncertain Variables (A Smart Composite Wing With Rectangular Cutout)

Forecasting of Supply Interruption Rate with Widespread Damage to the Water Supply Network

Toshiaki Chiba[1] and **Yoji Shimizu**[2]

[1] Professor, Department of Civil Engineering, Fukuyama University, Gakuen-cho, Fukuyama, Hiroshima, 729-02, JAPAN
[2] Engineer, Mitsubishi Construction Co., Tokyo, JAPAN

Abstract. The simple forecasting model of the supply interruption rate with widespread damage to the water supply network has been proposed by Hoshiya et al. In this model, the supply interruption rate is estimated by the linear multiple regression model. However, there are some unstable conditions in the calculation of multiple regression analysis. Since the intensity of importance of nodes and links is used as an explanatory variable, it is necessary to get a proper estimation of its degree of importance. Therefore, this study aims to consider its instability and to consider how to evaluate the intensity of importance.

Since the degree of importance is a vague variable, the other object of this study is to discuss the applicability of fuzzy theories to the forecasting model of the supply interruption rate. The examined fuzzy theories are the fuzzy quantification theory, type I and type II.

Keywords. Water supply network, supply interruption rate, degree of importance, fuzzy quantification theory

1 Introduction

There have been several simulation models for the evaluation of lifeline supply functions. These models so far developed are very complicated, and they require a long computational time due to a network inter- linkage analysis and so on.

Recently, to reduce the complexity, Hoshiya and others proposed a simple forecasting model, [1]-[3], for the seismic damage restoration of city water supply functions. In this model, the supply interruption rate is estimated by the linear multiple regression model. Therefore, it is easy to estimate its rate for any damage patterns at any time.

In the above model, however, there are some unstable conditions in the calculation of multiple regression analysis. Since the degree of importance of nodes and links is used as an explanatory variable, it is necessary to get a proper estimation of its levels. However, it is difficult to express these levels quantitatively. Therefore, this study aims to consider the subject of instability and to consider how to evaluate degrees of importance of nodes and links.

Harada and others, [4], discussed the estimation method of the importance of gas networks by using the shortest route from the supply plant to the demand node. To determine the intensity of importance of nodes and links of the water supply network, the Dijkstra method that determines the shortest route is also applied in this paper.

There are vague variables such as degrees of importance of nodes, links and so on. Therefore, the other object of this study is to discuss the applicability of fuzzy theories to the forecasting model of supply interruption rate. The fuzzy theories used are the fuzzy quantification theory, type I and type II.

The water supply network in the city of Fukuyama was used for the numerical demonstrations in this paper.

2 Outlines of the Supply Interruption Rate Model

The forecasting model for the damage restoration considered in this paper is a simple model, which is expressed with the mesh (overlaid mesh) where the node and link components exist. An actual network is not used directly, and overlaid meshes that represent the concerned area with the mesh of *500m* × *500m* are used for the calculation. In this study, the water supply network of Fukuyama City is used for the numerical examples, and its network is shown in Fig. 2.1. In this figure, only the links with diameters over *200mm* are drawn. The overlaid meshes are represented with the mesh numbers shown in Fig. 2.2. This model can incorporate any restoration strategies. Please refer to references [1] or [2] for details.

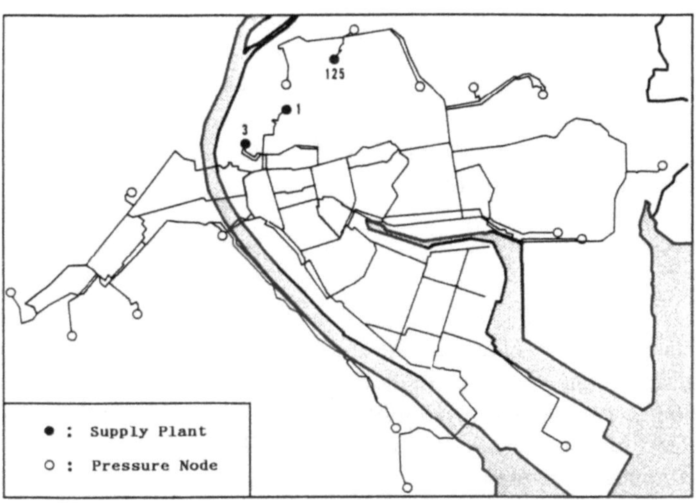

Fig. 2.1. Fukuyama City water supply network

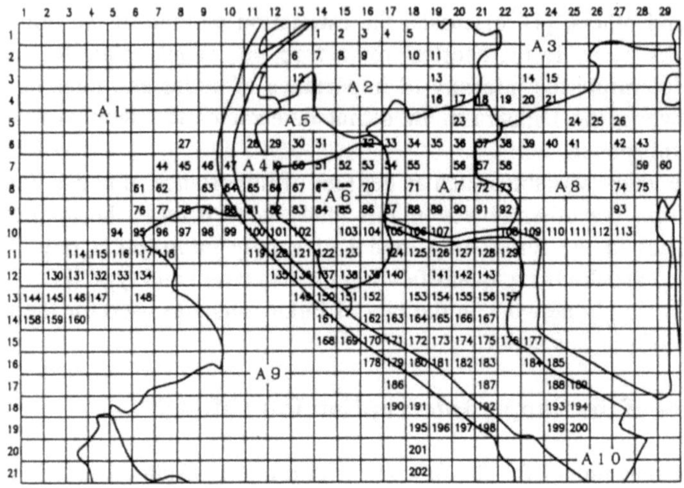

Fig. 2.2. Overlaid meshes (*500m* × *500m*) and supply area

The supply interruption rate F_t represents the degree of the functional supply interruption of the network at time *t*. It is obtained directly from the equation

$$F_t = b_0 + b_1 \cdot D1_t + b_2 \cdot D2_t + b_3 \cdot D3_t \ , \qquad (1)$$

where *t* represents the time considered, rather than from the network inter-linkage analysis and so on. When the water supply stops completely, F_t will become *1.0*, and F_t is *0.0* when water is supplied in normal circumstances. $D1_t$, $D2_t$ and $D3_t$ show the physical damage rate, the physical damage distribution rate, and the damage rate of important facilities, respectively.

The physical damage rate $D1_t$ in Eq. 1 can be represented by the ratio of the total damaged meshes to the total number of overlaid meshes of the water supply network and given by the equation

$$D1_t = \frac{1}{N}\sum_{i=1}^{N} WRC_t^i \ , \qquad (2)$$

where the superscript *i* represents the mesh number, *N* is the total number of overlaid meshes in the water supply network, and WRC_t^i is the completion index of restoration in the damage restoration model. $WRC_t^i=1$ when there is damage in mesh *i*, and for the meshes with no damage, $WRC_t^i=0$. Eqs. 1 and 2 are the same definitions as in reference [1].

The state of the damage distribution of the water supply network is represented by the physical damage distribution rate $D2_t$. The physical damage rate $D1_t$ could not represent the damage of important facilities such as water purification plants, distribution points, pumping points, and so on. The damage of important facilities, therefore, is given by the damage rate of important facilities $D3_t$. The definitions of $D2_t$ and $D3_t$ will be given in sections' 3.3 and 3.2.

The supply interruption rate F_t is calculated at each supply area in the concerned region. The partial regression coefficients, $b_{0,}$ b_1, b_2 and b_3 in Eq. 1, are determined from the multiple regression analysis. The values of the dependent variables F_t in this regression analysis are given by the network inter-linkage analysis or the data from the empirical judgments of experts. In the use of the experts' knowledge, the data F_t are obtained by asking the possible degrees of water supply interruptions that have occurred in a given system inflicted by certain earthquake damage. After determining the coefficients, $b_{0,}$ b_1, b_2 and b_3, from the multiple regression analysis, we can get the water supply interruption rate at any time and with any $D1_t$, $D2_t$ and $D3_t$. Therefore, this model is extremely simple, since the complicated network inter- linkage or pipeline capacity analyses are avoided.

However, when the multiple regression analysis is carried out following references [1] or [2], we encounter unstable conditions during the analysis. The regression coefficients become negative or the multiple correlation coefficients become unacceptably low. These unstable conditions may come from the absence of a redundant link when a certain node or link is damaged. In this case, many nodes in series may exist without water supply due to detour links. Moreover, the unreasonable estimations of the physical damage distribution rates and the damage rates of important facilities lead to unstable conditions in the multiple regression analyses. Therefore, this research is aiming at improvements in the estimation of the physical damage distribution rate and damage rate of important facilities by the evaluation of the degree of importance of nodes and links.

3 Estimation of Degree of Importance and Supply Interruption Rate

There is an instability in the multiple regression analysis mentioned previously. Therefore, it is attempted to improve the damage rate of important facilities $D3_t$ and the damage distribution rate $D2_t$ by considering the degree of importance of nodes and links hereafter.

3.1 Application of the Shortest Route and Degree of importance

In the Dijkstra method, the shortest route is evaluated only with the length of the link. Therefore, the size of the link diameter is disregarded. The size of the reservoir (supply plant) exhaust is disregarded in this method, too. However, a link with a large diameter may be more important than one with a small diameter. The degree of importance of a link that is connected to the supply plant with much exhaust may also be higher than that of links that are connected to the small supply plant.

Therefore, the length of a link with a small diameter should be evaluated longer, and the length of a link that is connected to the supply plant with a small amount of exhaust should also be evaluated longer. The lengths of the links are reevaluated based on the maximum diameter of the link and on the maximum amount of the supply plant. From these points of view, the link length L_e is estimated by the equation

$$L_e = L_r \cdot \left(2 - \frac{E_r}{E_{\max}}\right) \cdot \left(2 - \frac{D_r}{D_{\max}}\right), \tag{3}$$

where L_e is the estimated link length, and L_r is a real length of each link. E_r is the real exhausting amount of each supply plant, and E_{max} is the maximum exhausting amount of supply plants. D_r is the real diameter of each link, and D_{max} is the maximum diameter of links in the concerned area. The estimation method of the link length can be considered in several ways. In this paper, the link length is estimated to be a maximum of twice the real length. In the following numerical calculations, the estimated link lengths L_e are used.

In the estimation of the degree of importance of the network by the Dijkstra method, it is interpreted that the links or nodes with high passage frequencies are the important links or nodes in the network. This passage frequency is the number of times that water passes through the link or node along the shortest routes from one supply plant to each node. The passage frequencies of each node from three supply plants to each node are determined. These frequencies are summed up for each node and then normalized by the maximum number of frequencies. These normalized frequencies are shown in Fig. 3.1.

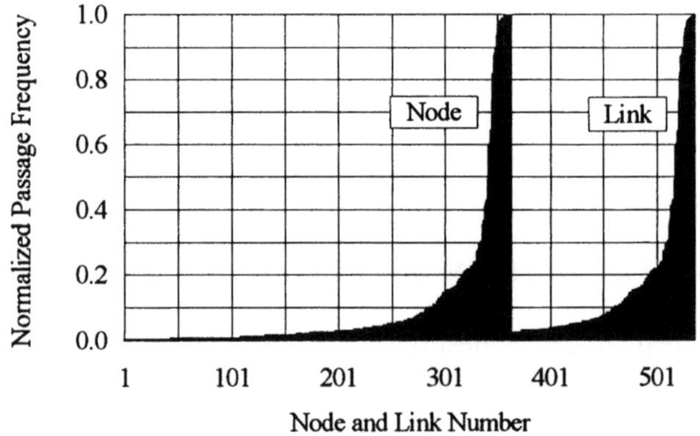

Fig. 3.1. Normalized passage frequencies of nodes and links

The degrees of importance of nodes or links were estimated to five levels as shown in Table 3.1 by using the normalized frequencies. Fig. 3.2 shows the results of degrees of importance of nodes. The degrees of importance of links are omitted here because it can be easily understood from Fig. 3.2.

Table 3.1. Importance level of node and link

Level of Importance : I_l	Normalized Passage Frequency : N_p
1 (Low Importance)	$N_p < 0.1$
2	$0.1 \leqq N_p < 0.2$
3	$0.2 \leqq N_p < 0.4$
4	$0.4 \leqq N_p < 0.7$
5 (High Importance)	$0.7 \leqq N_p$

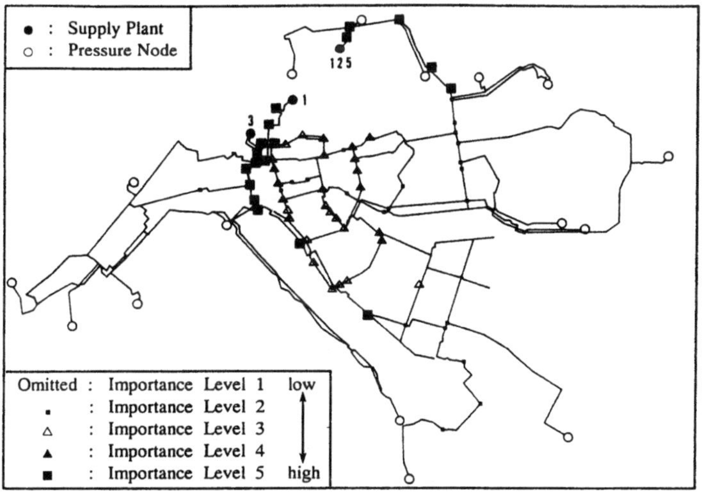

Fig. 3.2. Levels of importance of nodes by the Dijkstra method

On the other hand, carrying out the pipe network analysis by assuming damage in each node, the supply interruption rates of each node were determined. Then, the degrees of importance of nodes are estimated to five levels by using these supply interruption rates as shown in Fig. 3.3. The degrees of importance shown in Fig. 3.3 can be considered reasonably correct. As seen in Fig. 3.2, that was determined by the Dijkstra method directly, there are some important nodes located far from the supply plant. Moreover, the figure shows that some nodes located near the supply plant have the lower degrees of importance. These results are not accepted. This implies that the usage of the Dijkstra method should be improved.

The supply area for each supply plant is generally fixed up in the network systems. The water supply network systems in the city of Fukuyama have three supply plants. The supply area that is determined by the pipe network analysis is shown in Fig. 3.4. Since the supply area shown in Fig. 3.4 is determined by the pipe network analysis, the area can be considered as the actual supply area.

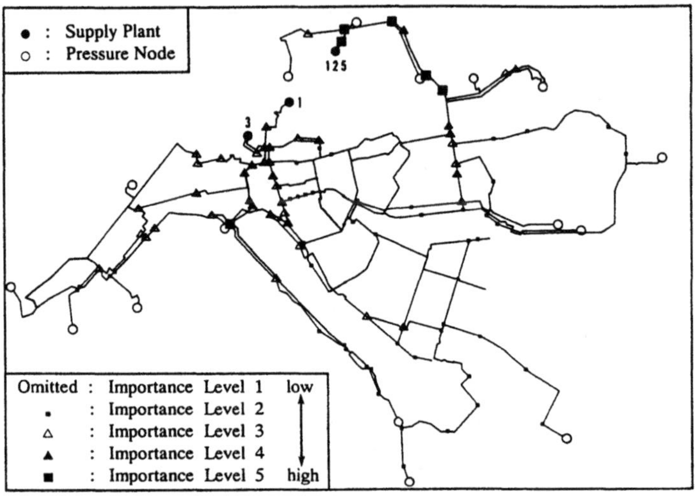

Fig. 3.3. Level of importance of nodes by pipe network analysis

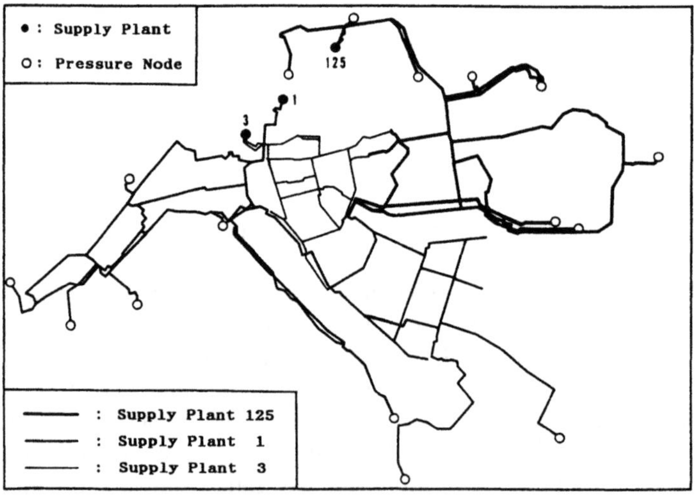

Fig. 3.4. Supply areas by pipe network analysis

Now, it is assumed that the supply plant for the certain demand node is the supply plant that has the shortest route among the routes from its demand node to the three supply plants. Fig. 3.5 shows the supply area that is estimated with the assumption mentioned above. Comparing these two figures, Fig. 3.4 and Fig. 3.5, both figures show almost the same supply areas. Therefore, estimating the shortest route by the Dijkstra method with the estimated pipeline length in Eq. 3, the supply area for each supply plant can be estimated easily without doing the pipe network analysis. Therefore, the passage frequencies represent the frequencies in the supply area of each supply plant hereafter.

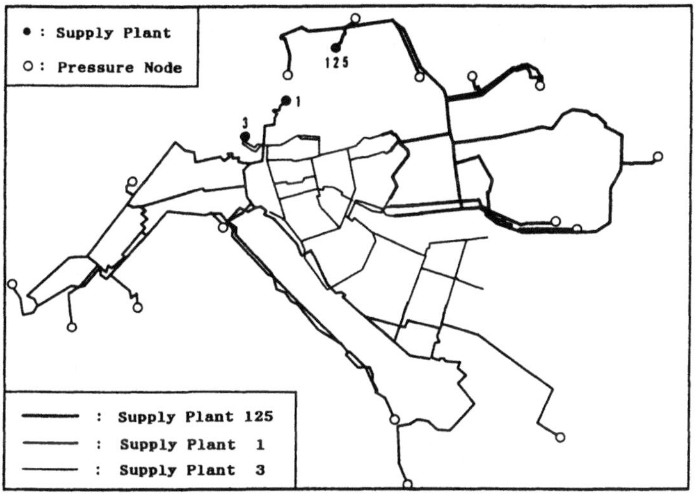

Fig. 3.5. Supply areas using the shortest route mrthod

The degree of importance of each node is estimated to five levels by using the normalized passage frequency with the maximum frequency. To get the importance level of the nodes, Table 3.1 is also used here. The higher numbers show the more important nodes. Then, we get the level of importance of the Fukuyama City water supply network as shown in Fig. 3.6. There are no big differences between Fig. 3.6 and Fig. 3.3. Therefore it can be considered the actual degree of importance. Almost the same degree of importance can be seen in the two. Therefore, we can say that the method proposed here that uses the shortest route gives the proper results.

Using the results of degrees of importance for the nodes and links, the degrees of importance of the overlaid meshes can be estimated. For instance, the highest level of importance of the nodes or links included in a mesh is the level of the degree of importance of the mesh. Thus, we can estimate the degree of importance of all overlaid meshes as shown in Fig. 3.7.

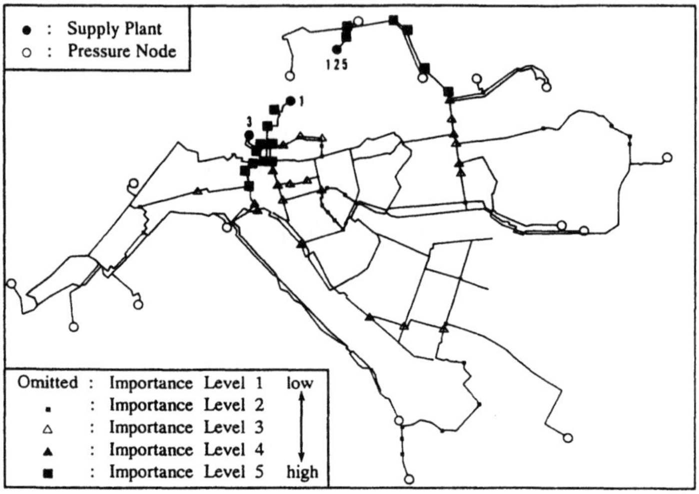

Fig. 3.6. Level of importance of nodes by the proposed method

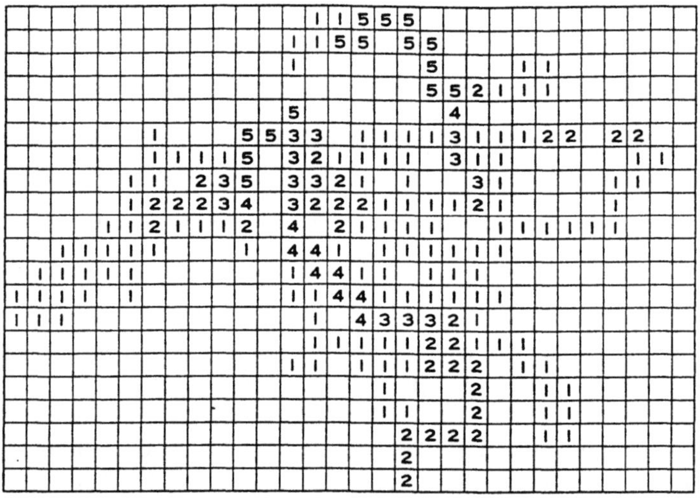

Fig. 3.7. Levels of importance of meshes

3.2 Estimation of the Damage Rate of Important Facilities: $D3_t$

The damage rate of important facilities should be an explanatory variable that is independent of the physical damage rate ($D1_t$: damaged meshes / all overlaid meshes). Therefore, the damage rate of important facilities should be a ratio of the damage of important facilities to the whole damage. To be an independent explanatory variable, the damage rate of important facilities $D3_t$ is evaluated here by the equation

$$D3_t = \frac{N_c}{N_{\max}} , \tag{4}$$

where N_c represents the converted number of damaged spots. One spot of damage in the mesh with the degree of importance of level I_l is converted into the number of damaged spots of the lowest level 1. Though the conversion rate of one spot of damage of level I_l is not determinate, its rate is assumed I_l^2 in this paper. Therefore, one spot of damage in the mesh of the level I_l is assumed to be I_l^2 spots of damage of the level 1. N_{max} in the denominator is a converted number of spots of damage if all damage occurs in the meshes of the highest level 5. Therefore, the damage rate of important facilities $D3_t$ becomes unity when all the damage really occurs in the meshes of the highest level 5.

3.3 Evaluation of Damage Distribution Rate: $D2_t$

The physical damage distribution rate represents the scattering degree of damage in the pipeline network. The damage scattering may have an influence on the water supply interruption rate and on the restoration strategy. The damage distribution rate is defined in reference [1] as the ratio of total distance between damaged meshes to the total distance between overlaid meshes. This definition is somewhat difficult to explain physically. This is also an explanatory variable that depends directly on the number of spots of damage. Therefore, it should be improved.

The damage distribution rate $D2_t$ is proposed by using the polar moment of inertia in the equation

$$D2_t = \frac{I_{p.d}}{I_{p.\max}} , \tag{5}$$

where $I_{p.d}$ is the polar moment of inertia with respect to the axis that passes through the centroid of the damaged meshes. $I_{p.max}$ is the maximum polar moment of inertia of the meshes that have the same amount of damaged meshes in the numerator. Since the polar moment of inertia represents a variance, this definition is easy to understand. This rate is independent of the physical damage rate. Therefore, a more suitable multiple regression analysis can be expected.

4 Application of the Fuzzy Quantification Theory

In the multiple regression analysis, there are some previously mentioned unstable conditions. Since the degree of importance of nodes and links is used as an explanatory variable, it is necessary to get a proper estimation of its level. The estimation method of this level is already mentioned in chapter 3. However, these levels are the vague variables, and it is still difficult to express them quantitatively. Therefore, by changing the viewpoint, the fuzzy theory is applied to the forecasting of water supply interruption rate.

4.1 Fuzzy Quantification Theory, type I

To get the influence of the damage in the central part of the city, the *29* types of damage patterns were analyzed by using the fuzzy quantification theory, type I. It is assumed that these damage patterns have no damage on the pipelines in series.

The objective variable (external criterion) is the water supply interruption rate of these damage patterns in this analysis. The explanatory variables (category) are the damage rate $D1_t$, the damage distribution rate $D2_t$, and the damage rate of important facilities $D3_t$.

The fuzzy group indicates a focal fuzzy event specified by its membership function. This membership function is represented by the downtown supply interruption rate in this analysis. The water supply interruption rate that is considered the downtown damage rate was determined from this analysis.

Considering the fuzzy group, the estimation of the supply interruption rate F_t is given by the equation

$$F_t = -0.023 + 1.36 \cdot D1_t + 14.28 \cdot D2_t + 5.33 \cdot D3_t \, . \qquad (6)$$

On the other hand, the water supply interruption rate F_t without considering the downtown damage rate becomes

$$F_t = -0.013 + 2.26 \cdot D1_t + 22.14 \cdot D2_t + 2.67 \cdot D3_t \, . \qquad (7)$$

Comparing with Eqs. 6 and 7, the water supply interruption rate with consideration of the downtown damage rate is influenced by $D3_t$ much more than without consideration of the downtown damage rate.

The partial correlation coefficients that represent the correlation's between external criteria and the explanatory variables are shown in Table 4.1. The partial correlation coefficients on $D1_t$ show no big difference between the two cases as shown in Table 4.1. The damage ratio $D1_t$ represents the ratio of the damaged meshes to whole meshes, and the damage locations are given at random. Therefore, the partial correlation coefficient on the damage rate $D1_t$ with consideration of the downtown damage rate may not differ to that without consideration of the downtown effect.

The partial correlation coefficient on $D2_t$ that considers the downtown effect shows negative correlation as shown in Table 4.1. The downtown network is very complicated, and there are many nodes and links in one mesh. Therefore, the interruption rate does not rise up easily due to the existence of detours, even if the damage occurs on the downtown network. The partial correlation coefficient may have become negative due to these reasons.

There are many demand nodes and detours in the downtown mesh. On the other hand, there are few demand nodes and detours in the surrounding area. Therefore, it is better to improve so that the sizes of the mesh have almost the same number of nodes and links in a mesh.

Table 4.1. Partial and multiple correlation coefficient, fuzzy quantification theory, type I

	Partial Correlation Coefficient			Multiple Correlation Coefficient
	$D1_t$	$D2_t$	$D3_t$	
fuzzy group, considered	0.517	− 0.206	0.610	0.757
fuzzy group, not considered	0.443	0.149	0.257	0.926

The partial correlation on $D3_t$ that considers the downtown damage rate is higher than that of the other cases. This means that downtown is apt to receive the effect of the damage of the important facilities.

The multiple correlation coefficients that represent the correlation between the observed data (the external criteria) and the estimated values are also shown in Table 4.1. Considering the downtown area, the multiple correlation coefficient is not acceptably high. However, the influence of the downtown or surrounding damage can be understood by the fuzzy quantification theory, type I. The effect of a certain factor to the supply interruption rate can be examined by changing the fuzzy group. Therefore, the fuzzy quantification theory, type I, is useful for examining certain effects or getting the data for making up the forecasting model of the supply interruption rate.

4.2 Fuzzy Quantification Theory, Type II

As mentioned previously, the supply interruption rates have to take a value between *0* and unity. However, it takes the value less than *0* or greater than unity due to the multiple regression model. Then the fuzzy quantification theory, type II, will be used for considering the supply interruption rate.

The external criteria are chosen as the supply interruption rate and the supply rate, and the items as $D1_t$, $D2_t$ and $D3_t$. After analyzing the quantification theory, we get the following equation

$$Y_s = 227.352 \cdot D1_t + 555.855 \cdot D2_t + 92.240 \cdot D3_t \qquad (8)$$

for the sample values Y_s.

The partial correlation coefficients are shown in Table 4.2. The value of the partial correlation on $D1_t$ is high, leading to the high correlation between the supply interruption rate and $D1_t$.

The relationships between the determined supply interruption rate(membership function of the external criterion) and the sample values are shown in Fig. 4.1. From this figure, when the sample value exceeds *20*, the supply interruption rate becomes *0.4* or more and shows a high interruption rate. When the sample values are less than *5*, there is almost no damage to the water supply.

Table 4.2. Partial correlation coefficient, fuzzy quantification theory, type II

Item	Partial Correlation Coefficient
$D1_t$	0.131
$D2_t$	0.049
$D3_t$	0.053

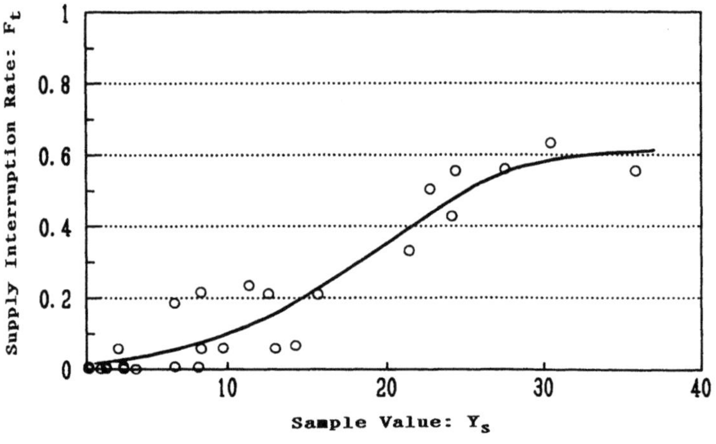

Fig. 4.1. Relationships between F_t and Y_s by fuzzy quantification theory, type II

The estimation of the supply interruption rate can be done as follows. The sample value Y_s is determined by substituting the values of $D1_t$, $D2_t$ and $D3_t$ into Eq. 8. Then, the supply interruption rate is the value that corresponds to the sample value Y_s. However, because the supply interruption rate is considerably different even in the case of the same level of Y_s as shown in Fig. 4.1, it is difficult to get the proper supply interruption rate. However, the method of using this fuzzy theory has an advantage that the rate does not exceed the range [0– 1].

The appropriate estimation of the degree of importance of nodes and links of the water supply network was done by the determination of the shortest route. Moreover, it was shown to be able to get the water supply area for each water supply plant by using this method. Since the method of the determination of the shortest route is simple, this method is sufficiently applicable.

The damage rate of the important facilities was estimated properly. The degrees of importance of nodes and links were estimated, and they were evaluated in the five levels of importance. Moreover, the estimation method of the damage distribution rate was proposed by using the polar moment of inertia. This rate can be a proper explanatory variable for the multiple regression analysis, because it is independent of the physical damage rate. However, the instabilities in the multiple regression analyses are not yet completely improved. Though this problem still remains.

In the application of the fuzzy quantification theory, it was demonstrated that the fuzzy quantification theory, type I is the useful method for examining the effect of a certain factor. However, in the fuzzy quantification theory, type II, the suitable demonstration was not shown, because the external criteria and items were not well chosen.

Acknowledgments

Mr. Maruyama of the Fukuyama City Waterworks Bureau is gratefully acknowledged for his cooperation in offering the data and the program for the pipe network analysis.

References

1. Ohno, H. & Hoshiya, M., An expert-based forecasting model for water supply systems on aseismic performance, Proc. of JSCE, No.422/I-14, 1990.10, pp.353-360, in Japanese.
2. Hoshiya, M., Ohno, H. & Chiba, T., A practical model for forecasting water supply functions at time of the earthquake, 5-th Int. Conf. on Structural Safety & Reliability, San Francisco, USA, Vol.1, 1989.8, pp.685-692.
3. Hoshiya, M., Ohno, H. & Chiba, T., A practical forecasting model for seismic damage restoration of a water supply system, 2-nd East Asian Conf. on Structural Engineering & Construction, Chiang Mai, Thailand, Vol.2, 1989.1, pp.1497-1502.
4. Harada, T. et al., Elementary study on the importance degree of lifeline network system, Annual Meeting of JSCE,I,1992.9,pp.1484-1485,in Japanese.
5. Watada, J. & Tanaka, H., Fuzzy quantification methods, 2-nd IFSA Congress, Tokyo, 1987.6, pp.20-25.

Realibility of Silo Ring Under Lognormal Stochastic Pressure Using Stochastic Interpolation

Ove Ditlevsen[1], Claus Christensen[1], and Søren Randrup-Thomsen[1]

[1]Department of Structural Engineering, Technical University of Denmark, Lyngby, Denmark

Abstract A Gaussian stochastic silo load model for circular cells has been formulated and used for reliability analysis in two previous papers. In order to avoid negative stresses orthogonal to the wall the possibility of modeling the wall stresses by a non-Gaussian field such as the lognormal field is considered herein. Several fundamental problems by such modeling are revealed. These problems concern the mandatory nonnegativity properties of the covariance structure under certain equilibrium restrictions. However, the investigations also show that the distribution assumption has a substantial influence on the reliability results in spite of the intuitive (but unjustified) reliance on the validity of the central limit theorem for integrals of random fields.

For the practical reliability analysis of the silo wall it is necessary to make a discretization of the random field into another random field defined in terms of a finite, but not too large, set of correlated random variables. It turns out to be essential for the discretization replacement to base it on stochastic interpolation in a particular form. In order to gain insight into the problem the mechanical part of it is kept simple by solely considering the analysis for a ring imagined to be cut out by two consecutive horizontal planes an infinitesimal distance apart at a place far away from regions of the silo with end effects on the internal forces.

The reliability analysis is made with respect to the initial formation of a yield hinge among a finite set of cross-sections. Moreover, a reliability analysis with respect to the development of a collapse mechanism in the ring is considered. The influence of the correlation length (scale of fluctuation) on the reliability is illustrated for both types of limit states. All reliability calculations are made by the standard methods FORM, SORM using the DNV (Det Norske Veritas) program PROBAN.

Keywords. Covariance of equilibrium field, lognormal field, random load field, reliability, silo pressure, stochastic finite elements.

1 Introduction

Silo pressure measurements made during the emptying period show that the wall pressure varies significantly in a random way both along the wall surface and in time. Continuum mechanical models have hitherto failed to describe these fluctuations that seem to be very dependent on imperfections of the silo wall. Therefore, adoption of a suitable empirical random field model for the wall pressure is presently more fertile. However, the choice of such a model is not at all trivial because the stress field in the silo medium is in equilibrium at any time.

Clearly, for the case of no other external forces than the gravity forces the total shear force is zero for any horizontal cut through the silo. Equilibrium consistent covariance functions for the random silo load field can therefore quite reasonably be constructed by requiring that the total system of horizontal stresses, by which the medium acts on the silo wall, is in equilibrium in any horizontal plane. For a circular cylindrical silo a family of equilibrium consistent correlation functions have been derived by Ditlevsen and Munch-Andersen[1] under the assumption that the random stress field is homogeneous and isotropic with respect to the covariance structure.

This equilibrium stress field covariance function is used herein as the basis for various reliability analysis calculations on a ring element imagined to be cut out of a circular concrete silo by two consecutive horizontal planes an infinitesimal distance apart at a place far away from regions of the silo with end effects on the internal forces.

Since physically there can be no tensile stresses between the silo medium and the silo wall it will herein be required that the mathematical model of the silo wall pressure satisfies the condition that the stresses orthogonal to the wall are positive everywhere. This implies that the random pressure field along the wall ideally should be modeled as a non-Gaussian field. The most straightforward model is to let the stress $p(\theta)$ orthogonal to the silo wall as function of the angle θ be a homogeneous lognormal field such that $(\log p(\theta), q(\theta))$ becomes a homogeneous Gaussian field where $q(\theta)$ is the shear stress field along the wall, see Fig. 1.1.

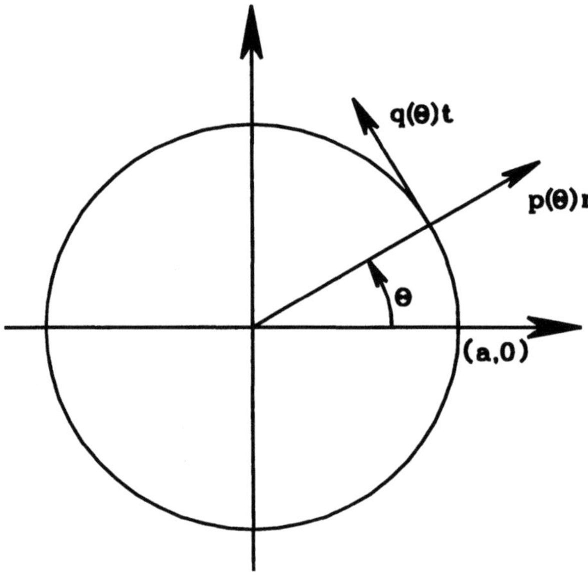

Fig. 1.1. Horizontal cross-section of silo medium body with symbol definitions.

This paper investigates the special problems that arise from using a lognormal silo pressure field for the reliability analysis of the silo ring. The main difficulty concerns the mandatory condition of having nonnegativity of the covariance structure together with the requirement of equilibrium along any horizontal perimeter of the silo, a difficulty which is present for most non-Gaussian field models that could be suggested for silo pressures.

The equilibrium condition difficulties are not present if $(p(\theta),\ q(\theta))$ is modeled as a homogeneous Gaussian field. Adopting the equilibrium covariance function family derived by Ditlevsen and Munch-Andersen[1] each single realization of the field will then be in equilibrium. Therefore the homogeneous field model obtained by censoring away from the Gaussian field all those realizations that are not positive everywhere will be a non-Gaussian field that satisfies the equilibrium condition. Unfortunately, the joint probability distributions of this censored Gaussian field are not known. Therefore it excludes itself as the basis for first or second order reliability analysis (FORM or SORM).

On the other hand, due to the averaging effect when calculating the internal forces in the silo ring the unphysical property of the Gaussian field of possibly giving tensile stresses at some minor parts of the wall may not influence the calculated reliability much given that the tensile situation is suitably rare.

In short, the silo load modeling situation is characterized by a conflict between the equilibrium requirement and the requirement of not having tensile stresses along the wall. Herein the last requirement is taken literally by adopting the lognormal field model. In return this modeling raises several problems with the equilibrium requirement and the proper definition of the covariance function of the field.

These problems make it interesting to study the lognormal load field model because such a study reveals inconvenient, but not necessarily serious inconsistency properties that can also be present for other types of random fields when it is required that the fields must satisfy certain restrictive conditions. For example, that certain integrals of the fields are given with specified properties.

2 Equilibrium covariance structure

It is required in the following that the total system of stresses $(p(\theta), q(\theta))$ along a horizontal perimeter of the circular silo is in equilibrium, that is, it is required that the three equilibrium equations

$$\int_0^{2\pi} q(\theta)d\theta = 0 \tag{1}$$

$$\int_0^{2\pi} [p(\theta)\cos\theta - q(\theta)\sin\theta]d\theta = 0 \tag{2}$$

$$\int_0^{2\pi} [p(\theta)\sin\theta + q(\theta)\cos\theta]d\theta = 0 \tag{3}$$

are satisfied, see Fig. 1.1. For a homogeneous field the covariance function matrix may be expanded into the Fourier series

$$\begin{bmatrix} C_p(\theta) & C_{pq}(\theta) \\ C_{qp}(\theta) & C_q(\theta) \end{bmatrix} = \begin{bmatrix} \frac{a_0}{2} & 0 \\ 0 & \frac{c_0}{2} \end{bmatrix} + \sum_{n=1}^{\infty} \begin{bmatrix} a_n\cos(n\theta) & b_n\sin(n\theta) \\ -b_n\sin(n\theta) & c_n\cos(n\theta) \end{bmatrix} \tag{4}$$

where

$$a_n = \frac{1}{\pi}\int_0^{2\pi} C_p(x)\cos(nx)dx, \quad c_n = \frac{1}{\pi}\int_0^{2\pi} C_q(x)\cos(nx)dx \tag{5,6}$$

$$b_n = \frac{1}{\pi}\int_0^{2\pi} C_{pq}(x)\sin(nx)dx \tag{7}$$

Nonnegative definiteness is ensured if and only if $a_n \geq 0$ and $a_n c_n - b_n^2 \geq 0$ for all $n \in \{0, 1, ...\}$. It can be shown by elementary calculations[1] that the constraints from the equilibrium equations (1) to (3) on the covariance functions $C_p(\theta), C_{pq}(\theta) = -C_{qp}(\theta), C_q(\theta)$ simply are

$$c_0 = 0, \quad a_1 = b_1 = c_1 \geq 0. \tag{8,9}$$

Thus any admissible equilibrium covariance function can be defined on the basis of eqs. (4) to (9). However, it is not easy to directly make a reasonable choice of the Fourier coefficients and therefore one might be lead to suggest several of these to be estimated from the measured data. It is more attractive to establish a reasonable covariance model where very few parameters are required to be estimated from the data. A set of reasonable, admissible covariance functions with such properties can be obtained by assuming that the randomly fluctuating horizontal stress tensor components correspond to biaxial stress equilibrium at any point of the horizontal cross-section through the silo medium body. Then $(p(\theta), q(\theta))$ can be obtained on the basis of a random Airy stress function. This reduces the problem to the simpler problem of choosing a suitable correlation function for a scalar two-dimensional homogeneous and isotropic random field without any other restrictions than that this function must be nonnegative definite and at least 4 times differentiable. Following this idea Ditlevsen and Munch-Andersen[1] established the equilibrium covariance functions

$$C_p(\theta) = 12(\frac{\sigma\alpha}{a^2})^2\frac{1}{3}e^{-2\alpha(1-\xi)} \cdot[1+2\xi^2-4\alpha(1-\xi^2)(1+2\xi)+4\alpha^2(1-\xi^2)^2] \tag{10}$$

$$C_q(\theta) = 4(\frac{\sigma\alpha}{a^2})^2 e^{-2\alpha(1-\xi)} \cdot[-1+2\xi^2+4\alpha(1-\xi-2\xi^2+2\xi^3)+4\alpha^2(1-\xi^2)(-1+2\xi-\xi^2)] \tag{11}$$

$$C_{pq}(\theta) = \sqrt{C_p(0)C_q(0)}\frac{2}{\sqrt{3}}e^{-2\alpha(1-\xi)}\sin\theta \cdot[-\xi+\alpha(3+\xi-4\xi^2)+2\alpha^2(-1+\xi+\xi^2-\xi^3)] \tag{12}$$

in which $\xi = \cos\theta$, α and σ are parameters, and a is the radius of the silo ring. The variances are $C_p(0) = 12(\sigma\alpha/a^2)^2$ and $C_q(0) = 4(\sigma\alpha/a^2)^2$.

Assuming that the field is Gaussian the parameters α and σ can be estimated by maximum likelihood principles from data obtained by pressure cell measurements of $p(\theta_1),...,p(\theta_n)$ at the discrete positions $\theta_1,...,\theta_n$. An example is given in Munch-Andersen et al[2] where 24 measured data sets were used for the estimation. These measurements were made with barley in the Swedish Karpalund silo (45 m high, 7 m diameter concrete silo). That the covariance functions defined by eqs. (10) to (14) for the estimated parameter values are reasonably behaving as the basis for stochastic interpolation between the discrete pressure values $p(\theta_1),...,p(\theta_n)$ can be judged subjectively by inspecting the graphs of the linear regression $\hat{E}[p(\theta) \mid p(\theta_1),...,p(\theta_n)]$ as functions of θ using the given observation sets for $p(\theta_1),...,p(\theta_n)$ in the graphs, Munch-Andersen et al[2]. Contrarily to judgmentally drawn interpolation graphs the linear regression $\hat{E}[p(\theta) \mid p(\theta_1),...,p(\theta_n)]$

together with the linear regression of $q(\theta)$ on $p(\theta_1), ..., p(\theta_n)$ automatically satisfy the equilibrium conditions (1) to (3) given that the covariance functions correspond to equilibrium. The proof of this is simply that if the field is assumed to be Gaussian then the linear regressions coincide with the conditional expectations. Since all the sample functions of the Gaussian field satisfy the equilibrium equations (1), (2), (3) with probability one given that the mean function and the covariance function satisfy the equilibrium requirements then any conditional expectation also satisfies the equilibrium conditions.

The ring beam under Gaussian field line load orthogonal to the perimeter has previously been investigated by Hørte, Bjerager and Skjong[3] with respect to reliability analysis illustrating stochastic finite element techniques under use of the DNV computer program PROBAN. However, their analysis was made without requiring internal equilibrium of the load field, that is, without any restrictions on the correlation structure. Instead equilibrium is restored by assuming that the ring beam is supported by soft springs in the plane of the ring. This spring reaction support eliminates the need for accurate stochastic interpolation between node points of the stochastic finite element mesh but also it makes the analysis less relevant to silo reliability problems.

3 Lognormal wall pressure field

Since the shear stress field $q(\theta)$ along the silo wall must be symmetrically distributed with respect to zero it is consistent with this property to assume that $q(\theta)$ is a Gaussian field. The study herein concerns the consequences of assuming that the homogeneous equilibrium stress field $(p(\theta), q(\theta))$ by transformation to the field $(\log p(\theta), q(\theta))$ becomes a homogeneous Gaussian field. It is first noted that

$$E[\log p] = \log\{E[p]\} - \frac{1}{2}\log(1 + V_p^2), \mathrm{Var}[\log p] = \log(1 + V_p^2) \qquad (13,14)$$

where $V_p = \sqrt{C_p(0)}/E[p]$ is the coefficient of variation of the stress field orthogonal to the silo wall, while the covariance function matrix of the field $(\log p(\theta), q(\theta))$ is

$$\begin{bmatrix} \mathrm{Cov}[\log p(\theta_1), \log p(\theta_2)] & \mathrm{Cov}[\log p(\theta_1), q(\theta_2)] \\ \mathrm{Cov}[q(\theta_1), \log p(\theta_2)] & \mathrm{Cov}[q(\theta_1), q(\theta_2)] \end{bmatrix}$$
$$= \begin{bmatrix} \log[1 + \dfrac{C_p(\theta)}{C_p(0)} V_p^2] & C_{pq}(\theta)/E[p] \\ C_{qp}(\theta)/E[p] & C_q(\theta) \end{bmatrix} \tag{15}$$

provided the matrix on the right hand side is a nonnegative definite matrix function of θ. Fig. 3.1 shows $\log[1 + V_p^2\ C_p(\theta)/C_p(0)]/\log(1+V_p^2)$ as a function of θ for a set of values of V_p. It is seen that for sufficiently large values of V_p the function gets values smaller than -1 which is not allowed for a correlation function. This problem exists for any value of $V_p > 0$, not just for large values. In fact, numerical calculations seem to indicate that among the Fourier coefficients of $\log[1 + V_p^2\ C_p(\theta)/C_p(0)]$ there will always be negative coefficients. Thus it seems that there exists no lognormal random field with the covariance function $C_p(\theta)$. However, this difficulty can be eliminated by defining the covariance function of $\log p(\theta)$ as the trigonometric polynomial

$$\tilde{C}_{\log p}(\theta) = \frac{1}{2}\max\{d_0, 0\} + \sum_{n=I}^{J} d_n \cos(n\theta) \tag{16}$$

where the coefficients $d_I, d_{I+1}, ..., d_J$ are obtained from the Fourier expansion

$$\frac{1}{2}\begin{bmatrix} d_0 & 0 \\ 0 & 0 \end{bmatrix} + \sum_{n=1}^{\infty} \begin{bmatrix} d_n \cos(n\theta) & \dfrac{b_n}{E[p]}\sin(n\theta) \\ -\dfrac{b_n}{E[p]}\sin(n\theta) & c_n \cos(n\theta) \end{bmatrix} \tag{17}$$

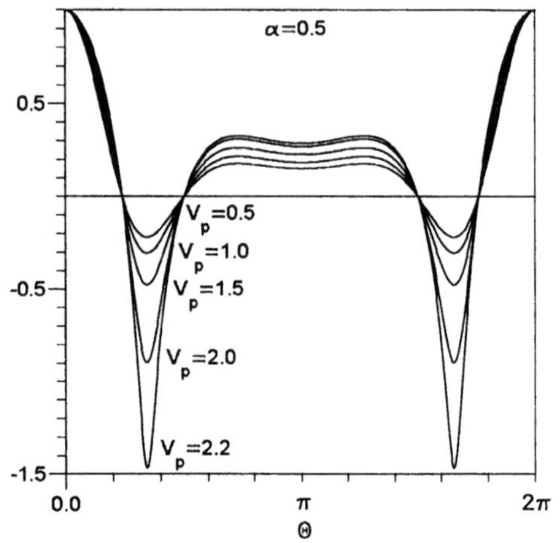

Fig. 3.1. Graphs of $\log\{1+\frac{C_p(\theta)}{C_p(0)}V_p^2\}/\log(1+V_p^2)$ for different values of V_p.

of the right hand side of eq. (15), see eq. (4). We have

$$d_n = \frac{1}{\pi}\int_0^{2\pi} \log[1+\frac{C_p(x)}{C_p(0)}V_p^2]\cos(nx)dx \tag{18}$$

while b_n and c_n are given by eqs. (7) and (6) respectively. The integers I and J are determined as the smallest and largest positive integer respectively such that $d_n \geq 0$ for all $n \in \{I, I+1, ..., J\}$. Moreover, in order to ensure non-negativity the coefficient b_n in eq. (17) is replaced by

$$\tilde{b}_n = b_n \min\{1, E[p]\sqrt{\frac{d_n c_n}{b_n^2}}\}. \tag{19}$$

Back-transformation to the modified covariance function matrix for $(p(\theta), q(\theta))$ gives

$$\begin{bmatrix} \tilde{C}_p(\theta) & \tilde{C}_{pq}(\theta) \\ \tilde{C}_{qp}(\theta) & \tilde{C}_q(\theta) \end{bmatrix} = \begin{bmatrix} E[p]^2\{\exp[\tilde{C}_{\log p}(\theta)]-1\} & \sum_{n=I}^{J}\tilde{b}_n \sin(n\theta) \\ -\sum_{n=I}^{J}\tilde{b}_n \sin(n\theta) & \sum_{n=I}^{J} c_n \cos(n\theta) \end{bmatrix} \tag{20}$$

which is ensured to be a nonnegative definite function matrix.

However, these covariance modifications violate the equilibrium conditions. In fact, given that $(\log p(\theta), q(\theta))$ is a Gaussian field such that the field $(p(\theta), q(\theta))$ satisfies the equilibrium conditions (2) and (3) then these conditions become nonlinear in the Gaussian field $(\log p(\theta), q(\theta))$. But this is in obvious conflict with the properties of the family of multidimensional Gaussian distributions. Thus $(p(\theta), q(\theta))$ can only satisfy the equilibrium conditions with probability zero. The search for a covariance function structure of $(p(\theta), q(\theta))$ that satisfies the equilibrium conditions will therefore never be successfull. This implies that in order for $(\log p(\theta), q(\theta))$ to exist as a Gaussian field there must necessarily be a balancing non-zero shear force in any horizontal cross-section of the silo wall. For realistic values of the coefficient of variation of $p(\theta)$ the covariance modifications to obtain eq. (20) are expected to be small so that only very small horizontal shear forces are needed to establish equilibrium. However, the modifications are necessary in order to avoid serious inconsistencies such as negative variances. In the reliability calculations the loss of equilibrium of $(p(\theta), q(\theta))$ must be compensated in a suitable way as explained later.

4 Limit state problem for reinforced concrete silo ring

In this paper the complicated limit state problem for the entire circular silo shell will not be studied. As a much simpler substitute consider a ring cut out of the circular reinforced concrete silo by two consecutive horizontal planes an infinitesimal distance apart. The ring is assumed to be subjected to the load stress field $(p(\theta), q(\theta))$ as indicated in Fig. 1.1 with mean $E[p] = 36.1$ kPa and coefficient of variation $V_p = 39.8\%$, Munch-Andersen et al[2]. The silo wall thickness is 20 cm and the reinforcement is placed symmetrically with respect to the concrete cross-section axis. The distance between the two reinforcement layers is 12 cm (center to center). The total reinforcement area is 1000 mm^2/m. Both the concrete compression and the reinforcement tensile yield strength are assumed to have lognormal distributions. The mean values are 30 MPa and 480 MPa respectively while the coefficients of variation are 0.15 and 0.05 respectively.

The yield condition for the cross-section is only relevant for the case of a tensile normal force. Yielding is assumed to occur whenever the one or the other of two linear combinations of the normal force N and the bending moment M takes the value zero. Specifically the limit state with respect to cross-section yielding is assumed to have the form

$$g(N,M) = \min\{1 - (\frac{N}{N_y} + \frac{M}{M_y}), 1 - (\frac{N}{N_y} - \frac{M}{M_y})\} = 0 \qquad (21)$$

where no yielding corresponds to $g(N,M) > 0$. The tensile yield normal force N_y corresponds to pure tension while the yield moment M_y corresponds to pure bending. Due to the symmetry of the cross-section M_y is the same for positive and negative bending. Assuming that the concrete has zero tensile strength the random strengths N_y and M_y are obtained by the standard concrete plastic cross-section carrying capacity formulae.

It is assumed that the elastic solution for the internal forces of the silo ring is valid before the occurrence of the first passage of the limit state $g(N,M) = 0$ in any cross-section. The elastic solution for the normal force $N(\theta)$ and the moment $M(\theta)$ at the position θ is

$$\frac{\pi}{a} N(\theta) = \pi \int_0^\theta [p(v)\sin(\theta - v) - q(v)\cos(\theta - v)]dv$$

$$- \int_0^{2\pi} \cos(\theta - u) \int_0^u \{p(v)\sin(u - v) + q(v)[1 - \cos(u - v)]\} dv\, du \qquad (22)$$

$$\frac{2\pi}{a^2} M(\theta) = 2\pi \int_0^\theta \{p(v)\sin(\theta - v) + q(v)[1 - \cos(\theta - v)]\} dv$$

$$- \int_0^{2\pi} [1 + 2\cos(\theta - u)] \int_0^u \{p(v)\sin(u - v)$$

$$+ q(v)[1 - \cos(u - v)]\} dv\, du. \qquad (23)$$

There is no exact solution to the problem of calculating the probability of cross-section yielding in the ring, that is, of calculating the probability that $(N(\theta), M(\theta))$ satisfies $g(N(\theta), M(\theta)) < 0$ for some $\theta \in [0, 2\pi[$.

An upper bound on the probability of yielding is the sum of the probability that $g(N(0), M(0)) < 0$ and the expected number of outercrossings of the set $g(N, M) > 0$ in the interval $\theta \in [0, 2\pi[$. A lower bound on the probability of yielding is obtained by calculating the probability that $g(N(\theta_i), M(\theta_i)) < 0$ for some $\theta \in \{\theta_1, ..., \theta_n\}$ where $\theta_1, ..., \theta_n$ are the positions of a finite number of cross-sections. Solely this lower bound calculation will be considered herein.

5 Numerical integration by stochastic discretization (finite elements)

Except possibly if $(p(\theta), q(\theta))$ is assumed to be a Gaussian field there is no other practicable way to deal with the distributional properties of the 2n-dimensional random vector $(N(\theta_1), M(\theta_1), N(\theta_2), M(\theta_2), ..., N(\theta_n), M(\theta_n))$ than to apply numerical integration in eqs. (22) and (23). For this purpose $(p(v), q(v))$ in the integrands is replaced by a random field $(\tilde{p}(v), \tilde{q}(v))$ completely defined as a linear function of a $(2m-2)$-dimensional random vector $(\boldsymbol{X}, \boldsymbol{Y}) = (X_1, ..., X_m, Y_1, ..., Y_{m-2})$ determined such that the two fields $(p(v), q(v))$ and $(\tilde{p}(v), \tilde{q}(v))$ become approximately statically equivalent over an angular section of size $\omega = 2\pi/m$. Moreover the field $(\tilde{p}(v), \tilde{q}(v))$ is constructed such that it satisfies the equilibrium conditions (1) to (3).

The random variables X_i and Y_i are defined by

$$X_i \sin\omega = a\int_0^\omega \{[p(i\omega + v) + p(i\omega - v)]\sin(\omega - v) + [q(i\omega + v) - q(i\omega - v)][\cos^2\frac{\omega}{2} - \cos(\omega - v)]\}dv \tag{24}$$

$$Y_i = \frac{a}{2}\int_0^\omega [q(i\omega + v) + q(i\omega - v)]dv \tag{25}$$

$i = 1, ..., m$. The random variable X_i is seen to be the size of the force acting orthogonal to the ring at $\theta = i\omega$ while Y_i is the size of the tangential force acting at the same point when it is supposed that the load field $(p(v), q(v))$ acts indirectly on the ring through m simply supported beams each curved as the ring and spanning over the angular intervals $[0,\omega]$, $[\omega, 2\omega]$,...,$[(m-1)\omega, m\omega]$ respectively, and with equal tangential reactions at the two supports.

Since the shear stress field $q(\theta)$ satisfies the equilibrium condition (1) it follows from the definition of Y_i by eq. (25) that $Y_1, ..., Y_m$ for m even satisfies the two linearly independent conditions $Y_1 + Y_3 + ... + Y_{m-1} = 0$ and $Y_2 + Y_4 + ... + Y_m = 0$. Therefore only the first $m-2$ of $Y_1, ..., Y_m$ are included in the vector $\boldsymbol{Y}$. A further reduction of the dimension of $(\boldsymbol{X}, \boldsymbol{Y})$ cannot be made because the linear equilibrium conditions (2) and (3) as emphasized previously cannot be satisfied for the lognormal stress field $p(\theta)$.

Next, a random field $(\hat{p}(v), \hat{q}(v))$ is defined as the linear regression

$$\begin{bmatrix}\hat{p}(v)\\ \hat{q}(v)\end{bmatrix} = \hat{E}[\begin{bmatrix}p(v)\\ q(v)\end{bmatrix} \Bigg| \begin{bmatrix}\boldsymbol{X}\\ \boldsymbol{Y}\end{bmatrix}] = \begin{bmatrix}E[p]\\ 0\end{bmatrix}$$
$$+\mathrm{Cov}[\begin{bmatrix}p(v)\\ q(v)\end{bmatrix}, [\boldsymbol{X}'\,\boldsymbol{Y}']]\mathrm{Cov}[\begin{bmatrix}\boldsymbol{X}\\ \boldsymbol{Y}\end{bmatrix}, [\boldsymbol{X}'\,\boldsymbol{Y}']]^{-1}(\begin{bmatrix}\boldsymbol{X}\\ \boldsymbol{Y}\end{bmatrix} - E[\begin{bmatrix}\boldsymbol{X}\\ \boldsymbol{Y}\end{bmatrix}]). \tag{26}$$

Using that $\hat{E}[(\boldsymbol{X}, \boldsymbol{Y}) \mid (\boldsymbol{X}, \boldsymbol{Y})] = (\boldsymbol{X}, \boldsymbol{Y})$ and that both X_i and Y_i are linear functionals of $(p(v), q(v))$ it follows that eqs. (24) and (25) remain valid with $(\hat{p}(v), \hat{q}(v))$ substituted for $(p(v), q(v))$. In this particular sense the field $(\hat{p}(v), \hat{q}(v))$ is statically equivalent to the field $(p(v), q(v))$.

The elements of the mean $E[\boldsymbol{X}]$ are all equal and given by

$$E[X_i] = 2aE[p](1 - \cos\omega)/\sin\omega \tag{27}$$

while $E[\boldsymbol{Y}] = \boldsymbol{0}$. The elements of the covariance matrix of $(\boldsymbol{X}, \boldsymbol{Y})$ as well as the elements of $\mathrm{Cov}[p(v), [\boldsymbol{X}'\,\boldsymbol{Y}']]$ and $\mathrm{Cov}[q(v), [\boldsymbol{X}'\,\boldsymbol{Y}']]$ are defined through the rather lengthy formulae written out in the appendix.

The aforementioned statical equivalence between ($\hat{p}(v)$, $\hat{q}(v)$) and ($p(v)$, $q(v)$) implies that ($\hat{p}(v)$, $\hat{q}(v)$) just as ($p(v)$, $q(v)$) slightly violates the equilibrium conditions (1) to (3). This lack of equilibrium must be compensated by a shear stress field in the horizontal cross-section of the silo wall. A simple way to establish equilibrium is to add a random sinusoidal normal pressure to $\hat{p}(v)$. By direct substitution into eqs. (1) to (3) it is seen that the equilibrium conditions are satisfied by the field ($\tilde{p}(v)$, $\tilde{q}(v)$) defined by

$$\tilde{p}(v) = \hat{p}(v) - \frac{1}{\pi}\int_0^{2\pi} [\hat{p}(\theta)\cos(\theta - v) - \hat{q}(\theta)\sin(\theta - v)]\, d\theta \tag{28}$$

$$\tilde{q}(v) = \hat{q}(v). \tag{29}$$

After the replacement in eqs. (22) and (23) of the random function pair ($p(v)$, $q(v)$) by the linear combination ($\tilde{p}(v)$, $\tilde{q}(v)$) of the at most $2m - 3$ random variable elements of $(\boldsymbol{X}, \boldsymbol{Y})$, noting that the coefficients to these random variables are deterministic functions of v, ordinary numerical integration by the trapezoidal formula gives $(N(\theta), M(\theta))$ as a pair of linear combinations of the elements of $(\boldsymbol{X}, \boldsymbol{Y})$ for each $\theta \in \{\theta_1, \ldots, \theta_n\}$.

6 Reliability analysis with respect to first yielding among n cross-sections

All subsequent calculations can easily be made by use of a standard reliability analysis computer program based on first and second order reliability methods (FORM or SORM). For the calculations made herein the DNV (Det Norske Veritas) program PROBAN has been used. Specifically $m = 8$ $(\omega = \pi/4)$ is chosen as an example and only the cross-sections placed at the discretization points are considered for analysis. The input to PROBAN is the two-mode limit state given by eq. (21) specified to be valid for each cross-section, the random variable characterizations of the concrete compression strength and the reinforcement yield strength (both strengths are assumed to be constant over the entire silo ring structure), and the set $(\boldsymbol{X}, \boldsymbol{Y})$ of random variables.

It follows from eq. (25) that the random variables $Y_1, ..., Y_{m-2}$ are jointly Gaussian. However, the entire set of random variable $(X_1, ..., X_m, Y_1, ..., Y_{m-2})$ has an unknown joint distribution because $p(v)$ is a lognormal field. Due to the central limit theorem the joint distribution is Gaussian asymptotically as $\alpha \to \infty$. In the opposite limit $\alpha \to 0$ all the random variables $X_1, ..., X_m$ become identical and lognormal. Herein the joint distribution of $(\boldsymbol{X}, \boldsymbol{Y})$ is for each given α approximated by a Nataf distribution, Der Kiureghian et al[4], using the Hermite approximation based on the first four marginal central moments to define the marginal transformation, Winterstein[5]. Formulae for the second central moments (the variances) are given in the Appendix. The third and fourth moments are difficult to obtain by analytical calculations. Herein the central moments are therefore obtained by simulation of the integrand in eq. (24) for a suitably dense discretization using the trapezoidal formula for the integration. Fig. 6.1 shows the skewness and the kurtosis of X_i as functions of α. Control of the discretization accuracy and the sufficient sample size is made by keeping the ratio of the estimated variance of X_i and the variance obtained from eq. (A.2) within an error limit of ± 0.02. This accuracy was obtained by a sample size of 50,000.

For $\alpha = 0$ the distribution of X_i becomes lognormal with coefficient of variation equal to $V_p = 0.398$. Then with $\omega = 1 + V_p^2$ the skewness is $(\omega + 2)(\omega - 1)^{1/2} \simeq 1.26$ and the kurtosis is $\omega^4 + 2\omega^3 + 3\omega^2 - 3 \simeq 5.94$, Johnson and Kotz[6]. Due to the asymptotic Gaussianity for $\alpha \to \infty$ the skewness must approach 0 and the kurtosis must approach 3 as $\alpha \to \infty$. However, it is seen from the simulation results in Fig. 6.1 that this convergence is not noticeable even for α as large as 30. The drop in both skewness and kurtosis down to a local minimum at $\alpha \simeq 4$ is explainable by the influence of the negative part of the correlation function of $\log p(\theta)$, Fig. 6.1 (right).

The joint distribution of the random vector $(\boldsymbol{X}, \boldsymbol{Y})$ is consequently defined in the input to PROBAN as a Nataf distribution specified by the mean vector as given by eq. (27) and $E[\boldsymbol{Y}] = 0$, the covariances as given by eqs. (A.2), (A.3), (A.4), the skewness and kurtosis of X_i, and the skewness 0 and kurtosis 3 for Y_i.

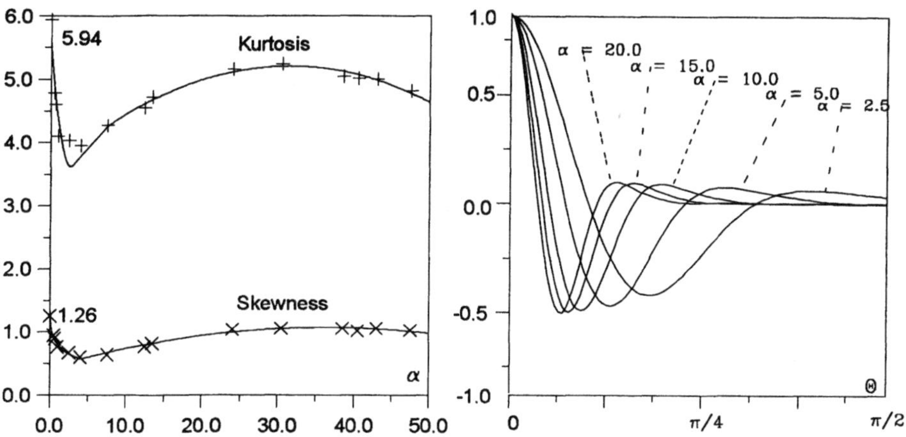

Fig. 6.1. Left: Skewness $E[(X_i - E[X_i])^3]/D[X_i]^3$ and kurtosis $E[(X_i - E[X_i])^4]/D[X_i]^4$ as function of the correlation parameter α for $m = 8$, $E[p] = 36.1$ kPa and $D[p] = 14.4$ kPa. Right: Correlation functions $\tilde{C}(\theta)/\tilde{C}(0)$ for p for different values of α. The negative parts explain the nonmonotonic behavior of the skewness and the kurtosis of X_i as functions of α.

The full thick curve in Fig. 6.2 (left) shows the reliability index β with respect to yield hinge formation at any given cross-section among the considered cross-sections as a function of the correlation parameter α under constancy of the product $\alpha\sigma$, eqs. (10), (11), except for a correction of the covariance matrix (20) by a factor that makes $\tilde{C}_p(0)$ equal to $C_p(0)$. This factor is about 1.02 for $\alpha = 20$. The reliability index is defined as $\beta = -\Phi^{-1}(p)$ where Φ is the standard normal distribution function and p is the probability of the relevant adverse event. For comparison the thin curve shows the reliability index obtained if the skewness is put to 0 and the kurtosis is put to 3 so that the Nataf distribution becomes the Gaussian distribution. It is seen that there is a high sensitivity of the calculated reliability index to the distribution assumption made for the random silo pressure field. This example is a serious warning against too of a hasty reference to the central limit theorem as a justification for adopting the Gaussian field assumption.

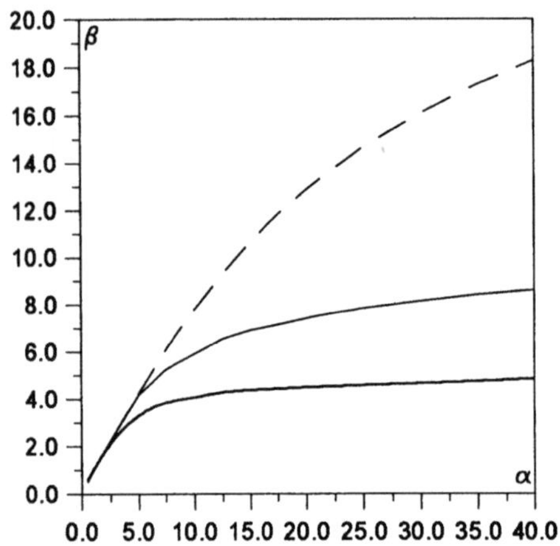

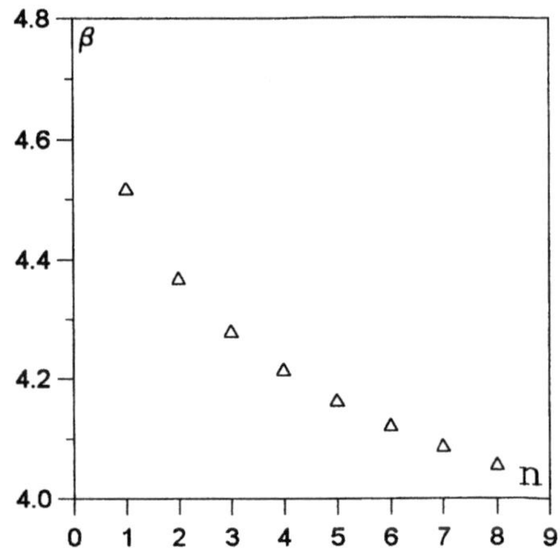

Fig. 6.2. Left: Reliability index β with respect to yield hinge formation at any preselected cross-section as function of the correlation parameter α (inversely proportional to the correlation length of the load field). Full thick curve corresponds to p being lognormal and q Gaussian, full thin curve corresponds to (p,q) being Gaussian in both cases with the same corrected covariance functions $\tilde{C}_p$, $\tilde{C}_q$, $\tilde{C}_{pq}$, eq. (20). Dotted curve corresponds to (p,q) being Gaussian but with the original equilibrium covariance functions C_p, C_q, C_{pq}, eqs. (10) to (12). The results are obtained by SORM. Right: Reliability index β with respect to yield hinge formation at some cross-section among n preselected cross sections. The correlation parameter is $\alpha = 20.9$.

The dotted curve in Fig. 6.2 (left) also corresponds to (p,q) being Gaussian but with the covariance functions changed to the equilibrium covariance functions (10) to (12). It is seen that the direct effect on the reliability index of changing C to $\tilde{C}$ is very large except for the smaller values of α. It should be noted, however, that the correlation lengths as defined by $1/\alpha$ (micro scale) are not comparable when changing from C to $\tilde{C}$. Moreover, since the correlation functions take both positive and negative values there will also be difficulties when attempting to compare the results by reference to one of the standard macro scale correlation length definitions of the type of an integral norm of the correlation function matrix. This topic is subject to further research.

Also the series system effect is considered, that is, the variation of the reliability index with respect to yield hinge formation at any cross-section among n given cross-sections as a function of n. The results are shown in Fig. 6.2 (right).

It will be shown later that if the limit state does not correspond to cross-section yielding but to the formation of a collapse mechanism in the silo ring then the reliability index gets values in the very high reliability range. This fact shows that the occurrence of cross-section yielding is less critical for the safety of the silo structure.

7 Load variation at the most central points

Fig. 7.1 shows examples of the most critical "equivalent" silo wall stress intensity distributions along the ring perimeter as they are determined by the linear regression (26) upon substitution of $\boldsymbol{X}, \boldsymbol{Y}$ corresponding to the most central points (the design points) on the limit state surface , eq. (21), for the cross-section at $\theta = 0$ and calculated by PROBAN.

8 Reliability analysis with respect to collapse

Due to the combination of rotation and elongation in a yielding cross-section it is sufficient for the occurrence of plastic collapse of the silo ring that three yield hinges 1, 2, 3 are formed, Fig. 8.1 (left). The arch 3-1 of the ring is kept fixed. Relative to this fixed arch 3-1 the arch 1-2 rotates with the angular velocity ω_1 and translates in the tangential direction with the velocity v_1. Relative to the arch 1-2 the arch 2-3 rotates with the angular velocity ω_2 and translates in the tangential direction with the velocity v_2. Relative to the arch 2-3 the arch 3-1 rotates with the angular velocity ω_3 and translates in the tangential direction with the velocity v_3. A kinematically admissible velocity field, that is, a velocity field that satisfies both the geometrical compatibility of the ring and the associated flow rule, Fig. 8.1 (right), is obviously obtained by setting

$$\omega_1 = kv_1,\ \omega_2 = -kv_2,\ \omega_3 = k'v_3,\ -k \leq k' \leq k \qquad (30,31,32)$$

(associated flow rule), and, Fig. 8.1 (left),

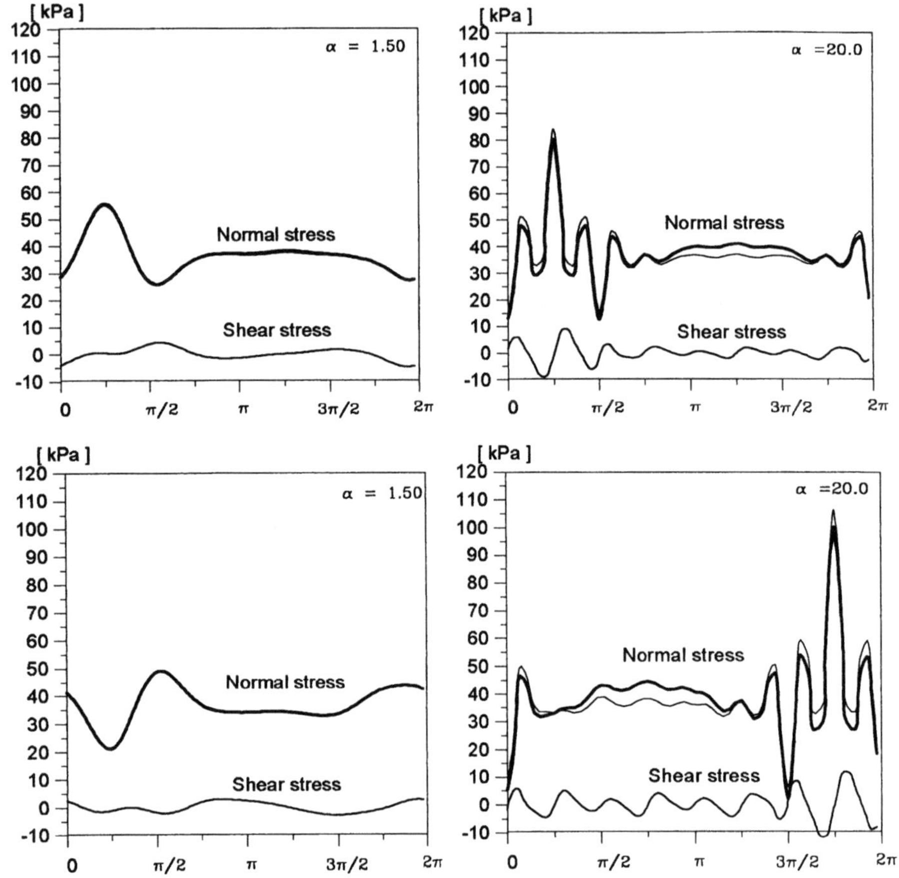

Fig. 7.1. The critical silo wall stress intensities $\tilde{p}(\theta)$, $\hat{p}(\theta)$, and $\tilde{q}(\theta) = \hat{q}(\theta)$ along the perimeter corresponding to the most central point on the limit state surface for the correlation parameter $\alpha = 1.5$ (left) and $\alpha = 20$ (right). The thick curves correspond to the equilibrium corrected intensities $\tilde{p}(\theta)$, $\tilde{q}(\theta)$, eqs. (28), (29). The thin curves correspond to the linear regression $\hat{p}(\theta)$, eq. (26). The two top/bottom figures correspond to the 1./2. failure modes numbered in the order as in eq. (21).

$$\omega_1 + \omega_2 + \omega_3 = 0 \tag{33}$$

$$v_2 \sin\varphi_1 + v_3 \sin\varphi_2 + \omega_3(y_1 - y_3) + \omega_2(y_2 - y_3) = 0 \tag{34}$$

$$v_1 + v_2 \cos\varphi_1 + v_3 \cos\varphi_2 + \omega_3(x_3 - x_1) + \omega_2(x_3 - x_2) = 0 \tag{35}$$

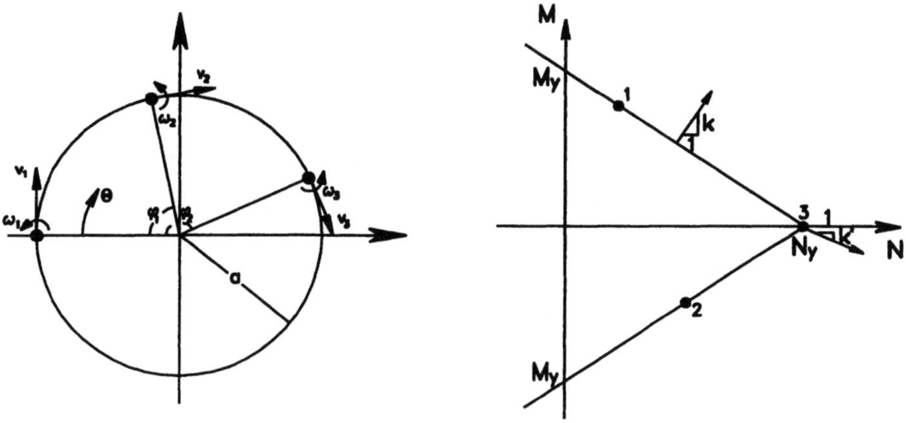

Fig. 8.1. Left: Yield hinge model for silo ring. Right: Yield condition, eq. (21), with indication of the associated flow rule (the normality condition).

(geometrical compatibility). Setting $\omega_1 = 1$, $v_1 = 1/k$ the angular velocities ω_2, ω_3 and the relative tangential velocities v_2, v_3 obtained by solving eqs. (31) to (35) become uniquely given in terms of the positions of the three yield hinges.

The random safety margin is given by the difference $D - W$ between the plastic dissipation

$$D = M_y(\mid \omega_1 \mid + \mid \omega_2 \mid) + N_y v_3 \tag{36}$$

and the external work

$$W = a(a\omega_1 + v_1)\int_0^{\varphi_2} p(\theta)\sin\theta d\theta + a(a\omega_2 + v_2)$$

$$\int_{\varphi_1}^{\varphi_2} p(\theta)\sin(\theta - \varphi_1)d\theta \quad + a\int_0^{\varphi_2} q(\theta)[v_1\cos\theta - a\omega_1(1 - \cos\theta)]d\theta$$

$$+a\int_{\varphi_1}^{\varphi_2} q(\theta)\{v_2\cos(\theta - \varphi_1) - a\omega_2[1 - \cos(\theta - \varphi_1)]\}d\theta \tag{37}$$

As in the previous reliability analysis with respect to first cross-section yielding the random field $(p(\theta), q(\theta))$ is next replaced by the field $(\tilde{p}(\theta),\ \tilde{q}(\theta))$ followed by ordinary numerical integration. The following results are obtained for the discretization with $m = 8$. By variation of the position of the yield hinges 2 and 3 keeping 1 fixed the collapse mechanism with the smallest reliability index is obtained. The full curve in Fig. 8.2 shows the obtained reliability index as function of the correlation parameter α. The fluctuations of the results are partly due to numerical errors. The large fluctuations in the range of small α-values may be caused by the negative correlation function parts, Fig. 6.1 (right). It is seen that the reliability index is in a much higher range of reliability than for the case of first yield hinge formation.

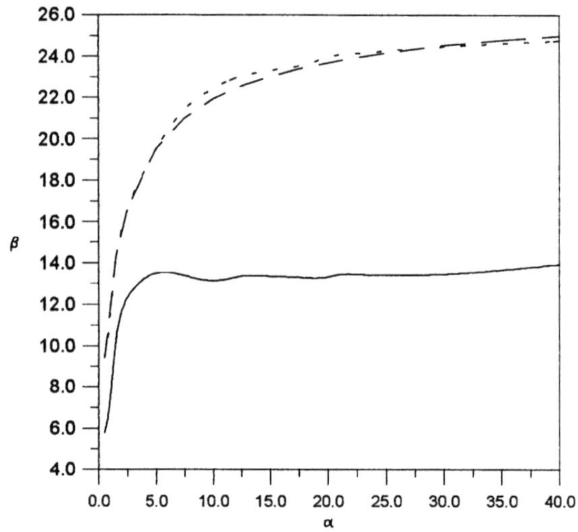

Fig. 8.2. Reliability index with respect to formation of collapse mechanism in silo ring with one yield hinge at fixed position as function of the correlation parameter α. Full curve corresponds to p being lognormal and q Gaussian, dotted curve (long bars) corresponds to (p, q) being Gaussian in both cases with the same corrected covariance functions $\tilde{C}_p$, $\tilde{C}_q$, $\tilde{C}_{pq}$, eq. (20). Dotted curve (short bars) corresponds to (p, q) being Gaussian but with the original equilibrium covariance functions C_p, C_q, C_{pq}, eqs. (10) to (12).

As in the reliability analysis with respect to first yield hinge formation the effect of changing the lognormal distribution to the normal distribution is also investigated here. It is seen from Fig. 8.2 that there is the same large sensitivity to the distribution assumption as observed in Fig. 6.2 (left) while the effect of changing from C to $\tilde{C}$ (Gaussian case) is small.

There is an infinity of possible critical mechanisms corresponding to letting the position of yield hinge 1 be at any place of the perimeter. Together these mechanisms make up a series system problem. Due to the common equilibrium load field it is expected that there will be high correlation between any two safety margins of the series system implying that the series system effect is expected to be of minor importance. This effect is not pursued in further detail in this paper.

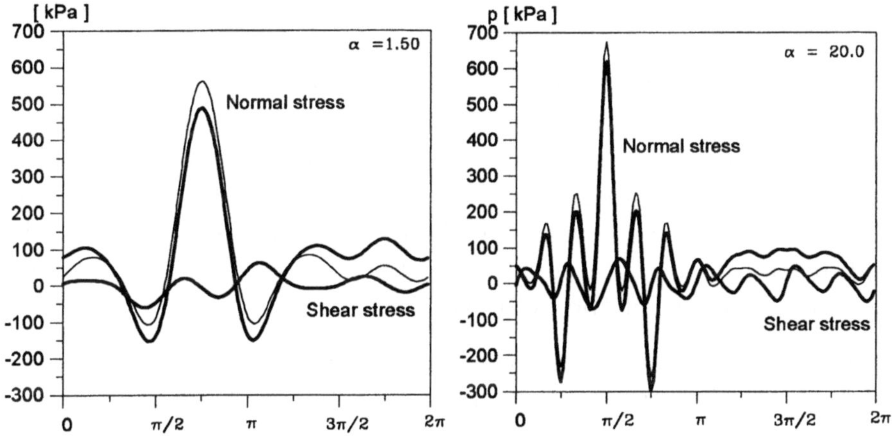

Fig. 8.3. The critical silo wall stress intensities $\hat{p}(\theta)$, $\tilde{p}(\theta)$, and $\hat{q}(\theta) = \tilde{q}(\theta)$ along the perimeter corresponding to the most central point on the collapse mechanism limit state surface for $\alpha = 1.5$ and $\alpha = 20$ respectively. The thick curves correspond to the equilibrium corrected intensities $\tilde{p}(\theta)$, $\tilde{q}(\theta)$, eqs, (28), (29). The thin curves correspond to the linear regression $\hat{p}(\theta)$, eq. (26).

As in Fig. 7.1 two different cases of the most critical "equivalent" silo load distribution $(\tilde{p}(\theta), \tilde{q}(\theta))$ along the ring perimeter are shown in Fig. 8.3 as they are obtained corresponding to the most central point of the limit state defined by the collapse criterion $D = W$. The negative values of $\tilde{p}$ along some minor parts of the perimeter are not in conflict with the lognormal assumption because $\hat{p}$ in the formula (28) for $\tilde{p}$ is defined as the linear regression of p on $(\boldsymbol{X}, \boldsymbol{Y})$ and not as the conditional mean of p given $(\boldsymbol{X}, \boldsymbol{Y})$.

9 Conclusions

The special problems that characterize the reliability analysis of a silo ring structure subject to an internal almost equilibrium load field $(p(\theta), q(\theta))$ of non-Gaussian type have been investigated. In particular the stress field $p(\theta)$ orthogonal to the circular perimeter has been taken to be lognormal while the corresponding shear stress field $q(\theta)$ tangential to the perimeter has been taken to be Gaussian. The following conclusions can be drawn from the investigation:

1. The considered non-Gaussian load field $(p(\theta), q(\theta))$ cannot possess exact internal equilibrium.

2. Starting from a given equilibrium covariance structure C of $(p(\theta),\ q(\theta))$ the requirement that $(\log p(\theta), q(\theta))$ is Gaussian necessitates some modifications of the spectral representation of the given equilibrium covariance structure in order to maintain nonnegativity of the covariance structure of $(\log p(\theta), q(\theta))$. This modification leads to a covariance structure $\tilde{C}$ giving almost internal equilibrium of the field $(p(\theta), q(\theta))$.

3. The necessary discretization for the reliability computations is based on a particular stochastic finite element formulation in which the direct load field is replaced by another load field defined as the linear regression on a discretized node force loading defined according to the principle of indirect loading to be statically equivalent to the direct load field.

The following conclusions 4 to 7 are all based on the results in Figs. 4 and 9:

4. For the given family of covariance functions the micro scale correlation length parameter $1/\alpha$ has for the smaller values of α a substantial influence on the reliability index of the considered concrete silo ring both with respect to first yielding of a cross-section and with respect to formation of a collapse mechanism in the silo ring. The reliability increases with increasing α.

5. Under the assumption that the load field $(p(\theta),\ q(\theta))$ is Gaussian the calculated reliability indices with respect to yield hinge formation are very sensitive to the change of the covariance struc-

ture from C to $\tilde{C}$. Such strong sensitivity is not observed for the reliability indices with respect to plastic collapse.

6. For the given covariance structure $\tilde{C}$ the calculated reliability indices are very sensitive to whether the distribution of the silo pressure field is assumed to be lognormal or to be Gaussian in spite of what should be intuitively expected when relying on the usual argument of the central limit theorem applied to integrals of non-Gaussian fields.

7. For the covariance structure $\tilde{C}$ and $p(\theta)$ being lognormal the calculated reliability index increases slowly with α except for small values of α in contrast to what is the case if $p(\theta)$ is Gaussian. This speaks in favor of standardizing the lognormal field rather than the Gaussian field as a codified silo pressure model for reliability investigations.

8. It is noted that the reliability analysis of the cut out ring is a gross simplification of the reliability analysis of the cylindrical silo shell structure. The existence of reserve carrying capacity in the vertical direction is judged to be beneficial for the reliability at least with respect to plastic collapse.

9. There is a need for further research on the probabilistic properties of integrals of non-Gaussian fields in relation to the central limit theorem in particular when the correlation functions take both positive and negative values.

Acknowledgement

This work has been financially supported by the Danish Technical Research Council.

References

1. Ditlevsen, O., & Munch-Andersen, J., Stochastic silo load model. I: Correlation theory. Submitted to *J. Eng. Mech.*, ASCE, 1993.

2. Munch-Andersen, J., Ditlevsen, O., Christensen, C., & Randrup-Thomsen, S., Stochastic silo load model. II: Data analysis and reliability applications. Submitted to *J. Eng. Mech.*, ASCE, 1993.
3. Hørte, T., Bjerager, P., & Skjong, R., Finite element reliability method, FERM. A.S Veritas Research Report No. 90-2062, 1990.
4. Der Kiureghian, A., & Liu, P-L., Structural reliability under incomplete probability information. *J. Eng. Mech.*, ASCE, **112**, 1986, 85-104.
5. Winterstein, S.R., Nonlinear vibration models for extremes and fatigue. *J. Eng. Mech.*, ASCE, **114**, 1988, 1772-1790.
6. Johnson, N.L., & Kotz, S., *Continuous Univariate Distributions.* I. Wiley, New York, 1970.

APPENDIX

The elements of the covariance matrix $\text{Cov}[\boldsymbol{X}, \boldsymbol{X}']$ are given by

$$
\begin{aligned}
&\text{Cov}[X_i, X_j](\sin^2 \omega)/a^2 \\
&= \int_0^\omega \int_0^\omega \Big\{ \tilde{C}_p[(i-j)\omega + (u-v)] + \tilde{C}_p[(i-j)\omega - (u-v)] \\
&\quad + \tilde{C}_p[(i-j)\omega + (u+v)] + \tilde{C}_p[(i-j)\omega - (u+v)] \Big\} \\
&\quad \sin(\omega - u)\sin(\omega - v)\, dudv \\
&+ \int_0^\omega \int_0^\omega \Big\{ \tilde{C}_q[(i-j)\omega + (u-v)] + \tilde{C}_q[(i-j)\omega - (u-v)] \\
&\quad - \tilde{C}_q[(i-j)\omega + (u+v)] - \tilde{C}_q[(i-j)\omega - (u+v)] \Big\} \\
&\quad \Big\{ \cos^4(\frac{\omega}{2}) - \cos^2(\frac{\omega}{2})[\cos(\omega - u) + \cos(\omega - v)] \\
&\quad + \cos(\omega - u)\cos(\omega - v) \Big\}\, dudv
\end{aligned}
$$

$$+\int_0^\omega \int_0^\omega \Big\{ -2\tilde{C}_{pq}[(i-j)\omega + (u-v)] + 2\tilde{C}_{pq}[(i-j)\omega - (u-v)]$$

$$+2\tilde{C}_{pq}[(i-j)\omega + (u+v)] - 2\tilde{C}_{pq}[(i-j)\omega - (u+v)] \Big\}$$

$$\sin(\omega - u)[\cos^2(\frac{\omega}{2}) - \cos(\omega - v)]\, dudv.$$

$$\text{(A.1)}$$

By the substitution $x = u - v$, $y = u + v$ these double integrals can be reduced to single integrals:

$$\mathrm{Cov}[X_i, X_j]\frac{\sin^2 \omega}{a^2}$$

$$= \frac{1}{2}\int_0^\omega \Big\{ \tilde{C}_p[(i-j)\omega + x] + \tilde{C}_p[(i-j)\omega - x] \Big\}$$

$$\Big[2\sin x - x\cos(2\omega - x) + 2(\omega - x)\cos x - \sin(2\omega - x) \Big]\, dx$$

$$+ \frac{1}{2}\int_\omega^{2\omega} \Big\{ \tilde{C}_p[(i-j)\omega + x] + \tilde{C}_p[(i-j)\omega - x] \Big\}$$

$$\Big[\sin(2\omega - x) - (2\omega - x)\cos(2\omega - x) \Big]\, dx$$

$$+ \int_0^\omega \Big\{ \tilde{C}_q[(i-j)\omega + x] + \tilde{C}_q[(i-j)\omega - x] \Big\}$$

$$\Big\{ (\cos x + 2\cos^4 \frac{\omega}{2})(\omega - x) + 2\cos^2 \frac{\omega}{2}[\sin x - \sin(\omega - x)]$$

$$+ \cos\omega \sin(\omega - x) - [\frac{1}{2}\cos(2\omega - x) + \cos^4 \frac{\omega}{2}]\, x$$

$$- 2\cos^2 \frac{\omega}{2} \sin(\omega - x) - \frac{1}{2}\sin x \Big\}\, dx$$

$$-\frac{1}{2}\int_{\omega}^{2\omega}\Big\{\tilde{C}_q[(i-j)\omega+x]+\tilde{C}_q[(i-j)\omega-x]\Big\}$$

$$\Big\{\Big[\cos(2\omega-x)+2\cos^4\frac{\omega}{2}\Big](2\omega-x)$$

$$-(4\cos^2\frac{\omega}{2}-1)\sin(2\omega-x)\Big\}\,dx$$

$$-\int_0^{\omega}\Big\{\tilde{C}_{pq}[(i-j)\omega+x]-\tilde{C}_{pq}[(i-j)\omega-x]\Big\}\Big\{4\cos^2\frac{\omega}{2}\cos\omega$$

$$+2\cos^2\frac{\omega}{2}[1-2\cos(\omega-x)]-2\cos^2\frac{\omega}{2}\cos x$$

$$+2(\omega-x)\sin x+x\sin(2\omega-x)\Big\}\,dx$$

$$+\int_{\omega}^{2\omega}\Big\{\tilde{C}_{pq}[(i-j)\omega+x]-\tilde{C}_{pq}[(i-j)\omega-x]\Big\}$$

$$\Big\{2\cos^2\frac{\omega}{2}[1-\cos(2\omega-x)]-(2\omega-x)\sin(2\omega-x)\Big\}\,dx. \tag{A.2}$$

The elements of the covariance matrix $\mathrm{Cov}[\boldsymbol{Y},\boldsymbol{Y}']$ are given by

$$\mathrm{Cov}[Y_i,Y_j]\frac{4}{a^2}=\int_0^{\omega}\int_0^{\omega}\Big\{\tilde{C}_q[(i-j)\omega+(u-v)]+\tilde{C}_q[(i-j)\omega-$$

$$(u-v)]+\tilde{C}_q[(i-j)\omega+(u+v)]+\tilde{C}_q[(i-j)\omega-(u+v)]\Big\}\,dudv$$

$$=\int_0^{2\omega}(2\omega-x)\Big\{\tilde{C}_q[(i-j)\omega+x]+\tilde{C}_q[(i-j)\omega-x]\Big\}\,dx. \tag{A.3}$$

while the elements of the cross-covariance matrix $\mathrm{Cov}[\boldsymbol{X},\boldsymbol{Y}']$ are given by

$$\mathrm{Cov}[X_i, Y_j]\frac{2\sin\omega}{a^2} =$$

$$\int_0^\omega \int_0^\omega \Big\{ -\tilde{C}_{pq}[(i-j)\omega + (u-v)] - \tilde{C}_{pq}[(i-j)\omega - (u-v)]$$

$$-\tilde{C}_{pq}[(i-j)\omega + (u+v)] - \tilde{C}_{pq}[(i-j)\omega - (u+v)] \Big\}$$

$$\sin(\omega - u)\, du dv$$

$$+\int_0^\omega \int_0^\omega \Big\{ \tilde{C}_q[(i-j)\omega + (u-v)] - \tilde{C}_q[(i-j)\omega - (u-v)]$$

$$+\tilde{C}_q[(i-j)\omega + (u+v)] - \tilde{C}_q[(i-j)\omega - (u+v)] \Big\}$$

$$\left[\cos^2\frac{\omega}{2} - \cos(\omega - u)\right] du dv$$

$$= \int_0^\omega \Big\{ \tilde{C}_{pq}[(i-j)\omega + x] + \tilde{C}_{pq}[(i-j)\omega - x] \Big\}$$

$$[2\cos\omega - 1 - \cos x]\, dx$$

$$+\int_\omega^{2\omega} \Big\{ \tilde{C}_{pq}[(i-j)\omega + x] + \tilde{C}_{pq}[(i-j)\omega - x] \Big\}$$

$$[\cos(2\omega - x) - 1]\, dx$$

$$+\int_0^\omega \Big\{ \tilde{C}_q[(i-j)\omega + x] - \tilde{C}_q[(i-j)\omega - x] \Big\}$$

$$\left[x\cos^2\frac{\omega}{2} - \sin x\right] dx$$

$$+\int_{\omega}^{2\omega}\left\{\tilde{C}_q[(i-j)\omega+x]-\tilde{C}_q[(i-j)\omega-x]\right\}$$
$$\left[(2\omega-x)\cos^2\frac{\omega}{2}-\sin(2\omega-x)\right]dx. \tag{A.4}$$

The elements of Cov $\left[\begin{array}{c} p(v) \\ q(v) \end{array}\right], [\boldsymbol{X}'\ \boldsymbol{Y}'])$ are given by

$$\mathrm{Cov}[p(v), X_i]\frac{\sin\omega}{a} =$$
$$\int_0^{\omega}[\tilde{C}_p(i\omega-v+u)+\tilde{C}_p(i\omega-v-u)]\sin(\omega-u)\,du$$
$$+\int_0^{\omega}[\tilde{C}_{pq}(i\omega-v+u)-\tilde{C}_{pq}(i\omega-v-u)] \tag{A.5}$$
$$[\cos^2\frac{\omega}{2}-\cos(\omega-u)]\,du$$

$$\mathrm{Cov}[q(v), X_i]\frac{\sin\omega}{a} =$$
$$\int_0^{\omega}[\tilde{C}_{pq}(-i\omega+v-u)+\tilde{C}_{pq}(-i\omega+v+u)]\sin(\omega-u)\,du$$
$$+\int_0^{\omega}[\tilde{C}_q(i\omega-v+u)-\tilde{C}_q(i\omega-v-u)]$$
$$[\cos^2\frac{\omega}{2}-\cos(\omega-u)]\,du$$
$$\tag{A.6}$$

$$\mathrm{Cov}[p(v), Y_i]\frac{2}{a} = \int_0^{\omega}[\tilde{C}_{pq}(i\omega-v+u)+\tilde{C}_{pq}(i\omega-v-u)]\,du \tag{A.7}$$

$$\mathrm{Cov}[q(v), Y_i]\frac{2}{a} = \int_0^{\omega}[\tilde{C}_q(i\omega-v+u)+\tilde{C}_q(i\omega-v-u)]\,du. \tag{A.8}$$

Improved Bounds on First-Passage Times for Scalar Gaussian Processes

S. Engelund[1], R. Rackwitz[1], C. Lange[2]

[1] *Technical University of Munich, Lehrstuhl für Massivbau, Arcisstrasse 21, 80290 München, Germany.*
[2] *Hochschule für Technik und Wirtschaft Dresden, Fachbereich Informatik/Angewandte Mathematik Friedrich List-Platz 1, 01069 Dresden, Germany*

Abstract. Several methods for the calculation of first excursion probabilities of Gaussian processes are discussed. Two of them are based on factorial moments of the number of crossings into the failure region including a new method based on a Gram-Charlier series expansion. Numerical methods for the evaluation of factorial moments are investigated. On the basis of these methods two examples are studied. The examples show that the new method converges faster towards the exact solution than Rice's "in- and exclusion" series. However, it is difficult to establish when bounds are obtained. Further it is shown that for engineering applications the Poisson assumption as modified by Ditlevsen such that the initial conditions are taken into account provide excellent results in almost all cases.

Keywords. Gaussian processes, First-passage, Bounds, Factorial moments

1 INTRODUCTION

In many engineering applications it is necessary to determine the reliability of structural components subject to Gaussian process loading. Then the probability that the structural component enters, during a given time interval, a critical state (failure) must be determined. Let failure be defined by $S = \{g(X(t)) \leq 0\}$ where X(t) denotes a Gaussian process. The probability of failure in the interval $[0;T]$ then is

$$P_f(T) = 1 - P\left[g(X(t)) > 0, \quad \forall t \in [0;T]\right] \tag{1}$$

No simple analytical results for his problem exist. In most cases the computation of the failure probability uses the outcrossing approach together with the assumption of independent outcrossings. Then the number of outcrossings follows a Poisson distribution and

$$P_f(T) \approx 1 - \exp\left[-E\left[N^+(T)\right]\right] \tag{2}$$

where $E[N^+(T)]$ is the mean number of crossings of $X(t)$ into the failure domain S during the interval $[0;T]$ for regular streams of crossings.In the stationary case this is $E[N^+(T)] = \upsilon^+T$. Cramer and Leadbetter[1] have shown that eq. (2) is asymptotically correct for increasing levels. However, for low levels the error made by assuming independent crossings can be large depending on the correlation structure of the process. Further, it is in general not possible to quantify the error made by the Poisson assumption.

If a number of time-independent variables **R** also are present the total failure probability is

$$P_f(T) = \int_\Omega P_f(T|\boldsymbol{r}) f_{\boldsymbol{r}}(\boldsymbol{r}) d\boldsymbol{r} \tag{3}$$

where **R** is defined in Ω. Significant contributions to $P_f(T)$ may come from cases where $P_f(T|\boldsymbol{r})$ is large. For this type of problem it is, therefore, important that the method for the calculation of $P_f(T|\boldsymbol{r})$ is sufficiently accurate even for low reliability levels. Numerous improvements for eq. (2) exist. In the stationary case Ditlevsen[2] suggests by taking the initial conditions into account

$$P_f(T) \approx 1 - \left(1 - P_f(0)\right) \exp\left[-\frac{\nu^+ T}{1 - P_f(0)}\right] \tag{4}$$

where ν^+ denotes the outcrossing intensity. In some cases this simple modification provides a remarkable improvement over eq. (2).

Madsen and Krenk[3] solve an integral equation which governs the first passage probability function. The equation contains an unknown kernel which can be approximated in different ways (see e.g. Nielsen and Sorensen[4] and Bernard and Shipley[5]).

Especially for narrow-band processes, a number of other approximations exist. Yang and Shinozuka[6] obtained good results based on the assumption that the extreme points follow a Markov process. Another successful method is to consider the outcrossing of the envelope process rather than the actual process (see e.g. Krenk, Madsen and Madsen[7]). For vector processes it is difficult to define a suitable envelope process and the method is less accurate (Veneziano[8]). For narrow band processes Yang and Shinozuka[6] also investigated the clump size approximation where the upcrossing intensity is reduced in order to take into account that upcrossings tend to occur in clumps. More recent studies for first passage times for narrow band processes have been performed by Langley[9] and Ditlevsen and Lindgren[10].

The scope of the present work is to compare a number of general methods for the calculation of upper and lower bounds of the outcrossing probability. In particular, we will investigate only those methods which do not set out from a narrow-band assumption. Special attention will be given to the method of Lange[11] because this method is less well-known and, possibly the most accurate.

2 BOUNDS ON FIRST PASSAGE TIME

A method to determine bounds on the outcrossing probability was already given by Rice[12] in the form of the so called "in- and exclusion" series, given and derived in Lange[11] in the form.

$$P_f(T) = \sum_{i=1}^{\infty} \frac{(-1)^i}{i!} m_i \tag{5}$$

where m_i denotes the i'th factorial moment of the number of outcrossings in $[0;T]$. The series provides upper and lower bounds upon truncation after an odd or even term, respectively. The computational effort involved in evaluating $P_f(T)$ according to this method, however, is extensive. Further an increasing number of terms has to be taken into account for increasing T. Bolotin[13] gave a similar series based on the moments of the number of outcrossings $E[N^+(T)]$, $E[(N^+(T))^2]$... . This series, can be regarded as a rearrangement of eq. (5) in terms of moments. Bolotin performed a few numerical studies indicating that the bounds are rather wide for large values of $E[N^+(T)]$.

Lange[11] considered the discrete random variable N^+ describing the number of crossings into the failure region in $[0;T]$ of the continuous stochastic process $X(t)$. Let p_k denote the probability of k crossings in $[0;T]$. Evidently, the survival probability is

$$p_0(T) = P\left[g(X(t)) > 0 \quad \forall t \in [0;T]\right] \tag{6}$$

The distribution function $F_N(k)$ of N^+ is generally unknown, but a Gram- Charlier series expansion (type B) can be used to approximate the unknown probabilities of a discrete random variable with known factorial moments m_i by a Poisson distributed random variable Y with parameter m_1 (see e.g. Kendall and Stuart[14]).

$$p_k = w_k \sum_{i=0}^{\infty} q_i \, Q(i,k) \tag{7}$$

where

$$w_k = \frac{m_1^k}{k!}\exp(-m_1) \tag{8}$$

are the probabilities $w_k = P(Y = k)$. The so-called Gram-Charlier polynomials can be obtained from

$$Q(i,k) = \frac{d^i}{dm_1^i} w_k \qquad (i = 1,2,\ldots) \tag{9}$$

and the coefficients q_i are given by

$$q_i = \frac{m_1^i}{i!}\sum_{i=0}^{\infty} p_k\, Q(i,k). \tag{10}$$

It can be shown that

$$q_i = \sum_{j=0}^{\infty}(-1)^j \frac{m_1^j}{j!(i-j)!}(m_{i-j} - m_1^{i-j}). \tag{11}$$

It is further seen that $Q(i,0) = (-1)^i$. From eq. (7) it then follows

$$p_0 = \exp(-m_1)\sum_{i=0}^{\infty} q_i(-1)^i \tag{12}$$

and the following approximations for the survival probability up to the fourth order are obtained

$$p_{0,1} = \exp(-m_1) \tag{13}$$

$$p_{0,2} = \exp(-m_1)\left(1 + \frac{1}{2}\left(m_2 - m_1^2\right)\right) \tag{14}$$

$$p_{0,3} = \exp(-m_1)\left(1 + \frac{1}{2}\left(m_2 - m_1^2\right) - \frac{1}{6}\left(m_3 - m_1^3\right) + \frac{m_1}{2}\left(m_2 - m_1^2\right)\right) \tag{15}$$

$$p_{0,4} = \exp(-m_1)\left(1 + \frac{1}{2}\left(m_2 - m_1^2\right) - \frac{1}{6}\left(m_3 - m_1^3\right) + \frac{m_1}{2}\left(m_2 - m_1^2\right)\right.$$
$$\left. + \frac{1}{24}\left(m_4 - m_1^4\right) - \frac{m_1}{6}\left(m_3 - m_1^3\right) + \frac{m_1^2}{2}\left(m_2 - m_1^2\right)\right) \tag{16}$$

It is well-known that for high levels the outcrossings are independent. This implies that $m_i = m_1{}^i$. Eq (12) then reduces to $p_0 = \exp(-m_1)$. Eq. (12), therefore, is correct in the limit. It is also evident that the outcrossings follow a Poisson distribution when $1 >> m_1 >> m_2 >> ... >> m_i$ i.e. the Poisson assumption is correct for low values of T. Further under the condition

$$q_{n_0} \geq q_{n_o+1} \geq q_{n_o+2} \geq ... > 0 \tag{17}$$

the error of approximation for p_0 is smaller than

$\exp(-m_1)\left|q_{n_o'+1}\right|$ for $n_0 > 0$ and

$$\exp(-m_1)\sum_{i=1}^{2n} q_i(-1)^i \qquad (2n > n_0) \tag{18}$$

are upper bounds for p_0 and

$$\exp(-m_1)\sum_{i=1}^{2n+1} q_i(-1)^i \qquad (2n+1 > n_o) \tag{19}$$

are lower bounds for p_0. If the inequality signs in eq. (17) were changed then eqs. (18) and (19) would provide lower and upper bounds respectively. Even if the condition eq. (17) is not fulfilled, convergence of eq. (12) can be established for a wide class of stochastic processes (see Lange[11])

3 EVALUATION OF FACTORIAL MOMENTS

Cramer and Leadbetter[1] have shown that the factorial moments of the number of crossings of a scalar stochastic process of the constant level ξ are

$$m_i = \int_0^T\int_0^T \cdots \int_0^T\int_0^\infty\int_0^\infty \cdots \int_0^\infty \dot{x}_1\dot{x}_2 \ldots \dot{x}_i \, f(\xi,\dot{x}_1,\xi,\dot{x}_2 \ldots \xi,\dot{x}_i)\, d\dot{x}_1 d\dot{x}_2 \ldots d\dot{x}_i dt_1 dt_2 \ldots dt_i \tag{20}$$

where f denotes the $2i$-dimensional normal density function of $\left(X(t);\dot{X}(t)\right)$.

The moments of crossings of a constant level up to the 4'th order have been studied by Gaganov[15] for special Gaussian scalar processes. Miroshin and Zvetkov[16] calculated moments of crossings of a linear function by a Gaussian process. In the general case, however, analytical solutions are unknown. Lange[11] evaluated the factorial moments by simulating the process and counting the number of upcrossings. Such a method, however, is not efficient for high levels. It is also not possible to use a simple numerical method as e.g. Gauss-integration because the dimension of the integral in general is too large.

Within the field of reliability analysis a number of different methods for the evaluation of integrals of the type (20) have been suggested. The integral can be solved by nested FORM/SORM (Hohenbichler[17]), as a Laplace integral (Breitung[18]) or as the sensitivity measure of an associated parallel system (Madsen[19], Hagen[20]). All these methods locate an important region where the boundary of the integral can be approximated by a linear or quadratic term. However, if the correlation function of the considered process has a periodic term a large number of important regions exist (multiple β-points). Therefore, these methods generally cannot be used. For further discussion it is convenient to rewrite the integral eq. (20) as

$$m_i = \int_0^T \int_0^T \dots \int_0^T \nu^+(t_1, t_2 \dots t_i)\, dt_1 dt_2 \dots dt_i \tag{21}$$

where $\nu^+(t_1,t_2..t_i)$ denotes the joint crossing intensity following from eq. (20). The factorial moments can also be evaluated by conditional sampling, where for each random sample of $(t_1,t_2..t_i)$ the crossing intensity is determined by one of the three mentioned methods (see e.g. Ayyub and Chia[21]). The multiple β-points problem is thus avoided. Then, it is important that the computational effort involved in the evaluation of ν^+ is small. Unfortunately, serious problems have been met with either method. In order to determine the n'th order crossing intensity by the sensitivity method it is necessary to calculate the n'th order derivative of a failure probability determined by a FORM/SORM method. Derivatives of order higher than one are in general difficult to calculate and all but sufficiently accurate. The amount of effort involved in performing a SORM analysis or evaluating a Laplace integral directly increases rapidly with the dimension of the problem. In conclusion, conditional sampling, even if performed in an adaptive manner, does not offer an efficient alternative. The most efficient numerical method for the evaluation of the factorial moments is simple importance sampling (see e.g. Rubinstein[22]). This method has the advantage that it is relatively insensitive towards the number of dimensions (see Engelund and Rackwitz[23]). In the following examples this method has been used. The samples of $(t_1,t_2..t_i)$ are equally distributed. The samples of $(\dot{x}_1, \dot{x}_2 \dots \dot{x}_i)$ are generated from a normal distribution with independent variables, centered in the point which gives the largest contribution to ν^+. It is believed that the sampling distribution still can be improved for special cases having regard to the correlation structure of the process.

4 INITIAL CONDITIONS

Consider a stationary Gaussian process $X(t)$ which without loss of generality can be assumed to have zero mean and unity standard deviation. The autocorrelation function is denoted by $\rho(\tau)$. For low levels the probability that the process initially exceeds the level ξ is large. This is taken into account by calculating the failure probability $P_f(T)$ as

$$P_f(T) = 1 - (1 - P_f(0))\, p_0(T \mid X(0) < \xi)\, P(X(0) < \xi) \tag{22}$$

where $P(X(0) < \xi)$, the probability that the process does not initially exceed the level ξ, is a simple time-invariant reliability problem. $p_0(T \mid X(0) < \xi)$ is determined by investigating the conditional process. For the condition $X(0) = x_0$ the covariance function $c(t)$ and the mean value function $\mu(t)$ of the conditional process are

$$\mu(t) = \rho(t)\, x_0 \tag{23}$$

$$c(t_1; t_2) = \rho(t_1 - t_2) - \rho(t_1)\rho(t_2) \tag{24}$$

This is a non stationary Gaussian process. The factorial moments of the number of outcrossings in $[0;T]$ where $X(0) < \xi$ can now be calculated as

$$m_i\left(T \mid X(0) < \xi\right) = \int_{-\infty}^{\xi} m_i\left(T \mid X(0) = x_0\right) f_{x_o}(x_0)\, dx_0 \tag{25}$$

where $m_i(T \mid X(0) = x_0)$ can be evaluated by eq. (20). The variable x_0 has a truncated normal distribution. This implies that in order to determine the factorial moments of the conditional process (and thereby the survival probability) it is necessary to solve an integral of dimension $2i + 1$. Fortunately, this does not offer serious problems in view of the method used to solve the integral. For small values of T this improvement corresponds to the improvement to the Poisson approximation suggested by Ditlevsen[2] (eq. (4)).

5 EXAMPLE 1

In the first example the first passage probability of a scalar stationary Gaussian process over the constant threshold ξ within the time interval [0;T] is studied. The stochastic process has mean value $\mu(t) = 0$ and autocorrelation function

$$\rho(\tau) = \exp(-\tau^2 / t_o^2)$$

The standard deviation is one. The same process has already been investigated by Bolotin[13] and Lange[11]. It is convenient to introduce

$$c_i = \frac{p_{0,i}}{p_{0,1}}.$$

The probability ratio c_i expresses the relative difference between $p_{0,i}$ and the survival probability determined on basis of the Poisson assumption with the initial condition taken into account.

In figure 5.1, 5.2 and 5.3 the failure probability, the probability ratios and Rice's bounds are shown for $\xi/\sigma = 1.0$, $\xi/\sigma = 2.0$ and $\xi/\sigma=3.0$ respectively.

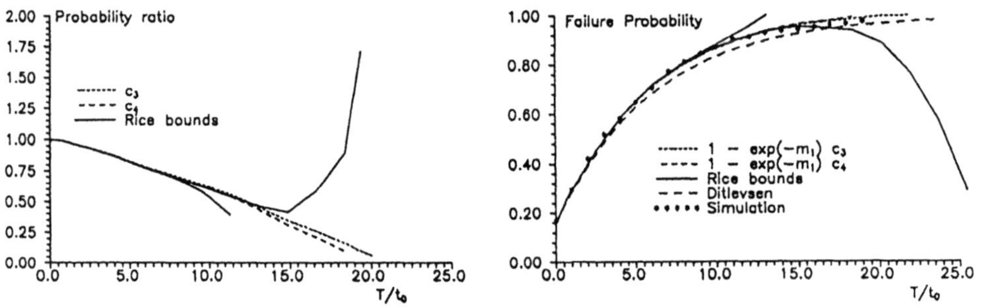

Figure 5.1: Probability ratios and failure probabilities for $\xi/\sigma = 1.0$.

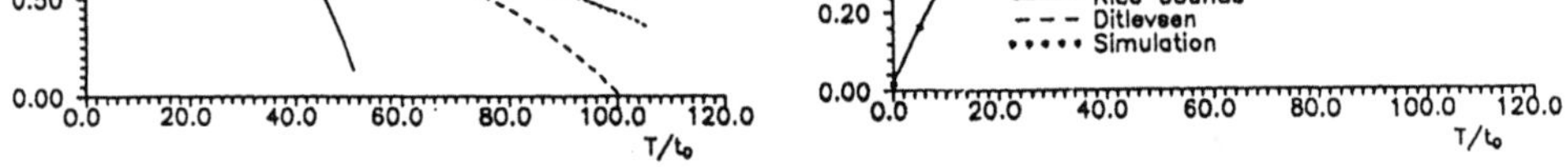

Figure 5.2: Probability ratios and failure probabilities for $\xi/\sigma = 2.0$.

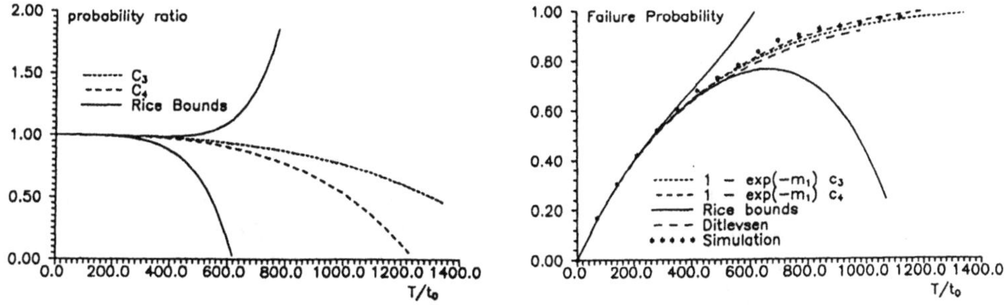

Figure 5.3: Probability ratios and failure probabilities for $\xi/\sigma = 3.0$.

At first it should be mentioned that the numerical results reported by Bolotin[13] and Lange[11] only in part could be reproduced. In the figures 5.1-5.3 it is seen that for small values of T the probability ratios are close to one for all methods. This implies that for small T and/or small number of exits the Poisson assumption with the initial conditions taken into account is very accurate. It is further shown that the Poisson approximation leads to better results for increasing levels as expected. Consideration of the initial conditions according to eq. (25), in fact, improves the results significantly for larger failure probabilities. As T increases Rice's bounds become very wide for all levels. For large values of T the estimated failure probabilities obtained by eq. (12) are much closer together than Rice's bounds. However, also for this method the number of factorial moments which are needed to obtain good estimates of the failure probability increases with increasing T. Further since the correction factors c_3 and c_4 both become negative for high values of T it is evident that the condition eq. (13) is not always fulfilled. This is, however, only true for very low reliability levels. Nevertheless, it is unfortunate that it is not possible to determine whether c_3 and c_4 are bounds because even higher factorial moments would be required in order to asses the validity of eq. (13). From the simulation (dots) it can be concluded that eq. (12) with the third or fourth factorial moment included yield rather accurate results, even sligthly more accurate than eq. (4) for all levels. It is finally noted that the Poisson assumption becomes unconservative for all levels and times in this example.

6 EXAMPLE 2

In the second example we consider a simple linear oscillator with Gaussian white noise excitation. The response $X(t)$ is a stationary Gaussian process with zero mean, unity standard deviation and autocorrelation function

$$\rho(\tau) = \exp(-\varsigma\omega_0|\tau|)\left(\cos\omega_d\tau + \frac{\varsigma}{\sqrt{1-\varsigma^2}}\sin\omega_d\tau\right)$$

where ω_0 denotes the eigenfrequency, ς the damping ratio, $\omega_d = \omega_0\sqrt{1-\varsigma^2}$ and $\tau = t_1 - t_2$. The standard deviation is σ. Failure is defined for $g(X(t)) = \xi - X(t) \leq 0$ where ξ is a constant. In figure 6.1 the probability ratios and Rice's upper and lower bounds are shown for $\xi/\sigma = 2.0$, $\zeta = 0.05$ and $\omega_0 = 1.0$.

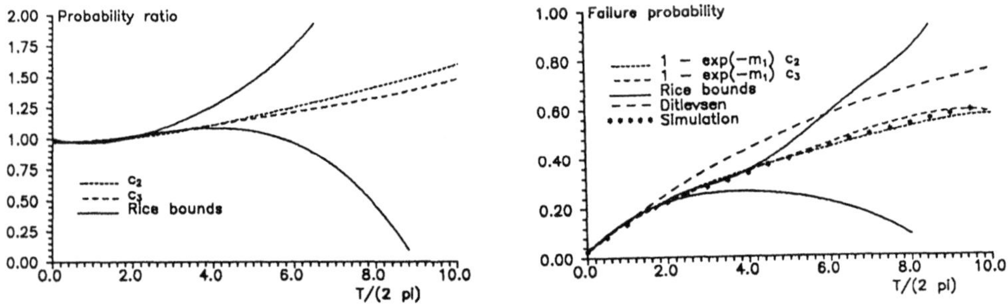

Figure 6.1: Probability ratios and failure probability.

For this example the outcrossings are highly correlated and occur in clusters. Therefore $\exp(-m_1)$ is a lower bound for the survival probability. This implies that all probability ratios are larger than 1.0. Clearly, this example represents exactly the case where one of the narrow-band approximations would work.

For this example only the first three factorial moments have been calculated. The same conclusions as for the first example can be made. However, the example also demonstrates that it is difficult to evaluate higher factorial moments if the correlation function is not very well-behaved. Especially for large values of T the numerical effort required to evaluate the factorial moments is excessive.

7 SUMMARY AND CONCLUSION

The methods by Rice and in this paper, eq. (12), for calculating upper and lower bounds on the outcrossing probability are reviewed. Some problems in numerical implementation are discussed. For two examples the results obtained by these methods have been compared. The examples show that the method proposed in this paper converges faster towards the exact solution than Rice's "in- and exclusion" series. However, not under all circumstances the method by Lange provides bounds on the outcrossing probability and it is difficult to determine whether bounds are obtained since this involves the calculation of higher factorial moments. Further work is necessary in order to establish better conditions for upper and lower bounds. The proposed method can also be used for vector processes. In that case the factorial moments can be determined by importance sampling. It has been shown that for engineering purposes where the failure probability normally is small the simple improvement of the Poisson assumption suggested by Ditlevsen, which allows the initial conditions to be taken into account is sufficiently accurate. Finally the Poisson assumption cannot be improved easily. Any improvement based on higher order moments of crossings is numerically rather involved. This also applies for the suggested Gram-Charlier expansion.

ACKNOWLEDGMENTS

This work is supported by Deutsche Forschungsgemeinschaft.

REFERENCES

1. Cramer, H., Leadbetter, M. R., Stationary and Related Stochastic Processes, Wiley, New York, 1967.
2. Ditlevsen, O., Duration of Gaussian Process visit to Critical Set, DCAMM Report No. 282, Technical University of Denmark, 1984.
3. Madsen, P. H., Krenk, S., An Integral Equation Method for the First-Passage Problem in Random Vibration, *Journ. Eng. Mech.*, 106, 1984, pp. 674-679.
4. Nielsen, S. R. K., Sørensen, J. D., Probability of Failure in Random Vibration, *Journ, Eng. Mech.*, 114, 1986, pp. 1218-1230.
5. Bernard, M. C., Shipley, J. W., The First Passage Problem for Stationary Random Structural Vibration, *Journal of Sound and Vibration*, 24, 1972, pp. 121-132.
6. Yang, J.-N., Shinozuka, M., On the First Excursion Probability in Stationary Narrow-Band Random Vibration, *Journ. Appl. Mech.*, 38, 1971, pp. 1017-1022.
7. Krenk, S., Madsen, H. O., Madsen, P. H., Stationary and Transient Response Envelopes, *Journ. Eng. Mech.*, 109,1983, pp. 263-278.
8. Veneziano, D., Envelopes of Vector Processes and their Crossing Rates, *Annals of Probability*, 7, 1979, pp. 62-74.
9. Langley, R. S., A Firat Passage Approximation for Normal Stationary Processes, *Journal of Sound and Vibration*, 122, 1988, pp. 261-275.
10. Ditlevsen, O., Lindgren, G., Empty Envelope Excursions in Stationary Gaussian Processes, *Journal of Sound and Vibration*, 1987.

11. Lange, C., First Excursion Probabilities for Low Threshold Levels by Differentiable Processes, Reliability and Optimization of Structural Systems '91, Proceedings of the 4th IFIP Conference Munich, Eds.: R. Rackwitz, P. Thoft-Christensen, pp. 263-278.
12. Rice, S. O., Mathematical Analysis of Random Noise, in: Selected Papers on Noise and Stochastic Processes, Ed.: N. Wax, Dover Publications, New York, 1954.
13. Bolotin, V. V., Wahrscheinlichkeitsmethoden zur Berechnung von Konstruktionen, VEB Verlag für Bauwesen, Berlin, 1981.
14. Kendall, M., Stuart, A., The Advanced Theory of Statistics, Macmillan Publishing Co., New York, 1977.
15. Gaganov, G. A., Ob ozenkach momentov tschisla perestscheni nulevo urovnja Gaussovskim stationarym prozessom, Vestnik LGU, 1973, 1, pp. 136-138.
16. Miroshin, P. N., Zvetkov, V. I., Asymptotika tretjo momenta tchisla peresetschini prjyomi kt + a Gaussovskim stationarym prozessom, Vestnik LGU, 1977, 1, pp. 90-97.
17. Hohenbichler, M., Expectations by FORM, Internal Report, Technical University Munich, 1983.
18. Breitung, K., Asymptotic Approximations for The Outcrossing rates of Stationary Gaussian Vector Processes, University of Lund, Dept. of Math. Statistics, 1984.
19. Madsen, H. O., Sensitivity Factors for Parallel Systems, in: Miscalleaneous Papers in Civil Engineering, Danish Engineering Academy, Ed.: G. Mohr, 1992.
20. Hagen, Ø., Conditional and Joint Failure Surface Crossing of Stochastic Processes, *Journ. Eng. Mech.*, 1992, 118, pp. 1814-1839.
21. Ayyub, B. M., Chia, C.-Y., Generalized Conditional Expectation for Structural Reliability Assessment, *Structural Safety*, 11, 1992, pp. 131-148.
22. Rubinstein, R. Y., Simulation and the Monte Carlo Method, Wiley, New York, 1981.
23. Engelund, S., Rackwitz, R., A Benchmark Study on Importance Sampling Techniques in Structural Reliability, Accepted for publication in *Structural Safety*.

Reliability Assessment via Log-Likelihood Maximization

Lucia Faravelli

Department of Structural Mechanics, University of Pavia,
via Abbiategrasso 211, I27100, Pavia (Italy)

Abstract. A general approach for assessing the reliability of nonlinear stochastic mechanical systems under stochastic conditions is presented. It combines the skeleton of an asymptotic reliability method, that can be used in the original space of the variables and provides a good estimate of the probability of failure, with an extended response surface scheme by which any limit state can be put in analytical form.

Keywords. Asymptotic theory, log-likelihood function, reliability, response surface

1 Introduction

Attention is focused on the reliability assessment of nonlinear mechanics systems subjected to stochastic external excitations. Nonlinearity arises from geometrical considerations and/or material properties. Practical applications in engineering of nonlinear mechanics (both static and dynamic), however, are characterized by numerical analyses. Moreover, the external excitations due to natural hazards must be modeled as stochastic processes, and spatially-distributed geometrical and mechanical parameters are correctly idealized by random fields. This prevents one from a direct application of standard reliability methods. A technique which couples iterated response-surface analytical approximations with asymptotic theory was therefore proposed in [5]. In this paper the operative features and the efficiency of this approach, which can be used in the original space of the random variables and gives a good estimate of the probability of failure, are illustrated with reference to hysteretic systems.

2 Governing Relations

This paper is mainly devoted to mechanical systems whose input-output relationship is governed by a numerical operator like a general-purpose finite-element code. The most direct approach to such problems is via Monte-Carlo simulation which requires the evaluation of the response for many sets of input values. A useful alternative to Monte Carlo simulation is to regard the numerical operator as a black box and to optimize the plan of experiments to be carried out. An approach based on the response surface methodology (r.s.m.) and able to provide an analytical approximation of the system limit state was proposed in [8]. An asymptotic theory to calculate the probability of failure of the system against any limit state assigned in analytical form is also available [4]. The reliability approach which makes use of the response surface methodology (r.s.m.) in conjunction with the asymptotic approach is illustrated.

2.1 The r.s.m. Scheme

Let x_j $(j = 1, ..., J)$ be a random design variable of a mechanical system. Any response variable y is a random variable being a function of the input variables. Let this function be unknown from an analytical point of view. Group in $\mathbf{x}_v$ the vector of the random variables of interest, and let some design variables x_j be functions of the spatial coordinates (random fields). When the continuous system is discretized, these functions become random vectors. Any vector $\mathbf{x}_{sj}$, expressing the spatial variability of the design variable x_j, can be written in one of the two forms

$$\begin{aligned} \mathbf{x}_{sj} &= x_{svj}\mathbf{x}_{ssj} \quad , \\ \mathbf{x}_{sj} &= x_{svj} + \mathbf{x}_{ssj} \quad . \end{aligned} \tag{1}$$

In equation (1), x_{svj} is the central value of $\mathbf{x}_{sj}$, and $\mathbf{x}_{ssj}$ denotes the deviations of $\mathbf{x}_{sj}$ from the central value. The scalar quantity x_{svj} has the nature of a random variable. The second factor on the r.h.s., $\mathbf{x}_{ssj}$, is a random vector with a unit or zero central value. When x_j is a stochastic function of any parameter t in $(0, T)$, equation (1) still holds provided that the l.h.s. represents the vector resulting from the discretization of the range $(0, T)$. The variables $\mathbf{x}_v$ and $\mathbf{x}_{sv}$ are mapped in the standardized space $\mathbf{z}$, and a suitable transformation $Y = Y(y)$ of the response could also be introduced. The extended response surface model is based on the relationship [7]:

$$Y = f[\mathbf{z}, \boldsymbol{\theta}] + \epsilon(\mathbf{x}_{ss}) \quad , \tag{2}$$

with $\boldsymbol{\theta}$ being model parameters; the random term ϵ collects the lack of fit and the randomness of neglected terms as in classical experiment theory. The same framework can be reached by selecting a suitable subset of the x_j by appropriate sensitivity analyses.

The dependence of Y on $\mathbf{z}$ is conveniently described by a second order polynomial relationship

$$\begin{aligned} Y &= \theta_0 + \mathbf{z}^T \boldsymbol{\theta}_1 + \mathbf{z}^T \Theta \mathbf{z} + \epsilon_{ss} = \\ &= \mathbf{z}_1^T \boldsymbol{\theta} + \epsilon_{ss} = \tilde{\mathrm{Y}} + \epsilon_{ss} \quad , \end{aligned} \tag{3}$$

where $\tilde{\mathrm{Y}}$ denotes the estimate of the response variable achieved by the response surface model. The coefficients θ_0, $\boldsymbol{\theta}_1$, Θ, which play the role of $\boldsymbol{\theta}$ in equation (2), can be computed by the regression analysis of the results obtained in numerical experiments appropriately planned [10]. In equation (3), the term ϵ_{ss} takes into account the model error ϵ, the effects of the vectors $\mathbf{x}_{ssj}$ of the deviations from the spatial averages, and their interactions.

Once the parameters of model (3) are evaluated, its optimization can be pursued by the following scheme:

- several experiments at the same coordinates $\mathbf{z}$ are repeated with different simulated vectors $\mathbf{x}_{ssj}$. For each experiment, different response values are found, but the estimate $\tilde{\mathrm{Y}}$, obtained by equation (3), is always the same. This error of the model is called the *pure error*. It is measured by the corresponding standard deviation s_ϵ.

- experiments are now conducted by changing both the coordinates $\mathbf{z}$ and the vectors $\mathbf{x}_{ssj}$. Again discrepancies between Y and $\tilde{\mathrm{Y}}$ are found. This error of the model is called the *lack of fit* error. It is measured by the standard deviation s_ℓ which includes the pure error and the lack of fit of the second-order polynomial model.

- any transformation Y of y is better than another when the corresponding ratio:

  $$T = s_\ell / s_\epsilon \tag{4}$$

  is closer to 1 [1]. This criterion states that the optimal response surface is characterized by a space-averaged error (s_ℓ) very close to the local ensemble error (s_ϵ).

Note that the global pure error has to be almost constant over the region under consideration. Otherwise the analyzer can conceive a piece-wise polynomial relationship $Y(\mathbf{z})$ in different regions of the space of the variables.

2.2 The Asymptotic Procedure

Let $g(\mathbf{x})$ be a limit state function in analytical form which indicates the state of the structure when the random vector $\mathbf{x}$ (which groups the vectors $\mathbf{x}_v$ and $\mathbf{x}_{sv}$) has the realization $x_1, ..., x_n$. If $g(\mathbf{x}) > 0$, the structure is safe, if $g(\mathbf{x}) \leq 0$, it is unsafe. If $p(\mathbf{x})$ is the joint probability density function

(JPDF) of the random vector $\mathbf{x}$, the probability of failure $P(F)$, with the failure $F = \{\mathbf{x}; g(\mathbf{x}) \leq 0\}$, is then

$$P(F) = \int_{g(\mathbf{x})\leq 0} p(\mathbf{x})d\mathbf{x} \quad .$$

The standard reliability methods, which are used to calculate this probability $P(F)$, are the so called FORM (First Order Reliability Methods) and SORM (Second Order Reliability Methods) procedures. The basic idea of these methods is to transform all the random variables into Gaussian variables as shown in [9] and to derive an approximation for its probability content $P(F)$ by simple geometric arguments: the point $\mathbf{x}^0 \in F$ with minimal distance to the origin is determined; this point must lie on the limit state surface $B = \{\mathbf{x}; g(\mathbf{x}) = 0\}$. At this point a Taylor expansion of $g(\mathbf{x})$ is made.

A better estimate of the probability of failure can be obtained by an asymptotic procedure, as proved in [4]. For this purpose, consider the Gaussian PDF $p(\mathbf{u})$ and the so called log-likelihood function defined as

$$\ln(p(\mathbf{u})) = -n/2\ln(2\pi) - |\mathbf{u}|^2/2 \quad .$$

The minimization of the distance $|\mathbf{u}|$ corresponds to the maximization of the log-likelihood. Asymptotic approximations can be derived also for non-normal random vectors with PDF $p(\mathbf{x})$ and log-likelihood $l(\mathbf{x}) = \ln(p(\mathbf{x}))$. The point $\mathbf{x}^*$ is calculated for which

$$l(\mathbf{x}^*) = \max_{\mathbf{x}\in F} l(\mathbf{x}) \quad .$$

Then, a suitable expansion of the log-likelihood is made and the following approximation is derived [4]:

$$P(F) \sim (2\pi)^{(n-1)/2} \frac{p(\mathbf{x}^*)}{|\nabla l(\mathbf{x}^*)|\sqrt{|det\mathbf{H}^*|}} \quad , \tag{5}$$

where ∇ denotes the gradient operator.
Here the matrix $\mathbf{H}^*(\mathbf{x}^*)$ has the form

$$\mathbf{H}^* = \mathbf{P}\mathbf{H}\mathbf{P}^{\mathbf{T}} - \nu\nu^{\mathbf{T}} \quad ,$$

with

1) $\nu = |\nabla l(\mathbf{x}^*)|^{-1}\nabla l(\mathbf{x}^*)$, the unit vector normal to the surface;

2) $\mathbf{P} = \mathbf{I_n} - \nu\nu^{\mathbf{T}}$ ($\mathbf{I}_n$ being the n-dimensional identity matrix);

3) $\mathbf{H} = (l_{,ij}(\mathbf{x}^*) - \frac{|\nabla l(\mathbf{x}^*)|}{|\nabla g(\mathbf{x}^*)|} g_{,ij}(\mathbf{x}^*))$, $i,j = 1,\ldots,n$, the Hessian of $l(\mathbf{x})$ in local coordinates (suffix j after a comma denotes the derivatives with respect to x_j).

An important advantage of the method of log-likelihood approximation is the fact that it can be used also for dependent random variables, if only the joint PDF $p(\mathbf{x})$ and its log-likelihood $l(\mathbf{x}) = \ln(p(\mathbf{x}))$ are known. In this case one calculates the point of maximum likelihood, then the first and second derivatives of the likelihood function at this point are computed and inserted into the approximation equation. Here, no conditional distribution function, as needed in the Rosenblatt transformation, must be computed. Further, we can calculate directly from the log-likelihood function in the original space importance and sensitivity factors [3].

3 The Reliability Assessment Method

When the limit state function has no analytical form, it can be expressed by the extended response surface model (2) using the equation

$$G(\mathbf{z}|\mathbf{x}_{ss}) = F_R(\mathbf{z}_v, \mathbf{z}_{sv}, \boldsymbol{\theta}) + e(\mathbf{x}_{ss}) \quad , \tag{6}$$

where $\mathbf{z}_v$ is a vector of standardized random variables, $\mathbf{z}_{sv}$ is a vector of standardized random central values, and $\mathbf{x}_{ss}$ is the deviation from the central values; e takes into account the error.
The reliability assessment procedure consists of the following steps.

Consider equation (6) written in the central region of all the variables in order to identify the actual location of $G(\ )$ in the standardized $\mathbf{z}$ space. A linear polynomial form is preliminarly selected for the response surface. An appropriate fractional replicate of a factorial design centered at the origin of the space $\mathbf{z}$ defines the experiment plan [8]. This first response surface application (where the contribution of the quantities $\mathbf{x}_{ss}$ is neglected) leads to an estimation of the design point $\mathbf{z}^0_{(1)}$ and of the maximum likelihood point $\mathbf{z}^*_{(1)}$ for the linear model.

The sensitivity factors of each variable z_j can be estimated from the gradient of the log-likelihood function at the point $\mathbf{z}^*_{(1)}$ due to the fact that the probability of failure satisfies in first approximation the relation

$$\partial P(F)/\partial z_i \sim (\partial l(\mathbf{z}^*)/\partial z_i) P(F) \quad .$$

No substantial error is generally introduced by considering the random variables with low sensitivity factors as fixed and equal to their means in the further calculations. Alternatively, they can still be regarded as random variables but grouped into the vector $\mathbf{x}_{ss}$.

A better estimation is then pursued by a sequence of repeated response surface applications. Each of them consists of several deterministic structural analyses for different input data. They form an experiment design conceived for both fitting second-order polynomial response surface and incorporating the terms $\mathbf{x}_{ss}$. At the i-th step, the resulting approximation for the response function is given by the equation

$$G^{(i)}_{\mathbf{V}}(\mathbf{v}_{(i)}) = F_R(\mathbf{v}_{(i)}, \boldsymbol{\Theta}_{(i)}) + \epsilon_{(i)}(\mathbf{x}_{ss}) \quad ,$$

where

$$\mathbf{v}_{(i)} = \mathbf{z} - \mathbf{z}^x_{(i)} \quad . \tag{7}$$

By equation (7), the origin is shifted along the vector from the origin to the relevant design point $\mathbf{z}^0_{(i-1)}$. The convergence to the actual value is pursued by selecting a point on the line between this point and the origin: this defines $\mathbf{z}^x_{(i)}$.

Practical reasons suggest that one should work, rather than in the space $\mathbf{v}$, in a particular space $\mathbf{w}$, obtained from it by rotation. The new first axis w_1 coincides with the normal to the function $G(\ ) =$ const at the selected point. This iterative procedure stops when convergence on the distance of the design point from the origin is reached. A scheme of the iterative procedure is provided in figure 1. At each step one also computes the current estimate of the maximum-likelihood $\mathbf{z}^*_{(i)}$ by the algorithm of constrained nonlinear optimization proposed by Schittkowski [11]. If required, then, a last iteration can make use of an experiment plan centered at the estimated maximum likelihood point.

It is worth noting that the Schittkowski algorithm [11] generates a sequence of quadratic programming subproblems which are to be solved successively (sequential quadratic programming method). At each step of the sequence, the Lagrangian function associated with the likelihood function is approximated by a quadratic form, while the quadratic constraint $G(\ ) = 0$, is linearized.

4 Numerical Example

Though the proposed approach can be applied to any complex structure, a simple nonlinear plane frame is considered in order to illustrate its operative features. Nevertheless, it contains all the complication of a large hysteretic system.

4.1 Mechanical Modeling

Consider a nonlinear plane frame subjected to a base excitation. It is discretized into perfectly elastic finite elements (see figure 2: all the data are in example 5.16 of reference [6]). Let the inelastic deformations $\mathbf{q}$ be concentrated at critical regions with hysteretic constitutive laws given [6] by the equation

$$V_i = (1 - \alpha_{2i})\alpha_{1i} z_i + \alpha_{1i}\alpha_{2i} q_i \quad , \tag{8}$$

where V_i is the bending moment at i-th section. The auxiliary variable z_i obeys the endochronic Bouc-Wen model [2,12]

$$\dot{z}_i = \alpha_{3i}\dot{q}_i - \alpha_{4i} z_i |\dot{q}_i| |z_i|^{\alpha_{6i}-1} - \alpha_{5i}\dot{q}_i |z_i|^{\alpha_{6i}} \quad , \tag{9}$$

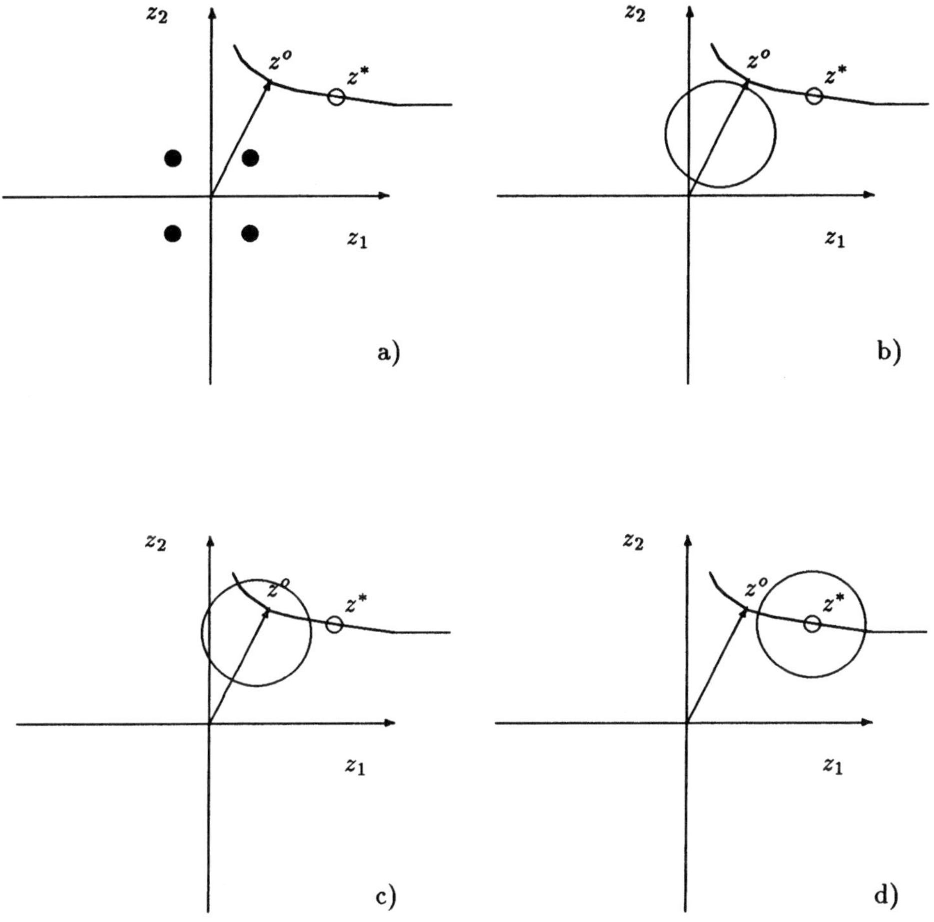

Figure 1: a) Factorial design for fitting a preliminary approximation of the response surface; b) and c) subsequent shifted composite central designs during the iterative procedure of reliability assessment; d) final composite central design in the estimated point of maximum likelihood

a

b

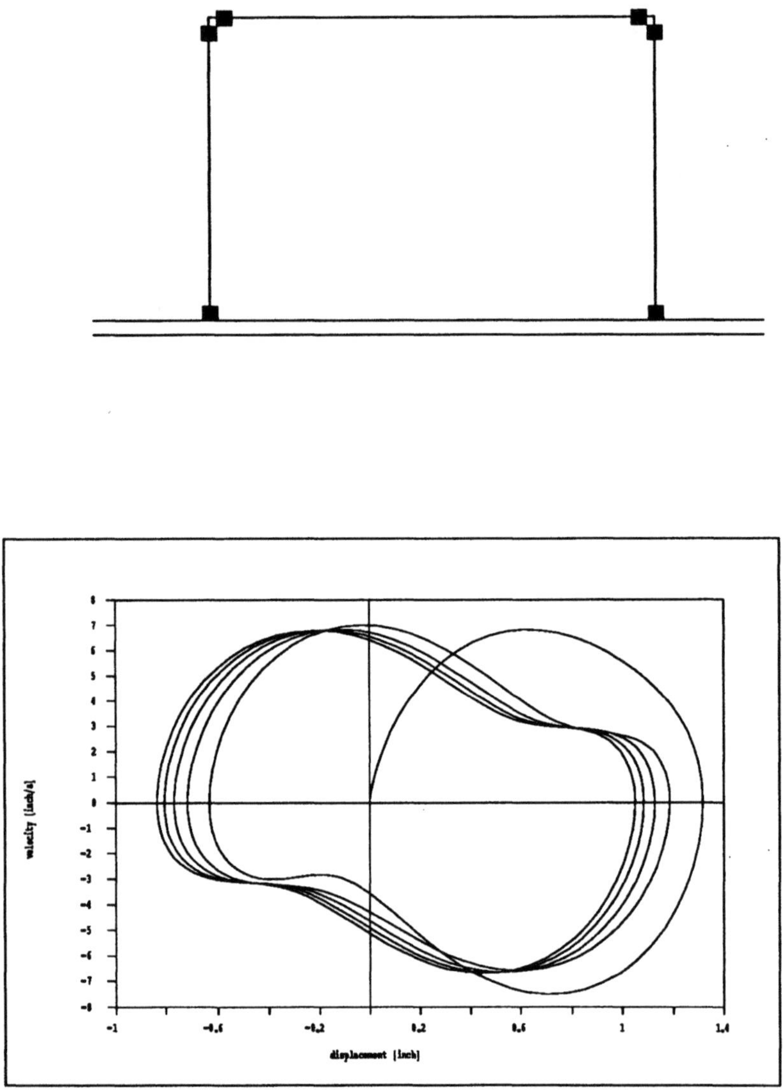

Figure 2: a) The hinged frame model considered in the numerical example; b) phase-space representation of the transient structural response from rest under a sine excitation: T=5s.

where α_{1i} to α_{6i} are system parameters controlling the hysteresis shape of the system. The previous relationships are illustrated for the transient motion under a sine excitation of duration $T = 5$s in figure 3.
Equations (8) and (9) can be assembled over all the critical regions to form the equations

$$\begin{aligned} \mathbf{S} &= \mathbf{A}_1\mathbf{z} + \mathbf{A}_2\mathbf{q} \quad , & (10)\\ \dot{\mathbf{z}} &= \dot{\mathbf{q}} - \mathbf{z}^T\mathbf{A}_4|\dot{\mathbf{q}}| - |\mathbf{z}|^T\mathbf{A}_5\dot{\mathbf{q}} \quad , & (11)\end{aligned}$$

where $\mathbf{A}_1$, $\mathbf{A}_2$, $\mathbf{A}_4$, $\mathbf{A}_5$ are diagonal matrices. The nonlinear behaviour of the structure can be approached by the imposed rotation method. It considers the structural response as the sum of the linear elastic response under the external load and the linear elastic response due to the inelastic deformations regarded as distorsions. Then, the internal actions in the critical regions can be expressed as

$$\mathbf{S} = \mathbf{E}_1\mathbf{u} + \mathbf{E}_2\mathbf{q} \quad . \tag{12}$$

Moreover let the vector of restoring force $\mathbf{r}$ be linearly related to $\dot{\mathbf{u}}$; it follows that

$$\mathbf{r} = \mathbf{E}_3\mathbf{S} + \mathbf{C}\dot{\mathbf{u}} \quad , \tag{13}$$

where $\mathbf{E}_1$, $\mathbf{E}_2$, $\mathbf{E}_3$ are structural matrices; $\mathbf{C}$ is a classical damping matrix. From the equality of the r.h.s of equations (10), (12) one has

$$\mathbf{z} = \mathbf{A}_1^{-1}(\mathbf{E}_1\mathbf{u} + (\mathbf{E}_2 - \mathbf{A}_2)\mathbf{q}) \quad . \tag{14}$$

By putting equation (14) into the l.h.s of equation (11) and its derivative into its r.h.s, one obtains

$$\mathbf{Q}_1\dot{\mathbf{u}} + \mathbf{Q}_2\dot{\mathbf{q}} = \mathbf{I}\dot{\mathbf{q}} - (\mathbf{Q}_1\mathbf{u} + \mathbf{Q}_2\mathbf{q})^T\mathbf{A}_4|\dot{\mathbf{q}}| - |\mathbf{Q}_1\mathbf{u} + \mathbf{Q}_2\mathbf{q}|^T\mathbf{A}_5\dot{\mathbf{q}} \quad , \tag{15}$$

where $\mathbf{Q}_1 = \mathbf{A}_1^{-1}\mathbf{E}_1$ and $\mathbf{Q}_2 = \mathbf{A}_1^{-1}(\mathbf{E}_2 - \mathbf{A}_2)$.
After some algebra equation (15) becomes

$$\begin{aligned} \dot{\mathbf{q}} &= (\mathbf{u}^T\mathbf{B}_1 + \mathbf{q}^T\mathbf{B}_2 + \mathbf{I} - \mathbf{Q}_2)^{-1}\mathbf{Q}_1\dot{\mathbf{u}} \qquad \text{if} \qquad \dot{\mathbf{q}} > 0 \quad , & (16)\\ \dot{\mathbf{q}} &= (\mathbf{u}^T\mathbf{B}_3 + \mathbf{q}^T\mathbf{B}_4 + \mathbf{I} - \mathbf{Q}_2)^{-1}\mathbf{Q}_1\dot{\mathbf{u}} \qquad \text{if} \qquad \dot{\mathbf{q}} < 0 \quad , &\end{aligned}$$

with the appropriate expressions for $\mathbf{B}_1$, $\mathbf{B}_2$, $\mathbf{B}_3$ and $\mathbf{B}_4$ given by the equations

$$\begin{aligned} \mathbf{B}_1 &= -(\mathbf{Q}_1\mathbf{A}_4 \pm \mathbf{Q}_1\mathbf{A}_5) \quad ,\\ \mathbf{B}_2 &= -(\mathbf{Q}_2\mathbf{A}_4 \pm \mathbf{Q}_2\mathbf{A}_5) \quad ,\\ \mathbf{B}_3 &= -(-\mathbf{Q}_1\mathbf{A}_4 \pm \mathbf{Q}_1\mathbf{A}_5) \quad ,\\ \mathbf{B}_4 &= -(-\mathbf{Q}_2\mathbf{A}_4 \pm \mathbf{Q}_2\mathbf{A}_5) \quad .\end{aligned}$$

In the previous expressions the sign of the second addendum in the r.h.s. depends on the sign of $(\mathbf{Q}_1\mathbf{u} + \mathbf{Q}_2\mathbf{q})$.

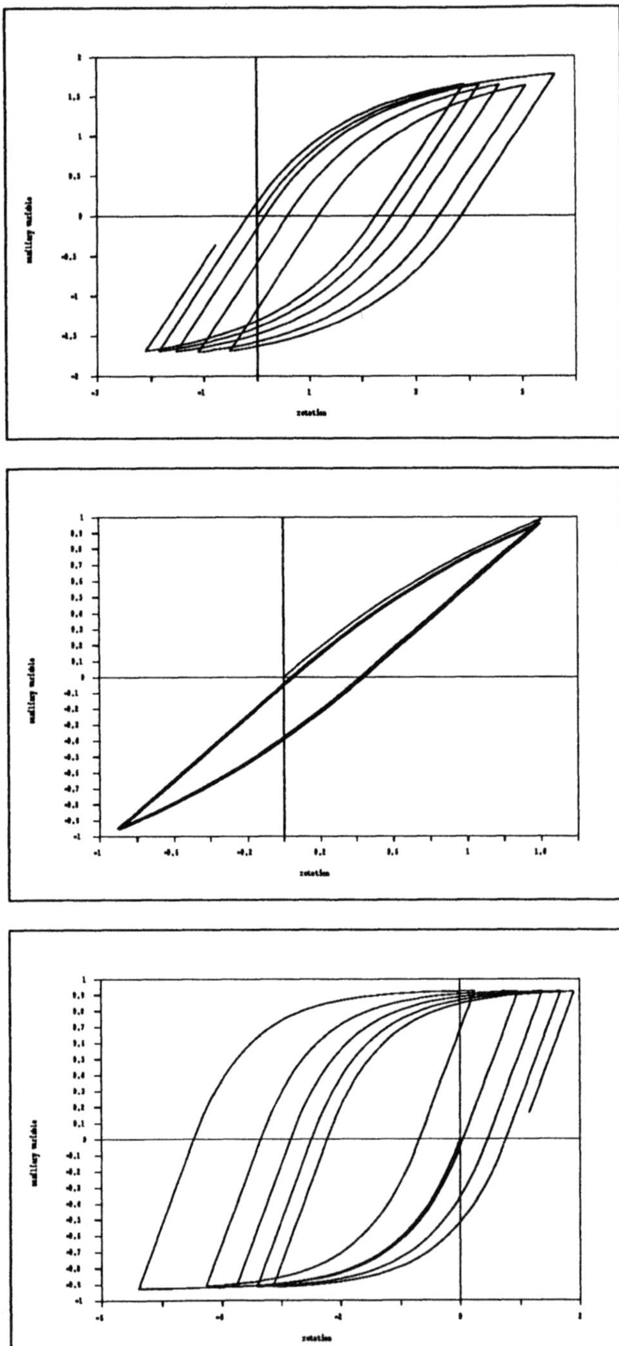

Figure 3: Rotation auxiliary variable relationships for a sine excitation (T=5s): a) bottom column; b) top column; c) beam

Table I. Probabilisitc definition of the random variables of the numerical example.

Symbol	Distribution	Mean Value	Standard dev.	Correl.
η_1 η_2	Joint extreme value distribution	1.00	0.1	0.5
R	Lognormal	12.1	3.347	-

Therefore the dynamics of the system may be described by equation (16) together with the equation of motion. By putting $\mathbf{d}_1 = \mathbf{u}$, $\mathbf{d}_2 = \dot{\mathbf{u}}$, and $\mathbf{d}_3 = \mathbf{q}$, it is written

$$\begin{aligned} \dot{\mathbf{d}}_1 - \mathbf{d}_2 &= 0 \quad , \\ \dot{\mathbf{d}}_2 - \mathbf{M}^{-1}\mathbf{C}\mathbf{d}_2 + \mathbf{M}^{-1}\mathbf{E}_3\mathbf{E}_1\mathbf{d}_1 + \mathbf{M}^{-1}\mathbf{E}_3\mathbf{E}_2\mathbf{d}_3 &= \mathbf{M}^{-1}\mathbf{f}(t) \quad . \end{aligned} \tag{17}$$

4.2 Stochastic Modeling

The probabilistic definition of the random variables is summarized in Table I. In particular the hardening and the yielding moment at each section are modeled by the equations

$$\begin{aligned} \boldsymbol{\alpha}_2 &= \boldsymbol{\alpha}_2^0 \eta_1 \quad , \\ \mathbf{M}_y &= \mathbf{M}_y^0 \eta_2 \quad , \end{aligned} \tag{18}$$

where $\boldsymbol{\alpha}_2^0$ and $\mathbf{M}_y^0$ are vectors of nominal values, and η_1 and η_2 are random variables with joint extreme distribution of type B.

The base excitation $f(t)$ is characterized by a ground acceleration modeled as a filtered white-noise. The power spectral density function of the white-noise is $200/(2\pi)$ sqi/s^3; the values of the filter parameters are given in [6].

4.3 Limit State Definition

With reference to a transient of duration $T = 5$s, let the performance function be assigned in terms of a low-cycle fatigue limit state. The criterion in [6] is adopted. For any potential plastic hinge, a function of the dissipated energy at the end of the excitation and of the maximum rotation achieved during the excitation is computed. The maximum of the values computed at different locations is then compared with a random target value R.

Table II. Maximum likelihood point estimates $\mathbf{z}^*_{(i)}$ and probability approximations for the dynamic example. The first step consider a linear response surface.

Iteration	l_{ott}	z_1	z_2	z_3	ϵ	$P_F.10^5$
1 (linear)	-	-0.648	- .511	-2.89	-	-
2	10.63	-0.625	- .476	-2.25	-1.60	.83
3	10.69	-0.520	- .436	-2.26	-1.58	1.22

4.4 Iterative Response Surface and Numerical Results

The proposed response surface scheme starts with a linear polynomial and by building a response plan of experiments centered at the origin of the standardized space $\mathbf{z}$. The governinig equations (16) and (17) are the standard differential equations which are integrated numerically. For this purpose one first updates $\mathbf{u}$ and $\mathbf{q}$; this makes possible to compute $(\mathbf{Q}_1\mathbf{u} + \mathbf{Q}_2\mathbf{q})$ and its sign fully defines equation (16) and hence $\dot{\mathbf{q}}$. The first iteration provides an estimation of the design point and of the maximum likelihood point.

A second order polynomial form is now considered, and the composite plan of experiments is adopted. The reference system is rotated in such a way that the first axis coincides with the straight line joining the origin with the design point. The iteration process moves the center of the experiment plan in $\mathbf{z}^0_{i-1}/2$.

The results of the iteration process are given in Table II. Some input-output time histories were also simulated in order to validate the results achieved by the technique discussed in this paper. The histogram of the ratio between the random load effect r and the random resistance R (see Table I) is given in Figure 4.

5 Conclusions

For any performance of nonlinear structures subjected to stochastic excitation, the system reliability can be assessed by a maximum likelihood algorithm coupled with an iterative use of an extended response surface method.

Several aspects of the way the supporting numerical experiments are selected must still be optimized. Nevertheless, the procedure is very efficient and accurate and, unlike other standard reliability methods, offers the possi-

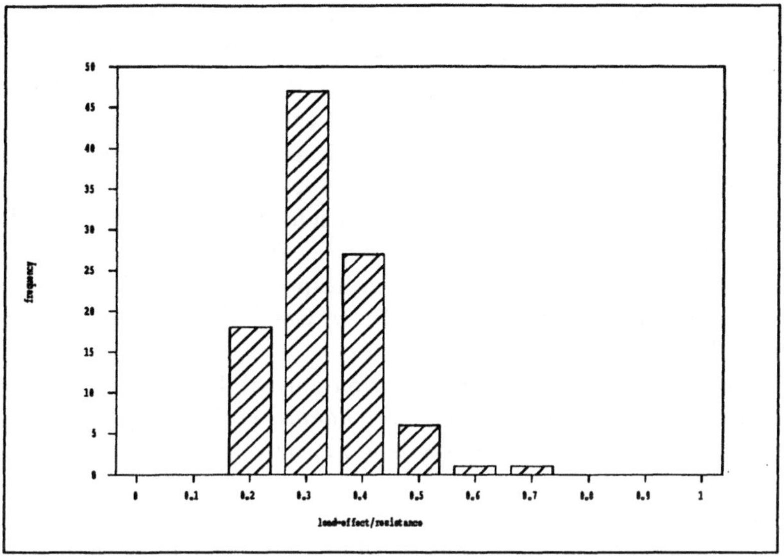

Figure 4: Histogram of the ratio between the random load effect r and the random resistance R (see Table I). A sample of size 100 has been simulated.

bility of incorporating the uncertainty arising from stochastic processes and random field idealizations.

6 Acknowledgement

The results illustrated in this paper were obtained within a research supported by the Innovative Technology Committee of the Italian National Research Council (CNR): contract number 93.03398.CT11.

7 References

1. Bohm F. & Bruckner-Foit A., On Criteria for Accepting a Response Surface Model, *Probabilistic Eng. Mechanics*, 7 (3), (1992), 183-190.
2. Bouc R., Forced Vibrations of a Mechanical System with Hysteresis, *4th. Conf. on Non-linear Oscillations*, Prague, 1967.
3. Breitung K., Parameter Sensitivity of Failure Probabilities, *Reliability and Optimization of Structures*, ed. Der Kiureghian A. & Thoft-Cristensen P., Springer Verlag, Berlin, 1990, 43-51.

4. Breitung K., Probability Approximations by Log Likelihood Maximization, *Journal of Engineering Mechanics*, **117**, (1991), 457-477.
5. Breitung K. & Faravelli L., Log-likelihood Maximization and Response Surface in Reliability Assessment, *Nonlinear Dynamics*, (1993).
6. Casciati F. & Faravelli L., *Fragility Analysis of Complex Structural Systems*, Research Studies Press, Taunton, 1991.
7. Faravelli L., A Response Surface Approach for Reliability Analysis, *Journal of Eng. Mechanics*, **115 (12)**, (1989), 2763-2781.
8. Faravelli L., Structural Reliability via Response Surface, *Nonlinear Stochastic Mechanics*, ed. Bellomo N. & Casciati F., Springer Verlag, Berlin, 1992, pp. 213-224.
9. Hohenbichler M. & Rackwitz R., Non-normal Dependent Vectors in Structural Safety, *Journ. of Eng. Mechanics*, **107**, (1981), 1227-1241.
10. Petersen R.G., *Design and Analysis of Experiments*, M.Decker Inc., New York, 1985.
11. Schittkowski K., NLPQL: a Fortran Subroutine Solving Constrained Nonlinear Programming Problems, *Annales of Operation Research*, **5**, (1985), 485-500.
12. Wen Y.K., Equivalent Linearization for Hysteretic Systems Under Random Excitation, *Journal of Applied Mechanics*, **47**, (1980), 150-154.

Reliability Analysis of Nonlinear 3-D Frames with PR Connections by SFEM

Liwei Gao and Achintya Haldar
Department of Civil Engineering & Engineering Mechanics
University of Arizona, Tucson, Arizona, U. S. A.

Abstract. An efficient stochastic finite element-based procedure is proposed for the reliability analysis of three-dimensional frames with partially restrained (PR) connections. The method is based on the Advanced First-Order Second-Moment reliability analysis procedure. The Richard model is used to describe the flexible behavior of connections. The assumed stress field approach is used in the finite element formulation to compute the nonlinear structural response and the corresponding response gradients. The material properties, geometry, connection parameters and external loads are considered to be random variables. The failure criteria of the structure are expressed in terms of the ultimate and serviceability limit state functions. The proposed method directly provides information on the structural reliability.

Keywords. Finite element, structural reliability, nonlinear analysis, probability of failure, reliability index, partially restrained, semi-rigid

1 Introduction

The use of the finite element method to evaluate the reliability of complicated structures where the limit state functions are not available in closed form has become very common. Gao and Haldar[1] proposed an efficient stochastic finite element method (SFEM) to evaluate the reliability of two dimensional (2-D) and three dimensional (3-D) nonlinear structures. However, these studies did not consider the uncertainties in the connection behavior.

The importance of the flexible end restraint provided by semi-rigid connections has long been recognized. The flexible behavior of a connection is commonly described by the relationship between the moment, M, transmitted by the connection and the relative angle of rotation between connecting members, θ. In practice, this moment-relative rotation relationship is nonlinear over the complete range. The uncertainties in the connection behavior come from uncertainties in the manufacturing and assembling processes and modeling uncertainties. Deterministic predictions of connection behavior, based on either empirical formulations or single test data, are likely to overestimate the values of strength and stiffness of connections[2]. Thus, uncertainties in the connection behavior should be considered explicitly in a structural reliability analysis. Mahadevan and Haldar[3] attempted to consider the uncertainties in connection behavior for linear analyses. Since the moment-rotation curves of connections are nonlinear over the complete loading range, the uncertainties in connection behavior should be considered in any nonlinear analysis. In this study, the four-parameter Richard model is used to described the M-θ curves of connection elements.

2 FEM of 3-D Frames with PR Connections

Details of the assumed stress-based nonlinear FEM formulation can be found in the literature[4,5,6]. Only the essential steps required for the SFEM formulation are discussed very briefly below.

The commonly used linear iterative strategy for solving nonlinear structural problems can be expressed as:

$$K^{(n)}\Delta D^{(n)} = F^{(n)} - R^{(n-1)} \quad , \tag{2.1}$$

where $K^{(n)}$, $\Delta D^{(n)}$, and $F^{(n)}$ are the tangent stiffness matrix, displacement vector, and external load vector at the nth iteration, respectively, and $R^{(n-1)}$ is the internal force vector at the (n-1)th iteration. The tangent stiffness matrix of a beam-column element using the assumed stress method can be expressed as:

$$K = A^{t}_{\sigma do} A^{-1}_{\sigma\sigma} A_{\sigma do} + A_{ddo} \quad , \tag{2.2}$$

where $A^{-1}_{\sigma\sigma}$ is the elastic property matrix, $A^{t}_{\sigma do}$ is the transformation matrix, and A_{ddo} is the geometric stiffness matrix. The internal force vector can be expressed as:

$$R = -A^{t}_{\sigma do} A^{-1}_{\sigma\sigma} R_{\sigma} + R_{do} \quad , \tag{2.3}$$

where R_{do} is the homogeneous part of the internal nodal force vector.

Since $\boldsymbol{K}$ and $\boldsymbol{R}$ can be expressed in explicit form, geometric nonlinear problems can be solved very efficiently using Eq. 2.1.

For numerical analyses, an ordinary beam-column element is used to represent a flexible connection except that its stiffness needs to be updated continuously. The stiffness and the current Young's modulus of an element are functions of the relative rotation θ. The stiffness of a connection can be represented by the following expression:

$$K(\theta) = E(\theta) \cdot I / l \quad , \qquad (2.4)$$

where $E(\theta)$ is the current Young's modulus as a function of θ. Knowing $K(\theta)$, the moment of inertia I, and the element length l, the current Young's modulus $E(\theta)$ can be evaluated by Eq. 2.4. Almost all of the experimental and analytical methods described so far consider only in-plane properties of flexible connections. Celikag and Kirby[7] reported that out-of-plane behavior of flexible connections should not be ignored. Therefore, both major- and minor-axis rotational flexibility of connections are considered in the present work. By substituting the corresponding parameters in Eq. 2.4, one can easily obtain the $E(\theta)$ values for both axes.

Richard's four parameter moment-rotation model[8], as shown in Fig. 2.1, is chosen to represent the semi-rigid connection behavior because of its generality and applicability to any type of connection. This model represents observed experimental data very well and is convenient for implementation in a computer program. Its four parameters are derived from a rational interpretation of response.

With the Richard model, the moment-rotation (M-θ) relationship of the connection is represented by the following expression:

$$M = \frac{(K - K_p)\theta}{\left(1 + \left|\frac{(K - K_p)\theta}{M_0}\right|^N\right)^{\frac{1}{N}}} + K_p\theta \quad , \qquad (2.5)$$

where M is the connection moment, θ is the connection rotation, K is the initial or elastic stiffness, K_p is the plastic stiffness, M_0 is the reference moment and N is the curve shape parameter. These parameters are shown in Fig. 2.1.

By using the Richard model, the tangent stiffness of the M-θ curve, $K(\theta)$, can be shown to be:

$$K(\theta)=\frac{dM}{d\theta}=\frac{(K-K_p)\theta}{\left(1+\left|\frac{(K-K_p)\theta}{M_0}\right|^N\right)^{\frac{N+1}{N}}}+K_p \quad . \tag{2.6}$$

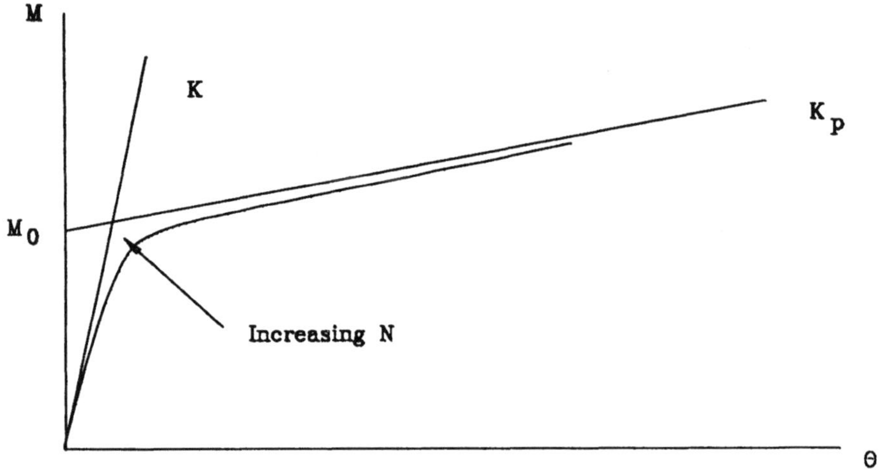

Figure 2.1 Definition of M-θ Curve by Richard Model

3 Evaluation of Gradients

The Advanced First-Order Second-Moment method (AFOSM) is used in this study to estimate the reliability of structures. In this context, G(**x**,**u**,s) is defined as the limit state function, where vector **x** denotes the set of basic random variables pertaining to a structure (e.g. loads, material properties and structural geometry), vector **u** denotes the set of displacements involved in the limit state function and vector **s** denotes the set of load effects (except the displacement) involved in the limit state function (e.g. stresses, internal forces). The displacement **u** can be expressed as $\boldsymbol{u}=\boldsymbol{QD}$, where **D** is the global displacement vector and **Q** is a transformation matrix. In general, **x**, **u** and **s** are related through a transformation which is available in an algorithmic sense, e.g., a finite element code. For a specified failure criterion, the limit state function G(**x**,**u**,s) >

0 defines the safe state, G(x,u,s) < 0 defines the failure state, and G(x,u,s) = 0 denotes the limit state surface.

For reliability computations, it is convenient to transform the variables **x** into the standard normal space **Y**= *Y*(**x**) such that the elements of **Y** are statistically independent and have a standard normal distribution. The limit state surface in the standard normal space is replaced with the tangent plane at the point with minimum distance from the origin, denoted the design point, and the reliability index $\beta = \sqrt{y^t y}$ is the minimum distance from the origin. To find the minimum distance point(s), **y** can be formulated as a constrained optimization problem. For nonlinear problems, this constrained optimization problem does not have a closed-form solution, and the following iterative scheme can be followed to obtain the design point:

$$y_{i+1} = \left(y_i^t \alpha_i + \frac{G(y_i)}{|\nabla G(y_i)|} \right) \alpha_i \quad , \tag{3.1}$$

where

$$\nabla G(y) = \left[\frac{\partial G(y)}{\partial y_1}, \ldots, \frac{\partial G(y)}{\partial y_n} \right]^t \quad ,$$

and α_i is the unit vector normal to the performance surface having direction away from the origin. It can be estimated as:

$$\alpha_i = -\frac{\nabla G(y_i)}{|\nabla G(y_i)|} \quad . \tag{3.2}$$

The implementation of the AFOSM in the reliability estimation requires the evaluation of the gradient of the limit state function $\nabla G(y_i)$. By the chain rule of differentiation, this gradient can be shown to be:

$$\nabla G = \left[\frac{\partial G}{\partial s} J_{s,x} + \left(Q \frac{\partial G}{\partial u} + \frac{\partial G}{\partial s} J_{s,D} \right) J_{D,x} + \frac{\partial G}{\partial x} \right] J_{y,x}^{-1} \quad , \tag{3.3}$$

where $J_{i,j}$ are the Jacobians of transformation, and all other variables were described earlier. The three gradients on the right-hand side of Eq. 3.3 are easy to compute since G(**x**,**u**,**s**) is an explicit function of **x**, **u** and **s**. Because of the triangular nature of the transformation, $J_{y,x}$ and its inverse are also easy to compute. If **s** is not an explicit function of the basic random variables **x**, then $J_{s,x} = 0$. The Jacobians of the transformation $J_{s,D}$ and $J_{D,x}$, however, are not easy to compute since **s**, **D** and **x** are implicit functions of each other. The adjoint method[9] is used here to compute the product of the second term in Eq. 3.3 directly rather than evaluating its constituent parts. This method is shown to be very accurate and efficient.

With the adjoint method, an auxiliary vector **λ** can be introduced such that

$$\lambda^t K = Q\frac{\partial G}{\partial u} + \frac{\partial G}{\partial s} J_{s,D} \quad . \tag{3.4}$$

It can be shown that:

$$\lambda^t K J_{D,x} = \left(Q\frac{\partial G}{\partial u} + \frac{\partial G}{\partial s} J_{s,D} \right) J_{D,x} = \lambda^t \left(\frac{\partial F}{\partial x} - \frac{\partial R}{\partial x} \right) \quad , \tag{3.5}$$

where $\partial F/\partial x$ is easy to obtain if the external load is not affected by the structural response, and $\partial R/\partial x$ can be derived from Eq. 2.3. $J_{D,x}$ can be obtained by the method described by Gao and Haldar[1]. Substituting Eq. 3.5 into Eq. 3.3, ∇ G can be obtained.

The practical implementation of the SFEM-based reliability analysis needs to be done in a computer environment. The amount of memory required in the computer grows rapidly with the number of basic random variables. Sensitivity analysis[3,10] can be used to reduce the number of basic random variables. Among the structural parameters, only the Young's modulus E, area A, Poisson ratio μ, yield stress F_y and the moments of inertia of a cross section J, I_2 and I_3, along with the external load **F**, are considered as basic random variables in this study.

The uncertainties of connection behavior should be addressed at this stage. In practice, any connection model tries to represent the connection behavior in a statistical way since the most precise representation of the parameters in the model is obtained through curve-fitting the model directly to the experimental data. Therefore, the deterministic moment-rotation curves do not account for the scatter of connection behavior. In this study, the scatter of connection behavior is considered by considering all four parameters in the Richard model as the basic random variables, as shown in Fig. 3.1.

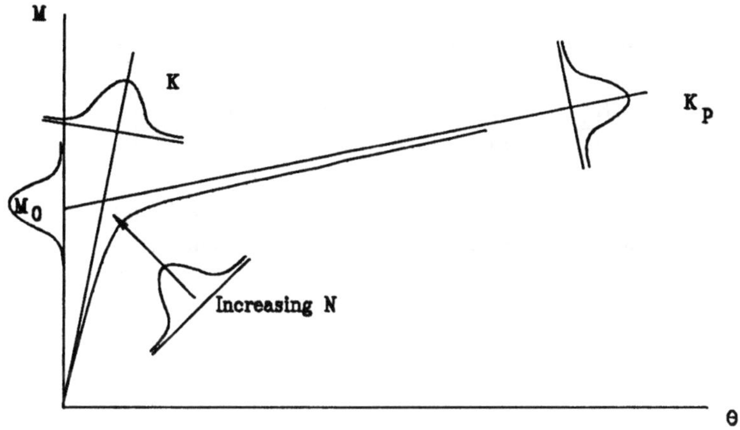

Figure 3.1 Stochastic Parameters in Richard Model

In order to calculate $\partial R/\partial x$, Eq. 2.3 can be written as

$$R = -A^{t}_{\sigma do} R_{A\sigma} + R_{do} \quad , \tag{3.6}$$

where

$$R_{A\sigma} = \begin{Bmatrix} n - \dfrac{EAH}{l} \\ m_1 - \dfrac{GJ}{l}({}^2\theta^{\bullet}_1 - {}^1\theta^{\bullet}_1) \\ {}^1m_2 + \dfrac{2EI_2}{l}(2\,{}^1\theta^{\bullet}_2 + {}^2\theta^{\bullet}_2) \\ {}^2m_2 - \dfrac{2EI_2}{l}({}^1\theta^{\bullet}_2 + 2\,{}^2\theta^{\bullet}_2) \\ {}^1m_3 + \dfrac{2EI_3}{l}(2\,{}^1\theta^{\bullet}_3 + {}^2\theta^{\bullet}_3) \\ {}^2m_3 - \dfrac{2EI_3}{l}({}^1\theta^{\bullet}_3 + 2\,{}^2\theta^{\bullet}_3) \end{Bmatrix} \quad . \tag{3.7}$$

Since $\boldsymbol{R}_{do}$ and $\boldsymbol{A}^{t}_{\sigma do}$ are not functions of basic variables, the derivative of $\boldsymbol{R}$ with respect to $\boldsymbol{x}$ can be expressed as:

$$\frac{\partial \boldsymbol{R}}{\partial \boldsymbol{x}}\bigg|_{D,\sigma} = -\boldsymbol{A}^{t}_{\sigma do}\frac{\partial \boldsymbol{R}_{A\sigma}}{\partial \boldsymbol{x}} \quad . \tag{3.8}$$

For a beam-column element, the following expression is available:

$$\frac{\partial \boldsymbol{R}_{A\sigma}}{\partial x} = \begin{bmatrix} \frac{\partial \boldsymbol{R}_{A\sigma}}{\partial E} & \frac{\partial \boldsymbol{R}_{A\sigma}}{\partial A} & \frac{\partial \boldsymbol{R}_{A\sigma}}{\partial \mu} & \frac{\partial \boldsymbol{R}_{A\sigma}}{\partial J} & \frac{\partial \boldsymbol{R}_{A\sigma}}{\partial I_2} & \frac{\partial \boldsymbol{R}_{A\sigma}}{\partial I_3} & \boldsymbol{0} \end{bmatrix} \quad . \tag{3.9}$$

Details of the calculation of each item in Eq. 3.9 can be found elsewhere[1]. For connection elements, the following items should be considered:

$$\frac{\partial R_{A\sigma}}{\partial x} = \begin{bmatrix} \frac{\partial R_{A\sigma}}{\partial K} & \frac{\partial R_{A\sigma}}{\partial K_p} & \frac{\partial R_{A\sigma}}{\partial M_o} & \frac{\partial R_{A\sigma}}{\partial N} & [0] \end{bmatrix} \quad . \tag{3.10}$$

Eq. 2.4 and Eq. 2.6 can be used to calculate some of the terms in Eq. 3.10 as discussed below:

$$\frac{\partial K(\theta)}{\partial K} = \frac{1 - Na^N}{\left[1 + a^N\right]^{\frac{2N+1}{N}}} \quad , \tag{3.11}$$

$$\frac{\partial K(\theta)}{\partial K_p} = \frac{-(N+2)a^N - 1}{\left[1 + a^N\right]^{\frac{2N+1}{N}}} + 1 \quad , \tag{3.12}$$

$$\frac{\partial K(\theta)}{\partial M_o} = \frac{(N+1)a^N}{\left[1 + a^N\right]^{\frac{2N+1}{N}}} \frac{(K - K_p)}{M_o} \quad , \tag{3.13}$$

$$\frac{\partial K(\theta)}{\partial N} = \left(-\frac{a^N(1+N)\log a}{N(1+a^N)^{\frac{1+2N}{N}}} + \frac{-N + (1+N)\log(1+a^N)}{N^2(1+a^N)^{\frac{1+N}{N}}} \right)(K - K_p) \quad , \tag{3.14}$$

where

$$a = \left| \frac{(K - K_p)\theta}{M_O} \right|$$

Denoting K, K_p, M_O and N as the four parameters representing the major-axis rotational flexibility of a connection, the components in Eq. 3.10 can be expressed as:

$$\frac{\partial R_{A\sigma}}{\partial K} = \left[0 \quad 0 \quad 2\frac{\partial K(\theta)}{\partial K}\left(2^1\theta_2^* + {}^2\theta_2^*\right) \quad -2\frac{\partial K(\theta)}{\partial K}\left({}^1\theta_2^* + 2^2\theta_2^*\right) \quad 0 \quad 0 \right]^t ,$$

$$\frac{\partial R_{A\sigma}}{\partial K_p} = \left[0 \quad 0 \quad 2\frac{\partial K(\theta)}{\partial K_p}\left(2^1\theta_2^* + {}^2\theta_2^*\right) \quad -2\frac{\partial K(\theta)}{\partial K_p}\left({}^1\theta_2^* + 2^2\theta_2^*\right) \quad 0 \quad 0 \right]^t ,$$

$$\frac{\partial R_{A\sigma}}{\partial M_0} = \left[0 \quad 0 \quad 2\frac{\partial K(\theta)}{\partial M_0}\left(2^1\theta_2^* + {}^2\theta_2^*\right) \quad -2\frac{\partial K(\theta)}{\partial M_0}\left({}^1\theta_2^* + 2^2\theta_2^*\right) \quad 0 \quad 0 \right]^t ,$$

$$\frac{\partial R_{A\sigma}}{\partial N} = \left[0 \quad 0 \quad 2\frac{\partial K(\theta)}{\partial N}\left(2^1\theta_2^* + {}^2\theta_2^*\right) \quad -2\frac{\partial K(\theta)}{\partial N}\left({}^1\theta_2^* + 2^2\theta_2^*\right) \quad 0 \quad 0 \right]^t .$$

Similarly, if K', K'_p, M'_0 and N' are the four parameters describing the minor-axis rotational flexibility of a connection, the above four expressions become :

$$\frac{\partial R_{A\sigma}}{\partial K'} = \left[0 \quad 0 \quad 0 \quad 0 \quad 2\frac{\partial K'(\theta)}{\partial K'}\left(2^1\theta_3^* + {}^2\theta_3^*\right) \quad -2\frac{\partial K'(\theta)}{\partial K'}\left({}^1\theta_3^* + 2^2\theta_3^*\right) \right]^t ,$$

$$\frac{\partial R_{A\sigma}}{\partial K'_p} = \left[0 \quad 0 \quad 0 \quad 0 \quad 2\frac{\partial K'(\theta)}{\partial K'_p}\left(2^1\theta_3^* + {}^2\theta_3^*\right) \quad -2\frac{\partial K'(\theta)}{\partial K'_p}\left({}^1\theta_3^* + 2^2\theta_3^*\right) \right]^t ,$$

$$\frac{\partial R_{A\sigma}}{\partial M'_0} = \left[0 \quad 0 \quad 0 \quad 0 \quad 2\frac{\partial K'(\theta)}{\partial M'_0}\left(2^1\theta_3^* + {}^2\theta_3^*\right) \quad -2\frac{\partial K'(\theta)}{\partial M'_0}\left({}^1\theta_3^* + 2^2\theta_3^*\right) \right]^t ,$$

$$\frac{\partial R_{A\sigma}}{\partial N'} = \left[0 \quad 0 \quad 0 \quad 0 \quad 2\frac{\partial K'(\theta)}{\partial N'}\left(2^1\theta_3^* + {}^2\theta_3^*\right) \quad -2\frac{\partial K'(\theta)}{\partial N'}\left({}^1\theta_3^* + 2^2\theta_3^*\right) \right]^t .$$

By now, all quantities required for the computation of $\nabla G(y)$ in Eq. 3.3 are available in very simple explicit forms. Eq. 3.1 can now be used to compute the design point, and the corresponding reliability index can be obtained. To improve the stability of the nonlinear finite element analysis, the Newton-Raphson method with arc-length procedure is used in this study.

4 Limit State Functions

Once the response of a structure has been calculated through structural analysis, it must be judged from the point of view of performance and safety. This judgment may be expressed in different forms according to the type of response parameters (strength or serviceability) under consideration. Both the strength and serviceability limit state functions must be considered. According to the *Load and Resistance Factor Design* (LRFD)[11], the component strength limit state can be defined by the following two interaction equations:

$$\frac{P_u}{\phi\ P_n} \geq 0.2$$
$$\frac{P_u}{\phi\ P_n} + \frac{8}{9}\left(\frac{M_{ux}}{\phi_b M_{nx}} + \frac{M_{uy}}{\phi_b M_{ny}}\right) \leq 1.0 \tag{4.1}$$

$$\frac{P_u}{\phi\ P_n} < 0.2$$
$$\frac{P_u}{2\phi\ P_n} + \left(\frac{M_{ux}}{\phi_b M_{nx}} + \frac{M_{uy}}{\phi_b M_{ny}}\right) \leq 1.0 \tag{4.2}$$

where ϕ and ϕ_b are the resistance factors, P_u is the required tensile/compressive strength, P_n is the nominal tensile/compressive strength, M_u is the required flexural strength, and M_n is the nominal flexural strength.

If δ and δ_{limit} are the actual and the limit displacement components, respectively, the serviceability limit state can be defined as:

$$\delta/\delta_{limit} \leq 1.0 \quad . \tag{4.3}$$

5 Numerical Example

To illustrate the applicability of the proposed method, a three-dimensional frame, shown in Fig. 5.1, is considered. The geometric dimensions of the frame are shown in the figure. Section W16 x 67 is used for all members of the frame. The frame is analyzed by using the stochastic finite element algorithm discussed in the paper. The probabilistic descriptions of basic variables of the structure which are required for this analysis are listed in Table 5.1.

Both the strength and serviceability limit states are considered in this example. Eq. 4.1-4.3 are used for this purpose. According to the UBC code, the drift at the top of column should not be more than h/400 where h is the height of the frame. Thus, for this example, δ_{limit} is 0.45 inches.

First, the frame is analyzed assuming its members are all connected rigidly. The results of linear and nonlinear analyses of the frame for strength limit state are listed in Table 5.2 and Table 5.3, respectively. Table 5.4 and Table 5.5 list the corresponding results for the serviceability limit state. The reliability index of the strength limit state in the nonlinear case is higher than that in the linear case. The reliability index of the serviceability limit state is almost the same in both cases.

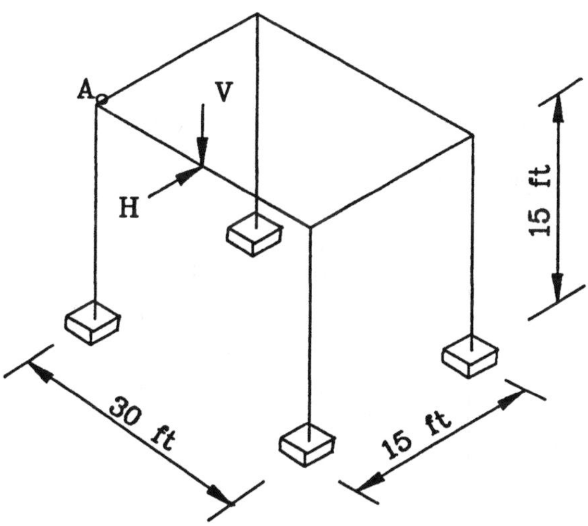

Figure 5.1 3-D Frame Structure in Example

The flexibility of connections is considered next. For the simplicity of the discussion, only one connection A, as shown in Fig. 5.1, is considered to be partially restrained, while all the others are rigid. For illustration purposes only, the major axis flexible behavior of connection A is considered. The M-θ curve for this connection is shown in Fig. 5.2. The probabilistic descriptions of the four parameters of the Richard's model to represent connection A are listed in Table 5.6. The three-dimensional frame with one flexible connection is again analyzed by using the proposed algorithm. Results for both the strength and serviceability limit states are listed in Tables 5.7 and 5.8, respectively. For this example, the reliability index corresponding to the strength limit state is much higher than that of the serviceability limit state.

Table 5.1 Description of Basic Random Variables In Example

Random Variables	Mean	C.O.V.	Distribution
E (ksi)	29000	0.06	Lognormal
A (in^2)	19.7	0.05	Lognormal
I_2 (in^4)	954	0.05	Lognormal
I_3 (in^4)	119	0.05	Lognormal
Z_x (in^3)	130	0.05	Lognormal
Z_y (in^3)	35.5	0.05	Lognormal
F_y (ksi)	37.8	0.10	Lognormal
V (kips)	27.0	0.15	Lognormal
H (kips)	7.5	0.15	Lognormal

Table 5.2 Rigid Frame, Linear analysis, Strength limit state (β=2.57)

Random Variables	Initial Point	Final Point	Sensitivity Index
E (ksi)	29000.00	28947.31	-0.0001
A (in^2)	19.70	19.67	-0.0041
I_2 (in^4)	954.00	952.83	0.0002
I_3 (in^4)	119.00	118.85	0.0002
Z_x (in^3)	130.00	127.18	-0.1609
Z_y (in^3)	35.50	34.69	-0.1700
F_y (ksi)	37.80	31.69	-0.6678
V (kips)	27.00	32.22	0.4897
H (kips)	7.50	9.02	0.5093

Considering the example of the three-dimensional frame with rigid connections and with one flexible connection, several important observations can be made. Considering the rigid frame and the strength limit state design criterion, the reliability index for the linear analysis appears to be smaller than that for the nonlinear analysis. In this particular case, the nonlinear analysis represents a safer design. When the serviceability limit state is considered, the reliability indexes for the linear and nonlinear analyses are almost identical. However, in this case, the strength limit state is controlling the design. Furthermore, both limit states give very similar reliability indexes, indicating a very efficient design.

Table 5.3 Rigid Frame, Nonlinear analysis, Strength limit state (β=2.76)

Random Variables	Initial Point	Final Point	Sensitivity Index
E (ksi)	29000.00	27218.43	-0.3723
A (in^2)	19.70	19.66	-0.0069
I_2 (in^4)	954.00	934.27	-0.1425
I_3 (in^4)	119.00	116.42	-0.1496
Z_x (in^3)	130.00	127.31	-0.1426
Z_y (in^3)	35.50	34.74	-0.1488
F_y (ksi)	37.80	31.99	-0.5880
V (kips)	27.00	32.67	0.4903
H (kips)	7.50	8.88	0.4359

Table 5.4 Rigid Frame, Linear Analysis, Serviceability (β=2.89)

Random Variables	Initial Point	Final Point	Sensitivity Index
E (ksi)	29000.00	27216.93	-0.3562
A (in^2)	19.70	19.68	0.0006
I_2 (in^4)	954.00	950.58	-0.0162
I_3 (in^4)	119.00	114.12	-0.2813
V (kips)	27.00	26.65	-0.0043
H (kips)	7.50	10.89	0.8909

Table 5.5 Rigid Frame, Nonlinear Analysis, Serviceability (β=2.88)

Random Variables	Initial Point	Final Point	Sensitivity Index
E (ksi)	29000.00	28317.35	-0.1274
A (in^2)	19.70	19.68	0.0002
I_2 (in^4)	954.00	951.98	-0.0161
I_3 (in^4)	119.00	114.36	-0.2680
V(kips)	27.00	26.66	-0.0039
H(kips)	7.50	11.17	0.9549

Table 5.6 Description of Stochastic Parameters of Connection A

Random Variables	Mean	C.O.V.	Distribution
K (k-in/rad.)	633000.00	0.15	Normal
K_p (k-in/rad.)	53000.00	0.15	Normal
M_0 (k-in)	1000.00	0.15	Normal
N	0.70	0.05	Normal

Table 5.7 Frame with PR Connection, Strength (β=3.45)

Random Variables	Initial Point	Final Point	Sensitivity Index
K (k-in/rad.)	633000.00	632966.13	-0.0001
K_p (k-in/rad.)	53000.00	52999.68	-0.0000
M_0 (k-in)	1000.00	999.96	-0.0001
N	0.70	0.70	-0.0000
E (ksi)	29000.00	26471.33	-0.4328
A (in^2)	19.70	19.65	-0.0079
I_2 (in^4)	954.00	925.93	-0.1661
I_3 (in^4)	119.00	115.44	-0.1693
Z_x (in^3)	130.00	126.17	-0.1663
Z_y (in^3)	35.50	34.44	-0.1690
F_y (ksi)	37.80	35.23	-0.3390
V (kips)	27.00	36.10	0.5866
H (kips)	7.50	9.55	0.4910

Table 5.8 Frame with PR Connection, Serviceability (β=2.57)

Random Variables	Initial Point	Final Point	Sensitivity Index
K (k-in/rad.)	633000.00	628995.25	-0.0164
K_p (k-in/rad.)	53000.00	52966.52	-0.0016
M_0 (k-in)	1000.00	995.35	-0.0121
N	0.70	0.70	-0.0033
E (ksi)	29000.00	28118.45	-0.1842
A (in^2)	19.70	19.68	0.0008
I_2 (in^4)	954.00	951.72	-0.0089
I_3 (in^4)	119.00	116.60	-0.1464
V (kips)	27.00	26.72	0.0016
H (kips)	7.50	10.76	0.9717

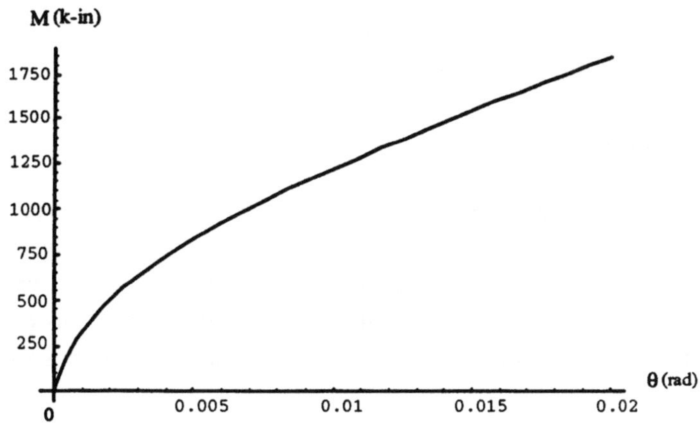

Figure 5.2 The M- θ curve of connection A

The situation changes significantly when even one partially restrained connection is considered. For the strength limit state, the reliability index goes up. This is expected since the redistribution of moments is to be expected due to the presence of the flexible connection. However, the reliability index for the serviceability limit state goes down significantly. This indicates that the serviceability limit state controls the design in this case. This behavior is totally opposite to the case where all the joints are rigid. Furthermore, the reliability indexes for both the limit states are quite different, indicating that the design is no longer efficient.

The examples considered here clearly indicate the implications of considering the flexibility of connections may change the controlling limit state , and the design could be inefficient.

6 Conclusions

An efficient SFEM is proposed for the reliability analysis of 3-D nonlinear frames with PR connections using the AFOSM method. The assumed stress field approach is used in computing the nonlinear structural responses and the corresponding response gradients. The comparison of the reliability indexes between the rigid frame and the frame with PR connections is stressed in the numerical example. Both nonlinearity and PR connections have considerable influence on the structural reliability. When PR connections are considered, the serviceability limit state may be more critical than the strength limit state. For structures with flexible connections, the assurance of structural performance considering serviceability limit state is highly recommended.

Acknowledgments

This paper is based upon work partly supported by the National Science Foundation under Grants No. MSM-8896267. Financial support received from the American Institute of Steel Construction (AISC), Chicago, is appreciated. Any opinions, findings, conclusions, or recommendations expressed in this publication are those of the writers and do not necessarily reflect the views of the sponsors.

References

1. Gao, L. and Haldar, A. "Reliability analysis for nonlinear structures," Proceedings of *ISUMA'93 The Second International Symposium on Uncertainty Modeling and Analysis*, 587-594, April, 1993.
2. Rauscher, T. R. and Gerstle, K. H. "Reliability of rotational behavior of framing connections," *Engineering Journal / American Institute of Steel Construction*, **29**(1), 1992, 12-19.
3. Mahadevan, S. and Haldar, A. "Efficient algorithm for stochastic structural optimization," *J. Struct. Engrg.*, ASCE, **115**(7), 1989, 1579--1598.
4. Haldar, A. and Nee, K.M. "Elasto-plastic large deformation analysis of PR steel frames for LRFD," *Comput. Struct.*, **34**(5), 1988, 811--823.
5. Kondoh, K. and Atluri, S.N. "Large deformation, elasto-plastic analysis of frames under non-conservative loading, using explicitly derived tangent stiffness based on assumed stress," *Comput. Mech.*, **2**(1), 1987, 1--25.
6. Shi, G. and Atluri, S.N. "Elasto-plastic large deformation analysis of space--frames," *Int. J. for Numerical Methods in Engineering*, **26**, 1988, 589--615.
7. Celikag, M. and Kirby, P. A. "Standardised method for measuring three dimensional response of semi-rigid joints", *Workshop on Connections - Strength and Design of Steel Structures*, Cachan, France, May 1987, 203-210.
8. Richard, R. M. and Abbott, B.J., "Versatile elastic-plastic stress-strain foumula," *J. Engng. Mech.*, ASCE, **101**(EM4), Aug., 1975, 511-515
9. Ryu, Y.S., Harian, M., Wu, C.C. and Arora, J.S. "Structural design sensitivity analysis of nonlinear response," *Comput. Struct.*, **21**(1/2), 1985, 245--255.
10. Der Kiureghian, A., and Ke, J-B. "The stochastic finite element method in structural reliability," *Probabilistic Engrg. Mech.* **3**(2), 1988, 83--91.
11. American Institute of Steel Construction, *Manual of Steel Construction: Load and Resistance Factor Design*, Chicago, Illinois, 1986.

Structural Reliability Optimization

Philippe Geyskens[1]

Department of Civil Engineering, Katholieke Universiteit Leuven, Belgium

Abstract. Maximum likelihood points play a predominant role in structural reliability computation. The point of maximum likelihood (or the failure point) is the solution of a nonlinear mathematical programming problem with a nonlinear equality constraint (NEP). Sequential quadratic programming (SQP) is regarded as one of the best procedures to solve an NEP [1, 2]. The different computational strategies on which the SQP method is based are described. Second-order reliability analysis methods need second-order derivative information of the likelihood and the limit state surface at the optimal point. An optimization-based algorithm for the calculation of the reduced Hessian of the Lagrangian is discussed.

Keywords. Structural reliability, nonlinear equality constrained program, optimization, second-order reliability method

1 Structural Reliability Approximation

Studying the asymptotic behavior of Laplace integrals, Breitung [3] proves that

$$\Pr(F) \approx \frac{(2\pi)^{(n-1)/2} f_{\underline{X}}(\mathbf{x}^*)}{||\nabla_{\mathbf{x}}\ell(\mathbf{x}^*)||\sqrt{|\det[\mathbf{Z}(\mathbf{x}^*)^T\mathbf{H}(\mathbf{x}^*)\mathbf{Z}(\mathbf{x}^*)]|}} \quad , \qquad (1)$$

where $\ell(\mathbf{x})$ is the log likelihood $\log(f_{\underline{X}}(\mathbf{x}))$, $\mathbf{x}^* = \max_{\{\mathbf{x}:g(\mathbf{x})=0\}} \ell(\mathbf{x})$ is the point of maximum likelihood (PML), and

$$\{\mathbf{H}(\mathbf{x}^*)\}_{i,j} = \left(\frac{\partial^2 \ell(\mathbf{x}^*)}{\partial x_i \partial x_j} - \frac{||\nabla_{\mathbf{x}}\ell(\mathbf{x}^*)||}{||\nabla_{\mathbf{x}} g(\mathbf{x}^*)||}\frac{\partial^2 g(\mathbf{x}^*)}{\partial x_i \partial x_j}\right) \quad i,j = 1,\ldots,n \quad . \qquad (2)$$

The expression in equation (1) is an asymptotically exact ($\Pr(F) \rightarrow 0$) approximation for the structural component reliability integral

$$\Pr(F) = \int_{g(\mathbf{x}) \leq 0} f_{\underline{X}}(\mathbf{x})\, d\mathbf{x} \quad , \qquad (3)$$

where $f_{\underline{X}}(\mathbf{x})$ is the PDF of the basic random variables $\underline{X}$, and $F = \{\mathbf{x} : g(\mathbf{x}) \leq 0\}$ is the failure domain. The columns of $\mathbf{Z}(\mathbf{x}^*)$ form an orthogonal basis of the $n-1$ dimensional subspace $\mathcal{Z}$, tangential to the limit state surface $g(\mathbf{x}) = 0$ in the PML $\mathbf{x}^*$: $\mathcal{Z} = \{\mathbf{y} : \mathbf{y}^T[\nabla_{\mathbf{x}} g(\mathbf{x}^*)] = 0\}$.

If all basic variables are standard-normal, the failure point $\mathbf{u}^*$ is equal to the PML $\mathbf{x}^*$. Equation (1) reduces to the asymptotic SORM approximation [4]

$$\Pr(F) \approx \hat{\mathrm{P}}_{2u} = \Phi(-\beta)[\prod_{i=2}^{n}(1+\beta\kappa_i)]^{-1/2} \quad , \qquad (4)$$

where κ_i, $i = 2, \ldots, n$ are the principal curvatures of the limit state surface $g(\mathbf{x}) = 0$, and $\beta = ||\mathbf{u}^*||$. In this particular case, the eigenvectors of $\mathbf{Z}(\mathbf{x}^*)^T\mathbf{H}(\mathbf{x}^*)\mathbf{Z}(\mathbf{x}^*)$ form the basis of the tangent subspace $\mathcal{Z}$.

2 The Structural Reliability Optimization Problem

The point of maximum likelihood (PML) for a structural component reliability problem (3) is the solution of a nonlinear mathematical programming problem with a single, nonlinear equality constraint (NEP)

$$\begin{cases} \underset{\mathbf{x}\in\mathbb{R}^n}{\text{maximize}} & \ell(\mathbf{x}) \\ \text{subject to} & g(\mathbf{x}) = 0 \end{cases} \quad . \qquad (5)$$

The objective $\ell(\mathbf{x})$ of the program (5) is the log likelihood function of the basic random variables $\ell(\mathbf{x}) = \log(f_{\underline{X}}(\mathbf{x}))$. The functional constraint $g(\mathbf{x}) = 0$ is the limit state function of the structural reliability problem. The set constraints $\mathbf{x} \in \mathcal{X} \subseteq \mathbb{R}^n$, i.e. constraints delimiting the domain in which the log likelihood $\ell(\mathbf{x})$ exists, are taken care of by a suitable coordinate transformation [5]. The multiple constraint case [6] occurs in e.g. system problems.

In the standard-normal domain, the failure point $\mathbf{u}^*$ is the solution of $\min_{\{\mathbf{u}:G(\mathbf{u})=0\}} ||\mathbf{u}||$ [7], or, equivalently,

$$\begin{cases} \underset{\mathbf{u}\in\mathbb{R}^n}{\text{maximize}} & -\frac{1}{2}\mathbf{u}^T\mathbf{u} \\ \text{subject to} & G(\mathbf{u}) = 0 \end{cases} \quad , \qquad (6)$$

where $\mathbf{u} = \mathbf{T}_{SN}(\mathbf{x})$, and $\mathbf{T}_{SN}$ is a suitable mapping to the standard-normal domain [8]. Program (6) is the standard-normal equivalent of (5), since the standard-normal log likelihood is $\ell(\mathbf{x}) = C - \frac{1}{2}\mathbf{u}^T\mathbf{u}$.

3 Sequential Quadratic Programming

3.1 The Quadratic Subproblem

According to the optimality conditions [1, 2] the local maximum of an NEP is an unconstrained local maximizer of the Lagrangian function $L(\mathbf{x}, \lambda) =$

$\ell(\mathbf{x}) - \lambda g(\mathbf{x})$, projected on the null space $\mathcal{Z}$ of the constraint at the optimum $\mathbf{x}^*$. The Lagrangian objective is replaced by its second-order Taylor expansion. Projection on the null space is obtained by introducing a linear constraint [9]. The linear constraint gradually forces the iterate $\mathbf{x}_k$ onto $g(\mathbf{x}) = 0$. Consequently, the quadratic program (QP)

$$\begin{cases} \underset{\mathbf{d}\in\mathbb{R}^n}{\text{maximize}} & [\nabla_{\mathbf{x}}\ell(\mathbf{x}_k)]^T\mathbf{d} + \frac{1}{2}\mathbf{d}^T\,\mathbf{H}_k\,\mathbf{d} \\ \text{subject to} & [\nabla_{\mathbf{x}}g(\mathbf{x}_k)]^T\mathbf{d} + g(\mathbf{x}_k) = 0 \end{cases} \quad , \tag{7}$$

with $\mathbf{d}_k = \mathbf{x} - \mathbf{x}_k$, approximates the NEP at iterate $\mathbf{x}_k$. The negative definite matrix $\mathbf{H}_k$, represents the Hessian of the Lagrangian function (Section 3.2). Successive quadratic approximation (7) produces a sequence of ascent directions $\{\mathbf{d}_k, k \geq 1\}$ of the NEP, with

$$\mathbf{d}_k = \frac{\mathbf{H}_k^{-1}[\nabla_{\mathbf{x}}g]([\nabla_{\mathbf{x}}g]^T\mathbf{H}_k^{-1}[\nabla_{\mathbf{x}}\ell] - g(\mathbf{x}_k))}{[\nabla_{\mathbf{x}}g]^T\mathbf{H}_k^{-1}[\nabla_{\mathbf{x}}g]} - \mathbf{H}_k^{-1}[\nabla_{\mathbf{x}}\ell] \quad , \tag{8}$$

where all gradients $\nabla_{\mathbf{x}}$ are evaluated in iterate $\mathbf{x}_k$. Solution $\mathbf{d}_k$ is a well-defined maximizer of the QP, provided the Hessian matrix $\mathbf{H}_k$ is negative definite. Consider the standard-normal optimization problem (6). The objective is quadratic. Hence, sequential linearization of the constraints leads to a QP with solution (8)

$$\mathbf{d}_k = \frac{\nabla_{\mathbf{u}}G(\mathbf{u}_k)}{||\nabla_{\mathbf{u}}G(\mathbf{u}_k)||^2}\{[\nabla_{\mathbf{u}}G(\mathbf{u}_k)]^T\mathbf{u}_k - G(\mathbf{u}_k)\} - \mathbf{u}_k \quad . \tag{9}$$

Equation (9) is the formulation of HL-RF algorithm [10, 11]. The HL-RF algorithm (9) is known to diverge [12]. Convergence properties can be enhanced by introducing step length control [13, 14, 15] (Section 3.3).

The solution $\mathbf{d}_k$ of the QP (7), formulated at the trial point $\mathbf{x}_k$, can be expressed as a sum of two orthogonal components $\mathbf{d}_k = [\nabla_{\mathbf{x}}g]d_Y + \mathbf{Z}_k\mathbf{d}_Z$. The columns of $\mathbf{Z}_k = \mathbf{Z}(\mathbf{x}_k)$ form an orthogonal basis, spanning the null space. Introducing the decomposition of $\mathbf{d}_k$ in (7), the following equations emerge:

$$[\nabla_{\mathbf{x}}g]^T\mathbf{d}_k = [\nabla_{\mathbf{x}}g]^T[\nabla_{\mathbf{x}}g]d_Y = -g(\mathbf{x}_k) \quad ; \tag{10}$$

$$\mathbf{Z}_k^T\mathbf{H}_k\mathbf{Z}_k\mathbf{d}_Z = -\mathbf{Z}_k^T\{[\nabla_{\mathbf{x}}\ell] + \mathbf{H}_k[\nabla_{\mathbf{x}}g]d_Y\} \quad . \tag{11}$$

The null-space component $\mathbf{d}_Z$ is the solution of an unconstrained quadratic program (11) in $\mathcal{Z}$. Equation (10) is a true Newton step solving the nonlinear equation $g(\mathbf{x}) = 0$. An implementable solution [16] of the QP (7) solves the linear system (11). If $\mathbf{Z}_k$ is known, the null-space component $\mathbf{d}_Y$ is the solution of a symmetrical negative definite system of $n-1$ equations. Matrix $\mathbf{Z}_k$ can be found by QR-factorization [16, 17] of $[\nabla_{\mathbf{x}}g]$.

3.2 The Hessian of the Quadratic Program

Matrix $\mathbf{H}_k$ in equation (7) stands for the Hessian of the Lagrangian function. If the true Hessian of the Lagrangian $\mathbf{H}_k = \mathbf{H}(\mathbf{x}_k) = [\nabla^2_{\mathbf{x}}\ell(\mathbf{x}_k) - \lambda_k \nabla^2_{\mathbf{x}} g(\mathbf{x}_k)]$ is used, the SQP method transforms into a Newton iteration algorithm. Unfortunately, the reduced Hessian of the Lagrangian function $\mathbf{Z}_k^T \mathbf{H}_k \mathbf{Z}_k$ in (11) can become indefinite. Moreover, second-order derivatives often require considerable computational expense. Aforementioned problems are avoided by using a quasi-Newton approximation for $\mathbf{H}_k$ [18]. The Hessian matrix is gradually approximated using gradient information gathered during the optimization sequence $\{\mathbf{x}_k, k \geq 1\}$. Powell [19] opts for the BFGS updating formula, due to its superior performance in unconstrained optimization [20],

$$\mathbf{H}_{k+1} = \mathbf{H}_k + \frac{\mathbf{H}_k \mathbf{s}_k \mathbf{s}_k^T \mathbf{H}_k}{\mathbf{s}_k^T \mathbf{H}_k \mathbf{s}_k} - \frac{\mathbf{y}_k \mathbf{y}_k^T}{\mathbf{y}_k^T \mathbf{s}_k} \quad , \qquad (12)$$

where $\mathbf{s}_k is the step \mathbf{x}_{k+1} - \mathbf{x}_k$, and with $\mathbf{y}_k$ the associated difference in gradients of the objective. The BFGS update has important hereditary properties. If started with a symmetric matrix $\mathbf{H}_0$, the sequence $\{\mathbf{H}_k, k \geq 1\}$ consists of symmetric matrices only. All matrices in the sequence will be negative definite, provided $\mathbf{H}_0$ is negative definite and $\mathbf{y}_k^T \mathbf{s}_k < 0, \forall k \geq 1$, i.e. $\mathbf{s}_k$ is strictly an ascent step.

Since $\mathbf{H}_k$ is supposed to approximate the Hessian of the Lagrangian function, $\mathbf{y}_k$ in (12) should be equal to the change in gradient of the Lagrangian function

$$\mathbf{y}_k \leftarrow \mathbf{y}_L = \nabla_{\mathbf{x}}\ell(\mathbf{x}_{k+1}) - \nabla_{\mathbf{x}}\ell(\mathbf{x}_k) - [\nabla_{\mathbf{x}} g(\mathbf{x}_{k+1}) - \nabla_{\mathbf{x}} g(\mathbf{x}_k)]\hat{\lambda} \quad , \qquad (13)$$

where $\hat{\lambda}$ is the best approximation of λ^* available. However, $\mathbf{y}_k^T \mathbf{s}_k < 0$ cannot be guaranteed for the aforementioned choice of the gradient change. Powell [19] therefore suggests using

$$\mathbf{y}_k = \theta \mathbf{y}_L + (1-\theta)\mathbf{H}_k \mathbf{s}_k \quad , \qquad (14)$$

with

$$\theta = \begin{cases} 1 & \text{for} \quad \mathbf{y}_L^T \mathbf{s}_k \leq 0.2 \mathbf{s}_k^T \mathbf{H}_k \mathbf{s}_k \\ \dfrac{0.8 \mathbf{s}_k^T \mathbf{H}_k \mathbf{s}_k}{\mathbf{s}_k^T \mathbf{H}_k \mathbf{s}_k - \mathbf{y}_L^T \mathbf{s}_k} & \text{for} \quad \mathbf{y}_L^T \mathbf{s}_k > 0.2 \mathbf{s}_k^T \mathbf{H}_k \mathbf{s}_k \end{cases} ,$$

and $\mathbf{y}_L$, the change in gradient of the Lagrangian, defined in (13). This definition of $\mathbf{y}_k$ forces $\mathbf{H}_k$ to be negative definite and subsequently guarantees that $\mathbf{d}_k$, the solution (8) of the QP at iteration step k, is an ascent direction of the NEP.

3.3 The Merit Function

The SQP formulation relies on the first- and second-order optimality conditions of the NEP. These conditions only hold in the vicinity of the maximum point $\mathbf{x}^*$. As such, the solution $\mathbf{d}_k$ of the QP (7) is to be considered as a search direction along which a step α_k is taken: $\mathbf{x}_{k+1} = \mathbf{x}_k + \alpha_k \mathbf{d}_k$. The step length α_k is chosen so that the next iterate $\mathbf{x}_{k+1}$ produces a significant increase of the log likelihood $\ell(\mathbf{x})$, while inducing negligible constraint violation. The compliance to aforementioned objectives is measured using a merit function $\psi(\mathbf{x}(\alpha))$, with $\mathbf{x}(\alpha) = \mathbf{x}_k + \alpha \mathbf{d}_k$. Schittkowski [21, 22] proposes an augmented Lagrangian merit function

$$\psi(\mathbf{x}(\alpha)) \doteq \ell(\mathbf{x}(\alpha)) - \mu g(\mathbf{x}(\alpha)) - \frac{1}{2}\rho[g(\mathbf{x}(\alpha))]^2 \quad . \tag{15}$$

The Lagrange multiplier μ of the augmented Lagrangian (15) is treated as an additional unknown in the search $\mu = \mu(\alpha) = \lambda_k + \alpha(\nu_k - \lambda_k)$ with λ_k the current estimate of the Lagrange multipliers and ν_k the Lagrangian multipliers of the QP at iterate k.

An appropriately chosen merit function $\psi(\mathbf{x}(\alpha))$ has the optimal steplength as an unconstrained maximizer $\alpha_k = \alpha^* > 0$. A univariate optimization scheme is necessary to find to local optimum of $\psi(\mathbf{x}(\alpha))$, nearest to the origin $\alpha = 0$. It is frequently argued that the line search need not be exact, as long as a 'sufficient increase' in the merit function $\psi(\mathbf{x}(\alpha))$, and consequently in the constrained objective, is obtained. The increase in the value of the merit function is considered to be sufficient if following criteria are satisfied:[1]

$$|\psi'(\mathbf{x}(\alpha_k))| \le \eta \psi'(\mathbf{x}(0)) \qquad 0 \le \eta < 1 \quad , \tag{16}$$

$$\psi(\mathbf{x}(\alpha)) \ge \psi(\mathbf{x}(0)) - \xi \alpha_k \psi'(\mathbf{x}(0)) \qquad 0 < \xi < \frac{1}{2} \quad , \tag{17}$$

with $\psi'(\mathbf{x}(\alpha)) = [\nabla_{\mathbf{x}} \psi(\mathbf{x} + \alpha \mathbf{d}_k)]^T \mathbf{d}_k$ the directional derivative of the merit function ψ. Condition (16) weakens the original optimality requirement: 'exact' optimization is obtained for $\eta = 0$. The second condition (17) guarantees a sufficient increase of function $\psi(\mathbf{x}(\alpha))$. Methods for exact optimization can easily be adapted to search for an α_k complying to (16).

4 Hessian reconstruction for the NEP

4.1 Conjugate directions and n-step convergence

Given a symmetric $n \times n$ matrix $\mathbf{B}$, a set of vectors $\{\mathbf{p}_1, \mathbf{p}_2, \ldots, \mathbf{p}_k\}$ is called $\mathbf{B}$-conjugate, if $\mathbf{p}_j^T \mathbf{B} \mathbf{p}_i = 0$ for $i \neq j$. A set of $\mathbf{B}$-conjugate vectors $\mathbf{p}_i$, $i = 1, \ldots, k$ is linearly independent, provided $\mathbf{B}$ is definite. If a set of n conjugate directions $\mathbf{p}_i$ can be found, then $\mathbf{P}^T \mathbf{B} \mathbf{P} = \mathbf{D}$, with $\mathbf{p}_i$ the ith column of matrix $\mathbf{P}$, $\mathbf{D} = \text{diag}\{d_1, d_2, \ldots, d_n\}$, and $d_i = \mathbf{p}_i^T \mathbf{D} \mathbf{p}_i$. Given $\mathbf{D}$ and $\mathbf{P}$, $\mathbf{B}$ is obtained by solving $[\mathbf{PD}^{-1}\mathbf{P}^T]\mathbf{B} = \mathbf{I}$. The singular value decomposition of $\mathbf{PD}^{-1/2}$ immediately produces the eigenvectors and eigenvalues of $\mathbf{B}$ [23].

It can be shown that a Broyden-class quasi-Newton iteration, applied to an unconstrained quadratic programming problem $\max_{\mathbf{x}\in\mathbb{R}^n} \frac{1}{2}\mathbf{x}^T\mathbf{B}\mathbf{x} + \mathbf{b}^T\mathbf{x}$, converges to a maximum in at most n steps. The search directions generated by such an algorithm are proven to be $\mathbf{B}$-conjugate. Due to the linear independence of the $\mathbf{B}$-conjugate search directions $\mathbf{d}_i$, the Hessian $\mathbf{B}$ will be exactly reproduced by the BFGS update $\mathbf{H}_k$ (12) if and only if exactly n steps are necessary to determine the optimum [1, 2].

4.2 Convergence of the Hessian Update

The n-step convergence theorem of the BFGS update (Section 4.1) may give the impression that a quasi-Newton algorithm is likely to produce a matrix sequence $\{\mathbf{H}_k, k \geq 1\}$ that will converge to the Hessian $\mathbf{H}(\mathbf{x}^*)$ of the optimization problem, or, equivalently, to the reduced Hessian of the Lagrangian $\overline{\mathbf{H}} = \mathbf{Z}(\mathbf{x}^*)^T\mathbf{H}(\mathbf{x}^*)\mathbf{Z}(\mathbf{x}^*)$ associated with the constrained optimization problem. For convergence analysis purposes, it is indeed frequently assumed [24] that the optimization objective is closely approximated by its second-order Taylor expansion in the proximity of the maximum point $\mathbf{x}^*$. In depth studies [18, 25] show that convergence statements based on assumed quadratic behavior are not generally applicable. Moreover, full reconstruction of the reduced Hessian of the Lagrangian is generally not implied. Approximation of the latter Hessian is only obtained along the search directions $\{\mathbf{d}_k, k \geq 1\}$ as $k \to \infty$ [18, 9]. Given certain conditions on the sequences $\{\mathbf{s}_k, k \geq 1\}$ and $\{\mathbf{y}_k, k \geq 1\}$, inducing convergence of $\{\mathbf{H}_k, k \geq 1\}$ can be proven [26]. Again, the latter conditions are, in general, not satisfied by optimization algorithms and convergence of the matrix sequence is subsequently difficult to prove.

4.3 The iterative curvature method

Der Kiureghian and De Stefano [12] prove that the search directions $\{\mathbf{d}_k,\ k \geq 1\}$ of the HL-RF algorithm (9) converge ($k \to \infty$) to the principal axis of the limit state surface $G(\mathbf{u}) = 0$, provided the sequence of iterates $\{\mathbf{u}_k, k \geq 1\}$ converges to the failure point $\mathbf{u}^*$. The proof relies on the quadratic nature of the standard-normal likelihood and on assumption that the limit state surface is adequately approximated by a parabola at the failure point $\mathbf{u}^*$.

Once the failure point and the principal axis are determined with the requested accuracy (after N iteration steps), the principal curvature is computed using gradient information of the last two iteration steps

$$\kappa_i = \frac{\text{sign}\{\boldsymbol{\alpha}_{N-1}^T(\mathbf{u}_N - \mathbf{u}_{N-1})\}}{\|\mathbf{u}_N - \mathbf{u}_{N-1}\|} \arccos(\boldsymbol{\alpha}_{N-1}^T\boldsymbol{\alpha}_N) \quad , \qquad (18)$$

with $\boldsymbol{\alpha}_k$ the negative unit gradient vector of the limit state surface at the iteration point $\mathbf{u}_k$. By restarting the procedure in a subspace orthogonal to the previously found principal axes, a descending sequence of the principal curvatures (and corresponding axes) is constructed. Equation (18) is based on the mean-value theorem. The quasi-Newton condition, on which the Hessian updating (Section 3.2) is founded, is an immediate consequence of the same theorem.

4.4 The conjugate Gram-Schmidt iteration

Algorithms for the reconstruction of the Hessian information, except for the straightforward application of finite differences, determine a set of $\overline{\mathbf{H}}$-conjugate vectors. Knowing that the eigenvectors are one particular conjugate set, the ICM (Section 4.3) falls under the same category. If a set of $\overline{\mathbf{H}}$-conjugate directions $\mathbf{p}_i$ can be found, matrix $\overline{\mathbf{H}}$ can easily be computed (Section 4.1). The previous discussion (Section 4.2) demonstrates that a BFGS iteration will, in general, not produce $n-1$ linearly independent $\overline{\mathbf{H}}$-conjugate directions.

The conjugate direction/gradient algorithm for optimization is a specific implementation of the conjugate Gram-Schmidt (CGS) iteration [27]. In general the CGS iteration transforms a set of linearly independent vectors $\mathbf{y}_1, \mathbf{y}_2, \ldots, \mathbf{y}_n$ into a set of $\mathbf{B}$-conjugate vectors $\mathbf{p}_1, \mathbf{p}_2, \ldots, \mathbf{p}_n$, with $\mathbf{B}$ a definite square matrix, by iteratively applying

$$\mathbf{p}_{k+1} = \mathbf{y}_{k+1} - \sum_{j=1}^{k} \frac{\mathbf{p}_j^T \mathbf{B} \mathbf{y}_{k+1}}{\mathbf{p}_j^T \mathbf{B} \mathbf{p}_j} \quad \text{for} \quad k = 1, \ldots, n-1 \quad , \quad (19)$$

with $\mathbf{p}_1 = \mathbf{y}_1$. Rewriting (19), following matrix formulation of the CGS iteration is obtained

$$\mathbf{p}_{k+1} = \mathbf{C}_k^T \mathbf{y}_{k+1} \quad \text{for} \quad k = 0, 1, \ldots, n-1 \quad , \quad (20)$$

with

$$\mathbf{C}_k = \mathbf{C}_{k-1} - \frac{\mathbf{B}\mathbf{p}_k \mathbf{p}_k^T}{\mathbf{p}_k^T \mathbf{B} \mathbf{p}_k} \quad , \quad (21)$$

and with $\mathbf{C}_0 = \mathbf{I}$. The CGS iteration consists of selecting a vector $\mathbf{y}_{k+1}$, independent of the previously determined $\mathbf{B}$-conjugate vectors $\mathbf{p}_i$, $i = 1, \ldots, k-1$, preceded by a rank-one update of the conjugacy matrix $\mathbf{C}_{k-1}$.

Applying the CGS iteration (20) in the null space effectively reconstructs the reduced Hessian of the Lagrangian function $\overline{\mathbf{H}}$. Given the PML $\mathbf{x}^*$ and the Lagrange multiplier λ associated with the equality constraint, the mean-value theorem, applied to the reduced gradient of the Lagrangian function produces

$$\mathbf{Z}(\mathbf{x}^*)^T \nabla_{\mathbf{x}} L(\mathbf{x}^* + \mathbf{Z}(\mathbf{x}^*)\mathbf{y}_Z, \lambda^*) = \overline{\mathbf{H}}\mathbf{y}_Z \quad \text{for} \quad \|\mathbf{y}_Z\| \to 0 \quad , \quad (22)$$

with $\mathbf{y}_Z$ a null space perturbation vector, and given that $\nabla_{\mathbf{x}} L(\mathbf{x}^*, \lambda^*) = 0$. The vector $\overline{\mathbf{H}}\mathbf{p}_k$, entering the conjugate rank-one update (21) is thus known.

Efficient reliability methods require the principal curvatures of the limit state surface $g(\mathbf{x}) = 0$ Hence, the eigenvectors and eigenvalues of the reduced Hessian of the limit state surface $\overline{\mathbf{G}} = \mathbf{Z}(\mathbf{x}^*)^T \nabla^2_{\mathbf{X}} g(\mathbf{x}^*) \mathbf{Z}(\mathbf{x}^*)$ are needed. The CGS iteration is no longer applicable as such, since $\overline{\mathbf{G}}$ need not be definite. Nevertheless, elementary linear algebra allows for the efficient reconstruction of the second-order derivative information of the limit state surface. Matrix $\overline{\mathbf{G}}$ is a constituent component of the reduced Hessian of the Lagrangian function

$$\overline{\mathbf{H}} = Z(\mathbf{x}^*) = \mathbf{Z}(\mathbf{x}^*)^T [\nabla^2_{\mathbf{X}} \ell(\mathbf{x}_k) - \lambda^* g(\mathbf{x}^*)] \mathbf{Z}(\mathbf{x}^*) = \overline{\mathbf{L}} - \| \nabla_{\mathbf{X}} \ell \| \overline{\mathbf{G}}. \quad (23)$$

The spectral decomposition of $\overline{\mathbf{G}}$ can be determined by elementary linear algebra, assuming that the (reduced) Hessian of the log likelihood $\overline{\mathbf{L}} = \mathbf{Z}(\mathbf{x}^*)^T [\nabla^2_{\mathbf{X}} \ell(\mathbf{x}^*)] \mathbf{Z}(\mathbf{x}^*)$ is explicitly known. The CGS iteration, followed by the singular value decomposition of $\mathbf{PD}^{1/2}$ [23], with $\mathbf{D} = \text{diag}\{\mathbf{p}_i^T \overline{\mathbf{H}} \mathbf{p}_i\}$ and $\mathbf{P}$ the matrix of $\overline{\mathbf{H}}$-conjugate directions $\mathbf{p}_i$, produces the eigenmatrix $\mathbf{U}$, i.e. the matrix with the eigenvectors as columns, and eigenvalues $\mathbf{S} = \text{diag}\{\sigma_i\}$ of $\overline{\mathbf{H}} \Rightarrow \overline{\mathbf{H}} = \mathbf{U}^T \mathbf{S} \mathbf{U}$. Then, given (23), $\| \nabla_{\mathbf{X}} \ell \| \mathbf{U}^T \overline{\mathbf{G}} \mathbf{U} = \mathbf{U}^T \overline{\mathbf{L}} \mathbf{U} - \mathbf{S}$ holds. Hence, the scaled eigenvalues v_i of $\mathbf{U}^T \overline{\mathbf{L}} \mathbf{U} - \mathbf{S}$ are the principal curvatures of the limit state surface, i.e. $\kappa_i = -v_i / \| \nabla_{\mathbf{X}} \ell \|$ for $i = 1, \ldots, n-1$. The principal directions are the columns of the $(n-1) \times (n-1)$ matrix $\mathbf{UQ}$, with $\mathbf{U}$ and $\mathbf{Q}$ the eigenmatrix of $\overline{\mathbf{H}}$ and $\mathbf{U}^T \overline{\mathbf{L}} \mathbf{U} - \mathbf{S}$, respectively. The aforementioned procedure minimizes the influence of the numerically unstable subtraction and scaling operations.

5 Examples

5.1 The nearly quadratic limit state

Given are n standard-normal basic random variables $\underline{X}$. The limit state function is

$$g(\mathbf{x}) = \beta - x_1 - \sum_{i=2}^{n} \delta_i [\cosh(x_i) - 1] \quad , \quad (24)$$

with parameters β and δ_i, $i = 2, \ldots, n$. The standard-normal log likelihood is quadratic: $\ell(\mathbf{x}) = -\frac{1}{2}\mathbf{x}^T\mathbf{x}$. If a quadratic limit state is chosen, the reduced Lagrangian function is also quadratic. The mean-value theorem (22) always applies (i.e. the asymptotic condition drops), and the CGS iteration encounters little challenge in reconstructing the reduced Hessian of the Lagrangian function. Hence, a nearly quadratic limit state function is selected. Given that the quadratic approximation holds in the neighborhood of the PML ($\cosh(x) = 1 + x^2/2 + x^4/4! + \ldots$), the proposed benchmark is intended to test the numerical stability of the construction of the $\overline{\mathbf{H}}$-conjugate set, rather than the accuracy of the mean-value theorem. The principal curvatures κ_i, $i = 1, \ldots, n-1$ of the limit state surface at the point of maximum likelihood $\mathbf{x}^*$ are given by $\kappa_i = -\delta_{i+1}$. The principal directions of the limit state surface

$g(\mathbf{x}) = 0$ at $\mathbf{x}^*$ coincide with the coordinate axes $x_2, x_3, \ldots, x_n$. Parameter β enters the reduced Hessian of the Lagrangian function $\overline{\mathbf{H}}$ through the Lagrangian multiplier $\lambda^* = \beta$. Hence, the value of β restricts the choices of the curvature parameters δ_i, $i = 2, \ldots, n$. If $\exists i : \delta_i \leq -\beta^{-1}$, the reduced Hessian of the Lagrangian becomes indefinite, and $\mathbf{x}^*$ is a saddle point instead of a true optimum. Parameter β is (arbitrarily) set to $\beta = 5$. The parameters δ_i, $i = 2, \ldots, n$ control the curvature of the limit state surface.

i	$\lvert\delta_{i+1}\rvert$	$\lvert\kappa_i\rvert$	i	$\lvert\delta_{i+1}\rvert$	$\lvert\kappa_i\rvert$	i	$\lvert\delta_{i+1}\rvert$	$\lvert\kappa_i\rvert$
1	.199	.1990	5	1.90	1.900	8	1.93	1.930
2	.198	.1980	6	1.91	1.910	9	1.94	1.940
3	.196	.1960	7	1.92	1.920	10	1.95	1.950
4	.195	.1950						

Table 1. nearly quadratic problem with closely spaced curvatures

Table 1 compares the analytical solutions with the numerical results for a limit state surface characterized by two groups of closely spaced curvature values. The numerical results reproduce the true curvatures up to the 8th significant digit. Conjugate direction methods are sensitive to the condition of matrix $\overline{\mathbf{H}}$ [2]. The larger the condition number: $K(\overline{\mathbf{H}})$, with $K(\overline{\mathbf{H}}) = \max\{|\sigma_i|\}/\min\{|\sigma_i|\}$, and σ_i, $i = 1, \ldots, n-1$ the eigenvalues of $\overline{\mathbf{H}}$, the more likely the CGS method will produce biased results. The influence of the conditioning of $\overline{\mathbf{H}}$ on the accuracy of the CGS iteration is analyzed using parameters δ_i

$$\begin{array}{rclllll} \delta_i &=& \delta_{i-1} + 0.001 & \text{for} & i = 2, \ldots, 6 & \text{and} & \delta_1 = -0.2 \\ \delta_j &=& \delta_{j+1} - 0.01 * D & \text{for} & j = 7, \ldots, 10 & \text{and} & \delta_{11} = 1.05 * D \end{array} .$$

As parameter D increases, the conditioning of matrix $\overline{\mathbf{H}}$ deteriorates. The effect of the increasing D is shown in Table 2.

D	$K(\overline{\mathbf{H}})$	$\lvert\delta_2\rvert$	$\lvert\kappa_1\rvert$	$\lvert\delta_3\rvert$	$\lvert\kappa_2\rvert$	$\lvert\delta_{10}\rvert$	$\lvert\kappa_9\rvert$	$\lvert\delta_{11}\rvert$	$\lvert\kappa_{10}\rvert$
1E0	1E3	.199	.1990	.198	.1980	1.040	1.040	1.050	1.050
1E1	1E4		.1991		.1981	10.40	10.40	10.50	10.50
1E2	1E5		.1993		.1978	104.0	105.2	105.0	175.4

Table 2. the accuracy for ill-conditioned problems

The results indicate that a problem, or its associated Hessian $\overline{\mathbf{H}}$, needs to be severely ill-conditioned before aberrant curvature values are produced. Further tests show that the CGS iteration is insensitive to the choice of the initial conjugacy direction $\mathbf{p}_1$, provided the condition of $\overline{\mathbf{H}}$ is within acceptable limits.

The dimensionality n of the problem is also expected to affect the accuracy of the CGS iteration. The iteration is indeed a sequential procedure in which numerical approximation errors will propagate [2]. The influence of the dimension is demonstrated using the parameter set

$$\begin{array}{lllll} \delta_i & = & \delta_{i-1} + 0.001 & \text{for} \quad i = 2, \ldots, 6 & \text{and} \quad \delta_1 = -0.2 \\ \delta_i & = & 1.0 & \text{for} \quad i = 6, \ldots n-5 & \\ \delta_j & = & \delta_{j+1} - 0.1 & \text{for} \quad j = n-4, \ldots, n-1 & \text{and} \quad \delta_n = 10.5 \end{array}$$

n	$\lvert\delta_2\rvert$	$\lvert\kappa_1\rvert$	$\lvert\delta_3\rvert$	$\lvert\kappa_2\rvert$	$\lvert\delta_{n-1}\rvert$	$\lvert\kappa_{n-2}\rvert$	$\lvert\delta_n\rvert$	$\lvert\kappa_{n-1}\rvert$
11	.199	.1990	.198	.1980	10.4	10.40	10.5	10.50
21		.1990		.1980		10.40		10.50
51		.1990		.1980		10.40		10.50
101		.1990		.1980		10.40		10.50

Table 3. the accuracy for increasing dimension n

The results in Table 3 show that the eigenvalues / curvatures predicted by the CGS iteration are almost insensitive to the dimensionality n of the problem, provided $\overline{\mathbf{H}}$ is reasonably conditioned. The previous analyses demonstrate that the CGS method is an accurate and efficient method for predicting the curvatures of the limit state surface. The applicability of the method depends on the conditioning of the problem. This is typical for any optimization-based Hessian reconstruction scheme. The condition of $\overline{\mathbf{H}}$ also influences the performance and the accuracy of the optimization algorithms themselves [2]. Nevertheless, efficient scaling heuristics [1, 2] can favorably influence the condition of the problem.

5.2 The uncertain primary-secondary system

In engineering structures, systems consisting of light, secondary components connected to heavier primary structures are frequently encountered (e.g. piping in a building, dashboard in a car). The two-degree-of-freedom primary-secondary system, provides a relevant model for such coupled systems [28]. The system is characterized by masses m_p and m_s, spring stiffnesses k_p and k_s, natural frequencies $\omega_p = (k_p/m_p)^{1/2}$ and $\omega_s = (k_s/m_s)^{1/2}$, and damping ratios ζ_p and ζ_s. Subscripts p and s refer to the primary and secondary oscillators, respectively. Igusa and Der Kiureghian [28] show that the mean-square relative displacement response of the secondary system to a white-noise base excitation is given by

$$\mathrm{E}(x_s^2) = \frac{\pi S_0}{4\zeta_s\omega_s}\left[\frac{\zeta_a\zeta_s}{\zeta_p\zeta_s(4\zeta_a^2 + \theta^2) + \gamma\zeta_a^2} \frac{(\zeta_p\omega_p^3 + \zeta_s\omega_s^3)\omega_p}{4\zeta_a\omega_a^4}\right] \quad , \tag{25}$$

with S_0 the intensity of the white-noise, $\gamma = m_p/m_s$ the mass ratio, $\omega_a = (\omega_p + \omega_s)/2$ and $\zeta_a = (\zeta_p + \zeta_s)/2$ the average frequency and damping ratio, respectively, and with $\theta = (\omega_p - \omega_s)/\omega_a$ the tuning parameter. Tuning and other interaction effects make the peak response of the simplified primary-secondary system strongly sensitive to the dynamic properties of its constituent components. The influence of uncertainties on the latter properties on the reliability of the second oscillator is investigated. Der Kiureghian and De Stefano [12] consider the limit state function

$$g(F_s, S_0, m_p, m_s, k_p, k_s, \zeta_p, \zeta_s) = F_s - k_s p \left[\mathrm{E}(x_s^2))\right]^{1/2} \quad , \qquad (26)$$

where p denotes the deterministic peak factor $p = 3$. The primary-secondary system fails when the differential peak displacement of the secondary oscillator induces a force, exceeding the (random) force capacity F_s of the spring. A mechanical model (25) relates the response $E(x_s^2)$ to the basic random variables $\underline{X} = [F_s, S_0, m_p, m_s, k_p, k_s, \zeta_p, \zeta_s]^T$. The stochastic characteristics [12] of the independent basic random variables are given in Table 4. The log normal variables are transformed to the standard-normal domain.

VAR	distribution	mean	COV	VAR	distribution	mean	COV
F_s	log normal	100.	0.1	S_0	log normal	15.0	0.1
m_p	log normal	1.00	0.1	m_s	log normal	0.01	0.1
k_p	log normal	1.00	0.2	k_s	log normal	0.01	0.2
ζ_p	log normal	0.05	0.4	ζ_s	log normal	0.02	0.5

Table 4. Stochastic characteristics of the primary-secondary system

Both HL-RF and the implemented SQP method find the same PML (or failure point). Using the modified HL-RF iteration algorithm including an Armijo-type line search [1, 2] on a merit function [13, 14], 38 iterations are necessary to converge to the optimal point. For an identical choice of initial values and convergence criterion, the implemented SQP only takes 10 steps to converge to the PML. Since HL-RF and SQP are similar, this faster convergence has to be mainly credited to a better choice of the merit function and of the line search algorithm, given in equations (16) and (17). The substantial gain in computation time justifies the implementation of the best possible step length control algorithm.

Curvature values obtained by ICM and the CGS method are listed in Table 5. The exact reference values are reported also reproduced [12]. Only the three main curvature values are listed. The remaining curvatures are approximately zero. The values listed in column $k_{\kappa_i}^{\mathrm{ICM}}$ refer to the number of iterations the ICM effectively needs to determine the particular curvature κ_i. Column $k_{\mathrm{tot}}^{\mathrm{CGS}}$ reproduces the total number of iterations necessary to determine all principal curvatures. Detailing the number of iterations in function of the curvatures is not relevant. The mean-value iterations of the CGS method are performed on a arbitrary conjugate set, other than the principal directions of the limit state surface.

CURV	EXACT	ICM	$k_{\kappa_i}^{ICM}$	CGS	k_{tot}^{CGS}
κ_1	8.132	8.134	76	8.134	53
κ_2	0.178	0.178	11	0.178	-
κ_3	0.0270	0.0269	11	0.0270	-

Table 5. curvature of the limit state surface

As is to be expected, both the ICM and the CGS method determine the curvatures of the limit state surface exactly. The ICM needs 78 HL-RF iteration steps to converge to the first principal curvature. Even if the steps necessary to converge to the PML are subtracted, 40 iterations are needed to obtain convergence to κ_1. This clearly indicates that the limit state surface is everything but quadratic, except in a very small neighborhood of the PML. The convergence of the ICM to the principal curvatures of the limit state is strongly tributary to the quadratic behavior of the limit state surface. The ICM has to comply to two convergence criteria. First, it must have reached the area in which the limit state is approximately quadratic. Only then will convergence of the mean-value iteration be possible. The CGS method is less affected by this non-quadratic behavior, since it applies mean-value iterations with the PML as its base. Nevertheless, the average number of iterations necessary to determine the value of the Hessian of the Lagrangian along each conjugate direction is quite large ($\overline{k}_{\mathbf{p}_i} \approx 8$).

Notwithstanding the previous remark, the ICM has an interesting advantage. Since it reconstructs the reduced Hessian of the limit state starting with the largest eigenvalue (in absolute value), its iterations can be stopped the moment a zero curvature value is obtained. The CGS approach does not allow for partial reconstruction. On the other hand, partial reconstruction is only useful in a standard-normal context, since zero principal curvatures have no contribution to the SORM estimator (equation 4). This is not the case when the structural reliability analysis is performed in the original domain of basic variables. Even if a specific CGS reconstruction sequence could be identified, it is impossible to define a criterion to stop the CGS iteration.

6 Summary and Conclusions

The search for the failure points, i.e. the points of maximum likelihood in the standard-normal domain, is often considered to be a specific structural reliability problem. Even though it is identified as a nonlinear optimization problem, little reference is made to the mathematical programming theory. Moreover, the commonly used (modified) HL-RF algorithm method remains restricted to structural reliability analysis in the standard normal domain.

In this paper, the most general formulation of the mathematical program, identifying the points of maximum likelihood, is given. It is independent of the distribution of the basic variables. The standard-normal optimization problem is shown to be a special case of the general structural reliability program. Sequential quadratic programming, in conjunction with the Powell-BFGS quasi-Newton update and the line search on an augmented Lagrangian merit function, is generally accepted to be the most successful optimization method for solving nonlinear programming problems with a nonlinear equality constraints. As such, it is perfectly suited to solve the general structural reliability program. Successive solution of an approximating quadratic program with linear equality constraint produces a sequence of ascent directions for the original problem. Line search along the ascent direction is essential for fast convergence of the optimization algorithm. The (modified) HL-RF method is a basic implementation of a sequential quadratic approximation procedure.

Efficient methods for structural reliability computation not only require the point of maximum likelihood (PML), they additionally use second-order derivative information of the log likelihood and the limit state function at the optimal point. The conjugate Gram-Schmidt (CGS) iteration is introduced. It constructs a set of vectors, conjugate to the Hessian of the Lagrangian function at the optimum, by elementary matrix updates and repeated application of the mean-value theorem. The Hessian of the limit state function is subsequently found by simple linear algebra. The CGS iteration is an extension of the iterative curvature method, built around the HL-RF optimization algorithm. Its mechanism does not rely on any quadratic assumption. It is therefore applicable to non standard-normal problems with nonlinear limit states. No restart of the iterative optimization is necessary. Given that any optimization based reconstruction of the Hessian of the Lagrangian invariably uses the conjugacy principle, it is the most general Hessian reconstruction method available. Implementing the proposed conjugacy iteration requires minimal coding effort, since it makes profitable use of the optimization modules. The method compares favorably with a finite difference approach as far as computational expenses are concerned. Benchmarks prove that the CGS iteration is remarkably accurate, if the Hessian of the Lagrangian is reasonably conditioned.

References

1. P.E. Gill, W. Murray, and M.H. Wright. *Practical Optimization.* Academic Press, New York, USA, 1981.
2. D.G. Luenberger. *Linear and Nonlinear Programming.* Addison-Wesley, Reading, Massachusetts, USA, second edition, 1984.
3. K. Breitung. Probability approximations by loglikelihood maximization. *Journal of Engineering Mechanics, ASCE,* 117(3):457–477, 1991.

4. K. Breitung. Asymptotic approximation for multinormal integrals. *Journal of Engineering Mechanics, ASCE*, 110(3):357–366, 1984.
5. M.A. Maes, K. Breitung, and D.J. Dupuis. Asymptotic importance sampling. *Structural Safety*, 12:167–186, 1993.
6. Ph. Geyskens. *Optimization and Quadrature Algorithms for Structural Reliability Analaysis in the Original Variable Domain.* PhD thesis, Department of Civil Engineering, Katholieke Universiteit Leuven, Leuven, Belgium, 1993.
7. M. Shinozuka. Basic analysis of structural safety. *Journal of Structural Engineering, ASCE*, 109(3):721–740, 1983.
8. M. Hohenbichler and R. Rackwitz. Non-normal dependent vectors in structural safety. *Journal of the Engineering Mechanics Division, ASCE*, 107(6):1227–1238, 1981.
9. W. Murray. Linear algebra and optimization. Lecture Notes, Stanford University, Stanford, CA, 1993.
10. A.M. Hasofer and N.C. Lind. Exact and invariant second-moment code format. *Journal of the Engineering Mechanics Division, ASCE*, 100(1):111–121, 1974.
11. R. Rackwitz and B. Fießler. Structural reliability under combined random load sequences. *Computers & Structures*, 9:489–494, 1978.
12. A. Der Kiureghian and M. De Stefano. Efficient algorithm for second-order reliability analysis. *Journal of Engineering Mechanics, ASCE*, 117(12):2904–2923, 1991.
13. P-L. Liu and A.Der Kiureghian. Optimization algorithms for structural reliability analysis. Technical Report UCB/SEMM-86/09, Department of Civil Engineering, University of California at Berkeley, Berkeley, California, USA, 1986.
14. P-L. Liu and A. Der Kiureghian. Optimization algorithms for structural reliability. *Structural Safety*, 9(3):161–177, 1991.
15. T. Abdo and R. Rackwitz. A new beta-points algorithm for large time-invariant and time-variant reliability problems. In A. Der Kiureghian A. and P. Thoft-Christensen, editors, *Reliability and Optimization of Structural Systems '90, Proceedings of the 3rd IFIP WG7.5 Conference, Berkeley, CA*, pages 1–11, Berlin, Germany, 1990. Springer-Verlag.
16. P.E. Gill and W. Murray. Numerically stable methods for quadratic programming. *Mathematical Programming*, 14:349–372, 1978.
17. G.H. Golub and C.F. Van Loan. *Matrix Computations.* The John Hopkins University Press, Baltimore, Maryland, USA, second edition, 1989.
18. J.E. Dennis and J.J. Moré. Quasi-newton methods, motivation and theory. *SIAM Review*, 19(1):46–89, 1977.
19. M.J.D. Powell. *The Convergence of Variable Metric Methods for Nonlinearly Constrained Optimization Problems*, pages 27–63. Academic Press, New York, USA, 1978.
20. L.C.W. Dixon. *The Choice of Step Length, a Crucial Factor in the Performance of Variable Metric Algorithms*, chapter 10, pages 149–170. Academic Press, London, UK, 1972.
21. K. Schittkowski. The nonlinear programming method of wilson, han and powell with an augmented lagrangian type line search function - part i: Convergence analysis. *Numerische Mathematic*, 38:83–114, 1981.

22. K. Schittkowski. On the convergence of a sequential quadratic programming method with an augmented lagrangian line search function. *Mathematische Operationsforschung und Statistic, series Optimization*, 14(2):197–216, 1983.
23. R.P. Brent. *Algorithms for Minimization without Derivatives*. Prentice-Hall, Englewood Cliffs, New Jersey, USA, 1973.
24. D.P. Bertsekas. *Constrained Optimization and Lagrange Multiplier Methods*. Academic Press, New York, USA, 1982.
25. P.T. Boggs, J.W. Tolle, and P. Wang. On the local convergence of quasinewton methods for constrained optimization. *SIAM Journal of Control and Optimization*, 20(2):161–171, 1982.
26. P.T. Boggs and J.W. Tolle. Convergence propoerties of a class of rank-two updates. Technical Report UNC/OR TR-90/16, Department of Operations Research, University of North Carolina at Chapel Hill, Chapel Hill, North Carolina, USA, 1992.
27. M.R. Hestenes. *Conjugate Directions Methods in Optimization*. Springer-Verlag, New York, USA, 1980.
28. T. Igusa and A. Der Kiureghian. Dynamic characterization of two-degree-of-freedom equipment-structure systems. *Journal of Engineering Mechanics, ASCE*, 111(1):1–19, 1985.

Response Functions and System Reliability of Bridges

Michel Ghosn [1], Fred Moses [2] and Nikhil Khedekar [3]

[1] The City College of the City University of New York, NY 10031.

[2] Department of Civil Engineering, The University of Pittsburgh, PA 15261.

[3] Department of Civil Engineering, Case Western Reserve University, Cleveland, OH 44106.

Abstract. Redundancy of a structural system is defined as the capability of the system to carry additional loads after the failure of one or more of its components. Research studies to include redundancy in the U.S. bridge design and evaluation codes are underway. These efforts are based on calibrating system factors that could be added to the checking equations. The purpose of these system factors is to "reward" redundant designs by allowing savings in required member capacities, and "penalize" nonredundant designs by requiring their members to be more conservative. The calibration process uses a target redundancy index as the safety criterion that bridges should satisfy. The redundancy index is defined as the difference between the reliability index of the bridge system and the reliability index of the members. This paper illustrates the method proposed to calculate the redundancy indices of typical bridges. The member safety index is calculated, as traditionally done, using the results of an elastic analysis of the bridge. The system reliability index accounts for the nonlinear behavior of the bridge system and the redistribution of the applied loads after the failure of a member. In this paper, the response function (response surface) method is used to calculate the system safety indices of typical multi-beam steel and prestressed concrete bridges. System factors are then calibrated to reflect the level of redundancy that different bridge configurations possess.

Keywords. Redundancy, bridges, system reliability, response surface, response function, system factor, code calibration

1 Introduction

Current bridge specifications deal with individual components ignoring the global behavior due to the interlocking of these components into one structural system. This paper proposes a method to bridge the gap between a component by component design and the system effect. This goal is achieved by introducing system factors that can be used to modify the member capacities of a system as a function of its level of redundancy. These system factors provide a method to insure that highway bridges will provide a minimum level of system safety when the bridge is intact or after a component's sudden failure.

A bridge system is safe if: a) It provides a reasonable safety level against first member failure. b) It does not produce large deformations under regular traffic conditions. c) The risk that it will reach its ultimate system capacity under extreme loading conditions is very small. d) It is able to carry some traffic loads after damage or the loss of a main load-carrying member. Therefore, the limit states that should be checked to insure adequate bridge redundancy and system safety are defined as:

a) Member failure This is a traditional check of individual member safety using elastic analysis and the capacity of individual members.

b) Serviceability limit state This is defined as a maximum live load displacement in a main longitudinal member equal to the span length/200 accounting for the nonlinear behavior of the bridge system.

c) Ultimate limit state This is defined as the ultimate capacity of the bridge system or the formation of a collapse mechanism for bridges that have high levels of member ductilities such as steel bridges with a positive bending failure mode. For other types of bridges, the ultimate limit state is defined in terms of the loads that will cause heavy damage rendering the bridge unsafe for truck traffic.

d) Damaged condition limit state This is defined as the ultimate capacity of the bridge system after the complete removal of one main load carrying component from the structural model.

1.1 Definition of Redundancy

Redundancy is defined as the capability of a bridge structure to continue to carry load after the failure of one of its members. A comparison between the safety index of one member and those of the safety indices for the ultimate capacity, the serviceability limit state, and the damaged condition will then provide a measure of bridge redundancy. A redundancy index is defined as the difference between the reliability index of the bridge system and the reliability index of the members. The three redundancy indices corresponding to the three system limit states b,c, and d are defined as

$$\rho_u = \beta_{ult.} - \beta_{member},$$
$$\rho_s = \beta_{serv.} - \beta_{member},$$
$$\rho_d = \beta_{damaged} - \beta_{member}. \qquad (1)$$

The redundancy indices give measures of the additional safety provided by the complete bridge system compared to the nominal safety obtained when a traditional check of individual members is performed. A reliability analysis was performed for a number of typical bridge configurations for steel, prestressed concrete, and reinforced concrete bridges. The analysis revealed that bridge systems that are known to have redundant configurations produced a redundancy index ρ_u equal to 1.0 or higher; ρ_s equal to -1.0 or higher; and ρ_d equal to -0.5 or higher. These values are used as the target values that bridge systems should satisfy to insure adequate levels of redundancy.[1]

Bridges with redundancy indices below the minimum target values should be penalized by requiring that their members be more conservatively designed than required by current standards. On the other hand, bridges that satisfy the redundancy index requirement will be rewarded by permitting less conservative designs. This could be achieved through the application of system factors which are proportional to the level of compliance with the recommended criteria.

To calculate the system safety indices and subsequently the redundancy indices of highly nonlinear structural systems, the most efficient procedure consists of using the response function (surface) method. The next section explains how this method is applied in practice to calculate the redundancy of prestressed concrete and steel multi-girder bridges.

2 Reliability Model for Highway Bridge Systems

The reliability of a structural system can be easily calculated when the functional relationship between member strengths and the response of the system is available for all possible failure modes. For complex structures such as bridge systems with highly nonlinear behavior that may involve member unloading, finding a closed form functional relationship is impractical. In this case, numerical simulations provide the most reasonable approaches. One efficient numerical simulation technique that can be used to calculate the reliability of bridge systems is the response function method. [2,3]

The method consists of first developing a nonlinear finite element F.E. program to model the deterministic behavior of bridges. The results of the deterministic F.E. analysis at predetermined realizations of the random variables are used in a regression analysis to obtain a response function. The response function gives a relationship between the response of the bridge and the random variables which for bridge systems include the member resistances and the dead load.

The incremental F.E. analysis is performed under the effect of loads simulating the configuration of typical trucks. In this paper, the representative trucks are chosen to have the configuration of the AASHTO HS-20 trucks[4]. The loads of these vehicles are then incremented until the limit states are reached. The response is expressed in terms of the load factor (LF) that will cause the violation of the different limit states. LF gives the multiplier of the AASHTO HS-20 vehicle that will produce system failure. Two limit states are recognized for intact bridges: The ultimate capacity and the serviceability limit. In addition, to insure the safety of the public after possible damage to a bridge member due to fatigue failure or a brittle loss in a member's capacity, bridges should be able to sustain these damages and still operate albeit at reduced capacity. Therefore, the reliability of bridges under damaged conditions should also be investigated.

The ultimate capacity limit is defined as the maximum possible truck load that can be applied on the structure before it collapses and/or permanent heavy damage ensues. In these calculations ultimate limit is reached when one of the following conditions occurs: a) A mechanism is formed, and the stiffness matrix becomes unstable. b) A very large displacement is obtained for a very small increment in load. c) A predetermined maximum plastic hinge rotation in a main longitudinal member is reached. At this point the member completely unloads, and the bridge is assumed to have reached its ultimate capacity. If a longitudinal member reaches its maximum plastic hinge rotation, major damage to the member is normally observed, and the member is no longer capable of carrying any load. Under certain circumstances some bridges may still be able to carry some additional load after the first member reaches its maximum plastic hinge rotation, however, the excessive visible damage that occurs when the maximum plastic hinge rotation is reached renders the bridge non-operational. Therefore, the maximum plastic hinge rotation is used in this study as one of the criteria to define the ultimate system capacity.

The response function gives the relationship between the response of the bridge, the member capacities, and the dead load. In addition to carrying its dead load, any bridge should be able to carry the maximum truck load expected to be applied on it in its lifetime without reaching its ultimate capacity. This maximum expected lifetime load is a function of the number of trucks that cross the bridge, the positions of the trucks on the bridge deck, the weight of the trucks, the distribution of the weight to the individual axles, and the axle configuration. In addition, the load will be a function of the dynamic impact. The design life of a bridge is normally equal to 75 years. Therefore, a 75-year exposure period is used herein for both the ultimate limit state and the first member failure limit state. In the latter case, the 75-year exposure period is used to be consistent with the basis for current code development efforts.[5] The maximum expected lifetime load can be expressed in terms of equivalent AASHTO HS-20 truck loads. For example, LL_{75} is defined as the number of HS-20 vehicles needed to produce the same load effect as the maximum expected lifetime load. Table 2.1 gives expected LL_{75} values and their coefficients of variation (COV) for simple span bridges of different span lengths for two-lane loadings. The mean values shown in table 2.1 are based on the work of Nowak.[6]

Regular traffic conditions are defined as recurrent loads expected to be regularly applied on the bridge. To determine the loads expected under regular traffic conditions, a two-year exposure period is used. The two-year exposure period is chosen because it corresponds to the biennial mandatory bridge inspection period. Thus, a two-year exposure period is used for the serviceability condition to reflect the fact that if a bridge has serviceability problems, these will be noticed during the bridge inspection process. Similarly, the two-year exposure period is used for the damaged bridge condition because, even if a damage (such as a fatigue fracture) goes unnoticed for a short period of time, it is bound to be discovered during inspection. LL_2 is defined as the number of AASHTO's HS-20 trucks needed to produce the same load effect as that expected under regular truck traffic conditions. Table 2.1 also gives the expected two-year load factors expressed as multipliers of AASHTO's HS-20 loads and their coefficients of variation.

Table 2.1. Mean and COV of applied loads as function of AASHTO load effect

Span length (ft)	LL_{75}	LL_2	COV
45	3.33	3.00	25%
60	3.43	3.09	23%
80	3.61	3.26	20%
100	4.07	3.44	20%

The reliability calculations are performed using a first order second moment (FORM) program based on the response function obtained from the nonlinear F.E. analysis and the live load factors. The safety margin can be expressed as

$$Z = f(R_i, D) - LL ,$$
$$Z = LF - LL , \tag{2}$$

where f () is the response function expressed in terms of the different member capacities R_i and the dead load D. LF is the load factor obtained from the nonlinear analysis, and LL is the applied live load factor.

The response function f() is obtained using a regression fit through the response obtained from the F.E. analysis at specific realizations of the random variables R_i and D. In a first stage, the realizations are chosen at points close to the mean values or close to the nominal (or design) values of the random variables R_i and D. FORM will provide a first estimate of the reliability index and values of R_i and D at the "failure point". To improve on the first estimate of the reliability index, the response function f() is expanded at a new set of realizations close to the failure point. This process can be repeated until convergence.

3 Nonlinear Behavior of Bridge Systems

A program for the nonlinear analysis of bridges (NONBAN) was developed to construct the response function by performing deterministic analyses of the bridge structural system. NONBAN uses a grillage analysis to model the main longitudinal and transverse members of the bridge. The concrete deck slab is modeled as transverse beam elements. The nonlinear behavior and the ductility of the concrete and steel members are modeled using moment versus plastic rotation curves. [1]

For concrete members, moment versus rotation curves are obtained using a combination of analytical and empirical methods according to the following steps: a) Moment versus curvature relationships are plotted for every concrete section using each section's geometry and its material properties. b) Given the moment-curvature plots and assuming that the plastic curvature is constant over the length of the plastic hinge, moment versus rotation curves are obtained by multiplying the plastic curvature by the length of the plastic hinge. c) The maximum plastic hinge rotation that each section can sustain is obtained using Naaman's equations.[7] When the plastic rotation of a section reaches this maximum value, the section is assumed to break and will not be able to carry load producing sudden and complete unloading. These steps provide a description of the nominal behavior of prestressed concrete members. Nowak [6] found that the bias in estimating the moment capacity of prestressed members is equal to 1.05 with a coefficient of variation equal to 7.5%. The estimate of the maximum plastic hinge rotation has a bias equal to 1.4 and a COV equal to 15%. Fig. 3.1 shows a typical moment versus rotation curve for a prestressed concrete member.

Considerable research on the inelastic behavior and the ductility of steel and composite bridge members took place since the mid-1960's. Schilling [8] summarizes the results of experimental studies on steel ductilities. These studies resulted in a number of moment-rotation curves that give the relationship between the applied moment and the plastic rotation of simple and continuous steel beams. Schilling's curves give the total plastic rotation over the finite length in which yielding occurs. Thus, the total plastic rotation is assumed to be concentrated in a small region representing an angular discontinuity, and the rest of the girder is otherwise assumed to be elastic. The curves that Schilling proposed were derived from experimental data, and they account for: Steel yielding including the effect of residual stresses; the spread of the yielding along the length as the loading progresses; cracking or local crushing of the slab; permanent distortion of the cross sectional shape; and any other factors that cause permanent rotations. Different curves are proposed to describe the nominal behavior of composite and non-composite, compact and non-compact sections in positive and negative bending. The bias in estimating the capacity of steel members is 1.08 with a COV of 14%.[6] Fig. 3.2 illustrates the typical behavior of steel bridge members represented by their moment versus rotation curves.

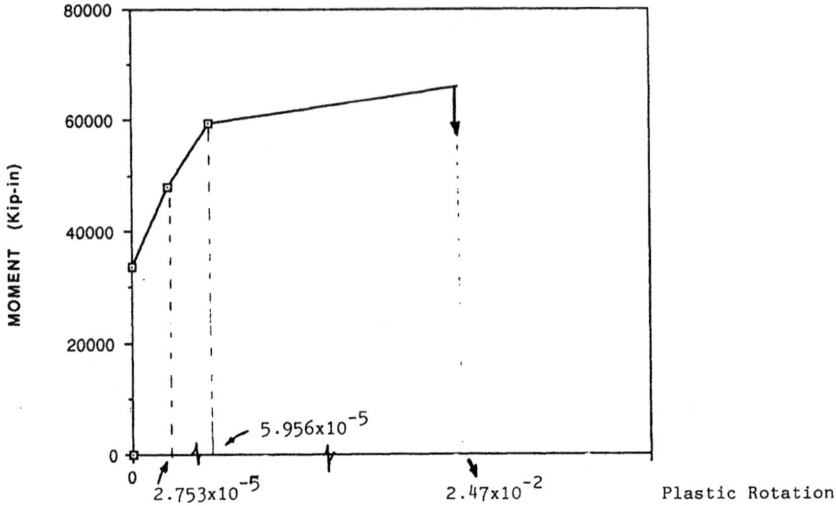

Fig. 3.1. Typical moment vs. plastic rotation curve for prestressed concrete girders

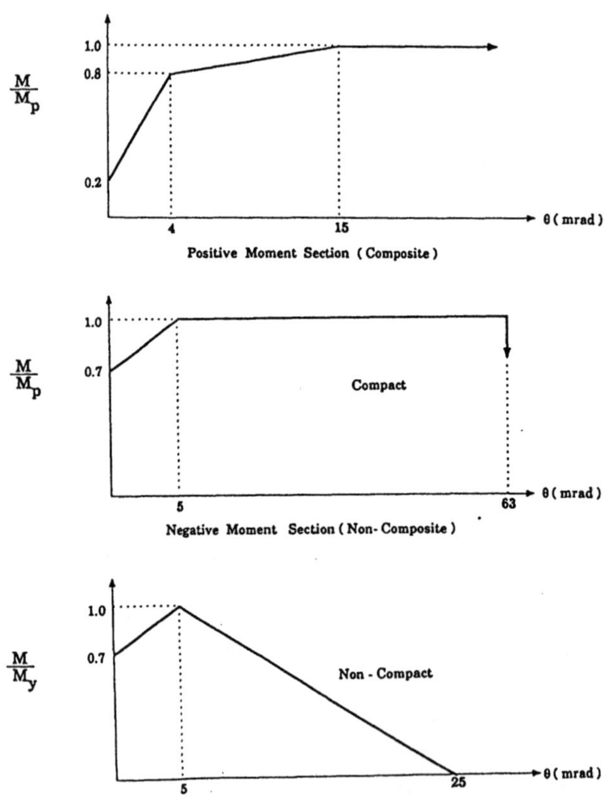

Fig. 3.2. Typical moment versus plastic rotation curves for steel girders

4 Redundancy of a Prestressed Concrete Bridge

The cross section of the six-girder simply-supported prestressed concrete bridge is shown in Fig. 4.1. The concrete deck is 7 in thick. This bridge was designed to exactly satisfy AASHTO's LFD criteria. The longitudinal members which are at 8 ft center to center are type IV AASHTO girders with 4.80 in^2 of grade 270 prestressing steel at an effective depth of 57 in. from the top of the slab. The effective prestressing force is equal to 726 kips. The section's concrete strength is 5000 psi while the slab's strength is 3000 psi. The moment versus rotation curve for this section was shown in Fig. 3.1. The nominal value for the maximum plastic hinge rotation obtained is $\theta_{p\ max}$ = 0.0247 rad. This assumes that the length of the plastic hinge L_p is 43.8 in.

In a first stage, a linear elastic analysis is performed for two AASHTO HS-20 trucks applied on the bridge as shown in Fig. 4.1. The total moment effect of the two HS-20 trucks is 3050 kip-ft. The most heavily loaded member is the external girder G1. G1 carries a live load $L_{HS\text{-}20}$ = 945 kip-ft constituting 31% of the total live load. The dead load moment (D) was found be 1970 kip-ft. Using the results of the elastic analysis, the safety margin against the failure of the most heavily loaded member is given by

$$Z = R_1 - D - L_{1,75} \ , \tag{3}$$

where R_1 is the member capacity of girder G1 with a nominal value equal to 5900 kip-ft, a bias equal to 1.05, and a COV equal to 7.5%. D is the dead load effect with a nominal value of 1970 kip-ft, a bias equal to 1.05, and a COV of 9%. $L_{1,75}$ is the effect of the maximum expected truck loads in the 75 year lifespan on the most heavily loaded member. This produces a mean load effect $L_{1,75}$ equal to 1788 kip-ft with a COV equal to 20%. Assuming that all the variables follow lognormal distributions, the reliability analysis produced a β_{member} equal to 3.60.

The incremental nonlinear analysis performed at the nominal values produces a plot of load factors versus displacements as shown in Fig. 4.2. The calculations shown were performed at the nominal values of the resistances, the dead load, and the maximum plastic hinge rotation. A maximum deflection of 6.00 in., constituting the span length/200, was obtained when the load factor was 3.94. This load factor is defined as LF_s. The load was increased until one girder reached a plastic hinge rotation equal to the maximum allowable rotation of 0.0247 rad. The girder that reached this maximum plastic rotation first was the external girder G1. At this point the bridge would undergo extensive damage and it is herein assumed that the ultimate capacity is reached. The load factor at which this maximum plastic hinge rotation was reached is defined as LF_u. For this structure LF_u was found to be equal to 6.42. Several similar incremental analyses were also performed after varying one random variable at a time by 10% from its nominal value. The values of LF_u and LF_s obtained for the various cases considered are shown in table 4.1.

Table 4.1. Sensitivity of results of incremental analysis to variation in input data

Realizations	LF_u	LF_s
Nominal Values	6.42	3.94
R_1 + 10%	6.61	4.19
R_1 - 10%	6.17	3.75
R_2 + 10%	6.59	4.13
R_2 - 10%	6.20	3.81
R_3 + 10%	6.56	4.07
R_3 - 10%	6.24	3.86
R_4 + 10%	6.51	4.02
R_4 - 10%	6.30	3.90
R_5 + 10%	6.49	3.97
R_5 - 10%	6.32	3.93
R_6 + 10%	6.47	3.94
R_6 - 10%	6.36	3.94
$\theta_{p\ max}$ + 10%	6.62	3.94
$\theta_{p\ max}$ - 10%	6.20	3.94
D + 10%	6.11	3.72
D - 10%	6.69	4.16

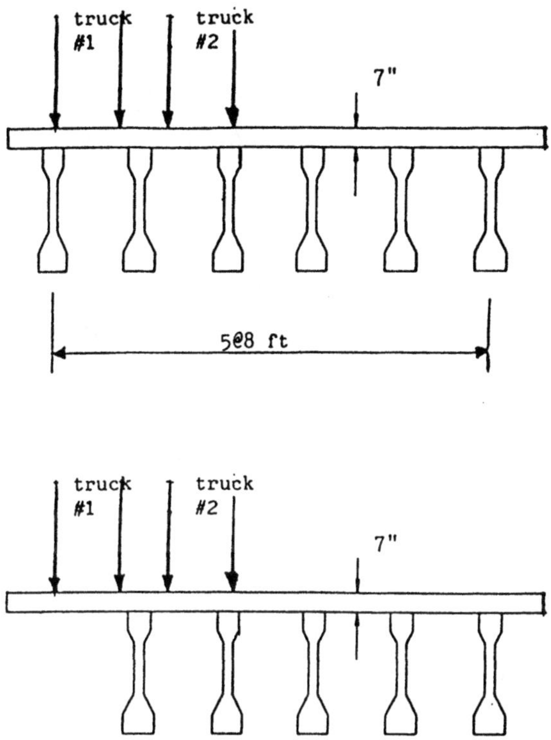

Fig. 4.1. Layout of prestressed concrete bridge

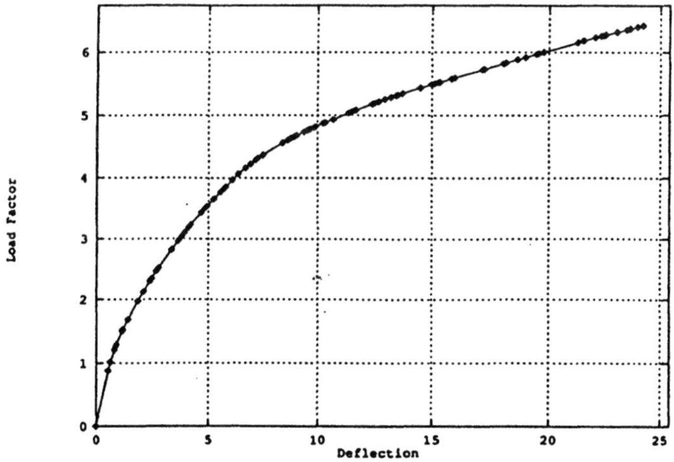

Fig. 4.2. Load factor versus deflection plot for prestressed concrete bridge

The linear fit produced the following equations:

$$LF_u = -1.0 + 31.01\times10^{-6} R_1 + 27.28\times10^{-6} R_2 + 22.29\times10^{-6} R_3 + 15.19\times10^{-6} R_4 + 11.60\times10^{-6} R_5 + 7.95\times10^{-6} R_6 + 85\, \theta_{p\,max} - 123.48\times10^{-6} D \quad , \qquad (4)$$

$$LF_s = 0.50 + 30.94\times10^{-6} R_1 + 22.50\times10^{-6} R_2 + 14.77\times10^{-6} R_3 + 8.44\times10^{-6} R_4 + 2.81\times10^{-6} R_5 + 0.0\, R_6 + 0.0\, \theta_{p\,max} - 93.19\times10^{-6} D \quad , \qquad (5)$$

where R_i are the resistances of members given in kip-in. D is the maximum dead load on one member given in kip-in, and $\theta_{p\,max}$ is the maximum plastic hinge rotation that the member can sustain given in radians. The system safety index obtained for the ultimate limit state is 6.01 assuming full correlation between the members' resistances and 6.56 assuming that the members are uncorrelated. The calculations for the ultimate limit state assume a maximum load corresponding to a 75 year exposure period. The reliability calculation for the serviceability limit state used a two-year exposure period and produced system reliability indices of 3.79 for correlated member resistances and 4.20 for uncorrelated resistances. The most likely failure points obtained from the reliability calculations for the four cases considered are shown in table 4.2.

Table 4.2. Most likely failure points for prestressed concrete bridge

Variable	Ultimate R correlated	Ultimate R uncorrelated	Serviceability R correlated	Serviceability R uncorrelated
R_1	6.08×10^4	7.00×10^4	6.43×10^4	6.97×10^4
R_2	6.08×10^4	7.04×10^4	6.43×10^4	7.09×10^4
R_3	6.08×10^4	7.11×10^4	6.43×10^4	7.20×10^4
R_4	6.08×10^4	7.21×10^4	6.43×10^4	7.30×10^4
R_5	6.08×10^4	7.27×10^4	6.43×10^4	7.40×10^4
R_6	6.08×10^4	7.32×10^4	6.43×10^4	7.47×10^4
D	2.86×10^4	2.86×10^4	2.74×10^4	2.77×10^4
$\theta_{p\,max}$	2.64×10^{-2}	2.65×10^{-2}	3.45×10^{-2}	3.46×10^{-2}

Repeating the process by expanding the response function at the failure points rather than the points corresponding to the nominal values, produced a system safety index for the ultimate limit state equal to 6.13 assuming full correlation between the members' resistances and 6.86 assuming that the members are uncorrelated. On the third iteration the uncorrelated condition produced a safety index of 6.94. On the second iteration, the reliability calculation for the serviceability limit state produced system reliability indices of 3.80 for correlated member resistances. Since the resistances of bridge members are expected to be correlated, only the full correlation case will be considered in the remainder of this study.

To estimate the capacity of the bridge to sustain load under damaged conditions, the same incremental analysis is repeated assuming that the external girder G1 was completely removed from the model. This simulates extreme damage to the girder due to severing of the prestressing strands because of fatigue failure or due to a collision by a truck, a ship, or debris. Girder G1 was chosen as the damaged girder in this example because it was the most critical member. Girder G1 was the most heavily loaded member and it was the first member to fail when performing the incremental analysis of the intact structure. Fig. 4.1 shows a cross section of the damaged model with the loading pattern used in this analysis. The critical loading pattern in this case was the same as that of the intact bridge; this however may not necessarily be the case for other bridge configurations. Different loading patterns should be checked when performing the incremental analysis. The HS-20 loads applied on the damaged bridge are incremented until the plastic hinge rotation of any remaining member reaches the limiting value of 0.0247 rad. This occurred at a load factor LF_d equal to 5.52. In this case, the member that failed was member G2. Several similar incremental analyses were also performed after varying one random variable at a time by 10% from its nominal value. The linear fit produced the following equation

$$\begin{aligned} LF_d = 0.09 &+ 29.11 \times 10^{-6} R_2 + 20.11 \times 10^{-6} R_3 + 13.22 \times 10^{-6} R_4 \\ &+ 7.45 \times 10^{-6} R_5 + 1.83 \times 10^{-6} R_6 \\ &+ 87.85\, \theta_{p\,max} - 77.93 \times 10^{-6} D \quad . \end{aligned} \tag{6}$$

The system safety index obtained for the ultimate limit state is 6.15 assuming full correlation between the members' resistances. The reliability calculation for the damaged conditions used a two-year exposure period.

Assuming full correlation between member resistances, the redundancy indices obtained for the prestressed concrete bridge are: $\rho_u = 2.53$, $\rho_s = 0.20$, and $\rho_d = 2.55$.

5 Redundancy of a Multi-Girder Steel Bridge

The second example assumes a simple span steel bridge with 4 parallel main girders. The span is 80 ft long, and the spacing between the girders is 8.33 ft. The total width of the bridge is 33.25 ft, and the slab deck is 7 in. thick. Fig. 5.1 gives the cross section and the elevation of this bridge as given in reference [9].

The bridge girders are designed to act compositely to produce a maximum plastic moment capacity M_p equal to 4791 kip-ft. The maximum dead load moment is given as 852 kip-ft per girder or a distributed load of 1.065 kip/ft. Two AASHTO HS-20 trucks are placed side-by-side as shown in Fig. 5.1. The analysis revealed that the first member will fail at a load factor equal to 5.0. The calculation of the member safety index followed the same approach used for the calculation of the safety index of prestressed concrete members. The bias associated with the resistance of steel members is 1.08, and the coefficient of variation is 14%. The bias for the dead load is assumed to be 1.0, and the coefficient of variation is 4%.[6] The linear analysis produced a safety index against the failure of the first member equal to 4.23.

Using a nonlinear structural model, the bridge is analyzed for two HS-20 trucks side-by-side in one span. The loads are incremented, and the maximum deflection of the main longitudinal girders is obtained for every load step. Fig. 5.2 shows a plot of the load factor versus the maximum deflection. A maximum deflection equal to span length/200 (6 in) is reached when the load factor is 3.78; this is defined as the load factor for the serviceability limit state LF_s. The loading is continued until a mechanism is formed and the bridge is not capable of carrying any additional load. The mechanism formed at a load factor equal to 7.90. This is defined as the load factor for the ultimate limit state LF_u. By varying one random variable at a time by 10% from its nominal value and performing a linear regression through the values obtained for LF_u and LF_s, the response function are obtained as

$$LF_u = -0.14 + 4.38 \times 10^{-4} \ (R_1 + R_2 + R_3 + R_4) - 1.40 \ D \quad , \qquad (7)$$

$$LF_s = 1.07 + 3.13 \times 10^{-4} R_1 + 3.13 \times 10^{-4} R_2 + 8.35 \times 10^{-5} R_3 + 0.0 \ R_4 - 0.467 \ D \quad , \qquad (8)$$

where the R_i's are the nominal resistances in kip-ft, and D is the dead load in kip/ft.

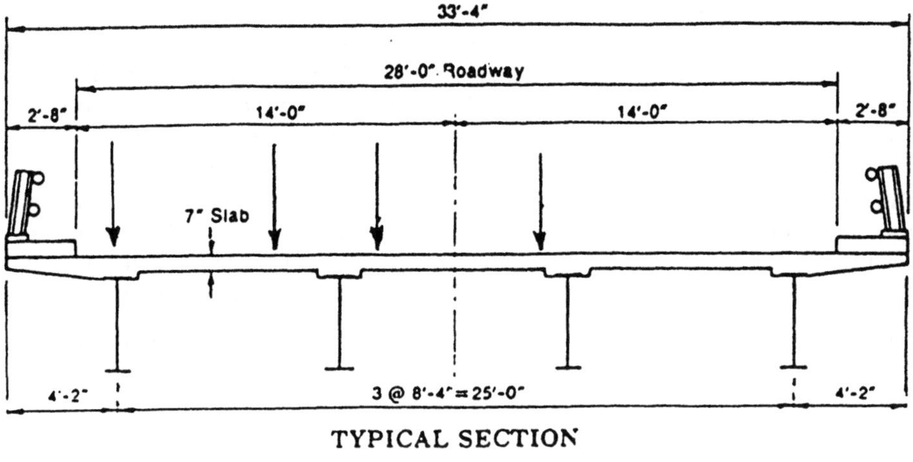

Fig. 5.1. Layout of steel multi-girder bridge

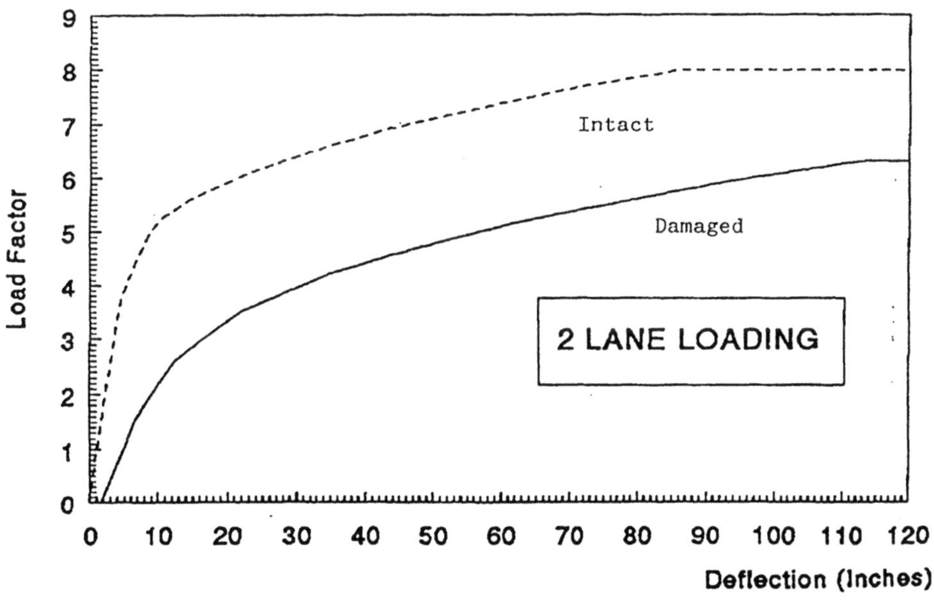

Fig. 5.2. Load factor versus deflection for steel bridge

To simulate possible damage to the most heavily loaded member G1, a 14 ft girder section is completely removed from the structural model. The section is removed from the middle of the span at the point of maximum positive bending. The incremental analysis of the damaged bridge is repeated with the same HS-20 vehicles positioned immediately on top of the removed section. This loading pattern constitutes the most critical loading for the assumed damaged condition. The incremental analysis of the damaged bridge produced a plot of load factors versus deflections as shown in Fig. 5.2. A mechanism forms when the load factor reaches the value of 6.15. This load factor obtained for the damaged bridge condition is defined as LF_d. After varying the variables around the nominal values and performing the regression analysis, the response function obtained becomes

$$LF_d = -0.10 + 4.38 \times 10^{-4} (R_1 + R_2 + R_3) - 1.40 \, D . \quad (9)$$

Assuming full correlation between member resistances, the first order reliability calculations produced safety indices of 5.31 for the ultimate limit state, 4.24 for the serviceability limit state, and 4.19 for the damaged condition. On the second iteration where the expansion of the response function is performed at the failure points, the safety indices become 5.37, 4.20, and 4.24 respectively. This produces the redundancy indices $\rho_u = 1.14$, $\rho_s = -0.03$, and $\rho_d = -0.01$.

6 Calibration of System Factors

To account for the redundancy of common type bridges, the framework proposed in this study consists of applying system factors ϕ_s to penalize nonredundant bridges or to reward redundant configurations. A system factor ϕ_s gives a measure of the level of redundancy of a bridge. If ϕ_s is less than 1.0, it indicates that the bridge under consideration has an unacceptably low level of redundancy. A system factor greater than one indicates that the level of redundancy is acceptable. A bridge that has a system factor less than 1.0, should be penalized by requiring that the load factor LF_1 corresponding to the failure of the main longitudinal members be increased by a factor equal to $1/\phi_s$. Bridges that have a system factor ϕ_s greater than 1.0 may be rewarded by allowing that the load factor of their main longitudinal members be decreased by $1/\phi_s$.

The ϕ_s factor is a scaling function that is applied in the case of redundant bridges to decrease their required member capacities, and in the case of nonredundant bridges to increase the required member capacities. Applying ϕ_s on the load factor that will cause the first member failure LF_1 will also result in the change of the values of the load factors for the ultimate limit state LF_u, the load factor for the serviceability limit state LF_s, and the load factor for the damaged condition LF_d. If a bridge is nonredundant and a system factor $1/\phi_s$ (where ϕ_s is less than 1.0) is applied to increase the values of LF_1, LF_u, LF_s, and LF_d, this will increase the values of β_{member}, β_{ult}, β_{serv}, and $\beta_{damaged}$. Thus, applying the system factor ϕ_s will not improve the redundancy of a nonredundant bridge but will simply penalize nonredundant designs by requiring that their members be more conservatively designed; hence, they will provide higher safety levels. The proposed approach will also reward redundant designs by allowing their members to have lower safety levels.

For example, the prestressed concrete bridge with 6 girders produced the redundancy indices ρ_u = 2.53, ρ_s = 0.20, and ρ_d = 2.55. These values are compared to the target values of 1.0, -1.0, and -0.5. This bridge satisfies the target redundancy indices. Also, even if β_{member} is decreased by 1.53 i.e from 3.60 to a value of 2.07, the target system safety indices will still be satisfied. This is achieved when the resistance is reduced from 5900 kip-ft to 4780 kip-ft. This will result in a change in LF_1 and in the live load margin R-D on the order of 40%.

The steel bridge with 4 girders produced the redundancy indices ρ_u = 1.14, ρ_s = -0.03, and ρ_d = -0.01. These values are compared to the target values of 1.0, -1.0, and -0.5. This bridge satisfies the target redundancy indices. Also, even if β_{member} is decreased by 0.14 to a value of 4.09, the target system safety indices will still be satisfied. This is achieved when the resistance is reduced from 4791 kip-ft to 4650 kip-ft. Or when LF_1 is reduced from 5.0 to 4.82. The system factor applied on LF_1 for this case is on the order of 1.18.

7 Conclusions

This paper illustrates the application of the response function method to find the reliability indices of bridges systems. The results are then used to compute the redundancy index for typical prestressed concrete and multi-girder steel bridges. When bridges designed according to current specifications are found to satisfy the target redundancy indices, these bridges are rewarded by allowing their members to be less conservative than required. On the other hand bridges that do not satisfy the target redundancy indices should be penalized by requiring a more conservative design.

8 Acknowledgments

The authors are grateful to the assistance provided by Dr. Y Liu formerly a research associate at Case Western Reserve University, Mr. Jian Ming Xu and Mr. Ge Chen both research assistants at The City College of New York. The contributions of Mr. Charles Schilling are also gratefully acknowledged. This work was sponsored by the American Association of State Highway and Transportation Officials, in cooperation with the Federal Highway Administration, and was conducted in the National Cooperative Highway Research Program which is administered by the Transportation research Board of the National Research Council. The opinions and conclusions expressed or implied in this paper are those of the authors. They are not necessarily those of the Transportation Research Board, the National Research Council, or the Federal Highway Administration, American Association of State Highway and Transportation Officials, or of the individual states participating in the National Cooperative Highway Research Program.

9 References

1. Ghosn, M. and Moses F., Redundancy in Highway Bridge Superstructures, Draft Final Report, NCHRP 12-36, September 1993.
2. Moses, F., Khedekar, N, and Ghosn, M., System Reliability of Redundant Structures Using Response Functions, *ICOSSAR 93*, Innsbruck, Austria, August 1993.
3. Augusti, G., Baratta, A. and Casciati, F., *Probabilistic Methods in Structural Engineering*, Chapman Hall, New York, 1984.
4. *Standard Specifications for Highway Bridges*, AASHTO, Washington DC, 1989.
5. Kulicki, J.M., Development of Comprehensive Bridge Specification and Commentary, NCHRP 12-33, Draft Report, March 1993.
6. Nowak, A.S., Calibration of LRFD Bridge Design Code, NCHRP project 12-33, May 1992.
7. Naaman, A.E., Harajali, M.H., and Wight, J.K., Analysis of Ductility in Partially Prestressed Concrete Flexural Members, *PCI Journal*, V. 31, No. 3, May-June 1986.
8. Schilling, C.G., A Unified Autostress Method, Report 51, American Iron and Steel Institute, November 1989.
9. *U.S. Steel Highway Bridge Design Manual*. US Steel, Pittsburgh, PA.

A Survey of Probabilistic Service Life Prediction Methods for Structures

R. A. Heller[1] and S. Thangjitham[1]

[1]Department of Engineering Science and Mechanics
Virginia Polytechnic Institute and State University
Blacksburg, Virginia

Abstract. This paper presents an overview of some proposed probability-based service life prediction methods. Primary concern is focused on the methods that are of interest in service life assessment of an outdoor structure. The survey includes methods for characterizing random load histories, for modeling material deterioration, and for evaluating both instantaneous and progressive probabilities of failure. A numerical example, showing the service life and reliability of a viscoelastic structure under outdoor environmental loadings, is given.

Keywords. Structures, service life, probability of failure, random loads, environmental loads, material degradation, fatigue, creep fracture, stress-strength interference

1 Introduction

Engineers have been designing and building structures since our ancestors climbed out of the trees. Though many ancient structures such as Roman aqueducts, the colosseum of Rome, and the pyramids of Giza are still partially standing, structural failures have been and still are quite common. Engineers have used their best know how as well as the most appropriate materials in their designs but structures still fail.

More recent examples include the collapse of two identical school gymnasiums in North Carolina under an unexpectedly heavy snow load, the collapse of a highway bridge under an overloaded truck, another highway bridge on the New England Turnpike, and the Point Pleasant-Gallopolis bridge on the Ohio River both due to stress-corrosion-fatigue, the demise of the Alexander Klieland Oil drilling platform in the North Sea due to rough seas combined with catastrophic crack propagation, and the Aloha Airlines 737 that lost the upper part of its fuselage due to fatigue cracking.

While the causes of these disasters are varied, there is a commonalty among them. All were caused by unexpected loads combined with deteriorated structural performance. Additionally, all of them should have been retired long before they failed. Most of them were not designed for a specific life span or the service life was not established with the statistical variability of the imposed loads, of the service environment, or of their effects on the mechanical properties of the materials.

Modern probabilistic service life prediction methods[1–3] take into consideration all of these factors and require that the structure be retired from service when its reliability drops below acceptable levels. The decision of what an acceptable level is depends on the uses of the structure and is usually dictated by public authorities and code writers.

In the following the most frequently used models and methods that describe the contributing causes of reliability degradation and the time dependent probability of failure will be discussed. These include models of stochastic loads, environmental effects on fatigue, creep, aging, stress-corrosion and crack propagation as well as techniques for the calculation of failure rates and failure probabilities utilizing *first passage*, *first-order-second-moment* (FOSM), and *weakest-link* approaches.

2 Load History Modeling

The development of accurate service life prediction procedure requires complete information on the time varying states of induced stresses and strains. Due to the fact that general service loading histories are lengthy and random in nature, the development of an accurate but nevertheless concise method for describing such histories is deemed necessary. The method is required not only to be able to preserve all relevant loading events found in the original history but also to contain a minimum number of model parameters.

2.1 Random Load Histories

The methods of modeling random load histories for service life analysis are divided into two main categories[4]– one that models only the relevant extreme events (e.g., peaks and valleys) and the other that models the complete history. Because the methods of modeling only the extreme events discard all intermediate points that exist between peaks and valleys, the number of data points can be greatly reduced. The *rain-flow matrix*[5] and *to-from matrix*[6] methods of random fatigue load modeling, for examples, fall in this category.

Modeling a complete random load history requires a continuous dynamic model. In this approach, the model parameters are obtained by considering statistical properties of the given history. The statistical analysis of observed data drawn from a time varying process is termed *time series analysis*[7]. Time series analysis can be performed in both the time and frequency domains.

2.1.1 Time Domain Analysis

A class of models referred to as *autoregressive moving average* (ARMA) is used to represent the random observed data in the time domain.[8,9] There has been a considerable amount of research work on the application of ARMA models in engineering analysis.[10,11] Specific applications have been focused in the fields of earthquake,[12,13] ocean,[14,15] and wind[16,17] engineering.

There are two components of an ARMA model: (1) the autoregressive part and (2) the moving average part. The autoregressive part represents the dependence of the output variable, $x(t)$, (variable under observation) on its own past, while the moving average part represents the dependence of the output process on the past values of the input process, $a(t)$. For example, the following is a pth order autoregressive and qth order moving average model, denoted as ARMA(p,q)[8,9]:

$$x_t - \phi_1 x_{t-1} - \phi_2 x_{t-2} - \ldots - \phi_p x_{t-p} = a_t - \theta_1 a_{t-1} - \theta_2 a_{t-2} - \ldots - \theta_q a_{t-q} \quad (1)$$

where $\phi_i; i = 1, 2, \ldots, p$, and $\theta_i; i = 1, 2, \ldots, q$, are the autoregressive and moving average parameters, respectively. In the above model, the input $a(t)$ is assumed to be an independently and identically distributed random process, i.e., $a(t)$ itself is considered to be non-autoregressive.

The ARMA processes can also be conveniently presented in the frequency domain. In this case, the one-sided power spectral density, $W(f)$, corresponding to the ARMA (p,q) model is given by[8,9]

$$W(f) = 2\sigma_a^2 \frac{\left|1 - \sum_{k=1}^{q} \theta_k e^{-2ki\pi f}\right|^2}{\left|1 - \sum_{k=1}^{p} \phi_k e^{-2ki\pi f}\right|^2}; \qquad 0 \leq f \leq \frac{1}{2} \quad (2)$$

where σ_a^2 is the variance of $a(t)$, f is the linear frequency, and $i = \sqrt{-1}$.

2.1.2 Frequency Domain Analysis

The stochastic characteristics of a stationary random process can be completely described by the variation of sinusoids of different frequencies. This can be accomplished by performaning spectral analysis[18] to obtain the corresponding *power spectral density* (PSD) function, $W(\omega)$, of the time series.

$$W(\omega) = \frac{1}{2\pi} \int_{-\infty}^{\infty} R(\tau) e^{-i\omega\tau} d\tau \quad (3)$$

where the autocorrelation function $R(\tau)$ of the given random process $x(t)$ is defined as

$$R(\tau) = E[x(t)x(t+\tau)] \quad (4)$$

for which $E[\cdot]$ is the expectation operator.

However, by utilizing the *Fast Fourier Transform* (FFT) algorithm, Eq. (3) can be digitally transformed directly from the measured data.[19,20] Examples for frequency domain description of random load histories are found in References 21, 22, and 23.

2.2 Environmental Loads

In the study of probabilistic service life of structures subjected to variations of an outdoor environment, the time varying induced stresses and strains are of interest. The random nature of the environmental loads, such as air temperature and solar radiation, are well known. Energy received by the structure at a given time can not be predicted with certainty but only within statistical variation. Its random characteristics vary significantly depending on a given location and time of the year.

2.2.1 Air Temperature Model

The variation of atmospheric temperature is directly influenced by the angular incidence of solar radiation and distance between the sun and the earth.[24] As a result, air temperature at a given altitude decreases with increasing latitude. This change, however, is not uniform. Air temperature along a given parallel also varies widely. This reflects the effect of the irregular composition of land and sea on the earth's surface. The vertical atmospheric temperature also decreases with increasing altitude.

Near the earth's surface, air temperature exhibits periodic variations from high to low peaks. These variations can be generally differentiated into two parts— deterministic trend and random variation. The deterministic trend represents the average weather conditions accounting for the mean and the typical periodic variations, such as diurnal and annual changes. For a given location, these variations can be predicted with fair accuracy. The random variation results from the instantaneous change in weather conditions such as local cloud formation and weather front movements.

As a result, the following model for air temperature, $T_a(t)$, at an instant of time t is proposed:[25,26]

$$T_a(t) = T_m + T_d \cos(\omega_d t + \phi_d) + T_y \cos(\omega_y t + \phi_y) + N_a(t) \qquad (5)$$

where T_m is the long term mean temperature, (T_d, ϕ_d) and (T_y, ϕ_y) are the amplitudes and phase angles corresponding to daily ($\omega_d = 2\pi/24$ rad/hr) and yearly ($\omega_y = 2\pi/8760$ rad/hr) temperature variations, respectively, and N_a is the random zero-mean temperature noise. The values for these site dependent constants can be evaluated by the method of least-square regression[27] of the metrorological data.[28]

2.2.2 Solar Radiation Model

The earth's daily rotation about its tilted axis and the annual revolution around the sun account for the distribution of solar radiation over the earth's surface, changing the length of hours of daylight and darkness. The diurnal variation of solar radiation is obvious. At night, the surface is on the shadow side of the earth, therefore, no solar radiant energy is received. During the day, however, the solar radiation intensity on the surface gradually increases with time before attaining its daily peak in the mid afternoon and decreases with time approaching zero in the evening. To describe this actual change, the following model for daily variation of solar radiation input on the nth day, $H_n(t)$, is proposed:[29]

$$H_n(t) = \begin{cases} 0 & 0 \le t \le t_{1n} \\ h_n \sin^p \pi \left(\frac{t - t_{1n}}{t_{2n}} \right) & t_{1n} \le t \le t_{2n} \\ 0 & t_{2n} \le t \le 24 \end{cases} \tag{6}$$

where h_n is the daily peak of solar radiation input, t_{1n} and t_{2n} are hours from midnight to sun rise and hours from sun rise to sun set (daylight), and p is shape constant.

The daily values for h_n, t_{1n}, and t_{2n} are obtained from the following expressions

$$h_n = h_m + h_y \cos(24\omega_y n - \phi_h) \tag{7}$$

$$t_{1n} = t_{1m} + t_{1y} \cos(24\omega_y n - \phi_1) \tag{8}$$

$$t_{2n} = t_{2m} + t_{2y} \cos(24\omega_y n - \phi_2) \tag{9}$$

where ω_y $(= 2\pi/8760$ rad/hr) is the annual frequency, ϕ is the phase angle, while the subscripts m and y indicate the long term mean value and annual amplitude, respectively. The values for these constants and shape parameter p vary from location to location and can be evaluated from the metrorological data.[28]

2.3 Wind Velocity and Ocean Waves

Wind refers to horizontal transport of air masses as a consequence of large scale differences in air pressure. Wind near the earth's surface is strongly affected by friction with the underlying surface and by thermal convection if the surface is warmer than the air.[30] These effects decrease the wind speed and produce turbulence. Wind turbulence increases with the roughness of the terrain as well as with the degree of temperature contrast between the warm surface and the cooler air above. As a consequence, winds are stronger at sea than on land for the same meteorological conditions.

Because a minimum of zero wind velocity exists, wind velocity is not evenly distributed.[27] It has been shown[31] that wind velocity has a form of the Rayleigh distribution.

Irregular random waves are regarded[32] as an infinite sum of sinusoidal waves with random amplitudes, A_i, and phase angles, ϕ_i. The surface deflection, η, measured from the mean water level is expressed as

$$\eta(x,t) = \sum_{i=1}^{\infty} A_i \cos(\lambda_i x - \omega_i t + \phi_i) \tag{10}$$

where λ_i is the wave number and ω_i is the frequency of the ith wave.
Frequency and wave number are related through the Airy equation[33]

$$\omega_i^2 = g\lambda_i \tanh \lambda_i h \tag{11}$$

where g and h are the acceleration of gravity and water depth respectively. The phase angle is assumed to be uniformly distributed between 0 and 2π and the amplitude may be expressed in a small frequency range, $\Delta\omega$, around ω_i in terms of the amplitude spectrum S_i, as

$$\frac{1}{2}A_i^2 = S_i \Delta\omega. \tag{12}$$

According to Bretschneider,[34] S_i may be written in terms of two constants, A and B:

$$S_i = \frac{A}{\omega^5} e^{-B\omega^{-4}} \tag{13}$$

with

$$A = 0.296\overline{H}^2 \left(\frac{2\pi}{\overline{T}}\right)^4 , \quad B = 0.461 \left(\frac{2\pi}{\overline{T}}\right)^2 \tag{14}$$

where $\overline{H}$ and $\overline{T}$ are the average wave height and average period, respectively.
The horizontal velocity, u, and acceleration, a, of the water, based on the Airy theory are given respectively by

$$u = \sum_{i=1}^{\infty} T_u A_i \cos(\lambda_i x - \omega_i t + \phi_i) \tag{15}$$

$$a = \sum_{i=1}^{\infty} T_a A_i \sin(\lambda_i x - \omega_i t + \phi_i) \tag{16}$$

where T_u and T_a are defined as

$$T_u = \omega_i \frac{\cosh \lambda_i y}{\sinh \lambda_i h}, \qquad T_a = -\omega_i^2 \frac{\cosh \lambda_i y}{\sinh \lambda_i h} \tag{17}$$

with (x, y) are the coordinates of a water particle.

3 Material Deterioration

Throughout the component's entire service life, many factors may deteriorate the performance of the material. Continuously subjected to cyclic stress may result in cumulative damage that reduces the strength and changes the stiffness of materials. Exposing them to environmental loading may cause chemical reaction that may result in the degradation of material properties, while temperature changes will induce viscoelastic variations.

3.1 Aging

Aging is a result of chemical degradation in material properties. Aging effects are responsible for hardening and softening as well as for reducing the strength and strain capacities of the material. It may occur in both loaded and unloaded states.

On the basis of extensive experimental results, Layton[35] found that aging effects on the mechanical properties of viscoelastic materials under constant temperature T can be expressed as

$$G'(T,t) = \eta(T,t)G(T,t) = [1 - \beta(T)\log(t)]\,G(T,t) \tag{18}$$

where G and G' represent the current unaged and aged values of material properties (i.e., modulus, strength, and strain capacities), respectively, η is the time and temperature dependent aging factor. The temperature dependent reduction coefficient is given by

$$\beta(T) = A\exp\left(-\frac{B}{T}\right) \tag{19}$$

where the material constants A and B are experimentally evaluated.

It is evident from Eq. (19) that aging is accelerated more rapidly at higher temperature. Variable temperature aging may be calculated based on the equivalent time concept suggested by Cost.[36] Considering an aging process under a constant temperature T_1 for a period of time t_1, the aging factor becomes

$$\eta = 1 - \beta_1\log(t_1) \tag{20}$$

where β_1 is evaluated at T_1. This same aging factor may also be obtained at a different temperature T_2 with an equivalent time, t_e, as

$$\eta = 1 - \beta_2\log(t_e) \tag{21}$$

where β_2 is evaluated at T_2.

Hence, the equivalent time during which the same aging factor is reached at temperature T_2 becomes equal to

$$t_e = t_1^{\beta_1/\beta_2}. \tag{22}$$

If the aging is now continued at T_2 for an additional time Δt, the total aging time, t_2, becomes

$$t_2 = t_e + \Delta t \tag{23}$$

and the aging factor at the end of t_2 is obtained as

$$\eta = 1 - \beta_2\log(t_2). \tag{24}$$

3.2 Time and Temperature Dependent Viscoelastic Properties

It is a well known fact that the modulus and strength of visco-elastic materials vary with both temperature and rate of loading. Such materials are frequently represented as several Maxwell bodies in a parallel arrangement.[37] The variations of the modulus in such cases is given by the relaxation modulus

$$E(t,T) = E_\infty + E'(t,T) \tag{25}$$

where E_∞ is the equilibrium modulus as $t \to \infty$ and $E'(t,T)$ is expressed in terms of a Prony Series as[36]

$$E'(t,T) = \sum_{i=1}^{M} E_i e^{-t/\tau_i a_T} \tag{26}$$

with E_i and $\tau_i; i = 1,2,\ldots,M$ are the elastic moduli and relaxation times of M parallel Maxwell elements, respectively, and a_T is the viscoelastic shift function based on the Williams-Landell-Ferry (WLF) equation[38] given by

$$\log a_T = \frac{-C_1(T - T_R)}{C_2 + (T - T_R)}. \tag{27}$$

Here C_1 and C_2 are constants and T_R is a reference temperature.

For periodic processes, the relexation modulus is Fourier transformed into the frequency and temperature dependent complex modulus

$$E(\omega,T) = E'(\omega,T) + iE''(\omega,T) \tag{28}$$

where $E'(\omega,T)$, the storage modulus and $E''(\omega,T)$, the loss modulus are respectively represented as

$$E'(\omega,T) = E_\infty + \sum_{i=1}^{M} \frac{E_i\omega^2 a_T^2 \tau_i^2}{1 + \omega^2\tau_i^2 a_T^2} \tag{29}$$

and

$$E''(\omega,T) = 2\pi \sum_{i=1}^{M} \frac{E_i\omega a_T \tau_i}{1 + \omega^2\tau_i^2 a_T^2} \tag{30}$$

where ω is the excitaion frequency.

The strength R, of such materials is also affected by time and temperature changes and may be expressed as a power function

$$R(t,T) = R_0 \left(\frac{t}{a_T}\right)^{-\alpha} \tag{31}$$

where the parameters R_0 and α are experimentally obtained.

3.3 Cumulative Damage

The linear cumulative damage rule proposed independently by Palmgren[39] and Miner[40] has been applied intensively in the aerospace industry for the study of fatigue damage.[41] According to this rule, the damage, d_i, produced in a unit of time spent at a particular stress level, σ_i, is inversely proportional to the time, t_{fi}, required to produced failure in the material at that stress level. Therefore

$$d_i = \frac{1}{t_{fi}} \tag{32}$$

and the damage, D_i, resulting from the total time, t_i, spent at σ_i is the sum of the damage per unit of time

$$D_i = \frac{t_i}{t_{fi}}. \tag{33}$$

When a mixture of N stress levels is present, the total damage, D, becomes

$$D = \sum_{i=1}^{N} D_i = \sum_{i=1}^{N} \frac{t_i}{t_{fi}}. \tag{34}$$

The linear cumulative rule assumes that damage, under variable stresses, accumulates until the total damage reaches unity.

Experiments conducted by Bills[42] indicate that the linear cumulative damage rule may be useful in the service life study of viscoelastic materials. It has further been shown that a power function can be used in describing the relationship between the applied constant stress and reduced time to failure of the viscoelastic materials.

$$\frac{t_f}{a_T} = C\sigma^{-B} \tag{35}$$

where C and B are material constants, t_f is time to failure, a_T is the temperature dependent viscoelastic shift function, and σ is the applied stress. Similar relations are used for the evaluation of cumulative creep damage.

Recently, a new phenomenological probabilistic model of cumulative damage has been proposed.[43] In this model, it is assumed that the damage is a finite state-discrete time imbedded Markoff process.

3.4 Creep and Shrinkage

Many engineering materials creep under long-term loading. Typical expression for creep strain $\epsilon(t)$ at the current time t produced by a unit constant stress σ that has been applied since time t' is given by[44]

$$\epsilon(t) = \epsilon^0(t) + \sigma J(t, t') \tag{36}$$

where $\epsilon^0(t)$ is the stress-independent inelastic strain and $J(t,t')$ is the time dependent creep compliance function.

The simplest form for creep compliance function of concrete at constant moisture and thermal state may be described by power curves of load duration $(t-t')$ and by inverse power curves of age t'. This leads to the double power law[44]

$$J(t,t') = \frac{1}{E_0} + \frac{\phi_1}{E_0}(t'^{-m} + \alpha)(t-t')^n \tag{37}$$

where E_0 is the asymptotic modulus, ϕ_1, α, m, and n are material constants.

For some materials, temperature has a major influence on creep. To describe the creep curves at various constant temperatures, the time and temperature dependent creep compliance function is defined as[44]

$$J(t,t',T) = \frac{1}{E_0} + \frac{\phi_T}{E_0}(t_e'^{-m} + \alpha)(t-t_e')^{n_T} \tag{38}$$

in which

$$t_e' = \int \beta_T(t')dt' \tag{39}$$

is called the reduced age which represents the effect of temperature on the rate of aging. In the above equations, the material constants ϕ_T, n_T, and β_T are functions of temeratuure and defined such that at reference temperature $\phi_T = \phi_1, n_T = n$, and $\beta_T = 1$.

For variable stresses, a cumulative creep damage relation similar to Eq. 14 has been proposed by Freudeulthal[45]

$$D_c = \frac{t_j}{t_{cj}} \tag{40}$$

where t_{cj} is the time to creep failure at the jth stress level and t_j is the time at that level. An interactive damage rule combining the effects of creep, D_c, at the mean stress and fatigue, D_f, produced by the alternating portion of the stress cycle has also been proposed as

$$D = D_f + D_c = \sum_{i=1}^{M} \frac{t_i}{t_{fi}} + \sum_{j=1}^{k} \frac{t_j}{t_{cj}}. \tag{41}$$

Based on such a linear cumulative damage rule failure takes place when $D = 1$.

When used in a probabilistic service life analysis it is assumed that the initial strength of the material is linearly reduced by cumulative damage and aging[46] so that the reduced strength, R_r, at time t becomes

$$R_r(t) = R(t,T)\,\eta(T,t)\,[1-D] \tag{42}$$

Shrinkage strain is usually small and is somewhat less important than creep strain under long-term loading. A typical expression for shrinkage strain for concrete is given by[44]

$$\epsilon_s(t,t_0) = \frac{t-t_0}{f_c+(t-t_0)}\epsilon_u^s \tag{43}$$

where t is the current time (age), t_0 is the age at the completion of curing, f_c is the material constant, and ϵ_u^s is the ultimate shrinkage strain after $t \to \infty$.

3.5 Fatigue Crack Growth

When a structural component is subjected to cyclic loading, fatigue failure is expected. Fatigue cracks start at the surface of the material in the direction perpendicular to the main principal stress. In addition to component material properties and geometry parameters, fatique crack growth also depends on the cyclic stress amplitude, the mean stress ratio, and the environment. Generally, the fatigue crack growth rate is written as a function of the stress intensity factor range, ΔK, and the stress ratio, R, such as

$$\frac{da}{dN} = f(\Delta K, R) \tag{44}$$

where a is the crack length and N is the number of stress cycles.

A number of empirical and theoretical crack rate relations in the form of Eq. (44) have been proposed. Among the more well-known are:

Paris-Erdogan's relation:[47]

$$\frac{da}{dN} = C(\Delta K)^n \tag{45}$$

Forman's equation:[48]

$$\frac{da}{dN} = C\frac{(\Delta K)^n}{(1-R)K_{IC}-\Delta K} \tag{46}$$

Walker's formula:[49]

$$\frac{da}{dN} = C\left[\frac{\Delta K}{(1-R)^{1-m}}\right]^n \tag{47}$$

where C, n, and m, are material constants, and K_{IC} is the material fracture toughness.

Under the variable amplitude loading condition, numerical integration of Eq. (44) cycle by cycle leads to the crack length after i load applications[50]

$$a_i = a_{i-1} + \Delta a_i \tag{48}$$

where Δa_i is the incremental crack length caused by the ith load cycle.

Probabilistic aspects of fatigue life prediction have been focused on assessing the probability of failure which is defined by the chance that the current crack size, a_i, will exceed the expect critical crack size, a_c. References 51 and 52 provide a collection of recent research articles in this area.

4 Probabilistic Service Life Analysis

In the probabilistic service life study of a structural component subjected to time-varying applied loads, the component's reliability or chance of success at a given time t is of interest. In order for the component to perform satisfactorily at time t, it must have survived up to that time. It is, therefore, a problem of conditional probability.

4.1 Approximate Method for Statistics

When Y is a function of n random variables such as

$$Y = g(X_1, X_2, \ldots, X_n), \tag{49}$$

the application of the method of transformation to obtain the probability density function of Y, $f_Y(y)$, may not be possible, depending on the nature of the function g. It is, however, possible to calculate the first two statistical moments (mean, $E[Y]$, and variance, $E[(Y - E[Y])^2]$) of random variable Y approximately if the moments of the random variables $X_1, X_2, \ldots, X_n$ are known. This can be accomplished via a Taylor series expansion.[53,54,55] The accuracy of the approximation, therefore, depends on the order of terms neglected in the series.

For a set of uncorrelated random variables $X_1, X_2, \ldots, X_n$, the second-order approximation gives

$$E[Y] = g(E[X_1], E[X], \ldots, E[X]) + \frac{1}{2}\sum_{i=1}^{n} \frac{\partial^2 g}{\partial X_i^2} E[(X_i - E[X_i])^2] \tag{50}$$

$$\begin{aligned} E[(Y - E[Y])^2] = &\sum_{i=1}^{n} \left(\frac{\partial g}{\partial X_i}\right)^2 E[(X_i - E[X_i])^2] \\ &+ \sum_{i=1}^{n} \frac{\partial g}{\partial X_i}\frac{\partial^2 g}{\partial X_i^2} E[(X_i - E[X_i])^3] \end{aligned} \tag{51}$$

where the derivatives are evaluated at the expected values of the random variables. The expressions for correlated random variables can be found in Reference 53.

4.2 Instantaneous Probability of Failure

The instantaneous probability of failure is defined as the chance that the component fails to perform satisfactorily within a given specific time interval, Δt. Depending on the characteristic time of the problem at hand, Δt may be defined such that the instantaneous hourly, daily, or yearly probability of failure is obtained.

4.2.1 Stress-Strength Interference

In the stress-strength interference method, failure of a structural component is said to occure if the applied stress, S, exceeds the strength, R, of the material, i.e., when $R \leq S$. Consequently, the probability of failure, p_f, is defined as[56]

$$p_f = \text{prob}\,[R \leq S] = \text{prob}\,[R - S \leq 0] \tag{52}$$

(In the case of crack propagation problem: $p_f = \text{prob}\,[a \leq a_c]$ where a_c is the critical crack length).

When both stress and strength are statistically distributed with joint density function, $f_{R,S}(r,s)$, and marginals $f_R(r)$ and $f_S(s)$, the probability of failure is calculated by

$$p_f = \int_{-\infty}^{\infty} F_{R|S}(s) f_S(s) ds = \int_{-\infty}^{\infty} L_{S|R}(r) f_R(r) dr \tag{53}$$

where $F_{R|S}(s)$ is the conditional distribution of the strength for given stress and $L_{S|R}(r)$ is the conditional probability of exceedence of the stress for given strength.

For the case where R and S are independent normal variates, the probability of failure becomes

$$p_f = 1 - \Phi\left[\frac{\mu_R - \mu_S}{\sqrt{\sigma_R^2 + \sigma_S^2}}\right] \tag{54}$$

where (μ_R, σ_R) and (μ_S, σ_S) are the means and standard deviations for the strength R and induced stress S, respectively, and $\Phi\,[\cdot]$ is the standard normal distribution function.

4.2.2 First-Order Second-Moment Method

The first-order second-moment (FOSM) method evaluates the probability of failure, p_f, of a structural component by considering only the first two moments, i.e., the means and variances of of the design random variables, $X_1, X_2, \ldots, X_n$. The formulation is based on the assumption that design variables are statistically independent. It involves the linearization of the limit-state (performance) function

$$g(X_1, X_2, \ldots, X_n) = 0 \tag{55}$$

at the *most probable failure point* $(\widehat{x}_1, \widehat{x}_2, \ldots, \widehat{x}_n)$ on the failure surface.[57]

The reliability is expressed in terms of the safety index, β, defined as

$$\beta = \frac{m_g}{s_g} \tag{56}$$

where m_g and s_g are the first-order approximations of the mean and variance of the function g evaluated at the most probable failure point. In terms of the reduced coordinates of the most probable failure point $(\widehat{y}_1, \widehat{y}_2, \ldots, \widehat{y}_n)$, m_g and s_g are given as

$$m_g = -\sum_{i=1}^{n} \widehat{y}_i \frac{\partial g}{\partial Y_i}, \qquad s_g^2 = \sum_{i=1}^{n} \left(\frac{\partial g}{\partial Y_i} \right)^2 \tag{57}$$

where $Y_i = (X_i - \mu_{X_i})/\sigma_{X_i}$ for which μ_{X_i} and σ_{X_i} are the mean and standard deviation of X_i, respectively, and the derivatives $\partial g/\partial Y_i$ are evaluated at $(\widehat{y}_1, \widehat{y}_2, \ldots, \widehat{y}_n)$.

When the function g is linear in design variables $X_1, X_2, \ldots, X_n$ and these variables are statistically independent and normally distributed, the safety index, β, relates to the probability of failure, p_f, by [57]

$$\beta = \Phi^{-1}[1 - p_f] \qquad \text{and} \qquad p_f = \Phi[-\beta] \tag{58}$$

where $\Phi[\cdot]$ is the standard normal distribution function.

In the case when the probability distributions of the design variables are not normal, the one-to-one relation between the probability of failure, p_f, and the safety index, β, given by Eq. (58) becomes invalid. To retain this relation for non-normal variables, an equivalent normal transformation method is applied. For correlated design variables, the application of the FOSM method requires that the original reduced variables be transformed to a set of uncorrelated reduced variables, $Z_1, Z_2, \ldots, Z_n$. This can be accomplished by the method of orthogonal transformation.[57]

For nonlinear performance functions, the most probable failure point is not, *a priori*, known. In this case, an iterative algorithm proposed by Rackwitz[58] is utlized.

Several approaches have recently been introduced to facilitate the calculation of the probability of failure for the case when the performance function is evaluated by numerical methods. Among these are: *Monte Carlo simulation*,[59] *response surface*,[60–62] *importance sampling*,[63–64] and *fast probability integration*.[66]

4.3 Failure Rate Function

The time dependent failure rate function or hazard function, $\lambda(t)$, is defined in terms of the probability density, $f(t)$, and reliability, $L(t)$, functions[67] as

$$\lambda(t) = \frac{f(t)}{L(t)} \tag{59}$$

where failure rate may be constant, decreasing, or increasing function of time t. Constant failure rates are associated with *chance* failures, decreasing failure rates represent *break-in* processes or early failures, and increasing failure rates are required to describe *wear-out* processes such as fatigue, aging, and other degradation of material properties.

4.4 Progressive Probability of Failure

The most common approach for assessing the time dependent reliability problem is through employing the failure rate function, Eq. (59). In this case, $\lambda(t)dt$ is interpreted as the conditional probability that a component that has survived to time t will fail in the next short time interval dt. Hence, from Eq. (59) and utilizing the relation defined as

$$f(t) = -\frac{dL(t)}{dt} \tag{60}$$

the reliability of the component at time t may be evaluated by integrating the failure rate function

$$L(t) = \exp\left[-\int_0^t \lambda(\tau)d\tau\right] = \exp\left[-\Lambda(t)\right] \tag{61}$$

where $\Lambda(t)$ is the cumulative failure rate up to time t.

For the problem in which the instantaneous probabilities of failure are defined over a specific time interval (e.g., daily probability of failure), the cumulative failure rate, $\Lambda(n)$, after n time intervals have passed is obtained by replacing the integral with the summation up to the nth interval and the time increment dt with the interval duration Δt, becomes[27]

$$\Lambda(n) = \sum_{k=1}^{n} \frac{p(k)}{L(k-1)}\Delta t \tag{62}$$

with $L(0) = 1$.

Consequently, the progressive probability of failure on the nth interval, $P(n)$, becomes

$$P(n) = 1 - L(n) = 1 - \exp\left[-\Lambda(n)\right]. \tag{63}$$

4.5 Service Life by First Passage Probability

The upcrossing rate for a stationary normal process $x(t)$, with zero mean, variance σ_x^2, and bounded second spectral moment $\sigma_{\dot{x}}^2$, of a barrier curve $u(t)$ is given by[68]

$$\nu^+(t) = \frac{1}{2\sigma_x}\phi\left[\frac{u(t)}{\sigma_x}\right]\left\{2\sigma_{\dot{x}}\phi\left[\frac{\dot{u}(t)}{\sigma_x}\right] + \dot{u}(t)\left(2\Phi\left[\frac{\dot{u}(t)}{\sigma_x}\right] - 1\right)\right\} \tag{64}$$

where $\phi[\cdot]$ and $\Phi[\cdot]$ are the density and distribution functions of the standard normal variate, respectively.

Assuming the upcrossing event is a Poisson process, the probability distribution, $F_{T_c}(T_c)$, of the first passage time, T_c, i.e., the probability that at least one upcrossing occurs in a given interval $(0, T_c)$, becomes

$$F_{T_c}(T_c) = \text{prob}\,[T_c \leq t] = 1 - \exp\left[-\int_0^{T_c} v^+(t)\,dt\right] \tag{65}$$

where Eq. (65) results in an exponential distribution function.

In the service life prediction problem, the barrier $u(t)$ may be taken to represent the difference between the mean aged and damaged strength, $R'(t)$, and deterministic stress, $S_m(t)$, as[69]

$$u(t) = R'(t) - S_m(t) \tag{66}$$

Similarly, if the crossing rate is calculated for a given time interval, Δt, the (discrete) distribution for the first passage time after n time intervals have elapsed becomes

$$F_{T_c}(n) = 1 - \exp\left[-\sum_{i=1}^{n} v^+(i)\Delta t\right] \tag{67}$$

where $v^+(i)$ is the upcrossing rate for the ith interval.

5 A Numerical Example

Tactical solid propellant rocket motors stored under field conditions are subjected to environmental temperature variation and, as a consequence, to variable thermal stresses. These stresses may, during the service life of the motor, exceed the cumulatively damaged strength of the propellant resulting in cracks that could induce failure upon firing. Because both temperature and material properties are random variables, service life calculations have been based on probabilistic considerations,[25,26] i.e., motor life is terminated when the progressive probability of failure which increases with elapsed time exceeds a predetermined value. For the purpose of thermal stress analysis, solid propellant motors are modeled as long, concentric, layered cylinders consisting of a hollow propellant layer bonded to a thin steel case as shown in Fig. 5.1. Both layers are assumed to be isotropic materials.

Air temperature is considered as a random process. The deterministic part represents average weather conditions accounting for the mean and the typical seasonal and daily variations.[27] For a given location, these variations can be predicted with fair accuracy. The superimposed random noise results from the instantaneous change in weather conditions such as local cloud formation and weather front movements. Figure 5.2 shows the typical hourly air temperature readings at Phoenix, Arizona. Table 5.1 provides values for

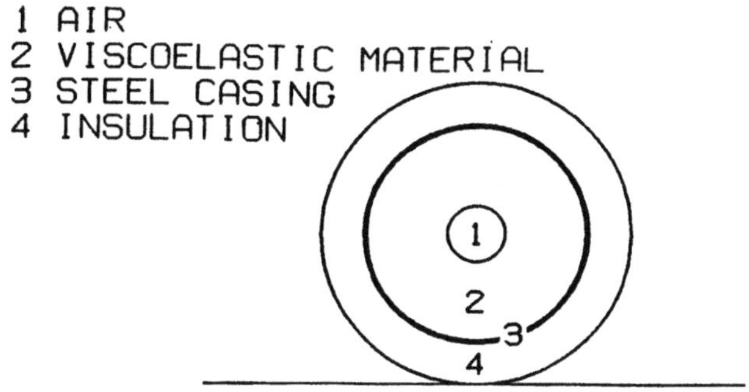

Figure 5.1 Configuration of the Structure.

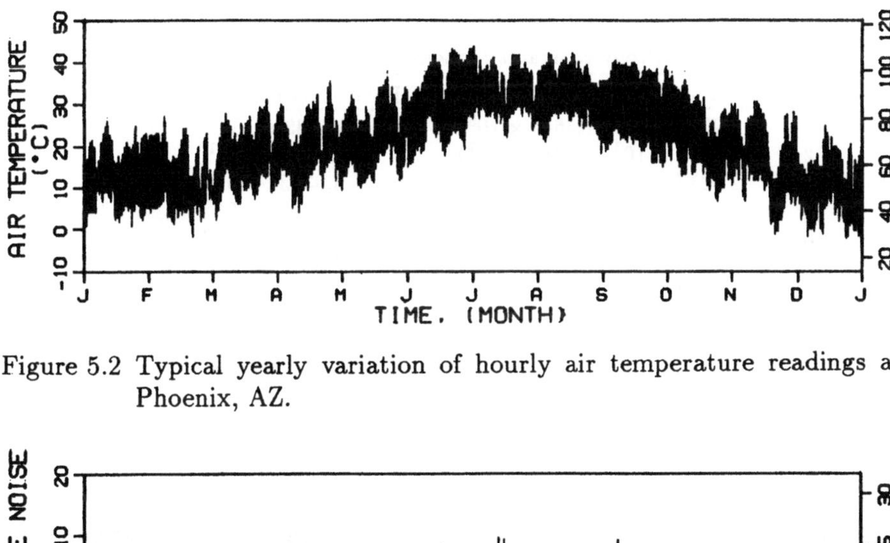

Figure 5.2 Typical yearly variation of hourly air temperature readings at Phoenix, AZ.

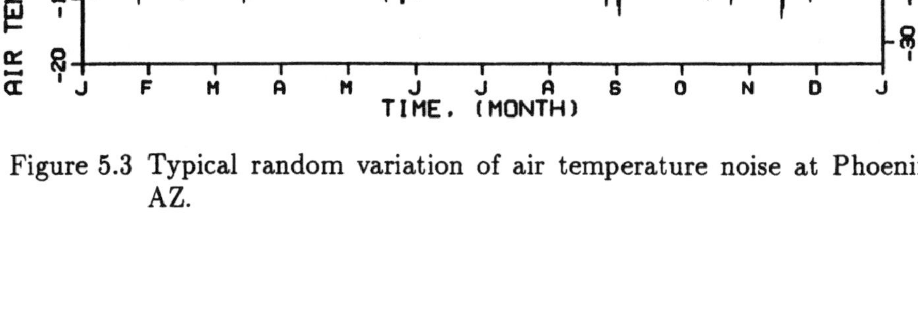

Figure 5.3 Typical random variation of air temperature noise at Phoenix, AZ.

Table 5.1. Characteristics of air temperature at Phoenix, AZ.

Mean, °C (°F)		23.05 (73.490)
Standard deviation, °C (°F)		9.53 (17.154)
Maximum value, °C (°F)		48.30 (118.940)
Harmonic components		−1.70 (28.940)
Period, hour	Amplitude, °C (°F)	Phase angle, hour
8760	10.68 (19.224)	4920
24	6.41 (11.538)	16
4380	1.46 (2.628)	1464
12	1.32 (2.376)	2

Table 5.2. Characteristics of air temperature noise at Phoenix, AZ.

Maximum value, °C (°F)	14.89 (26.81)	29.0
Minimum value, °C (°F)	−21.06 (−37.91)	−20.0
Standard deviation °C (°F)	3.48 (6.28)	5.66
Coefficient of skewness	−0.3115	0.6073

Table 5.3. Prony constants for tensile relaxation modulus

i	Modulus, E_i, psi (N/m^2)	Relaxation time, τ_i, hour
1	1.9789×10^4 (1.3644×10^8)	3.3333×10^{-12}
2	7.9896×10^3 (5.5088×10^7)	3.3333×10^{-10}
3	2.5217×10^3 (1.7378×10^7)	3.3333×10^{-8}
4	1.1526×10^2 (7.9472×10^6)	3.3333×10^{-6}
5	7.3456×10^2 (5.0648×10^6)	3.3333×10^{-4}
6	2.0414×10^2 (1.4975×10^6)	3.3333×10^{-2}
∞	2.8184×10^2 (1.9433×10^6)	

Table 5.4. Thermo- and elastomechanical properties

Layer	1	2	3
Conductivity, κ			
(Btu/ft-°-hr)	0.0145	0.7641	14.6
(W/m-°K)	0.02508	1.3215	25.25
Diffusivity, γ			
(in^2/hr)	106.282	4.351	49.04
(cm^2/hr)	685.672	28.064	316.308
Modulus of elasticity, E			
(psi)		281.84	30×10^6
(N/m^2)		1.9433×10^6	2.068×10^{11}
Coefficient of thermal expansion, α			
(in/in/°F)		5.95×10^{-5}	6.50×10^{-6}
(cm/cm/°C)		3.31×10^{-5}	1.17×10^{-5}
Poisson's ratio, v		0.49	0.25

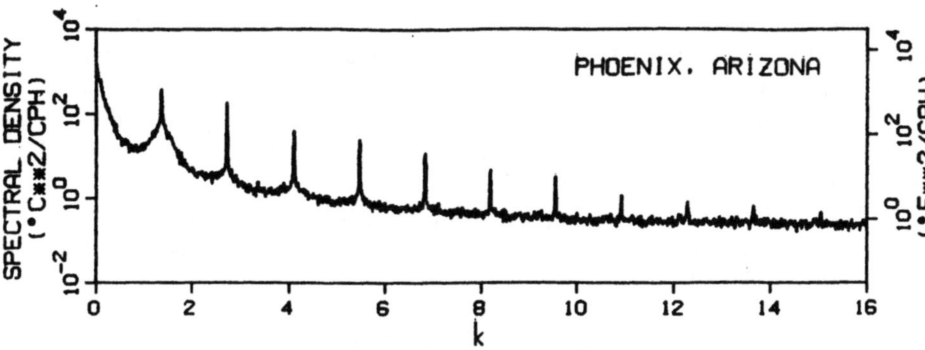

Figure 5.4 Smoothed spectral density function for air temperature noise at Phoenix, AZ. ($\omega_k = 2\pi k/32768$ rad/hr)

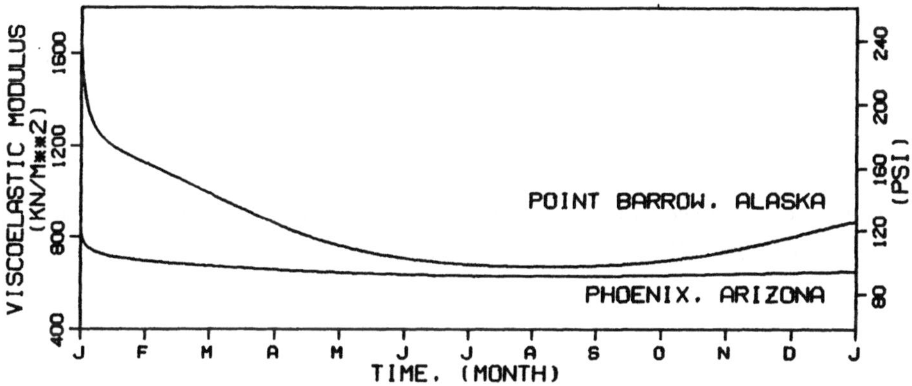

Figure 5.5 Daily variations of viscoelastic modulus during the first year of loading at Phoenix, AZ, and Point Barrow, AK.

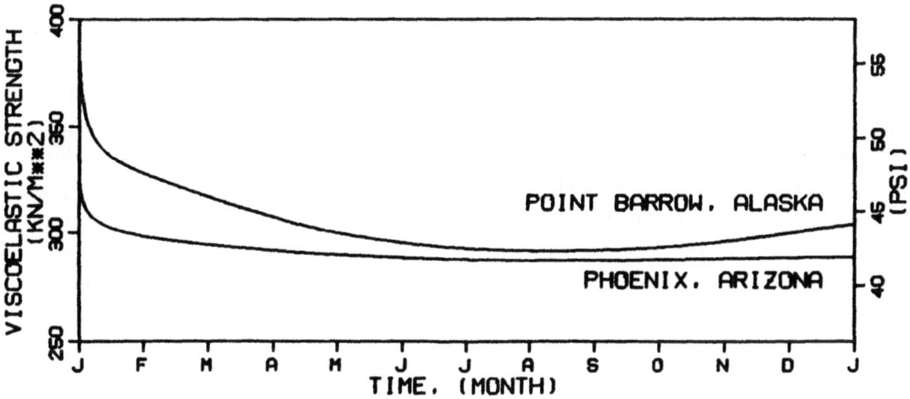

Figure 5.6 Daily variations of viscoelastic strength during the first year of loading at Phoenix, AZ, and Point Barrow, AK.

amplitude and phase of the first four dominant frequencies, Eq. (5), together with the statistical characteristics for the corresponding air temperature process. The typical random variation of the hourly air temperature noise at Phoenix is shown in Fig. 5.3, while the resulting smoothed power spectral density is plotted in Fig. 5.4. Table 5.2 gives the statistical descriptions of the corresponding air temperature noise.

The Fourier heat conduction equation for the jth layer of a system of long cylinders, with no input variation along the length and without the presence of heat source, is written as[70]

$$\nabla^2 T_j(r,t) = \frac{1}{\gamma_j}\frac{\partial T_j(r,t)}{\partial t} \tag{68}$$

where ∇^2 indicates the Laplacian operator, $T_j(r,t)$ is the temperature distribution function, γ_j is the thermal diffusivity, and r and t are the radial coordinate and time, respectively.

For an axially symmetric sinusoidal surface temperature input, the solution of Eq. (68) is sought in the following form:

$$T_j(r,t) = F_j(r,\omega)e^{i\omega t} \tag{69}$$

where $F_j(r,\omega)$ is the complex frequency response function for the temperature within the jth layer and is given as[71]

$$F_j(r,\omega) = C_1^j \mathrm{Br}\,(a_j r) + C_2^j \mathrm{Kr}\,(a_j r) \tag{70}$$

where $a_j = \sqrt{\omega/\gamma_j}$, ω is the loading frequency, C_1^j and C_2^j are the complex constants which must be determined by imposing the appropriate boundary and interface conditions, and $\mathrm{Br}(\cdot)$ and $\mathrm{Kr}(\cdot)$ are the Kelvin functions of zero order.

Figure 5.5 compares the daily variations of viscoelastic modulus during the first year of loading at two locations which have significant different weather conditions. Phoenix, Arizona represents a relatively warm environment and Point Barrow, Alaska represents a relatively cold environment. Similar comparison also shown in Fig. 5.6 for daily variations of viscoelastic strength.

The aging factor for the motor can be readily calculated by the equivalent time concept, Eq. (22). Plots showing comparisons of the variations of aging factors for viscoelastic modulus and for viscoelastic strength in the first year at Phoenix and Point Barrow are shown in Figs. 5.7 and 5.8, respectively.

For a plane stress problem, the elastic compatibility equation for the jth layer can be expressed in terms of a stress function $\phi_j(r,t)$ as[72]

$$\nabla^2\left[\nabla^2\phi_j(r,t) + 2\beta_j T_j(r,t)\right] = 0 \tag{71}$$

where $\beta_j = \alpha_j E_j/2(1-\nu_j)$ and α_j, ν_j, and E_j are the linear coefficient of thermal expansion, Poisson's ratio, and modulus of elasticity for the jth layer,

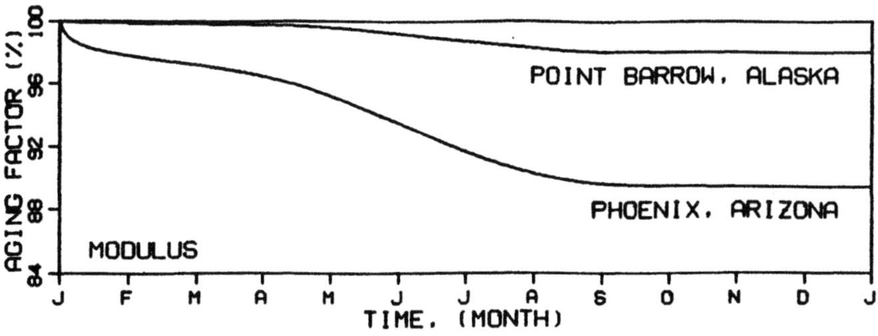

Figure 5.7 Variations of aging factor for viscoelastic modulus in the first year at Phoenix, AZ, and Point Barrow, AK.

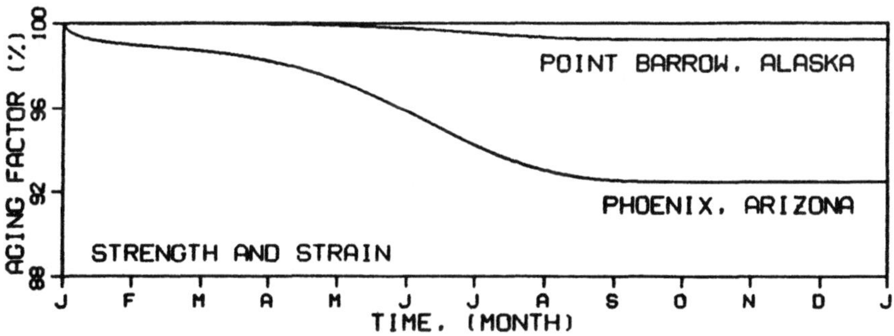

Figure 5.8 Variations of aging factor for viscoelastic strength and strain capacities in the first year at Phoenix, AZ, and Point Barrow, AK.

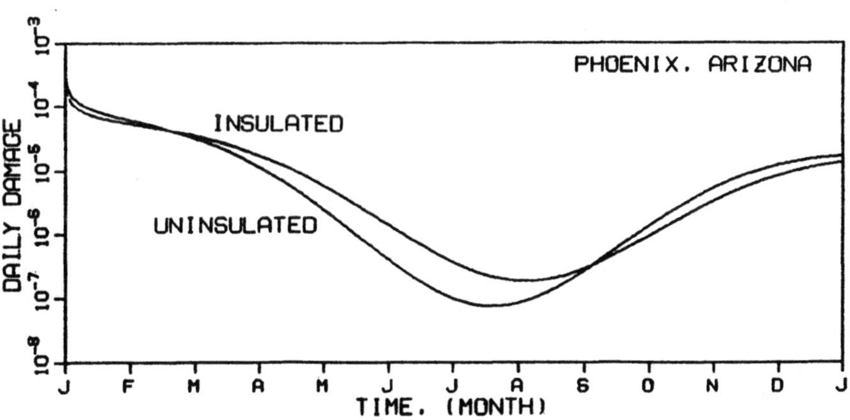

Figure 5.9 Yearly variations of daily damage for uninsulated and insulated cylinders at Phoenix, AZ.

respectively. For the type of temperature distribution governed by the Fourier heat conduction equation, the following form of stress function is assumed.

$$\phi_j(r,t) = f_j(r,\omega)e^{i\omega t} \tag{72}$$

where $f_j(r,\omega)$ is the complex frequency response for the stress function.

For an axisymmetric heating problem, the response function is given as[71]

$$\begin{aligned} f_j(r,\omega) =& K_1^j + K_2^j \ln(r) + K_3^j r^2 + K_4^j r^2 \ln(r) \\ &+ 2\beta_j \left[\int^r F_j(\rho,\omega)\rho \ln(\rho) d\rho - \ln(r) \int^r F_j(\rho,\omega)\rho d\rho \right] \end{aligned} \tag{73}$$

where $K_i^j; i = 1,2,3,4$, are complex integration constants which must be evaluated by imposing the appropriate boundary and interface conditions on stresses and displacements.

In terms of the resulting stress function, the radial and tangential stress components, $\sigma_{rr}^j(r,t)$ and $\sigma_{\theta\theta}^j(r,t)$, within the jth layer are readily calculated by the following relations:[73]

$$\sigma_{rr}^j(r,t) = s_{rr}^j(r,\omega)e^{i\omega t} = \frac{1}{r}\frac{\partial f_j(r,\omega)}{\partial r}e^{i\omega t} \tag{74}$$

$$\sigma_{\theta\theta}^j(r,t) = s_{\theta\theta}^j(r,\omega)e^{i\omega t} = \frac{\partial^2 f_j(r,\omega)}{\partial r^2}e^{i\omega t} \tag{75}$$

where $s_{rr}^j(r,\omega)$ and $s_{\theta\theta}^j(r,\omega)$ are the complex frequency response functions for the radial and tangential stress components, respectively.

Because the propellant is considered viscoelastic material, the stress-strain relation is given by[37]

$$S(t) = \int_{-\infty}^{t} E(t-\tau,T)d\epsilon(\tau) \tag{76}$$

where $S(t)$ and $\epsilon(t)$ are the time dependent stress and strain components, respectively, and $E(t,T)$ is the time and temperature dependent viscoelastic relaxation modulus, Eq. (25). For the steady-harmonic temperature input, the response stresses may be obtained via the correspondence principal[37] together with the frequency dependent complex modulus, Eq. (28). The viscoelastic constants E_i and τ_i for the propellant are shown in Table 5.3. Table 5.4 gives the dimensions and thermo- and elastomechanical properties of the structure.

Due to the characteristics of the environmental temperature, the response stresses in the propellant are random quantities. Furthermore, it was found in this case that the most significant stress component in the structure is the tangential stress at the bore ($r = r_1$). Consequently, the maximum induced stress is written in the form

$$S(t) = S_m(t) + S_N(t) \tag{77}$$

where $S_m(t)$ and $S_N(t)$ are the deterministic and random stress components, respectively. The deterministic stress component is due to the long-term mean, yearly, and daily variations of the environmental temperature, while the random stress component mainly accounts for the hourly fluctuation of the air temperature noise.

The variance of the maximum induced stress, σ_S^2, can be obtained via the use of the random process theory as[74]

$$\sigma_S^2 = 2\int_0^\infty W_a(\omega)\,|H_S(\omega)|^2\,d\omega \tag{78}$$

where $W_a(\omega)$ is the one-sided spectral density function of the air temperature noise and $H_S(\omega)$ is the complex frequency response function for the maximum induced stress.

The damage produced on the nth day is calculated by considering the variation of the damaging stress, $S_D(t)$, in the form

$$S_D(t) = S_m(n) + S_d\cos(\omega_d t + \phi_d) \tag{79}$$

where $S_m(n)$ and S_d are the daily mean and daily amplitude of the damaging stress (i.e., tangential stress at the bore), while ω_d and ϕ_d are the daily frequency and phase angle, respectively. Therefore, in a given day for which a fraction of time t_l is spent at stress level S_l, the linear damage rule, Eq. (34), results in

$$D(n) = \sum_{l=1}^{L}\frac{t_l t_d}{t_{fl}} \tag{80}$$

where $D(n)$ is the corresponding damage on the nth day, t_{fl} is the time to failure at stress level S_l, t_d = 24 hours, and L is the total number of time increments considered. For continuously varying stress level, the daily damage can be canculated based on the density function concept[75]

$$D(n) = \frac{t_d S_d^B}{\pi C a_T}\int_{-\pi/2}^{\pi/2}\left(\frac{S_m(n)}{S_d} + \sin\theta\right)^2 d\theta \tag{81}$$

where the damage rule of Eq. (35) is assumed. In this example, the damage parameters $C = 1.4211 \times 10^{16}$ for S in psi (5.49138×10^{49} for S in N/m^2) and $B = 8.75$ are used.

The statistical parameters (mean and variance) for the resulting cumulative damage can now be evaluated by Eqs. (50) and (51). Yearly mean variations of daily damage for both uninsulated and insulated motors are shown in Fig. 5.9, while the corresponding mean cumulative damages are shown in Fig. 5.10.

For a given day, Eqs. (50) and (51) can be used to calculate the approximate values for means and variances of the maximum induced viscoelastic stress and of the viscoelastic strength of the motor. The daily probability of

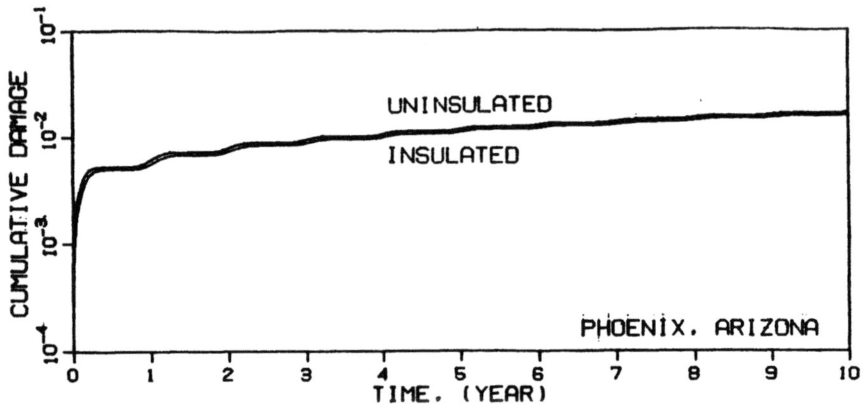

Figure 5.10 Variations of cumulative damage for uninsulated and insulated cylinders at Phoenix, AZ.

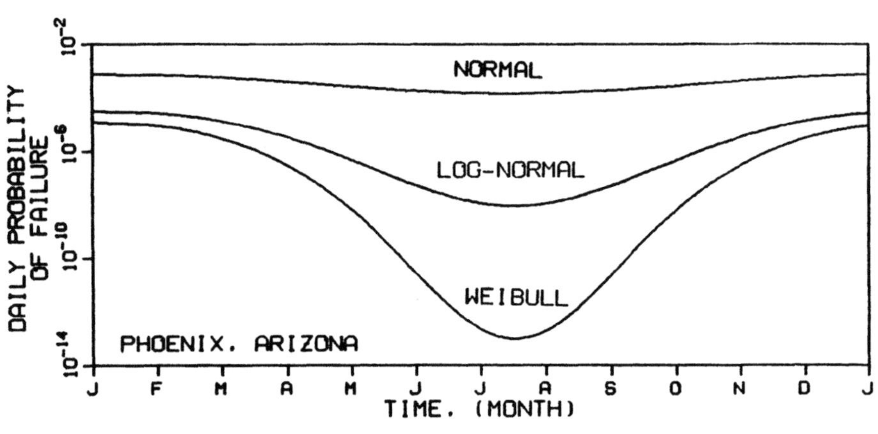

Figure 5.11 Daily probability of failure due to excessive daily maximum stress for uninsulated cylinder at Phoenix, AZ.

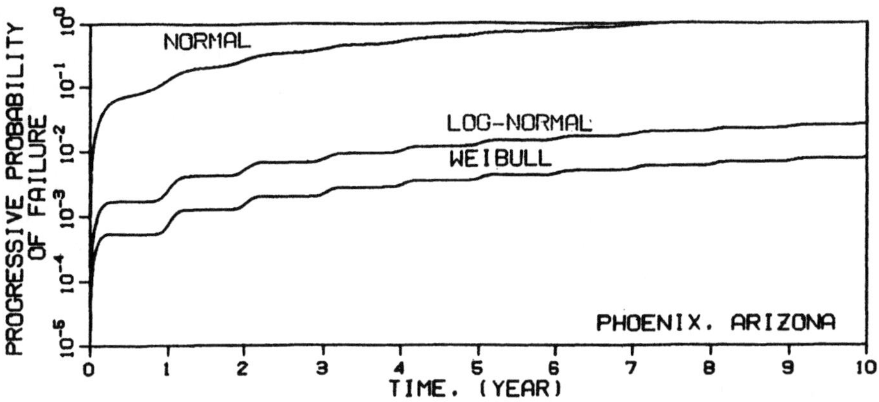

Figure 5.12 Progressive probability of failure due to excessive daily maximum stress for uninsulated cylinder at Phoenix, AZ.

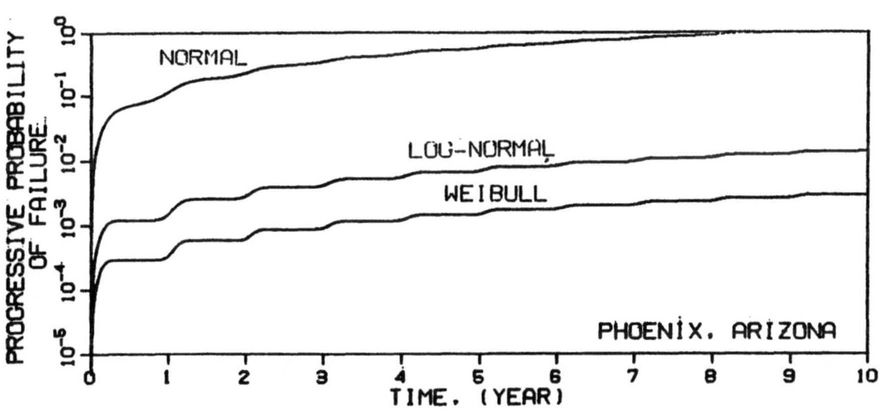

Figure 5.13 Progressive probability of failure due to excessive daily maximum stress for insulated cylinder at Phoenix, AZ.

failure due to the environmentally induced stress can then calculated based on the concept of interference of the distributions of stress and strength, Eq. (53). Figure 5.11 shows the variations of daily probability of failure due to excessive daily maximum stress for an uninsulated motor at Phoenix, AZ. Depending on the types of probability density function assumed for the design variables (stress, S, and strength, R), a significant difference in the predicted daily probabilities of failure may be found. Figures 5.12 and 5.13 show the variations of progressive probability of failure, Eq. (63), due to excessive daily maximum stress for uninsulated and insulated motors at Phoenix, AZ, respectively.

6 Concluding Remarks

The forgoing discussion has indicated some of the techniques available for the calculation of progressive probability of failure for structures in which time dependent degradation of material properties and changes in applied stresses produce variable probabilities.

The structure in the described example has only a single failure mode. The problem becomes more complicated when multiple load path structures are considered. In those cases reliability block diagrams or "fault trees" are constructed before the above techniques are applied to individual components. Discussion of such structures is, however, not within the scope of the present paper.

References

1. Haviland, R. P., *Engineering Reliability and Long Life Design*, D. Van Nostrand, New York, 1964.
2. Bolotin, V. V., *Prediction of Service Life for Machines and Structures*, ASME Press, New York, 1989.
3. Masters, L. W., ed., *Problems in Service Life Prediction of Bulding and Construction Materials*, NATO ASI Series E: Applied Sciences - No. 95, Martinus Publishers, the Netherland, 1985.
4. Čačko, J., Bílý, M., and Bukoveczky, J., *Random Processes: Measurement, Analysis, and Simulation*, Elsevier, Amsterdam, 1988.
5. Khosrovaneh, A. K. and Dowling, D. E., Fatigue Loading History Reconstruction Based on the Rainflow Technique, *International Journal of Fatigue*, Vol. 12, No. 2, 1990, pp. 99-106.
6. Haibach, E., Fischer, R., Schütz, W., and Hück, M., A Standard Random Loading Sequence of Gaussian Type Recommended for General Application in Fatigue Testing: Its Mathematical Background and Digital Generation, *Proceedings of the International Conference on Fatigue Testing and Design*, Vol. 2, SEE, London, April, 1976, pp. 29.1-29.21.
7. Box, G. E. P. and Jenkins, G. M., *Time Series Analysis, Forecasting and Control*, Holden-Day, San Francisco, 1976.

8. Priestley, M. B., *Spectral Analysis and Time Series, Volume 1: Univariate Series*, Academic Press, Florida, 1981.
9. Pandit, S. M. and Wu, S. M., *Time Series and System Analysis with Applications*, John Wiley and Sons, new York, 1983.
10. Samaras, E., Shinozuka, M., and Tsurui, A., ARMA Representation of Random Processes, *Journal of Engineering Mechanics*, ASCE, Vol. 111, No. 3, 1985, pp. 449-461.
11. Spanos, P. D. and Mignolet, M. P., Recursive Simulation of Stationary Multivariate Random Processes-Part 2, *Journal of Applied Mechanics*, ASME, Vol. 109, 1987, pp. 681-687.
12. Polhemus, N. W. and Cakmak, A. S., Simulation of Earthquake Ground Motions Using ARMA Models, *Earthquake Engineering and Structural Dynamics*, Vol. 9, 1981, pp. 343-354.
13. Kozin, F., Autoregressive Moving Average Models of Earthquake Records, *Probabilistic Engineering Mechanics*, Vol. 3, No. 2, 1988, pp. 58-63.
14. Houmb, O. G. and Overvik, T., Some Applications of Maximum Entropy Spectral Estimation to Ocean Waves and Linear Systems Response in Waves, *Applied Ocean Resources*, Vol. 3, No. 4, 1981, pp. 154-162.
15. Spanos, P. D., ARMA Algorithms for Ocean Wave Modeling, *Journal of Energy Resource Technology*, ASME, Vol. 105, September, 1983, pp. 300-309.
16. Reed, D. A. and Scanlan, R. H., Time Series Analysis of Cooling Tower Wind Loading, *Journal of the Engineering Mechanics Division*, ASCE, Vol. 109, No. 2, 1983, pp. 538-554.
17. Li, Y. and Kareem, A., ARMA Systems in Wind Engineering, *Probabilistic Engineering Mechanics*, Vol. 5, No. 2, 1009, pp. 50-59.
18. Bendat, J. S. and Piersol, A. G., *Engineering Application of Correlation and Spectral Analysis*, John Wiley, New York, 1980.
19. Cooley, J. W., Lewis, P. A., and Welch, P. D., The Fast Fourier Transform and Its Applications, *IEEE Transactions on Education*, Vol. 12, No. 1, 1969, pp. 27-34.
20. Enochson, L. D. and Otnes, R. K., Programming and Analysis for Digital Time Series Data, *Shock and Vibration Monograph Series*, U.S. Department of Defense, Navy Publication and Printing Service Office, Washington, D.C., 1968.
21. Shinozuka, M. and Jan, C.-M., Digital Simulation of Random Processes and Its Applications, *Journal of Sound and Vibration*, Vol. 25, 1972, pp. 111-128.
22. Heller, R. A., Thermal Stress as a Narrow Band Random Load, *Journal of the Engineering Mechanics Division*, ASCE, Vol. 102, No. EM5, October, 1976, pp. 787-805.
23. Wirsching, P. H. and Light, M. C., Fatigue under Wide Band Random Stresses, *Journal of the Structural Division*, ASCE, Vol. 106, No. ST7, July, 1980, pp. 1593-1607.
24. Brooks, E. M., *Climate*, Encyclopedia of America, Americana Corp., 1970.

25. Heller, R. A., Kamat, M. P., and Singh, M. P., Probability of Solid Propellant Motor Failure Due to Environmental Temperature, *Journal of Spacecraft and Rockets*, Vol. 16, No. 3, 1979, pp. 140-146.
26. Heller, R. A. and Singh, M. P., Thermal Storage Life of Solid Propellant Motors, *Journal of Spacecraft and Rockets*, Vol. 20, No. 2, 1983, pp. 144-149.
27. Thangjitham, S. and Heller, R. A., Test and Evaluation of Dynamic Properties of Solid-Propellant Rocket Motor Materials, Volume I-Random Environmental Thermal Load Models and Their Effects on the Service Life of Rocket Motors, Technical Report CR-RD-86-8, *USAMICOM*, Redstone Arsenal, Alabama, October, 1986.
28. Anon., *Hourly Solar Radiation-Surface Meteorological Observations*, National Oceanic and Atmospheric Administration, Vol. I & II, TD-9724, 1978.
29. Thangjitham, S. and Heller, R. A. Thermal Stresses in a Rocket Motor under Environmental Loading, *Journal of Spacecraft and Rockets*, AIAA, Vol. 23, No. 5, 1986, pp. 519-526.
30. Donn, W. L., *Meteorology*, 3rd ed., McGraw-Hill, New York, 1965.
31. Chou, K. C. and Corotis, R. B., Generalized Wind Speed Probability Distribution, *Journal of Engineering Mechanics Division*, Transactions of ASCE, Vol. 109, No. 1, February, 1983, pp. 14-29.
32. Dawson, T. H., *Offshore Structure Engineering*, Prentice Hall, N.J., 1983.
33. Dean, R. G. and Daynuple, R. A., *Water Wave Mechanics for Engineers and Scientists*, Prentice Hall, N.J., 1984.
34. Bretschneider, C. L., Overwater Wind and Wind Forces, in *Handbook of Ocean and Underwater Engineering*, Chapter 12, J. J. Meyers, et al., eds., McGraw-Hill, New York, 1969.
35. Layton, L. H., *Chemical Aging Studies on ANB-3066 and TH-H1011 Propellants*, Final Technical Report, AFRPL-TR-74-16, 1974.
36. Cost, T. L., Computer Simulation of Solid Rocket Motor Service Life for Thermal Loads, in *Thermal Stresses in Severe Thermal Environments*, D. P. H. Hasselman and R. A. Heller, eds., Plenum Press, New York, 1980, pp. 431-445.
37. Christensen, R. M., *Theory of Viscoelasticity*, Academic Press, New York, 1971.
38. Ferry, J. D., *Viscoelastic Properties of Polymers*, John Wiley & Sons, New York, 1970.
39. Palmgren, A., Ball and Roller Bearing Engineering, translated by G. Palmgren and B. Ruley, *SKF Industries, Inc.*, Philadelphia, PA, 1945, pp. 82-83.
40. Miner, M. A., Cumulative Damage in Fatigue, *Journal of Applied Mechanics*, ASME, Vol. 12, September, 1945, pp. a159-164.
41. Freudenthal, A. M., and Heller, R. A., On Stress Interaction in Fatigue and a Cumulative Damage Rule, *Journal of the Aero/Space Science*, Vol. 26, No. 9, 1959, pp. 431-442.

42. Bills, W. K., Jr. and Wiegand, J. H., The Application of an Integrated Structural Analysis to the Prediction of Reliability, *Annals of Reliability and Maintainability*, 1970, pp. 514-526.
43. Bogdanoff, J. L. and Kozin, K., *Probabilistic Models of Cumulative Damage*, John Wiley & Sons, New York, 1985.
44. Bažant , Z. P., Mathematical Models for Creep and Shrinkage of Concrete, in *Creep and Shrinkage in Concrete Structures*, Z. P. Bažant and F. H. Wittmann, eds., John Wiley & Sons, New York, 1982.
45. Freudenthal, A. M., Aspects of Fatigue Damage Accumulation at Elevated Temperature, *Proceedings of International Conference on Mechanisms of Fatigue*, Acta Metallurgica, Vol. 77, July, 1963.
46. Heller, R. A. and Donat, R. C., Fatigue Failure of a Redundant Structure, *ASTM-STP 404*, American Society for Testing and Materials, 1964, pp. 55-66.
47. Paris, P. C. and Erdogan, F., A Critical Analysis of Crack Propagation Laws, *Journal of Basic Engineering*, Trans. ASME, Series D, Vol. 85, 1963, pp. 528-534.
48. Forman, R. G., Kearney, V. E., and Engle, R. M., Numerical Analysis of Crack Propagation in Cyclic Loaded Structures, *Journal of Basic Engineering*, Trans. ASME, Series D, Vol. 89, 1967, pp. 459-465.
49. Walker, E. K., The Effect of Stress Ratio During Crack Propagation and Fatigue for 2024-T3 and 7075-T6 Aluminum, in *The Effect of Environment and Complex Load History on Fatigue Life*, ASTM-STP 462, 1970, pp. 1-14.
50. Heller, R. A. and Thangjitham, S., The Effects of Variable Temperature on Fracture, *Arch. Mech.*, Vol. 39, 1987, pp.395-409.
51. Heller, R. A., ed., *Probabilistic Aspects of Fatigue*, *ASTM STP 511*, American Society for Testing and Materials, 1972.
52. Provan, J. W., *Probabilistic Fracture Mechanics and Reliability*, Martinus Nijhoff, 1987.
53. Hahn, G. J. and Shapiro, S. S., *Statistical Models in Engineering*, John Wiley & Sons, New York, 1967.
54. Benjamin, J. R. and Cornell, C. A., *Probability, Statistics and Decision for Civil Engineers*, McGraw-Hill, New York, 1970.
55. Ang, A, H-S. and Tang, W. H., *Probability Concepts in Engineering Planning and Design: Volume I, Basic Principles*, John Wiley & Sons, New York, 1975.
56. Freudenthal, A. M., Garrelt, J. M., and Shinozuka, M., The Analysis of Structural Safety, *Journal of Structural Division*, Transactions of the ASCE, Vol. 92, No. ST1, 1966, pp. 267-325.
57. Ang, A, H-S. and Tang, W. H., *Probability Concepts in Engineering Planning and Design: Volume II, Decision, Risk, and Reliability,* John Wiley & Sons, New York, 1984.
58. Rackwitz, R., Practical Probabilistic Approach to Design, *Bullentin 112*, Comite European du Beton, Paris, France, 1976.

59. Moses, F. and Kinser, D. E., Analysis of Structural Reliability, *Journal of the Structural Division*, ASCE, Vol. 93, No. ST5, October, 1967.
60. Chamis, C. C., Probabilistic Structural Analysis Methods for Space Propulsion System Components, Probabilistic Engineering Mechanics, Vol. 2, No. 2, 1987, pp. 100-110.
61. Faravelli, L., A Response Surface Approach for Reliability Analysis, *Journal of Engineering Mechanics*, ASCE, Vol.115, No. 12, 2763-2781.
62. Heller, R. A. and Janajreh, I. M., The Reliability of Rocket Motors Subjected to Static and Dynamic Loads, Technical Report CR-RD-PR-92-4, *USAMICOM*, Redstone Arsenal, AL, April, 1982.
63. Fujita, M., Schall, G., and Rackwitz, R., Time Variant Component Reliabilities by FORM and SORM and Updating by Importance Sampling, *Proceedings of ICASP 5*, Vancouver, Vol. 2, 1987, pp. 520-527.
64. Schueller, G. I. and Stix, R., A Critical Appraisal of Methods to Determine Failure Probabilities, *Structural Safety*, Vol. 4, 1987, pp. 293-309.
65. Bjerager, P., Probability Integration by Directional Simulation, *Journal of Engineering Mechanics*, ASCE, Vol. 114, No. 8, 1988, pp. 1285-1301.
66. Wu, T. and Wirsching, P. H., Algorithm for Structural Reliability Estimation, *Journal of Engineering Mechanics*, ASCE, Vol. 113, No. 9, 1987, pp. 1319-1336.
67. Bazovsky, I., *Reliability Theory and Practice*, Prentice Hall, Englewood Cliffs, NJ, 1961.
68. Cramer, H. and Leadbetter, M. R., *Stationary and Related Stochastic Processes*, John Wiley and Sons, New York, 1967.
69. Zibdeh, H. S. and Heller, R. A., Environmental Thermal Stress as a First Passage Problem, in *Random Vibration–Status and Recent Developments*, I. Elishakoff and R. H. Lyon, eds., Elsevier, Amsterdam, 1986, pp. 537-553.
70. Carslaw, H. S. and Jaeger, J. C., *Conduction of Heat in Solids*, 2nd ed., Oxford University Press, London, 1976.
71. Thangjitham, S., Heller, R. A., and Singh, M. P., Frequency Response Functions for Thermal Stresses in Multilayered Cylindrical Structures, *Journal of Thermal Stresses*, Vol. 9, 1986, pp. 133-150.
72. Boley, B. A. and Weiner, J. H., *Theory of Thermal Stress*, John Wiley and Sons, Inc., New York, 1960.
73. Timoshenko, S. P. and Goodier, J. N., *Theory of Elasticity*, McGraw-Hill, New York, 1970.
74. Newland, D. E., *Random Vibrations and Spectral Analysis*, Longman, London, 1975.
75. Heller, R. A. and Singh, M. P., Allowable Strength of Viscoelastic Materials Under Variable Thermal Loads, in *Thermal Stresses in Severe Thermal Environments*, D. P. H. Hasselman and R. A. Heller, eds., Plenum Press, New York, 1980, pp. 491-502.

Prediction Control of Structures by ARMA Modeling

M.Hoshiya[1] and Y.Saito[2]

[1] Department of Civil Engineering, Musashi Institute of Technology, 1-28-1 Tamazutsumi,Setagaya-ku,Tokyo 147 Japan.
[2] Technical Research Institute, Maeda Corporation, 1-39-16 Asahicho, Nerima-ku, Tokyo 179 Japan.

Abstract. A basic analysis is carried out on identification, prediction, and control of a single degree of freedom system. First, the responses of the system excited by an active control device installed on the system are effectively utilized to identify the dynamic properties of the system which is modeled by a multi-variate ARMA model. Then, general modes of an instantaneous optimal prediction control rule are formulated in terms of the identified components of the coefficient matrix of the ARMA model and the weights included in the control objective function. Based on the formulation, a neural network is derived whose links have physically meaningful weights.

Keywords. Prediction control, ARMA model, Identification

1 Introduction

In the design of an optimal regulator, control force is generally determined such that a quadratic performance index becomes minimum. When applied to civil engineering structures under earthquake excitation or strong wind turbulence force, however, there arise the following critical issues even for a simple structure:

(a) Since excitation forces and corresponding responses are unknown apriori, instantaneous optimal control must be taken up as a method relevant to the problems.

(b) Time delay of control force due to the time spent for evaluation of control force and needed for sending signal to actuate control device must be avoided. To circumvent the delay, prediction of input and responses is necessary in the on-line real time process.

(c) Structures are subject to many uncertainties, and their structural properties must be identified apriori in order that control force is rationally computed.

Although the classical control procedures are well established within the scope of the optimal regulator problem, instantaneous optimal control remains to be solved for basic matters of system identification and prediction, and on clarification of mutual relationships among different control rules that meet with the above mentioned requirements.

This study addresses itself to the abovementioned issues and investigates a single degree of freedom linear system excited by earthquake ground motion and controlled by an actuator installed on the system. First, the system is represented by an ARMA model[1)2)], and a theoretical formulation is derived to identify the dynamic properties of the system by exciting the system by the actuator. Then, with the identified ARMA model, together with an AR representation[3)4)] of earthquake excitation, response displacement and velocity of the system as well as the earthquake excitation are predicted in recursive forms so that a procedure of identification, prediction, and control is established on the basis of an instantaneous optimal control rule[5)6)]. Finally, a sensitivity analysis is carried out for the effect of identification and prediction errors of the system control. Also demonstrated is an equivalent neural network model[7)] stemmed from the formulation, whose links retain physically meaningful weight values.

2 System Identification

A SDOF linear system is represented by a multi-variate ARMA model. This ARMA model is identified with observed responses of the system excited by an actuator which will be used in the instantaneous optimal control. The identified coefficient matrices of the ARMA model are theoretically related with physical parameters such as mass, spring constant, and viscous damping coefficient. Thus, all structural properties are to be identified.

The governing equation of the SDOF system which is subject to earthquake excitation and control force may be expressed by the state vector equation

$$\dot{\mathbf{X}}(t) = \mathbf{A}\mathbf{X}(t) + \mathbf{B}[u(t) - m\ddot{g}(t)] \quad , \tag{1}$$

where

$$\mathbf{X}^T(t) = [x(t) \;\; \dot{x}(t)]$$

$$\mathbf{A} = \begin{bmatrix} 0 & 1 \\ -k/m & -c/m \end{bmatrix} \qquad \mathbf{B} = \begin{bmatrix} 0 \\ -1/m \end{bmatrix}$$

m : mass
k : spring constant
c : viscous damping coefficient
u (t) : control force
$\ddot{g}$ (t) : earthquake acceleration

Discretization of eq(1) with time interval Δ leads to the multi-variate ARMA equation

$$X_{k+1} = \Phi X_k + \Gamma\,[u_k - m\ddot{g}_k] \quad , \tag{2}$$

where k stands for discrete time $t=k\Delta$,
where

$$\Phi = I + A\Delta + \frac{1}{2!}A^2\Delta^2 + \ldots + \frac{1}{r!}A^r\Delta^r + \ldots = \begin{bmatrix}\Phi_{11} & \Phi_{12}\\ \Phi_{21} & \Phi_{22}\end{bmatrix} \quad ,$$

and

$$\Gamma = (I + \frac{1}{2!}A\Delta + \frac{1}{3!}A^2\Delta^2 + \ldots + \frac{1}{r!}A^{r-1}\Delta^{r-1} + \ldots)\,B\Delta = \begin{bmatrix}\Gamma_1\\ \Gamma_2\end{bmatrix} \quad ,$$

since it is proved that eigenvalues Λ and eigenvector Z of matrix A and eigenvalues Λ^* of matrix Φ are mutually related[1)] through the equations

$$A\,Z = Z\,\Lambda \qquad \Phi Z = Z\,\Lambda^* \qquad \Lambda^* = \exp(\Lambda\,\Delta) \quad . \tag{3}$$

So if Φ is identified, matrix A is determined from solving the eigenvalue ploblem of Φ. If vector Γ is also identified, then matrix B is determined by the following equation

$$B = A\,(\Phi - I)^{-1}\,\Gamma \quad . \tag{4}$$

It is noted that since matrices A and B include physical parameters such as m, k, and c, all necessary properties of the system are determined.

Now, in order to identify Φ and Γ, the Kalman filter is applied to the state vector and observation vector equations, which yields respectively

$$[\Phi_{11}\,\Phi_{12}\Gamma_1\,\Phi_{21}\,\Phi_{22}\Gamma_2]^T_{k+1} = [\Phi_{11}\,\Phi_{12}\Gamma_1\,\Phi_{21}\,\Phi_{22}\Gamma_2]^T_k \quad , \tag{5}$$

and

$$\begin{bmatrix}x_{k+1}\\ \dot{x}_{k+1}\end{bmatrix} = \begin{bmatrix}x_k & \dot{x}_k & u_k & 0 & 0 & 0\\ 0 & 0 & 0 & x_k & \dot{x}_k & u_k\end{bmatrix}\begin{bmatrix}\Phi_{11}\,\Phi_{12}\,\Gamma_1\,\Phi_{21}\Phi_{22}\,\Gamma_2\end{bmatrix}^T_k + W_k \quad , \tag{6}$$

where equation(6) is obtained from eq(2) provided that $\ddot{g}_k=0$, and W_k is a white noise vector.

3 Instantaneous Optimal Control

The following performance index is employed for the instantaneous optimal control of the discrete system expressed by eq(2):

$$J_k = \frac{1}{2}\left[\mathbf{X}^T_{k+1}\mathbf{Q}\mathbf{X}_{k+1} + ru^2_k + 2\left(-m\ddot{g}_k\right)\mathbf{L}^T\mathbf{P}\mathbf{X}_{k+1}\right] \quad , \tag{7}$$

where

$$\mathbf{L} = \begin{bmatrix}0\\1\end{bmatrix} \quad , \quad \mathbf{P} = \begin{bmatrix}0 & 0\\0 & \alpha\end{bmatrix} \quad , \quad \mathbf{Q} = \begin{bmatrix}k & 0\\0 & m\end{bmatrix} \quad .$$

This performance index is basically identical to such an index as proposed by Sato et al.[8] that includes an additional term of effective input energy in the index of an open-closed loop control by Yang[5)6)]. In the equation(7), $\mathbf{Q}$, r , and $\mathbf{P}$ represent respectively the weighting coefficients to govern energy distribution among the first term of kinetic and potential energy, the second term of control force energy, and the third term of effective input energy. The control force may be determined such that the performance index becomes minimum under a constraint equation(2). So doing, we obtain the following control force consisting of feed back and feed forward terms:

$$u_k = \mathbf{G}_{FB}\mathbf{X}_k + G_{FF}\ddot{g}_k \quad , \tag{8}$$

where

$$\mathbf{G}_{FB} = -\left[r + \Gamma^T\mathbf{Q}\Gamma\right]^{-1}\Gamma^T\mathbf{Q}\Phi \quad , \tag{9}$$

and

$$G_{FF} = \left[r + \Gamma^T\mathbf{Q}\Gamma\right]^{-1}\Gamma^T\left[\mathbf{Q}\Gamma + \mathbf{P}^T\mathbf{L}\right] m \quad . \tag{10}$$

Equations(9) and (10) indicate that the feed back gain G_{FB} and the feed forward gain G_{FF} are known apriori if the system is identified in advance as described in the preceding section. However, in order to determine the control force u_k at time t=kΔ, we must know the responses $\mathbf{X}_k$ and input $\ddot{g}_k$ at the stage of t=(k-1)Δ. As for $\mathbf{X}_k$, we use equation(2) as

$$\mathbf{X}_k = \Phi\mathbf{X}_{k-1} + \Gamma\left[u_{k-1} - m\ddot{g}_{k-1}\right] \quad . \tag{11}$$

The prediction of $\ddot{g}_k$ is carried out by modeling the input acceleration by an AR representation[4)] as

$$\ddot{g}_k = \alpha_1\ddot{g}_{k-1} + \alpha_2\ddot{g}_{k-2} + \alpha_3\ddot{g}_{k-3} + \alpha_4\ddot{g}_{k-4} + v_k \quad , \tag{12}$$

where v_k is a white noise.

The unknown coefficients α_i are to be evaluated recursively during the control process by applying the Kalman filter to the state vector equation and the observation equation respectively. This produces the equations

$$\left[\alpha_1\ \alpha_2\ \alpha_3\ \alpha_4\right]^T_{k-1} = \left[\alpha_1\ \alpha_2\ \alpha_3\ \alpha_4\right]^T_{k-2} \quad , \tag{13}$$

and

$$\ddot{g}_{k-1} = \left[\ddot{g}_{k-2}\ \ddot{g}_{k-3}\ \ddot{g}_{k-4}\ \ddot{g}_{k-5}\right]\left[\alpha_1\ \alpha_2\ \alpha_3\ \alpha_4\right]^T_{k-1} + v_{k-1} \quad , \tag{14}$$

where v_{k-1} is a white noise.

In equation(14), $\ddot{g}_{k-1}$ to $\ddot{g}_{k-5}$ are already observed at time (k-1) Δ, and once α_i are identified, $\ddot{g}_k$ at time k Δ is predicted from equation(12).

A special case is emphasized in which the mass of the system is a known constant. Then, excitation terms $u_{k-1} - m\ddot{g}_{k-1}$ in equation(11) are to be known at time (k-1) Δ. Therefore, using them as the excitation, the remaining parameters k and c may be identified in a manner similar to the procedure of the preceding section. If this process is integrated into the sequential real time control algorithm, we can deal with a case in which the structural stiffness and damping are time dependent. The adaptive algorithm to determine the control force u_k at time k Δ based on the identification of the system at time (k-1) Δ and on the prediction of X_k and $\ddot{g}_k$ is schematically illustrated in Fig. 3.1. At the stage of t=(k-1) Δ, the responses X_k and input $\ddot{g}_k$ are predicted by eq.(11) and (12), and the feed back gain G_{FB} and the feed forward gain G_{FF} are determined from the Φ and Γ which are identified using the observations. Therefore, the control force u_k at time k Δ is determined from the prediction and identified values.

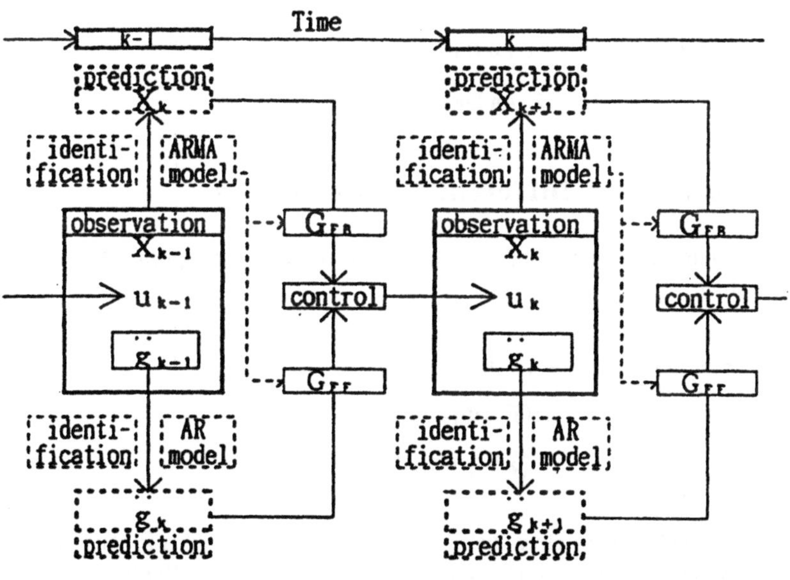

Fig.3.1. Schematical Procedure of Identification, Prediction, and Control

The proposed procedure may be also interpreted as a neural network model as in Fig.3.2.. It should be noted that every link appearing in the network retains physically meaningful weight values.

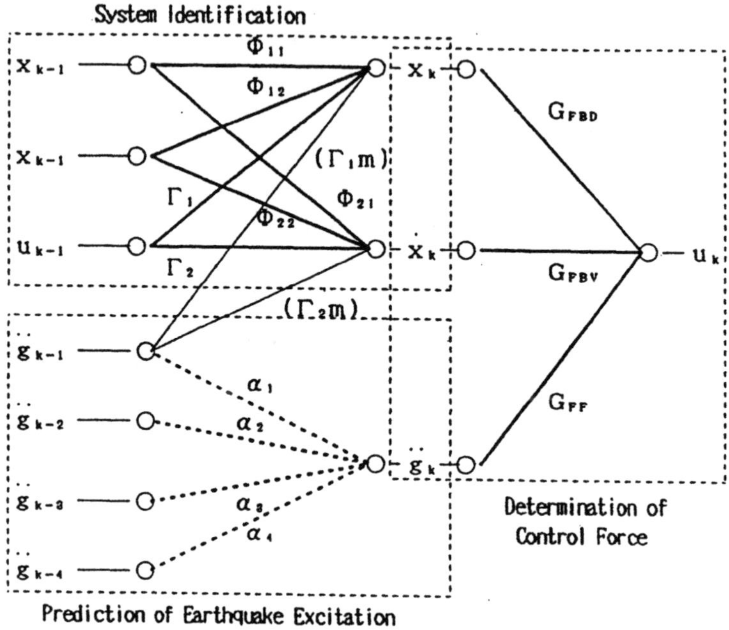

Fig.3.2. An Equivalent Neural Network

4 Comparison of Different Control Rules

The control force of equation(8) is restated in terms of components as shown in the equation

$$u_k = \frac{-1}{r + k\Gamma^2_1 + m\Gamma^2_2} \Big[\left(\Gamma_1 k\Phi_{11} + \Gamma_2 m\Phi_{21}\right) x_k + (\Gamma_1 k\Phi_{12} + \Gamma_2 m\Phi_{22})\, \dot{x}_k - \left(k\Gamma^2_1 + m\Gamma^2_2 + \alpha\Gamma_2\right) m\ddot{g}_k \Big] \quad (15)$$

From equation(15), it is clear that the weighting component α for the effective input energy regulates the feed forward gain relatively against the feed back gain of the first and second terms in the numerator, whereas the weighting component r for the control force energy that appears in the denominator regulates the magnitude of the control force.

Introducing a parameter β to relate α and r as

$$\alpha = \beta * r / \Gamma_2 \quad , \quad (16)$$

different control rules will be discussed in the following paragraphs.

If the third term is dropped out, equation(15) becomes (a)the closed loop control by Yang[5)6)], whereas if $\beta = 0$, it becomes (b)the open-closed loop control proposed again by Yang[5)6)]. The latter case(b) means that the ratio of the feed back gain to the feed forward gain is fixed. By varying the value of α in equation(15), different control rules are designed which are namely (c)hybrid feed back and feed forward control. The value $\beta = 1$ indicates that the third term of the feed forward control force is always equal to the input excitation force. In this control(d), the excitation force is perfectly counterbalanced with the feed forward control force, and no response will be expected. So there arises no feed back control force. As r becomes smaller, however, the value of the feed back gain increases. Finally, if $\beta = (r - k\Gamma_1^2 - m\Gamma_2^2)/2r$, the feed forward term of equation(15) becomes half of the input excitation force. This is (e)a control rule that counterbalances with half of input force.

In order to visualize the above discussions and the mutual relationships between different control rules (a) to (e), numerical analyses are carried out for a system given in Table 4.1. and excited by an El Centro earthquake-NS component acceleration record.

Table 4.1. A SDOF System

mass	(m)	1.0 (ts²/cm)
stiffness	(k)	39.478(t/cm)
natural period	(T)	1.0 (sec)
damping coefficient	(h)	2.0 (%)

The results are shown in Fig.4.3. through 4.8. where maximum values of the control force u_k and maximum response displacement are comparatively given as functions of r . Fig.4.3. and 4.4. show results based on true input and responses whereas all other figures are obtained based on the prediction of input force. The symbols used are summarized in Table 4.2.

Table 4.2. Different Control Rules

Control Rules	Symbols
(a)closed loop by Yang	—■—
(b)open-closed loop by Yang	—◆—(α=0.0)
(c)hybrid feed back and feed forward control	—◇—(α=0.1) —▲—(α=0.5) —△—(α=1.0) —●—(α=2.0)
(d)feed forward control fixed for $\beta=1$	—□—
(e)feed forward control fixed for $\beta=(r-k\Gamma_1^2-m\Gamma_2^2)/2r$	—○—

It is observed that since the closed loop control(a) is dependent upon only the state vector $\mathbf{X}_k$, it is naturally not influenced by the prediction errors of input force (compare the curves(— ■ —) in Fig.4.3. and 4.5., and in Fig.4.4. and 4.6.). The open-closed loop control(b) approaches the closed loop control as r increases and approaches the control rule(d) as r decreases (see the curves (—◆—)). In the hybrid feedback and feed forward control(c), curves become the open-closed loop control(b) when r is large, but they approach the control(d) when r decreases. When r satisfies with $r = \alpha \Gamma^2$, this control rule coincides with the control(d). If r furthermore decreases beyond this critical point, the control will diverge. As for the control(d), the response is always zero if no prediction error is involved. However , if some errors exist in the input prediction, the input force can not be counterbalanced with the feed forward control force, and the difference will be shared by the feed back control. The control(e) is always between the open-closed loop control(b) and the control(d).

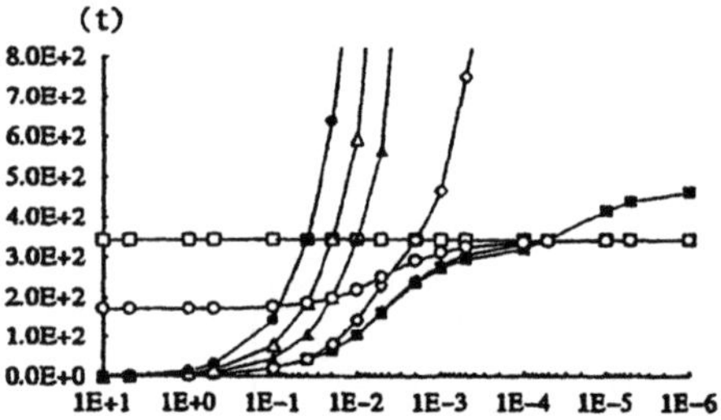

Fig.4.3. Maximum Control Force
(True Input)

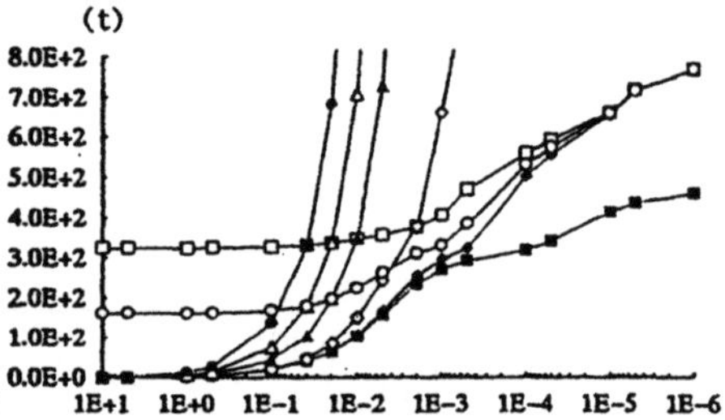

Fig.4.5. Maximum Control Force
(Prediction Input)

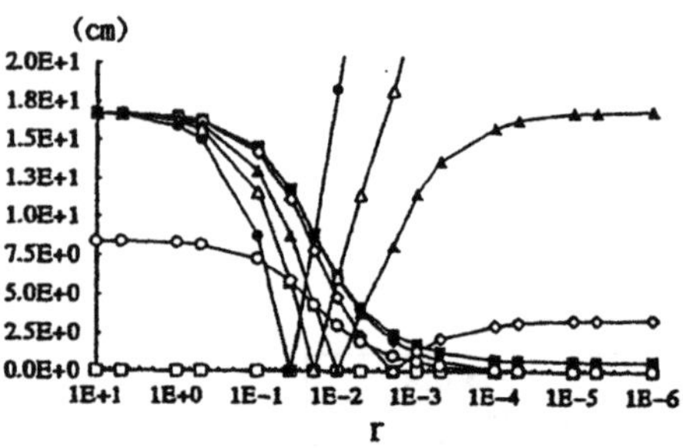

Fig.4.4. Maximum Response
Displacement

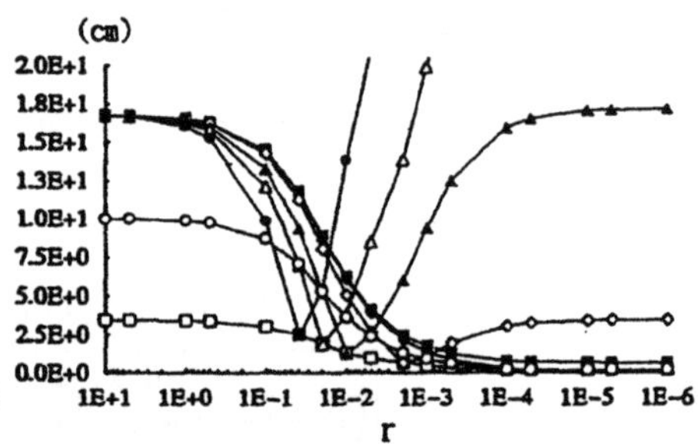

Fig.4.6. Maximum Response
Displacement

Fig.4.7. and 4.8. show results in the case that the control device has an upper bound of the control capability of 150t. So it is observed that the maximum values of the control force are always less than or equal to 150t (see Fig.4.7.). On the contrary, the maximum values of the response displacements are increased greater than those in the case of having no upper bound of the control force capacity (compare Fig.4.6. and 4.8.). As for the hybrid feed back and feed forward control(c), however, the responses caused by the surplus control force are apt to be restrained (see Fig.4.8.).

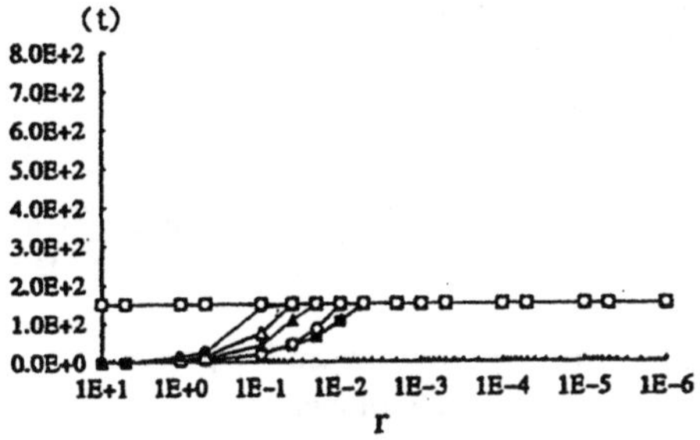

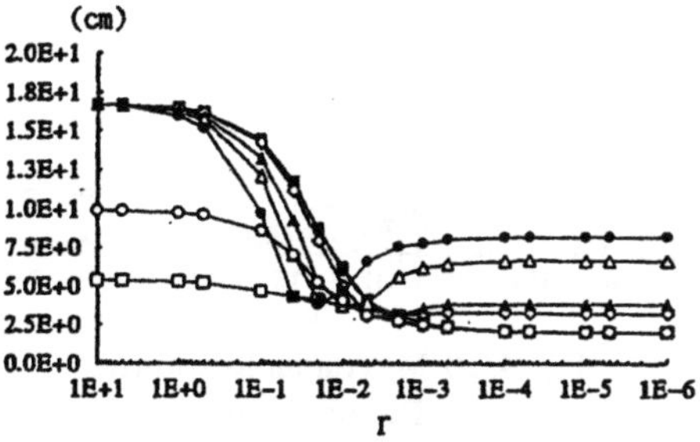

Fig.4.7. Maximum Control Force (Prediction Input and Limit of Control Force)

Fig.4.8. Maximum Response Displacement (Prediction Input and Limit of Control Force)

Fig.4.9. through 4.12. show results including identification error for the mass. They are obtained based on the true input in order to observe the effect of only identification error. First, the error of $\pm$ 50% of the mass is considered in the analysis. As for the closed-loop control(a), since the control forces increase in the case of +50% error, the responses decrease compared to the case of no prediction error(see the curves(— ■ —) in Fig.4.3.,4.4. and 4.9.,4.10.). In the case of -50% error, the responses increase as the control forces decrease(see the curves(— ■ —) in Fig.4.3.,4.4. and 4.11.,4.12.). The critical points of the control(c) where the responses become minimum, shift to the left side in the case of +50% error and to the right side in the case of -50% error (see Fig.4.10. and 4.12.). The control(d) and the control(e) are naturally influenced greatly by the identification error for the mass, because the feed forward control force is influenced by the error, even if the input acceleration is true.

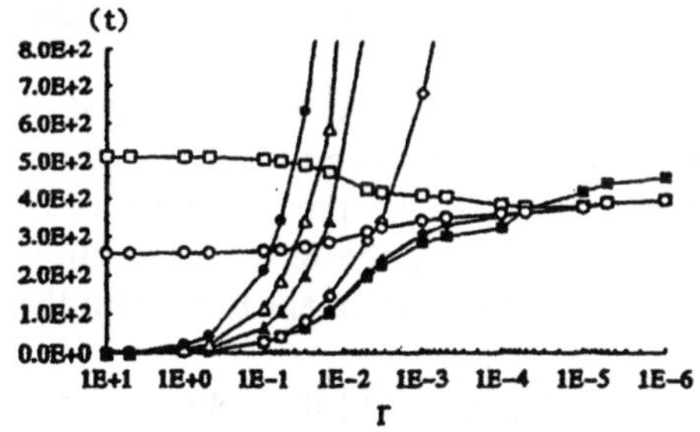

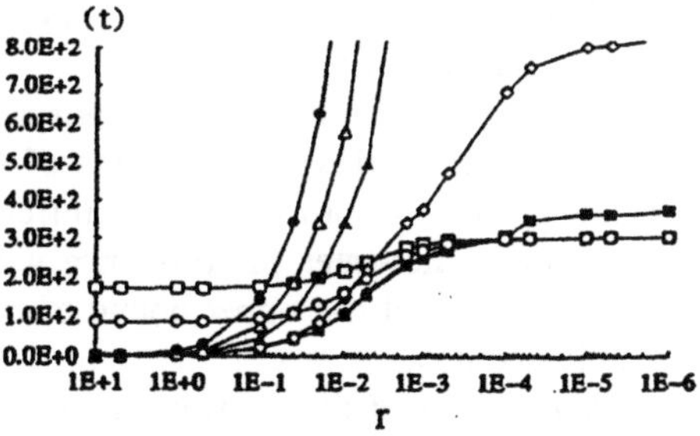

Fig.4.9. Maximum Control Force (True Input and +50% error of the mass)

Fig.4.11. Maximum Control Force (True Input and -50% error of the mass)

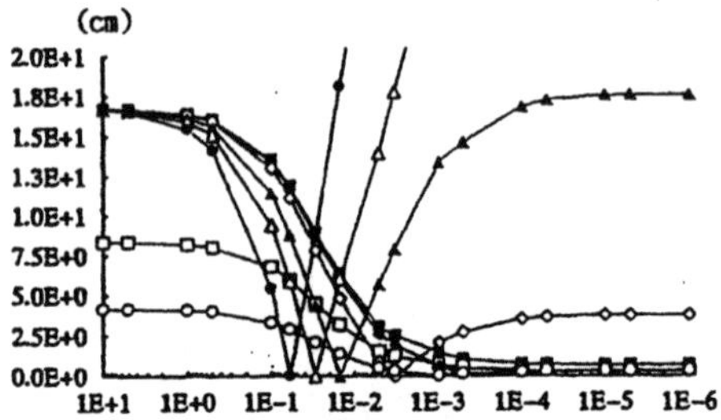

Fig.4.10.Maximum Response Displacement (True Input and +50% error of the mass)

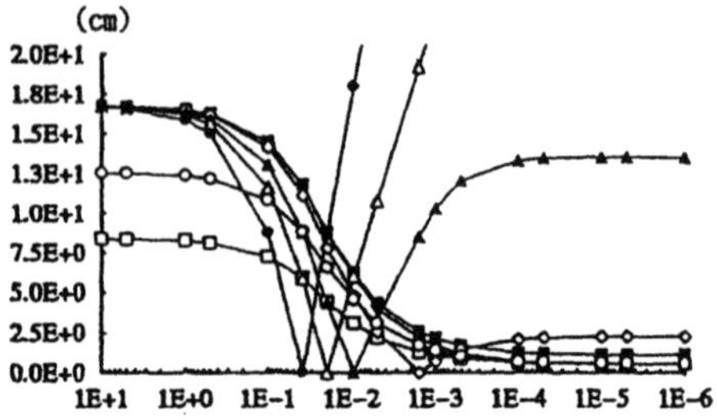

Fig.4.12.Maximum Response Displacement (True Input and -50% error of the mass)

Similar analyses are carried out on the effect of identification errors for the spring constant and the viscous damping coefficient, and it is found that they scarcely influence the effect of control. This is mainly due to the fact that these parameters have no connection with the input force, and furthermore the prediction of the state vector is not influenced by them as far as the error is within the range of $\pm 50\%$ of the true values.

5 Concluding Remarks

A basic analysis is carried out on the identification, prediction and control of a single degree of freedom system which is modeled by a multi-variate ARMA model. The following conclusions are drawn:

(1) Utilizing the responses of the system excited by an active control device installed on the system, all structural properties of the system can be identified.

(2) Based on this formulation, a neural network is derived whose links have physically meaningful weights.

(3) General modes of an instantaneous optimal control rule are formulated in terms of the identified components of the control performance index.

(4) The feed forward control is influenced greatly by the prediction errors of input force, however the influence will be decreased by the feed back control. The existence of the upper bound of control force capacity is apt to restrain the surplus control.

(5) The effects of the control are influenced greatly by the identification error of the mass, however scarcely by the other properties.

Finally, this study is performed by the second author, incorporated with theoretical suggestion by the first author.

6 References

1. PI, Y.L., & Mickleborough, N.C., Time domain modal identification of a linear vibrating structure. UNICIV Report No.R-249, The University of New South Wales, Apr., 1988.
2. Maruyama, O., Aizawa, J. & Hoshiya, M., Identification of dynamic properties of existing structures by multivariate ARMA representation. Proc. Japan Society of Civil Engineers, No.416, Apr., 1990, pp.439-447 (in Japanese).
3. Hoshiya, M. & Maruyama, O., Identification of an autoregressive model for multivariate one-dimensional nonstationary gaussian random processes. Proc. Japan Society of Civil Engineers, No.416, Apr., 1990, pp.349-355 (in Japanese).
4. Sato, T., Toki, K. & Hashimoto, M., Predictive control of seismic response of structure with time delay. Proc. Japan Society of Civil Engineers, NO.428, Apr., 1991, pp.193-202 (in Japanese).
5. Yang, J.N., Control of tall buildings under earthquake excitation. Journal of the Engineering Mechanics Division, ASCE, Vol.108, No.EM5, Oct., 1982, PP.833-849.
6. Yang, J.N., Akbarpour, A. & ghaemmaghami, P., New optimal control algorithms for structural control. journal of the Engineering Mechanics Division ASCE, Vol.113, No9, Sep., 1987, pp.1369-1386.
7. Sato, T., Toki, K. & Hashimoto, M., Active structual response control with self-learning mechanism. Proc. Japan Society of Civil Engineers, NO.471, Jul., 1993, pp.115-124 (in Japanese).
8. Sato, T., Toki, K. & Sugiyama, K., Optimal control of seismic response of structure. Proc. Japan Society of Civil Engineers, NO.416, Apr., 1990, PP.191-200.

Reliability Analysis Methods for Linear Uncertain Dynamic Systems

Yaacob Ibrahim[1] and Loh Han Tong[2]

[1] Dept. of Industrial and Systems Engineering, National University of Singapore, 10 Kent Ridge Crescent, Singapore 0511
[2] Dept. of Mechanical and Production Engineering, National University of Singapore, 10 Kent Ridge Crescent, Singapore 0511

Abstract. Reliability of simple linear uncertain systems subjected to Gaussian excitation can be estimated using First- and Second-Order Reliability Methods (FORM/SORM). These methods require the design point which can be obtained by solving a constrained optimization problem. In structural reliability the Rackwitz-Fiessler (RF) optimization algorithm is commonly used. However, for dynamic problems, this routine appears to have a convergence problem. This is examined in greater detail for a one degree-of-freedom (dof) and two dof systems. Alternative methods which make use of the RF routine in conjunction with some known solvers are proposed. This strategy greatly improves the convergence rate.

Keywords. Reliability, dynamic systems, uncertainty, optimization

1 Introduction

Recent research and progress has been made in developing techniques to evaluate the response and performance of an uncertain system subjected to a stochastic excitation. Several researchers have considered uncertainty in system parameters (Igusa and Kiureghian, 1988; Madsen and Moghtaderi; 1987; Soong and Chen, 1987; Wen and Chen, 1987) and uncertainty in load parameters (Grigoriu, 1986) in evaluating the system responses and reliability. Uncertainties in special systems such as primary-secondary system (e.g., Ibrahim et al., 1989) and tuned mass dampers (e.g., Jensen et al., 1992) have also received attention.

This paper presents results of a study of simple linear uncertain systems subjected to stochastic excitation. The reliability is estimated using the First- and Second-Order Reliability Methods (FORM/SORM). These methods provide a good estimate by approximating the failure surfaces with simpler surfaces which are made to pass through one point on the failure surface. This point is called the design point which is obtained by solving a constrained optimization problem. A study of a commonly used algorithm known as the Rackwitz-Fiessler algorithm is made, and alternative procedures are recommended.

2 Reliability Analysis of Linear Uncertain Systems

Consider an uncertain system subject to a Gaussian excitation, and let $\mathbf{Z}$ be the vector of uncertain parameters. Reliability, defined as p_s, is the probability that a system response stays within a safe domain for a duration of time, i.e.,

$$p_s = P[| X(t) | \leq r]; \quad 0 \leq t \leq \tau , \tag{1}$$

where $X(t)$ is a system response, r is the allowable threshold, and τ is the reference time. The probability of failure p_f is equal to $1.0 - p_s$ and can be written as

$$p_f = \int_{\mathbf{Z}} p_f(\mathbf{z}) f(\mathbf{z}) d\mathbf{z} , \tag{2}$$

where $p_f(\mathbf{z})$ is the conditional failure probability given the system parameters, and $f(\mathbf{z})$ is the joint probability density function of the random vector $\mathbf{Z}$. The above relation can only be used effectively if the conditional failure probability can be obtained easily. The case of a linear system subject to Gaussian input is one instance in which the conditional failure probability due to first crossing can be obtained easily. In such a case, Eq. 2 can be estimated using schemes such as Monte Carlo simulation or numerical integration. These methods tend to be inefficient and expensive since a large number of conditional random vibration analyses are needed for an accurate estimate. Alternative methods based on the First- and Second-Order Reliability Methods (FORM/SORM) (Madsen et al., 1986) have been developed by several researchers (Igusa and Kiureghian, 1988; Madsen and Moghtaderi, 1987; Wen and Chen; 1987). In these methods, the performance of the system is expressed in terms of a g-function defined as $g(\mathbf{Z}, S)$, where S is the critical value of some desired system response. The probability of failure is, therefore, given by the equation

$$p_f = P[g(\mathbf{Z}, S)] \leq 0; \quad 0 \leq t \leq \tau \tag{3}$$

and is estimated using FORM/SORM. An example of a g-function would be

$$g(\mathbf{z}, s) = s - Y(t); \quad 0 \leq t \leq \tau\ , \tag{4}$$

where $Y(t)$ is some system response (e.g. maximum stress or displacement), and s is the allowable (critical) value of the system response, which can be either random or deterministic. The advantage of using FORM/SORM is that the number of conditional random vibration analyses needed to estimate p_f can be considerably less than that required by Monte Carlo simulation. Igusa and Kiureghian (1988) employed FORM/SORM to investigate the response of an uncertain linear system subjected to stationary excitation. Also, analytical equations for the partial derivatives of the g-function were developed. Madsen and Moghtaderi (1987) proposed a formulation of the g-function that can be used (with simple modifications) with available time-invariant routines in the FORM/SORM algorithm. Wen and Chen (1987) proposed an alternative g-function which is equivalent to Madsen and Moghtaderi's (1987) formulation, i.e. for the same problem similar results will be obtained with both methods. The proposed g-function is

$$g(\mathbf{u}) = u_{n+1} - \Phi^{-1}\left[p_f(\mathbf{T}^{-1}(\mathbf{u}))\right]\ , \tag{5}$$

where $\mathbf{u}$ is the vector of unit independent normal random variables, u_{n+1} is a unit independent normal auxiliary random variable, n is the number of uncertain parameters, $\mathbf{T}$ is the transformation matrix of $\mathbf{Z}$ to $\mathbf{U}$, and $\Phi()$ is the cumulative distribution function of a standard normal variate. The g-function in Eq. 5 is used in this study.

2.1 First- and Second-Order Reliability Methods

First- and Second-Order Reliability Methods (FORM/SORM) estimate the reliability or probability of failure by approximating the failure surface with either a hyperplane (hence FORM) or a second-order surface (hence SORM). These approximations are done in the space of unit independent normal random variables. The approximate surfaces are made to pass through one point on the failure surface known as the ***design point***. It is the point closest to the origin which has the highest density on the failure surface. The distance to this point is defined as β which is also known as the ***reliability index*** (Madsen et al., 1986). β can be found by solving the constrained optimization problem given by the equations

$$\begin{aligned} \text{Minimize :} \quad & \beta(\mathbf{u}) = \sqrt{\textstyle\sum_i^n u_i^2} \\ \text{such that :} \quad & g(\mathbf{u}) = 0 \ . \end{aligned} \tag{6}$$

The FORM estimate of p_f is simply $\Phi(-\beta)$. The SORM estimate can be approximated using Tvedt's equation (Madsen et al., 1986). The curvatures at the design point are needed in this case.

For dynamic systems, the number of random parameters will be $n+1$ if the g-function given in Eq. 5 is used. With this g-function the optimization problem given above can be re-cast into an unconstrained problem. Such problems are much easier to solve.

3 Rackwitz-Fiessler (RF) Algorithm for Reliability Analysis

An algorithm that has been used successfully in conjunction with FORM/SORM for structural reliability analysis is the Rackwitz-Fiessler (RF) algorithm (Madsen et al., 1986). It is a gradient-based iterative algorithm which is inexpensive and known to converge quickly to the design point. In the algorithm, the g-function is successively linearized at a sequence of points which will eventually converge to the design point. The sequence of points is given by

$$\mathbf{u}^{i+1} = \left[\mathbf{u}^{i'}\mathbf{v}^i\right]\mathbf{v}^i + \frac{g(\mathbf{u}^i)}{|\nabla g(\mathbf{u}^i)|}\mathbf{v}^i \ , \tag{7}$$

where $\mathbf{u}^i$ is the value of $\mathbf{u}$ at the i-iteration, the superscript $'$ means transpose, $\nabla g(\mathbf{u})$ is the gradient vector, and

$$\mathbf{v}^i = \frac{\nabla g(\mathbf{u}^i)}{|\nabla g(\mathbf{u}^i)|} \ . \tag{8}$$

This algorithm has been known to encounter some problems of convergence (Ibrahim et al., 1989; Liu & Kiureghian, 1991). The following two examples demonstrate the convergence problem of the RF algorithm. In implementing the algorithm, a convergence criterion of $1E-06$ was used, and the gradients were evaluated numerically using a two-sided finite difference scheme. A maximum of 25 iterations was fixed, and the origin was used as the starting solution for all cases.

3.1 Example 1

Consider a one dof linear system in which the system parameters, natural frequency ω_0 and damping ratio ς, are uncertain. The parameters are assumed to be independent and uniformly distributed. Three different combinations of parameters are considered as shown in Table 1.

Case No.	ω_0 (rads/s)	ς
1	$U[\pi, 2\pi]$	0.045
2	1.5π	$U[0.02, 0.07]$
3	$U[\pi, 2\pi]$	$U[0.02, 0.07]$

Table 1. Combinations of Parameter Values

The system is subjected to a stochastic ground excitation which is modeled as a stationary Gaussian white noise process with zero mean and a one-sided power spectral density (psd) of intensity G_0. The conditional reliability for upcrossing is approximated by

$$p_s(\mathbf{z}, \tau) = \exp(-2.0\nu(r, \mathbf{z})\tau) \ , \tag{9}$$

where $\nu(r, \mathbf{z})$ is the stationary conditional mean crossing rate for threshold r, and this is given by

$$\nu(r, \mathbf{z}) = \frac{\omega_0}{2.0\pi} \exp\left\{-r^2 \frac{2\varsigma\omega_0^3}{\pi G_0}\right\} \ . \tag{10}$$

For this example G_0 is set to 1.0, and τ is fixed at 20.0 seconds. Results for various thresholds r are shown in Table 2. When convergence is achieved, the number of iterations N_i and the reliability index β are recorded.

	Case 1		Case 2		Case 3	
r	β	N_i	β	N_i	β	N_i
1.0	-0.3026	4	0.9982	4	-0.2605	4
1.5	0.7227	4	No convergence		0.6217	4
2.0	1.5818	7	No convergence		1.2575	7
3.0			No convergence			
4.0			No convergence			
5.0			No convergence			

Table 2. Results of RF for One-Dof System

For low thresholds, the routine converges to the design point very quickly. A plot of the g-functions at various values for case 1 at $r = 1.5$ is shown in Fig. 1. The axis U_1 refers to the uncertain parameter (in this case it is the natural frequency). While the axis U_{n+1} refers to the additional parameter arising from using the g-function given in Eq. 5. The linearization points are also shown, i.e., points marked #1, #2, and #3. Notice that the design point is located at a surface which is almost linear. For the non-convergence cases, typically, the linearization points would oscillate between two points with a cycle of about 3 to 4 iterations. This behaviour is shown in Fig. 2 for case 1 for a threshold of 3.0. From the sequence of the linearization points, the oscillatory behaviour of the algorithm is evident, and this could be due to the existence of curvatures at the design point.

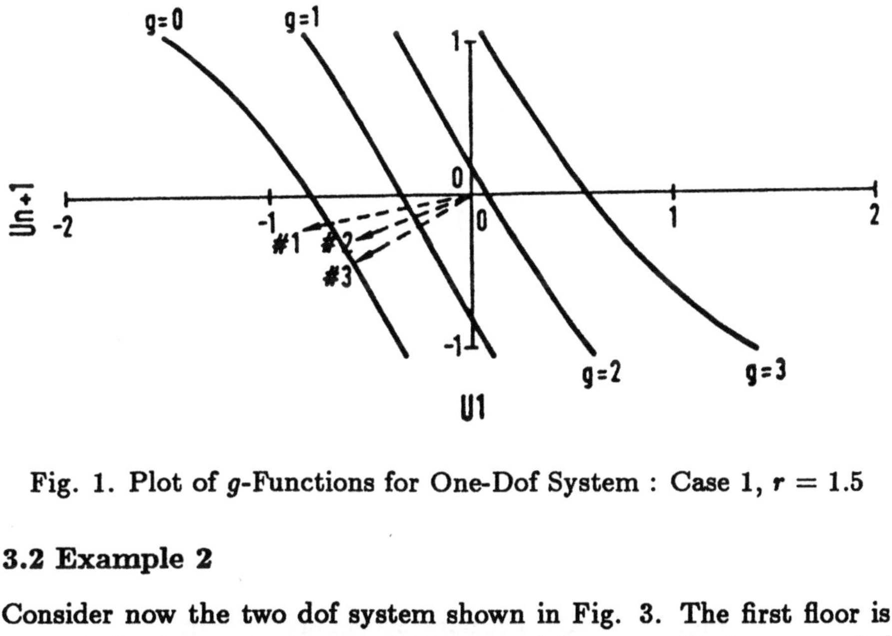

Fig. 1. Plot of g-Functions for One-Dof System : Case 1, $r = 1.5$

3.2 Example 2

Consider now the two dof system shown in Fig. 3. The first floor is a one dof primary system, and the second floor constitutes a one dof secondary system. The reliability is defined as the probability that the interstory displacement of the second floor does not exceed a threshold, i.e.

$$p_s = P[|\, \Delta_2(t) \,| \leq r]; \quad 0 \leq t \leq \tau \;, \tag{11}$$

where

$$\Delta_2(t) = X_2(t) - X_1(t) \;. \tag{12}$$

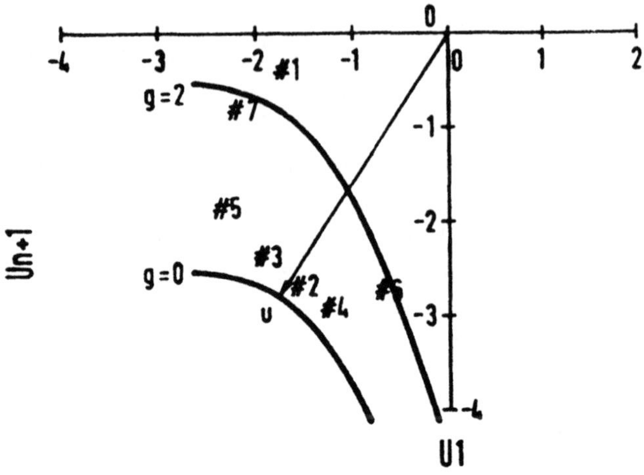

Fig. 2. Plot of g-Functions for One-Dof System : Case 1, $r = 3.0$

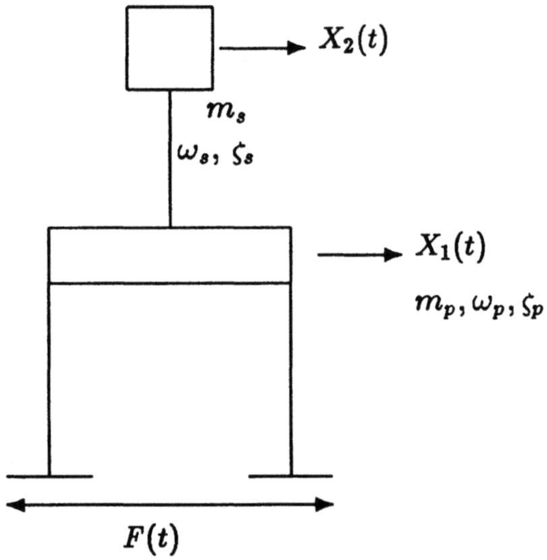

Fig. 3. 2-DOF Primary-Secondary System

The conditional mean crossing rate $\nu_\Delta(\mathbf{z}, r)$ is given by (Ibrahim et al., 1989)

$$\nu_\Delta(\mathbf{z}, r) = \frac{1}{\sigma_{\Delta(t)}} \phi(\frac{r}{\sigma_{\Delta(t)}}) \left[\sigma' \phi \left\{ \frac{\mu'}{\sigma'} \right\} + \mu' \Phi \left\{ \frac{\mu'}{\sigma'} \right\} \right] , \qquad (13)$$

where

$$\mu' = \frac{rE\left[\Delta(t)\dot{\Delta}(t)\right]}{\sigma^2_{\Delta(t)}} , \tag{14}$$

$$\sigma'^2 = \sigma^2_{\dot{\Delta}(t)} - \frac{\left\{E\left[\Delta(t)\dot{\Delta}(t)\right]\right\}^2}{\sigma^2_{\Delta(t)}} , \tag{15}$$

$$\phi(x) = \frac{1}{\sqrt{2\pi}} \exp(-\frac{x^2}{2}) , \tag{16}$$

and

$$\Phi(y) = \int_{-\infty}^{y} \phi(x)dx . \tag{17}$$

The parameters are assumed to be independent and uniformly distributed. Three different combinations of parameter values were considered as shown in Table 3.

Parameters	Case No. 1	Case No. 2	Case No. 3
ω_p(rads/sec)	$U[0.65\pi, 3.35\pi]$	$U[0.65\pi, 3.35\pi]$	$U[0.65\pi, 3.35\pi]$
ω_s(rads/sec)	2π	$U[0.65\pi, 3.35\pi]$	$U[0.65\pi, 3.35\pi]$
ζ_p	0.04	0.04	$U[0.01, 0.07]$
ζ_s	0.04	0.04	$U[0.01, 0.07]$

Table 3. Combinations of Parameter Values for Two Dof System

A mass ratio (ratio of the mass of the secondary system to that of the primary system) of 0.01 was used. The system is subjected to a non-stationary zero-mean Gaussian white noise excitation, $F(t)$, which is modelled as

$$F(t) = a(t)W(t); \quad 0 \leq t \leq \tau , \tag{18}$$

where $a(t)$ is a deterministic modulating function given by

$$a(t) = 2.32\left[\exp(-0.09t) - \exp(-1.49t)\right] . \tag{19}$$

$W(t)$ is a stationary zero-mean Gaussian white noise with a one-sided psd set at 1.0. τ is fixed at 20 seconds. The mean and variances of the conditional responses (which are Gaussians) were obtained using the method developed in Ibrahim et al., (1989).

Results for various thresholds are shown in Table 4. From the results, clearly there are problems of convergence. A plot of the failure surface for Case 1 at threshold 8 is given in Fig. 4. As in the one dof case, the existence of curvatures around the design point is perhaps a contributing factor to the oscillatory behaviour.

	Case 1		Case 2		Case 3	
r	β	N_i	β	N_i	β	N_i
1.0	-4.9009	4	-4.9009	4	-4.9009	4
2.0	-3.4441	4	-3.4441	4	-3.4441	4
3.0	No convergence		-0.1777	12	-0.1773	8
4.0	-0.1524	5	-0.9663E-01	21	-0.9619E-01	6
5.0	-0.5246E-02	4	-0.2133E-02	3	-0.2038E-02	3
6.0			No convergence			
8.0			No convergence			
10.0			No convergence			

Table 4. Results of RF Algorithm for Two Dof System

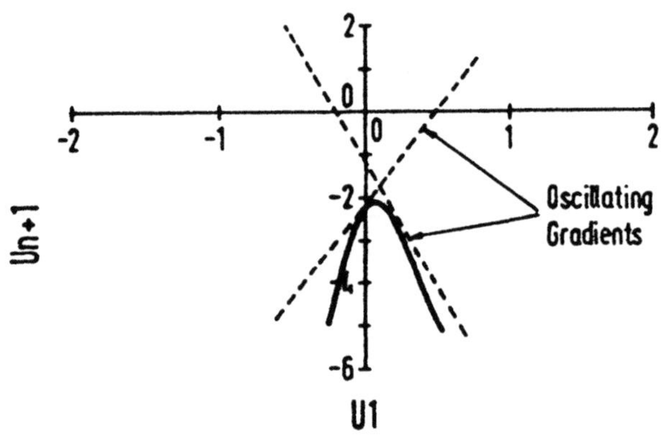

Fig. 4. Plot of Failure Surface for Two-Dof System : Case 1, Threshold = 8.0

3.3 Observations

Clearly the RF algorithm has some convergence problems. For those cases which did not converge, it was observed that the points of linearization were oscillating, i.e., lack of convergence is accompanied by oscillatory behaviour. Thus, even if the maximum number of iterations is increased, convergence will not take place. It was also observed that when there was convergence, the β value will tend to decrease towards the global minimum. Hence, an important indicator of lack of convergence is the tendency for the β value to increase after decreasing for a few iterations. Using different starting solutions did not help in the convergence. Starting solution at the point where all the u-values were equal to -1.0 gave almost identical results as those reported above.

For the above two examples, the problem generally occurs at high threshold. This observation is important since the approximation used in computing the conditional reliability is valid only for high thresholds (Nigam, 1983). Thus, the RF algorithm, which works well for low threshold, cannot be used for the more accurate cases of high thresholds.

4 Alternative Methods

In order to guarantee convergence for all cases and thresholds, a more robust solver is needed. There are available routines in the International Mathematics and Statistics Library (IMSL) which have been used to overcome this problem (Ibrahim, 1992). However, here we are interested to find a cheap and quick method of estimating the reliability. From the results of the two examples given earlier, it was evident that the RF routine tended to oscillate about the design point when it did not converge. This fact can be used to combine, for example, two solvers. The RF routine can be used to identify the approximate location of the design point. A second solver which needs such information can then be triggered to home in quickly to the design point. The choice of the second solver depends upon the information available from the RF routine. Two different solvers are considered. The first is a simple method known as the Golden Section Search (GSS) and the second the Broyden-Fletcher- Goldfarb-Shanno (BFGS) algorithm. The GSS is a line search technique between two points on the objective function. While BFGS requires a good starting point for rapid convergence. Thus, the information from RF routine on the location of the design point is relevant for these methods.

4.1 Golden Section Search (GSS) Method

The Golden Section Search (GSS) is an one-dimensional interval reduction search technique that successively reduces the interval of uncertainty within which the optimum lies. The rate of reduction is the Golden Ratio ($\approx$ 1.618) which is derived from the Fibonacci sequence.

The GSS is initiated after two points delimiting the initial interval of uncertainty are given. The search is along the straight line joining these two points and the interval can be reduced to any small acceptable value. This procedure is very well known and is described in any optimization textbook (e.g., Luenberger, 1984).

4.2 Broyden-Fletcher-Goldfarb-Shanno (BFGS) Method

The BFGS is a Quasi-Newton method that uses second-order derivative information to assist in the convergence to a local minimum (Gill & Wright, 1981). Quasi-Newton methods avoid the expensive cost of calculating exact second-order derivative by approximating the Hessian or its inverse with the information accumulated from previous iterations using only first-order derivatives. For positive definite quadratic functions, it can be shown that the convergence to the optimum takes at most n iterations, where n is the number of variables. In a general objective function, this ideal situation does not exist; however, near the minimum, the objective function will behave like a positive definite quadratic function. Hence, an initial point given near the minimizer will lead to rapid convergence. In approximating the Hessian or its inverse, update formulas are used, and the different formulas lead to different algorithms for the family of Quasi-Newton methods. It is essential to preserve symmetry and positive definiteness of the matrix in the updating process to ensure that the search will be carried out in a direction of descent. The BFGS update formula keeps the Hessian approximation symmetric and positive definite, and computational experience has shown good convergence rates.

The switching from RF to GSS or BFGS can either be defined apriori or adapted to the problem at hand. In the latter case the switching is triggered when oscillation is detected. Thus, by simply monitoring the iterations one is able to incorporate an automatic switch into the routine. In the former case, a fixed number of the RF algorithm iterations is pre-determined. Here, based on the study, an adaptive switch was used. It was found that oscillations tend to occur after about 3 to 5 iterations. And as mentioned earlier, another indicator of lack of convergence is the increasing value of β. Using these two facts, the switch is made after 3 iterations and only if the β value is increasing. Note also that GSS and BFGS solve the unconstrained optimization problem mentioned earlier.

In a preliminary study it was found that BFGS is very sensitive to the starting solution. A weighted average of the last two points obtained from the RF algorithm is used as a starting solution for the BFGS routine. The weights are computed based upon the distances of the two points, where more weights are assigned to points closer to the failure surface.

4.3 Results of Using GSS and BFGS for One Dof System

Results for the one dof system are shown in Table 5. The number of function evaluations N_f is monitored. This is used as a basis of comparison between the various methods.

	RF-GSS		RF-BFGS		IMSL's DUMINF	
Threshold	β	N_f	β	N_f	β	N_f
3.0	2.4477	100(79)[1]	2.373	41(20)	2.374	24
4.0	3.6160	104(62)	3.436	64(22)	3.436	32
5.0	4.5787	85(64)	4.446	41(20)	4.437	37

[1] = Number of function evaluations required by GSS or BFGS shown in brackets

Table 5. Results for One Dof System : Case 3

Between GSS and BFGS, the latter requires much less function evaluations. Solutions obtained from using IMSL's DUMINF routine have been included to serve as a check on the accuracy of the two solvers. This routine solves the unconstrained optimization problem using essentially the BFGS method. For this routine the origin is used as a starting solution. The results compared quite well with DUMINF results with BFGS being the faster routine.

4.4 Results of Using GSS and BFGS for Two Dof System

The GSS routine worked quite well for case 1, but failed to converge for cases 2 and 3. GSS is a line minimization routine. The search for the minimum is along the line joining the two initial bounding points. Thus, if the bounding points do not contain the global minimum, GSS will return a local optimum. For this reason, GSS worked well only for one-dimensional and regular curved surfaces. For cases 2 and 3, where there are two or more uncertain parameters, GSS can be expected to fail. Thus, for the two dof system, the BFGS routine is preferred. The results of this are given in Table 6.

	Case 1		Case 2		Case 3	
Threshold	β	N_f	β	N_f	β	N_f
6	0.7676	37	0.3068	74	0.2582	89
8	2.0529	41	0.7992	81	0.6838	83
10	3.1173	48	1.1706	75	1.0048	97
12	4.0799	46	3.9578	63	3.9299	71

Table 6. Results for Two Dof System Using RF-BFGS

As mentioned earlier, it was found in this study that BFGS and also IMSL's DUMINF routine are sensitive to the starting solution. In the former case, the RF routine is used to give a starting point which gave convergence in all cases. But there is no guarantee that this solution is globally optimal. In order to test this, a comparison is made between the results of RF-BFGS and DUMINF. Case 2 was chosen as a test case at two thresholds; 6 and 8. In the case of DUMINF, 100 randomly chosen starting solutions were used. The results were then checked for that which gave the smallest distance. Table 7 gives results of this comparison.

	DUMINF			RF-BFGS		
Threshold	U^*	β	N_f	U^*	β	N_f
	-0.1764			-0.1773		
6	-0.2139	0.3068	102	-0.2148	0.3068	74
	-0.1313			-0.1287		$(53)^1$
	-0.5066			-0.4987		
8	-0.5472	0.8004	85	-0.5344	0.7992	81
	-0.2908			-0.3233		$(53)^1$

U^* = Coordinates at Design Point

^{1} = Number of function evaluations taken by BFGS

Table 7. Comparison Between DUMINF and RF-BFGS for Case 2 of Two Dof System

From the results, it would appear that RF-BFGS requires less function evaluations. However, the more interesting observation is that for DUMINF, at threshold 6 only 6 of the 100 starting solutions gave the absolute minimum and for threshold 8 only 1 gave the absolute minimum. Hence, RF-BFGS is more robust and less sensitive to the starting solution.

5 Conclusion

The Rackwitz-Fiessler (RF) algorithm used to estimate the reliability index was shown to have convergence problems for linear uncertain systems. This problem was encountered even in very simple cases of one uncertain parameter. However, it was observed that the non-convergence occurred with oscillations about the design point. This fact was used to design an alternative approach to overcome the convergence problem. Essentially, it consists of using first RF algorithm to locate approximately the design point, and then switching to another solver to home in quickly on to the design point. Two solvers were tested; one based on the Golden Section Search (GSS) method and the other on the Broyden-Fletcher-Goldfarb-Shanno (BFGS) method. It was shown that the latter was more robust, arriving at the design at a much faster rate that GSS. The switching from the RF routine to either solvers was made by monitoring the oscillations. It was found that at about 3 to 5 RF iterations, oscillations would have taken place. Also oscillations tend to occur with changing β values. Based on these observations an adaptive switch was developed.

The strength of the RF routine is in its ability to locate approximately very quickly the design point. This hence gave a very good starting solution which can be used for another solver. Here, the BFGS solver is recommended for its efficiency and robustness in arriving at the design point.

6 References

1. Gill, P.E., & Wright, M. H., *Practical Optimization*, Academic Press, 1984.

2. Grigoriu, M., Structural Response to Uncertain Seismic Excitations *Journal of Structural Engineering, ASCE,* **112** (6) (1986) 1355 - 1365.

3. Ibrahim, Y., Reliability Analysis of Linear Dynamic Systems Using Optimization Techniques. *International Conference on Optimization: Techniques and Applications*, eds. K.H. Phua, C.M. Wang, W.Y. Yeong, T.Y. Leong, H.T. Loh, K.C. Tan, and F.S. Chou, Singapore, June 3-5 (1992), pp. 664 - 671.

4. Ibrahim, Y., Grigoriu, M. & Soong, T. T., Random Vibration and Reliability Analysis of Primary-Secondary Structural Sys-

tems. Report No NCEER-89-0031, State University of New York at Buffalo, 1989.

5. Igusa, T., & Kiureghian, A. D., Response of Uncertain Systems to Stochastic Excitation *Journal of Engineering Mechanics, ASCE,* **114** (5) (1988) 812 - 832.

6. DUMINF. *International Mathematics and Statistics Library*, IMSL, Inc., U.S.A., (1987) 802 - 806.

7. Jensen, H., Setareh, M., & Peek, R., TMDs For Vibration Control of Systems with Uncertain Parameters *Journal of Structural Engineering, ASCE,* **118** (12) (1992) 3285 - 3296.

8. Liu, P.-L., & Kiureghian, A. D., Optimization Algorithms for Structural Reliability *Structural Safety*, **9** (1991) 161 - 177.

9. Luenberger, D. G., *Linear and Nonlinear Programming*, 2nd. ed., Addison-Wesley, 1984.

10. Madsen, H. O., Krenk, S. & Lind, N. C., *Methods of Structural Safety*, Prentice-Hall, Inc., Englewood Cliffs, N. J., 1986.

11. Madsen, H. O., & Moghtaderi-Zadeh, M., Reliability of Plates Under Combined Loading. *Marine Structural Reliability Symposium*, SNAME, Arlington, Virginia, 1987, pp. 185 - 191.

12. Nigam, N. C., *Introduction to Random Vibrations*, MIT Press, 1983.

13. Soong, T. T. & Chen, Y., Stochastic Response Sensitivities of Secondary Systems to Primary Structural Uncertainties. *Symposium on Stochastic Structural Dynamics*, Urbana, Illinois, 1988.

14. Wen, Y. K. & Chen, H-C., On Fast Integration for Time Variant Structural Reliability, *Probabilistic Engineering Mechanics*, **2** (3) (1987) 156 - 162.

Improved Formula for Upper Bound of Probability of Failure Based on Means and Variances

Masahiro Ichikawa

Department of Mechanical and Control Engineering
University of Electro-Communications
Chofu-city, Tokyo, Japan

Abstract. In relation to practical structural reliability analysis, this paper is concerned with the formula $P_{fu}=(4/9)(f_C^2\eta_R^2+\eta_S^2)/(f_C-1)^2$ for the upper bound of the probability of failure based on means and variances of strength R and stress S, where f_C is the central factor of safety, and η_R and η_S are coefficients of variation of R and S, respectively. This formula was previously derived using Camp-Meidell inequality under the assumptions (a) the distribution of $Z(=R-S)$ is continuous and unimodal, and (b) the mode of Z is equal to the mean of Z. By formulating and solving an optimization problem, the present paper reveals that this formula holds even when only (a) is assumed. This finding extends largely the applicability of this formula in practical analysis of structural reliability.

Keywords. Reliability, probability of failure, upper bound, optimization problem, linear programming

1 Introduction

In structural reliability analysis, one often encounters such a case in which the distributions of the strength R and the stress S are unknown except for their means and variances. In this case, along with using safety index[1,2], it is useful to evaluate the upper bound of the probability of failure $P_f=Pr(R\leq S)$ as noted by Mischke[4]. Applying Tchebysheff inequality, a formula for the upper bound of the probability of failure, P_{fu}, based on means and variances was derived by Thien and Massoud[4] in 1974 as given by the equation

$$\left.\begin{aligned} P_{fu} &= \frac{f_C^2[(1+\eta_R^2)(1+3\eta_S^2)+\lambda(\lambda-2)(1+\eta_S^2)^2]}{[\lambda f_C(1+\eta_S^2)-1]^2} \\ \lambda &= \frac{f_C[(1+\eta_R^2)(1+3\eta_S^2)/(1+\eta_S^2)]-1}{f_C(1+\eta_S^2)-1} \end{aligned}\right\}, \tag{1}$$

where (μ_R, σ_R^2, η_R) and (μ_S, σ_S^2, η_S) are (mean, variance, coefficient of variation) of R and S, respectively, and $f_C(=\mu_R/\mu_S)$ is the central factor of safety. By a somewhat different application of Tchebyseff inequality, a more powerful formula was derived in 1983 by the present author[5] as

$$P_{fu} = \frac{f_C^2\eta_R^2+\eta_S^2}{(f_C-1)^2+f_C^2\eta_R^2+\eta_S^2}. \tag{2}$$

In deriving Eq.(2) no assumption was made on the distributions of R and S. A Tchebyseff bound of P_f was also discussed by Veneziano[6] for multi-dimensional cases in relation to reliability index.

Applying the Camp–Meidell inequality, a still more powerful formula was derived in 1983 by the present author[5] as given by the equation

$$P_{fu} = \frac{4}{9}\cdot\frac{f_C^2\eta_R^2+\eta_S^2}{(f_C-1)^2}. \tag{3}$$

However, use of the Champ–Meidell inequality requires one to assume (a) the distribution of $Z(=R-S)$ is continuous and unimodal, and (b) the mode of Z is equal to the mean of Z. The assumption (b) may be satisfied only in special cases, hence, practical applicability of Eq.(3) is limited. In the present paper, we attempt to obtain a formula for P_{fu} under the assumption (a) alone without assuming (b).

2 Formulation of the Problem

Calculation of the upper bound of P_f is equivalent to finding such a distribution of $Z(=R-S)$ that makes the probability $Pr(R\leq S)$ largest. We first consider the case where no assumption is made on the distributions of R and S. Let $f(Z)$ be the probability density function of Z. The upper bound P_{fu} may be obtained as a solution of the optimization problem

$$\int_{-\infty}^{0} f(Z)dZ \rightarrow \max \tag{4a}$$

subjected to:

$$\left.\begin{aligned}&\int_{-\infty}^{\infty} f(Z)\,dZ = 1\\&\int_{-\infty}^{\infty} Zf(Z)\,dZ = \mu_Z\\&\int_{-\infty}^{\infty} Z^2 f(Z)\,dZ = \mu_Z^2+\sigma_Z^2\end{aligned}\right\}, \tag{4b}$$

where μ_Z and σ_Z^2 are given by $\mu_Z=\mu_R-\mu_S$ and $\sigma_Z^2=\sigma_R^2+\sigma_S^2$ when R and S are independent. This problem can be reduced approximately to a linear programming problem. First, replace the infinite regions $0\leq R<\infty$ and $0\leq S<\infty$ by finite regions $0\leq R\leq R_{max}$ and $0\leq S\leq S_{max}$, where R_{max} and S_{max} are sufficiently large finite values. Then, $Z(=R-S)$ varies within $-S_{max}\leq Z\leq R_{max}$. Let the variable X be $X=Z+S_{max}$ and $G(X)$ be the distribution function of X. Using partial integrations, Eqs.(4a) and (4b) are rewritten as

$$G(S_{max}) \rightarrow \max \tag{5a}$$

subjected to:

$$\left.\begin{aligned}&G(0) = 0, \quad G(R_{max}+S_{max}) = 1\\&\int_0^{R_{max}+S_{max}} G(X)\,dX = R_{max}-\mu_Z\\&\int_0^{R_{max}+S_{max}} XG(X)\,dX = \frac{1}{2}(R_{max}^2-\mu_Z^2-\sigma_Z^2)+S_{max}(R_{max}-\mu_Z)\end{aligned}\right\}. \tag{5b}$$

Divide the range $0\leq X\leq R_{max}+S_{max}$ into n intervals with the width of $d=(R_{max}+S_{max})/n$, and let g_i be defined as

$$g_i = G(id)-G((i-1)d) \qquad i=1,2,\cdots,n. \tag{6}$$

It is evident that $g_i>0$. Using the trapezoidal rule for integration in Eq.(5b), the system of Eqs.(5a) and (5b) is reduced to the following linear programming problem

$$g_1+g_2+\cdots\cdots+g_k \rightarrow \max \tag{7a}$$

subjected to:

$$\left.\begin{aligned}&\sum_{i=1}^{n} g_i = 1\\&\sum_{i=1}^{n} i g_i = k+\frac{1}{2}+\frac{\mu_Z}{d}\\&\sum_{i=1}^{n} i^2 g_i = k^2+k+\frac{1}{2}+\frac{(2k+1)\mu_Z}{d}+\frac{\mu_Z^2+\sigma_Z^2}{d^2}\end{aligned}\right\}, \tag{7b}$$

where k is the number of intervals within the region $0\leq X\leq S_{max}$ i.e. $-S_{max}\leq Z\leq 0$, and given by $k=S_{max}/d=nS_{max}/(R_{max}+S_{max})$.

Next, we consider the case when the distribution of Z is assumed to be continuous and unimodal. When the mode of Z is located within the i_0th interval, the above assumption is expressed as

$$\left.\begin{array}{ll} g_i - g_{i+1} \leq 0 & (1 \leq i \leq i_0 - 1) \\ g_i - g_{i+1} \geq 0 & (i_0 \leq i \leq n-1) \end{array}\right\}. \tag{7c}$$

Let $P_{fu}(i_0)$ be the solution to the system of Eqs.(7a), (7b), and (7c) for a given value of i_0. Then, the maximum of $P_{fu}(i_0)$ with respect to i_0 gives the upper bound of the probability of failure, P_{fu}, under the assumption that the distribution of Z is continuous and unimodal. In order to obtain P_{fu}, it is necessary to solve the system of Eqs.(7a) to (7c) by changing i_0 in a systematic manner.

3 Preliminary Calculations

In order to see if the linear programming approach gives a reasonable result, we first solve the system of Eqs.(7a) and (7b), and compare the solution with Eq.(2). Calculations were carried out for the case of η_R=0.05 and η_S=0.2. f_C was changed as f_C=2, 4, 6, 8, and 10. μ_S was given as μ_S=5(in an arbitrary unit). It should be noted that the value of μ_S does not affect the solution. Thus, values of $\mu_R(=f_C\mu_S)$, μ_S, $\sigma_S(=\mu_R\eta_R)$ and $\sigma_S(=\mu_S\eta_S)$ were determined. From these values, μ_Z and σ_Z^2 were determined. R_{max} and S_{max} were tentatively chosen based on a criterion $R_{max} \simeq \mu_R + 4\sigma_R$ and $S_{max} \simeq \mu_S + 4\sigma_S$. n was chosen as n=60 after some trials.

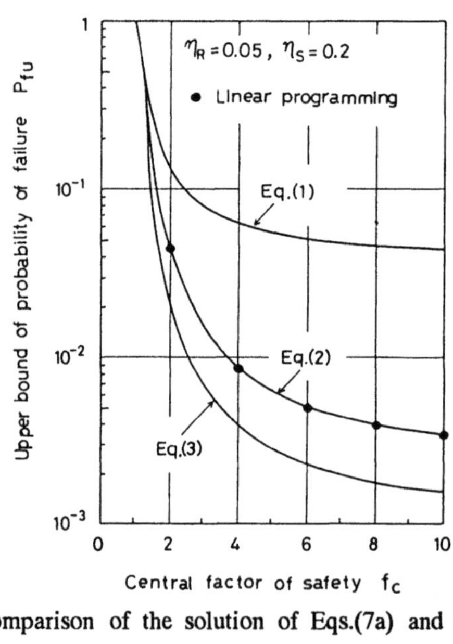

Fig. 3.1. Comparison of the solution of Eqs.(7a) and (7b) with Eq.(2)

The result is shown in Fig. 3.1. It is seen that the solution of Eqs.(7a) and (7b) is in agreement with Eq.(2). This indicates that the upper bound of the probability of failure can be obtained by solving the linear programming problem. Eqs.(1) and (3) are also shown in Fig.1 for comparison.

4 Evaluation of Upper Bound of Probability of Failure

Now, we proceed to evaluate the upper bound of the probability of failure under the assumption that the distribution of $Z(=R-S)$ is continuous and unimodal. In Section 3, n=60, $R_{max} \simeq \mu_R+4\sigma_R$, and $S_{max} \simeq \mu_S+4\sigma_S$ were used. Since the condition of being continuous and unimodal is now imposed, we newly study how large n, R_{max}, and S_{max} should be.

Influence of n was first studied. For several sets of (f_C, η_R, η_S), the system of Eqs.(7a) to (7c) was solved for various values of n ranging from n=10 to n=200. R_{max} and S_{max} were given by $R_{max} \simeq \mu_R+4\sigma_R$ and $S_{max} \simeq \mu_S+4\sigma_S$. For each value of n, i_0 in Eq.(7c) was changed in a wide range to find the maximum value of $P_{fu}(i_0)$ with respect to i_0. It was found that a stable value of P_{fu} could be obtained when n$\gtrsim$60.

Next, influence of R_{max} and S_{max} was studied. Taking n=200, the system of Eqs.(7a) to (7c) was solved for various sets of (R_{max}, S_{max}). It was found that a stable value of P_{fu} could be obtained if $R_{max} \gtrsim \mu_R+4\sigma_R$ and $S_{max} \gtrsim \mu_S+4\sigma_S$.

Now, calculations for various sets of parameter values were carried out to obtain the upper bound of the probability of failure, P_{fu}, under the assumption that the distribution of $Z(=R-S)$ is continuous and unimodal. n was taken to be 200, and R_{max} and S_{max} were determined based on the criterion $R_{max} \simeq \mu_R+4\sigma_R$ and $S_{max} \simeq \mu_S+4\sigma_S$. Calculations were made for 15 cases of (f_C, η_R, η_S) =(2, 0.05, 0.05), (2, 0.05, 0.2), (2, 0.05, 0.4), (2, 0.2, 0.05), (2, 0.2, 0.2), (2, 0.2, 0.4), (2, 0.4, 0.05), (2, 0.4, 0.2), (2, 0.4, 0.4), (4, 0.05, 0.2), (6, 0.05, 0.2), (6, 0.4, 0.4), (8, 0.05, 0.2), (10, 0.05, 0.2), (10, 0.4, 0.4).

In order to show how $P_{fu}(i_0)$ varies with i_0, $P_{fu}(i_0)$ was plotted against $(Z_{MO}-\mu_Z)/\sigma_Z$ in Fig. 4.1 for some sets of (f_C, η_R, η_S), where Z_{MO} is the mode of Z. As seen in Fig. 4.1, $P_{fu}(i_0)$ takes a maximum at a certain value of $(Z_{MO}-\mu_Z)/\sigma_Z$. This maximum gives P_{fu} under the assumption that the distribution of Z is continuous and unimodal.

Figs. 4.2 and 4.3 show P_{fu} in two ways. It is seen that the solution of linear programming problem is almost equal to Eq.(3) in some cases, and lower than Eq.(3) in other cases. But, an overall tendency is not clear from Figs. 4.2 and 4.3. In order to find an overall tendency, the solution of the linear programming problem was plotted in Fig. 4.4 against $\eta_Z(=\sigma_Z/\mu_Z)$, the coefficient of variation of Z. It is seen that the solution of the linear programming problem is almost equal to Eq.(3) when η_Z is small, and somewhat lower than Eq.(3) when η_Z is large. This implies that Eq.(3) can still be used even when it is only assumed

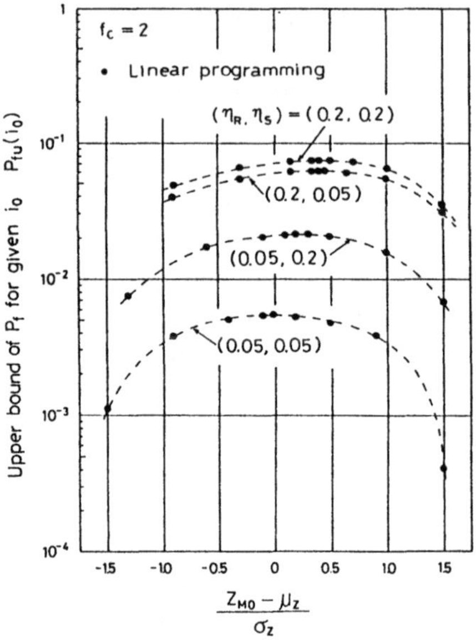

Fig. 4.1. Variation of $P_{fu}(i_0)$ with the relative location of the mode Z_{MO} with respect to the mean μ_Z

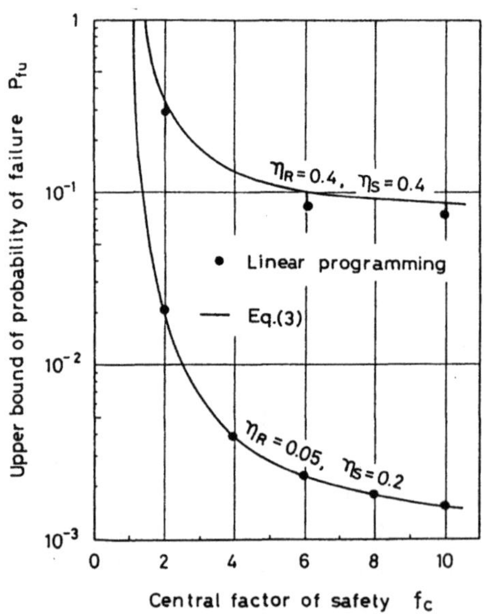

Fig. 4.2. P_{fu} as plotted against f_C

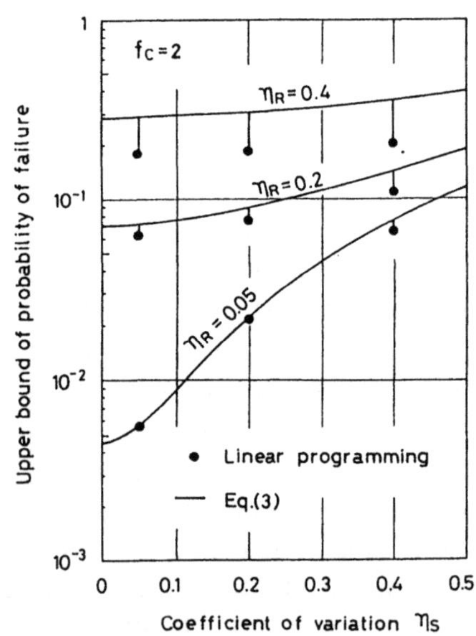

Fig. 4.3. P_{fu} as plotted against η_S

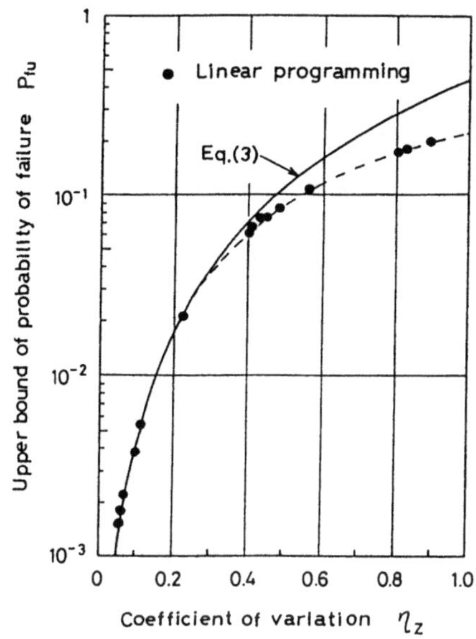

Fig. 4.4. Solution of Eqs.(7a) to (7c) as plotted against η_Z

that the distribution of Z is continuous and unimodal. That is, Eq.(3) does notrequire the assumption of $Z_{MO}=\mu_Z$.

The dotted curve in Fig. 4.4 shows the equation

$$P_{fu}=\frac{4}{9}\cdot\frac{\eta_Z^2}{1+\eta_Z^2}. \tag{8}$$

It is seen that the solution of the linear programming problem coincides with Eq.(8) quite well within the present range of calculation.

5 Discussion

As mentioned before, Eq.(3) was derived previously using the Camp–Meidell inequality under the assumptions: (a) the distribution of Z is continuous and unimodal, and (b) the mode of Z is equal to the mean of Z, i.e, $Z_{MO}=\mu_Z$. On the other hand, P_{fu} was obtained in the present study under the assumption of (a) alone without assuming (b). One might consider that Eq.(3) should give the value lower than the present solution, since use of the assumption (b) implies more informations. However, the actual result is opposite as shown in Fig.5. This is due to that Eq.(3) was derived using the Camp–Meidell inequality which is not very sharp as may be understood from the derivation of this inequality.

6 Conclusions

In relation to practical structural reliability analysis, the upper bound of the probability of failure $P_f = Pr(R \leq S)$ based on the means and the variances of the strength R and the stress S was evaluated by formulating and solving an optimization problem under a mild assumption that the distribution of $Z(=R-S)$ is continuous and unimodal. The principal conclusions obtained are as follows:

(1) The solution of an optimization problem revealed that Eq.(3) still holds even when it is only assumed that the distribution of $Z(=R-S)$ is continuous and unimodal.

(2) Since the previous derivation of Eq.(3) using the Camp–Meidell inequality requires an additional assumption that the mode of Z is equal to the mean of Z, which may not be satisfied in usual, the present finding extends largely practical applicability of Eq.(3).

(3) Within the present range of calculation, P_{f_u} can be well expressed by Eq.(8).

Acknowledgement

This study was partly supported by grant–in–aid for scientific research, Japanese Ministry of Education, Science and Culture.

References

1. Cornell, C. A., A probability–based structural code, ACI Journal, **66**(1969)974–85.
2. Hasofar, M. A. and Lind, N. C., Exact invariant second moment code format, J. Eng. Mech. Div., Proc. ASCE, **100–EM1**(1974)111–21.
3. Mischke, C., A method of relating factor of safety and reliability, J. Eng. Ind., Trans. ASME, **B92**(1970)537–42.
4. Thien, M. D. and Massoud, M., On the relation between the factor of safety and reliability, J. Eng. Ind., Trans. ASME, **B96**(1974)853–57.
5. Ichikawa, M., New formula for upper bound of probability of failure, Reliability Eng., **5**(1983)173–80.
6. Veneziano, D., New index of reliability, J. Eng. Mech., Proc. ASCE, **105–EM2**(1979)277–96.

Stochastic Simulation of Initiation and Early Growth of Small Cracks in Creep-Fatigue

Takayuki Kitamura[1], Naoya Tada[1], and Ryuichi Ohtani[1]

[1] Department of Engineering Science,
Kyoto University,
Kyoto 606-01, Japan

Abstract. The creep-fatigue failure of polycrystalline steels and alloys is caused by multiple intergranular cracks which begin to initiate from the early stage of life. The initiation and the growth of small cracks show random behavior due to the local inhomogeneity of materials. In order to investigate the mechanics hidden behind the stochastic nature, a method of inverse analysis based on a numerical simulation is proposed. As a result, the characteristics of the failure process is successfully drawn out in terms of the cracking behavior by the analyses.

Keywords. creep-fatigue, small cracks, crack initiation, crack growth, numerical simulation

1 Introduction

The failure of steels and alloys in creep-fatigue at high temperatures is, in general, brought about by multiple small cracks which often begin to initiate along grain boundaries on smooth surfaces in the early stages of fatigue life[1-3]. The essence of the mechanics is hidden behind the stochastic nature since the microstructural inhomogeneity conspicuously affects the cracking process[4].

Recent development of computer technology provides a powerful tool for analyzing many kinds of stochastic phenomena. The authors have proposed stochastic models of initiation and growth of small creep-fatigue cracks and the method of numerical simulation by means of a Monte Carlo method[5-8]. Although the computation has given fruitful results for understanding the intricate failure process, it is still difficult to deduce the mechanics of initiation and growth of actual cracks only from the model proposed.

In this study, discussions are made on a procedure to extract inversely or inductively the characteristics of the mechanics of initiation and growth of small cracks in creep-fatigue on the basis of the experimental observation.

2 Experimental Observation on Initiation and Growth of Small Cracks

For polycrystalline steels and alloys with fine grains, intergranular surface cracks initiate at the unit of about one grain boundary facet. Although most of them are arrested at the grain boundary triple junction, the rest grow discretely by the unit of one facet after the temporal arrest[9]. On the other hand, a small crack initiates on a grain boundary to propagate continuously along it for steels and alloys with large grain size[3]. The cracks propagate at high rates along grain boundaries, and are arrested or decelerated when they hit triple points or sharp bends in grain boundaries. In any case, the junction of grain boundaries causes the fluctuation of the growth rate for small cracks.

The circles in Fig.2.1 show the number of cracks per unit area (mm^2) initiated along grain boundaries on the surface of a smooth specimen of a 304 stainless steel, of which average grain diameter is 40 μ m, in creep-fatigue with a slow-fast strain waveform at 923K (650C) in a vacuum environment[4]. The total strain rage is 1% and the failure life is 388 cycles. The number of cracks initiated per unit cycle is readily obtained as the slope of the density except the late stage where the frequent coalescence of cracks reduce the density. The cracks, of which length is about 20 μ m, initiate from less than 10% of the failure life to the final stage. The distribution of crack initiation life is attributed to the inhomogeneity of resistance and driving force for crack initiation. The crack growth rate is plotted against the half length of surface crack in Fig.2.2. The growth of small cracks is characterized by huge scatter whereas it is well-known that the fluctuation of the growth rate for large cracks is within a factor of two[10]. The distribution of crack

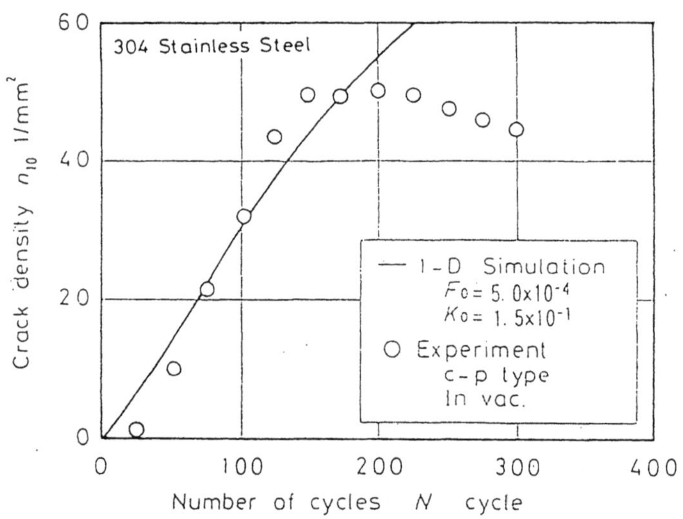

Fig.2.1. Change in crack density with increasing number of fatigue cycles. Crack density is defined as the number of cracks per unit area.

length at the early stage of life is shown in Fig.2.3, which results from the randomness of the crack growth rate as well as the initiation time.

3 Modelling of Initiation and Growth of Small Cracks

In order to deal with the random nature of small cracks due to the microstructural inhomogeneity, the authors have proposed stochastic models. The Monte Carlo technique is used to simulate the process of initiation and growth of intergranular cracks. In one of the models[5], the acceleration and the deceleration of crack growth along grain boundaries is directly modelled as illustrated in Fig.3.1. The intergranular crack grows at alternately increasing and decreasing rates between the upper and lower bounds of the rates. The lower bound appears at the grain boundary triple points "A" and "C" and the upper one does at the midway "B". The bounds of the growth rates are given as random numbers. Although the random growth is successfully reproduced by the model, it is insufficient to analyze the failure since the initiation is not included.

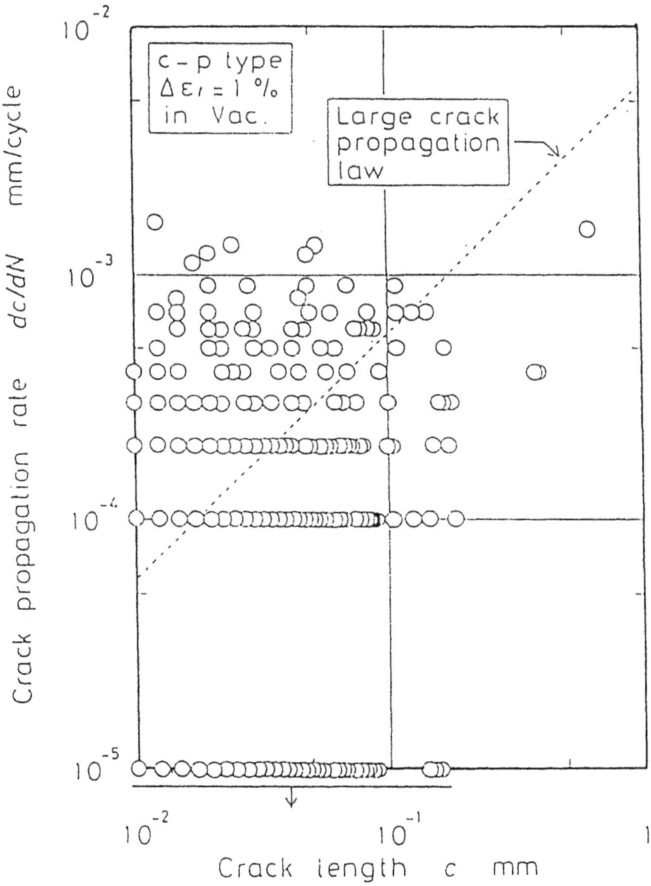

Fig.2.2. Growth rates of intergranular small cracks (experimental results).

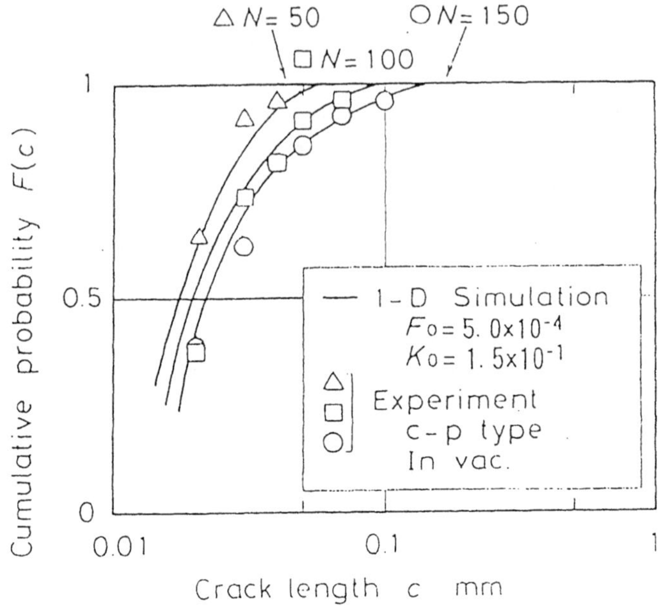

Fig.2.3. Cumulative probability of crack length at the early stage of fatigue life.

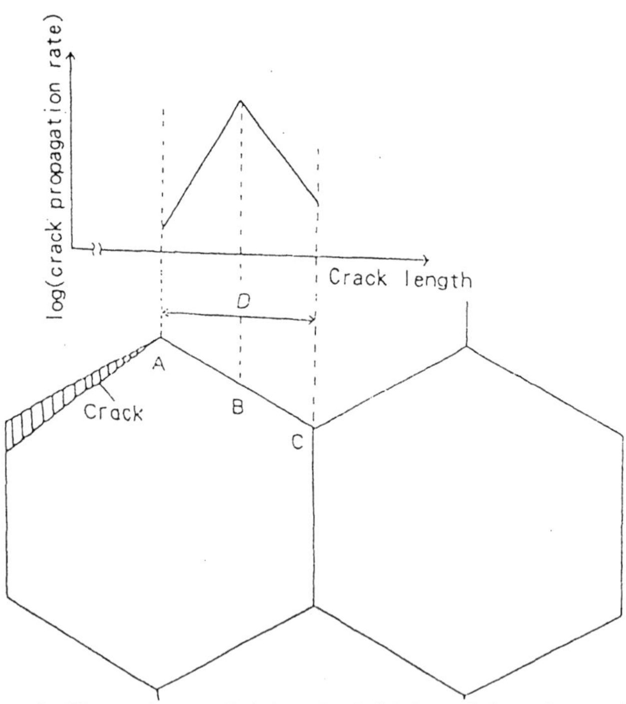

Fig.3.1. Schematic illustration explaining the initial model on the stochastic growth of microstructurally small cracks.

Another model has been developed to simulate not only the growth but also the initiation at once[6,8]. The model which is illustrated in Fig.3.2 is summarized as follows;

(1) Grain boundaries between two adjacent triple points are projected and linked on a straight line perpendicular to the applied stress axis, and the lengths are given as random numbers of a normal distribution.

(2) The grain boundaries have their own fracture resistance, R, which are random variables.

(3) Driving force for crack initiation is imposed on each grain boundary. The fracture resistance of each grain boundary, R, decreases by the magnitude of the driving force in every fatigue cycle.

(4) When the residual resistance reaches zero, the grain boundary is assumed to fracture and to yield an intergranular crack.

(5) The driving force for a grain boundary on either side of a cracked boundary increases under the influence of stress intensity, and it is given as a function of crack length.

(6) When the resistance adjacent to the crack reaches zero, the grain boundary is assumed to fracture and to exhibit a discrete propagation.

Though the resistance and the driving force are essentially independent random variables, the crack initiation and growth is mathematically governed by the convolution of the two kinds of the variables. Therefore, in this study, only the resistance is assumed to be represented by random numbers and the driving force is treated as a deterministic variable. The mathematical equivalence is proved elsewhere[11]. For simplicity, the resistance is assumed to have fictitious uniform random numbers between zero and unity in this study. The dependence of driving force on the crack length is formulated as

$$D(c)=Fo+K(c/d)=Fo+Ko\cdot c, \qquad (1)$$

where Fo is the driving force for crack initiation (when c=0), $Ko \cdot c$ is the additional driving force for crack growth, c is the half crack length and d is the average of grain boundary length. The model consists of only two random numbers (the grain boundary length and the resistance) and two deterministic variables (Fo and Ko).

Figure 3.3 shows simulated results by means of a computer for some arbitrary magnitudes of driving forces, Fo and Ko. The figures reveal that the crack density is sensitive to the magnitude of Fo whereas the distribution of crack length is controlled by that of Ko. This implies that Fo and Ko govern the initiation and the growth of small cracks, respectively[8].

The simulation result is compared with the experimental observation. The magnitudes of driving forces, Fo and Ko, are determined by a trial and error as discussed in the next section in detail. The crack density simulated is shown by a solid line in Fig.2.1. Up to the middle stage of life, the calculation agrees with the experimental result though it does not after then. The distribution of crack length simulated at the early stage of life is shown by solid lines in Fig.2.3 which reveals successful agreement. These suggest the applicability of this simple model to

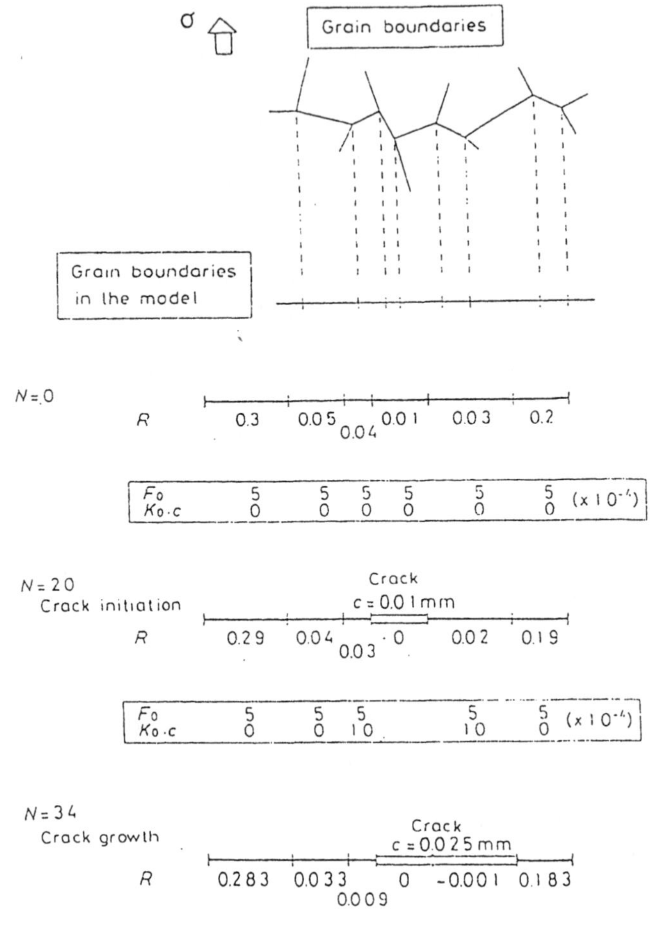

Fig.3.2. Schematic illustration explaining the one-dimensional model used in this study on the stochastic behavior of initiation and growth of microstructurally small cracks.

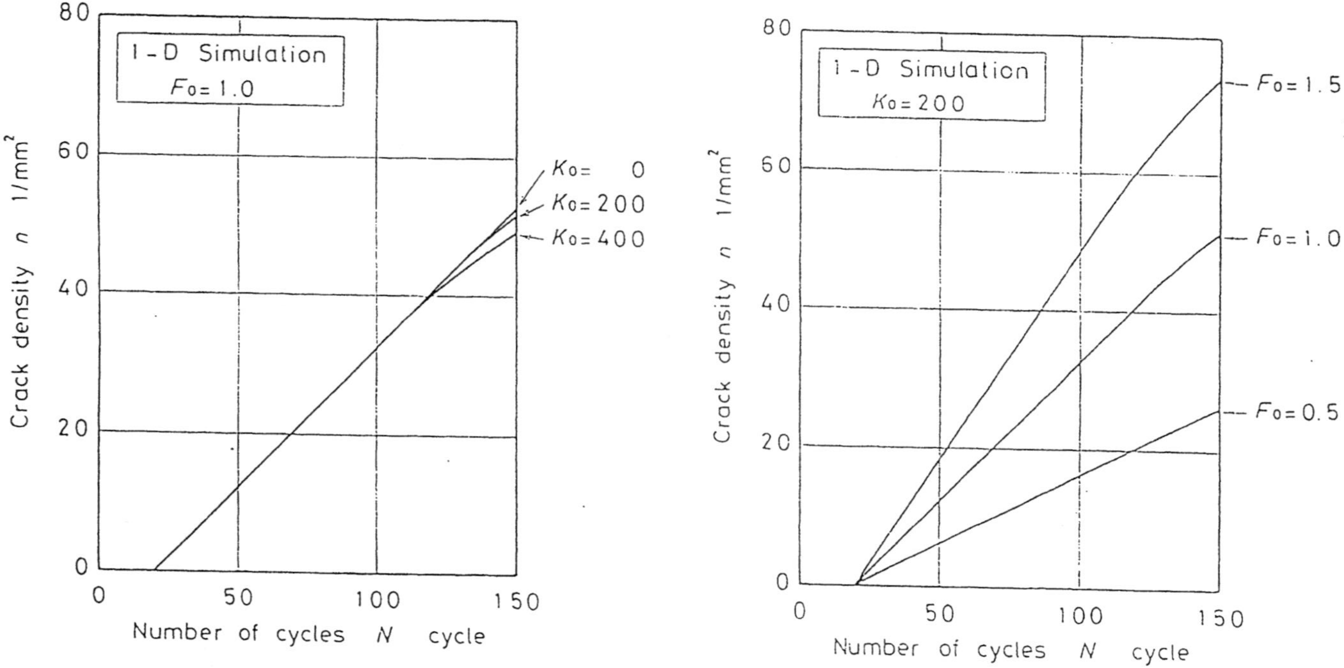

(a) Effect on crack density.

Fig.3.3. Effect of the fracture driving force on the initiation and the growth of small cracks.

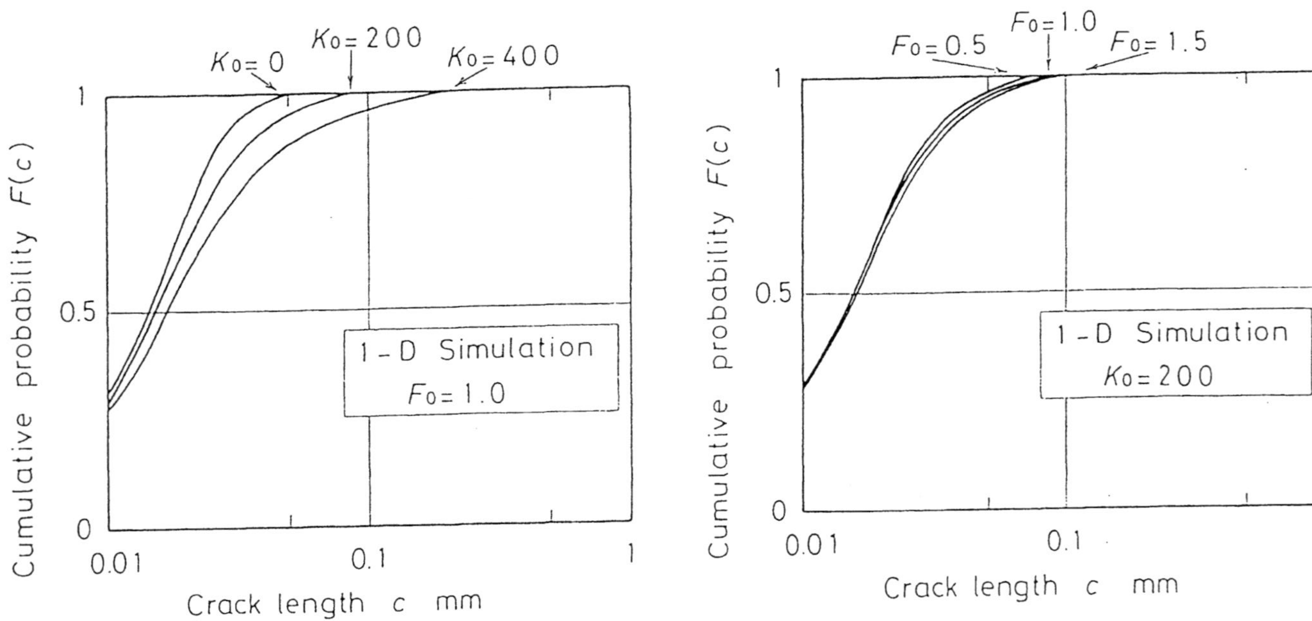

(b) Effect on cumulative probability of crack length.

Fig.3.3. continued.

reproduce the random cracking process at the early stage. The simulation conducted here holds enough validity for the investigation on the random nature of small cracks, as the prediction of damage process at the early stage of failure life is important in the application to an actual component. Since the disagreement at the later stage is attributed to the crack coalescence and the stress relaxation near large cracks, it is necessary to refine the model for improving the simulation at the later stage on the basis of three-dimensional analyses as described in Section 5.

4 Extraction of Stochastic Nature by Inverse Analyses

The authors have investigated the growth of large creep-fatigue cracks. As their growth rates show little fluctuation, they can be evaluated on the basis of the J-integral concept of nonlinear fracture mechanics[10]. In an infinite body, the growth rate is proportional to the crack length. Assuming that the growth rate of the large crack corresponds with the average in the fluctuation of the small crack growth rate, the magnitude of Ko can be determined in terms of the J-integral.

On the other hand, the magnitude of Fo could be discussed on the basis of the creep cavity growth or grain boundary sliding as well. However, it is generally difficult to determine the precise value of Fo because a functional relation between Fo and grain boundary structure and energy, micromechanics of grain deformation at the surface of polycrystals, environmental effect and so on have never been clarified.

An inverse analysis of small cracks is proposed here to extract the characteristics of the random failure process by using the numerical simulation. Fitting the simulation results to the experimental data by trial and error, an optimum set of magnitudes of Fo and Ko are determined. As the magnitudes of Fo and Ko are representative of the driving forces for the initiation and the growth of small cracks, respectively, their magnitudes are presumed through the inverse calculation by the following procedure.

(1) The simulation described in Section 3 is carried out using a provisional set of values of Fo and Ko.

(2) If the crack density obtained from the simulation is higher than that from the experiment, the magnitude of Fo is reduced, and vice versa.

(3) If the average crack length obtained from the simulation is longer than that from the experiment, the magnitude of Ko is reduced, and vice versa.

(4) The simulation is carried out again using the modified set of Fo and Ko.

(5) Repeat (2) through (4) until the simulation agrees with the experimental observation.

Figure 4.1 is an example showing the change in the values of Fo and Ko during a repeated calculation for the inverse analyses of the experimental data shown in Figs.2.1 and 2.3. Values of 0.0001 and 0.2 are chosen as the initial values of Fo

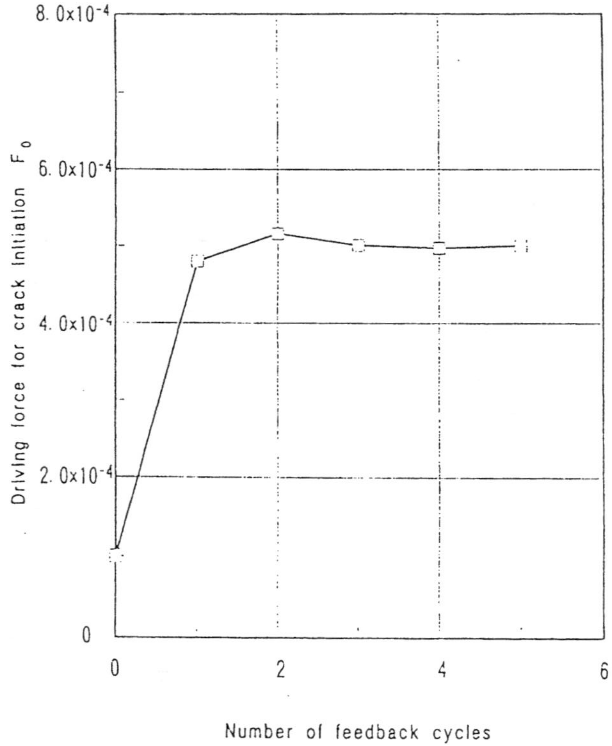

(a) Convergence of Fo.

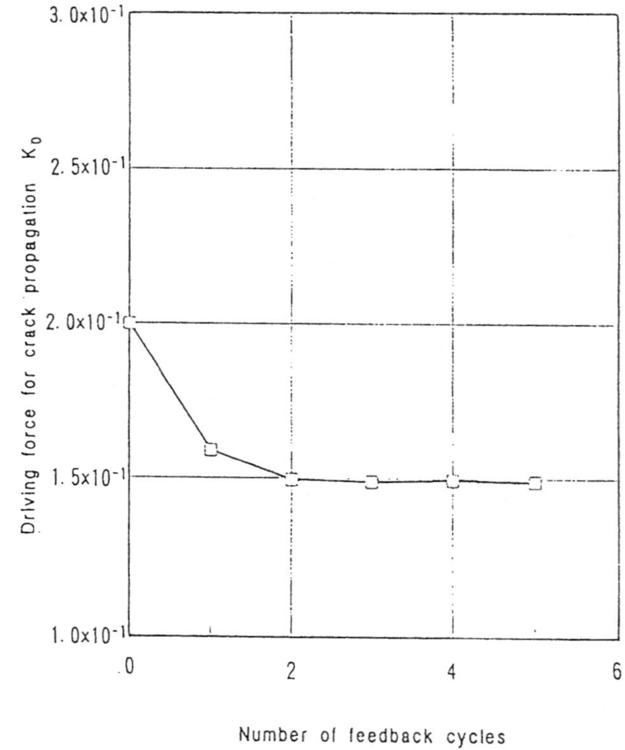

(b) Convergence of Ko.

Fig.4.1. Convergence of fracture driving force by repeated calculation in the inverse analyses.

and Ko. The convergence is attained in a few cycles. As shown by the solid lines in Figs. 2.1 and 2.3, the simulation statistically reproduces the distribution of multiple small cracks observed in the experiment.

Other sets of values of Fo and Ko for other creep-fatigue conditions are examined by the similar inverse calculation corresponding to the experiments on the same 304 stainless steel with slow tensile strain rate at 923K in air[4]. The effect of strain range and compressive strain rate on the magnitudes of Fo and Ko is plotted in Figs. 4.2 and 4.3, respectively. The driving force for crack initiation, Fo, shows stronger dependence on the strain range than that for crack growth, Ko. The compressive strain rate affects eminently on the magnitude of Fo while it does not on that of Ko. By analyzing these results, the characteristics hidden behind the actual randomness can be extracted and suggestive information on the cracking mechanism and mechanics can be obtained.

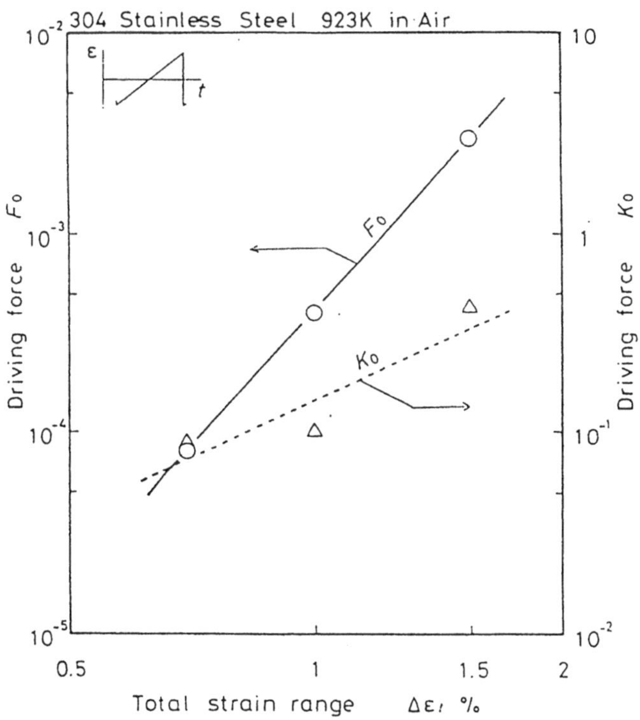

Fig.4.2. Dependence of fracture driving force on the total strain range in creep-fatigue of 304 stainless steel at 923K in air.

The failure process in creep-fatigue is characterized by the ratio, Ko/Fo [12]. For an equal value of Fo/Ko, the crack initiation and growth go through the same process with the same cracking pattern. A limited numbers of crack initiation and the successive growth of a single crack yields the failure due to a main crack in the case of the high magnitude of Ko/Fo, whereas the initiation of many distributed cracks brings about the failure due to their coalescence in the low magnitude of Ko/Fo. Figures 4.2 and 4.3 point out that the crack growth governs the life in lower strain ranges and slow-slow strain waveforms whereas the coalescence characterizes it in higher strain ranges and slow-fast strain waveforms.

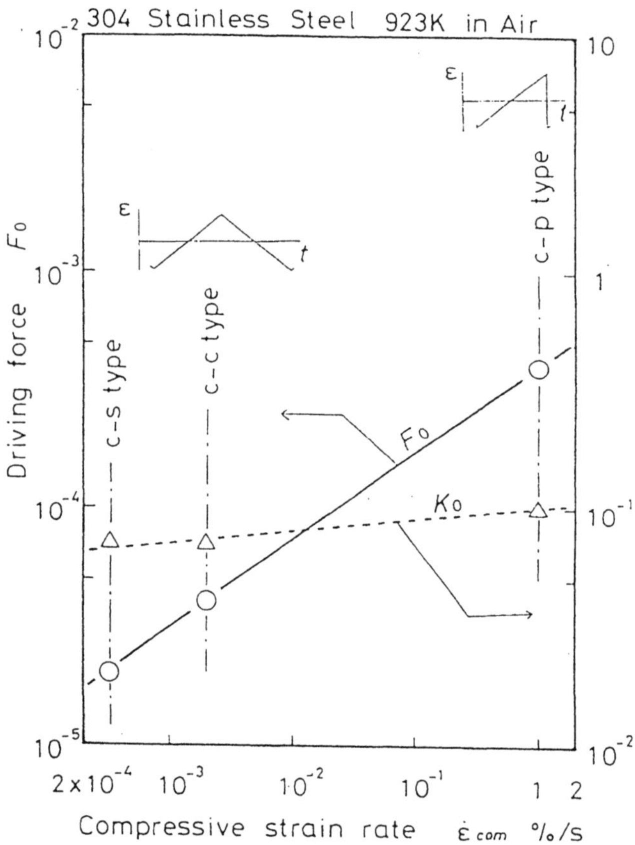

Fig.4.3. Dependence of fracture driving force on the compressive strain rate in creep-fatigue of 304 stainless steel at 923K in air.

5 Three-Dimensional Model

The present model is extended to a three-dimensional analysis in order to deal with the middle stage through the final one in the failure process[8]. The model is summarized as follows.

(1) Grain boundaries on the surface of a polycrystalline material are numerically constructed on the basis of an isotropic grain growth model. Nuclei of grains are generated at random points and time in a fixed plane, and expand at a constant speed until the fronts of two adjacent nuclei meet, where a grain boundary is formed. The inward inclination of each grain boundary facet is determined by theoretical distribution of an isotropic material.

(2) Each grain boundary facet on the surface has its own fracture resistance, R, with a uniform distribution between zero and unity.

(3) The fracture driving force for crack initiation is assumed to be a function of the normal stress to each facet, σ_n, as

$$D = F_0 \cdot (\sigma_n / \sigma)^m, \qquad (2)$$

where σ is the applied stress and m is a material constant. The fracture resistance of each grain boundary, R, decreases by F in every fatigue cycle.

(4) When the residual resistance reaches zero, the grain boundary is assumed to fracture and an intergranular crack is initiated there.

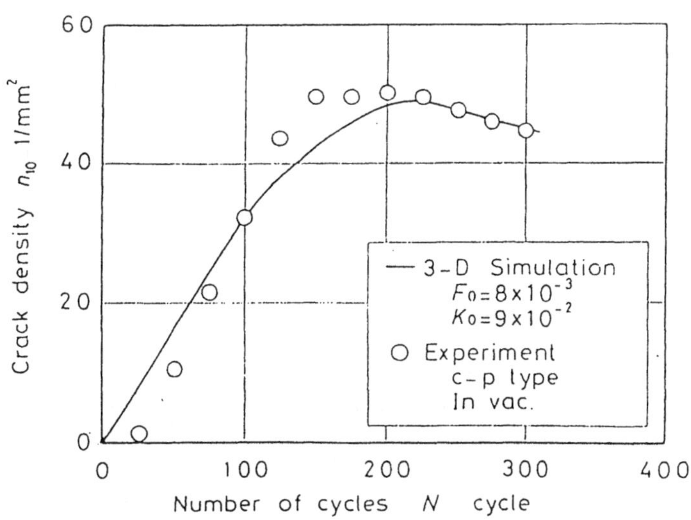

Fig.5.1. Crack density simulated on the basis of the three-dimensional model comparing with the experimental result.

(5) The driving force for a grain boundary on either side of a cracked grain boundary is given by

$$D(c)=F+K=Fo\cdot(\sigma_n/\sigma)^m+Ko\cdot c. \qquad (3)$$

(6) When the resistance adjacent to the crack reaches zero, the grain boundary is assumed to fracture and to exhibit a discrete propagation.

(7) No fracture driving force is assigned to the grain boundary facets within a circular area containing a pre-existent crack due to the stress relaxation.

(8) When two neighboring cracks happen to coalesce, the crack is assumed to become a large crack immediately.

Fitting the simulation with the experimental result, the calculated result of crack density is shown by a solid line in Fig.5.1. The decrease in the crack density at the latter half of fatigue is reproduced by the calculation, which displays remarkable improvement of the simulation as compared with Fig.2.1. The crack propagation rates and the distribution of crack length shown in Figs.5.2 and 5.3, respectively, are also in good accordance with the experimental observation (See experimental results in Fig.2.2 in comparison of Fig.5.2). Figure 5.4 shows a simulated image of distributed surface cracks comparing the experimental observation. These results prove the validity of the model until the final stage of creep-fatigue life. Following the procedure described in the previous section, the inverse analyses is available to the three-dimensional calculation for the determination of the magnitude of variables.

6 Conclusions

An investigation is made on the inverse analysis for extracting the characteristics of initiation and growth of microstructurally small cracks in creep-fatigue which exhibits complex mechanical behavior. As a result, the nature of cracking is successfully drawn out by the inverse calculation, and the information obtained gives us a clue for understanding the failure mechanism and mechanics. It also contributes to the damage evaluation of structural materials in practical components. The authors proposed a rational method for evaluating the remaining life of electric power plants on the basis of the numerical simulation where the driving force is determined inversely by the data obtained from the regular inspection[13].

Similar complexity is found on the multi-site cracking in low cycle fatigue, stress corrosion cracking, corrosion fatigue, etc[14-16]. For example, the growth of transgranular small cracks in room temperature fatigue is interfered by grain boundaries while it proceeded continuously at a high rate in grains[15]. For the precise prediction of failure life, their stochastic nature due to the microstructural inhomogeneity must be understood. The characteristics may be inversely extracted by the computer simulation based on a simple model as proposed in this paper.

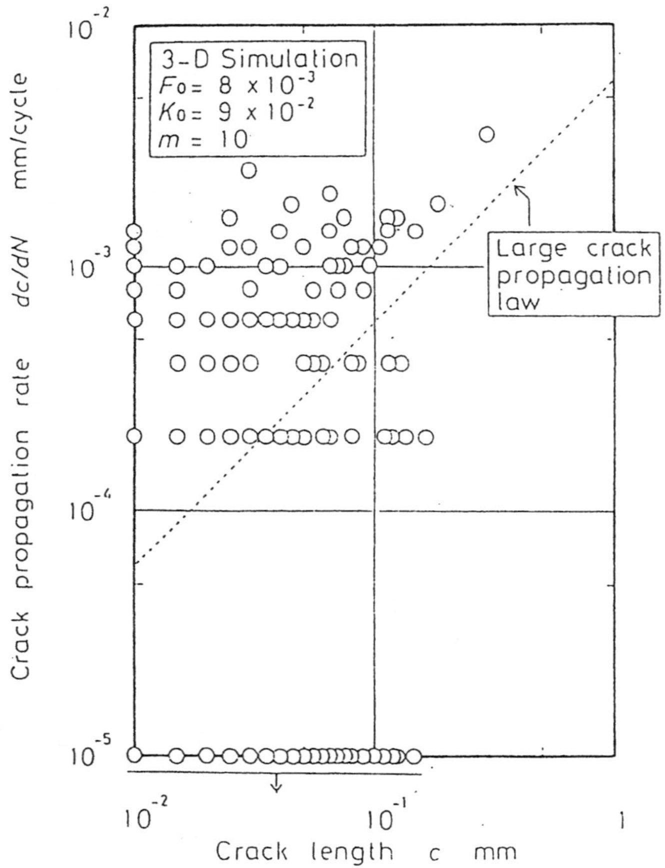

Fig.5.2. Crack propagation rate simulated on the basis of the three-dimensional model.

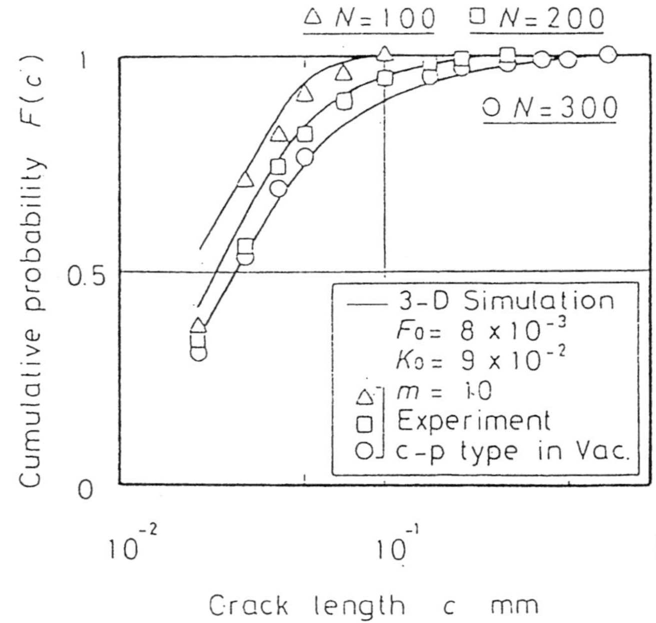

Fig.5.3. Cumulative probability of crack length simulated on the basis of the three-dimensional model comparing with the experimental result.

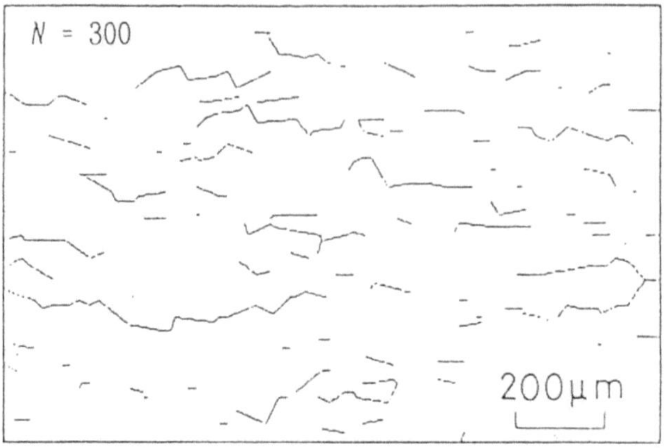

(a) Simulated image.

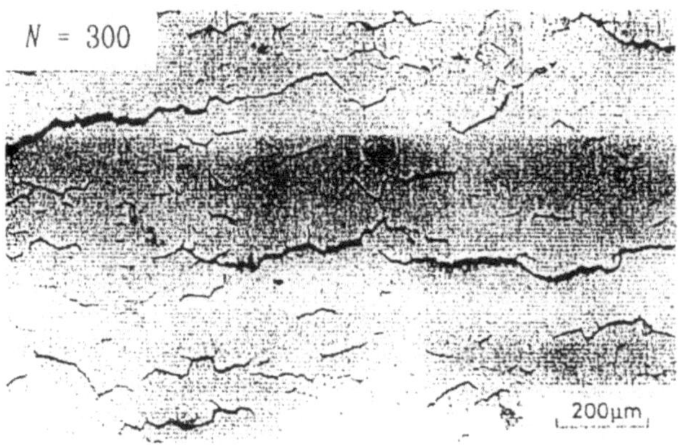

(b) Actual image.

Fig.5.4. Distributed surface cracks at N=300 in creep-fatigue of 304 stainless steel.

References

1. Ohtani, R., Okuno, M. and Shimizu, R., Microcrack growth in grain boundaries at the surface of smooth specimen of 316 stainless steel in high temperature creep, J. of Society of Materials Science, Japan, 31 (1982) 505-509, (in Japanese).

2. Ohtani, R., Kinami, T. and Sakamoto, H., Small crack propagation in high temperature creep-fatigue of 304 stainless steel, Trans. of JSME, Ser.A, 52 (1986) 1825-1830, (in Japanese).

3. Ohtani, R. and Kitamura, T., Behavior of initiation and growth of creep and creep-fatigue small cracks in austenitic stainless steels, in Proceedings of the 7th International Conference on Fracture, ed. K. Salama, K. Ravi-Chandar, D.M.R. Taplin and R. Pao, Pergamon Press, 1989, 213-221.

4. Ohtani, R. and Kitamura, T., Initiation and propagation of microstructurally small cracks under creep-fatigue condition, in Proceedings of the 4th International Conference on Creep and Fracture of Engineering Materials and Structures, ed. B. Wilshire and R.W. Evans, The Institute of Metals, London, UK, 1990, 791-802.

5. Ohtani, R. and Kitamura, T., Numerical simulation of microstructurally-short-crack propagation in creep, JSME International, 30 (1987) 1741-1749.

6. Ohtani, R., Kitamura, T. and Tada, N., Numerical simulation of initiation and early propagation of creep-fatigue small cracks based on a model of random fracture resistance of grain boundaries, in Proceedings of the 1989 ASME Pressure Vessels and Piping Conference, ed. C. Becht IV, R. Ohtani and S.Y. Zamrik, ASME, 1989, 123-127.

7. Ohtani, R., Kitamura, T. and Tada, N., Stochastic simulation of initiation and growth of small cracks in creep-fatigue condition, in Proceedings of the 4th International Conference on Fatigue and Fracture, 1990, 2143-2148.

8. Tada, N., Kitamura, T. and Ohtani, R., Monte Calro simulation of creep-fatigue small cracks based on a three-dimensional model of random fracture resistance of grain boundaries, Trans. JSME, Ser.A, 56 (1990) 708-714.

9. Kitamura, T. and Tada, N. Ohtani, R., To be published, Evaluation of creep fatigue damage based on initiation and growth of small cracks, European Structural Integrity Society Special Technical Publication, Behavior of Defects at High Temperatures, 1993.

10. Ohtani, R. and Kitamura, T., Characterization of high temperature strength of metals based on the mechanics of crack propagation, Current Japanese Materials Research Vol.3, ed. R. Ohtani, M. Ohnami and T. Inoue, Elsevier Applied Science, London, 1988, 65-90.

11. Kitamura, T., Ghosn, L.J. and Ohtani, R., Stochastic modeling of crack initiation and short-crack growth under creep and creep-fatigue conditions, Trans. ASME, J. of Applied Mechanics, 59 (1992) S35-S42.

12. Kitamura, T., Tada, N. and Ohtani, R., Characterization of creep-fatigue failure process based on stochastic behavior of small crack initiation and growth, in Proceeding of IUTAM Symposium on Creep in Structure, ed. M. Zyczkowski, Springer-Verlag, Berlin, 1991, 347-354.

13. Ohtani,R., Kitamura, T. and Tada, N., Methodology of remaining life assessment in high temperature applications based on a Monte Calro simulation of grain boundary cracking, in Proceedings of the 6th Mechanical Behavior of Materials, ed. M. Jono and T. Inoue, Pergamon Press, 1991, 205-212.

14. Kitagawa, H., Nakasone, Y. and Miyashita, S., Measurement of fatigue damage by randomly distributed small cracks data, ASTM STP 811, ed. J. Lankford, D.L. Davidson, W.L. Morris and R.P. Wei, 1983, 233-263.

15. Tokaji, K., Ogawa, T. and Harada, Y., Evaluation on limitation of linear elastic fracture mechanics for small fatigue cracks under rotating bending, J. of Society of Materials Science, Japan, 35 (1986) 394-400, (in Japanese).
16. Riesch-Oppermann, H., Bruckner-Foit, A., Muntz, D. and Winkler, T., To be published, Crack initiation, growth and interaction in thermal cyclic loading -- a statistical approach, Evaluation of creep fatigue damage based on initiation and growth of small cracks, European Structural Integrity Society Special Technical Publication, Behavior of Defects at High Temperatures, 1993.

Study of Three Reliability Methods for Fatigue Crack Growth

W. K. Liu*, Y. J. Lua†, Y. Chen*, and T. Belytschko*

* Department of Mechanical Engineering, Northwestern University, Evanston, IL 60208, USA

† Applied Research Associates, Inc., Southeast Division, Raleigh, NC 27615, USA

Abstract. In this paper, three reliability methods are presented to solve general fatigue crack growth problems: the first order approximation method (FAM), the Lagrange multiplier formulation (LMF) and a Monte Carlo Simulation (MCS) with exact sampling technique. With the help of the exact sampling technique used in MCS, we are able to determine the extremely low probabilities of failure efficiently and with high accuracy. A comparison of the three methods shows that the LMF is the most efficient method to solve the fatigue crack growth reliability problem. It is also shown that the crack growth direction law plays an important role in the curvilinear crack propagation problem.

Keywords. Fatigue reliability, curvilinear crack growth, Lagrange multiplier formulation, first order approximation, Monte Carlo simulation

1. Introduction

Probabilistic Fracture Mechanics (PFM) is a rapidly developing field in modern engineering and technology. Due to randomness in geometric defects, material properties, external loads and other uncertainties, deterministic approaches are often insufficient. Probabilistic fracture mechanics, which combines the fracture mechanics with stochastic methods, provides a useful tool to address these problems. The basic aspects of probabilistic mechanics and its application to the analysis of uncertainties in structural systems are summarized and recapitulated in the recent book by Liu and Belytschko[16]. The application of PFM to the reliability analysis of brittle fracture of a homogeneous material has been given by Besterfield et al.[1] based on the probabilistic finite element method. Lua et al.[17;19] developed a stochastic damage model to quantify the inherent statistical distribution of the fracture toughness of multi-phase brittle materials. The mixed boundary integral equation method coupled with the first order reliability method (FORM) has been developed by Lua et al.[18;20] for the reliability analysis of fatigue crack growth.

Given the complexity of the constitutive material model and failure mechanism, the combination of the finite element method with the theory of statistics and structural reliability has become a powerful tool for safety and reliability analysis[11;12;13;14;15]. A finite element method coupled with the first and second-order reliability methods (FORM and SORM) has been developed by Der Kiureghian and Ke[5] for linear structural problems and Liu and Der Kiureghian[10] for geometrically non-linear problems. A probabilistic finite element method in conjunction with FORM has been developed by Besterfield et al.[2] for the reliability analysis of the fatigue crack growth. The stochastic finite element method coupled with the response-surface approach has been used by Faravelli[8;9] to evaluate both the response uncertainty and reliability of a system with a high degree of mechanical complexity and nonlinearity. In addition to the extremely small range of failure probability, which is the major concern in reliability engineering, all or part of the Cumulative Distribution Function (CDF) is generated in Wu et al.[22] by an advanced algorithm based on the Fast Probability Integration (FPI). The performance of the FPI based on either mean-value methods or mean-value first-order methods has been demonstrated in Cruse et al.[4] through the reliability analysis of a turbine blade.

In this study, the stochastic boundary element method, which combines the mixed BIEM[18] with the FORM, is developed to study probabilistic fatigue crack growth. The response gradient, the key in FORM, is determined by two approaches: the first order approximation method (FAM) and Lagrange multiplier formulation (LMF). Finite difference is employed to determine the sensitivity of stress intensity factors. An iterative scheme based on the HL-RF method[6] is employed to determine the most probable failure point (or design point). In order to show the accuracy of the numerical algorithm, the cumulative distribution function of the fatigue life calculated by these two formulations is compared with the result obtained by Monte Carlo Simulation (MCS). With the help of this exact sampling technique, the extremely small value of probability of failure can be determined with a relatively small number of expensive simulations based on the mixed BIEM.

2. Stochastic Process in a Fatigue Crack Growth

Fatigue crack growth is a stochastic process. A crack will propagate along a curvilinear path from its initial crack size to its final size after a given number of load cycles. In the numerical simulation of curvilinear fatigue crack growth, both the crack growth direction and growth step are required at each intermediate step.

The fatigue crack growth step can be determined from the fatigue crack growth law. The fatigue crack growth law provides a mapping from the time domain to the spatial domain. While the problem can be solved in the spatial domain via this mapping, the fatigue process is truly a stochastic process. Furthermore, due to the step-wise solution procedure in propagating a crack from its initial length to the final length, the statistics of sequential points must rely on previous information. Based on the above observation, a fatigue crack propagation should be considered as a Markov process. To fully describe this stochastic process, in addition to the crack growth law, the crack growth direction law is also required.

2.1 Fatigue Crack Growth Law

In this study, the most popular Paris-Erdogan law[21] is used:

$$T = \int_{a_1}^{a_f} \frac{da}{D\left(\Delta\kappa_{eq}\right)^n}, \tag{1}$$

where a_1 is the initial crack length, a_f is the final crack length. D and n are fatigue parameters, and $\Delta\kappa_{eq}$ is the range of equivalent model I Stress Intensity Factor (SIF) in a cycle given by

$$\Delta\kappa_{eq} = \kappa_{eq}^{max} - \kappa_{eq}^{min}. \tag{2}$$

The quantities κ_{eq}^{min} and κ_{eq}^{max} are the minimum and maximum equivalent mode I SIFs associated with the minimum and maximum cyclic applied stress, respectively. If $\kappa_{eq}^{min} = 0$, Equation (2) reduces to

$$\Delta\kappa_{eq} = \kappa_{eq}^{max} = \kappa_{eq}. \tag{3}$$

The relationship among κ_{eq}, mode I and mode II SIFs (κ_I and κ_{II}) is represented by

$$\kappa_{eq} = \cos^3\frac{\theta}{2}\,\kappa_I - 3\cos^2\frac{\theta}{2}\sin\frac{\theta}{2}\,\kappa_{II}, \tag{4}$$

where θ is the crack growth direction angle.

The differential form of the Paris-Erdogan law can be written as

$$da = D\left(\Delta\kappa_{eq}\right)^n dT. \tag{5}$$

Because of the randomness in material properties, loading and crack geometry, the resulting probability density function (pdf) of da can be analyzed by Eq. (5) as shown in Fig. 1.

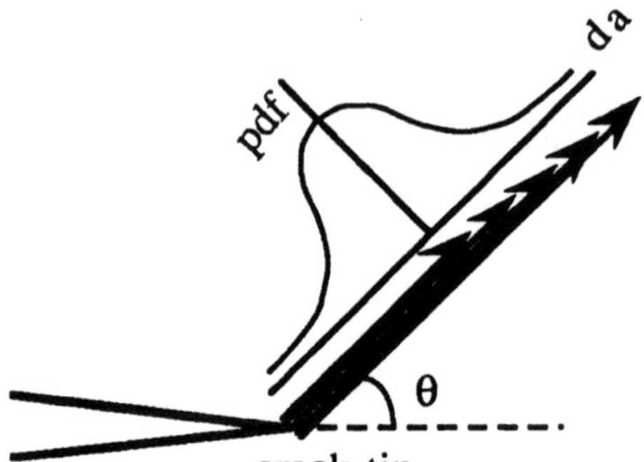

Fig. 1. Distribution of fatigue crack growth step size

2.2 Crack Growth Direction Law

The direction of the crack growth is also a random function. One common choice of crack growth direction law is the maximum circumferential tensile stress theory[7]:

$$\sin\theta\, \kappa_I + (3\cos\theta - 1)\, \kappa_{II} = 0, \tag{6}$$

where κ_I and κ_{II} represent mode I and mode II stress intensity factors, respectively.

As can be seen from Eq. (6), the randomness of material properties, loading and crack geometry also gives rise to the uncertainty in the crack growth direction (see Fig. 2).

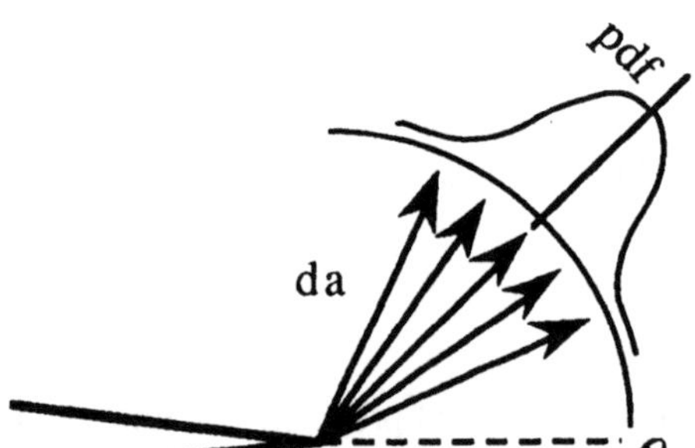

Fig. 2. Distribution of crack growth direction

Under the combined effects of random crack growth step and crack direction angle, the joint pdf of the crack tip position can be obtained. The uncertainty in the crack-tip position will be propagated as the crack grows from its initial length to the final length. In order to show the propagation of randomness, a rectangular plate with a single inclined edge crack subjected to both normal (σ) and shear (τ) loading is considered. The initial crack angle, initial crack size, fatigue parameters and external loadings are assumed as random parameters. The mixed BIEM[18] coupled with Monte Carlo simulation is applied to show the stochastic process. By performing fifty simulations (NHIS=50), the resulting random crack paths corresponding to three crack growth steps are shown in Fig. 3.

3. Curvilinear Fatigue Crack Reliability Analysis by Lagrange Multiplier Formulation

All random variables characterizing the fatigue crack propagation are classified into two groups: master random variables and slave random variables. The slave random variables are used to characterize the current geometric configuration of a growing crack. The master random variables which represent the primary problem parameters are given by a vector, i.e.,

$$\mathbf{b} = \left[\theta_0, y_0, a_1, a_f, D, n, \sigma\right], \tag{7}$$

where θ_0 is the initial crack angle, y_0 is the y-coordinate of the initial crack, a_1 is the initial crack length, a_f is the final crack length, D and n are fatigue parameters used in the fatigue law (see Eq. (1)), and σ is the applied stress.

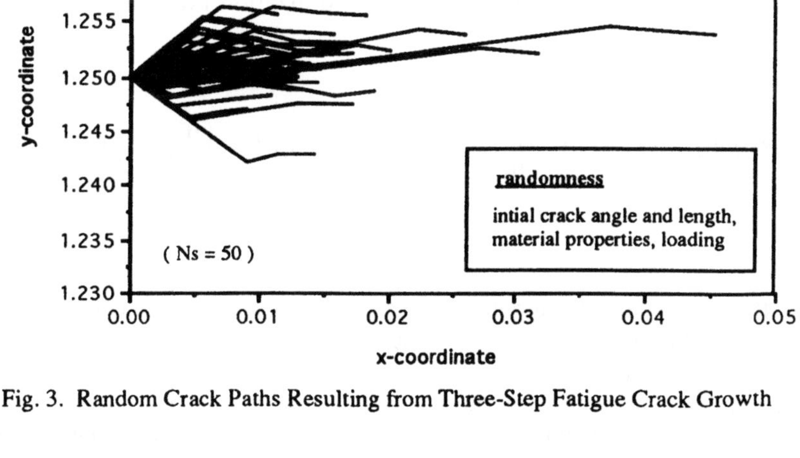

Fig. 3. Random Crack Paths Resulting from Three-Step Fatigue Crack Growth

The first order reliability method (FORM) can be formulated by the following optimization problem:

$$\text{minimize}\left(\mathbf{r}^{T}\mathbf{r}\right)^{1/2}, \tag{8}$$

where **r** is a set of standard, uncorrelated normal variables transformed from the master random vector **b**. The constraints posed to the optimization problem are:

(1) limit state function $$T - T_S = 0, \tag{9}$$

where T is the fatigue life determined by Eq. (1) and T_s is a given service life, and

(2) equilibrium equation $$\mathbf{K}(t)\,\boldsymbol{\delta}(t) = \mathbf{F}(t), \quad t_i \le t \le t_f, \tag{10}$$

where $\mathbf{K}(t)$ and $\mathbf{F}(t)$ are the generalized stiffness matrix and external force vector, respectively, and the generalized displacements are

$$\boldsymbol{\delta}(t) = \begin{bmatrix} \mathbf{d}(t) \\ \boldsymbol{\kappa}(t) \end{bmatrix}, \tag{11}$$

where $\mathbf{d}(t)$ is the nodal displacement vector and $\kappa(t)$ is the stress intensity vector. In Eq. (10), the initial and final fatigue time are given by t_i and t_f, respectively.

(3) crack growth direction law

$$Z(t) = \sin\theta(t)\,\kappa_I(t) + (3\cos\theta(t) - 1)\,\kappa_{II}(t) = 0, \quad t_i \le t \le t_f, \tag{12}$$

where $\theta(t)$ is the direction of crack growth measured from the current crack line. Equations (8)-(12) can be converted to an optimization problem by defining a Lagrange function, L, namely,

$$L(\mathbf{b},\boldsymbol{\mu}(t),\boldsymbol{\delta}(t),\gamma(t),\theta(t),\lambda) = \mathbf{r}^{T}\mathbf{r} + \int_{t_i}^{t_f} \boldsymbol{\mu}^{T}(t)\,[\mathbf{F}(t) - \mathbf{K}(t)\boldsymbol{\delta}(t)]\,dt + \int_{t_i}^{t_f} \gamma(t)\,Z(t)\,dt + \lambda(T - T_S), \tag{13}$$

where $\mu(t)$ is a vector of Lagrange multipliers for equilibrium, $\gamma(t)$ is a Lagrange multiplier of the crack direction law, and λ is the Lagrange multiplier for the performance function.

After applying the time-space mapping, Eq. (13) can be written in the spatial domain. In order to perform the integration in Eq. (13) numerically, the crack path is discretized into N-1 discrete elements by N discrete points. Applying necessary conditions of Eq. (13) at each discrete point, we have

$$\frac{\partial L}{\partial \mathbf{b}} = \frac{\partial}{\partial \mathbf{b}}\left[\mathbf{r}^{T}\mathbf{r}\right] + \sum_{i=1}^{N} \mu_i^{T} \frac{\partial}{\partial \mathbf{b}}\left[\mathbf{K}_i\delta_i - \mathbf{F}_i\right] + \sum_{i=1}^{N} \gamma_i \left(Z_i\right)_{,\mathbf{b}} + \lambda\, T_{,\mathbf{b}} = 0, \tag{14}$$

$$\frac{\partial L}{\partial \delta_i} = \mu_i^{T}\mathbf{K}_i + \sum_{k=1}^{N} \gamma_k \left(Z_k\right)_{,\delta_i} + \lambda\, T_{,\delta_i} = 0, \tag{15}$$

$$\frac{\partial L}{\partial \theta_i} = \sum_{i=1}^{N} \mu_i^{T} \frac{\partial}{\partial \theta_i}\left[\mathbf{K}_i\delta_i - \mathbf{F}_i\right] + \sum_{i=1}^{N} \gamma_i \left(Z_i\right)_{,\theta_i} + \lambda\, T_{,\theta_i} = 0, \tag{16}$$

$$\frac{\partial L}{\partial \mu_i} = \mathbf{K}_i\delta_i - \mathbf{F}_i = 0, \tag{17}$$

$$\frac{\partial L}{\partial \gamma_i} = Z_i = 0, \tag{18}$$

where $T_{,\mathbf{b}}$, $T_{,\delta_i}$, $T_{,\theta_i}$, $Z_{i,\mathbf{b}}$, Z_{i,δ_i} and Z_{i,θ_i} are the derivatives of equations (1) and (12) with respect to master and slave random variables, respectively. Equations (17) and (18) are simply equations of equilibrium and crack growth direction law.

Since $\mathbf{b}$, δ_i and θ_i are independent variables in the Lagrange function, the partial derivatives of equilibrium with respect to $\mathbf{b}$ and θ_i become

$$\frac{\partial}{\partial \mathbf{b}}\left[\mathbf{K}_i\delta_i - \mathbf{F}_i\right] = \frac{\partial \mathbf{K}_i}{\partial \mathbf{b}}\delta_i - \frac{\partial \mathbf{F}_i}{\partial \mathbf{b}}, \tag{19}$$

$$\frac{\partial}{\partial \theta_i}\left[\mathbf{K}_i\delta_i - \mathbf{F}_i\right] = \frac{\partial \mathbf{K}_i}{\partial \theta_i}\delta_i - \frac{\partial \mathbf{F}_i}{\partial \theta_i}. \tag{20}$$

Employing the first-order probabilistic finite element equation[12], i.e.

$$\frac{\partial \mathbf{K}_i}{\partial \mathbf{b}}\delta_i - \frac{\partial \mathbf{F}_i}{\partial \mathbf{b}} = -\mathbf{K}_i \frac{\partial \delta_i}{\partial \mathbf{b}}, \tag{21}$$

$$\frac{\partial \mathbf{K}_i}{\partial \theta_i}\delta_i - \frac{\partial \mathbf{F}_i}{\partial \theta_i} = -\mathbf{K}_i \frac{\partial \delta_i}{\partial \theta_i}, \tag{22}$$

and substituting these results into Eqs. (14)-(18), we have

$$L^{\beta} + \lambda L^{g} = 0, \tag{23}$$

where

$$L^{\beta} = \frac{\partial}{\partial \mathbf{b}}\left[\mathbf{r}^{T}\mathbf{r}\right] \tag{24}$$

$$L^{g} = T_{,\mathbf{b}} + \sum_{i=1}^{N} T_{,\kappa_i}{}^{T}\left(\frac{\partial \kappa_i}{\partial \mathbf{b}}\right) + \sum_{i=1}^{N}\left(-\frac{\dfrac{\partial \tilde{T}}{\partial \theta_i}}{\dfrac{d\tilde{Z}_i}{d\theta_i}}\right)\left(\frac{d\hat{Z}_i}{d\mathbf{b}}\right). \tag{25}$$

Note that $T_{,\kappa_i}$, and κ_{eq,θ_i} can be determined explicitly from Eqs. (1)-(4) as shown by Lua et al.[20]. Other terms in Eq. (25) are defined by

$$T = T\left(\mathbf{b}, Z_i(\kappa_i,\theta_i)\right) = \tilde{T}\left(\mathbf{b}, \tilde{Z}_i(\kappa_i(\theta_i),\theta_i)\right), \tag{26}$$

$$Z_i = Z_i\left(\kappa_i,\theta_i\right) = \tilde{Z}_i\left(\kappa_i(\theta_i),\theta_i\right) = \hat{Z}_i\left(\mathbf{b},\kappa_i(\mathbf{b})\right), \tag{27}$$

$$\frac{\partial \tilde{T}}{\partial \theta_i} = T_{,\theta_i} + T_{,\kappa_i}{}^{T}\left(\frac{\partial \kappa_i}{\partial \theta_i}\right), \tag{28}$$

$$\frac{d\hat{Z}_i}{d\mathbf{b}} = Z_{i,\mathbf{b}} + Z^{T}_{i,\kappa_i}\left(\frac{\partial \kappa_i}{\partial \mathbf{b}}\right), \tag{29}$$

$$\frac{d\tilde{Z}_i}{d\theta_i} = Z_{i,\theta_i} + Z_{i,\kappa_i}{}^{T}\left(\frac{\partial \kappa_i}{\partial \theta_i}\right), \tag{30}$$

where $Z_{i,\mathbf{b}} = 0$, and Z_{i,κ_i} is determined from the crack growth direction law (Eq. (11)). Unlike other terms, two terms, namely, $\partial\kappa_i/\partial\theta_i$ and $\partial\kappa_i/\partial\mathbf{b}$, in Eqs.(28)-(30) cannot be determined explicitly. The finite difference method is employed to calculate these derivatives. After determining these derivatives, an iterative numerical scheme based on the HL-RF method is used to locate the design point.

4. Curvilinear Fatigue Crack Reliability Analysis by the First Order Approximation Method

Different from the Lagrange Multiplier formulation (LMF), where the crack growth direction law is used as an explicit constraint in the formulation of the Lagrange function (see Eq. (13)), the crack growth direction law is satisfied implicitly in the first order approximation method (FAM). A detailed description of FAM can be found in Lua et al.[20].

In FAM, the line mapping is also applied to map a curvilinear crack path to a local coordinate system, ξ ($\xi \in [-1, +1]$). The mapping function is defined by

$$a = \frac{1}{2}\left[(a_f - a_1)\,\xi + (a_f + a_1)\right]. \tag{31}$$

By rewriting the fatigue law (Eq. (1)) and using the mapping function, we have

$$T(\mathbf{b}, \kappa_{eq}(\mathbf{b})) = \int_{-1}^{+1} f(\mathbf{b}, \kappa_{eq}(\mathbf{b}, \xi))\, d\xi, \tag{32}$$

where the function f in Eq. (32) is given by

$$f(\mathbf{b}, \kappa_{eq}(\mathbf{b}, \xi)) = \frac{J(\mathbf{b})}{D\left[\kappa_{eq}(\mathbf{b}, \xi)\right]^n}. \tag{33}$$

In Eq. (33), the Jacobian is derived from Eq. (31) and given by

$$J(\mathbf{b}) = \frac{1}{2}(a_f - a_1). \tag{34}$$

The performance function in $\mathbf{r}$-space can be written as

$$\tilde{g}(\mathbf{r}) = g(\mathbf{b}(\mathbf{r})) = T(\,\mathbf{b}(\mathbf{r}), \kappa_{eq}(\mathbf{b}(\mathbf{r}), \xi)) - T_s. \tag{35}$$

In the first order reliability method (FORM), the most critical step is the determination of the design point $\mathbf{r}^*$. This point is obtained by an iterative optimization algorithm which involves repeated computation of the performance function (Eq. (35)) and its gradient. The gradient of the performance function $\tilde{g}(\mathbf{r})$ can be expressed as

$$\frac{\partial \tilde{g}(\mathbf{r})}{\partial \mathbf{r}} = \frac{\partial g}{\partial \mathbf{b}}\frac{\partial \mathbf{b}}{\partial \mathbf{r}} = \frac{\partial T}{\partial \mathbf{b}}\frac{\partial \mathbf{b}}{\partial \mathbf{r}}. \tag{36}$$

The term $\frac{\partial \mathbf{b}}{\partial \mathbf{r}}$ in Eq. (36) can be determined from the probability transformation as shown by Liu and Der Kiureghian[10]. In order to ensure rapid convergence, an accurate determination of the response gradient ($\partial T/\partial \mathbf{b}$) is required. By applying the direct differentiation, we have

$$\frac{\partial T}{\partial \mathbf{b}} = \int_{-1}^{+1} \left\{ f_{,\mathbf{b}} + f_{,\kappa_{eq}} \left[\kappa_{eq,\kappa_I} \frac{\partial \kappa_I(\mathbf{b}, \xi)}{\partial \mathbf{b}} + \kappa_{eq,\kappa_{II}} \frac{\partial \kappa_{II}(\mathbf{b}, \xi)}{\partial \mathbf{b}} + \kappa_{eq,\theta} \frac{\partial \theta(\mathbf{b}, \xi)}{\partial \mathbf{b}} \right] \right\} d\xi, \tag{37}$$

where $\kappa_{eq,\theta}$, $f_{,b}$, $f_{,\kappa_{eq}}$, κ_{eq,κ_I}, and $\kappa_{eq,\kappa_{II}}$ can be determined explicitly as shown by Lua et al.[20]. Since κ_I, κ_{II}, and θ are implicit functions of both $\mathbf{b}$ and ξ in Eq. (37), the direct calculation of these response sensitivities, namely, $\partial\kappa_I/\partial\mathbf{b}$, $\partial\kappa_{II}/\partial\mathbf{b}$, and $\partial\theta/\partial\mathbf{b}$ are unfeasible in this case. The mixed BIEM coupled with the first order approximation method has been used by Lua et al.[20] to determine these response sensitivities.

It has been shown numerically that the contribution from the last term in Eq. (37) is very small. This is mainly because $\kappa_{eq,\theta} = 0$ along the curvilinear crack path. Thus, we can conclude that compared with LMF, the crack growth direction law is satisfied approximately via the deterministic solver in FAM.

5. Monte Carlo Simulation with Exact Sampling

Monte Carlo simulation (MCS) has been used extensively in evaluating the accuracy of both first and second order reliability methods. To enhance the efficiency of MCS in determining low probabilities of failure, various variance reduction techniques have been developed[22;3]. In this section, a new sampling scheme, called exact sampling, is presented.

In zero-one indicator-based MCS, the probability of failure can be simply expressed as

$$p_f = \int_{\mathbf{x} \in \mathbf{R}^n} I\,[g(\mathbf{x}) \le 0]\, f(\mathbf{x})\, d\mathbf{x}, \tag{38}$$

where p_f is the probability of failure, and I is an indicator function equal to one when the argument is true and zero otherwise. The probability of failure can be easily determined from the ratio of the total number of "1s" to the total number of simulation histories. As an alternative, the above procedure can be performed via a sorted outcome, i.e. sorted ascendantly by the values of the performance function. The probability of failure is equal to the order number of the breaking point, where the performance function changes the sign, divided by the total number of simulation histories, NHIS.

Since the deterministic solver based on the mixed BIEs[18] is computationally intensive for a fatigue crack growth problem, the exact sampling technique is developed to determine the small value of probability of failure with a relatively small number of expensive simulations based on the mixed BIEs. The exact sampling technique consists of two steps. In the first step, which is called pseudo-MCS (PMCS), the deterministic solver based on the mixed BIEs is replaced by a simplified approximation formula. During the process of PMCS, all the realizations of the simulated random variables along with the value of the performance function are stored and sorted ascendantly by the values of the performance function. In the neighborhood of the breaking point, an important sampling domain is constructed by an inverse mapping. All the samples located in this region will be used in the second step of the exact sampling method, where the mixed BIEs will be used to determine the value of the performance function. Since the size of the important sampling domain is much smaller than the total sample size, NHIS, the resulting computational effort is reduced considerably.

The key step in PMCS is the selection of an appropriate approximation formula. As has been derived, the simplified approximation formula is given by

$$T = \frac{\Delta a}{D \sigma^n} \exp\left(\bar{f}\, n + \bar{g}\right), \tag{39}$$

where T is the fatigue life, Δa is the step size of crack growth, σ is the applied stress, and $\bar{f}$ and $\bar{g}$ are geometric-dependent-only parameters determined from the linear interpolation. To insure the accuracy of the exact sampling technique, the important sampling domain centered at the breaking point has to be also selected. The lower and upper bound of the important sampling domain are given by

$$g^- = -T_s\left(\frac{\varepsilon_{max}}{1+\varepsilon_{max}}\right), \text{ and } g^+ = T_s\left(\frac{\varepsilon_{max}}{1+\varepsilon_{max}}\right), \tag{40}$$

respectively, where T_s is the service life specified in the performance function and ε_{max} is the maximum error tolerance. In this study, the tolerance, ε_{max}, is equal to 6%.

In order to get a probability estimate with good confidence, a proper sample size has to be selected in the MCS. As a rule of thumb[3], the necessary sample size, NHIS, is around $100/p_f$, where p_f is the probability of failure.

6. Numerical Example

In order to show the accuracy of both the Lagrange multiplier formulation (LMF) and the first order approximation method (FAM) for the curvilinear fatigue crack reliability analysis, an edge cracked plate under uniaxial fatigue loading is considered. As shown in Fig. 4, the initial crack length a_1, initial crack angle θ_0, external load σ, and fatigue parameters D and n are assumed as random variables. The statistical distribution along with the statistical parameters of these random variables are listed in Table 1. As shown in Table 1, the initial crack size a_1 has the largest dispersion (COV $\approx$ 60%). For the initial crack length a_1, a uniform with tail distribution is employed (see Fig. 5). The detection threshold, which is equal to 7.5×10^{-3} in (0.1905 mm) shown in Fig. 5, represents the lower limit of an inspection device to detect the presence of a small crack. Below the detection threshold, the probability density is assumed uniform; above the threshold the probability density decays linearly to zero, representing false negatives of the inspection technique. In addition to these random parameters, the plate geometry W (width) and L (length), initial crack location (x_0, y_0), final crack length a_f, and material properties are chosen as deterministic parameters and are given by

$$L = 2.5 \text{ in } (63.5 \text{ mm}),\ W = 2.0 \text{ in } (50.8 \text{ mm}),$$
$$x_0 = 0.0 \text{ in } (0.0 \text{ mm}),\ y_0 = 1.25 \text{ in } (31.75 \text{ mm}), \tag{39}$$
$$a_f = 0.5 \text{ in } (12.7 \text{ mm}),\ \mu = 0.8 \times 10^6 \text{ psi } (0.552 \text{ MPa}),\ \nu = 0.25,$$

where μ is the shear modulus and ν is the Poisson's ratio. Twenty-five crack growing steps are used to grow the crack from the initial crack length a_1 to the final length a_f. For the purpose of verifying the accuracy of both LMF and FAM, Monte Carlo simulation (MCS) coupled with the exact sampling technique with the sample sizes NHIS=100,000 and 500,000 is used to determine the middle range and tail region of the probability of failure, respectively. The Cumulative Distribution Functions (CDFs) of the fatigue life T obtained by both LMF and FAM for a series of values of T_s are presented in Figs 6 and 7 for the middle range and tail region of the probability of failure, respectively, along with the results generated from MCS. As shown in Fig. 6, the accuracy of both LAM and FAM is comparable for the middle range of probability of failure. In the tail region of the CDF curve (see Fig. 7), the LMF is more accurate than the FAM for the curvilinear fatigue reliability problem.

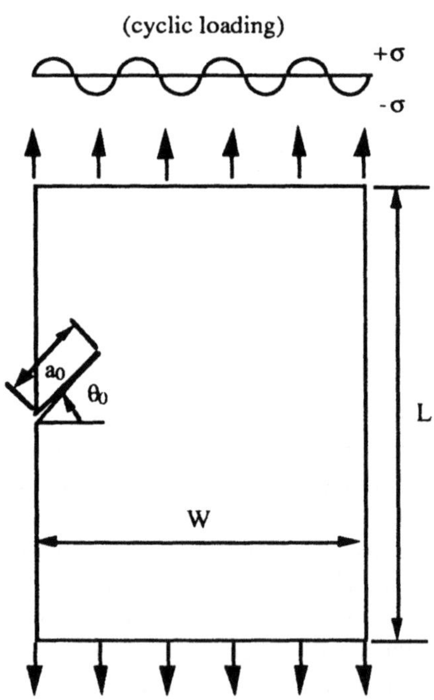

Fig. 4. An edge cracked plate under uniaxial fatigue loading

Table 1. Statistical parameters of random input variables

Random Parameters	Mean	Standard deviation	COV
θ_0 (Uniform)	45.0	0.4534	1.01%
a_i (Uniform with tail)	0.5833×10^{-2} in (0.1482 mm)	0.3584×10^{-2} in (0.09103 mm)	61.44%
D (Log-normal)	0.3770×10^{-9}	0.1885×10^{-10}	5.00%
n (Log-normal)	3.60	0.18	5.00%
σ (Normal)	11.0 ksi (75.845 MPa)	1.1 ksi (7.5845 MPa)	10.00%

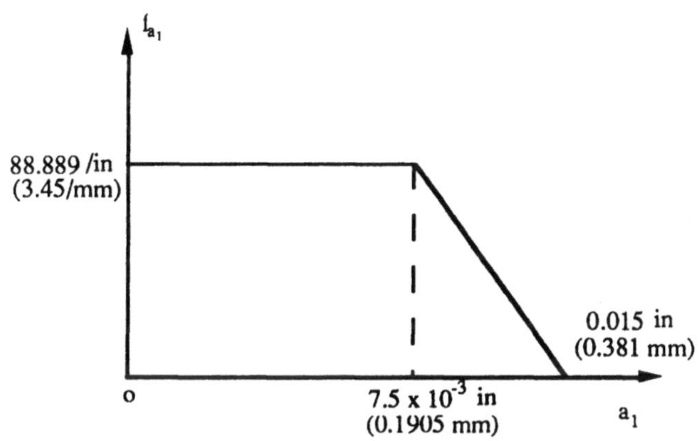

Fig. 5. Statistic distribution of the initial crack length a_1

Since the main difference between the LMF and FAM is in the way of treating crack growth direction law explicitly or implicitly, it can be concluded that the crack growth direction law plays an important role in determining small probabilities of fatigue failure. It can be expected that the effect of crack growth direction on the response gradient will be increased under mixed mode fatigue loading.

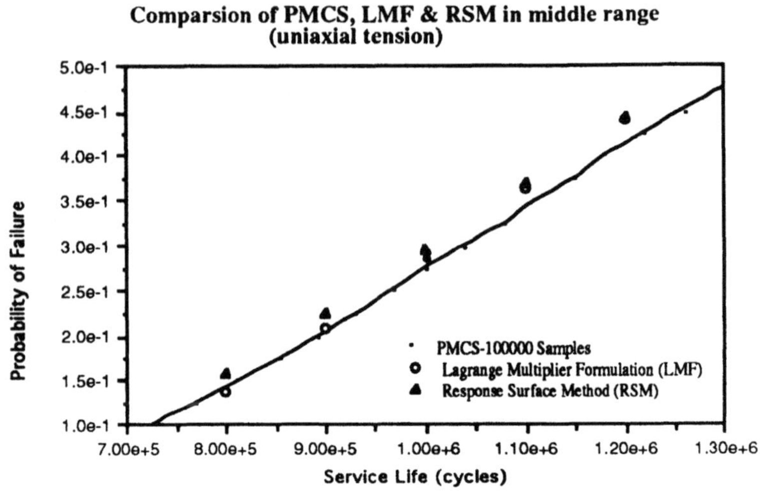

Fig. 6. Comparison of MCS, LMF and FAM in the middle range of p_f

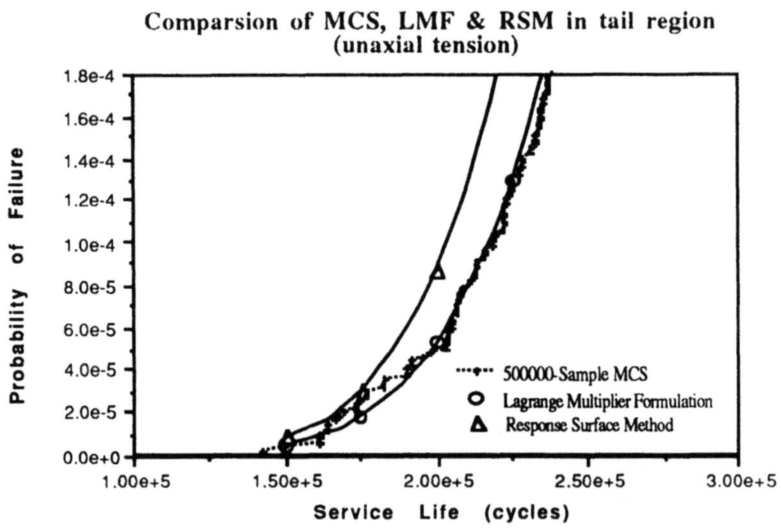

Fig. 7. Comparison of MCS, LMF and FAM in the tail region of p_f

7. Conclusions

The Lagrange multiplier formulation (LMF) has been shown to provide a more accurate evaluation of the response gradient when compared with the first order approximation method (FAM). This is mainly because the crack growth direction law has been incorporated in LMF directly and explicitly. The accuracy of using LMF in the calculation of response gradients is more pronounced for the problem with an extremely low value of failure probability. In order to achieve rapid convergence of an iterative scheme used in FORM, LMF has to be implemented in determining the response gradient.

MCS coupled with the exact sampling technique has been shown to be an efficient and accurate tool for calculating the probability of fatigue failure in the tail region. Since this technique heavily relies on the existence of a good approximation formula, its application to a general problem may not be feasible.

Acknowledgments

The support of NASA Lewis Grant No. NAG3-822 for this research and the encouragement of Dr. Christos Chamis are gratefully acknowledged. This work was also supported in part by the FAA-Center for Aviation System Reliability, operated by the Ames Laboratory, USDOE, for the Federal Aviation Administration under Contract No. W-7405-ENG-82 for work by Iowa State University and Northwestern University.

References

1. Besterfield, G. H., W. K. Liu, M. A. Lawrence, and T. Belytschko, "Brittle Fracture Reliability by Probabilistic Finite Elements," *J. Eng. Mech., ASCE,* **116** (1990) 642–659.
2. Besterfield, G. H., W. K. Liu, M. A. Lawrence, and T. Belytschko, "Fatigue Crack Growth Reliability by Probabilistic Finite Elements," *Comput. Methods Appl. Mech. Engr.*, **86** (1991) 297–320.
3. Bjerager, P., "Probability Integration by Directional Simulation," *J. Eng. Mech., ASCE,* **114** (1988) 1285–1302.
4. Cruse, T. A., Y.-T. Wu, B. Dias and K. R. Rajagopal, "Probabilistic Structural Analysis Methods and Applications," *Computer & Structure,* **30** (1988) 163–170.
5. Der Kiureghian, A. and B.-J. Ke, "The Stochastic Finite Element Method in Structural Reliability," *Probabilistic. Engr. Mech.*, **3** (1988a) 83–91.
6. Der Kiureghian, A. and P.-L. Liu, "Optimization Algorithms for Structural Reliability," *ASME Comp. Prob. Mech.*, **ADM-93** (1988b) 185–196.
7. Erdogan, F., and G. H. Sih, "On the Crack Extension in Plates under Plane Loading and Transverse Shear," *J. Basic Eng., ASME,* **85** (1963) 519–527.
8. Faravelli, L., "A Response Surface Approach for Reliability Analysis," *RILEM Symp on Stoch. Meth. in Mat. and Struct. Eng.*, 1986.
9. Faravelli, L., "Response Surface Approach for Reliability Analysis," *J. Eng. Mech., ASME* , **115** (1989) 2763–2781.

10. Liu, P.-L., and A. Der Kiureghian, "Finite Element Reliability of Geometrically Nonlinear Uncertain Structures," *J. Eng. Mech., ASCE,* **117** (1991) 1806–1825.

11. Liu, W. K., T. Belytschko and A. Mani, "Probabilistic Finite Element Methods for Nonlinear Structural Dynamics," *Comput. Methods Appl. Mech. Engr.*, **56** (1986a) 1–81.

12. Liu, W. K., T. Belytschko and A. Mani, "Random Field Finite Elements," *Int. J. Numer. Methods Eng.*, **23** (1986b) 1831–1845.

13. Liu, W. K., T. Belytschko and A. Mani, "Application of Probabilistic Finite Element Methods in Elastic/Plastic Dynamics," *ASME J. Eng. Ind.*, **109** (1987) 2–8.

14. Liu, W. K., G. H. Besterfield and T. Belytschko, "Transient Probabilistic Systems," *Comput. Methods Appl. Mech. Engr.*, **67** (1988a) 27–54.

15. Liu, W. K., G. H. Besterfield and T. Belytschko, "Variational Approach to Probabilistic Finite Elements," *J. Eng. Mech., ASCE*, **114** (1988b) 2115–2133.

16. Liu, W. K., and T. Belytschko, *Computational Mechanics of Probabilistic and Reliability Analysis*, Elme Press International, 1989.

17. Lua, Y. J., W. K. Liu and T. Belytschko, "A Stochastic Damage Model for the Rupture Prediction of a Multi-phase Solid, Part I: Parametric Studies," *Int. J. Fracture Mech.*, **55** (1992a) 321–340.

18. Lua, Y. J., W. K. Liu and T. Belytschko, "Elestic Interaction of a Fatigue Crack with a Micro-defect by the Mixed Boundary Integral Equation Method," *Int. J. Numer. Methods Eng.*, **36** (1993a) 2743–2759.

19. Lua, Y. J., W. K. Liu and T. Belytschko, "A Stochastic Damage Model for the Rupture Prediction of a Multi-phase Solid, Part II: Statistical Approach," *Int. J. Fracture Mech.*, **55** (1992b) 341–361.

20. Lua, Y. J. , W. K. Liu and T. Belytschko, "Curvilinear Fatigue Crack Reliability Analysis by Stochastic Boundary Element Method," *Int. J. Numer. Methods Eng.*, **36** (1993b).

21. Paris, P. C. and F. Erdogan, "A Critical Analysis of Crack Propagation Laws," *ASME J. Basic Engr.*, **85** (1963) 528–534.

22. Wu, Y.-T., H. R. Millwater, and T. A. Cruse, "Advanced Probabilistic Structural Analysis Method for Implicit Performance Functions," *AIAA Journal,* **28** (1990) 1663–1669.

Reliability-Based Tail Estimation

Marc A. Maes[1] and Karl Breitung[2]

1 Associate Professor, University of Calgary, Calgary, AB, Canada T2N 1N4
2 Schellingstr. 21, D-8000 Munich 40, Germany

Abstract. In reliability analysis a probabilistic model has two components. One is the mathematical description of its properties; this is part of probability theory. The other one is the description of its relation with reality, i.e. the data and the observations; this is part of statistical inference. Most research up to this date has focused on the problem of the mathematical description of the model, while the second aspect has been somewhat neglected.

One of the basic aspects of statistical inference in reliability theory is considered in this paper. It is the fact that in reliability problems we are interested mainly in the tail behavior of a probability distribution function. Most traditional statistical estimation procedures try to obtain a good fit of the theoretical distribution in the central part of the observed data, rather than in the tail part.

In the present paper, an alternative way of estimating the tail of a distribution is outlined. By weighting the approximation error with weights which are larger in the distribution tail, it is ensured that the fit of the approximate distribution is good in the tail region. The selection of the weights is based on a risk criterion appropriate for structural reliability analysis.

The method is described for the case of a generalized extreme value distribution as an approximating distribution, which has three distribution parameters. All three types of extreme value distribution are included in this distribution family. The type of the distribution is determined by the value of one of the distribution parameters.

The extreme value distribution is then fitted to the data using the weights which increase in the tail. By using a suitable form of a probability density function, this can be treated as a maximum likelihood estimation problem. This asymptotic distribution of the estimators or the Bayesian posterior distribution of the parameters can be found subsequently by the usual asymptotic expansion. The method is also shown to be well suited for large or medium-sized data sets and the tricky problem of selecting a cut-off point between the "central" part and the "tail" part of a data set is avoided.

Keywords. Structural Reliability Analysis, Tail Fitting, Extreme Values, Data Analysis

1 Introduction

In reliability-based design it is very common to use statistics with the aim of determining certain tail characteristics of random variables. This includes : (1) the computation of high, or low, quantiles of a random variable; (2) the determination of very small probabilities of exceedance associated with fixed levels; or (3) the accurate description of the tail of a probability distribution. The latter objective is critically important in structural reliability analysis. If one fails to model the tail behavior of the basic variables correctly, then the resulting reliability level is questionable.

In the present paper, several statistical methods for tail estimation are discussed and contrasted. A least squares approach is suggested as a practical way to deal with medium-size data sets; this method does not require the choice of a distribution type, it avoids the rather arbitrary selection of a cut-off point between the "central" part and the "tail" part of a data set, it is consistent with simple risk measures, and it yields an estimate of the associated model uncertainty.

2 Methods of Tail Estimation

Many statistical estimation procedures focus on estimating the central part of a probability distribution. In traditional statistical applications, the shape of the distribution in this region does indeed constitute the main subject of interest. This is, however, not the case in risk and reliability problems, where the interest often lies in the tail of a distribution. The use of a "central" method in conjunction with tail estimation amounts to estimating for a given parametric distribution family the values of the parameters (e.g. on the basis of maximum likelihood or a method of moments) and then using the tails of this estimated distribution in a subsequent analysis. As a result, the data in the central part of the distribution clearly have a dominating influence on the estimated distribution and the data in the tails which are usually few in number do not carry much weight. But, in order to get a good fit of the distribution tail, it is precisely these data that should have the most influence on the tail behavior.

In reliability analysis the main modeling effort does not concern the central part of the distribution, but one or both tails of the distribution, since failure events are generally caused by extreme behavior of random variables. A criterion for the quality of approximation of a distribution function F by another distribution G in the tail region is given by the concept of tail equivalence. If we consider two distribution functions $F(x)$ and $G(x)$, these distribution functions are called tail equivalent (at the right) if

$$\lim_{x \to \infty} \frac{1 - F(x)}{1 - G(x)} = 1. \tag{1}$$

This means that, asymptotically, the relative error of approximating the small exceedance probability $1-F(x)$ by $1-G(x)$ approaches zero as $x \to \infty$.

As shown by Castillo[1], the problem of approximating a distribution function in the tails is closely related to that of extreme values. The practical implication is that it becomes legitimate to fit an extreme value distribution to the tail of an arbitrary distribution, even though full asymptotic extreme value behavior is not achieved. Actually, even though most probability distributions commonly used in engineering belong to the Gumbel domain of attraction for extremes, it is often the case that the sample size n is not large enough to detect the Gumbel trend. In these situations a penultimate (or, pre-asymptotic) approximation to extremes has been suggested[1]; examples abound in literature of Weibull and Fréchet extreme value approximations, even though, strictly speaking, the Gumbel domain of attraction applies.

In order to capture this critical dependence of tail approximations on the sample size, n, while, at the same time, allowing for flexibility in the selection of the type of extreme value distribution, the best approach is to consider the generalized extreme value distribution (GEVD) defined by :

$$F(x|\theta) = \exp\left\{-\left[1-\frac{\gamma(x-\alpha)}{\beta}\right]^{\frac{1}{\gamma}}\right\}. \tag{2}$$

The parameter set θ contains three values: β is a positive constant and

$$-\infty < x < \alpha + (\beta/\gamma) \qquad \text{when } \gamma > 0$$

$$\alpha + (\beta/\gamma) < x < \infty \qquad \text{when } \gamma < 0. \tag{3}$$

The problem is then reduced to finding the parameter values of α, β, γ, so that a good tail approximation is obtained. The criterion for "good tail approximation" will be discussed subsequently. If $\gamma = 0$, we have a type I or Gumbel distribution whereas the case $\gamma < 0$ corresponds to a type II or Fréchet distribution and the case $\gamma > 0$ represents a type III or Weibull distribution.

3 Tail Approximation

To estimate the tail of a distribution in a more precise way, several methods have been proposed. A first method, by Maritz/Munro[2] assumes that the whole distribution and not only the tail has an extreme value form. Functions of order statistics are then used to estimate the parameters of this distribution. Therefore, this method cannot be considered a proper tail fitting method.

Weissman[3] and Boos[4] propose a method based on using only the higher order statistics to estimate the shape of the extreme value distribution. The problem here is that some relatively arbitrary cut-off number k must be determined; from a sample of n data only the k largest are used. All the remaining sample information is discarded. A further extension of this method is given by Hasofer and Wang[5]. They develop a test for the extreme value Type based on these extreme order statistics. The method yields satisfactory results only for very large data sets, and quantile estimation is possible only if the specified exceedance probability is greater then $1/n$, where n is the sample size.

4 A Risk-Based Least Squares Criterion

In setting up a suitable tail fitting procedure, we wish to accomplish the following objectives :

- to avoid the problem of choosing cut-offs and discarding large parts of the data set;
- to assign greater weights to data in the tail of interest, and less weight to data in the central and opposite parts of the data range;
- to respect the aforementioned principle of tail equivalence;
- to produce a method which is consistent with risk criteria used in structural reliability analysis and which can be interpreted visually;
- to produce estimates of the associated uncertainties.

Consider the simple case where reliability is governed by the behavior of a random variable X with distribution function $F(X)$, with failure **F** occurring at large values of X. From the point of view of risk, it is clear that in the upper tail region the error on -log [Pr(**F**)] (and not on Pr(**F**)) itself should be considered as a suitable measure of fit. This is consistent with the traditional use, in reliability-based design, of the reliability index β, a nonlinear scaling of Pr(**F**). Any fit $F(x)$ of the upper tail region should therefore be such that the error on -log (1 - $F(x)$) is minimized; this is also equivalent to minimizing the relative error $R(x)=1/(1-F(x))$ on the return period since :

$$\Delta\left(-\log(1-F(x))\right)=\Delta\left(\log R(x)\right)=\frac{\Delta(R(x))}{R(x)} \tag{4}$$

Let the empirical distribution function be described by the n plotting positions p_i

$= i / n + 1$ of the ordered sample $x_1 \leq x_2 \leq \ldots \leq x_n$. Then we intend to fit the GEVD $F(x \mid \theta)$ given by (2) to this empirical distribution following the above error control criterion (4); consequently the following expression needs to be minimized with respect to θ :

$$\sum_{i \in \mathbf{T}} \left[\frac{\frac{1}{1-F(x_i|\theta)} - \frac{1}{1-p_i}}{\frac{1}{1-F(x_i|\theta)}} \right]^2 = \sum_{i \in \mathbf{T}} \frac{1}{(1-p_i)^2} \left[F(x_i|\theta) - p_i \right]^2 \quad (5)$$

where **T** is the upper tail region. Following [1] this expression can also be approximated by

$$\sum_{\text{all } i} \frac{1}{(1-p_i)^2} \left[F(x_i|\theta) - p_i \right]^2 = \sum_{\text{all } i} w_i \left[F(x_i|\theta) - p_i \right]^2 \quad (6)$$

since for large n, the weights $w_i = (1-p_i)^{-2}$ are concentrated in the right tail of the distribution anyway, so that the entire data range may be considered. Note that the above derivation is entirely consistent with the principle of tail equivalence, since the left-hand side of (1) must tend as closely as possible, to the value of 1; this results in the least squares expression :

$$\sum_{i \in \mathbf{T}} \left[\frac{1 - F(x_i|\theta)}{1 - p_i} - 1 \right]^2 \quad (7)$$

which is identical to (6).

In what follows the minimization problem with general weights w_i will be treated rather than the specific values $w_i = (1-p_i)^{-2}$ obtained above. In the special case where F represents the Gumbel distribution, other possible choices for the weights w_i are discussed in [1]; most of these choices are found to result in inferior performance for the type of applications studied by the author.

5 Weighted Tail Estimation as a Maximum Likelihood Problem

The parametric family $F(x|\theta)$ of cdf's given by (2) and depending on a parameter vector θ is assumed. A weighted tail estimation is achieved by minimizing the following sum of weighted squared errors (SWSE) :

$$SWSE = \sum_{i=1}^{n} w_i \left[p_i - F(x_i|\theta) \right]^2 \quad (8)$$

with respect to θ. As discussed before, the weights w_i must increase as i approaches n.

If we assume that the model errors which represents the deviations from the best fit $F(x|\theta)$ are independent with :

$$\varepsilon(x_i) = F(x_i) - F(x_i|\theta) \tag{9}$$

$$\mathbf{E}(\varepsilon(x_i)) = 0 \tag{10}$$

$$\mathbf{E}(\varepsilon^2(x_i)) = \frac{\sigma^2}{w_i} \tag{11}$$

then the above heteroscedastic model can be viewed as a maximum likelihood problem for the following multivariate pdf :

$$f(x_1,\ldots,x_n|\theta) = \prod_{i=1}^{n} \frac{w_i^{\frac{1}{2}}}{\sqrt{2\pi}\sigma} \exp\left[-\frac{w_i}{2\sigma^2}\left(p_i - F(x_i|\theta)\right)^2\right]. \tag{12}$$

This approach is approximate since the correlation between the errors is neglected, but the use of a full instead of diagonal covariance structure in (12) has been found to change little to the least squares results.

It can be seen that the maximum likelihood estimators $\hat{\theta}$ of the k parameters θ_1, ..., θ_k which are obtained by maximizing the loglikelihood L associated with (12)

$$L = \text{constant} - \log \sigma^n - \sum_{i=1}^{n} \frac{w_i}{2\sigma^2}(p_i - F(x_i|\theta))^2 \tag{13}$$

are the same as the parameters resulting from the minimization of (8). The model error variance σ^2 can be found from the additional likelihood equation,

$$\frac{\partial L}{\partial \sigma} = 0 \quad \text{or} \quad -\frac{n}{\sigma} + \sum_{i=1}^{n} \frac{w_i}{\sigma^3}(p_i - F(x_i|\theta))^2 \tag{14}$$

so that, following (8),

$$\sigma^2 = \frac{1}{n} SWSE(\hat{\theta}) \tag{15}$$

and equation (11) becomes :

$$\mathbf{E}(\varepsilon^2(x_i)) = \frac{SWSE\ (\hat{\theta})}{nw_i} \tag{16}$$

6 Parameter Uncertainty

There remains the additional problem of determining the parameter uncertainty associated with the vector $\hat{\theta}$. If n is not too small, it may be appropriate to use the following large sample approximation to the posterior distribution of θ. Following Kendall and Stuart[6], and Berger[7], θ has a multivariate normal distribution with the maximum likelihood estimator $\hat{\theta}$ as mean vector and with a covariance matrix given by minus the inverse of the matrix formed by the second derivatives of the log-likelihood function $L(\boldsymbol{x}\,|\,\theta)$ at the point $\hat{\theta}$ having as its elements :

$$\frac{\partial^2 L}{\partial\theta_i \partial\theta_j} = -\frac{1}{2\sigma^2}\left[\frac{\partial^2 SWSE}{\partial\theta_i \partial\theta_j}\right]_{\theta=\hat{\theta}} =$$

$$\frac{1}{\sigma^2}\sum_{i=1}^{n} w_i\left[-\frac{\partial F(x_i|\theta)}{\partial\theta_j}\frac{\partial F(x_i|\theta)}{\partial\theta_l} + \left(p_i - F(x_i|\theta)\right)\frac{\partial^2 F(x_i|\theta)}{\partial\theta_l \partial\theta_j}\right] \tag{17}$$

The first and second derivatives of F with respect to the parameters $\theta = [\alpha\ \beta\ \gamma]^T$ associated with the GEVD (2) are given in Appendix A. The results of the parameter uncertainty analysis can now be used to determine probability (or, confidence) intervals on high quantiles or on small exceedance probabilities or return periods.

7 Example

In what follows we will focus on the problem of estimating high quantiles from various data sets. The GEVD weighted tail fitting approach is first contrasted with two straightforward tail estimation techniques. The first one consists of fitting a straight line to the - log (exceedance probability) points for the upper 15% portion of the ordered data. We will refer to this approach as the "exponential tail fit" since it amounts to fitting an exponential tail to the right-hand side of an empirical probability density, using simple least squares.

The second technique is based on) high quantile estimators[3]. Its use has been advocated by Hasofer and Wang[5] for the purpose of reliability analysis. Quantiles $\eta(\varepsilon)$ with a specified small exceedance probability ε are estimated on the basis of the k upper order statistics $X_{n-k+1}, \ldots, X_n$ from a sample of size n as follows :

$$\hat{\eta}(\varepsilon) = \left[\frac{1}{k}\sum_{i=n-k+1}^{n} X_i - X_{n-k+1}\right]\log(\frac{k}{n\varepsilon}) + X_{n-k+1}. \tag{18}$$

In assessing the merit of these different techniques, it is important to cover a wide range of different behaviors. To illustrate this point, we will limit ourselves to three samples with size $n = 50$ taken from a standard normal, a gamma ($\alpha = 2$, $\beta = 2$), and a lognormal (with standard normal logs) distribution, respectively. The first distribution is characterized by a lighter than exponential tail, the second one by a nearly exponential tail, while the third one has a heavier than exponential tail. Tail behavior of this kind can conveniently be described by the tail heaviness coefficient $h(\varepsilon)$ at a chosen exceedance probability ε :

$$h(\varepsilon) = -\varepsilon\left[f'/f^2\right]_{x=\eta(\varepsilon)} - 1 \tag{19}$$

where η is the corresponding quantile and f represents the density function. For all distributions in the domain of attraction of the Gumbel distribution, $h(\varepsilon)$ becomes zero as $\varepsilon \to 0$ or $\eta \to \infty$, zero tail heaviness being associated with the exponential density over its entire range (for all ε). For the examples with $n = 50$ considered here, a Gumbel plot shows sufficient curvature which explains the positive (heavy tails) or negative (light tails) heaviness coefficients in Table 1. The three techniques are used to compute a 95%, a 99%, and a 99.9% quantile (Table 1). Of these values, only the 95% quantile lies within the data range, since for $n = 50$, the largest empirical cdf is about 98%. Also included in Table 1, is an "exact" quantile of the parent cdf; because of the small sample size, it should be used for comparative purposes only if a whole set of samples were to be generated from that cdf and analyzed. However, it is useful in confirming the trends characteristic of the tail heaviness. It can be seen that a straightforward exponential tail fit (Method 2) is acceptable only if the tail heaviness coefficient is close to zero; as discussed before, this occurs either when the sample size is very large, or else when the tail behavior is sufficiently close to exponential. If Method 2 is used regardless, overestimation or underestimation of quantiles will result, depending on whether the data indicate light or heavy tails. The Weissman estimators are seen to have the opposite effect : they exaggerate the curvature in a Gumbel plot and generally perform poorly for this size of sample. Besides, the estimator seems to be rather sensitive to the selected k/n ratio. Further analysis on a large number of samples is to provide conclusive results about the performance of the proposed tail estimation techniques.

Table 1 : Comparison of High Quantiles

	Exceedance Probability		
	ε = 0.05	**ε = 0.01**	**ε =0.001**
LIGHT TAIL			
Tail heaviness h at 0.01	*- 0.12*		
1.GEVD Weighted Tail Fit	1.60	2.17	2.86
2. Exponential Tail Fit	1.62	2.52	4.21
3. Weissman *k/n* = 0.2	1.41	1.99	2.80
(Exact)	(1.65)	(2.33)	(3.09)
NEAR EXPONENTIAL TAIL			
Tail heaviness h at 0.01	*- 0.02*		
1.GEVD Weighted Tail Fit	9.63	12.43	15.72
2. Exponential Tail Fit	9.70	13.11	17.66
3. Weissman *k/n* = 0.2	8.92	10.93	13.82
(Exact)	(9.49)	(13.27)	(18.46)
HEAVY TAIL			
Tail heaviness h at 0.01	*+ 0.25*		
1.GEVD Weighted Tail Fit	5.69	11.96	29.39
2. Exponential Tail Fit	6.05	11.22	17.62
3. Weissman *k/n* = 0.2	5.16	7.98	12.03
(Exact)	(5.18)	(10.24)	(21.98)

8 Conclusions

In this paper the importance of using tail-specific estimation techniques in risk and reliability applications is emphasized. the proposed method is suited for tails of medium to small sized samples. Tail estimation is based on the principle that errors on [- log (failure probability)] should be minimized : this results in weighted least squares. The GEVD is selected in order to capture correctly the curvatures in Gumbel plots, as characterized by pre-asymptotic tail heaviness coefficients. It is pointed out how uncertainty estimates on quantiles and return periods may be derived, and how an asymptotic approximation of the posterior density of the parameters can be obtained.

References

1. Castillo, E., *Extreme Value Theory in Engineering*, Academic Press Inc., Boston, 1988.
2. Maritz, J.S. and Munro, A.H., On the Use of the Generalized Extreme Value Distribution in Estimating Extreme Quantiles, *Biometrics*, **23**, (March., 1967), 79-103.
3. Weissman, I., Estimation of Parameters and Large Quantiles Based on the *k* Largest Observations, *J. Am. Statistical Assn.*, **73**, **264**, (1978), 812-815.
4. Boos, D.D., Using Extreme Value Theory to Estimate Large Percentiles, *Technometrics*, **26**, **1**, (February, 1984), 33-39.
5. Hasofer, A.M. and Wang, Z., System Reliability Calculations Using Extreme Value Theory, *Proc. ICASP*, **1**, (1991), 41-47.
6. Kendall, M. and Stuart,A., *The Advanced Theory of Statistics*, Volume II, Fourth Edition, MacMillan, London, 1979.
7. Berger, J.O., *Statistical Decision Theory and Bayesian Analysis*, Second Edition, Springer, New York, 1985.

Appendix : Derivatives Needed in the Evaluation of Parameter Uncertainties

In equation (17), the first and second derivatives of the GEVD given by (2) with respect to α, β and γ are needed. The first derivatives are equal to :

$$\frac{\partial F}{\partial \alpha} = -F \cdot \frac{1}{\gamma} h(x)^{\frac{1}{\gamma}-1} \frac{\partial h}{\partial \alpha} \tag{20}$$

$$\frac{\partial F}{\partial \beta} = -F \cdot \frac{1}{\gamma} h(x)^{\frac{1}{\gamma}-1} \frac{\partial h}{\partial \beta} \tag{21}$$

$$\frac{\partial F}{\partial \gamma} = -F \cdot h(x)^{\frac{1}{\gamma}} \frac{\partial g}{\partial \gamma} \tag{22}$$

where :

$$h(x) = 1 - \frac{\gamma(x-\alpha)}{\beta} \tag{23}$$

$$F \equiv F(x) = \exp\left(-h(x)^{\frac{1}{\gamma}}\right) \tag{24}$$

$$\frac{\partial h}{\partial \alpha} = \frac{\gamma}{\beta} \qquad \frac{\partial h}{\partial \beta} = \frac{\gamma(x-\alpha)}{\beta^2} \tag{25}$$

$$g(x) = \frac{1}{\gamma} \ln h(x) \tag{26}$$

$$\frac{\partial g}{\partial \gamma} = \frac{-1}{\dfrac{\beta\gamma}{x-\alpha} - \gamma^2} - \frac{g(x)}{\gamma} \tag{27}$$

The second derivatives are :

$$\frac{\partial^2 F}{\partial \alpha^2} = -\frac{\partial F}{\partial \alpha}\frac{1}{\gamma}h(x)^{\frac{1}{\gamma}-1}\frac{\partial h}{\partial \alpha} - F\cdot\frac{1}{\gamma}\left(\frac{1}{\gamma}-1\right)h(x)^{\frac{1}{\gamma}-2}\left(\frac{\partial h}{\partial \alpha}\right)^2 \tag{28}$$

$$\frac{\partial^2 F}{\partial b^2} = -\frac{\partial F}{\partial b}\frac{1}{g}h(x)^{\frac{1}{g}-1}\frac{\partial h}{\partial b} - F\cdot\frac{1}{g}\left(\frac{1}{g}-1\right)h(x)^{\frac{1}{g}-2}\left(\frac{\partial h}{\partial b}\right)^2 - F\cdot\frac{1}{g}h(x)^{\frac{1}{g}-1}\left[\frac{-2g(x-a)}{b^3}\right] \tag{29}$$

$$\frac{\partial^2 F}{\partial \alpha \partial \beta} = -\frac{\partial F}{\partial \alpha}\frac{1}{\gamma}h(x)^{\frac{1}{\gamma}-1}\frac{\partial h}{\partial \beta} - F\cdot\frac{1}{\gamma}\left(\frac{1}{\gamma}-1\right)h(x)^{\frac{1}{\gamma}-2}\left(\frac{\partial h}{\partial \alpha}\right)\left(\frac{\partial h}{\partial \beta}\right) - F\cdot\frac{1}{\gamma}h(x)^{\frac{1}{\gamma}-1}\left(-\frac{\gamma}{\beta^2}\right) \tag{30}$$

$$\frac{\partial^2 F}{\partial \gamma^2} = -\frac{\partial F}{\partial \gamma}h(x)^{\frac{1}{\gamma}}\frac{\partial g}{\partial \gamma} - F\cdot h(x)^{\frac{1}{\gamma}}\left(\frac{\partial g}{\partial \gamma}\right)^2 - F\cdot h(x)^{\frac{1}{\gamma}}\left(\frac{\partial^2 g}{\partial \gamma^2}\right) \tag{31}$$

$$\frac{\partial^2 F}{\partial\alpha\partial\gamma} = -\frac{\partial F}{\partial\alpha}h(x)^{\frac{1}{\gamma}}\frac{\partial g}{\partial\gamma} - F\cdot\frac{1}{\gamma}h(x)^{\frac{1}{\gamma}-1}\left(\frac{\partial h}{\partial\alpha}\right)\left(\frac{\partial g}{\partial\gamma}\right) - F\cdot h(x)^{\frac{1}{\gamma}}\left(\frac{\partial^2 g}{\partial\alpha\partial\gamma}\right) \tag{32}$$

$$\frac{\partial^2 F}{\partial\beta\partial\gamma} = -\frac{\partial F}{\partial\beta}h(x)^{\frac{1}{\gamma}}\frac{\partial g}{\partial\gamma} - F\cdot\frac{1}{\gamma}h(x)^{\frac{1}{\gamma}-1}\left(\frac{\partial h}{\partial\beta}\right)\left(\frac{\partial g}{\partial\gamma}\right) - F\cdot h(x)^{\frac{1}{\gamma}}\left(\frac{\partial^2 g}{\partial\beta\partial\gamma}\right) \tag{33}$$

where :

$$\frac{\partial^2 g}{\partial\gamma^2} = +\left[\frac{\beta\gamma}{x-\alpha} - \gamma^2\right]^{-2}\left(\frac{\beta}{x-\alpha} - 2\gamma\right) + \frac{2g(x)}{\gamma^2} + \frac{1}{\gamma^2 h(x)}\frac{(x-\alpha)}{\beta} \tag{34}$$

$$\frac{\partial^2 g}{\partial\alpha\partial\gamma} = \left[\frac{\beta\gamma}{x-\alpha} - \gamma^2\right]^{-2}\frac{\beta\gamma}{(x-\alpha)^2} - \frac{1}{\beta\gamma h(x)} \tag{35}$$

$$\frac{\partial^2 g}{\partial\beta\partial\gamma} = \left[\frac{\beta\gamma}{x-\alpha} - \gamma^2\right]^{-2}\frac{\gamma}{(x-\alpha)} - \frac{(x-\alpha)}{\beta^2\gamma h(x)} \tag{36}$$

Some Developments in Directional Simulation for Time Variant Structural Systems

R.E. MELCHERS

Department of Civil Engineering and Surveying
The University of Newcastle, NSW, Australia 2308

ABSTRACT The mechanical response of a structural system subject to one or more load processes is described in reliability theory by a limit state (hyper-) surface. For complex systems this surface may not be known explicitly. In the formulation presented, a probabilistic structural analysis technique for structural system evaluation is employed to obtain the probabilistic but implicit limit state function as it exists in the load process space. Directional simulation is employed to estimate the outcrossing rate and the initial (zero time) probability of failure and hence to estimate the probability of failure at any subsequent time. Both continuous (Gaussian) and discrete pulse (Poisson) processes are considered. Some discussion of the relationship of the proposed method to importance sampling methods applied to multi-limit state structural systems is also given.

KEYWORDS. Probability; Outcrossing; Structural Systems; Failure; Monte Carlo; Sampling; Importance; Random Processes, Poisson, Gaussian

1. INTRODUCTION

Consider a problem in stochastic mechanics and let it be formulated in the space defined by all load processes. Also, let the mechanical response of a structural system be described by one or more limit state functions. However, since the member or component strengths are considered here not to be known with certainty, the limit state functions are uncertain and may be seen as probabilistically described "boundaries" in the space of the load processes. [As should readily be evident, this formulation may be generalised to include time dependent resistance or member properties in the "load" space[1] but this detail will be ignored herein for simplicity of exposition].

Further, let the limit state functions be assumed to have the property that they are not dependent on the path followed by the load process vector in reaching the limit state function (see Figure 1). Although this is a major assumption and one which certainly is not always justified, it is common with most techniques currently available. Load path dependence significantly complicates structural mechanics reliability assessment because of the need to consider explicitly all possible combinations of loading when each load is governed by a stochastic description. More generally, there are innumerable load realisations possible; only for some of these is the limit state function likely to be affected. In most cases the only realistic approach is to ignore load path dependency – alternatively some particular cases can be identified *a priori* as being critical for the analysis[2].

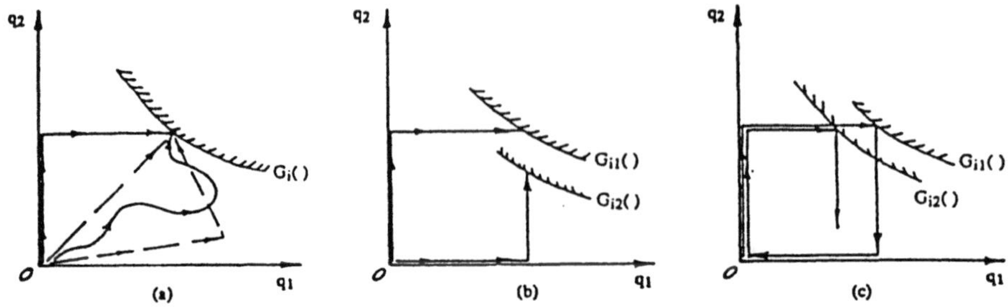

Fig. 1 Load Path Independence and Dependence

Unlike other techniques, for the present approach the limit state expressions for the description of the mechanics of the structural system need not be explicit. As will be seen, this is handled through the use of a method such as the "Truncated Enumeration" technique[3] or similar[4] but with (repeated) use of deterministic loadings rather than random variable loadings. This latter aspect has the immediate advantage of removing an important limitation of these methods, namely that their validity is in doubt when there is more than one random load or process acting on the system[2].

The time variant reliability problem formulated in the (load) process space was discussed earlier for purely continuous Gaussian processes[1] and later for mixed discrete (Poisson) processes having Gaussian amplitudes[5]. The present work considers both types of processes. The essential theoretical background is given in the first part of the paper, followed by an exposition of the way the Truncated Enumeration method is incorporated into the work. An example application is given next. Several observations and a comparison to importance sampling methods close the paper.

2. LOAD SPACE FORMULATION

Let the mechanical behaviour of a structural system be described by a number of deterministic parameters and by the k vector of random variables **X**. Also, let the system be acted on by m load processes **Q**(t). In general, the vector **X** could contain parameters of **Q**(t). In principle, failure of the structural system is defined by the first time violation of one or more limit states $G_i(\mathbf{q}, \mathbf{x}) = 0$, $i = 1, \ldots,$ sometime during the period $(0, t_L)$, where t_L is a known design life. The probability of failure is estimated from the k dimensional integral

$$p_f = \int_X p_f(x)\, f_X(x)\, dx \qquad (1)$$

where $f_X(\ )$ = the (presumed known) joint probability density function (PDF) in **X**. The term $p_f(\mathbf{x})$ represents the structural failure probability conditional on **X** = **x**. Unfortunately, often it is not possible to express $p_f(\mathbf{x})$ in analytic form, creating considerable difficulties in solution of equation (1).

One approach to the solution of (1) is to reformulate the problem in the load space[1]. Let the m vector **R** be the load capacity vector of the structural system, defined such that each component of **R** corresponds in direction and sense to one (and only one) component of **Q**. Evidently **R** is a direct function of the components of **X** which contribute to the overall structural strength. If, as here, no components of **X** are involved in describing **Q**(t), it follows that **R** = **R**(**X**) and that **Q**(t) is independent of **R**. This restriction is convenient but not necessary, since a conditional formulation could be developed to allow for it. Also, the precise specification of **Q**(t) has not yet been defined and is therefore not a limitation. It is possible, for example, to have a degree of dependence between various components of **Q** as might occur at different locations, such as on the legs of a jacket–type offshore platform as a result of wave loading.

With these preliminaries it follows that equation (1) may be rewritten in terms of **R** as

$$p_f = \int_{R} p_f(\mathbf{r})\, f_{\mathbf{R}}(\mathbf{r})\, d\mathbf{r} \tag{2}$$

where $p_f(\mathbf{r})$ = the probability of structural failure given the load capacity **R** = **r**, and $f_{\mathbf{R}}(\ )$ = the joint PDF of **R**. The multiple integration in (2) is m–fold. Since in most realistic structural reliability problems m is significantly less than k, the formulation in (2) has obvious advantages, provided $p_f(\mathbf{r})$ and $f_{\mathbf{R}}(\mathbf{r})$ can be derived. One way to implement the integration in (2) is by transformation to the spherical co–ordinate hyper-space in which case it is advantageous to have $f_{\mathbf{Q}}(\ )$ in a form such that the variances in each q_i direction are of similar magnitude. This can be achieved by a linear transformation, and ensures that the radius vector in the polar space **Q** has roughly equivalent implications for each Q_i. Fortunately, the loading space usually has this property, except, perhaps, for dead loads. However, as a rule, these should not be treated as stochastic processes and might be more appropriately considered part of the vector **X**.

Consider now a hyper-polar co–ordinate system centred at some point **C** = **c** in the m dimensional real space R^m (see Figure 2). The location of **c** is of some interest and importance. However, no obvious rules for its selection have been developed to date, apart from suggesting that the point be chosen to lie in the safe domain (as defined by the mechanics of the system) and to expose as much of the (generally non–convex and piecewise) inner envelope of limit state functions (for any given **X** = **x**) as possible when viewed from **C** = **c**. Moreover, since loads are often so specified that $Q_i \geq 0$, the point **C** = **c** = **0** is convenient. In general, however, **c** ≠ **0**. Let the ray from **c** to some point in **R** space be defined by **R** – **c** = S · **A**, where S ≥ 0 is a random variable with m degrees of freedom in R^m, such that s = 0 at **c** and is not necessarily independent of **A**, the random unit vector of direction cosines having a PDF given by $f_{\mathbf{A}}(\ )$. Hence for given **A** = **a**, the scalar S defines the magnitude of **R** relative to **c**. It is possible now to define the conditional PDF $f_{S|\mathbf{A}}(\ )$ on s, it is a function of **X** and hence **R** and it might depend also, in general, on **A**.

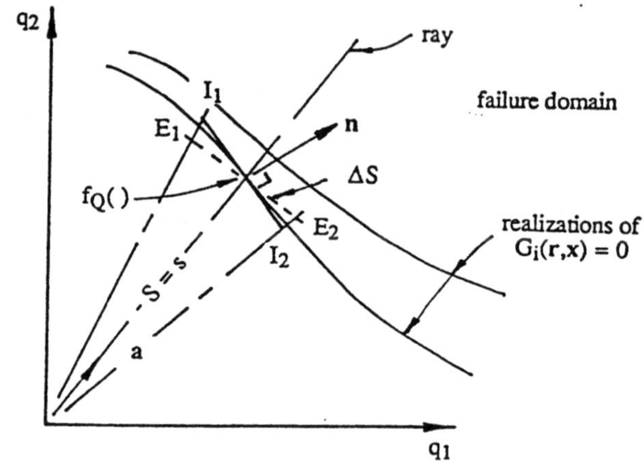

Fig. 2 Realisation of a Ray and of One Limit State Showing Also the Relationship to the Hyper Surface Element ΔS

If the limit state expression G( ) is unique and single-valued, there will be a unique mapping from **X** to **R** to S for **A** = **a**. In general, however, there will be a multiplicity of component and combined limit state functions. In this case there will be a unique mapping from **X** to **R** (and hence to S for a given direction) for each $G_i(\ )$, but obviously the reverse mapping is not unique.

Rewriting the conditional failure probability term in (2) in polar co–ordinates and changing the integration from over **R** to over **A** and S produces[1]

$$p_f = \int_{\substack{\text{unit}\\\text{sphere}}} f_{\mathbf{A}}(\mathbf{a}) \left[\int_S p_f(s|\mathbf{a})\, f_{S|\mathbf{A}}(s|\mathbf{a})\, ds \right] d\mathbf{a} \tag{3}$$

where $p_f(\ )$ depends both on $f_{\mathbf{Q}}(\ )$ and on the type of process used to represent each component of **Q**. The integral over the unit sphere in (3) may be replaced by the expected value, which may then be used with an importance sampling Monte Carlo estimate

$$p_f = E_\theta \left\{ \frac{f_{\mathbf{A}}(\mathbf{a})}{h_\theta(\theta)} \left[\int_S p_f(s|\mathbf{a})\, f_{S|\mathbf{A}}(s|\mathbf{a})\, ds \right] \right\} \tag{4}$$

where $h_\theta(\theta)$ = an importance sampling PDF from which samples θ are drawn. These are used to obtain **a** = **a**(θ) and so to evaluate (4). For the special (and important) case of constant probability density over the unit radius n dimensional hyper-sphere, having a surface area S_A, we have[6,7] $h_\theta(\theta)/f_A[a(\theta)] = S_A^{-1}$ = constant.

Under the condition that individual outcrossings are independent (Poisson) events, as is reasonable for highly reliable structural systems and for processes not narrow banded, the conditional failure probability $p_f(s|a)$ for a given structural life $[0 - t_L]$ can be bounded from above. For a given direction **A** = **a**, and for the limit state at S = s this bound becomes

$$p_f(s|\mathbf{a}) \leq p_f(0, s|\mathbf{a}) + [1 - p_f(0, s|\mathbf{a})] \cdot \left\{1 - \exp\left[-\int_0^{t_L} \nu_D^+(t, s|\mathbf{a})\, dt\right]\right\} \quad (5)$$

where $p_f(0, s|\mathbf{a})$ is the failure probability at time t = 0 and $\nu_D^+(t)$ is the outcrossing rate of the vector process **Q**(t) out of the same domain D. In what follows, the conditions S = s and **A** = **a** will be noted only if the context requires it.

As is well known, expression (5) may be simplified in various ways. If $\nu_D^+(\ )$ is independent of time (a common assumption) and if $p_f(0)$ is small relative to unity, then

$$p_f(s|\mathbf{a}) \leq p_f(0, s|\mathbf{a}) + [1 - p_f(0, s|\mathbf{a})] \cdot \left\{1 - \exp\left[-\int_0^{t_L} \nu_D^+(t, s|\mathbf{a})\, dt\right]\right\} \quad (6)$$

while for high reliability problems

$$p_f(s|\mathbf{a}) \leq p_f(0, s|\mathbf{a}) + \nu_D^+(s|\mathbf{a}) \cdot t_L \quad (7)$$

Both $p_f(0, s|\mathbf{a})$ and $\nu_D^+(s|\mathbf{a})$ may be estimated in the hyper-polar co-ordinate system. Evaluation of expressions (5) – (7) requires evaluation of both $\nu_D^+(\ )$ and $p_f(\ )$. This will now be considered.

3. EVALUATION OF MEAN OUTCROSSING RATE ν_D^+

3.1 Loads Modelled As Continuous Processes

For continuous processes, the (conditional) local outcrossing rate through an elementary domain boundary ΔS at point **R** = **r** on S_D in the structural resistance space is given by the well known generalised Rice formula. It can be written for the **A** = **a** direction with s on the domain boundary S_D (i.e., **R** = **r**)

$$\nu_D^+(s|\mathbf{a}) = E\left[\mathbf{n}(s|\mathbf{a}) \cdot \dot{\mathbf{Q}}(t)\right]^+ \cdot f_Q(\mathbf{r}) \cdot \Delta S_D \tag{8}$$

where E( ) = the expectation operator, with $E(\ )^+ = 0$ if E( ) is negative; $f_Q(\ )$ = the PDF of the vector load process; **n**( ) = the outward unit vector perpendicular to S_D at **S** = **s**; and $\dot{\mathbf{Q}}(t)$ = the (continuous) derivative of **Q**(t) assumed, as written, independent of **Q**(t) at any time t. Also ΔS_D is an elemental part of S_D.

Since the domain boundary in the **Q** space is uncertain and a function of the radial distance s, the local outcrossing rate at **A** = **a** will be conditional also on s

$$\nu_D^+\Big|_{\substack{\mathbf{A}=\mathbf{a}\\ S=s}} = E\left[\mathbf{n}\Big|_{\substack{\mathbf{A}=\mathbf{a}\\ S=s}} \cdot \dot{\mathbf{Q}}(t)\right]^+ \cdot f_Q(\mathbf{r}) \cdot \Delta S_D \tag{9}$$

where the elemental domain boundary ΔS_D remains constant. The interpretation of expression (9) as $\Delta S_D \to 0$ is straightforward – $f_Q(\mathbf{r})$ represents the "probability" of the load process being "exactly" at the point **r** in **Q** space, irrespective of the precise process trajectory in reaching this position, and $E[\ ]^+$ represents the expectation that the next "increment" in the process will have a component in a direction parallel to the outward normal to the limit state surface at **r**; i.e., $E[\ ]^+$ indicates whether the process is about to move into the "failure" domain.

Using expressions (3) and (5), the unconditional mean outcrossing rate becomes

$$\nu_D^+ = \int_{S_D} \nu^+(\mathbf{a})\,d(\Delta S) = \int_{\substack{\text{unit}\\ \text{sphere}}} f_A(\mathbf{a}) \int_0^\infty E\left[\mathbf{n} \cdot \dot{\mathbf{Q}}(t)\right]^+ \cdot f_Q(\mathbf{r})$$

$$\cdot f_{S|A}(s|\mathbf{a})\, s^{(n-1)} \left(\mathbf{a} \cdot \mathbf{n}\right)^{-1} ds\, da \tag{10}$$

where $s^{(m-1)}da$ = an elemental surface area perpendicular to **a** at s, and the scalar product $\mathbf{a} \cdot \mathbf{n}(\ )$ converts $s^{(m-1)}da$ to d(ΔS), representing the elemental surface area equivalent to ΔS with normal **n**( ) (see Figure 2).

In general, the outward unit normal vector **n** varies with s. It can be obtained directly from the expression for the jth limit state surface $G_j(\mathbf{r}, \mathbf{x}) = 0$ at **R** = **r** as $\mathbf{n} = \{n_i\}$, having components

$$n_i = \frac{\partial G_j}{\partial r_i} \left[\sum_k^n \left(\frac{\partial G_j}{\partial r_k}\right)^2\right]^{-1/2} \tag{11}$$

if the function G_j is known explicitly (e.g. Figure 2). If $G_j(\ )$ is not known explicitly, **n** may be (i) approximately estimated by numerical means (a computationally expensive technique) or (ii) essentially ignored by assuming **n** = 1 for all $\mathbf{A} = \mathbf{a}$. The latter approximation may be demonstrated by example to produce only a relatively small error for the part of the limit state function making most contribution to ν_D^+.

The restriction on independence between $\dot{Q}(t)$ and $Q(t)$ may be removed if directional sampling (i.e., over A) is used, provided $\mathbf{n} \cdot \dot{\mathbf{Q}}$ in $E(\ )^+$ is replaced by $\mathbf{n}(\mathbf{r}|\,\mathbf{a}) \cdot \dot{Q}(t)|\,Q(t)$. This is satisfied for $\mathbf{A} = \mathbf{a}$ by the directional sampling procedure[6]. Further, if Q(t) is a standard Gaussian process, the preceding discussion may be particularised and somewhat simplified. This case has been described by Ditlevsen *et al.*[8].

3.2 Loads Modelled As Poisson Pulse Processes

For Poisson pulse processes, the above expressions must be modified to account for the mean pulse rate ν_{mi} of each process. The basic theory has been provided by Breitung and Rackwitz[9] and formulated in hyper-polar space previously[5]. For Poisson pulse processes having a mean rate of arrival ν_{mi} and a mean duration μ_{md} such that $\mu_{md}\,\nu_{md} = 1$, the conditional local rate of the vector process Q(t) crossing out of the safe domain D through the elemental surface ΔS_D is given by

$$\nu_D^+\Big|_{\substack{A=a\\S=s}} = \sum_{i=1}^{n} \Big[\nu_{mi}\left\{F_i(r_{1i}) - F_i(r_{2i})\right\} \left\{1 - F_i(r_i)\right\} \cdot |n_i| \cdot f_{Q_{(i)}}(r_{(i)})\Big] \cdot \Delta S_D \qquad (12)$$

where $f_{Q_{(i)}}(\ )$ is the probability that the vector load process lies along r_{1i}, r_{2i} in Figure 3 with

$$\begin{Bmatrix} r_{1i} \\ r_{2i} \end{Bmatrix} = \begin{Bmatrix} \max \\ \min \end{Bmatrix} \left\{z;\ (r_1, r_2, \ldots, r_{i-1}, z, r_{i+1}) \in D\right\} \qquad (13)$$

where $\mathbf{n}$ is the outward unit normal given that $\mathbf{q}$ is at ΔS, the limit state surface, so that $\mathbf{q} = \mathbf{r}$, or $q_i = r_i$ for all i. $F_i(r_{1i})$ denotes the probability that $r_i < r_{1i}$ for the ith process, and $F_i(r_{1i}) - F_i(r_{2i})$ denotes the probability that q_i lies in D given that $\mathbf{q}$ is at ΔS (see Figure 3). For independent $Q_i(t)$ it follows that $f_{Q_i}(q_{(i)}) = f_Q(q)/f_{Q_i}(q_i) = f_Q(r)/f_{Q_i}(r_i)$ and for systems having high reliability $F_i(r_{1i}) - F_i(r_{2i}) \simeq F_i(r_{1i}) \simeq F_i(r_i)$. If the pulses always return to zero at the end of each pulse duration, the first term in (12) reduces to $F_i(r_i)$ for $n_i < 0$ and to $[1 - F_i(r_i)]$ for $n_i > 0$.

If the Poisson pulses are of the "mixed" type and "sparse", such that $\nu_{mi} \cdot \mu_{mi} \ll 1$, which means that most of the time each process is "off", their PDFs may be described as shown in Figure 4. In principle, expression (12) may be used to obtain the conditional outcrossing rate. However, a practical obstacle is that it is unlikely that the major contributors to probability and outcrossing will be accounted for in a simulation technique. The reason for this is that the probability contributions at zero load (see Figure 4) are unlikely to be detected by simulation since they occur at zero load precisely. Evidently, even large sample sizes may not detect a particular value. It is better, therefore, to separate out the various combinations of "off" and "on" states, and to evaluate each of these separately. With this approach, expression (12) becomes

$$\nu_D^+\Big|_{\substack{A=a \\ S=s}} = \sum_{j=1}^{ncomb} \sum_{i=1}^{m} \Big[\nu_i \left\{\left(p_i + q_i \bar{F}_i(r_i)\right) q_i \left[1 - \bar{F}_i(r_i)\right]\right\} \cdot |n_i| \cdot f_{Q(i)}(r(i))\Big] \tag{14}$$

where $\bar{F}_i(\ )$ is the improper CDF for the ith process, with corresponding improper PDF $\bar{f}_i(\ )$;
ν_i is the pulse rate conditional on the ith process being "on";
μ_{di} is the mean pulse duration for the ith process (i.e., when "on"), so that $\nu_i \mu_{di} = 1$;
and ν_{mi} is the mean pulse rate (unconditional) as before (for the whole process).
Note that $\nu_{mi} = \nu_i q_i$, where q_i = probability of the ith process being "on".

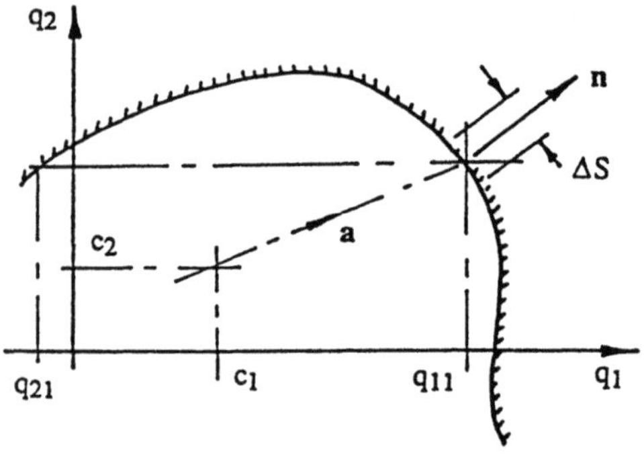

Fig. 3 Local Outcrossing Through ΔS

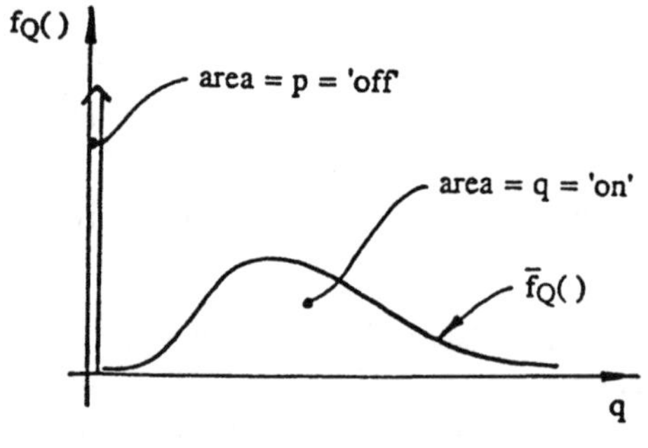

Fig. 4 Schematic "Mixed" Pulse Process

Also, the number of combinations of "on"/"off" states is given by

$$\text{ncomb} = \sum_{\ell=1}^{m} 2^{(\ell-1)} \tag{15}$$

By integrating over all s and **a**, expression (14) may be made unconditional

$$\nu_D^+ = \int_{\substack{\text{unit}\\ \text{sphere}}} f_{\mathbf{A}}(\mathbf{a}) \int_{S=0}^{\infty} \left[\sum_{j=1}^{\text{ncomb}} \sum_{i=1}^{m} \left[\nu_i \left\{ \left(p_i + q_i \overline{F}_i(r_i) \right) q_i \left[1 - \overline{F}_i(r_i) \right] \right] \right. \right.$$

$$\left. |n_i| \; f_{Q(i)}(r(i)) \right] \cdot f_{S|\mathbf{A}}(s|\mathbf{a}) \, s^{(n_j-1)} (\mathbf{a} - \mathbf{n})^{-1} \Big] \, ds \, da \tag{16}$$

where n_j is the number of "active" components of the load process vector **Q**(t) for the jth combination. As for continuous Gaussian processes, the term $s^{(n_j-1)}(\mathbf{a} \cdot \mathbf{n})^{-1}$ corrects the elemental surface area ΔS_D at radius s to the equivalent surface area on the unit sphere.

3.3 Mixed Load Models

When the load vector consists of both Poisson pulse process and continuous process components, as might occur in a structure under man induced live loads and natural live loads (e.g., wind, waves, etc.) respectively, the outcrossing rate can be determined from a combination of the above formulations. Combining (9) and (12) for highly reliable systems and for **A** = **a** and S = s (i.e., at the limit state surface S_D), yields the conditional outcrossing rate as

$$\nu_D^+\Big|_{\substack{A=a\\ S=s}} = \left\{ E\left[\mathbf{n} \cdot \dot{\mathbf{Q}}(t)\right]^+_{(i=1,\ldots,m_1)} + \sum_{j=1}^{m_2} \nu_{mj} \frac{F_j(r_j)\left[1 - F_j(r_j)\right]}{f_{Q_j}(r_j)} |n_j| \right\} \cdot f_{\mathbf{Q}}(\mathbf{r}) \, \Delta S_D \tag{17}$$

with m_1 components of the continuous process vector and m_2 independent pulse processes, the latter are assumed, realistically, independent from the continuous process vector.

The implications in terms of outcrossings as expressed by equation (17) may be visualised in two dimensions – see Figure 5. It is evident from the "typical" realisation shown that, for a given Poisson pulse, the continuous process can outcross regardless of the pulse process, provided it is assumed: (i) that the system is of high reliability so that the Poisson approximation for outcrossings of continuous process(es) remains valid for all values of the pulse process(es); and (ii) that the probability of a joint outcrossing may be neglected. If these assumptions are satisfied, the continuous vector process(es) can be considered separately, only within its own subspace, with outward normal unit vector $\mathbf{n}_1$ (i = 1, ..., n_1) being the normal to the limit state function defined in the m_1 subspace.

This is illustrated in Figure 6 for a two dimensional case, with q_1 being a pulse process and q_2 a continuous process. Similarly, the components of the outward normal unit vector n_j ($j = 1, ..., m_2$) exist only in the directions corresponding to the pulse process.

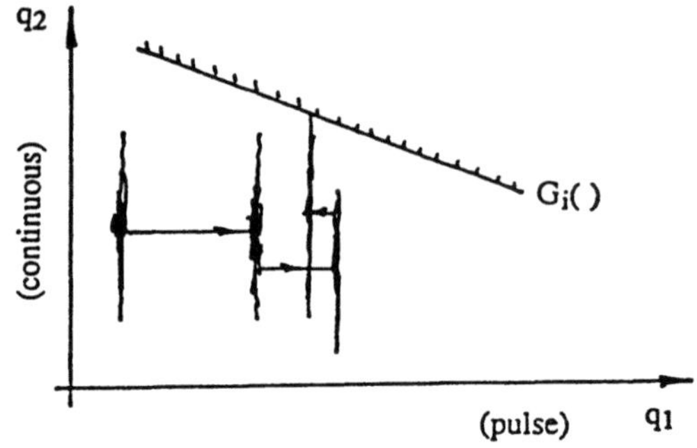

Fig. 5 Realisation of Continuous and Pulse Process Vector in Two Dimensions for a Given Limit State

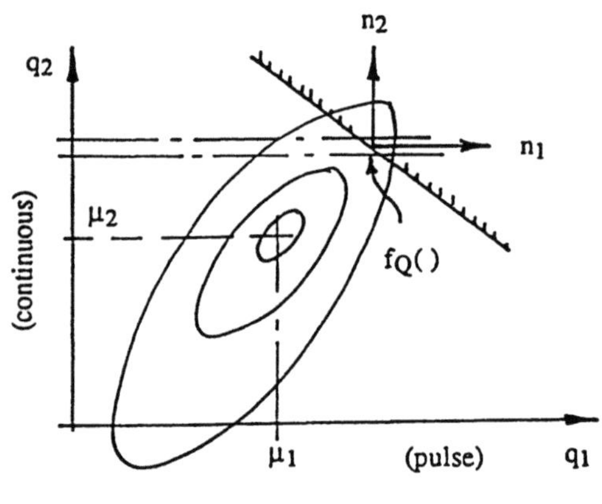

Fig. 6 Unit "Normal" Vectors for Outcrossings in Two Dimensional Process Space for the Processes Shown in Figure 4

The unconditional outcrossing rate for mixed load models may be obtained from expression (17) by integrating over **A** and S in a manner similar to that detailed earlier. Also, expression (17) may be particularised for mixed pulse processes of the type shown in Figure 4 by substituting expression (14) for the second term of (17), taking care to allow for $f_{Q(j)}(\ )$.

4. EVALUATION OF INITIAL FAILURE PROBABILITY $p_f(0)$

For the evaluation of the failure probability at t = 0, both the continuous and the pulse load processes **Q**(t) become random variables, and $p_f(s|\mathbf{a})$ is then the hyper volume under $f_Q(\ )$ in the region $G(\ ) \leq 0$ in the radial direction **A** = **a**. The resulting directional simulation approach has been discussed extensively for the special case of Q( ) being standard Gaussian and also for the more general situation[6,7,8,10]. In these expositions, use was made of successive approximations by closed hyper-polar limit state functions (hyper-spheres in the standard Gaussian case). An alternative exposition, leading to the same mathematical outcome, is to estimate the probability content using hyper-sectors of arbitrary span, as shown in Figure 2. An approximation for the hyper-sectors of finite size is obtained by using E_1, E_2 to approximate I_1, I_2. Clearly the approximation improves with increased sampling of **A**, thus leading to convergence in the same sense as applied for the closed hyper-polar limit state functions.

Using the segment approach, it follows that for polar co-ordinates in R^m, the volume element at radius ρ is $\rho^{(m-1)} \cdot d\rho \cdot da$ so that at radius S = s for given A = **a**

$$p_f(s|\mathbf{a}) = \int_s^\infty f_Q(\rho \cdot \mathbf{a} + \mathbf{c})\, \rho^{m-1}\, d\rho \tag{18}$$

which may be substituted directly into (3) or as $p_f(0, s|\mathbf{a})$ into (7). Also the unconditional initial failure probability can be obtained by combining (3) and (18)

$$p_f = \int_{\substack{\text{unit}\\\text{sphere}}} f_A(\mathbf{a}) \int_s \left[\int_s^\infty f_Q(\rho \cdot \mathbf{a} + \mathbf{c})\rho^{(m-1)} d\rho\right] f_{S|A}(s|\mathbf{a})\, ds\, da \tag{19}$$

In (19) the inner integral represents the volume under $f_Q(\ )$ in the failure region $G \leq 0$ for a given limit state location s along the radial direction **a**. The middle integral obtains the expected value of the inner integrals, using $f_{S|A}$ as the PDF. It is more efficient computationally to consider elemental parts of the inner integrand summed after weighting by $f_{S|A}$ (i.e., such that the order of integration is changed), thus (see also Figure 2)

$$p_f = \int_{\substack{\text{unit}\\\text{sphere}}} f_A(\mathbf{a}) \int_0^\infty \left[\int_0^\rho f_{S|A}(s|\mathbf{a})\, ds\right] f_Q(\rho \cdot \mathbf{a} + \mathbf{c})\rho^{(m-1)} d\rho\, da \tag{20}$$

Evidently, the inner integral is the cumulative distribution function (CDF) for S| **A**, so that

$$p_f = \int_{\substack{\text{unit}\\\text{sphere}}} f_A(\mathbf{a}) \left[\int_0^\infty F_{S|A}(\rho|\mathbf{a}) \cdot f_Q(\rho \cdot \mathbf{a} + \mathbf{c})\rho^{(m-1)} d\rho\right] da \tag{21}$$

which can be integrated numerically.

5. DETERMINATION OF $f_{S|A}$ OR $F_{S|A}$

The above formulations require the variation of the mechanical response of the structural system to be known along the radial direction s for given **A** = **a**. For any ***known*** limit state function $G_i(\mathbf{q}, \mathbf{x}) = 0$ and for given **A** = **a**, the conditional probability density function $f_{S|A}$ can be obtained in general by multiple integration or, for simple structures for which the limit state functions for the structure are known, explicit and differentiable, the technique described in Melchers[1] may be employed. However, this is not the general situation. For structural systems, and particularly for complex systems, the limit state function(s) may not be known for the system (as distinct from those for the elements or components). In other cases, such as those involving finite element analysis, the system limit state(s) may be known only implicitly. It is these last cases which are of particular interest here.

Even though the system limit state function(s) may not be available, it is useful for the present to suppose that such a function $G_i(\mathbf{q}, \mathbf{x}) = 0$ exists. By definition, at any point on $G_i(\ ) = 0$, the loading on the structure will be such that the structural resistance has been reached. More particularly, for any given direction **a**, the condition $G_i(\ ) = 0$ implies $\mathbf{Q} = \mathbf{R}(\mathbf{X})$ and for a point **q**, it implies $\mathbf{q} = \mathbf{r}(\mathbf{x})$. Since **X** is a vector of random variables at any such point in the load space, $G_i(\ ) = 0$ will be a random variable in **X**, and can be represented schematically as shown in Figure 7. Further, for given direction, or ray, **a**, the variation of $G_i = 0$ with s is required; this then gives the variation also of **R**(X), the structural strength since R_j corresponds to Q_j (j = 1, ..., m).

Consider now a point Z_1, say, at a given value of the radial distance, s, along the ray **A** = **a** and with the loading acting on the structure given by the vector value (**q**). For this vector to correspond to structural failure, the point Z_1 at (**q**) must be on the structural limit state. Given that the load space is bounded by structural strength limitations as expressed by the limit state functions, it follows that $p_{f|\mathbf{q},\mathbf{r}}(s) \rightarrow 1$ as $s \rightarrow \infty$ (see Figure 8). Also, for zero value loading, by definition structural failure under loading cannot occur; $p_{f|\mathbf{q},\mathbf{r}}(s) = 0$ at $s = 0$. Generalising, it follows that the probability of structural failure under loading (**q**) at point S = s along the ray represents the probability that the structural strength is less than s or

$$p_f(s)|_{\mathbf{a}} = \mathrm{Prob}(S < s)|_{\mathbf{A}=\mathbf{a}} = F_{S|\mathbf{a}}(s) \tag{22}$$

Hence, evaluating the structural failure probability along **a** for different values of s allows $F_{S|\mathbf{a}}(s)$, to be approximated. Also, differentiation produces $f_{S|\mathbf{a}}(s)$. This approach may be applied where the limit state equations for the structure are not known, and only mechanics criteria for the members or components are available.

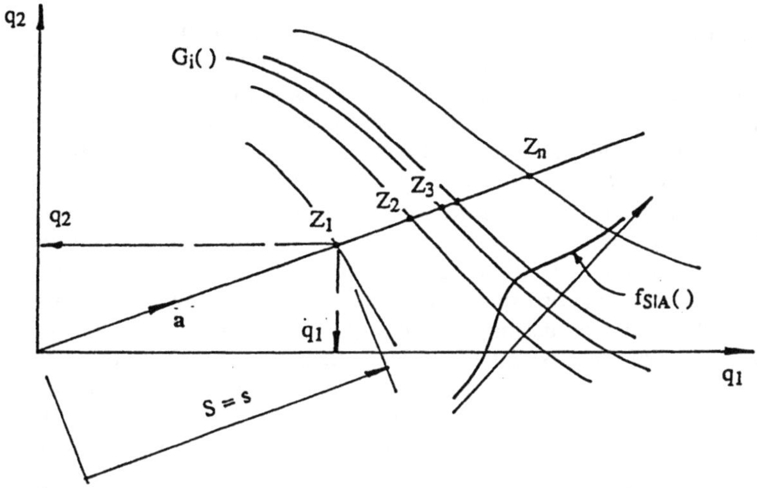

Fig. 7 Realisation of a Ray and Several Realisations of the Limit State Functions Showing Evaluating Points

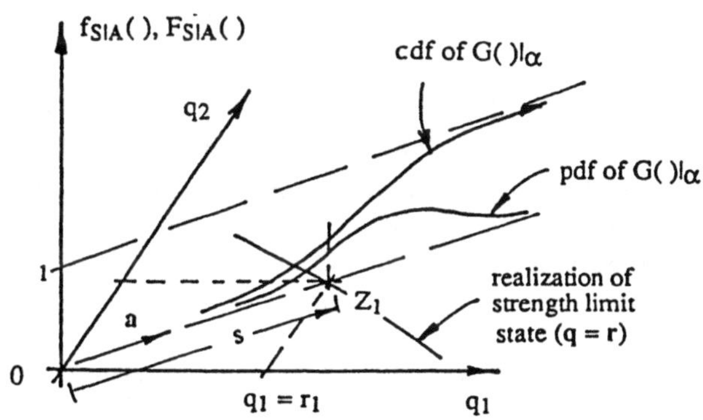

Fig. 8 Realisation of a Ray and Definition of $F_{S|\mathbf{a}}(\ )$ and $f_{S|\mathbf{a}}(\ )$

6. FRAMED STRUCTURAL SYSTEMS

The above concept may be used to obtain the probabilistic description of the system limit state function if $p_f(s)|_{\mathbf{a}}$ can be evaluated for all (s, a), or equivalently for systems at the limit state, for all **q** = **r**. As noted, this requires that the probability of failure for a probabilistic system can be evaluated under defined, deterministic, loading conditions. Evidently, this requirement can be met by an extensive range of methods developed to obtain the probabilistic response of uncertain systems, including finite element and structural dynamic analyses. For the present, attention will be confined to framed structural systems.

Earlier work[3,4] on reliability analysis of structural frames, dealt with systems subject to random variable loading(s). The methods determine the structural system failure probability by considering progressive member failure under given but uncertain loading, ideally through a program of enumeration of all possible modes of structural system failure. Some rules or algorithms are applied to discard, early in the analysis, those combinations of member failure unlikely to make a significant contribution to the overall estimate of the structural system failure probability. Only in the case of a one parameter loading situation (i.e., all loading governed by one random variable) have the techniques the possibility of validity[2]. Further, care is required where members have negative stiffness or "unloading" characteristics. There is also the interesting situation that the failure modes developed using these techniques, which all rely on repeated elastic analysis of the structure, may not correspond precisely to the failure modes which might be obtained by the common assumption of rigid plastic behaviour. As should be evident, nearly all of these potential problems disappear in the case of deterministic rather than random variable loading. This is the case implied by evaluation under loading at Z_1 in Figure 7, i.e., under the deterministic load vector ($\mathbf{q}$). Hence the value of $F_{S|\mathbf{A}}(s)$ at any radius $S = s$ may be obtained using a procedure such as that of Melchers and Tang[3] but with deterministic loading. In using such a method, all the possible structural system limit state functions are automatically generated and then considered for system failure probability estimation.

The integration shown in equation (21) for p_f can be carried out directly by evaluation of $f_{S|\mathbf{a}}(s)$ at each integration point(s), even though it is clear that the computational cost might be high. If it is likely that $F_{S|\mathbf{a}}(s)$ has regularity features for all $\mathbf{A} = \mathbf{a}$ of interest, it is possible to reduce the number of $F_{S|\mathbf{a}}(s)$ evaluations significantly through the use of curve fitting techniques. Evidently, it is important that low values of the $F_{S|\mathbf{A}}(s)$ function be approximated with greater accuracy than higher values, since the lower values correspond to greater contributions to ν_D^+ and $p_f(0)$.

For the integration required in equation (16) for the outcrossing rate, the probability density $f_{S|\mathbf{a}}(s)$ is required – this might be obtained relatively simply by differentiation if an explicit function has been established for $F_{S|\mathbf{a}}(s)$ as suggested above, or discrete values of $F_{S|\mathbf{a}}(s)$ may be used to estimate $f_{S|\mathbf{a}}(s)$ locally.

7. SOME OBSERVATIONS

In using the load process space for the formulation of the problem, unless "negative" or reversing loads are involved, the first quadrant in hyper-cartesian space R^m is the only space of real interest. "Negative" loads usually are not considered to be valid models – a different load process representation is required. If loads can reverse their direction, such as in the case of wave loading during a wave cycle, a space larger than the first quadrant is required. However, in other cases, such as service and wind loads, there is not usually a significant probability of the load acting in a direction diametrically opposite to the expected value. If there is, the probabilistic description needs to be improved in general. It follows that the structural system strength also needs be defined largely in the first quadrant or somewhat more generally if reversing loads or "negative" loads are possible.

Implicit in the formulation earlier in this paper was the assumption that the system limit state function(s) will be prescribed everywhere in the space of interest and that gaps do not exist, that is, for any arbitrary radial direction, s will be bounded with probability one. This follows from the fact that the materials constituting the structural system will each be described by a constitutive equation of some type, representing in each case the response of the material to external agencies (such as loads). It follows from structural mechanics theory that the constitutive equations fully describe the space of the external agencies. For ideal plastic structures the equivalent case is the plastic yield surface.

In terms of efficiency of simulation, the implication of the existence of a completely enveloping limit state surface is that, without further information, $f_{\mathbf{A}}(\ )$ need not be biased in a particular way through importance sampling techniques. It might be assumed uniformly distributed throughout the space of interest. However, $f_{\mathbf{A}}(\ )$ could be biased in those cases where sufficient information exists. Also, as already seen, for load processes having a significant zero component (see Figure 4), sampling efficiency (and the accuracy of the result) can benefit from biasing $f_{\mathbf{A}}(\ )$ appropriately. Further, algorithmic rules might be applied to ensure that directional samples are not taken too close to each other, to avoid repeating (almost) identical calculations along (slightly) different $\mathbf{A} = \mathbf{a}$ directions.

In integrating along a realisation ray $\mathbf{A} = \mathbf{a}$ to account for the uncertain limit state location, automatic allowance is made in the present procedure for the critical structural system strength (i.e., for the "critical" structural system limit state function, without evaluating it explicitly). This follows naturally from the way $f_{S|\mathbf{a}}(\ )$ is obtained in the present procedure using a structural system analysis which allows for strength uncertainty. In contrast, standard importance sampling approaches (for probability of failure evaluation) require that each structural system limit state function be known explicitly so that for each an appropriate selection of the importance sampling function can be made. As is well known, if this is not done for each individual limit state function, the outcome of the importance sampling analysis can be extremely poor and worse than the "crude" sampling approach. Moreover, the selection of the location and form of importance sampling distributions is an art. Most available techniques tend to use the points of maximum likelihood for the location of the distributions. Although for simple structural systems these points may not be difficult to estimate, in general some constrained optimisation scheme is required. Unfortunately, for a structural system there are, usually, many points of local maximum likelihood, one (at least) for each limit state function. It is likely, therefore, that the solution of an optimisation scheme implies considerable difficulty since: (i) *all* local points must be identified to ensure the critical one(s) have been identified; and (ii) the solution for each such point requires repeated evaluation of the limit state function (i.e., a structural analysis) and, for most optimisation routines, determination of the gradient at each such evaluation. It should be evident that this imposes considerable computational requirements.

In the technique described herein, multi-parameter optimisation for all local extremal points (i.e., the points of maximum likelihood) has been replaced by repeated structural analysis along a directional realisation of the possible load vector co–ordinate points (such as the points Z_i in Figure 7). Explicit identification of structural limit state functions is not required nor is a search scheme necessary. However, evaluations of the structural system capacity for given, deterministic, load vector outcomes biased to the lower tail are required, together with estimates of **n**, the local outward normal vector.

Whether the computational requirements are significantly in favour of one procedure or the other remains to be investigated, but it should be clear that the procedures proposed herein have two significant practical advantages: (i) structural limit state functions do not need to be known explicitly – a major advantage for complex systems; and (ii) identification of points of maximum likelihood as in importance sampling is not required. Both of these matters are of clear importance where the structural analysis is complex, such as in dynamic or finite element situations and for complex structures such as an offshore jacket type structure.

The approach described herein cannot deal with progressive failure of a structure, in the sense that under a given loading state the structure may deteriorate but not fail even though it is weakened against imposition of a later different loading state. As should be evident, the evaluation of the structural failure probability at Z_i (see Figure 7) implies a new structural analysis which makes no allowance for previous loading. Incorporation of progressive failure, except in a simplified form perhaps, is paramount to the study of systems which are load path dependent (see comments in relation to Figure 1).

8. EXAMPLE

In order to test the structural mechanics part of the work outlined herein, a very simple structural frame was selected for study (see Figure 9). The frame is composed of ideal elastic-plastic material having the strength properties X_1 = N(360, 54) and X_2 = N(400, 72). Two independent sparse Poisson processes model the loadings Q_1 and Q_2. They have Gaussian pulse heights Q_1 = N(100, 10) and Q_2 = N(50, 15) and occur with various intensities ν_{mi} and durations μ_{mi} (see Table 1).

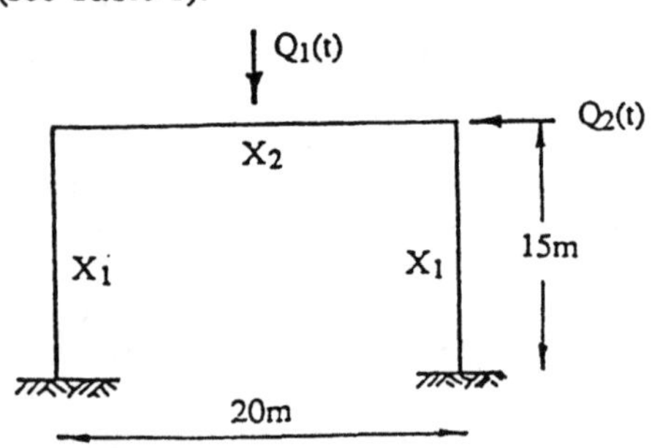

Fig. 9 Example frame

Results for this problem have been reported earlier using directly the known limit state functions for the problem[5]. In the present example, these limit state functions were ignored and recourse was had to the algorithm of Tang and Melchers[3] to (i) obtain the critical limit state functions for any set of load realizations and (ii) calculation of the probability of failure for that load realization.

Table 1 shows results obtained by the present approach compared to those obtained more directly by Melchers[5] It is seen that the results from the present approach somewhat underestimate, rather consistently, those obtained earlier. The present results were all obtained with only 80 radial direction samples.

Table 1. Typical Results

Poisson Process Loading Parameters				Outcrossing rate	
ν_{mi}	ν_{m2}	μ_{mi}	μ_{m2}	Present method	Melchers[5]
5	5	0.01	0.01	0.072	0.077
20	20	0.01	0.01	0.331	0.365
20	20	0.025	0.025	0.403	0.44
/year	/year	yr	yr		

The estimate for the outcrossing rate may be improved to yield close results (within 2%) by substantially increasing the amount of radical sampling, as shown in Figure 10. However, as Figure 11 shows, the coefficient of variation of the outcrossing rate estimate does not show significant improvement beyond about 80 radical samples. Of course, the computation times increase considerably with the number of samples - from 14.6 minutes for 80 to about 290 for 1600 radical samples. These high CPU times arise largely from repeated derivation of the limit state functions in the truncated enumeration technique. This is a matter currently under investigation with the aim to reduce unnecessary repeated computation.

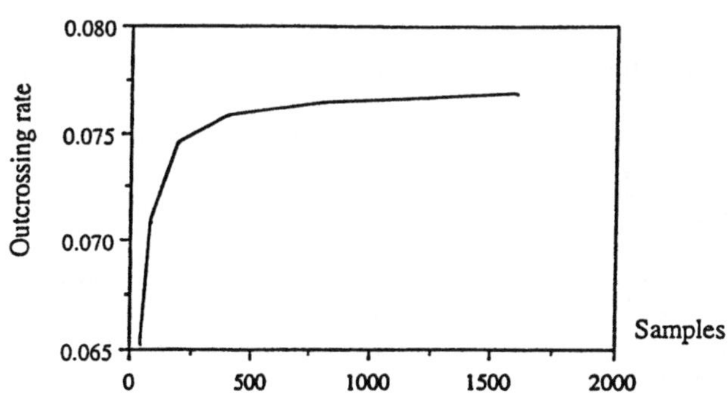

Fig. 10 Outcrossing rate estimate as a function of number of solid samples

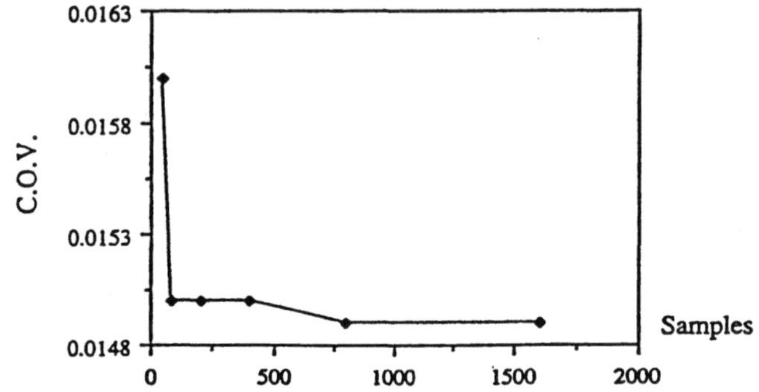

Fig. 11 Coefficient of variation of outcrossing rate as a function of the number of radical samples

9. CONCLUSION

When the structural reliability problem is formulated in the space of the load processes, it is possible to use a directional simulation approach to estimate (an upper bound of) the mean outcrossing rate and the zero time probability of failure. Existing truncated enumeration methods for the evaluation of conditional failure probabilities may be used to develop the probabilistic description of the probabilistic system. The feasibility of this approach was verified using a relatively simple example. It was argued also that the solution strategy obviates the need for explicit limit state functions for the system (although local member limit state functions are still required). Some remarks compare the present approach to importance sampling techniques for systems.

10. ACKNOWLEDGMENT The assistance of Dr Chun Qing Li in carrying out the numerical work for the example is gratefully acknowledged.

11. REFERENCES

1. Melchers, R.E., Load Space Formulation for Time Dependent Structural Reliability, *Journal Engineering Mechanics*, ASCE, **118**(5), (1992), 853–870.
2. Ditlevsen, O. and Bjerager, P., Methods of Structural Systems Reliability, *Structural Safety*, **3**(3/4), (1986), 195–229.
3. Melchers, R.E. and Tang, L.K., Dominant Failure Modes in Stochastic Structural Systems, *Structural Safety*, **2**, (1984), 127–143.
4. Thoft–Christensen, P. and Murotsu, Y., Application of Structural Systems Reliability Theory, Springer Verlag, Berlin, (1986).
5. Melchers, R.E., Load Space Reliability Formulation for Poisson Pulse Processes, (1993), (Submitted for Publication).
6. Ditlevsen, O., Melchers, R.E. and Gluver, H., General Multi Dimensional Probability Integration by Directional Simulation, *Computers and Structures*, **36**(2), (1990), 355–368.

7. Melchers, R.E., Radial Importance Sampling for Structural Reliability, *Journal Engineering Mechanics*, ASCE, **116**(1), 1990), 189– 203.
8. Ditlevsen, O., Bjerager, P., Olsen, R. and Hasofer, A.M., Directional Simulation in Gaussian Processes, *Probabilistic Engineering Mechanics*, 3(4), (1988), 207–217.
9. Breitung, K. and Rackwitz, R., Non Linear Combination of Load Processes, *Journal Structural Mechanics*, **10**(2), (1982), 145–166.
10. Bjerager, P., (1988), Probability Integration by Directional Simulation, *Journal of Engineering Mechanics*, ASCE, **114**(8), 1285–1302.
11. Vanmarcke, E.H., On the Distribution of the First Passage Time for Normal Stationary Random Processes, *Journal Applied Mechanics*, ASME, **42**(3), (1975), 215–220.

13. NOTATION

c	vector co–ordinates in **Q** space for origin of hyper polar co–ordinate system
D	domain
E( )	expectation operator
$f_X(\ )$	joint probability density function in **X**
$G_i(\ )$	limit state function (ith)
h( )	sampling probability density function
k	dimension of **X**
m	dimension of **Q**
n	direction cosines of unit outward normal
p_f	probability of failure (unconditional)
$p_f(x)$	conditional probability of failure
Q(t)	vector of load processes
R(x)	vector of structural resistance parameters corresponding to **Q**
S	radial direction (from point **c**)
S_A	surface area
S_D	surface of domain D
t	time
t_L	structure life time
X	vector of basic random variables
a	vector of direction cosines of hyper polar co–ordinate system
ΔS	elemental surface area
θ	vector of circumferential co–ordinates in hyper polar space
μ_X	vector of mean values for **X**
ν_D^+	outcrossing rate for domain D

Empirical Limit State Modelling in Reliability Analysis

Gunnar Mohr

Department of Civil Engineering, Engineering Academy of Denmark, 2800 Lyngby, Denmark

Abstract. Carrying capacity interaction diagrams are often established by using regression analysis on a given, usually small, set of interaction data. From a structural reliability point of view such regression analyses are not completely satisfactory. In this paper a different approach is followed.

First, a suitable (and preferably mechanically based) analytical expression of the failure function is formulated. The expression typically contains a number of unknown parameters. These are modelled as random variables whose joint distributional properties are obtained by use of Bayesian methods. Thus, a random set of failure functions is established. This set can be used directly in a modern reliability analysis. For the communication of the results of the analysis, it is often convenient not to have parameters without mechanical relevance in the interaction diagram expression. Therefore, special attention is given to the problems of eliminating these parameter random variables from the interaction diagram equation and transforming their randomness into model uncertainty randomness.

The ideas are exemplified by the examination of a small sample of interaction data related to the moment-normal force failure condition of a specific type of wall element.

1. Introduction.

We will consider specific types of structural elements which are subjected to a certain number of loads $\mathbf{S} = (S_1, S_2, \cdots, S_q)$. Actually $\mathbf{S}$ can also contain internal forces, moments, deformations, etc., coming from a given set of actions. They will however be denoted loads in the following. Let us assume that, due to the complexity of the elements, it is impossible to determine the failure criterion by a structural mechanics model. Thus we are not able to derive a failure function, and no resistance parameters are defined in an obvious way. The only objective information at hand is a small set of loads $(\mathbf{s}_1, \mathbf{s}_2, \cdots, \mathbf{s}_n)$ coming from measurements on n elements. The observed load vector $\mathbf{s}_i$ corresponds to occurrence of failure of element i. The purpose with the present analysis is to demonstrate how these data can be treated within the frame of structural reliability theory.

First, an analytical expression of the failure function has to be formulated. Even though it is not considered possible to derive a failure function directly from a mechanical model, at least some qualitative statements can be established sometimes. If possible, such information should be included in the subjective formulation of the failure function by giving this a suitable analytical expression. This expression will not only contain the loads **S** but also a number (k) of basic parameters (variables) $\mathbf{R} = (R_1, R_2, \cdots, R_k)$ representing the variability of the resistance of the elements. Moreover, a random vector consisting of p parameters $(A_1, A_2, \cdots, A_p)$ representing model uncertainty is supposed to enter the model. The problem now can be regarded as a reliability problem with the k+q+p basic variables $\mathbf{R}, \mathbf{S}, \mathbf{A}$, and we will first evaluate the joint distribution for $(\mathbf{R}, \mathbf{S}, \mathbf{A})$.

Let the failure function be denoted by g. Then the failure region is defined by

$$g(\mathbf{r}, \mathbf{s}, \mathbf{a}) \leq 0 \quad . \tag{1}$$

Each observation of loads by which failure occurred, denoted $\mathbf{s}_i$, is assumed to appear in the following manner. First, q-1 of the q loads are given previously chosen values (called $\mathbf{s}_i'$ corresponding to the i'th element), and then the remaining variable is increased until failure occurs. Hereby is observed a value s_i^*, which is an outcome of a conditional stochastic variable denoted S_i^*. We thus have one observation from each of the n (eventually different) conditional variables $S_1^*, S_2^*, \cdots, S_n^*$. The calculation should now continue in a conditional space. Indeed there is information in the fact that the structural element actually did survive the fixed load $\mathbf{s}_i'$ first put on it. We will however assume that the probability associated with this event is close to one (this can be ensured by a sensible planning of the experiment), and that we therefore can neglect the mentioned information and thus work with the unconditional variables **R** (and **A**).

A problem arises regarding whether the set $\{\mathbf{s}\}$ is independent of the load path leading to the points of failure. If there is no such independency, the choice of the fixed loads $\mathbf{s}_i$ will be important, and when planning the experiments, proper care with respect to this problem has to be taken.

The following basic relations are valid

$$g(\mathbf{R}, S_i^*, \mathbf{s}_i', \mathbf{A}) = 0 \quad , i = 1,2,\ldots,n \quad . \tag{2}$$

We now have to make an assumption concerning the joint distribution for **R** and will do this by assuming that **R** follows the normal distribution with expectation $E[\mathbf{R}] = \mathbf{m}$ and covariance matrix $\mathbf{c}$. The sample $(\mathbf{s}_1, \mathbf{s}_2, \cdots, \mathbf{s}_n)$ will give information about these distributional parameters. When estimating **m** and **c**, we are concerned with **statistical** uncertainty. This will be treated by considering **m** and **c** as Bayesian random variables. Following the Bayesian viewpoint, it is possible to establish the predictive joint distribution for $(\mathbf{R}, \mathbf{A})$. The **model** uncertainty related to the description of the failure surface by the failure function g is to some extent taken care of by introducing the variables **A** in the description. The uncertainty related to the choice of the functional relationship is not included herein.

2. Determination of the predictive joint distribution of $(\mathbf{R}, \mathbf{A})$.

Following a Bayesian formulation, the distributional parameters **m** and **c** are considered to be outcomes of random variables **M** and **Σ**. Moreover, a prior density

$f_{\mathbf{M},\Sigma,\mathbf{A}}(\mathbf{m},\mathbf{c},\mathbf{a})$ is assumed chosen. The posterior density is given by

$$f_{\mathbf{M},\Sigma,\mathbf{A}}(\mathbf{m},\mathbf{c},\mathbf{a}\,|\,\mathbf{s}^*) \propto L(\mathbf{m},\mathbf{c},\mathbf{a})\cdot f_{\mathbf{M},\Sigma,\mathbf{A}}(\mathbf{m},\mathbf{c},\mathbf{a}) \quad , \tag{3}$$

where L is the likelihood function

$$L(\mathbf{m},\mathbf{c},\mathbf{a}) = \prod_{i=1}^{n} f_{S_i^*}(s_i^*\,|\,\mathbf{m},\mathbf{c},\mathbf{a}) \quad . \tag{4}$$

Let $B_{s,i}$ be the set

$$B_{s,i} = \{\mathbf{r}\,|\,S_i^* \leq s\}$$

derived from

$$g(\mathbf{r},\mathrm{S}_\mathrm{i}^*,\mathrm{s}_\mathrm{i}',\mathbf{a}) = 0 \quad . \tag{2'}$$

Then

$$f_{S_i^*}(s\,|\,\mathbf{m},\mathbf{c},\mathbf{a}) = \frac{d}{ds}\int_{B_{s,i}} f_{\mathbf{R}}(\mathbf{r}\,|\,\mathbf{m},\mathbf{c})d\mathbf{r} \quad . \tag{5}$$

Finally the predictive density for $(\mathbf{R},\mathbf{A})$ is determined by

$$\begin{aligned} f_{\mathbf{R},\mathbf{A}}(\mathbf{r},\mathbf{a}\,|\,sample) &= \int_{\mathbf{m}}\int_{\mathbf{c}} f_{\mathbf{R},\mathbf{M},\Sigma,\mathbf{A}}(\mathbf{r},\mathbf{m},\mathbf{c},\mathbf{a}\,|\,sample) \\ &= \int_{\mathbf{m}}\int_{\mathbf{c}} f_{\mathbf{R}}(\mathbf{r}\,|\,\mathbf{m},\mathbf{c})\cdot f_{\mathbf{M},\Sigma,\mathbf{A}}(\mathbf{m},\mathbf{c},\mathbf{a}\,|\,sample) \quad . \end{aligned} \tag{6}$$

When used in an actual structure, the actions hereon are assumed to have known distributional properties. This in turn implies that this assumption also holds for the actual loads. Moreover, these will be independent of $(\mathbf{R},\mathbf{A})$. We thus have a standard reliability problem with the safety margin M_s given by

$$M_s = g(\mathbf{R},\mathbf{S},\mathbf{A}) \quad . \tag{7}$$

Suppose now that the problem is considered in connection with a safety method based on the design value format (/1/). According to (7), we can calibrate the distributional parameters for the loads to correspond to a wanted reliability index β. Moreover, we can determine design values $\mathbf{r}_\mathrm{d}$ and $\mathbf{a}_\mathrm{d}$ for $\mathbf{R}$ and $\mathbf{A}$. The safety margin can then be given the form

$$M_s = g(\mathbf{R},\mathbf{S};\mathbf{a}_d) \quad , \tag{8}$$

where **R** contains the resistance characteristics, and **S** is the load vector. Together with (8), the design values of **R** must be shown. In this formulation the distribution of **R** corresponding to (8), however, is unknown, which is not quite satisfactory. We will discuss this in more depth in the following example.

3. The special case with only 1 resistance parameter.

When the sample size is small, it seems reasonable to formulate the problem with a limited number of parameters. The case k=1 is easy to handle and is often supposed to give a sufficient description. Assume that the failure function is formulated in such a way that

$$r_2 > r_1 \;\Rightarrow\; g(r_2, \mathbf{s}, \mathbf{a}) > g(r_1, \mathbf{s}, \mathbf{a}) \quad all\; \mathbf{s}\,,\, \mathbf{a} \quad . \tag{9}$$

Then R can simply be called the resistance. If we further make the assumption that increasing values of S^* correspond to increasing resistance (an assumption which does not always hold and also is not vital for the proposed method), then formula (2') implicitly determines S_i^* as an increasing function h of R for every i and **a**

$$S_i^* = h_i(R) \quad . \tag{10}$$

The dependency of **a** is not explicitly shown. With σ denoting the standard deviation $D[R]$, we have

$$f_{S_i^*}(s\,|\,m,\sigma,\mathbf{a}) = f_R(h_i^{-1}(s)\,|\,m,\sigma)\cdot\frac{1}{h_i'(r_i)} \quad , \tag{11}$$

where $r_i = h_i^{-1}(s_i^*)$.

The likelihood function (4) becomes

$$L(m,\sigma,\mathbf{a}) = \prod_{i=1}^{n} f_R(r_i\,|\,m,\sigma)\cdot\prod_{i=1}^{n}\frac{1}{h_i'(r_i)} \quad . \tag{12}$$

With the following choice of prior density

$$f_{M,\Sigma,\mathbf{A}}(m,\sigma,\mathbf{a}) \propto \frac{1}{\sigma} \quad , \;\sigma > 0 \quad , \tag{13}$$

we get the posterior density

$$f_{M,\Sigma,\mathbf{A}}(m,\sigma,\mathbf{a}\,|\,sample) \propto \prod_{i=1}^{n}\frac{1}{\sigma}\phi(\frac{r_i - m}{\sigma})\cdot\prod_{i=1}^{n}\frac{1}{h_i'(r_i)}\cdot\frac{1}{\sigma} \quad . \tag{14}$$

The product $\prod_{i=1}^{n}\frac{1}{h_i'}$ depends not only on the sample, but also on **a**. This function is denoted $\psi(\mathbf{a})$.

Introducing the mean $\bar{r}$ and the standard deviation s_r through

$$\bar{r} = \frac{1}{n}\sum_{i=1}^{n} r_i \tag{15}$$

and

$$s_r^2 = \frac{1}{n-1}\sum_{i=1}^{n}(r_i - \bar{r})^2 \quad , \tag{16}$$

the posterior density (14) can be reformulated as

$$f_{M,\Sigma,\mathbf{A}} \propto \frac{1}{\sigma^n}\cdot\exp[-\frac{n-1}{2}(\frac{s_r}{\sigma})^2]\cdot\frac{1}{\sigma}\phi(\frac{m-\bar{r}}{\sigma/\sqrt{n}})\cdot\psi(\mathbf{a}) \quad . \tag{17}$$

This yields by integration

$$f_{\mathbf{A}}(\mathbf{a}\,|\,\text{sample}) \propto \frac{1}{s_r^{n-1}}\cdot\psi(\mathbf{a}) \quad . \tag{18}$$

The predictive density for $(R\,,\mathbf{A})$ can now be determined from formula (6)

$$\begin{aligned} f_{R,\mathbf{A}}(r\,,\mathbf{a}\,|\,\text{sample}) &= \psi(\mathbf{a})\cdot\int_\sigma\int_m \frac{1}{\sigma}\phi(\frac{r-m}{\sigma})\cdot\phi(\frac{m-\bar{r}}{\sigma/\sqrt{n}})\cdot\exp[-\frac{n-1}{2}(\frac{s_r}{\sigma})^2]\,dm\,d\sigma \\ &\propto \psi(\mathbf{a})\cdot\int_{\sigma=0}^{\infty}\frac{1}{\sigma^{n+1}}\exp[-\frac{1}{2\sigma^2}\{(n-1)s_r^2+\frac{n}{n+1}(r-\bar{r})^2\}]\,d\sigma \\ &\propto \psi(\mathbf{a})\cdot\frac{1}{s_r^n}\cdot\{1+\frac{n}{n^2-1}\cdot(\frac{r-\bar{r}}{s_r})^2\}^{-n/2} \quad . \end{aligned} \tag{19}$$

From this is seen that the conditional random variable

$$\frac{(R\,|\,\mathbf{A}=\mathbf{a})-\bar{r}}{s_r\sqrt{1+1/n}} \tag{20}$$

follows a t-distribution with n-1 degrees of freedom. This is a well known result (/2/) which reflects the statistical uncertainty related to the estimation of m and σ. The difference between (20) and f_R found from (19) is due to the model uncertainty which is included in (19). In the limiting situation when $n\rightarrow\infty$, the statistical uncertainty, as well as the model uncertainty, will vanish, since

$$\lim_{n\rightarrow\infty} D[M]=0 \quad , \tag{21}$$

which also holds for Σ and $\mathbf{A}$.

Up to this point, we have not been concerned with the uncertainty due to the measurement error related to the observations $\{s_i^*\}$. In this type of problem, it is not possible to separate the measurement error from the random variation of the resistance. The measurement error ε can only been taken into consideration by an a priori assignment of a distribution or at least a second moment representation for ε. In order to make an approximate quantitative evaluation of the influence of the measurement error, we let $S_{i,m}^*$ be defined by

$$S_{i,m}^* = S_i^* + \varepsilon_i \quad . \tag{22}$$

The index m corresponds to measured values. Assume that $E[\varepsilon_i]=0$ and that ε_i and S_i^* are uncorrelated. For simplicity we assume that $\mathbf{A}$ is constant. With R_m being the resistance calculated by measured values, it is easily seen that

$$E[R] \simeq E[R_m] \tag{23}$$

and

$$Var[R] \simeq \left(1 - \frac{Var[\varepsilon_i]}{Var[S^*_{i,m}]}\right) \cdot Var[R_m] \ . \tag{24}$$

Not taking the measurement error into consideration corresponds to substituting R by R_m. When $Var[\varepsilon_i]$ is small compared to $Var[S^*_{i,m}]$, neglecting the measurement error thus has an insignificant influence. If the model uncertainty is also involved by not considering **A** as being constant, the significance of the measurement error will be further reduced. In the following, we will assume that the measurement error is small in the above sense.

Having derived the joint density for $(R, \mathbf{A})$, equation (7) appears as a standard reliability problem. It is, however, inconvenient to end up with a formulation of the limit state function which involves parameters without any mechanical reference, and we will therefore transform the model uncertainty to the other variables. Load distributions are often defined in the relevant codes, and we prefer not to interfere with these predetermined distributions. Thus the model uncertainty is transformed (by a linear approximation) to the resistance parameter R only.

Consider the equation

$$g(R, \mathbf{S}, \mathbf{A}) = 0 \ . \tag{25}$$

When solved with respect to R, we obtain a function w such that

$$R = w(\mathbf{S}, \mathbf{A}) \tag{26}$$

$\Leftrightarrow$

$$R \approx w(\mathbf{S}, \mathbf{a}_0) + \sum_{i=1}^{p} w'_{a_i}(\mathbf{s}_0, \mathbf{a}_0) \cdot (A_i - a_{0i}) \ , \tag{27}$$

where $\mathbf{s}_0$ and $\mathbf{a}_0$ are appropriately chosen constants.

With

$$R^* = R - \sum_{i=1}^{p} w'_{a_i}(\mathbf{s}_0, \mathbf{a}_0) \cdot (A_i - a_{0i}) \ , \tag{28}$$

equation (27) can be rewritten as

$$g(R^*, \mathbf{S}, \mathbf{a}_0) = 0 \ , \tag{29}$$

which corresponds to the desired failure function. A suitable choice of $\mathbf{s}_0$ and $\mathbf{a}_0$ could be the design values found by (25) as previously described. The difference between (8) and (29) is that in (29) we know the distribution of R^*.

4. An example.

A bearing wall constructed of two alu-profiles with a plaster plate on each side is considered. The bearing capacity of the wall is examined in the following way. The wall is first loaded by a certain moment M and is then, in addition, given a normal force N until failure occurs. In this way, $n = 11$ observations (m_i, n_i) of the combined bearing capacity are obtained by the measuring on 11 walls. The actual 11 pairs are shown in figure 1. Due to the complexity of the wall, it is considered impossible to determine the failure criterion by a structural mechanics model,and we will therefore use the described method and establish a reliability model related to the specific type of wall elements.

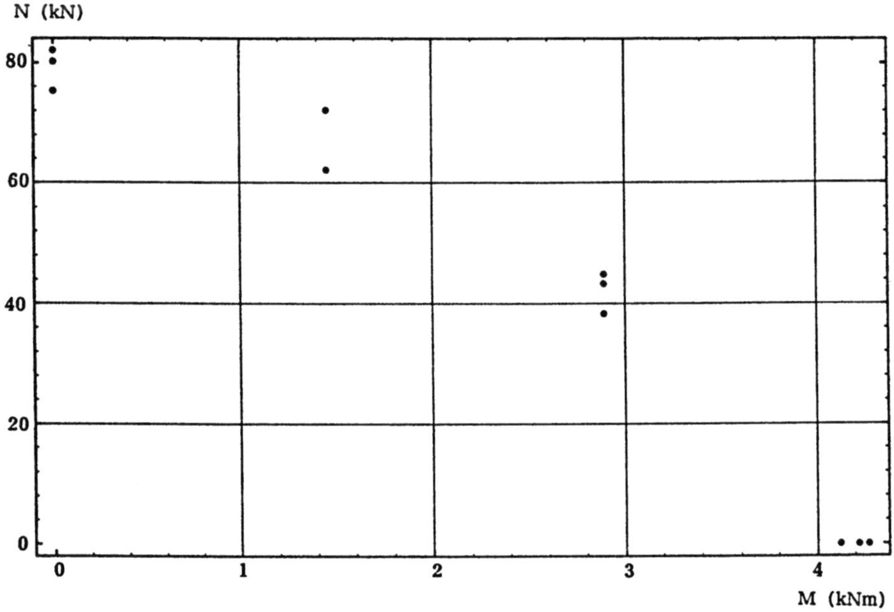

Fig. 1. Observations of combined loads leading to failure.

The following model is suggested for further analysis

$$\left(\frac{N}{N_0}\right)^{\alpha} + \left(\frac{M}{M_0}\right)^{\beta} = 1 \ , \qquad (30)$$

where

N_0 = bearing capacity by a normal load only
M_0 = moment capacity by a moment load only
(M,N) = pair of moment capacity and normal force capacity by combined loads
α , β = parameters to be estimated.

Moreover, the relation $M_0 = k{\cdot}N_0$ is included in the model. Here k is an unknown parameter which also is to be estimated. The carrying capacity R is now simply defined as

$$R = N_0 \quad .$$

Thus the function g in (1) is given by

$$g(r,n,m,\alpha,\beta,k) = 1 - (\frac{n}{r})^{\alpha} - (\frac{m}{k{\cdot}r})^{\beta} \quad . \tag{31}$$

With $s_i' = m_i$ and $s_i^* = n_i$, the function h_i defined according to (10) is determined by (31), i.e.

$$1 - (\frac{n_i}{r_i})^{\alpha} - (\frac{m_i}{k r_i})^{\beta} = 0$$

$\Leftrightarrow$

$$n_i = h_i(r_i) = r_i(1 - (\frac{m_i}{k r_i})^{\beta})^{1/\alpha} \quad . \tag{32}$$

From this relation the values r_i are found, $r_i = r_i(\alpha,\beta,k)$. The derivatives h_i' become

$$h_i'(r_i) = \frac{n_i}{r_i} + \frac{\beta}{\alpha}\cdot(\frac{n_i}{r_i})^{1-\alpha}\cdot(\frac{m_i}{k r_i})^{\beta} \quad . \tag{33}$$

These expressions apply when $i=4,5,6,7,8$. For $i=1,2,3$ we have (since $m_i = 0$) $n_i = r_i$. Thus $h_i' = 1$. For $i=9,10,11$, we have (since here $s_i' = n_i = 0$ and $s_i^* = m_i$) $m_i = k r_i$, which implies that $h_i' = k$. The function $\psi(\mathbf{a}) = \psi(\alpha,\beta,k)$ in (19) is then given as the product

$$\psi(\alpha,\beta,k) = \frac{1}{k^3}\cdot\prod_{i=4}^{8}(\frac{n_i}{r_i} + \frac{\beta}{\alpha}(\frac{n_i}{r_i})^{1-\alpha}\cdot(\frac{m_i}{k r_i})^{\beta})^{-1} \quad . \tag{34}$$

The predictive density (19) is now completely specified. Numerical calculations based on this formula yield the following second moment representation (with A ~ α, B ~ β and K ~ k) of (R, A, B, K)

Expectations:

$$(E[R], E[A], E[B], E[K]) = (79.1; 1.43; 1.57; 0.053)$$

Variances:

$$(Var[R], Var[A], Var[B], Var[K]) = (18.1; 0.208; 0.228; 2.76e^{-6}) \quad .$$

The correlation matrix becomes

$$\rho = \begin{bmatrix} 1 & 0.041 & -0.15 & -0.28 \\ 0.041 & 1 & -0.85 & 0.22 \\ -0.15 & -0.85 & 1 & 0.24 \\ -0.28 & 0.22 & 0.24 & 1 \end{bmatrix} .$$

The marginal density functions f_A , f_B , and f_K calculated from (19) are well approximated by log-normal distributions. The marginal density function f_R is well approximated by the normal-distribution. These approximations are shown in the figures 2-5 below. In the next section a reliability study is performed by use of the software package PROBAN. In these analyses the joint distribution $f_{A,B,R,K}$ is calculated using the Nataf-distribution and not according to (19). Reliability indices are found by FORM analysis.

5. Reliability analysis.

Consider the situation where the actions on the wall are due to a wind pressure (which causes a bending moment M) and a permanent load N on the top of the wall. More precisely, let M be the maximum moment within a reference period of 1 year. M is assumed to follow the Gumbel distribution

$$F_M(m) = exp(-exp(-\alpha(m-u))) \tag{35}$$

with constant coefficient of variation $V_m = 0.22$ corresponding to $\alpha u = 5.18$. (This is a recommended value in Danish Codes when no other information is available). The characteristic value m_c is defined as the 98%-fractile. This leads to the formula

$$m_c \approx \frac{5.18 - \ln 0.02}{\alpha} = \frac{9.09}{\alpha} . \tag{36}$$

The permanent load N is assumed to follow the log-normal distribution with coefficient of variation 0.10. The characteristic value is defined as being the mean $E[N]$.

According to the manufacturer, the characteristic moments of relevance are most frequently of magnitude ≈ 2.5 kN, corresponding to $\alpha=3.7$. If we choose a target reliability index $\beta=3.71$, this corresponds to a mean permanent load $E[N]=7.9$ kN. The $\beta-$point is found to be

$$(n_0, m_0, r_0, a_0, b_0, k_0) = (7.92\,;\,3.83\,;\,76.3\,;\,1.18\,;\,1.63\,;\,0.052) \ . \tag{37}$$

In order to evaluate R^* from (28), the derivatives w'_{a_i} must be found. By implicit differentiation of

$$(\frac{n}{r})^a + (\frac{m}{kr})^b - 1 = 0 \ ,$$

one finds

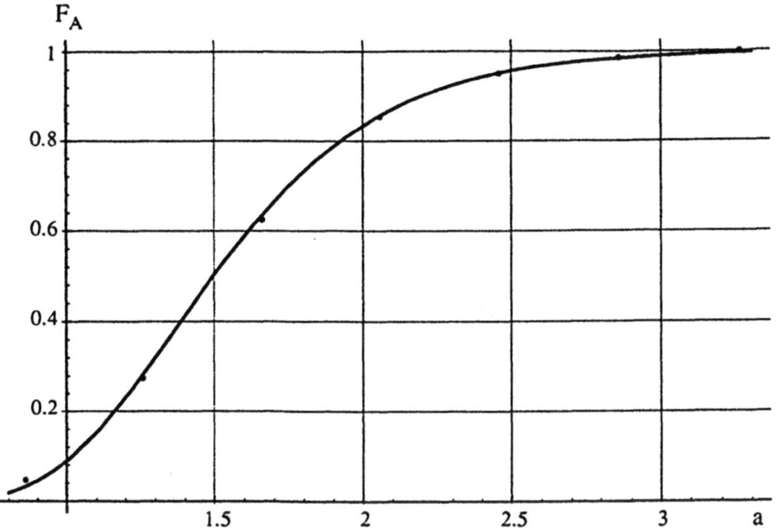

Fig. 2. Distribution function of A. Calculated values (points) vs. log-normal model distribution.

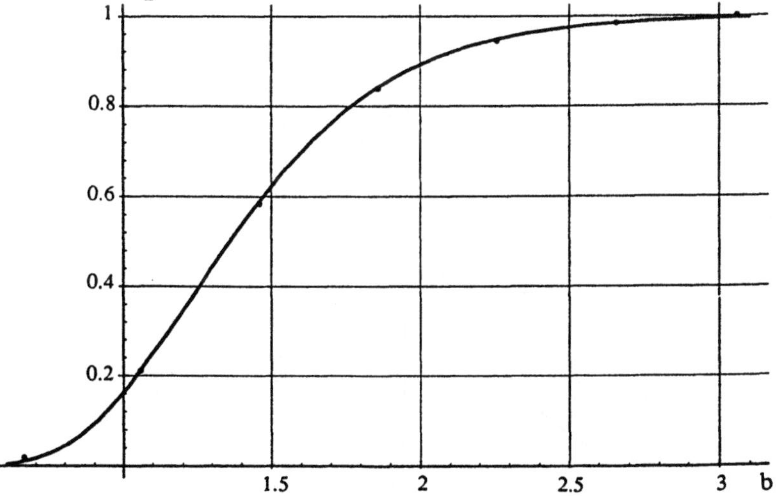

Fig. 3. Distribution function of B. Calculated values (points) vs. log-normal model distribution.

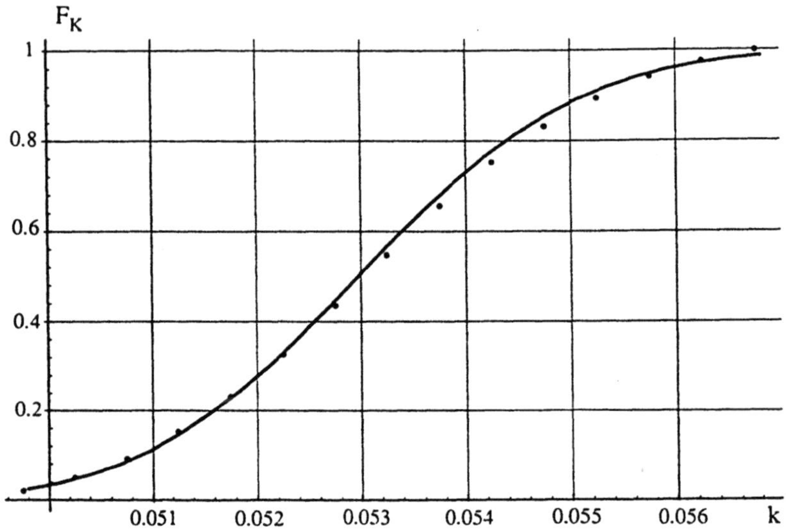

Fig. 4. Distribution function of K. Calculated values (points) vs. log-normal model distribution.

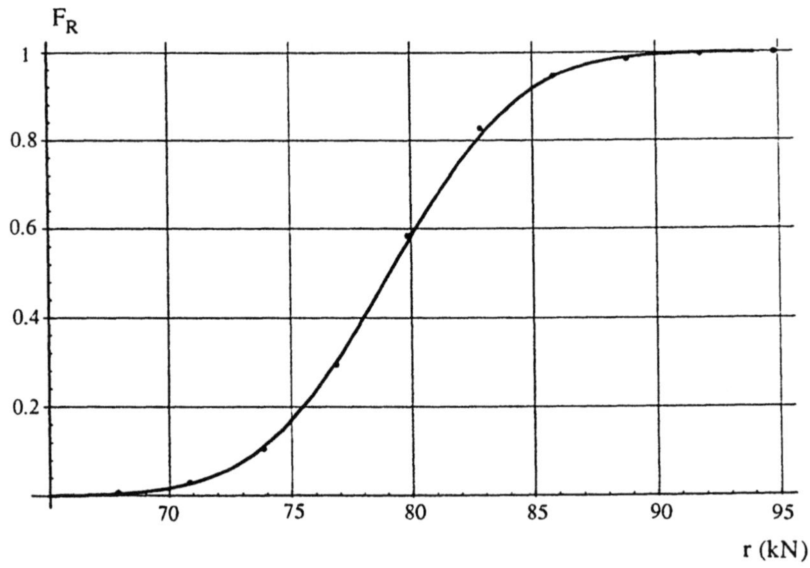

Fig. 5. Distribution function of R. Calculated values (points) vs. normal model distribution.

$$w_a' = \frac{(\frac{n_0}{r_0})^{a_0} \cdot ln(\frac{n_0}{r_0}) \cdot r_0}{(\frac{n_0}{r_0})^{a_0} \cdot a_0 + (\frac{m_0}{k_0 r_0})^{b_0} \cdot b_0} = -7.45 \quad , \tag{38}$$

$$w_b' = \frac{(\frac{m_0}{k_0 r_0})^{b_0} \cdot ln(\frac{m_0}{k_0 r_0}) \cdot r_0}{(\frac{n_0}{r_0})^{a_0} \cdot a_0 + (\frac{m_0}{k_0 r_0})^{b_0} \cdot b_0} = -1.92 \quad , \tag{39}$$

$$w_k' = -\frac{(\frac{m_0}{k_0 r_0})^{b_0} \frac{b_0}{a_0} r_0}{(\frac{n_0}{r_0})^{a_0} \cdot a_0 + (\frac{m_0}{k_0 r_0})^{b_0} \cdot b_0} = -61.37 \quad , \tag{40}$$

$$R^* = R + 7.45A + 1.92B + 61.37K - 15.1 \quad , \tag{41}$$

$$E[R^*] = 80.98\ kN \ , \quad D[R^*] = 4.99\ kN \quad ,$$

where $r = w(n, m, a, b, k)$. The distribution of R^* is determined by (19) and (41) but is not easy to handle. When approximated with a normal distribution,we will refer to the model as model II. We thus consider the following two models

Model I:

Safety margin: $$M_I = 1 - (\frac{N}{R})^A - (\frac{M}{KR})^B \tag{42}$$

Model II:

Safety margin: $$M_{II} = 1 - (\frac{N}{R^*})^{1.18} - (\frac{M}{0.052R^*})^{1.63} \tag{43}$$

R^* normally distributed with $(E[R^*]\,;D[R^*]) = (80.9\,;4.99)\ kN$

The reliability index corresponding to model II becomes $\beta_{II} = 3.71$ equal to the target value.

Finally we will compare model I and model II by calculations corresponding to the entire range of load combinations (i.e. with varying characteristic values). The results are shown in Table 1 below. The calculations are carried out with m_c equal to the values in column 1. Columns 2 and 3 show the calculated values of $n_c = E[N]$ corresponding to (42) and (43) respectively. Reliability index β equals 3.71 . Column 4 shows the β-values β_{II} calculated from (43) with n_c equal the values in column 2. The deviations between results from the two models are less than 2%. Thus the process of transforming the model uncertainty to the resistance variable R gives an insignificant error.

Table 1. Comparison between model I and model II.

m_c (kNm)	$n_{c,I}$ (kN)	$n_{c,II}$ (kN)	β_{II}
-	51.9	52.1	3.75
0.75	50.2	50.6	3.77
1.0	49.0	49.5	3.79
1.5	44.1	44.8	3.77
2.0	30.3	30.7	3.73
2.2	22.5	22.7	3.72
2.3	18.1	18.3	3.72
2.5	7.9	7.9	3.71
2.60	-	$m_{c,II} =$ 2.59 kNm	3.71

Figure 6 shows the interaction diagram between the characteristic values. The load combinations (m_c, n_c) on the curve are the characteristic values by which failure occurred. If the safety format of the relevant code is based on design values and partial coefficients γ, the interaction diagram can be given an alternative formulation. As an example,assume that $\gamma_N = 1$ and $\gamma_M = 1.3$. Curve (I) in figure 7 then shows the combinations of design values (m_d, n_d) leading to failure. With a suitably chosen design value r_d^* (here the value 66.6 kNm), this curve can be approximated by a curve (curve (II)) consistent with model II. The model is thus formulated as

Model II:

$$\text{Failure function:} \quad g(m,n,r) = 1 - \left(\frac{n}{r}\right)^{1.18} - \left(\frac{m}{0.052r}\right)^{1.63} \tag{44}$$

R normally distributed with $(E[R]\,;D[R]) = (80.9\,;4.99)\ kN$

Design value of resistance: $r_d^* = 66.6\,\text{kN}$. The model is valid for $m_d \in [1.8\,;3.3]$.

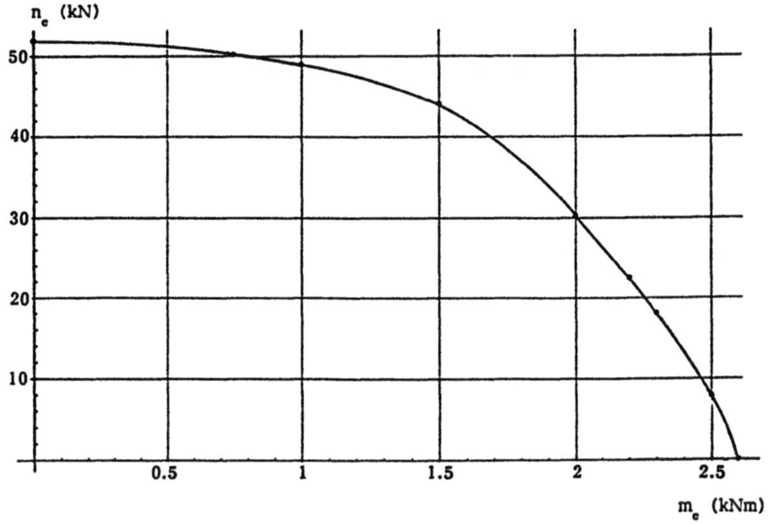

Fig. 6. Interaction diagram. Load combinations on the curve are the characteristic values corresponding to occurrence of failure.

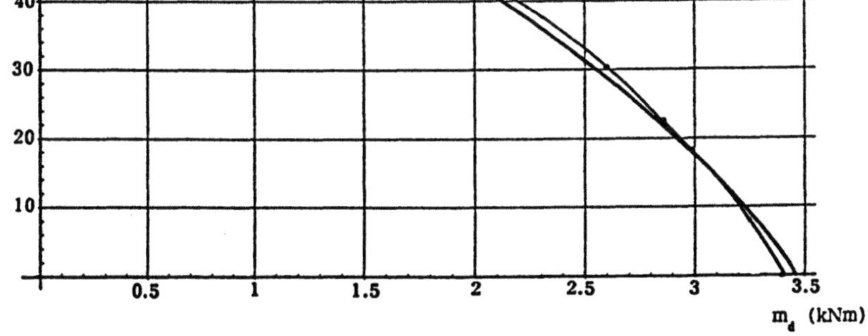

Fig. 7. Interaction diagram. Load combinations on curve (I) are the design values corresponding to occurrence of failure. Curve (II) is calculated from model II.

Acknowledgement.

The author wants to thank Ove Ditlevsen for his valuable comments and suggestions.

References.

1. Madsen, H.O., Krenk, S. & Lind, N.C., *Methods of Structural Safety*. Prentice Hall, Inc. New Jersey. 1986.

2. Ditlevsen, O., *Uncertainty Modeling*. McGraw-Hill Inc. New York. 1981.

Fatigue Specific Linearization Methods

A. Naess[1] and S. R. Winterstein[2]

[1] Faculty of Civil Eng., Norwegian Institute of Technology, N-7034 Trondheim, Norway
[2] Dept. of Civil Eng., Stanford University, CA 94305, USA

Abstract. The paper discusses the problem of linearization for the purpose of simplified calculation of fatigue damage in engineering structures. Specific linearization procedures are proposed that, for some structural models, may be better suited to deal with the problem of fatigue life calculation than the standard mean square stochastic linearization (MSSL). The proposed methods are based on minimizing higher order moments of the linearization error, and it is indicated how the optimal order is linked to the stress exponent of the S-N curve. It is shown by specific example studies that the proposed method may lead to substantial improvement over MSSL in estimating fatigue damage.

1 Introduction

The only method of almost general applicability for simplifying nonlinear dynamic models subjected to random loads is the method of statistical linearization[1]. For a broad range of nonlinear problems, the application of this method makes it possible to analyse the response statistics in an almost routine manner. For the obtained results to be of practical value, it is obviously necessary to supplement this kind of analysis by some means of verification. We shall not go into a detailed discussion of the rationale behind linearization methods here, but merely note that their properties make them singularly attractive as tools for practical engineering analyses of structural response statistics.

However, one of the basic short-comings of current linearization methods as general tools is their emphasis on mean square deviation errors between the given nonlinear model and the 'equivalent' linear one. This forces the linear model to focus on agreement at the level of second-order statistical moments, which is also borne out by experience. This is obviously not always the desired level where agreement is sought. Both for extreme response and fatigue analyses, attention is generally on agreement at higher statistical moments in such a way that emphasis is shifted to higher response levels.

The goal of the present paper is to investigate the possibility to extend the standard mean square statistical linearization (MSSL) method for the purpose of estimating fatigue damage in a more consistent manner than allowed by the MSSL method. As already mentioned, to achieve this, emphasis must usually be shifted to higher response levels in a manner consistent with the standard way of calculating accumulated fatigue damage in engineering systems subjected to random loads. As a rule, this implies the use of an S-N curve relationship together with a Palmgren-Miner linear damage accumulation law. It is generally recognized that the fatigue process in specific components of a structure is a very complex and not fully understood phenomenon. However, even if the practical procedures for calculating fatigue life that have been proposed as a rule fall far short of being accurate, they are still an indispensable tool for the engineer grappling with the fatigue design of real-life structures.

For a fatigue estimation procedure to be satisfactory as a design tool, disregarding its absolute accuracy, it is important that it has a sufficient degree of inner consistency in the sense that it captures the relative importance of the various elements that determine the fatigue life according to the given model. For engineering fatigue design, this requirement of consistency must be combined with the requirement of ease of application to produce what would be a satisfactory decision tool for this purpose.

The current status in this respect is that such a goal has not yet been fully achieved. As an effort in this direction, the present paper investigates the possibility of applying linearization methods. This particular idea has been around for some time, see e.g. Melchers and Ahammed[2]. But, to the authors' knowledge, a systematic study along such lines has not previously been undertaken.

A description of the proposed linearization methods and numerical procedures for fatigue damage calculation is given below. The theory is illustrated by application to a simplified structural model. Naess and Winterstein[3] have also applied the proposed methods to a quasistatic offshore structure subjected to Morison type wave forces.

2 Fatigue Damage Accumulation Model

The following relation is assumed between stress cycles with stress range S, and the number N of such stress cycles to fatigue failure

$$cNS^{m} = 1\,, \tag{1}$$

where c and m are material parameters. For simplicity, they are considered to be constants. Assuming a linear elastic material, to a stress range S there corresponds a displacement range R, where $S = qR$ for some material constant q. Eq (1) can then be rewritten as

$$kNR^{m} = 1\,, \tag{2}$$

where $k = cq^{m}$.

If the displacement response process is narrow band, it may be assumed that

$R = 2A$, where A denotes a random variable giving the displacement response peak. Let the PDF of A be denoted by $f_A(\cdot)$. By adopting a linear damage accumulation law, i.e. the Palmgren-Miner hypothesis, the mean damage accumulation rate D is then given by the expression

$$D = k\nu_0^+ \int_0^\infty (2a)^m f_A(a)\, da\,, \tag{3}$$

where ν_0^+ is the mean upcrossing rate of the mean response level.

If it is further assumed that the response process is Gaussian, A is Rayleigh distributed. Then one obtains the following well-known expression for the mean damage accumulation rate

$$D = k\nu_0^+ (2\sqrt{2}\,\sigma)^m \Gamma(1 + \frac{m}{2})\,. \tag{4}$$

Here σ is the standard deviation of the displacement response. $\Gamma(\cdot)$ denotes the gamma function[4]. Various correction procedures have been developed to account for non-narrow band stress processes. In offshore applications it is often the case that a non-narrow band stress process corresponds to a process with a multipeaked spectral density with, for example, energy at both the wave and structural frequencies. Wirsching and Light[5] have proposed a simple method to estimate the fatigue damage rate in this situation. It amounts to multiplying the right hand side of Eq (4) by a correction factor derived directly from the spectral density. A similar correction method has been proposed by Jiao and Moan[6]. For simplicity, such correction procedures will be left out of the subsequent discussion.

3 Nonlinear Dynamic Model and Stochastic Linearization Methods

For the subsequent development, the following multi-degree-of-freedom (MDOF) equation of motion describing the (generalized) displacement response of a nonlinear structure with r degrees of freedom, will be assumed

$$\boldsymbol{M}\,\ddot{\boldsymbol{X}} + \boldsymbol{g}(\dot{\boldsymbol{X}}) + \boldsymbol{h}(\boldsymbol{X}) = \boldsymbol{F}(t)\,. \tag{5}$$

Here $\boldsymbol{M}$ denotes an $r\times r$ generalized mass matrix. $\boldsymbol{g}(\dot{\boldsymbol{X}})$ and $\boldsymbol{h}(\boldsymbol{X})$ are r-dimensional nonlinear vector functions of (generalized) velocity and displacement, respectively. $\boldsymbol{F}(t) = (F_1(t),\dots,F_r(t))^T$ denotes the (generalized) excitation force vector, which is assumed to be a wide band, stationary Gaussian vector process. $\boldsymbol{X}(t) = (X_1(t),\dots,X_r(t))^T$ denotes the corresponding (generalized) displacement response vector.

It will be recognized from the subsequent developments that the methods described are particularly well adapted to deal with the case when the damping and restoring force terms can be represented by polynomial type expansions of the following generic kind,

$$q_i(x) = \sum_{j=1}^{r} \{q_{0i}^{j}\, sign(x_j) + q_{1i}^{j} x_j\} + \sum_{j=1}^{r}\sum_{k=1}^{r} q_{2i}^{jk} x_j |x_k| + \\ + \sum_{j=1}^{r}\sum_{k=1}^{r}\sum_{l=1}^{r} q_{3i}^{jkl} x_j x_k x_l + \dots \, . \tag{6}$$

Here $sign(\cdot)$ denotes the sign function, $q_{...}^{...}$ are suitable constants. For practical applications, these models should cover a large range of actual damping and restoring force behaviour. For simplicity, it has been assumed that $\boldsymbol{g}(\cdot)$ and $\boldsymbol{h}(\cdot)$ are antisymmetric functions, i.e. $\boldsymbol{g}(-\dot{X}) = -\boldsymbol{g}(\dot{X})$ and $\boldsymbol{h}(-X) = -\boldsymbol{h}(X)$.

To achieve the simplified procedure for calculating fatigue damage described in the previous section, the principal aim of the present work is to investigate the possibility to develop rational methods for replacing Eq (5) by the following linear, time-invariant dynamic model

$$\boldsymbol{M}\ddot{\boldsymbol{X}} + \boldsymbol{C}\dot{\boldsymbol{X}} + \boldsymbol{K}\boldsymbol{X} + \boldsymbol{d} = \boldsymbol{F}(t) \, . \tag{7}$$

Here $\boldsymbol{C}$ and $\boldsymbol{K}$ are constant $r{\times}r$ matrices. $\boldsymbol{d}$ is a constant r-dimensional vector thrown in to balance mean values, that is

$$\boldsymbol{d} = E[\boldsymbol{h}(\boldsymbol{X})] - \boldsymbol{K}\,E[\boldsymbol{X}] \, . \tag{8}$$

More specifically, the aim has been to determine the quantities $\boldsymbol{C}$, $\boldsymbol{K}$ and $\boldsymbol{d}$ so that the fatigue damage predicted by using Eq (4), is a reasonably accurate estimate of the fatigue damage that would be obtained from Eq (3) by using the exact expression for the PDF of displacement response amplitudes $f_A(\cdot)$.

As a guide in our effort to establish a linearization method for fatigue estimation, the obvious starting point is Eq (3). From this equation it is seen that the essential quantity is the m'th statistical moment of the displacement response amplitude A. Assuming that the underlying displacement response process $X(t)$ is narrow-band Gaussian, $f_A(\cdot)$ will be a Rayleigh PDF. By comparing the formulas giving the Gaussian and the Rayleigh PDF, it is recognized that, apart from a fixed multiplicative constant, an m'th order statistical moment of the amplitude corresponds to an $(m + 1)$'th order statistical moment of the absolute value of the underlying displacement response. Since the m-values to be met in practical fatigue calculations usually range from about 3 and upwards, it is realized that fatigue calculations are mostly concerned with 4'th order and higher statistical moments of the relevant displacement response. This statement immediately leads to the conclusion that a standard mean square stochastic linearization (MSSL) method, with its intrinsic focus on second-order statistical moments, cannot in general be expected to produce accurate estimates of fatigue damage. And the departure from the target value will typically increase with increasing values of m.

A method for replacing the nonlinear equation of motion by a linear one that may be more suitable for fatigue estimation, has been proposed by Naess[7], see also Naess et al.[8]. To describe it, the following notation is expedient, $|\boldsymbol{x}|^{\boldsymbol{n}} = \sum_{i=1}^{r} |x_i|^{n_i}$, where $\boldsymbol{n} = (n_1, \ldots, n_r)$, $n_i = 2,3,4,\ldots$. The stochastic linearization procedure to be used here for replacing the nonlinear equations of motion with a set of linear equations, is obtained by minimizing the following deviation or error measure with specified exponent vectors $\boldsymbol{m}$ and $\boldsymbol{n}$,

$$e_{(\boldsymbol{m},\boldsymbol{n})} = E[|\boldsymbol{g}(\dot{X}) - \boldsymbol{C}_{(\boldsymbol{m})}\dot{X}|^{\boldsymbol{m}}] + E[|\boldsymbol{h}(X) - \boldsymbol{K}_{(\boldsymbol{n})}X - \boldsymbol{d}(\boldsymbol{K}_{(\boldsymbol{n})})|^{\boldsymbol{n}}] \tag{9}$$

with respect to the matrices $\boldsymbol{C}_{(\boldsymbol{m})} = (c_{ij}^{(m_i)})$ and $\boldsymbol{K}_{(\boldsymbol{n})} = (k_{ij}^{(n_i)})$ so that

$$\boldsymbol{d}(\boldsymbol{K}_{(\boldsymbol{n})}) = E[\boldsymbol{h}(X)] - \boldsymbol{K}_{(\boldsymbol{n})}\, E[X] \tag{10}$$

and under the assumption that $X(t)$ is the solution of Eq (7) with $\boldsymbol{C}, \boldsymbol{K}$ and $\boldsymbol{d}$ replaced by $\boldsymbol{C}_{(\boldsymbol{m})}$, $\boldsymbol{K}_{(\boldsymbol{n})}$ and $\boldsymbol{d}(\boldsymbol{K}_{(\boldsymbol{n})})$, respectively. Hence, $X(t)$ is a stationary Gaussian vector process. This linearization method will be denoted by ESL($\boldsymbol{m},\boldsymbol{n}$), (ESL - Extended Stochastic Linearization). To simplify notation, $e_{(\boldsymbol{n})} = e_{(\boldsymbol{n},\boldsymbol{n})}$ and ESL($\boldsymbol{n}$) = ESL($\boldsymbol{n},\boldsymbol{n}$). We have described the extended stochastic linearization method in full generality. However, for practical fatigue calculations it may be suggested to use identical component values in both exponent vectors, i.e. $\boldsymbol{m} = \boldsymbol{n} = (n, n, \ldots, n)$. It may be noted that the standard stochastic linearization method[1], that is the MSSL, is obtained for $m_i = n_i = 2$.

From the previous discussion on the statistical moments involved in fatigue calculations, it can be anticipated that the optimal values of the exponent vectors in the error measure $e_{(\boldsymbol{m},\boldsymbol{n})}$ will be closely related to m-value of Eq (1).

In principle, the components of the exponent vectors $\boldsymbol{m}$ and $\boldsymbol{n}$ could be any real positive numbers, but for practical analysis it is convenient to restrict them to integer values. And, in fact, even odd integers are difficult to handle directly. Hence, from a practical, computational point of view, the candidates are the even integers $m_i, n_i = 2, 4, 6, \ldots$. However, from an expected smooth behaviour of the determined optimum parameters as a function of $\boldsymbol{m}$ and $\boldsymbol{n}$, a suitable interpolation procedure may provide optimum parameters for values of the exponent vector components in the intervals between the even integers.

Observing Eq (1), an apparently different alternative as a linearization strategy would be to base the linearization on the amplitude probability distributions. However, from the comments of the previous paragraph, and the earlier discussion on the statistical moments involved in fatigue calculations, it seems that no substantial improvement can be expected from such an approach as compared to the method introduced in this paper.

As for the case of the MSSL method, it is usually impractical to solve the problem of minimizing the right hand side of Eq (9) directly. The reason for this is that there is a coupling between the optimum 'equivalent' parameters $c_{ij}^{(m_i)}$ and $k_{ij}^{(n_i)}$ and statistical moments of $X(t)$ and $\dot{X}(t)$. This difficulty can be avoided by using the following iterative procedure. Initial estimates of the necessary statistical moments of $X(t)$ and $\dot{X}(t)$ are provided by applying the standard MSSL method, which is also in general iterative. These estimates are then used in the error measure $e_{(m,n)(m_i)}$ which can be minimized to produce numerical estimates of the parameters $c_{ij}^{(m_i)}$ and $k_{ij}^{(n_i)}$. With the resulting $C_{(m)}$ and $K_{(n)}$ substituted into Eq (7), new estimates of the statistical moments of $X(t)$ and $\dot{X}(t)$ can be provided. Thus an iterative loop is established that provides numerical estimates of $C_{(m)}$ and $K_{(n)}$ by minimizing $e_{(m,n)}$ and then uses Eq (7) to produce new estimates of the necessary statistical moments of $X(t)$ and $\dot{X}(t)$. When convergence is obtained, the desired numerical estimates of $C_{(m)}$ and $K_{(n)}$ are established. So far, the convergence properties of this procedure have not been fully tested.

4 Numerical Examples

As has already been discussed, the optimum linearization method to use in a given situation will expectedly depend on the S-N curve exponent m. However, little has been said about concrete criteria for choosing an optimum method. One would tend to argue that such criteria must be developed on the basis of particular case studies. For simplicity, we shall here restrict attention to a class of nonlinear SDOF systems characterized by the following (normalized) equation of motion

$$\ddot{X}(t) + 2\zeta\{\dot{X}(t) + \mu\dot{X}(t)^3\} + X(t) = \sqrt{4\zeta}\,N(t) . \tag{11}$$

Here μ is a positive constant. $N(t)$ is a zero-mean Gaussian white noise satisfying

$$E[N(t)N(t+\tau)] = \delta(\tau) , \tag{12}$$

where $\delta(\cdot)$ denotes Dirac's delta function. Note that the standard deviation of both X and $\dot{X}$ is equal to unity in the linear case ($\mu = 0$).

By applying the ENLE method, it can be shown that, for a certain parameter range, a good approximation to the PDF $f_A(\cdot)$ of the envelope $A = \{X^2 + \dot{X}^2\}^{1/2}$, is given as follows[1]

$$f_A(a) = C a \exp\{-\frac{a^2}{2} - \frac{3\mu a^4}{16}\} . \tag{13}$$

C is a normalization constant ensuring total probability equal to unity. To simplify the analysis, it is expedient in the case at hand to use the PDF of the envelope as a substitute for the PDF of the amplitude in Eq (3).

Application of the linearization method described above to the present case amounts to finding the parameter $\beta_{(n)}$ that minimizes the error measure

$$e_{(n)} = E[\,|\dot{X}_{(n)}(t)^3 - \beta_{(n)}\dot{X}_{(n)}(t)\,|^n] \tag{14}$$

for the appropriate value of n. Here $X_{(n)}(t)$ denotes the linearized and normalized displacement response. Since the optimum value of n is unknown a priori, we shall make calculations for a range of n-values. As previously discussed, the practical calculation of $\beta_{(n)}$ is carried out for the even integers. Let $\sigma^2_{(n)} = Var[X_{(n)}]$ and $\dot{\sigma}^2_{(n)} = Var[\dot{X}_{(n)}]$. Writing $\beta_{(n)} = b_{(n)}\dot{\sigma}^2_{(n)}$, it is found that in the present case, cf. Naess et al.[8], $b_{(2)} = 3.0$, $b_{(4)} = 6.27$ and $b_{(6)} = 9.54$. The corresponding expression for the variance in each case is given by the relation

$$\sigma^2_{(n)} = Var[X_{(n)}] = \frac{(1 + 4b_{(n)}\mu)^{1/2} - 1}{2b_{(n)}\mu}\,, \tag{15}$$

where the numerical identity $\dot{\sigma}_{(n)} = \sigma_{(n)}$ for the present example has been invoked.

Eq (4) now gives us the mean fatigue damage accumulation rate $D_{(n)}$ associated with $X_{(n)}(t)$ for each value of n. The expression is

$$D_{(n)} = k\nu_0^+(2\sqrt{2}\,\sigma_{(n)})^m\,\Gamma(1 + \frac{m}{2}) \tag{16}$$

for a suitable constant k. In this equation ν_0^+ denotes the mean zero-upcrossing rate of $X_{(n)}(t)$, which is the same for every n.

For comparison, the mean fatigue damage rate for the present example will be calculated by invoking Eqs (3) and (13). This leads to the expression

$$D = 2^m k\nu_0^+ C\int_0^\infty a^{m+1}\exp\{-\frac{a^2}{2} - \frac{3\mu a^4}{16}\}da\,. \tag{17}$$

For specified values of the damage exponent m and the parameter μ, an optimum value of n, denoted by n_{opt}, which in general is nonintegral, can be calculated by requiring that $D/D_{(n)} = 1$. This is done as follows. By combining Eqs (15) and (16) with the relation $D/D_{(n)} = 1$, an optimum value of the parameter $b_{(n)}$ can be calculated. Noting that $b_{(n)}$ shows a perfectly linear dependence on n for the values calculated, the corresponding optimum values of n can then be determined simply by linear interpolation. Results for the optimum values of n as a function of m for a range of μ-values are given in Fig 1. It is interesting to observe that n_{opt} shows an almost perfectly linear dependence on m for each value of μ. Also worth noting is the small spread in the values of n_{opt} over the different μ-values for each fixed m.

In Figs 2 - 5 are plotted the relative mean damage rate $D/D_{(n)}$ for suitable values of $n = 2, 4$ and 6 as a function of the parameter μ for four different values of the damage exponent m, viz. $m = 4, 8, 12$ and 16. As was to be expected, the

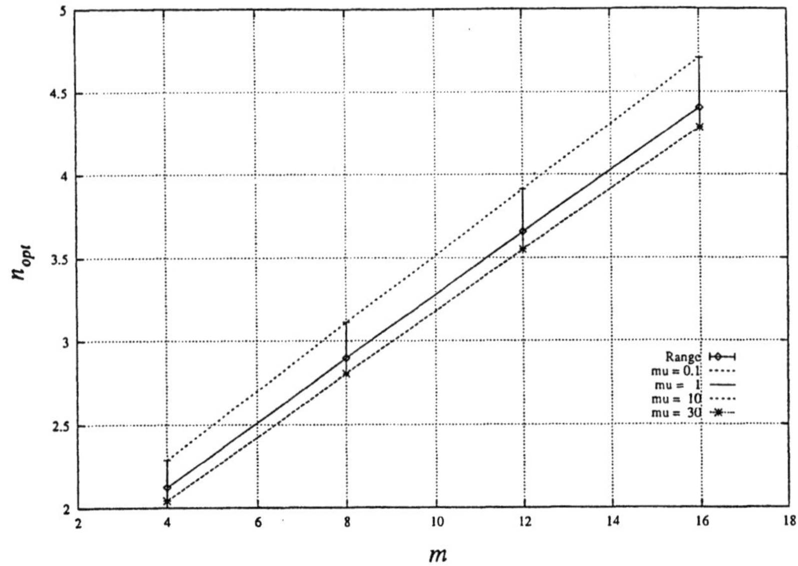

Figure 1 Plot of the optimum value of the linearization exponent n as a function of fatigue damage exponent m for various values of the nonlinearity parameter μ .----- , μ = 0.1; ———— , μ = 1; -----, μ = 10; - - -✱- - , μ = 30.

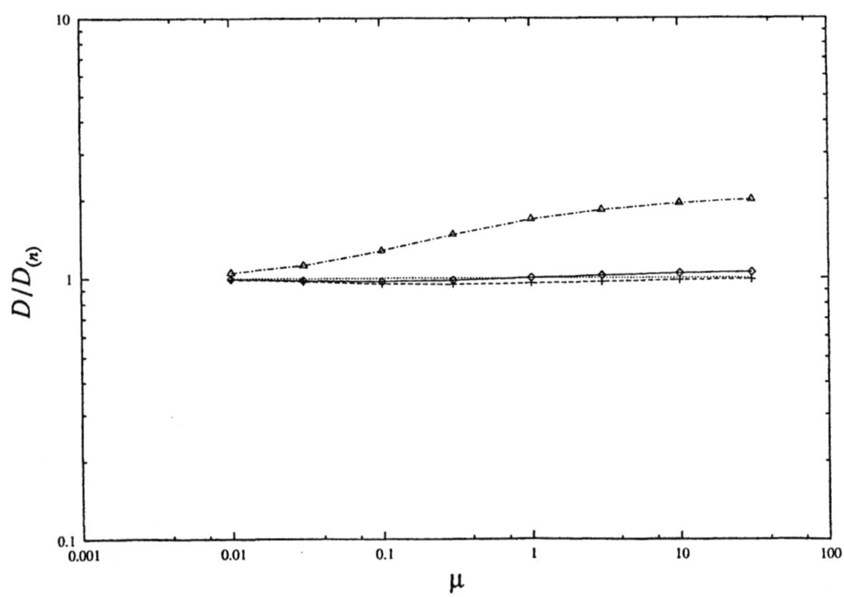

Figure 2 The relative mean damage rate $D/D_{(n)}$ as a function of the nonlinearity parameter μ for damage exponent m = 4. ---+--- , n = 2; -·-△-·- , n = 4; -·-×-·- , n = 6; ---◇--- , n_{opt} = 2.12. ·········· , $D/D_{(n)}$ = 1

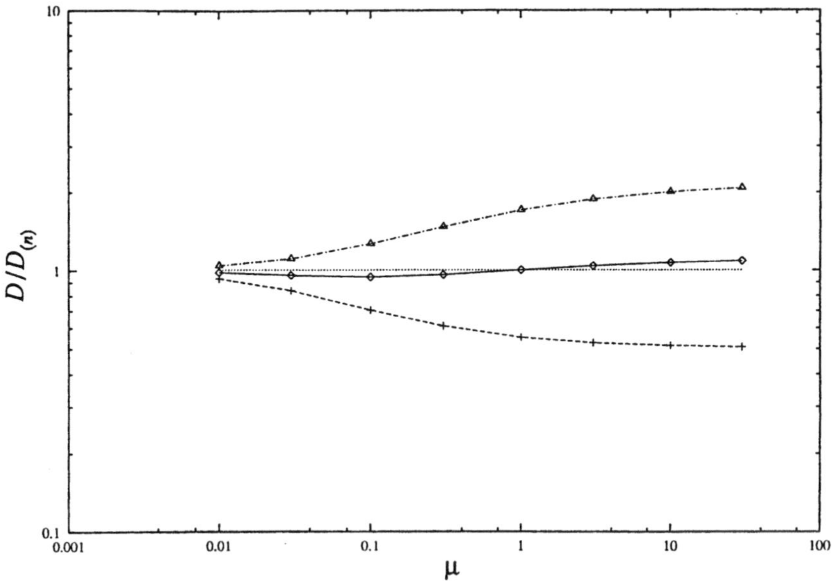

Figure 3 The relative mean damage rate $D/D_{(n)}$ as a function of the nonlinearity parameter μ for damage exponent m = 8. ---+--- , n = 2; -·-Δ-·- , n = 4; -·-×-·- , n = 6; ---◇--- , n_{opt} = 2.90. ·········· , $D/D_{(n)}$ = 1

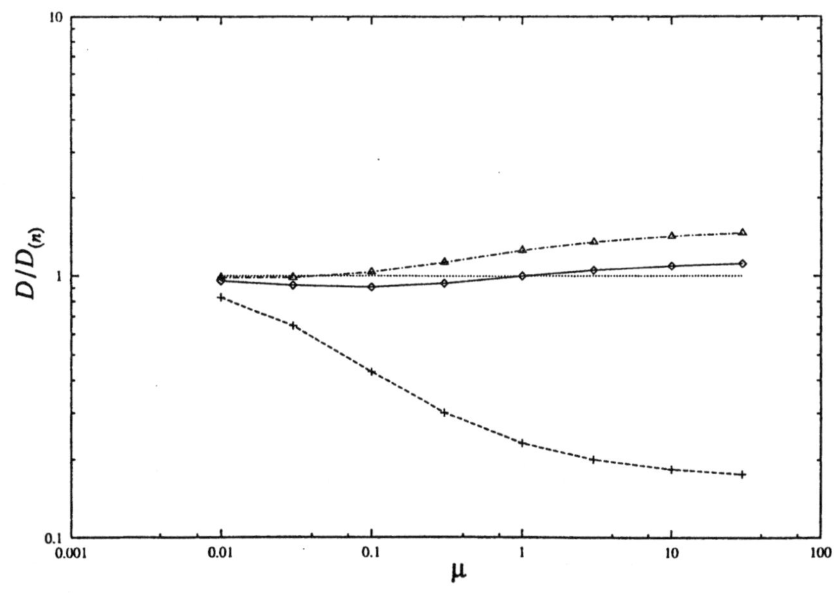

Figure 4 The relative mean damage rate $D/D_{(n)}$ as a function of the nonlinearity parameter μ for damage exponent m = 12. ---+--- , n = 2; -·-Δ-·- , n = 4; -·-×-·- , n = 6; ---◇--- , n_{opt} = 3.66. ·········· , $D/D_{(n)}$ = 1

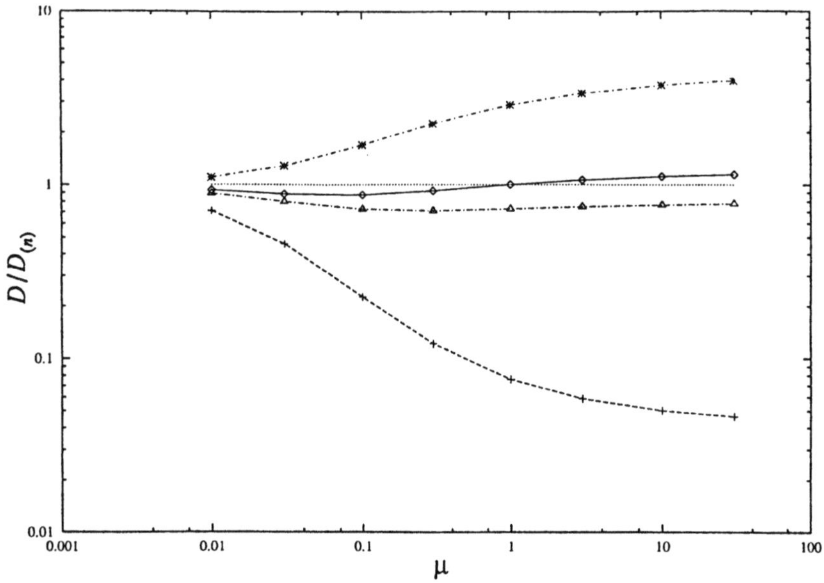

Figure 5 The relative mean damage rate $D/D_{(n)}$ as a function of the nonlinearity parameter μ for damage exponent m = 16. ---+--- , n = 2; -·-▵-·- , n = 4; -·-×-·- , n = 6; ---◇--- , n_{opt} = 4.40. ········· , $D/D_{(n)}$ = 1

figures clearly show the tendency of the MSSL method to produce increasingly conservative estimates of the fatigue damage with increasing values of m. Also plotted in Figs 2 - 5 are the results for $D/D_{(n_{opt})}$ where the value of n_{opt} corresponding to $\mu = 1$ has been chosen. In agreement with the comments made in connection with Fig 1, it is seen that the influence of μ on the ratio $D/D_{(n_{opt})}$ is not very significant. This indicates that for the specific type of nonlinearity studied here, a reasonably accurate practical rule for choosing n_{opt} can be set up, where n_{opt} only depends on the damage parameter m and not at all on the degree of nonlinearity in the model.

5 Conclusions and Comments

In this paper we have presented a study of the application of a class of linearization methods to the problem of calculating fatigue damage accumulation rates in nonlinear models. The proposed methods have been applied to the particular case of a linearly elastic structure whose displacement response is determined by a specific nonlinear equation of motion subjected to broad band Gaussian loading. For this example it has been demonstrated that among the class of proposed linearization methods there always exists one that can provide good agreement with fatigue damage rates calculated by a nonlinear method of analysis.

However, for the methods to be really useful as a practical tool for acceptably accurate fatigue damage calculations, it is necessary to develop ways of determining the optimum linearization method on the basis of a set of characteristics of the given structure and its expected loading in a straight-forward, routine manner. Whether such a goal is attainable cannot, of course, be decided within the scope of this limited study. The main purpose here has been to show that fatigue specific linearization methods may in fact be developed that can cope with the problem of fatigue damage calculations in nonlinear systems, leaving the question about its potential practical value as a general tool for accurate estimation largely unanswered.

Nevertheless, it seems that with a modest degree of development, the methods can be put to good use for a straight-forward assessment of the effect of nonlinearities on the fatigue life. For example, if such application discloses a dependence of fatigue life on the specific linearization method that is deemed too strong, other methods are available for a more elaborate and detailed study of the fatigue life.

Acknowledgements. This work was undertaken while the first author was staying at Stanford University on sabbatical leave. The second author was supported by ONR, Grant No. N00014-87-K-0475, and the Reliability of Marine Structures Program of Stanford University. The authors also gratefully acknowledge the assistance of A. Jha in carrying out the calculations in this paper.

References

1. Roberts, J.B. and Spanos, P.D., *Random Vibration and Statistical Linearization.* John Wiley & Sons, New York, 1990.
2. Melchers, R.E. and Ahammed, M., Equivalent Linearization and Fatigue Reliability Estimation for Offshore Structures, *Proc. 2nd Int. Offshore and Polar Engineering Conference* (ISOPE-92), San Francisco, California, 1992.
3. Naess, A. and Winterstein, S.R., Statistical Linearization as a Tool for Fatigue Damage Estimation, *Proc. 12th International Conference on Offshore Mechanics and Arctic Engineering* (OMAE-93), Glasgow, Scotland, 1993.
4. Abramowitz, M. and Stegun, I., *Handbook of Mathematical Functions*, Dover Publications, New York, 1965.
5. Wirsching, P.A. and Light, M.C., Fatigue under Wide Band Random Stresses, *Journal Struct. Div.*, ASCE, Vol. **106** (1980), pp. 1593-1607.
6. Jiao, G. and Moan, T., Probabilistic Analysis of Fatigue due to Combined Gaussian Load Processes. *Probabilistic Engineering Mechanics*, Vol. **5** (1990), No. 2, pp. 76-83.
7. Naess, A., Statistical Linearization Methods for Prediction of Extreme Response of Nonlinear Structures. *Proc. 6th Int. Conference on Structural Safety and Reliability* (ICOSSAR '93), Innsbruck, Austria, 1993.
8. Naess, A., Galeazzi, F. and Dogliani, M., Extreme response predictions of non-linear compliant offshore structures by stochastic linearization. *Applied Ocean Research*, Vol. **14** (1992), No. 2, pp. 71-81.

Stochastic Materials Design of Fibrous Composite Laminates

Hidetoshi Nakayasu[1]

[1]Department of Industrial Management, Osaka Institute of Technology,
5-16-1 Ohmiya Asahi-ku Osaka 535, Japan
E-mail: nakayasu@nak101.dim.oit.ac.jp

Abstract. One of the major objectives of this paper is to offer a practical tool for materials design of unidirectional composite laminates under in-plane multi-axial load. Design-oriented failure criteria of composite materials are applied to construct the evaluation model of probabilistic safety based on the extended structural reliability theory. Typical failure criteria such as maximum stress, maximum strain and quadratic polynomial failure criteria are compared from the viewpoint of reliability-oriented materials design of composite materials. The new design diagram which shows the feasible region on in-plane strain space and corresponds to safety index or failure probability is also proposed. These stochastic failure envelope diagrams which are drawn in in-plane strain space enable one to evaluate the stochastic behavior of composite laminate with any lamination angle under multi-axial stress or strain condition. Numerical analysis for graphite/epoxy laminate of T300/5208 is shown for the comparative verification of failure criteria under the various combinationsof multi-axial load conditionsand lamination angles. The stochastic failure envelopes of T300/5208 are also described in in-plane strain space.

Keywords. Stochastic materials design, composite materials, reliability, failure probability, safety index, first/second order reliability method, stochastic failure envelopes

1 Mechanical Behavior of Composites Laminates

1.1 In-plane Stress-Strain Relationship

When the coordinate system of a unidirectional composite laminate is defined as in Fig. 1.1, and it is assumed that the laminate is in a plane stress condition of linear elasticity without non-linearity, plasticity, or viscosity, the stress-strain relationships are expressed by

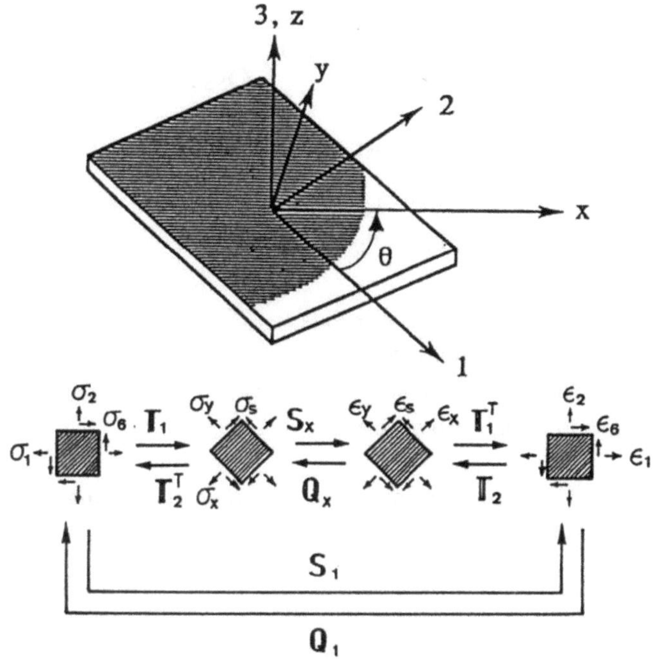

Fig. 1.1. Coordinates system and relation between stress and strain components of unidirectional composites laminates.

$$\sigma_x = Q_x \varepsilon_x,\ \sigma_1 = Q_1 \varepsilon_1,$$
$$\varepsilon_x = S_x \sigma_x,\ \varepsilon_1 = S_1 \sigma_1 \quad (1)$$

where

$$\sigma_x = \{\sigma_x\ \sigma_y\ \sigma_s\}^T, \quad \sigma_1 = \{\sigma_1\ \sigma_2\ \sigma_6\}^T$$
$$\varepsilon_x = \{\varepsilon_x\ \varepsilon_y\ \varepsilon_s\}^T, \quad \varepsilon_1 = \{\varepsilon_1\ \varepsilon_2\ \varepsilon_6\}^T \quad (2)$$

There are two kinds of coordinate systems in Fig.1.1, i.e., x-y-s coordinate system and 1-2-6 coordinate system. The former means the coordinate system based on fibrous direction, where x-axis corresponds to longitudinal direction of fiber, y to transverse direction and s to shear direction on x-y. On the other hand, the latter is the main coordinate system, where the 1 or 2 axis corresponds to normal loads direction and 6 to shear load direction on the laminate. The angle between the 1-axis and x-axis is called the fiber lamination angle and is a controllable angle. In this paper, the subscript x means x-y-s coordinate system and 1 means 1-2-6 system such as the subscript of the stress or strain vector in Eq.(2).

In Eq.(1), Q_x is in-plane stiffness matrix on x-y-s coordinate system and S_x is compliance matrix which is inverse matrix of the following matrix

$$\mathbf{Q}_x=\begin{bmatrix} Qxx & Qxy & 0 \\ & Qyy & 0 \\ sym. & & Qss \end{bmatrix},$$

$$Qxx=\eta Ex, \quad Qyy=\eta Ey,$$
$$Qxy=\eta \nu_x Ey, \quad Qss=Es,$$
$$\eta=(1-\nu_x^2 \, Ey/Ex)^{-1} \tag{3}$$

where

Ex :elastic modulus on x axis,
Ey :elastic modulus on y axis,
Exy :shear modulus on x-y axis, and
ν_x :Poisson ratio on x axis.

When θ is the lamination angle of a composite laminate, the coordinate transformation matrices

$$\mathbf{T}_1=\begin{bmatrix} m^2 & n^2 & 2mn \\ n^2 & m^2 & -2mn \\ -mn & mn & m^2-n^2 \end{bmatrix},$$

$$\mathbf{T}_2=\begin{bmatrix} m^2 & n^2 & mn \\ n^2 & m^2 & mn \\ -2mn & 2mn & m^2-n^2 \end{bmatrix} \tag{4}$$

enable one to transfer stress/strain components from the 1-2-6 system to the x-y-s system as

$$\sigma_x = \mathbf{T}_1\sigma_1, \quad \varepsilon_x=\mathbf{T}_2\varepsilon_1 \tag{5}$$

where m=cosθ, n=sinθ. On the other hand, the transformation of stiffness/compliance matrices can be easily shown using the orthogonal relationship $\mathbf{T}_1^{-1}=\mathbf{T}_2^T$, $\mathbf{T}_2^{-1}=\mathbf{T}_1^T$ such as

$$\mathbf{Q}_1=\mathbf{T}_2^T\mathbf{Q}_x\mathbf{T}_2, \ \mathbf{S}_1=\mathbf{T}_1^T\mathbf{S}_x\mathbf{T}_1 \ . \tag{6}$$

1.2 Failure Criteria of Unidirectional Composite Laminate

It is important to undertake further probabilistic studies in failure criteria, because of the criticality of failure criteria in material design and the necessity for better confidence in the predicted strength of a unidirectional composite laminate.

Though there are many kinds of research work concerned with the failure laws of a unidirectional composite laminate, they are divided broadly into two categories of treatment such as micromechanical research and macromechanical models for design. The former consists of

1)weakest link model[1,2],

2)fracture mechanics[3],

3)stochastic process and time-dependent failure[4],

4)interlaminar shear[5], and

5)residual stress effect[6].

The latter is the hypothesis law for design based on the fracture phenomena and it is expressed by a mathematical equation.

One of main objectives of the former is not design-oriented problems but to clarify the mechanism of the failure based on the assumption of idealized load condition and situation such as a longitudinal tensile load, zero lamination angle, no shearing state, and so on. In this paper, the latter treatment is selected for reliability formulation. Therefore, a comparative study of failure criteria is performed in probabilistic fields on the basis of six kinds of failure laws for a unidirectional composite laminate, 1)maximum stress law, 2)maximum strain law, and the quadratic polynomial failure criteria by 3)Tsai-Wu[7], 4)Chamis[8], 5)Hoffman[9], and 6)Hill[10] which belong to the latter category.

2 Basic Concept for Reliability Analysis

If the failure of a unidirectional composite laminate occurs by serial state or mode, the probability of failure of laminate is

$$Pf,s=P(Fs)=P(\cup Fi)=P(\cup\{gi(X)\leq 0\}). \tag{16}$$

On the other hand,if failure of laminate occurs by parallel state or mode,

$$Pf,p=P(Fp)=P(\cap Fi)=P(\cap\{gi(X)\leq 0\}) \tag{17}$$

where

Fs :failure event with serial state or mode
Fp :failure event with parallel state or mode
Fi :failure event of i-th element of the system
P(.) :occurrence probability of the event (.)
$\cup$:union of the set
$\cap$:product of the set
$g_i(\mathbf{X})$:limit state function of i-th element

and **X** is the random vector with uncertainty of design variables and environment parameters.

$$\mathbf{X} = \{ X_1, X_2, \ldots, X_n \}^T . \tag{18}$$

From the definitions of limit state function of i-th element, the state of i-th element is divided into three categories:

$$V = \{ g_i(\mathbf{X}) \leq 0 \}, \ S = \{ g_i(\mathbf{X}) > 0 \}, \ G = \{ g_i(\mathbf{X}) = 0 \} \tag{19}$$

V, S, and G expresses the failure domain,safety domain, and failure surface, respectively.

In general, since the composite laminate consists of several plies with different kinds of lamination angles, the probability of failure must be calculated by Eq.(17), it is very difficult to evaluate the probability of failure of a composite laminate with multiple plies of various orientations. In addition, because of the many kinds of failure modes generated by tensile, compressive, and shear loads, even the probability of failure of i-th ply must be calculated by Eq.(16). Therefore, the evaluation of the failure probability must be performed by using the combination of Eqs.(16) and (17), and it is a most complicated process to obtain the probability of failure analytically.

Fortunately, it is possible to identify the laminate as a one layer composite in the case of unidirectional laminate, since the lamination angles of each layer are the same. Furthermore, if the mechanism of failure of the ply is followed by only one mode or state such as quadratic failure law, and the continuity of the random vector is satisfied, the probability of failure is expressed by the simple form

$$P_f = P(g(\mathbf{X}) \leq 0) = \int_{D_g} f_{\mathbf{X}}(\mathbf{x}) d\mathbf{x} \tag{20}$$

where

$$f_{\mathbf{X}}(\mathbf{x}) = f_{X_1, X_2, \cdots X_n}(x_1, x_2, \cdots x_n),$$
$$d\mathbf{x} = dx_1 dx_2 dx_3 \cdots dx_n ,$$

Dg shows failure region corresponding to failure domain V by Eq.(19), and ∂D_g is called a failure surface which corresponds to G in Eq.(19).

Since the random vector **X** in Eq.(20) is composed of more than 13 random variables from the strength, load and modulus on each axis, lamination angle, and so on, even in this simple case of Eq.(20), it is difficult to solve Eq.(20) directly by a numerical method because of the accuracy of solution and computational efficiency.

On the other hand, if the mechanism of failure of the ply is effected by a multiple modal state such as a maximum stress or strain law, it is more complicated to obtain the probability of failure of the laminate since it must be calculated by Eq.(16).

3 Approximated Reliability Analysis and Safety Index

When the probability of failure must be evaluated by a high order dimensional integral such as Eq.(20), the first/second order reliability methods are powerful tools to obtain the probability integral. In this method, the multiple dimensional probability integral can be easily obtained using the standardizing transformation[11] of a random vector

$$X \leftrightarrow U: \quad X = T(U) \text{ or } U = T^{-1}(X) \,. \tag{21}$$

After the transformation from original X-space to standardized U-space, each safety index β_i , which is equivalent to the probability of failure through the standard normal distribution function F(.), is evaluated respectively according to the approximated failure surface on U-space.

3.1 Safety Index by First Order Reliability Method(FORM)[12]

If the failure surface at design point $\mathbf{u}^*$ in U-space is approximated by the linear equation

$$\partial D_{gu} \approx \{\mathbf{u}:\ \alpha^T(\mathbf{u}-\mathbf{u}^*) = 0\} = \{\ \mathbf{u}: l(\mathbf{u}) = \alpha^T\mathbf{u} + \beta = 0\ \} \tag{22}$$

from a Taylor series expansion until first order, the safety index and the probability of failure are derived from the equations

$$\beta_1 = |\mathbf{u}^*| = \min\{|\mathbf{u}^*|\} = -\alpha^T\mathbf{u}^* \quad \text{for } \{\mathbf{u}: g_u(\mathbf{U}) \le 0\} \tag{23}$$

where

$$\alpha = \mathrm{grad}\{g_u(\mathbf{u}^*)\} / |\mathrm{grad}\{g_u(\mathbf{u}^*)\}| \tag{24}$$

and

$$P_{f_1} \cong \Phi(-\beta_1) \tag{25}$$

where

∂D_{gu} :failure surface transformed into U-space
$\mathbf{u}^*$:design point or beta point on U-space
α :gradient of limit state function on transformed U-space
$\Phi(.)$:cumulative distribution function of standardized normal distribution
$|.|$:norm of vector.

3.2 Safety Index by Second Order Reliability Method(SORM)[13]

In this method, the safety index is calibrated with the curvature of the failure surface on the design point in U-space which is approximated by the Taylor series expansion until second order

$$\partial D_{q\mu} \approx \{ \mathbf{u}: q(\mathbf{u}) = \frac{1}{2}(\mathbf{u}-\mathbf{u}^*)\mathbf{H}(\mathbf{u}-\mathbf{u}^*) + \alpha^T(\mathbf{u}-\mathbf{u}^*) = 0 \} \tag{26}$$

Thus the relation[14]

$$P_{f_2} \approx \Phi(-\beta_1) \prod_{j=1}^{n-1} (1-\psi(-\beta_1)\kappa_j)^{-\frac{1}{2}} \tag{27}$$

$$\psi(.) = \phi(.)/\Phi(.) \tag{28}$$

is derived, where

κ_j :curvature of failure surface on design point.

$\mathbf{H}$:Hessian matrix

$\phi(.)$:probability density function of standardized normal distribution.

From Eq.(26) to Eq.(28),

$$\beta_2 = -\Phi^{-1}(P_{f_2}) \tag{29}$$

which is more accurate, when the limit state function is not linear about design point[15].

3.3 Safety Index by Importance Sampling Method (ISM)[16]

When the nonlinearity of the limit state function is strong, the safety index by FORM/SORM is not an appropriate tool for estimating the probability of failure of a composite laminate. For the verification of the safety index by FORM/SORM with a strongly nonlinear failure surface, the method of Monte Carlo Simulation is useful. In this method, the simulation is performed on the transformed space from U-space to V-space where

$$\mathbf{u}^* \rightarrow \mathbf{v}^* = \{0, \cdots, -\beta\}^T \tag{30}$$

and safety index is obtained as

$$\beta_3 = -\Phi^{-1}(P_{f_3}) \tag{31}$$

with

$$P_{f_3} = P_{f_2}\{\frac{1}{N}\sum_{i=1}^{n} \Phi(f(v_i))\} \;/\; \{\Phi(-\beta)\exp(-\frac{1}{2}\psi(-\beta)\sum_{j=1}^{n-1}\kappa_j V_{ij})\} \;. \tag{32}$$

4. Evaluation of the Limit State Function and Safety Index of a Unidirectional Composite Laminate

4.1 Evaluation by Maximum Stress or Strain Law

When the failure of a laminate is evaluated by a maximum stress or strain law, the mechanism of failure depends upon the multiple modal failure such as tensile/compressive failure in the fibrous direction, that in the transverse direction, and shear failure. The failure of a unidirectional laminate will occur, even if only one failure mode occurs. Therefore, the probability of failure may be evaluated by Eq.(16). In this case, the limit state function $g_{(.)}(\mathbf{X})$ in Eq.(16) is as follows.

4.1.1 Maximum Stress Law

$$g_x(\mathbf{X}) = \sigma_{L,t} - \sigma_x \quad (\sigma_x \geq 0), \text{ or } \sigma_{L,c} + \sigma_x \quad (\sigma_x < 0)$$
$$g_y(\mathbf{X}) = \sigma_{T,t} - \sigma_y \quad (\sigma_y \geq 0), \text{ or } \sigma_{T,c} + \sigma_y \quad (\sigma_y < 0)$$
$$g_s(\mathbf{X}) = \sigma_{LT} - |\sigma_s| . \tag{33}$$

4.1.2 Maximum Strain Law

$$g_x(\mathbf{X}) = \varepsilon_{L,t} - \varepsilon_x \quad (\varepsilon_x \geq 0), \text{ or } \varepsilon_{L,c} + \varepsilon_x \quad (\varepsilon_x < 0)$$
$$g_y(\mathbf{X}) = \varepsilon_{T,t} - \varepsilon_y \quad (\varepsilon_y \geq 0), \text{ or } \varepsilon_{T,c} + \varepsilon_y \quad (\varepsilon_y < 0)$$
$$g_s(\mathbf{X}) = \varepsilon_{LT} - |\varepsilon_s| . \tag{34}$$

From Eq.(16), the inequality

$$\max\{ P_{f,i} \} \leq P_f \leq \sum P_{f,i} \quad (i=x,y,s) \tag{35}$$

must be satisfied. When the relationship between the probability of failure and safety index

$$\Phi^{-1}(P_f) = -\beta, \qquad \Phi^{-1}(1 - P_f) = \beta \tag{36}$$

can be introduced into Eq.(36), the Eq.(35) can be rewritten as

$$-\Phi^{-1}\left(\sum P_{f,i}\right) \leq \beta \leq -\Phi^{-1}(\max\{ P_{f,i} \}) \quad (i=x,y,s) . \tag{37}$$

As a unidirectional composite laminate has strong orthotropic properties, the following relation can be easily recognized intuitively

$$P_{f(1)} << P_{f(2)} << P_{f(3)} \tag{38}$$

where $P_{f(.)}$ is the probability of failure in smallest order.

From the relation of Eq.(38)

$$\sum P_{f,i} \approx \max\{ P_{f,i} \} \quad (i=x,y,s) \tag{39}$$

yields the approximated estimation of safety index

$$\beta \approx \min\{ \beta_i \} \quad (i=x,y,s) \tag{40}$$

where

$$\beta_i = \Phi^{-1}(1- P_{f,i}) \quad (i=x,y,s) \tag{41}$$

is easily obtained by the approximated reliability analysis discussed before in Eqs.(23), (29), or (31).

4.2 Evaluation by Quadratic Polynomial Failure Criteria

Based on the expressions of Eq.(9) to (13) for the quadratic polynomial failure law, the approximated reliability method will be useful for the evaluation of the safety index since only one mode or state should be taken into account to evaluate the failure of a unidirectional composite laminate. In this case, the limit state functions for the identification of Eqs.(23), (29), (31) are formulated as follows:

4.2.1 On Stress Space

Hill's law, and Chamis' law:

$$g(\mathbf{X}) = 1 - \sigma_x^T \mathbf{F}_{A,x} \sigma_x \,. \tag{42}$$

Hoffman's law, and Tsai-Wu's law:

$$g(\mathbf{X}) = 1 - \sigma_x^T \mathbf{F}_{A,x} \sigma_x - \mathbf{F}_{B,x}^T \,. \tag{43}$$

4.2.2 On Strain Space

Hill's law, and Chamis' law:

$$g(\mathbf{X}) = 1 - \varepsilon_x^T \mathbf{G}_{A,x} \varepsilon_x \,. \tag{44}$$

Hoffman's law, and Tsai-Wu's law:

$$g(\mathbf{X}) = 1 - \varepsilon_x^T \mathbf{G}_{A,x} \varepsilon_x - \mathbf{G}_{B,x}^T \varepsilon_x \,. \tag{45}$$

5 Stochastic Failure Envelope

In a deterministic field, the threshold of the limit state equation from Eq.(9) to Eq.(13) creates failure envelopes which show safety regions of stress or strain.

This DFE(deterministic failure envelope) is defined by the equation

$$\mathrm{DFE} = \{ \varepsilon_1 : \varepsilon_1^T \mathbf{G}_{A,1}\varepsilon_1 + \mathbf{G}_{B,1}^T \varepsilon_1 = 1 \}. \tag{46}$$

On the other hand, in the probabilistic field the evaluation of the safety index of the in-plane stress and strain condition will suggest the envelope which corresponds to the safety index or the probability of failure. This envelope corresponding to the safety index will be a stochastic failure envelope (SFE) and formulated by the equation

$$\mathrm{SFE} = \{ \varepsilon_1 : \mathbf{P}(g(\mathbf{X}) \leq 0) = \Phi(-\beta) \approx P_f \} \tag{47}$$

where ε_1 is in-plane strain of unidirectional composite laminate.

Fig. 5.1 shows the aspectsof the relationship between a DFE and SFE. As the shape of the DFE is an ellipsoid in the x-y-s coordinate system, the SFE will be estimated as the ellipsoid whose contents are concerned with the safety index or the probability of failure.

Since it is impossible to evaluate the locus of the SFE of Eqs.(47) by an analytical method, the method for evaluating the SFE depends on a numerical search technique. The technique of searching the point $\varepsilon_1(\beta)=\{ \varepsilon_1 \; \varepsilon_2 \; \varepsilon_6 \}^T$,

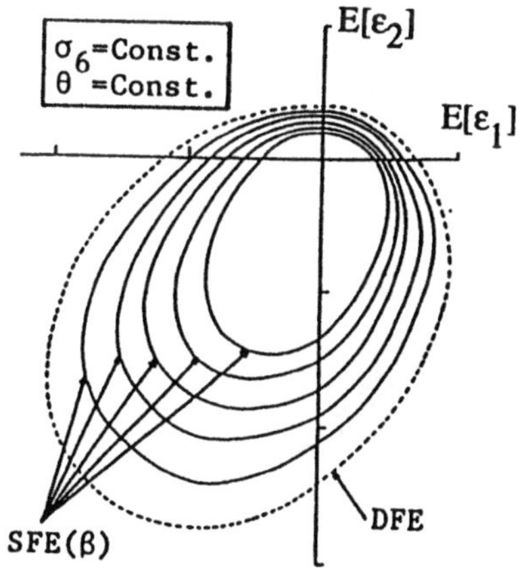

Fig. 5.1. Aspect of stochastic failure envelopes.

which is sequentially in conjunction with the value of safety index, is due to the numerical searching method in the ε_1-ε_2 plane with fixed ε_6. This computational procedure is so wasteful in time improvement of the searching technique has been required.

6 Numerical Analysis

Consider the load condition of a unidirectional composite laminate as shown in Fig.6.1 with

$$\sigma_1 = P_1^* / (B \cdot H), \quad \sigma_2 = P_2^* / (L \cdot H), \quad \sigma_6 = P_6^* / (B \cdot H) \tag{48}$$

where

$$P_1^* = P_1/SF_1, \quad P_2^* = P_2/SF_2, \quad P_6^* = P_6/SF_6 \tag{49}$$

and SF_1, SF_2, and SF_6 are safety factors in materials design. In the formulation of the above equations, the applied strains on the laminate are derived from the nominal loads by Eq.(49) and dimensional parameters through the safety factor. As the actual material design has many parameters with uncertainty such as load, strength, modulus, dimensions, and so on, all of these parameters must be evaluated in the probabilistic fields. Though, in the deterministic analysis, the results of the analysis with the specified strain conditions directly applied to a laminate coincide with those with the strain conditions derived from the many parameters such as loads, modulus, dimensions of the laminate and so on, they are not equivalent, exactly in the treatment of probabilistic field because of the dimensional unbias . That is the reason why the numerical analysis does not start from the specified stress which is directly applied to a laminate but from the many parameters through safety factors.

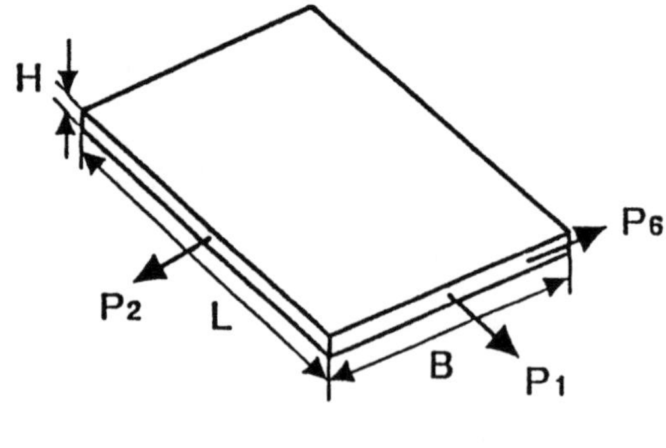

Fig. 6.1. Load conditions and dimensional parameters.

Table 6.1. Stochastic features of design variables and materials parameters as random variables.

Parameter	Expectation	Standard deviation	CV
$\sigma_{L,t}$	1500 (MPa)	150	0.10
$\sigma_{L,c}$	1500 (MPa)	180	0.12
$\sigma_{T,t}$	40 (MPa)	4.40	0.11
$\sigma_{T,c}$	246 (MPa)	19.68	0.08
σ_{LT}	68 (MPa)	4.08	0.06
k_{12}	Value calculated by Table 1	-	0.01
Ex	181000 (MPa)	9050	0.05
Ey	10300 (MPa)	515	0.05
Es	7170 (MPa)	358.5	0.05
νx	0.28 (MPa)	0.0028	0.01
θ	0 (deg)	0.1	-
B	30 (mm)	1.5	0.05
L	150 (mm)	7.5	0.05
H	2 (mm)	0.1	0.05
P_1	90000 (MPa)	4500	0.05
P_2	120000 (MPa)	600	0.05
P_6	4080 (MPa)	204	0.05

Table 6.1 shows the values of stochastic features of parameters and design variables for T300/5208, whose distribution types are respectively

$$\sigma_{L,t}\ ,\ \sigma_{L,c}\ ,\ \sigma_{T,t}\ ,\ \sigma_{T,c}\ ,\ \sigma_{LT} \rightarrow \text{ Log-normal distribution}$$

$$k_{12}\ ,\ E_x\ ,\ E_y\ ,\ E_s\ ,\ \nu_x\ ,\ B,\ L,\ H \rightarrow \text{ Normal distribution}$$

$$P_1\ ,\ P_2,\ P_6 \rightarrow \text{ Weibull distribution (2 parameters).}$$

Table 6.2 shows the calculation results for the off-axis loading condition and lamination angle $E[\theta]=0°$ for the various failure criteria. Fig.6.2 describes the behavior of the β_1 curve corresponding to various load conditions and lamination angles. Finally, SFE diagrams for T300/5208 based on the Tsai-Wu criterion and the proposed procedure are shown in Fig.6.3 in in-plane strain space.

7 Discussion

Many considerations can be derived from the calculation results of the safety index or the probability of failure in Table 6.2:

Table 6.2. Three kinds of safety indices and failure probabilities obtained by proposed method for various failure criteria.

(a)Case of unidirectional tensile load condition(σ_1=1007.18 MPa)

Criteria	β_1	β_2	β_3	Pf_1	Pf_2	Pf_3
TW(S)*1	3.0285	3.0404	3.0448	0.00123	0.00118	0.00116
HO(S)	3.0285	3.0403	3.0447	0.00123	0.00118	0.00116
HI(S)	3.0285	3.0410	3.0455	0.00123	0.00118	0.00116
CH(S)	3.0285	3.0415	3.0461	0.00123	0.00118	0.00116
MS	3.0285	3.0446	3.0496	0.00123	0.00117	0.00115
TW(N)*2	3.0285	3.0404	3.0448	0.00123	0.00118	0.00116
HO(N)	3.0285	3.0403	3.0447	0.00123	0.00118	0.00116
HI(N)	3.0285	3.0410	3.0455	0.00123	0.00118	0.00116
CH(N)	3.0285	3.0415	3.0461	0.00123	0.00118	0.00116
MN	3.0285	3.0446	3.0496	0.00123	0.00117	0.00115

*1 ·(S) designates the abbreviation of each criterion in stress space.
*2 ·(N) designates the abbreviation of each criterion in strain space.

(b)Case of multiaxial load condition (σ_1=604.31, σ_2=16.12, σ_6=27.40 MPa)

Criteria	β_1	β_2	β_3	Pf_1	Pf_2	Pf_3
TW(S)*1	3.0692	3.0473	3.0367	1.07×10^{-3}	1.15×10^{-3}	1.20×10^{-3}
HO(S)	2.4690	2.4555	2.4456	6.77×10^{-3}	7.03×10^{-3}	7.23×10^{-3}
HI(S)	4.0188	3.9388	3.9330	2.93×10^{-5}	4.10×10^{-5}	4.20×10^{-5}
CH(S)	6.8868	6.8049	6.8214	2.87×10^{-12}	5.08×10^{-12}	4.54×10^{-12}
MS	6.5339	6.5363	6.5239	3.22×10^{-11}	3.17×10^{-11}	3.44×10^{-11}
TW(N)*2	3.0692	3.0473	3.0367	1.07×10^{-3}	1.15×10^{-3}	1.20×10^{-3}
HO(N)	2.4690	2.4555	2.4456	6.77×10^{-3}	7.03×10^{-3}	7.23×10^{-3}
HI(N)	4.0188	3.9388	3.9330	2.93×10^{-5}	4.10×10^{-5}	4.20×10^{-5}
CH(N)	6.8868	6.8049	6.8214	2.87×10^{-12}	5.08×10^{-12}	4.54×10^{-12}
MN	6.9650	6.9660	6.9653	1.65×10^{-12}	1.64×10^{-12}	1.65×10^{-12}

*1 ·(S) designates the abbreviation of each criterion in stress space.
*2 ·(N) designates the abbreviation of each criterion in strain space.

1) From the comparative studies between β_1, β_2, and β_3, there are few significant differences among β_1, β_2 and β_3. These results suggest that the safety index β_1 from FORM is a sufficient measure to estimate the probability of failure of a composite laminate, since the limit state function of a composite laminate has stable behavior for searching for a beta point.

2) The results of the analysis in stress space are almost equivalent to those on strain space though each case has a different number of basic random variables. The reasons why both the results of the analysis in stress and strain space coincide are due to the small sensitivity of the 4 coefficients of modulus to safety indices which are additional random variables in strain space. The details of the reasons are discussed in the reference[17].

These results are interesting and important for materials design, since the reliability formulation in this paper is a fundamental study for the reliability analysis on multiple layer composite laminates with various lamination angles. In the advance, the evaluation of the failure probability must be performed using the combination of Eqs.(16) and (17) on the basis of system reliability theory, and it is a most complicated process to obtain the safety index. For these purposes, the two kinds of considerations discussed above will be recommended for constructing a mathematical model of reliability on multiple layer composite laminates for materials design, such as :

1) For the advanced analysis of reliability for a multiple layer composite laminate, the safety index β_1 determined by the formulation in this paper based on the fist order reliability method yields a sufficient measure for evaluating the probability of failure.

2) Since both the results of the analysis in stress and strain space coincide, though the number of random variables on strain space is greater than that in stress space, it is recommended that the reliability analysis must be performed in strain space because the limit state function has more stable behavior for searching for the beta point in strain space, and the investigation is possible with different lamination angles on strain space.

Secondly, comparisons of failure criteria can also be made on the basis of Fig.6.2 from the viewpoint of the behavior of the curves which show the relation between safety index and the value of the expectation of the lamination angle. There are six kinds of load combinations drawn in Fig.6.2 for the comparative study of the efficiency of the failure criteria. There are considerable differences under shear load and biaxis load conditions (Figs.6.2 (c)-(f)). It is seen from Figs.6.2(c) and (f) that the Hill's criterion is not sufficient under pure shear and multiple axis load conditions with compressive loads, since the effect of compressive strength has not been covered in the expression for the failure parameters.

As a total rough conclusion from the various considerations, the Tsai-Wu criterion is superior among the many criteria along with a design-oriented viewpoint in the probabilistic fields. In addition to the above, the Chamis, Hoffman and Hill criteria can be deduced from the Tsai-Wu criteria. The Tsai-Wu criteria reduces to the Hoffman criteria if

$$k_{12}(=F_{xy}^{*})=-\frac{1}{2}\sqrt{(\sigma_{T,t}\cdot\sigma_{T,c})/(\sigma_{L,t}\cdot\sigma_{L,c})}$$

(for T300/5280, k_{12}=-0.33), reduces to the Hill's criteria if

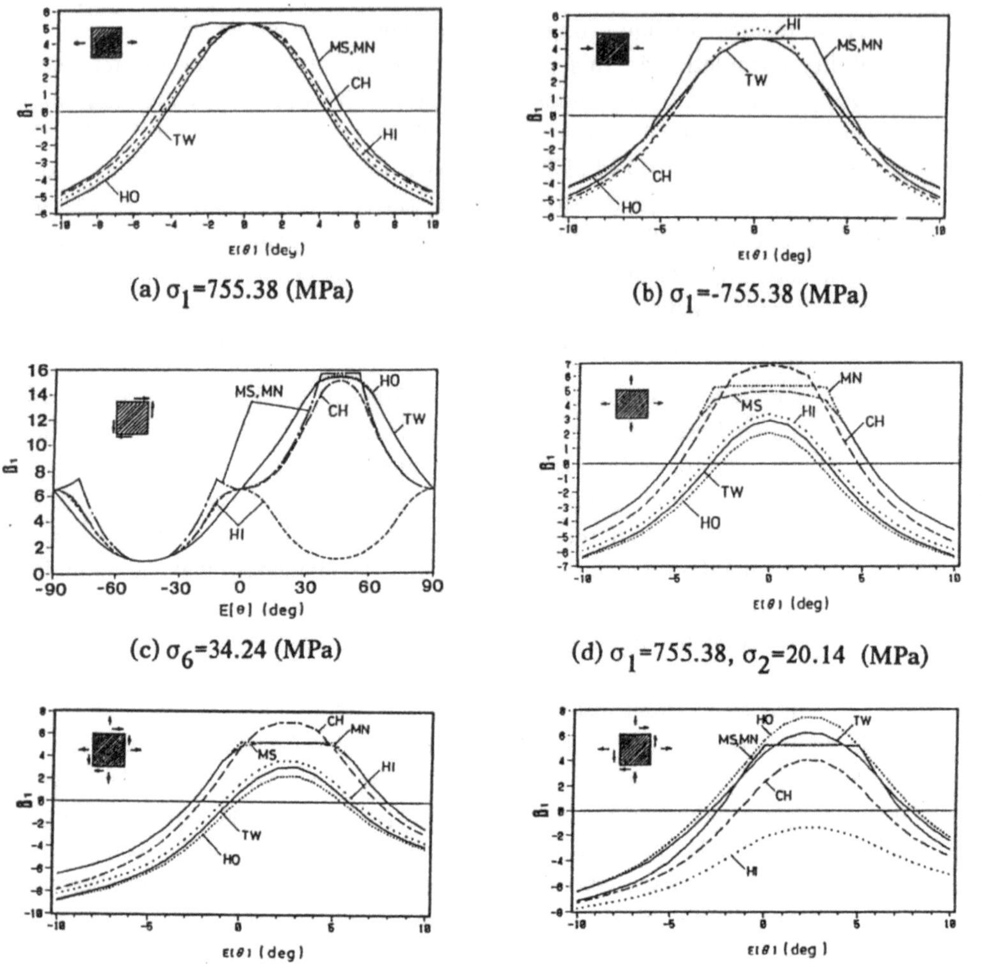

(a) σ_1=755.38 (MPa)

(b) σ_1=-755.38 (MPa)

(c) σ_6=34.24 (MPa)

(d) σ_1=755.38, σ_2=20.14 (MPa)

(e) σ_1=755.38, σ_2=20.14 , σ_6=34.24 (MPa)

(f) σ_1=755.38, σ_2=-40.29 , σ_6=34.24 (M

Fig.6.2. Curves which show the relation between safety indices and fiber orientation angle at different load conditions for the comparative study of various failure criteria.

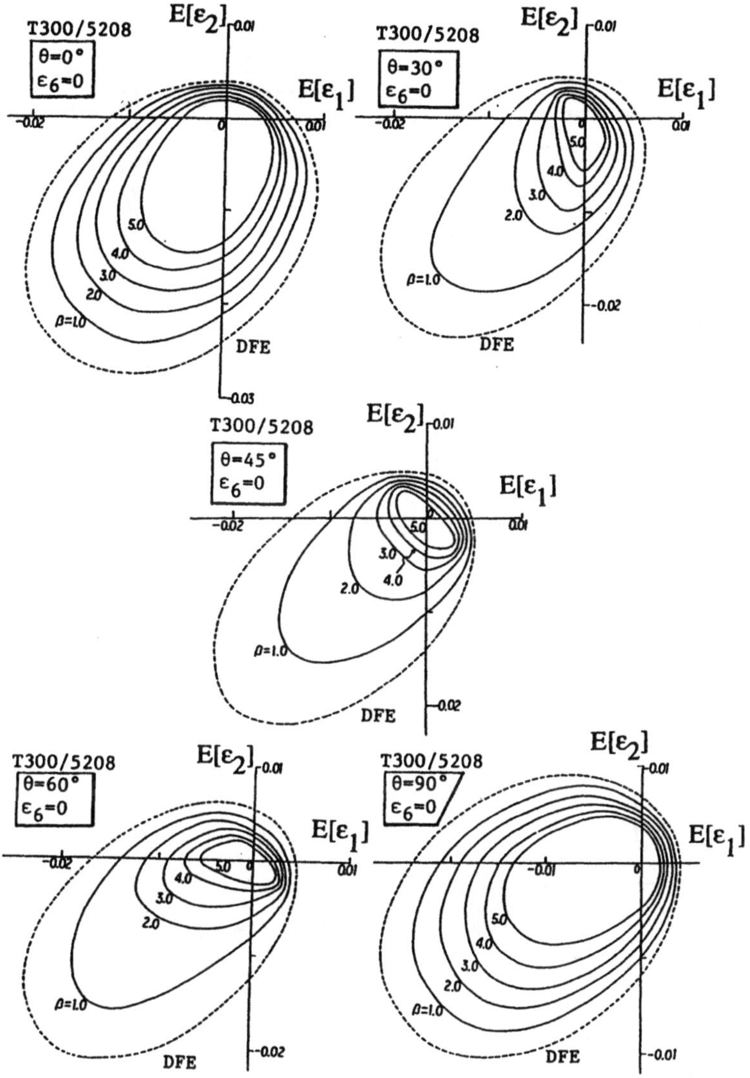

(a) SFE on specified orientation angles at $\varepsilon_6=0$

Fig.6.3. Stochastic Failure Envelopes for Graphite/Epoxy (T300/5208) based on Tsai-Wu's failure criterion.

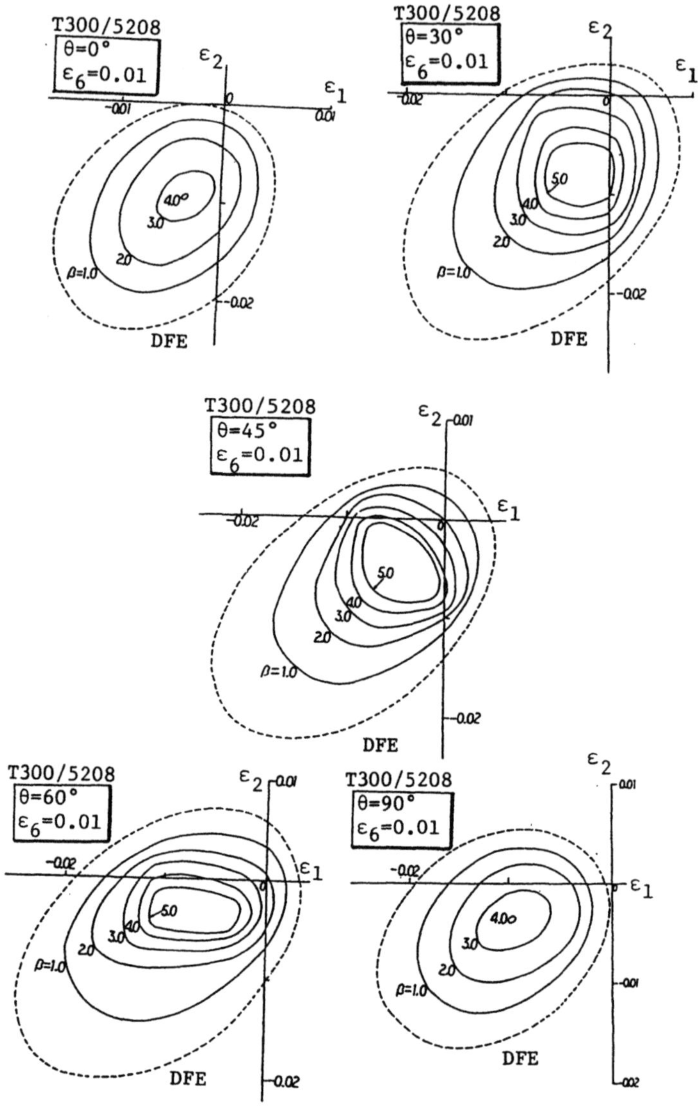

(b) SFE on specified orientation angles at ε_6=0.01

Fig. 6.3. Stochastic Failure Envelopes for Graphite/Epoxy (T300/5208) based on Tsai-Wu's failure criterion(continued).

$$\sigma_{L,c} = \sigma_{T,c} = \sigma_{T,t} \text{ and } k_{12} = -\sigma_{T,t}/2\sigma_{L,t}$$

(for T300/5208, k_{12}= -0.0133), and reduces to the Chamis criteria for appropriate values of Fxx, Fyy and k_{12} in different quadrants for loading. k_{12} for the Tsai-Wu criteria is -1/2 and that for the Chamis criteria varies from material to material (for T300/5208, k_{12}=-0.7). That is also a reason why the Tsai-Wu criterion is superior among the other criteria compared in this study.

Finally, the calculation result of the SFE in Fig.6.3 shows the locus in strain space which corresponds to the safety index. It is seen in Fig.6.3 that all of the results show the concentric ellipsoids.

8 Conclusions

A quantitative comparison of six different failure criteria has been studied in the probabilistic fields on the basis of the formulated model in conjunction with approximated structural reliability theory and lamination theory. The stochastic failure envelopes which show the feasible region is obtained by the presented model in in-plane strain space. For multiaxial composite laminates, however, the reliability analysis will be so complicated that some methods must be developed

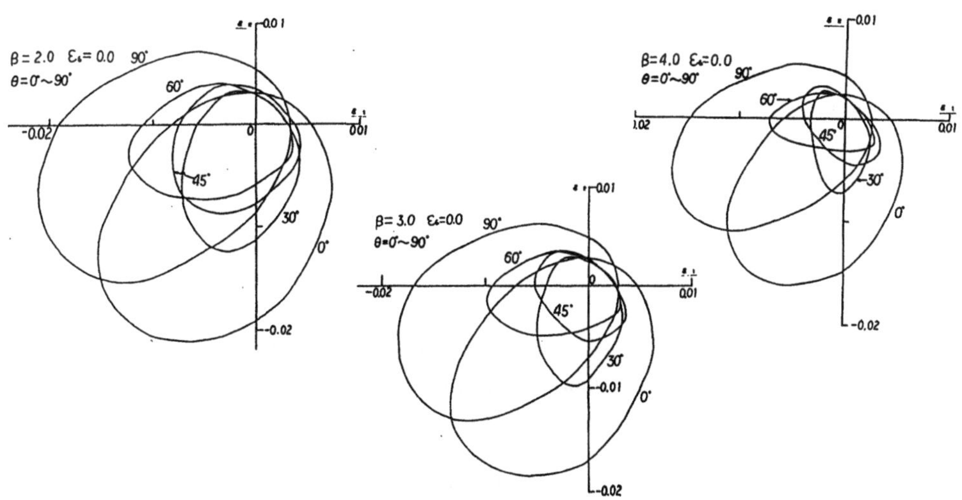

(c) SFE on various orientation angles at ε_6=0

Fig. 6.3. Stochastic Failure Envelopes for Graphite/Epoxy (T300/5208) based on Tsai-Wu's failure criterion(continued).

for this system reliability analysis. The first is a method for evaluating the failure path of a composite laminate where the sequential failure path of each layer of composite laminates is simulated by means of modulus degradation and safety indices under the offaxis load conditions. The second is a method of evaluating the effective failure path of multiple layer composite laminates in order to select the failure path whose occurrence probabilities are greater than others since the number of combinations of failure paths for multiple layer composite laminates is too large for evaluation.

Based upon these findings and the importances of the failure criteria and the stochastic failure envelopes proposed in this paper, it is necessary to undertake further analytical and experimental studies for the reliability-oriented materials design of composite laminates in the future.

Acknowledgement

This work was partly supported by the Science Research Fund of the Ministry of Education, Science and culture of Japan, and the Science Research Promotion Fund of Japan Private School Promotion Foundation.

References

1. Rosen,W., Tensile failure of fibrous composites, AIAA J. .**2** (1964) 1985-1991.
2. Zweben,C., Tensile failure of fiber composites, AIAA J. **6** (1968) 2325-2331.
3. Yang,J.N., Residual strength degradation model and theory of periodic proof tests for graphite/epoxy laminates, J. Composite Materials, **11** (1977) 176-203.
4. Coleman, B.D., Fox,T.G., General theory of stationary random sequences with application to the tacticity of polymers, J. Polymer Sci., **1** (1963) 3183-3197.
5. Fukuda, H., Chou, T.W., A Probabilistic theory for the strength of short fibre composites," J. Materials science, **16** (1981) 1088-1096.
6. Ishikawa, T., Strength and thermal residual stress of unidirectional composites," J. Composite Materials, **16** (1980) .40-52.
7. Tsai, S.W., Wu, E., M., A general theory of strength for anisotropic materials, J. Composite Materials, **5** (1971) 58-80.
8. Chamis, C.C., Failure criteria for filamentary composites, Composite Material:Testing and Design, ASTM STP 460, ASTM, 1970, pp.336-350.
9. Hoffman, O., The brittle strength of orthotropic materials, J. Composite Materials, **1** (1967) 200-206.

10. Hill, R., The Mathematical Theory of Plasticity, Clarendon, Oxford,1950, pp.317-331.
11. Rosenblatt,M., Remarks on a multivariate transformation, Annals of Mathematical Statistics, **23** (1952) 470-472.
12. Hochenbichler, M., Rackwitz,R., First-order concepts in system reliability, Structural Safety, **1** (1983) 177-188.
13. Fiessler,B., Neumann,H.J., Rackwitz,R., Quadratic limit state in structural reliability," J. Eng. Mech. Div., ASCE, **105** (1979) 661-676.
14. Breitung, K., Asymptotic approximations for multinormal integrals, J. Engng. Mech. Div., ASCE, **110** (1984) 357-366.
15. Madsen, H.O., Krenk,S., Lind, N. C., Methods of Structural Safety, Prentice-Hall, Inc., 1986, pp.65-69.
16. Fujita,M., Rackwitz,R., Updating first- and second- order reliability estimates by importance sampling, Proc. JSCE, No.392/I-9 (1988) 53-59.
17. Nakayasu, H., Maekawa, Z., Parameter sensitivity study of uni-directional fiber reinforced composite laminates subjected to off-axial loads, Trans. JSME, **40** (1991) 289-295(in Japanese).

Editors' Note: The discontinuity in the numbering of equations, from (6) to (16), does not alter the context of the paper.

Manufacturing Requirements from a Probabilistic Design Analysis Approach to Optimizing for Robust Design

G. E. Orient[1] and K. R. Rajagopal[1]
D. L. O'Connor[1], and J. F. Newell[1]

[1]Rocketdyne Division, Rockwell International Corp. Canoga Park CA 91303

Abstract. As new areas of technology and product development mature, commercial viability requires adopting and developing engineering tools that assure high system reliability while meeting performance goals. A quality called robustness is required to minimize sensitivity of response and durability to various uncertainties in a system. A requirement for successful development is that all the technical disciplines involved in defining and fabricating a product are included in the model. Probabilistic Design Analysis (PDA) is emerging as a powerful set of tools to quantify the variability of response as a result of variations in system parameters. Design evolution approaches such as the Taguchi method and optimization in concert with PDA provide the capability to obtain robust design through tailoring manufacturing parameters, design and operating environment. In the paper, software selection and additional development activity is described, and the application of the tools is illustrated.

Keywords. Robustness, probabilistic design analysis, design optimization

1 Introduction

The goal of designing robust components in space systems has originated from the need to design highly reliable components, as the cost consequence of failures are enormous. Further, reliability needs to be achieved through design practices so that costs associated with inspection and maintenance requirements are minimized. Past emphasis on minimum weight and maximum performance objectives resulted in products such as the Space Shuttle Main Engine (SSME) requiring labor intensive assembly, inspection and launch procedures. The current design challenge is to develop space systems that in addition to meeting performance constraints, exhibit tolerance to variability in manufacturing, operating and maintenance procedures.

Manufacturing technology influences the design in several ways. For example, consider the liner of a liquid rocket engine main combustion chamber that may be cooled by fuel flowing in rectangular channels in the liner. The specific configuration, performance, reliability and cost of this generic design is fundamentally influenced by manufacturing. Various manufacturing approaches to the cooled liner and the forward and aft manifold joining are considered. Also, the forward and aft manifolds may be welded from separate wrought subassemblies or they may be an integral casting. With respect to the liner, each process exhibits a unique range of capabilities such as the minimum and maximum wall thickness, attainable channel aspect ratio, material properties, characteristic failure concerns (bondline failure, thermal ratchetting, etc.) and the varying degree of controllability of each parameter. These technological bounds, in turn, influence design and performance through different coolant flow and resulting pressure requirement characteristics, life prediction via thermal, constitutive and damage properties determining maximum wall temperature, constitutive response and failure resistance. Furthermore, fabrication and operating costs, as well as inspection requirements are also closely related to manufacturing, completing the interdependence of all aspects of a product development.

2 Robustness

A design tolerant to variations in manufacturing, material and load environment variabilities is called robust, a property that needs to be further quantified. Considering the cost factor associated with inadequate performance and implementing extensive manufacturing control measures, achieving robust design must be the objective at the early phase of product design. Several choices of defining a robustness measure can be made. A classical definition, the signal to noise (mean to standard deviation) ratio is directly available from probabilistic assessment. Alternative definitions of robustness may include the norms of the non-normalized physical or the probabilistic sensitivity vector. In the design process, the variables are classified as control and noise variables. Since controlling the noise variables is typically costly or impossible, the objective of the robust design methodology is to arrive at the optimum robust design by changing the control variables only. For example, the variability of dimensions may be given by a manufacturing process, but the mean value is a controllable design parameter. If appropriate, an initial Taguchi analysis may be performed to examine various classifications of variables and to select an initial robust design. All these elements may be organized in a sequential manner, as shown in Figure 2.1, to form the base of a consistent methodology. Probabilistic variable characterization preceding the analysis steps include material property variations, manufacturing tolerances and capabilities, and load environment definition. Accurate characterization of variables is always a challenge due to insufficient data. Proper balance between accuracy and the level of analysis needs to be maintained. Since a valid model of the component needs to describe all aspects of manufacturing and operation, the robust design methodology requires closely tied analysis modules supplied by the various engineering disciplines. Once the modules are generated, design assessment can be made including sensitivity studies, probabilistic analysis and various levels of robust design search methods.

The software implementation needs to be flexible enough to permit various configurations of the tools, incorporation of external deterministic modules including databases, expert systems and deterministic analysis tools.

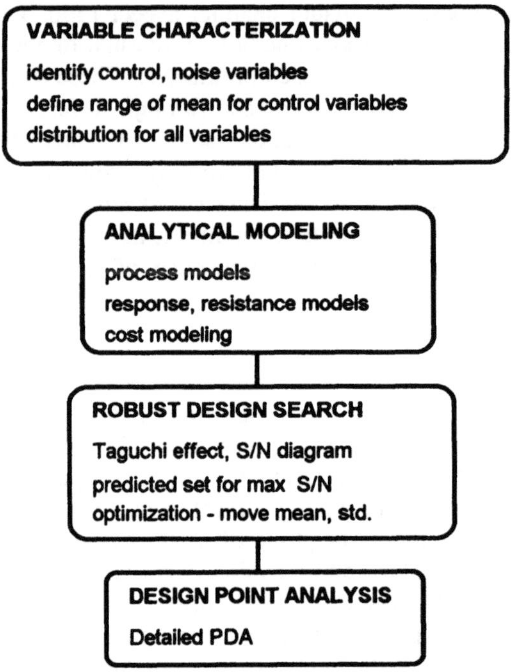

Figure 2.1. Robust design methodology

In the current work, various combinations of probabilistic, optimization and Taguchi type quality engineering methods are presented, and their interrelationship examined using examples derived from manufacturing process and design selection problems in aerospace engineering.

3 Analytical Approaches

3.1 Probabilistic Analysis

At a particular design point, Probabilistic Design Analysis (PDA) can be performed using a variety of reliability methods, if the random variables and their variations are quantified. Simulation approaches (Monte Carlo, Latin Hypercube Sampling, etc.) are often used to assess the reliability of mechanical systems. However, the complexity of a single deterministic analysis may reach a level where simulation approaches are no longer practical. Therefore, application of fast approximate reliability analysis methods such as MVFO (Mean Value First Order) and AMV (Advanced Mean Value) methods[1] is necessary. In recent years, several probabilistic analysis tools in the form of generic software have been developed[2, 3]. The probabilistic results reported in this study were obtained through the software tool, NESSUS[2]. In addition to the probabilistic methods, the package also provides a FEM and BEM modules designed for efficient perturbation evaluation.

In practice, a single deterministic analysis may consist of a sequence of analyses spanning across disciplines. For example, thermomechanical damage analysis includes material database handling, parametric solid modeling capable of handling large geometric variation, thermal analysis, mechanical analysis and advanced damage modeling accounting for various failure modes such as thermomechanical fatigue and flaw growth and plastic instability. Consequently, the deterministic analysis needs are much more general than the options that can be reasonably expected from a single application. NESSUS_EXT developed at Rocketdyne was designed to handle such situations. Geometric variability is best described by parametric solid modeling, a tool that is widely available in the industry (ANSYS, PATRAN, etc.). The efficiency of the NESSUS perturbation evaluation approach is lost, which may be partially offset by the capability of NESSUS_EXT to simultaneously evaluate perturbations on multiple networked workstations.

Although direct evaluation of response gives the most accurate results, building a simple influence function (IF) model is often desirable. In the current work, the model is a quadratic approximation that expresses the response variables as functions of random variables, as given in Equations (1)-(3).

$$\bar{r}_i = \sum_j a_{1ij}\bar{v}_j + a_{2ij}\bar{v}_j^2 \tag{1}$$

$$\bar{r}_i = \frac{R_i - R_i^{ref}}{R_i^{ref}} \quad \text{if } R_i^{ref} \neq 0,\ R_i - R_i^{ref} \quad \text{otherwise} \tag{2}$$

$$\bar{v}_j = \frac{V_j - V_j^{ref}}{V_j^{ref}} \quad \text{if } V_j^{ref} \neq 0,\ V_j - V_j^{ref} \quad \text{otherwise} \tag{3}$$

where R_j and R_j^{ref} are actual and reference response values and V_j and V_j^{ref} are actual and reference variable values. The coefficients a_{1ij}, a_{2ij} are related to the normalized change in response due to the normalized change in a variable, thus called influence coefficients. The original data that is fitted are obtained from direct evaluation of the external response such as a finite element model. The choice of response that is fitted may impact the accuracy of the model significantly. For example, it is advantageous to fit stress components, even though the effective stress may be required in the damage model. Similarly, the inverse of the dominant stress may give a better approximation in stress concentration problems when the independent variable is a local radius.

The influence model may be used to do Monte Carlo simulations, for convergence studies using AMV type analyses, for probabilistic evaluation of several different response types without re-running a full direct evaluation based probabilistic analysis for each response, or as a means of quick response evaluation for health monitoring purposes. The program called IF_GEN, uses the information stored in a NESSUS_EXT input file to generate a quadratic influence function model for all the computational variables.

In addition to distribution of the response, the sensitivity of the response is also calculated with respect to the independent random variables. Physical sensitivity is a measure of the slope of the response surface at the design point irrespective of the probabilistic description of the variables. A better measure of the importance of a variable is the probabilistic sensitivity factor that accounts for the randomness of the variables. For example, if a variable is tightly controlled, high physical sensitivity will not mean that the probabilistic sensitivity is high.

In order to compare the distribution function resulting from the probabilistic analysis with deterministic calculations, nS designs are automatically calculated $(n = 1,2,3)$. Using the signs of the physical sensitivities from the MVFO analysis, each variable is perturbed by $n\sigma$ in the direction to increase the response. For each σ level, the response is calculated deterministically and responses are stored in the log file. In practice, it is often found that the nS deterministic response corresponds to probabilistic response associated probability levels much higher than $n\sigma$ of the distribution function.

NESSUS_EXT consists of a control program written in UNIX C-shell language, and a number of utility codes written in FORTRAN. The program can be fully configured to suit application needs and network resources. Multiple layers of error checking procedures ensure proper execution even in the case of network connection failure. A network activity log file is created documenting on which node the perturbations were executed, network failures along with information relating the mode of failure (no connection, cannot start application, application failed while executing, etc.).

3.2 Optimization

Optimization in some form is part of most engineering activity, whether or not the procedure is formally laid out. For example, in product development, initial product design is conceived and either through testing or analysis the design is improved based on the observation of some objective and constraints. Manufacturing processes are also tailored to meet productivity, quality and cost objectives and constraints. Data reduction of material test data is a simpler application of optimization, whereby an analytical function is fitted over measurements by minimizing the difference between the model and data. A large set of these problems can be handled by the same numerical tools. The requirement for applicability is that the objective and the constraints corresponding to a set of design variables can be evaluated by a numerical model. The model may be based on physical principles such as heat transfer, fluid mechanics and solid mechanics, or it may be a multivariate curve fit, which is often used in modeling manufacturing processes.

The OPTIMIZE front-end has been designed for flexible and unified problem description irrespective of the complexities involved in evaluating the response. The optimization problem is stated via keywords allowing specification of the variable bounds, scanning resolution to find initial estimates, detailed control of gradient evaluation and optimization package. Forward and central differences are available for objective and constraint gradient evaluation. The gradient evaluation method can be specified uniformly, or evaluation method of one particular function with respect to a variable may be specified. Previous knowledge and/or intuition related to the response can aid selection of the gradient evaluation method, permitting minimization of the number of function evaluations. The philosophy of the OPTIMIZE tool is to utilize as much of the available resources as possible. Therefore, user defined modules that implement standard parametric engineering response evaluation tools are interfaced with well established optimizing subroutine packages resulting from years of research in the field. Currently, links to optimization routines from the IMSL mathematical subroutine package[4] or the DOT (Design Optimization Tool) package[5] are available. The optimizers may be 'daisy chained' up to 10 optimization stages by issuing subsequent optimizer specification commands. A two-step approach may be optimal for problems where accuracy is required, but the function evaluations are non-trivial. Three kinds of function evaluations are available: user defined FORTRAN routine, user defined executable, and user defined script. Gradients of the objective and constraints may be evaluated on multiple networked workstations using the task scheduling utility package developed for NESSUS_EXT. This way, optimization problems ranging from simple curve fitting to structural optimization involving coupled FEM analysis can be solved through a consistent interface providing sufficient choice of standard optimizers (IMSL, DOT) and control parameters. Extensive control over gradient evaluation and a keyword to allow complete convergence control of the individual optimizer packages assure that the analyst can tailor the problem statement to minimize the number of function evaluations.

3.3 Taguchi Analysis

Recent advances in quality engineering made by Taguchi are based on the recognition that in the manufacturing process, not all variables are controllable, or more realistically, only the combined effect of many independent variables may be known and measurable. Cost-effective tailoring of manufacturing processes requires concentrating on controllable variables and seeking a set of parameters resulting in a product least sensitive to variations in noise, thus inherently robust. Appropriate selection and grouping of variables is critical in the success of developing a robust product.

The method provides a formal framework to processing quantified responses once the signal and noise variables, often referred to as factors, are determined. The objective of the robust design methodology is to maximize the signal-to-noise ratio (S/N), a concept originated in classical signal processing theory. Recognizing that depending on the character of the hardware and selection of the response quantity, optimal design may be nominal, highest, or lowest response, Taguchi introduced appropriate signal-to-noise ratios. In the Taguchi methodology, orthogonal arrays are utilized to reduce the number of experiments or response evaluations necessary to predict which combination of factors will be most robust.

The current software tool implements these calculations in a flexible manner so that responses ranging from experimental data to results of complex numerical-analytical models may be processed. If extensive calculations are required to obtain response, the tool can make use of multiple networked workstations to evaluate the response matrix efficiently using the above mentioned parallel processing utility library. The TAGUCHI tool is designed to execute the data analysis part of the methodology. Careful selection of factors with special regards to additivity and independence remains the analyst's task. The following general description section does not describe all the details of a Taguchi analysis, many references exist[6].

The TAGUCHI tool can be used to process an array of responses obtained from experiment, or to invoke a numerical model to generate the response array preceding processing. A standard library of orthogonal arrays[7] is used to select combinations of variable levels to evaluate the response. Variables designated by names are grouped into control and noise variables. For each group, the smallest orthogonal array is selected whose number of columns is larger than the number of variables in the group. For example, if three levels of seven variables are defined in a group, then the $L_{27}\,(3)^{13}$ array is selected. The naming convention is $L_a\,(b)^c$ where a is the number of evaluations, b is the number of levels for each variable and c is the number of columns in the array. The structure of the orthogonal arrays is such that main effects and variable interactions are covered together. For both control and noise variables, the TAGUCHI tool will assign the available main effect columns to the first i variables ($i = \sqrt[b]{a}$), which minimizes interactions between those variables. The rest of the variables are assigned the first available unused columns. Thus, by ordering the variables, some prioritization is implied.

An important concept in the Taguchi framework is the quality loss function. The optimization goal may be maximizing (Larger Is Best), minimizing (Smaller Is Best) a response, or ensuring that the mean response is close to a target value

(Nominal Is Best). The expressions for signal-to-noise ratios have been defined for each quality loss scenario so that higher values indicate more robust design:

$$S/N = -10\log\left(\overline{\left(y^2\right)}\right) \quad \text{Smaller Is Best (SIB)} \tag{4}$$

$$S/N = 10\log\left(\frac{\bar{y}^2}{\sigma_{n-1}^2}\right) \quad \text{Nominal Is Best (NIB), + response} \tag{5}$$

$$S/N = 10\log\left(\frac{1}{\sigma_{n-1}^2}\right) \quad \text{Nominal Is Best (NIB), } \pm \text{ response} \tag{6}$$

$$S/N = -10\log\left(\overline{\left(1/y^2\right)}\right) \quad \text{Larger Is Best (LIB)} \tag{7}$$

where y_i are the responses and n is the number of noise states.

After obtaining the response array either from the input file or by running the response evaluator through combinations of variables dictated by orthogonal arrays, the data for the standard Taguchi effect and S/N diagrams are generated. Prediction to the best and the worst design is made using these diagrams and significance criteria provided in the input file. Denote the threshold value for effect by e and for signal-to-noise ratio by s, the prediction is made by selecting levels for each control variable by the logic illustrated in Table 3.3.1

effect	S/N	decision based on
$\Delta e > e$	$\Delta s > s$	S/N
$\Delta e > e$	$\Delta s < s$	effect
$\Delta e < e$	$\Delta s > s$	S/N
$\Delta e < e$	$\Delta s < s$	S/N

Table 3.3.1. Level selection logic

The quantity Δe is the difference between the highest and the lowest effect values at a given variable, Δs is the analogous difference in signal-to-noise ratio. Once the best case design is selected, the best case effect and S/N ratio are predicted by linear extrapolation. Worst case predictions are made analogously. In order to complete the Taguchi study, a confirmation experiment is required to verify the linear assumption. The confirmation must be done either experimentally or by running the analytical model with the predicted levels of the significant control variables and random numbers for the rest of the variables.

It is re-emphasized that definitions for signal-to-noise ratios are somewhat arbitrary and fundamentally different, reflecting the desired goal of the component. They were defined to be always positive and monotonic with robustness for good indication of ranking. Also, the orthogonal arrays minimize the number of tests as compared to full factorial evaluation, therefore the accuracy of the obtained statistics is not guaranteed. The objective is to capture trends and identify significant control variables.

Conceptually, the effect diagrams depict the change of mean response when a control variable is set to a level, while the rest of the control and noise variables are left random. Factors that dominate the effect diagram are called signal variables indicating that the response is strongly driven by them. Once the response table is complete, the signal-to-noise ratio associated with a particular level of a control variable is generated by evaluation at fixed sets of control variables and averaging these values over the sets that included the control variable at the given level. Alternatively, a compound S/N can be computed over all the variable sets that are have the particular control variable at the given level. This latter method may help to further isolate signal variables. The compound S/N ratio can also be evaluated by probabilistic tools by removing variables from the random variables, setting them at a given level and performing a probabilistic analysis. Once the full cumulative distribution function is obtained from an AMV analysis, for example, the signal-to-noise ratio appropriate to the selected quality loss function can be determined with better accuracy than that provided by the orthogonal arrays. A software tool called TAGUCHI_PROB was developed to implement this logic. The inputs to this module are a NESSUS input file posing the reliability problem with all the variables and a TAGUCHI input file classifying the random variables, providing criteria and quality loss function specifications. As shown in an example, the number of required function evaluations may be less than when using orthogonal arrays, depending on the nonlinearity of the response and the non-normality of the random variables.

4 Application Examples

The methodology is demonstrated on two examples, a Main Combustion Chamber (MCC) liner design study and a pressure vessel weld parameter selection problem. The MCC liner is a high heat flux component with inherent uncertainties in operating parameters, geometry and material properties even under nominal conditions. Robustness is defined as insensitivity of LCF damage against variabilities in operating conditions. Various disciplines such as heat transfer, material selection, engine balance and global performance model, stress analysis and resistance modeling modules are linked together to result in response evaluation that reflects the current set of probabilistic and deterministic parameters. Variabilities in manufacturing parameters manifest in variabilities in material fatigue resistance completing the coupling between control parameters. The weld design problem illustrates how manufacturing requirements are quantified through probabilistic and optimization analysis.

4.1 Liquid Propellant Rocket Engine Main Combustion Chamber Design

The physical component addressed in this example is the liner of a rocket engine main combustion chamber. The primary function of the liner is to provide the means of removing the heat generated by the combustion process, thereby protecting the structural shell often referred to as jacket. Various copper alloys, manufacturing processes and associated designs are considered.

name	description	group
PIN	coolant inlet pressure	engine balance
FLO	flowrate	"
PC	chamber pressure	"
TIN	inlet temperature	"
MR	mixture ratio	"
CHW	channel width	manufacturing
CHD	channel depth	"
WTHN	hot gas wall thickness	"
HGGF	hot gas coefficient factor	thermal model
COGF	coolant coefficient factor	uncertainty
CENH	coolant curvature enhancement	"
LNKF	liner thermal cond. factor	"
DUCT	ductility	fatigue properties
FTU	ultimate stress	"

Table 4.1. Random variables for liner example

Thermal deformation induced cyclic plasticity and creep are the predominant deformation modes. Consequently, low cycle fatigue damage due to duty cycle transients is the most important damage mechanism, although cyclic creep and plastic instability are also significant. At the current level of assessment the objective is preliminary design evaluation and only the LCF life is considered. For a complete evaluation of the designs, all failure modes need to be investigated before a selection can be recommended. The numerical model of the component comprises several engineering disciplines. A global engine balance model and influence function database was used to derive local load parameters such as pressures, flowrates and their variations. Manufacturing processes resulted in a set of tolerances defining geometric variabilities. Since the local strain response is sensitive to the thermal input, uncertainties in the thermal modeling were considered and quantified. Various manufacturing processes result in different forms of the alloys, significantly impacting their fatigue resistance. In the absence of measured fatigue curves, the most relevant tensile properties are ductility and ultimate strength that can be used to estimate full range fatigue curves via the Method of Universal Slopes (MUS). Statistics of ductility and ultimate stress were determined by fitting analytical distributions to measured data. Since the problem is driven by thermal strain, variability of local strain range due to variations in constitutive parameters are assumed to be negligible. The names of the variables and their grouping is given in Table 4.1.

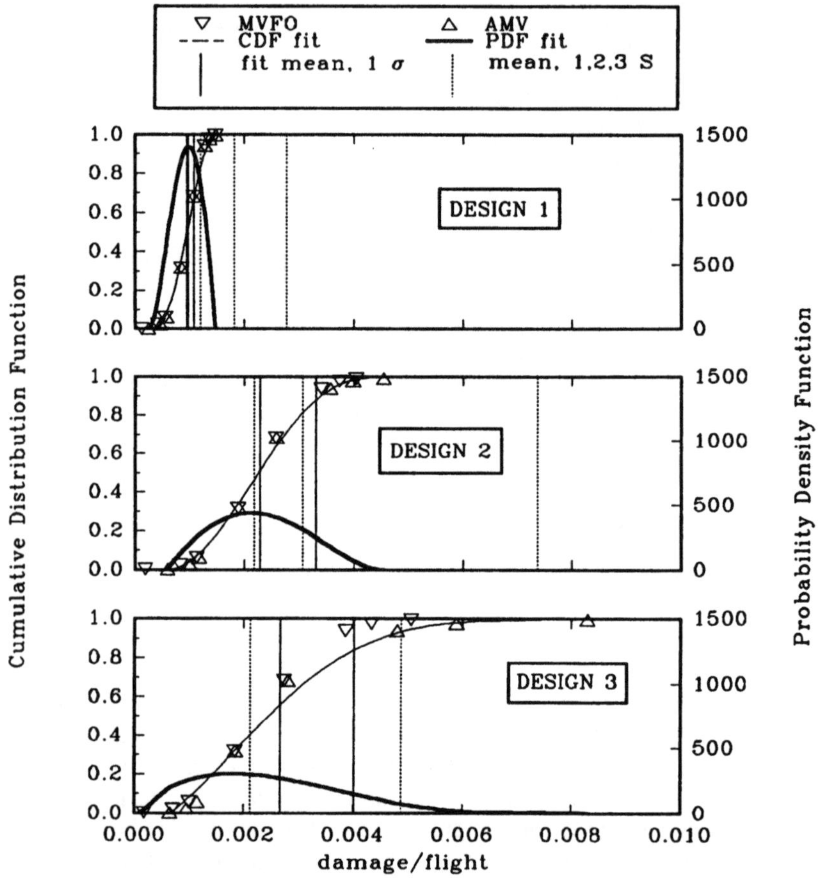

Figure 4.1. Effect of manufacturing and design on durability

The deterministic response evaluation module that was coupled with NESSUS_EXT included a local load evaluation module, a thermal code to calculate temperature distribution at steady-state, steps to evaluate local mechanical response with ANSYS and a damage module. Results of the probabilistic assessment of the designs based on the incurred LCF damage during a flight are given in Figure 4.1. Points of the cumulative distribution function calculated with Mean Value First Order and Advanced Mean Value First Order methods indicate that at the high probability levels the converged solution differs significantly from the MVFO results. Comparison of the three designs reveals that design 1 results in the lowest damage. In addition, a measure of robustness, the inverse coefficient of variation, μ / σ also is higher at that design as indicated by the shape of the probability density functions. It is worth mentioning that the maximum wall temperature distributions were also derived showing minor differences between the designs. The single most significant differentiator between the alternatives is ductility, which shows high capability resulting from material selection and the manufacturing process in design 1. For the LCF failure mode, design 1 is favorable from both design performance (life) and robustness points of view.

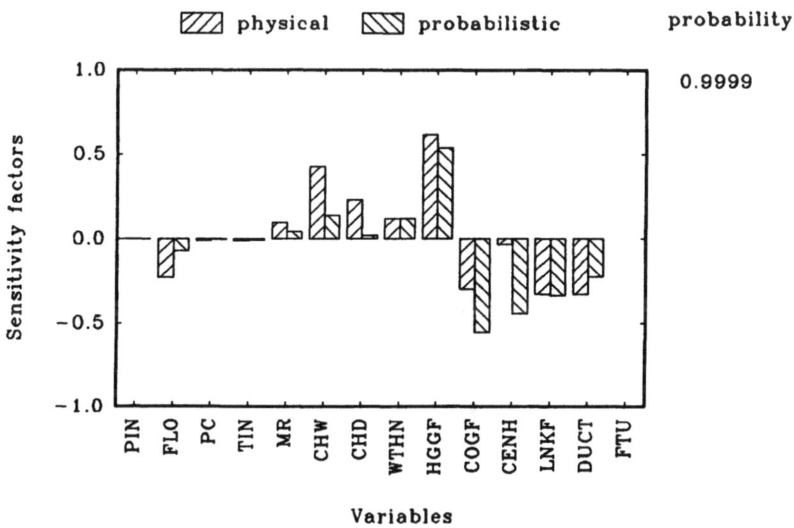

Figure 4.2. Sensitivity factors for Design 1

As mentioned earlier, an important result of the MVFO and AMV analyses is a set of sensitivity factors at each probability level. The physical sensitivity factor is a dimensionless representation of the gradient of the response surface, while the probabilistic sensitivities also reflect the distributions of the random variables. As illustrated in Figure 4.2, excluding modeling uncertainties, ductility has the highest probabilistic sensitivity factor. Some variables with high physical sensitivities such as the channel width are tightly controlled, therefore, their probabilistic sensitivity is low. Modeling uncertainties can be either significant or negligible from the physical point of view. However, since the variation in the variables associated with modeling uncertainties is often estimated with conservatism, they tend to have significant probabilistic sensitivity factors. These uncertainties were estimated by the same approach for all the designs, consequently, the ranking of the designs - the goal of the current study - is not impacted. Reliability measures and numerical measures of robustness may change by removing modeling uncertainties through additional methods development.

4.2 Weld Parameter Study

Consider a pressure vessel that is welded together from a spherical dome and a cylindrical shell, as shown in Figure 4.3.

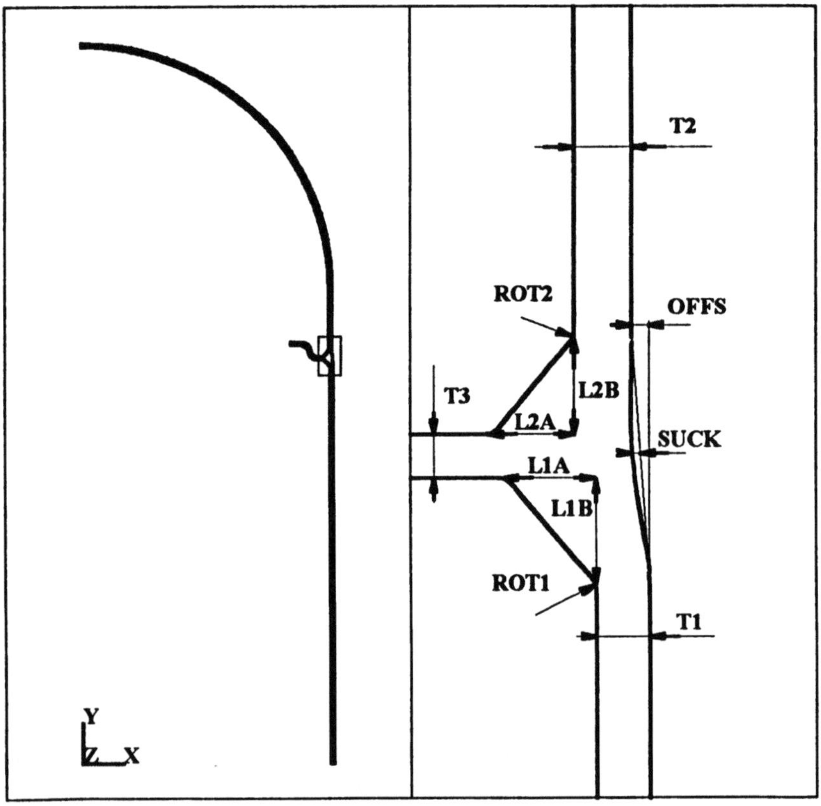

Figure 4.3. Geometric Variables

The objective of the current analysis is to quantify variability in the capability of the vessel to resist failure as indicated by the stress criterion, determine both best and worst case designs and to quantify sensitivity of the designs with respect to variabilities in key design parameters. As a result, effective control variables and dominant noise variables are identified, indicating directions for new design and weld technology development. Additionally, a measure of design robustness is evaluated to help in ranking alternative designs. While in the previous example geometric shape change was small, the weld parameters in the current example may result in drastically different geometric configurations.

The critical area of concern from a residual stress, crack initiation and crack propagation life is at the as-welded circumferential welds shown in Figure 4.3. The internal components of the pressure vessel prohibit heat treating the girth welds. The local stiffness change between the cylinder, welding and dome causes local bending moments and stress concentration.

Geometric design variables were identified and their limits were defined based on manufacturing tolerances, uncertainties and examination of existing hardware. Corresponding to the limits, all variables were characterized by truncated normal distributions. The thickness of the dome and cylinder shells are grouped into three ranges according to the availability of the material. Several response quantities were considered, including membrane and bending safety factors, proof and operating strain ranges. The axial safety factor is a measure of how far the cyclic stress state is from being fully reversed plastic. The following discussion is based on results from analyses using the axial safety factor defined from elastic deformation as

$$F.S._{ax}^{mem+bend} = \min\left(\frac{2FTY}{1.1\sigma_{ax,1}^{memb+bend}}, \frac{2FTY}{1.1\sigma_{ax,2}^{memb+bend}} \right) \tag{8}$$

where last subscript denotes location '1' (cylinder) or location '2' (dome).

This example illustrates the steps involved in selecting control variables, identifying a robust design and obtaining response distribution at the selected configuration. The Taguchi method was applied first to examine the variable space. The effect of weld curvature is assumed to be negligible, therefore the variable SUCK is readily classified as a noise variable. Although the effect of weld root radii and the weld offset are also non-negligible over the entire range of designs, they are impossible to control, therefore they will also be classified as noise variables. For the initial scoping the cylinder and the dome thickness are considered identical (T1=T2=T), even though the different fabrication processes yield slightly different ranges of thicknesses. The weld bead shape is somewhat controllable by weld technology control and by inspection/acceptance of the pressure vessel. Variables describing the weld angle will be taken as control variables along with the dome, cylinder and weld ring thicknesses. At this point it is not clear which variable will be truly a control factor, therefore all the candidate variables are placed in the control variable group. Using the appropriate orthogonal arrays, the effect and signal-to-noise ratios were calculated as indicated with lines with open symbols in Figure 4.4.

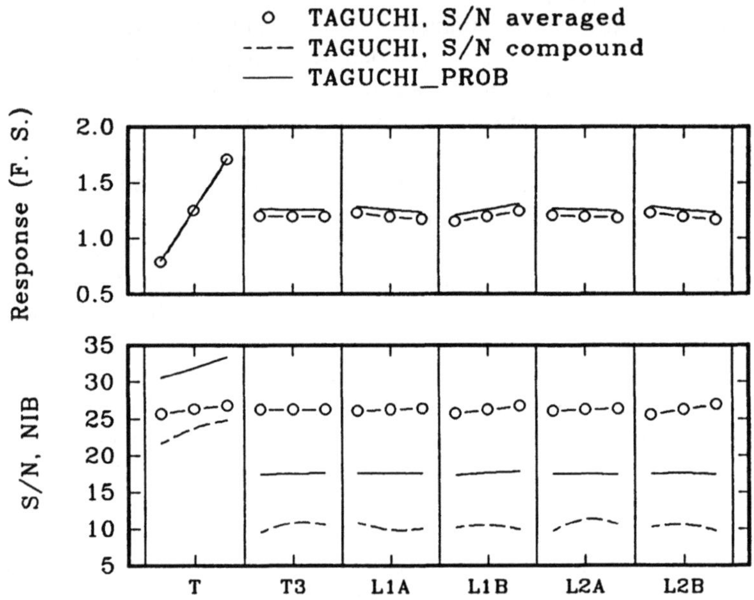

Figure 4.4. Taguchi effect diagram for the weld example

The effect diagram clearly shows that the shell thickness is a driving variable, while the rest of the factors produce various secondary effects. The compound S/N ratios were also calculated, further indicating the shell thickness as signal variable, as shown with dashed lines in Figure 4.4. The compound S/N ratios were also evaluated by probabilistic analysis using the MVFO approach and plotted with solid lines. The distributions derived from analyzing weld samples were used for variables determining weld geometry and for the welds ring thickness. Signal variables may be selected at any level, therefore, uniform distribution was used for the shell thickness. The accuracy of the values evaluated with the probabilistic approach are expected to be more accurate, due to the small sample size in the Taguchi response table. The figure indicates that the mean values are sufficiently accurate, but the standard deviation evaluated from the table is overestimated.

At this point, a decision may be made specifying the level of the signal variable at the highest value. In order to gain insight to the effect of different dome and cylinder thicknesses and to asses the changing response distribution with changing signal factor, three thickness groups were defined with appropriate distributions for the two shell thicknesses. Since the thickness variations are relatively small, a quadratic response function for all the component and linearized stresses and strains from elastic analyses was generated using IF_GEN, at two locations on opposite sides of the weld.

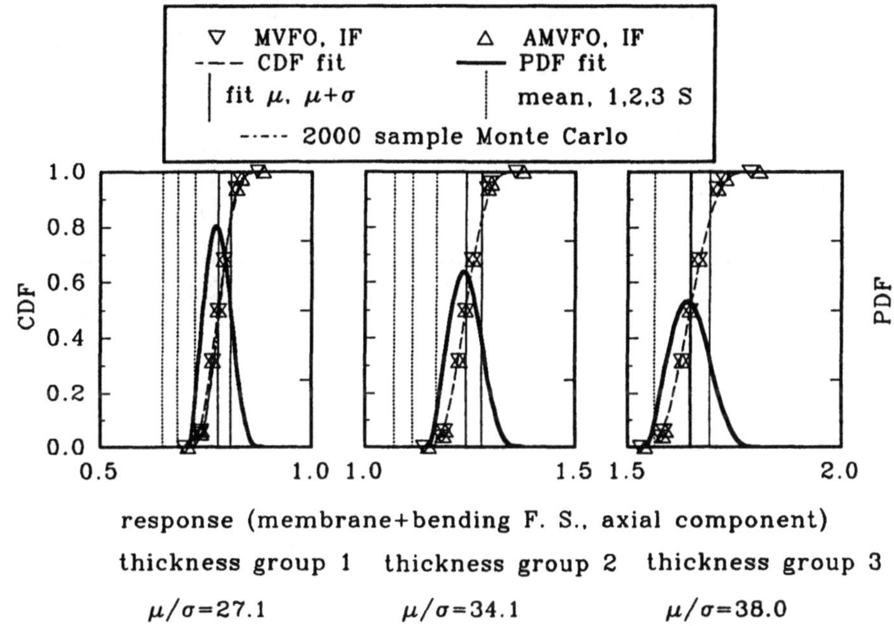

Figure 4.5. Response distribution

Probabilistic analysis was performed using NESSUS resulting in statistics of the response, physical and probabilistic sensitivity factors and the inverse of the coefficient of variation, a measure of robustness. Additionally, '*nS*' deterministic designs are also generated, as defined in Section 3.1.

A summary of the detailed distribution of the linearized membrane+bending factor of safety is shown in Figure 4.5. At thickness group 1, a Monte-Carlo analysis using the influence coefficient method was performed with 2000 samples to verify the accuracy of the AMV procedure for the current problem. In order to enhance accuracy, separate influence function models were fitted for each thickness group. As expected, increasing thickness results in better factor of safety. Additionally, the robustness measure is also increasing, although it appears to level off at the highest thickness. As indicated, thickness groups 2 and 3 meet the factor of safety criterion based of the current approximate analysis. It can be observed on Figure 4.5 that the *nS* configurations represent designs much worse than $n\sigma$ of the response distribution, indicating the conservative character of the deterministic designs. In all cases, the cylinder side of the weld was the most severe, which is inherent to the current range of the variables.

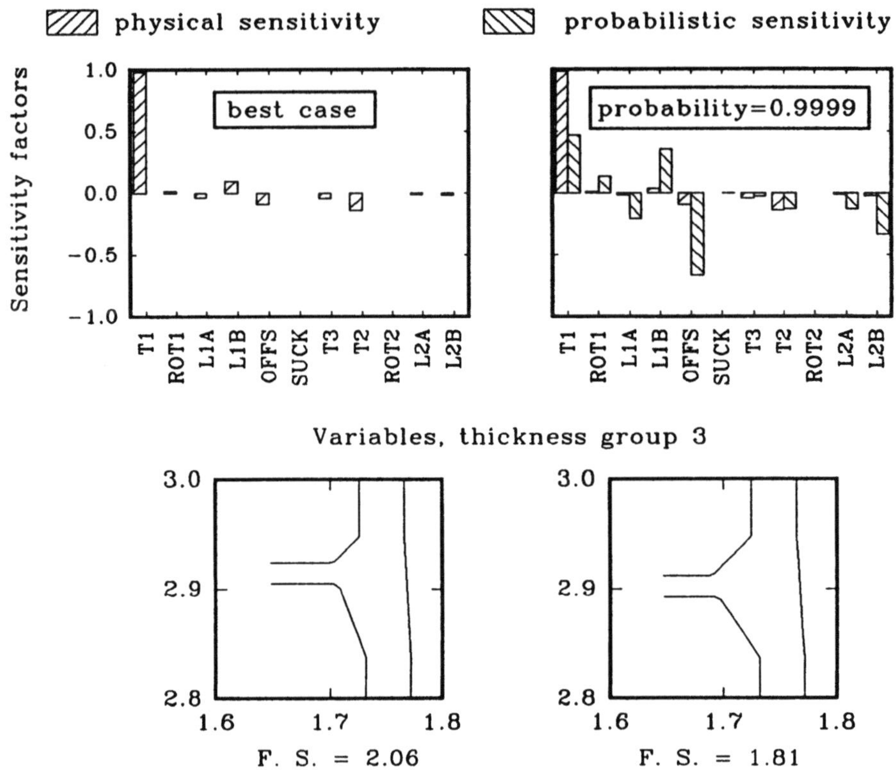

Figure 4.6. Sensitivity and weld configuration at the high capability end

The Taguchi results and the detailed probabilistic analyses consistently indicate thickness group 3 as the design point. Physical and probabilistic sensitivity factors are shown in Figures 4.6 and 4.7 at the high and at the low capability end for the selected configuration. Along with the sensitivity factors, the geometric configurations of the weld are also shown. The designs associated with the low and the high ends of the response distribution also served as starting points to determine the extreme designs. The OPTIMIZE front-end tool was used to arrive at the best/worst case designs. At the extreme designs probabilistic sensitivity factor is not available, since the configurations were obtained by optimization. The results agree with intuition suggesting that with the current mismatch of stiffness, the configuration with the most shallow weld angle allowed by the variable limit will perform best. Examination of the sensitivity factors reveals that the single most important variable is the cylinder thickness. Similarly to the previous example, probabilistic and physical sensitivity factors result in different ranking of the variables depending on the simultaneous effect of response sensitivity and the coefficient of variation associated with the random variables.

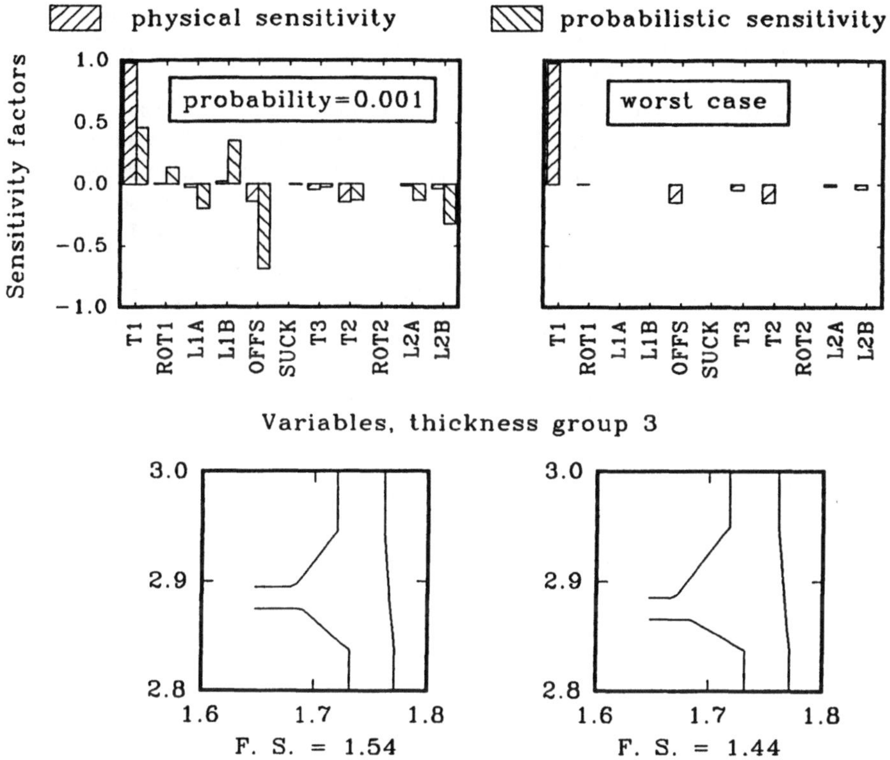

Figure 4.7. Sensitivity and weld configuration at the low capability end

Summary, Future Directions

Various design evaluation tools were examined, and the potential for their utilization in concerts were explored. As demonstrated, probabilistic design analysis is the main tool for evaluating a given configuration. Furthermore, approximate methods are preferred due to the complexity and computational intensity of a typical deterministic design evaluation. In order to evolve the design, scoping tools, such as Taguchi type methods and more rigorous design optimization tools may be useful. The next step in integrating probabilistic and optimization tools as well as robustness concepts adopted from Taguchi's work is to provide tools to link the modules together and develop some standard configurations.

1. Optimization of a design guided by probabilistic analysis - From the robustness stand point, optimization to maximum response is indeed finding maximum signal-to-noise ratio when Larger Is Best quality loss function is applied. This configuration was demonstrated in the second example where probabilistic analysis guided determination of extreme design by providing most probable configurations at the high and low probability.

2. Robustness assessment through probabilistic analysis - The TAGUCHI_PROB tool is one example of this configuration. The advantage of this approach is greater accuracy in the effects and the signal-to-noise ratios and possibly better efficiency in terms of the required number of realizations.

3. Replace the simple best case and worst case projection used in Taguchi type of analysis with constrained maximization of a signal-to-noise measure. - For example, an initial Taguchi analysis may be done to identify significant control variables. Assuming that the mean of the significant control variables may be shifted without changing the shape of the distribution, reliability or performance constraints may be included to limit acceptable configurations. For example, if the response is damage/mission, one may require that the probability of a certain damage level should not exceed a pre-set reliability limit. Using the set of significant control variables as design variables ($\underline{x}_{sel}$), and defining a robustness objective function, f, solve the following optimization problem:

$$\text{objective: maximize } f(\underline{x}_{sel}) \quad (9)$$

$$\text{constraint: } P(R > R_{\max}) < P_0 \quad (10)$$

$$\text{simple bounds: } \underline{x}^{lb} < \underline{x} < \underline{x}^{ub} \quad (11)$$

where $R_{\max}$ is the upper limit in response and P_0 is the maximum allowable probability for the upper limit to occur. The objective may be inverse coefficient of variation, or some compound objective accounting for cost, weight and assembly/maintenance time requirements. The resulting parameters will represent a configuration that meets reliability constraints and attains best signal-to-noise ratio.

The basic goal of the robust design technology is helping preliminary design development and ranking of configurations. Although the accuracy of the assumed/fitted distributions for the significant control variables may not be crucial, a balance must be found between credible characterization and the basic goal of the analysis. Since the methodology can encompass multiple disciplines and is best when applied early in the product development stage, it can impact manufacturing, material selection and design selection effectively.

Acknowledgment

The work reported herein was supported by the NASA task order titled Manufacturing Requirements from a Probabilistic Design Approach under the Development of Life Prediction Capabilities for Liquid Propellant Rocket Engines, contract (NAS3-26884). The support given by the NASA Contract Monitor Dr. Chris Chamis is gratefully acknowledged.

References

1. Wu, Y.-T., Millwater, H. R., Cruse, T. A.: Advanced Probabilistic Structural Analysis Method for Implicit Performance Functions, *AIAA Journal*, 28 (1990) 1663-1669.
2. Probabilistic Structural Analysis Methods - Annual Reports, NASA Lewis Research Contract NAS3-24390, Contractor Report, Southwest Research Institute, 1985 to 1991.
3. Probabilistic Analysis Proban-3, Veritas Sesam Systems, Houston, Texas.
4. MATH/LIBRARY ver. 2.0 IMSL INC., 1991.
5. DOT User's Manual VMA Engineering, 1989.
6. Phadke, S. M.: Quality Engineering Using Robust Design, Prentice Hall, 1989.
7. Taguchi, G. and Konishi, S.: Taguchi Methods, Orthogonal Arrays and Linear Graphs, American Supplier Institute, Inc., 1987.

Advances in Reliability Estimation of MDOF-Systems

H.J. Pradlwarter, G.I. Schuëller, and P.G. Melnik-Melnikov

Institute of Engineering Mechanics, University of Innsbruck, Technikerstr. 13, A-6020, Innsbruck, Austria

Abstract. The response surface methodology (RSM) and Monte Carlo simulation (MCS) appear to be the most general approaches to calculate failure probabilities of non-linear structural multi-degrees of freedom (MDOF) systems subjected to random dynamic loading. In an effort to reduce the number of time consuming simulations generally required for direct MCS, only a few of which are important for the assessment of failure rates, a new selective MCS technique named "Double & Clump" (D&C) is suggested. The main idea of this procedure is to increase for each time step the relatively low number of response samples containing higher mechanical energy. The efficiency of selective MCS is demonstrated by applying it to a 6-storey frame structure subjected to earthquake excitation. The comparison with results obtained by using the RSM is presented and discussed.

1. Introduction

In a recent review of methods currently available to analyse MDOF-systems under stochastic loading, their merits, limitations and potential to be used in engineering practice have been discussed[1]. From this study it becomes clear that only numerical procedures such as the Response Surface Method (RSM) and Monte Carlo simulation techniques - both in their direct and variance reduction forms - are capable of treating nonlinearities without restriction and systems of higher dimension with, moreover, reliable information on the tails. For this reason the present paper concentrates on the class of numerical procedures and their possibilities of future developments.

Since reliability analysis is one of the most important goals of utilizing the procedures of stochastic dynamics, the generality with respect to the type of nonlinearity - particularly in view of the non-restricted formulation of the limit state condition - represents a significant property. In other words, this becomes an important issue for cases where, in addition to the serviceability, the partial and total collapse, i.e. the ultimate failure condition, is to be analysed. Moreover, the method should be applicable to practical problems, i.e. systems of a larger scale to be modeled by finite element procedures as used in traditional, deterministic analysis. In this context the computational efficiency of the procedure becomes an issue in spite of the development of high speed computers. For this reason the development of "intelligent" simulation techniques becomes more and more important.

2. Response Surface Methodology

The RSM has been adapted to treat structural problems where the experiments are mainly replaced by structural analysis. One of the major advantages of the RSM applied to structural problems is its generality. Any available experimental and analytical methods can be utilized to establish an approximate relation between important parameters and the response surface. Hence, the treatment of highly nonlinear systems or systems with a large number of degrees of freedom does not pose any substantial difficulty. Due to this feature the RSM is well suited for solving structural reliability problems. In early developments the response was established in terms of response properties such as critical displacements and damage. On the basis of the obtained response surface, the reliability expressed by the failure probability has been obtained[e.g.2,3,4,5]. In a further development, the response surface has been established in terms of the so-called limit state function[6]. The limit state function divides the set of input variables in a safe domain (i.e. input sets leading to no failure) and an unsafe domain. The procedure is well established for nonlinear static problems and dynamic structural problems where the stochastic process can be represented sufficiently well by a few random variables. Only recently the procedure has been applied successfully in structural problems where the loading is represented by a stochastic process[7]. The RSM is utilized to determine points on the limit state, i.e. the set of input variables resulting in a transition from safe to unsafe states. The limit state function is then approximated by a second order polynomial[8],

$$\bar{g}(\mathbf{x}_j) = a_0 + \sum_{i=1}^{r} b_i \cdot x_{ij} + \sum_{i=1}^{r} \sum_{k=1}^{i} c_{ik} \cdot x_{ij} \cdot x_{kj} \quad , \tag{1}$$

where the regression coefficients a_0, b_i and c_{ik} are determined by a weighted least square procedure. An overbar is used in eq.(1) in order to indicate that $\bar{g}(\mathbf{x})$ is an approximation of the true limit state function $g(\mathbf{x})$. The efficiency of the procedure depends strongly on the number of random variables considered, i.e. the number of varying input parameters. In order to avoid an extensive numerical effort, the number of random variables should not exceed a certain number, e.g. twenty. By studying first the importance of all random variables involved, only the most significant random variables, i.e. those with the largest effect on the failure probability, are taken fully into account.

3. Selective Monte Carlo simulation

3.1. Introductory Remarks

Before discussing in detail the procedure for generating realizations important for the reliability assessment, some basic features of importance sampling are in order. In this procedure, the original density f(**x**), reflecting direct MCS, is modified by using a sampling distribution h(x). This can be done without distorting the original density f(x) by modifying the weight of the realization by 1/h(x). One can see from this feature that the density of realization can be increased, e.g. c times, provided the weight of the realization is decreased 1/c times. One can therefore conclude, that realizations in the low probability tails are only obtainable when the weight of these realizations is considerably decreased. Most important is the accurate representation of the non-Gaussian distribution defined by the simulated discrete state vectors. Assign to each realization a weight and normalize them such that they add up to one. These weights play then the role of discrete probabilities, and the cumulative distribution function CDF(x) is approximated by

$$\mathrm{CDF}(\mathbf{x};t) \cong \sum_{n=1}^{nSim} I\big[\mathbf{X}_n(t),\mathbf{x})\big]\cdot w_n(t) \tag{2}$$

where nSim denotes the sample size, and $I[\mathbf{X}_n,\mathbf{x}]$ is an indicator function. This function assumes the value 1 when each of the components $X_{nk} < x_k$, k=1(1)nState, of the n-th state vector $\mathbf{X}_n$ is smaller than the components of the vector **x**. Otherwise, the indicator function assumes 0, i.e. $I[\mathbf{X}_n,\mathbf{x}] = 0$. If direct MCS is applied, all weights equal 1/nSim. As a consequence, no estimates can be obtained for the region **x** with CDF(**x**) < 1/nSim or CDF(**x**) > 1.0-1/nSim.

Consider a system excited by a Markovian type loading process. This loading process may be continuous or discrete. For the special case of continuous white noise loading, the probability density function (PDF) is then governed by the Fokker-Planck equation. As an introduction to the following procedure, it is instructive to focus on the evolution of a single realization in time governed by the transition probability density function (TPD). For simplicity, the TPD is sketched in Fig.1 in a two-dimensional normalized phase plane determined by the displacement x and the velocity $\dot{x}$. In Fig.1 the drift as well the diffusion of the TPD can be recognized. The centre of the TPD is spiraling toward the origin due to the drift, whereas the TPD is spreading due to the diffusion.

For reliability analysis, the probability flow into the low probability region is of paramount interest. To approximate this flow, the following procedure denoted as "Double and Clump" (D&C) utilizes doubling of samples at certain time intervals. Doubling of state vectors (samples) at a certain time step simply means that an identical copy of the state vector is made. Let $\mathbf{x}_{k0}(t)$ and $w_{k0}(t)$ be the state vector to be doubled and the weight before doubling, respectively. Doubling implies the substitution of the pair ($\mathbf{x}_{k0}(t),w_{k0}(t)$) by two pairs ($\mathbf{x}_{k1}(t),w_{k1}(t)$) and ($\mathbf{x}_{k2}(t),w_{k2}(t)$) with

$$\mathbf{x}_{k1}(\tau) = \mathbf{x}_{k2}(\tau) = \mathbf{x}_{k0}(\tau) \tag{3}$$

$$w_{k1}(\tau) = w_{k2}(\tau) = \frac{1}{2} w_{k0}(\tau) \tag{4}$$

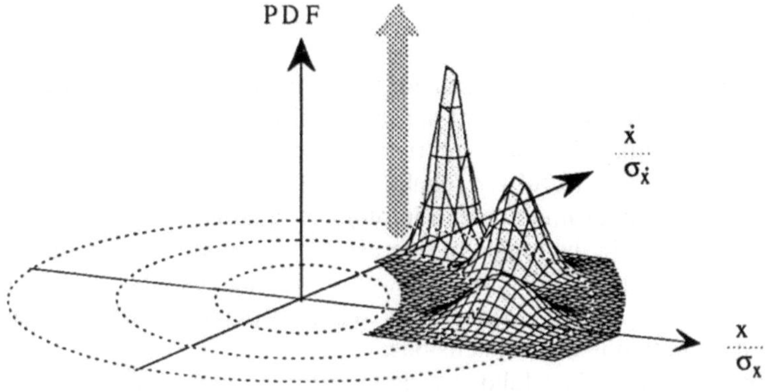

Fig.1:Evolution of the Transition Probability Density (TPD) with respect to time

Hence, the representation of the PDF by the sampling distribution is not altered at time t. The effect of doubling can be observed only at a time t > t. Starting with t = t, the loading increments are assumed to be independent in both doubled state vectors. Hence two different random paths in the state space will be observed due to the differences in loading for t > t. The effect of doubling at several consecutive time steps is shown in Fig.2, where the thickness of the paths indicate their respective weights. From Fig.2 it is obvious that the branching due to doubling approximates the TPD quite well and that the probability flow into the low probability range can be represented by a sequence of doubling. Due to the exponential growth of the sample size in each time step, however, this procedure is not practically feasible.

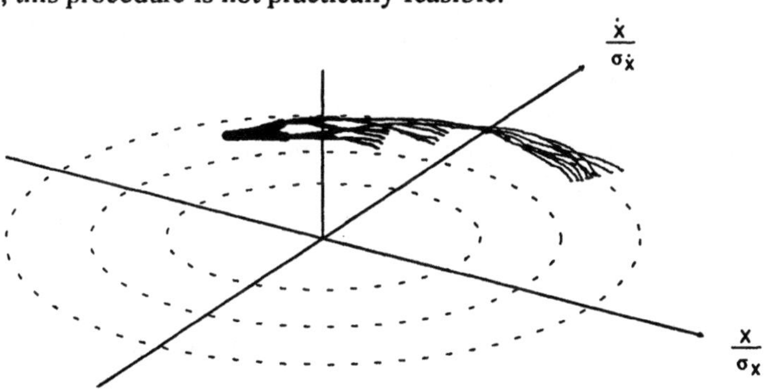

Fig.2: Approximation of the TPD by doubling

It is worth noting that the restriction to Markovian type of loading is not an indispensable requirement. There is no essential difficulty to incorporate other loading processes defined, for example, by an ARMA model. The only important requirement for this procedure is the independence of the loading increments.

3.2. Double & Clump

In this section, a procedure is presented, which maintains the basic advantages of doubling yet avoids the growth of the sample size. Doubling in each time step is not only infeasible, but also inefficient since only a tiny fraction of all generated samples are useful for reliability analysis. In order to enhance the efficiency, some type of importance sampling is required where the number of samples is increased in the region of interest and decreased in the region of less importance. Importance sampling, however, is not readily applicable for stochastic dynamic loading, since it is quite difficult to obtain a suitable sampling distribution in high dimensions.

The D&C procedure follows a quite different approach[9]. This approach requires the simultaneous simulation of all realizations, i.e. all simulated state vectors $y_k(t)$, k=1(1)nSim, must be available at a given time t. It is further assumed that a scalar function expressing the importance of each realization is available. For this purpose, an energy criterion is suggested in the following for quantifying the importance, $c_k(t)$, of each of these realizations defined by the state vectors y_k, k=1(1)nSim,

$$c_k(t) = \left(E_{k,pot}(t) + E_{k,kin}(t)\right) \cdot P_{k,exc}(t) \cdot w_k(t)^{\beta} \quad , \tag{5}$$

where $E_{pot}(t)$ denotes the potential energy of the mechanical system, $E_{kin}(t)$ denotes its kinetic energy, $P_{exc}(t) = f^T(t) \cdot \dot{x}(t)$ (f(t): external force, $\dot{x}$ (t): velocity of the response) denotes the input power by the external excitation and $w_k(t)$ is the weight of each particular realization. The parameter ß is used to control the importance of realizations in the tails. It is generally of the order of 0.1 - 0.5. Using a small value for ß, more samples in the tails of the distribution are generated; the accuracy of the procedure, however, decreases. The above criterion is useful for problems where the number of realizations in the tails of the stochastic response needs to be increased. This is the case for most practical reliability problems. However, depending on the problem, other suitable criteria may be used as well.

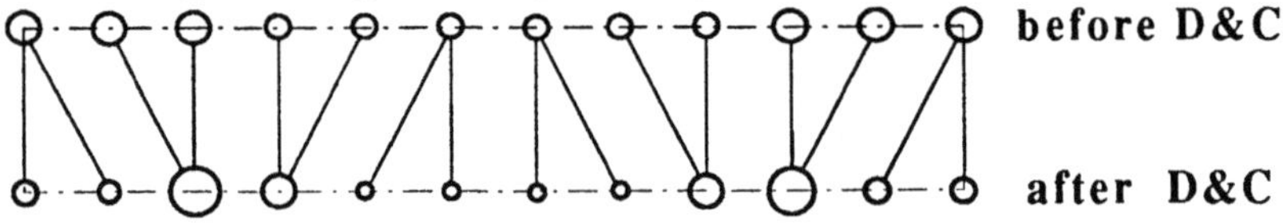

Fig.3: Sketch of the D&C procedure at a given time step

In a further step, the scalar values $c_k(t)$, $k = 1(1)$nSim which express the relative importance of all realizations of the stochastic response are then sorted. At a given time step, a certain number, e.g. M, of the most important realizations are doubled. In order to keep the sample size nSim constant, M pairs of realizations of less importance are then clumped together. This essential step of the D&C procedure is sketched in Fig.3, where an "I" indicates an important realization and "L" indicates a less important realization. The size of the circles indicates the weight of the realizations before and after D&C.

In order to be accurate, it is most important that the procedure affects the statistics of the samples only insignificantly. As pointed out above, doubling has no effect on the sample statistics. Clumping, however, does affect the sample statistics since it is impossible to maintain the same amount of information by less samples. The single state vector $\mathbf{x}_{j0}(t)$ resulting from clumping together a pair of state vectors $\mathbf{x}_{j1}(t)$ and $\mathbf{x}_{j2}(t)$ is selected as the mean, i.e.

$$w_{j0}(t) = w_{j1}(t) + w_{j2}(t) \tag{6}$$

$$\mathbf{x}_{j0}(t) = \frac{1}{w_{j0}(t)}\left[\mathbf{x}_{j1}(t) \cdot w_{j1}(t) + \mathbf{x}_{j2}(t) \cdot w_{j2}(t)\right] \tag{7}$$

Clumping to the mean value of the pairs ensures identical first moments. Due to effect of clumping, higher moments, however, are generally slightly underestimated. The main effort of the D&C procedure is actually devoted to keeping the effect on the sample statistics small. The key for keeping the distortion due to clumping small is the fact that the difference of the higher moments of the statistics depends mainly on the distance between the pairs of clumped state vectors expressed by

$$d_{ij}(t) = \sum_{k=1}^{nState} \frac{\left|x_{ki} - x_{kj}\right|}{x_{k,max} - x_{k,min}} \cdot 2^{\alpha(k)} \quad , \tag{8}$$

where the first index k denotes the component of the state vector, i and j denote two different realizations, and $x_{k,max}$ and $x_{k,min}$ are the maximum and minimum value of the n-th component among all nSim realizations. For avoiding problems associated with the large dimension of the state vector x(k), k=1(1)nState, only the most important components are considered. For this purpose, the array a(k) is introduced. Presently, the algorithm is restricted to less than 15 components with a(k) > 0. The factor $2^{a(k)}$ is then used in eq.(8) to express the respective importance of each component. Clearly, the sampling statistics are better preserved if the distances $d_{ij}(t)$ are smaller for all clumped pairs. To determine these pairs, however, is not a trivial task in a realistic case where the sample size is of the order nSim=1000. The introduction of an upper limit for the weight which a clumped realization can reach proved to be useful to stabilise the D&C procedure. This limit avoids an ever- decreasing sampling density in the region of highest probability and ensures a similar accuracy as MCS in this domain. The limit 2/nSim is presently used. Pairs which result in a larger weight than this limit are not considered for clumping.

Although clumping is optimized to keep the sampling statistics as close as possible to the origin, a very small bias of higher moments remains. To avoid any bias in the D&C procedure, all realizations are distorted slightly in such a way that the second moments of all components remain identical. Hence the statistics of the realization are affected only insignificantly by the D&C procedure.

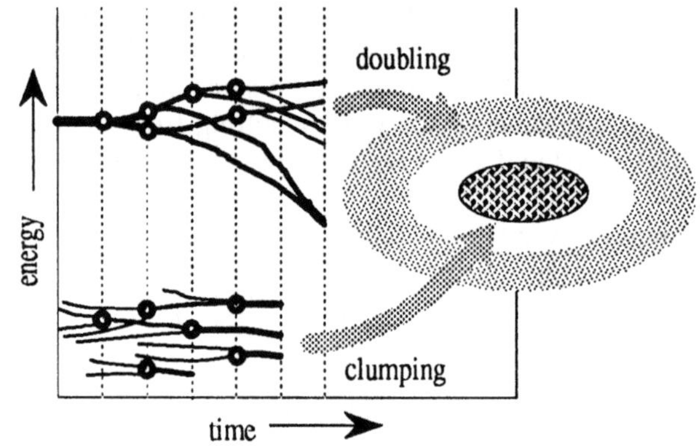

Fig.4 Sketch of the "Double & Clump" procedure in consecutive time steps

The evolution of the realizations in consecutive time steps is shown in Fig.4. Realizations with a high mechanical energy and power are doubled while realizations with less energy are clumped. Again, the thickness of the path indicates the weight which decreases after doubling and increases after clumping.

Finally, it is interesting to note that an algorithm with similar basic properties as the introduced "Double & Clump" has been applied successfully in nuclear physics[10] for solving neutron transport problems. This has been brought to the attention of the authors just recently.

4. NUMERICAL EXAMPLE

In the following, the procedure is now applied to a realistic MDOF-system, in particular to a 6-storey building, subjected to earthquake excitation. When evaluating the restoring forces of the steel columns, the P-Δ effect and elasto-plastic constitutive stress - strain relations have been considered. This example has been analysed in a previous paper[7] utilizing the Response Surface Method (RSM) and non-Gaussian equivalent linearization where all structural data, as well as the stochastic loading, are defined. The limit state is defined by specific amounts of accumulation of plastic deformations as shown in Fig.5. It is assumed that damage is a function of this drift expressed by the ratio γ_i of plastic displacement $d_{pl,i}$ to the onset of yielding displacement u_{yi} for the i-th floor :

$$\gamma_i = \frac{d_{pl,i}}{u_{yi}} \tag{9}$$

In ref.7 the exceedance probabilities of the drift level γ_i corresponding to 2, 3 and 5 were obtained using the RSM. The 6-storey frame was analysed for an earthquake excitation within a duration of 7.5 sec. The calculation has been carried out for 8 and 28 combinations of pairs of the eight most important variables in order to obtain additional points on the limit state function (response surface). Based on this surface, the exceedance probabilities (i.e. the probabilities of failure in this context) were obtained for the respective levels of accumulated drift γ_i = 2,3,5 as shown in Fig.6. The RSM results of Fig.6 are taken from ref.7.

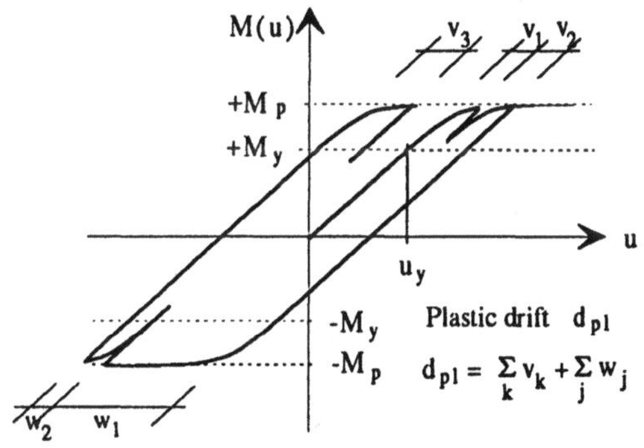

Fig.5: The accumulation of drift as damage indicator

For comparison, the response statistics obtained by direct MCS with nSim = 10000 samples and D&C procedure with nSim = 1600 samples are also shown in Fig.6. The results obtained by the RSM, D&C and direct MCS (DMC) reveal an excellent agreement. They also show the main drawback of direct MCS, which in this example (nSim=10000) can not assess the exceedance probabilities smaller than 10^{-4}. At the same time, these results demonstrate the main advantage of D&C, which can easily cover exceedance probabilities of the order 10^{-6} with the number of samples six times less (nSim=1600) than direct MCS. It can be seen that the results for the first floor with nSim = 10000 and nSim = 1600 fit quite well even in the low probability domain of order 10^{-4}-10^{-5}. It also shows that only the first and fifth floors experience considerable accumulated drift, whereas all others (for example - the third floor) remain almost unaffected by plastic deformations.

Finally it should be mentioned that the accuracy of the suggested D&C procedure has been tested extensively by comparing the results with known analytical solutions for special simple SDOF systems subjected to white noise excitation and linear MDOF systems[9,11]. It has been shown that the agreement with theoretical solutions fits quite well within the range 10^{-7} <CDF< $1.0\text{-}10^{-7}$.

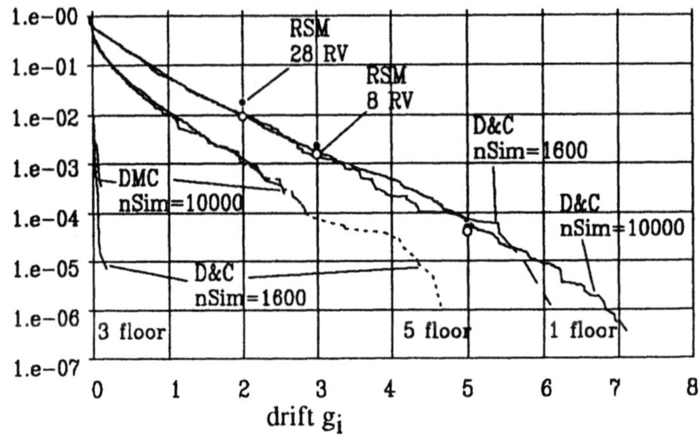

Fig.6: Exceedance probabilities of cumulative drift g_i in the first ,third and fifth floors after t =7.5 sec (RSM results are taken from ref.7)

DISCUSSION AND OUTLOOK

Based on the results presented in this paper, the following conclusions can be drawn:

(1) Reliability estimates of nonlinear structures subjected to dynamic random loading can be assessed by numerical procedures such as the Response Surface Methodology (RSM) and the presented "Double & Clump" procedure. Both methods have no restrictions with respect to mechanical modeling, and hence numerical procedures reflecting the current state of the art may be utilized.

(2) "Double & Clump" is most advantageous for random loading represented by a sequence of random variables as used in a Markov process and ARMA models.

(3) Further research on the D&C procedure may be helpful to enhance the accuracy and efficiency of the proposed method. One of the possible extensions is to drop the restriction of clumping two realizations or even to develop alternative procedures to avoid excessive growth of the samples due to doubling. The method requires also the development of objective selection criteria expressing the importance of the realizations. The rather heuristic energy criterion given in the paper works well, but it may not be the best choice.

ACKNOWLEDGEMENT

This work was partially supported by the Austrian Industrial Research Promotion Fund (FFF) under project No.6/636, which is gratefully acknowledged by the authors.

REFERENCES:

1. Schuëller,G.I., Pradlwarter H.J., Pandey M.D., Methods for Reliability Assessment of Nonlinear Systems under Stochastic Dynamic Loading - A Review, Proceedings of EURODYN'93, Balkema, 1993, pp.751-759.
2. Bucher, C.G., Bourgund, U., A fast and efficient response surface approach for structural reliability problems, *Structural Safety*, Vol. 7, No. 1, (1990), pp.57-66.
3. Faravelli, L., Response Surface Approach for Reliability Analysis, *Journal of Engineering Mechanics*, ASCE, Vol. 115, No. 12, (1989), pp.2763-2781.
4. Schuëller,G.I., Bucher, C.G., Bourgund, U., Ouypornprasert, W., On Efficient Computational Schemes to calculate Structural Failure Probabilities, *Probabilistic Engineering Mechanics*, Vol. 4, No 1, (1989), pp.10-18.
5. Pradlwarter, H.J., Schuëller,G.I., Jehlicka, P., Steinhilber, H., Structural Failure Probabilities of the HDR-Containment, *Nuclear Engineering and Design*, 128, (1991), pp.237-246.
6. Bucher, C.G., Chen,Y.M., Schuëller,G.I., Time variant reliability analysis utilizing response surface approach, in: Reliability and Optimization of Structural Systems 88, ed. P.Thoft-Christensen, Lecture Notes in Engr, 48, Springer Verlag, Berlin, 1989, pp.1-14.
7. Schuëller,G.I., Pradlwarter, H.J., Bucher, C.G., Efficient Computational Procedures for Reliability Estimates of MDOF Systems, *Int.J. Non-linear Mechanics*, Vol.26,No.6, (1991), pp.961-974.
8. Schuëller,G.I., Bucher, C.G., Computational Stochastic Analysis - A Contribution to the Software Development for the Reliability Assessment of Structures under Dynamic Loading, *Probabilistic Engineering Mechanics*, Vol.6, Nos 3 /4, (1991), pp.134-138.
9. Pradlwarter H.J., A Selective MC simulation Technique for Non-linear Structural Reliability, Proc. ASCE Speciality Conference, Denver, Colorado, July 8 -10, 1992, ASCE, N.Y. USA, 1992, pp.69-72.
10. Spanier J., Gelbard E.M., Monte Carlo Principles and Neutron Transport Problems, Addison-Wesley Publishing Company, Mass., USA, 1969.
11. Pradlwarter, H.J., Schuëller,G.I., Melnik-Melnikov, P.G., Reliability of MDOF-Systems, to appear in *Probabilistic Engineering Mechanics*, 1994.

Grain Boundary Failure and Geometrical Models of Creep Damage

H. Riesch-Oppermann [1] and A. Brückner-Foit [2]

[1] Kernforschungszentrum Karlsruhe, Institut für Materialforschung II, P.O.Box 3640, D-76021 Karlsruhe, FRG
[2] Universität Karlsruhe, Institut für Zuverlässigkeit und Schadenskunde im Maschinenbau, P.O.Box 3640, D-76021 Karlsruhe, FRG

Abstract. A geometrical model for the grain structure of a polycrystalline material forms the basis of a statistical model of creep damage. Damage is introduced into the grain boundary structure by failure of the grain boundary facets. Suitable relations such as creep potentials are used to relate damage parameters on a microscopic level to macroscopic quantities such as the creep strain rate and the creep life. Two examples with different materials are presented together with experimental and simulation results.

Keywords. Creep damage, creep potential, grain boundary failure, damage simulation, Stochastic Geometry, Dirichlet tesselation

1 Introduction

The paper intends to show how the scatter of the macroscopic creep damage can be related to the scatter of the microstructure of the material, which is an important task in the field of lifetime predictions based on micromechanical models of creep damage. The basic idea is to perform statistical simulations of creep damage on a microscopic level and to have then a parameter, which allows one to relate the microscopic damage to macroscopic quantities like the creep strain rate or the creep life. This parameter should also be related to experimental investigations of creep damage.

In the following, a specific class of creep damage is considered. Damage is assumed to be caused by the nucleation and growth of voids on grain boundaries.

In this case, the scatter in the microstructural area comes mainly from two sources. The first is the geometrical one, characterized by the scatter in grain size and orientation of the grain boundaries. The second source is damage-related and is characterized by the scatter of the input-variables for local damage evolution.

This leads to a simulation procedure consisting of two steps: In a first step, the grain boundary structure of the material, as it appears in metallographic sections, is simulated. In a second step, damage is introduced into the grain boundary structure.

The results of the simulation have then to be compared with experimental results in order to draw conclusions about the applicability of the selected models. Modifications and improvements of the models have to be made according to experiments.

It is shown how the simulation of the microstructure can be performed using methods of stochastic geometry, how the simulated microscopic damage can be incorporated in a macroscopic formulation and how simulated and experimental results fit together.

2 Creep damage by cavitating grain boundary facets

On a micromechanical scale, different mechanisms of creep damage can be identified [1]. An important creep damage mechanism is the formation and growth of voids on grain boundaries. In metals, voids nucleate mainly on grain boundaries which are oriented perpendicularly to the loading direction. Existing voids grow by stress directed diffusion of matter from the surface of the void into the grain boundary. The growth rate of the voids is proportional to the normal stress σ_b acting on the grain boundary.

Two limiting cases of void growth can be distinguished. If the stress σ_b acting on the grain boundary coincides with the remote applied stress σ_n, void growth is stress-controlled. This corresponds to a situation where we have 'isotropic damage' of the material with the amount of cavitation on each grain boundary being representative for the whole section of the material. For isolated cavitating facets, however, the surrounding grains, which have non-cavitated grain boundaries, may have a supporting influence on the cavitated grain boundary, which leads to a relaxation of the normal stress σ_b that controls void growth. If the stress on the grain boundary relaxes completely ($\sigma_b \simeq 0$) the cavitated facet acts as a microcrack [2]. The growth of voids is then mainly determined by the deformation of the adjacent grains and thus becomes strain-controlled. $\sigma_b \simeq 0$ corresponds to a situation where the nucleation of voids on the grain boundary is identical with the formation of a microcrack. In an advanced stage of damage, the isolated microcracks begin to interact and coalesce. Final failure occurs by formation of a dominating crack.

This damage mode, which has been observed in creep experiments, will be considered in this paper.

3 Relation between microscopic damage and macroscopic response

The presence of microcracks in a material affects its macroscopic deformation behaviour. Hutchinson [3] gives an expression for the creep strain rate of a material containing sparsely distributed (i.e. non-interacting) penny-shaped microcracks oriented perpendicularly to the loading direction. The additional deformation rate induced by the presence of microcracks is obtained in terms of a creep potential Φ which consists of a term Φ_0 related to the undamaged material and a term Φ_c indicating the presence of the microcracks. Specifically, one obtains

$$\Phi = \Phi_0 + N\ \Phi_c \quad ,$$

where N is the number of cracks per unit volume. The following expression for the creep strain rate is obtained for uniaxial loading and a material obeying NORTON's creep law with parameters B and n:

$$\dot{\varepsilon} = (1 + \rho)B\sigma^n \quad ,$$

where ρ denotes the damage-induced part of the creep strain rate and is related to the volume density of microcracks, N, by

$$\rho = \frac{n+1}{2\sqrt{1+\frac{3}{n}}}\ d^3\, N$$

for cracks with diameter d.

If the interaction of microcracks cannot be neglected, a self-consistent model, e.g. as introduced by Riedel [4], leads to the following relationship for the creep strain rate:

$$\dot{\varepsilon} = \frac{B\sigma^n}{1-\rho} \quad .$$

The parameter ρ, which plays the role of a damage parameter is related to experimental damage parameters, which can be determined by ultrasonic measurements or from metallographic sections of crept specimens [5].

The important feature of the damage parameter ρ is the fact that it links the microscopic damage and macroscopic creep behaviour of the material and that it is related to experimental damage parameters.

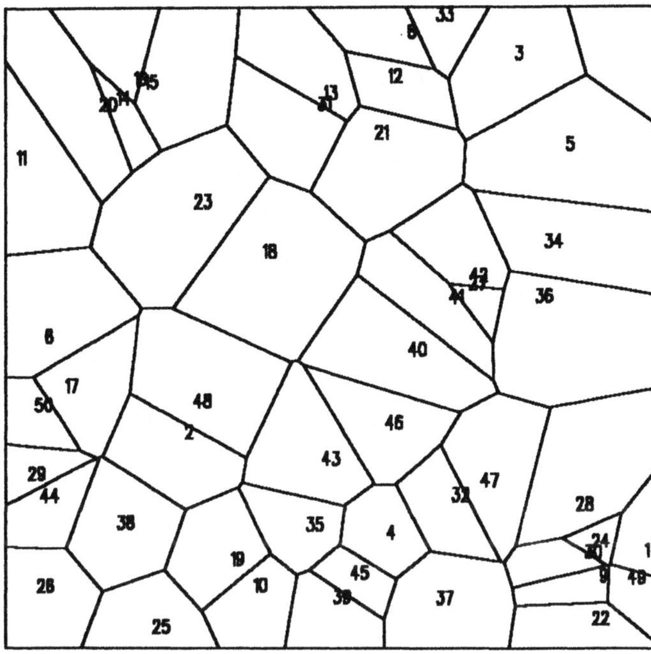

Figure 1. Dirichlet tesselation of a window containing 50 points. Labels of points centred at their respective locations

4 Simulation procedure for grain boundary structure and creep damage

For a given physical model, simulation can replace some part of the experimental investigations by calculations and may lead to the verification of the model under consideration. Using different physical models, simulation can give hints about how these models affect the results. Thus, it is possible to study the influence of specific variables which in turn may lead to deeper insight about the dominant mechanisms.

In our case, the simulation procedure consists of two steps: In a first step, the grain boundary structure as it appears in metallographic sections is simulated. In a second step, damage is introduced into the grain boundary structure.

If we assume that grains form by isotropic crystallization starting from germs in a metallic melt, grain boundaries of neighbouring grains are the midplanes of the respective germs. Mathematically, the germs are created as outcomes of a Poisson-point-process in a certain part of the plane which is called a window for convenience. Each point is associated with a cell consisting of all points whose distance to the respective point is less than to all other points of the window. This leads to a partition of the window into cells, the so-called Dirichlet tesselation (also known as POISSON-VORONOI-mosaic). The cells of the Dirichlet tesselation correspond to the grains of a metallographic section and the boundaries of the cells to the grain boundaries.

Here we idealize in the sense that we assume the locations of the germs to be in the plane of the metallographic section, which is not true but seems to be a reasonable assumption for the time being.

The Dirichlet tesselation is determined by use of an algorithm given by Green & Sibson [6]. The great advantage of this algorithm is that it allows one to include additional points into a given configuration, with the amount of work neccesary for this being almost independent of the number of existing grains. Figure 1 illustrates a Dirichlet tesselation with 50 points in a square window. Upon completion of the algorithm, we have for each grain a so-called contiguity list of adjacent grains and the coordinates of the grain boundaries which in a later stage can be used to model interaction effects of neighbouring grain boundaries which is normally a very complicated and time-consuming task.

The first step of the simulation is now completed. A random grain boundary structure is obtained which is based on randomly distributed germs in a window which can be thought of as part of a metallographic section. The characteristics of this grain boundary pattern depend on the characteristics of the underlying point process. Now, in a second step,

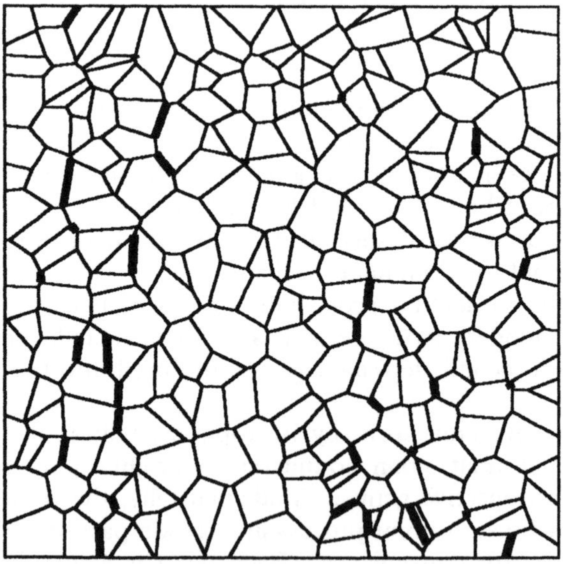

Figure 2. Simulated metallographic section by use of Dirichlet tesselation

damage has to be introduced into the grain boundary structure. A criterion for the failure of a grain boundary facet is necessary to model creep damage by formation of microcracks on a grain boundary. At the moment, this is not yet available. Damage is therefore modeled in a qualitative way. Experimental investigations of an austenitic steel show that the formation of microcracks occurs preferentially on grain boundaries which are oriented perpendicularly to the loading direction. This fact is used for the damage simulation of the grain boundary facets.

A failure probability which depends on the orientation of the facet relative to the loading direction is assigned to each facet. The dependencies of grain boundary failure on other parameters is included in the failure probability in terms of a reference value for the failure probability. This corresponds to the assignment of a certain 'damage level'. At the end of the simulation procedure, a list of cavitated and non-cavitated boundaries for each grain of the random cell structure is obtained.

5 Comparison between simulation and experiment

For the time being, damage is introduced into the simulated grain boundary configuration in a purely phenomenological way. Naturally, one can only expect a qualitative result from this approach. Figure 2 and Figure 3 show damage patterns obtained by simulation and from experiments, respectively. The metallographic section was obtained within investigations of creep damage of an austenitic steel AMCR 0033, which were performed at the JRC Ispra [5]. This kind of material shows damage by failure of

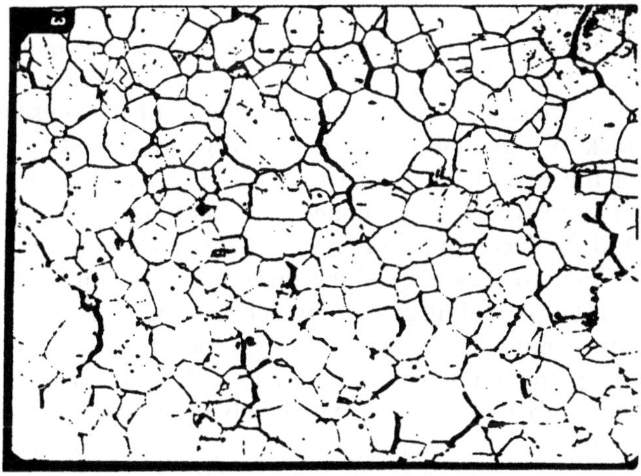

Figure 3. Metallographic section of the austenitic steel AMCR 0033

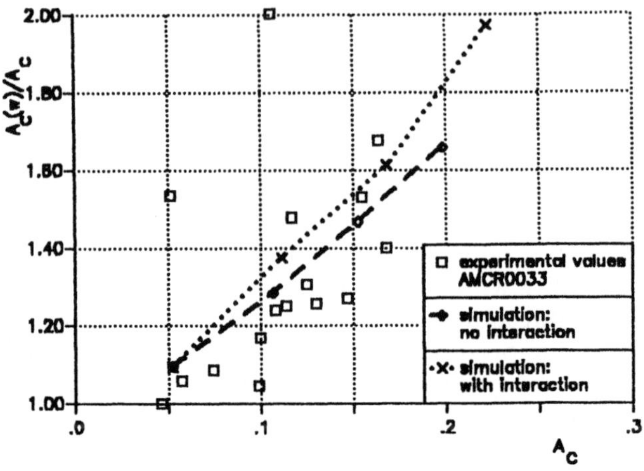

Figure 4. Assessment of interaction effects.

grain boundary facets, which are oriented perpendicularly to the loading direction. For the simulation of damage, no specific damage mechanism of the grain boundary facets is included. It is merely assumed that damage of the grain boundaries depends on the orientation with respect to the loading direction. Nevertheless, at a first glance, simulated and experimentally obtained damage patterns show a surprisingly good agreement.

Obviously, more quantitative measures are needed to characterize the differences and the common features of simulated and experimental patterns. Experimentally, damage is determined by line section methods which lead to an estimation of a damage parameter A_c, the fraction of cavitated grain boundaries along a test line on the surface of the metallographic section. The simulated patterns can be evaluated in the same way with the advantage that bias resulting from different treatments of simulated and experimental results can be avoided.

To assess damage interaction effects, that is, effects of the presence of an existing microcrack on the failure of adjacent grain boundary facets, a weighted form of the parameter A_c can be used. The weighted form $A_c(w)$ of the damage parameter A_c is obtained by counting a cavitated grain boundary n-fold, if the facet contributes to a microcrack which is made up of n facets. The fraction $A_c(w)/A_c$, which gives the mean number of facets per crack, can then be plotted against A_c, which is done in Figure 4. Here, damage interaction is modelled by attributing an increased value for the failure probability to grain boundary facets adjacent to existing microcracks, dependent on the number of facets of the already existing microcrack. Figure 4 shows an increase of $A_c(w)/A_c$, which may be due to two reasons: first, there is a purely random effect for the cavitation of facets adjacent to existing microcracks, and second, the increase of $A_c(w)/A_c$ may be due to

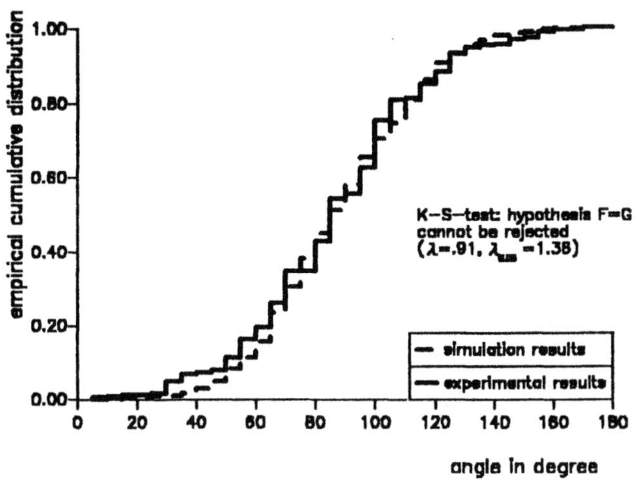

Figure 5. Empirical distribution of crack orientation with loading direction.

interaction effects. As can be seen, the scatter coming from the experimental results is quite large, so that it is not possible to draw any conclusions about interaction effects, but the qualitative agreement of experiment and simulation in the tendency of damage evolution is quite surprising.

Quantitative comparisons of simulated and experimentally obtained damage patterns may also be obtained with the help of empirical distributions of some selected quantities. First, the orientation of the cracks which initiate on the grain boundaries is considered. Figure 5 shows the empirical distribution function of the angle between a crack and the loading direction for both simulation and experiment. A statistical Kolmogorov-Smirnov-test shows the distributions of experimental and simulated crack orientations are compatible with each other and, hence, the model chosen for the orientation dependence is in agreement with the findings of the experiment.

Geometrical quantities of the simulated and experimental patterns are also the area of the grains and the length of the formed cracks. These quantities reflect the geometrical differences of the real grain structure and the POISSON-VORONOI-mosaic which is used as a geometrical model for the simulation. As there is no natural length scale in the simulation, the median of the grain area distribution is used to adapt the scale of the metallographic section obtained in the experiments and the POISSON-VORONOI-mosaic obtained in the simulation procedure. If we do this, Figure 6 shows that the simulation (DZ1P1000) tends to contain less grains with very small grain sizes and also less grains with relatively large grain sizes, which means that the distribution of simulated grain areas has small-

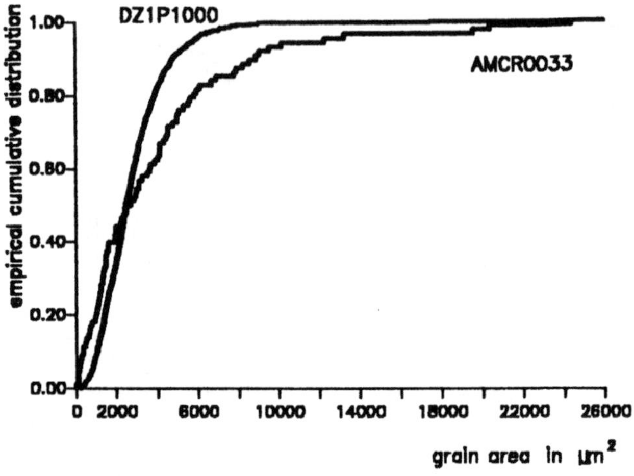

Figure 6. Empirical distribution of grain area. The length scale of the simulation is obtained by imposing identical values of the median for both experimental and simulated grain structures.

er variance than that of the experimental grain areas (AMCR0033). This is due to the fact that we simulate a 2-D mosaic, whereas the metallographic section is a section of a 3-D grain arrangement.

Figure 7 shows the empirical distribution function of the observed together with the simulated crack length. Here, we observe that the simulated crack length tends to be shorter than the experimentally observed one. Two reasons may explain this behaviour. The first reason may be the fact that very short cracks are difficult to find on metallographic sections so that there is a bias in the empirical distribution of the observed crack length. The second reason comes from a specific property of the used 2-D mosaic for the simulation, which has a relatively large fraction of very short edges leading to very short cracks. The first reason leads to a shift of the experimental distribution towards larger crack lengths, whereas the second reason leads to a shift of the simulated distribution function towards smaller crack length values.

Hence, while there are inherent differences in the geometrical properties of simulated and experimentally obtained grain structures, the orientation dependence of the underlying damage mechanism is well represented by the simulation model.

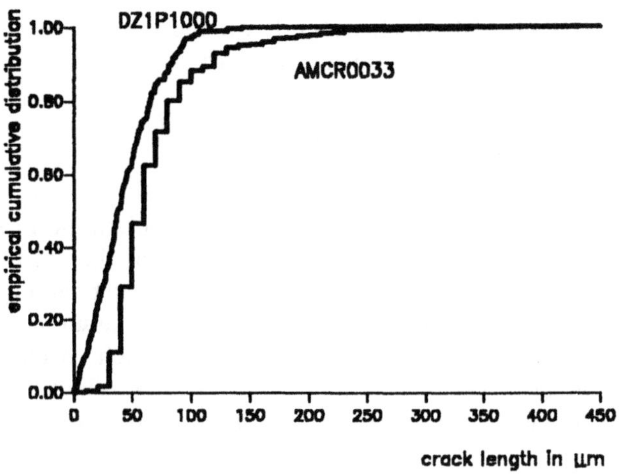

Figure 7. Empirical distribution of crack length.

6 Creep damage of ceramics: a tentative example

In the preceding part of the paper, damage introduced into the stochastic grain boundary structure was based on phenomenological arguments rather than on a micromechanical model. In the following, an example will be given where a micromechanical description of damage is used. The purpose of this example is to show that identical grain boundary configurations may lead to completely different damage patterns depending on the active mechanisms of damage.

The material under consideration is alumina subjected to compressive creep loading. Following Chan & Page [7], nucleation of cavities on grain boundaries occurs as the result of stress concentrations at grain boundary ledges due to discontinuous grain boundary sliding events. The nucleation of cavities is described by classical nucleation theory (see e.g. [4]). The stress concentrations which are necessary to form critical clusters of vacancies on the grain boundary result from two competitive processes: the evolution of stress concentrations when grain boundary sliding occurs by transformation of shear stresses along the grain boundaries into normal stresses on the grain boundary ledges and the relaxation of stress concentrations by grain boundary diffusion. Grain boundary sliding relaxes within a characteristic time, t_{BR}, whereas diffusion processes relax within a characteristic time, t_c. t_{BR} and t_c depend on grain boundary characteristics and diffusion parameters [7]. A time-dependent stress concentration factor q is given by Chan & Page [7] which depends on the characteristic times for grain boundary sliding, t_{BR}, and diffusion, t_c, with

$$q(t) = \frac{1}{2}\left\{\frac{\lambda}{h}[1 - \exp(-t/t_{BR})]\exp(-t/t_c) + c\right\}$$

where c depends on the kind of loading and on the orientation of the grain boundary, t is time and h, λ are height and spacing of the grain boundary ledges, which are assumed to be periodically distributed along the grain boundary. Nucleation of voids on a grain boundary is assumed to occur if the nucleation rate for the maximum of q, which occurs at time $t = t_{BR} \times \ln(1 - t_c/t_{BR})$, exceeds a certain critical value, which is arbitrarily set to 1. Since nucleation increases very rapidly with the stress level the

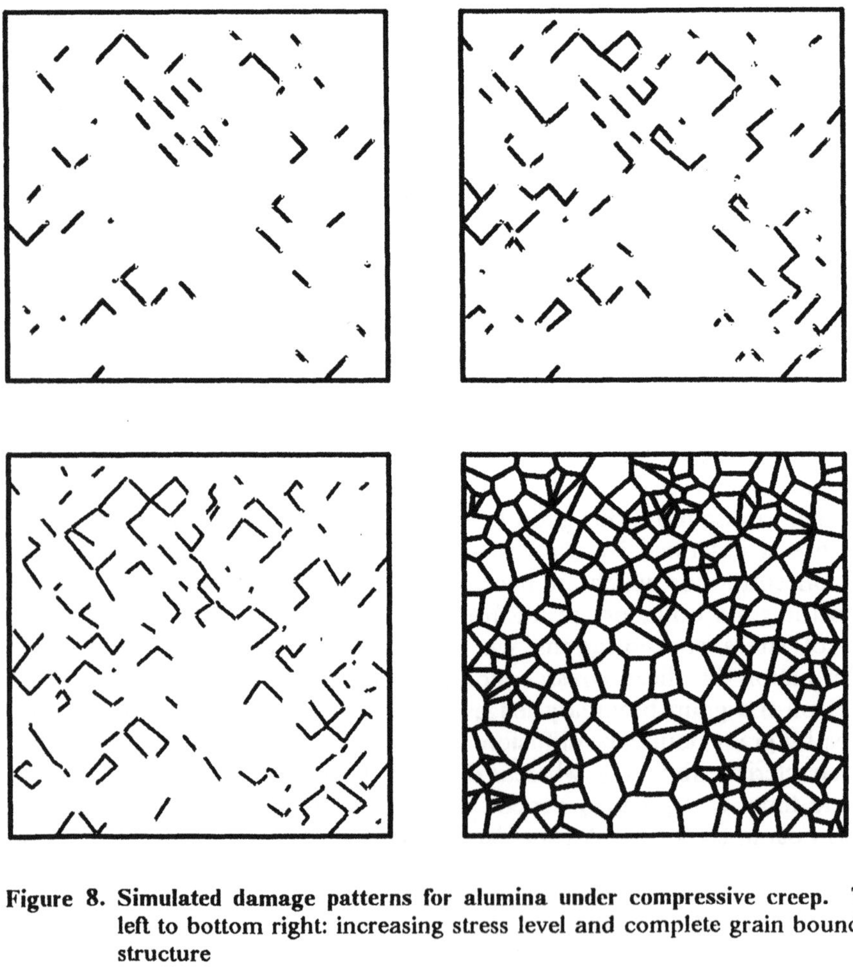

Figure 8. Simulated damage patterns for alumina under compressive creep. Top left to bottom right: increasing stress level and complete grain boundary structure

absolute value of the critical nucleation rate does not affect the results of the simulation qualitatively. The quantities for diffusion and grain boundary structure are treated as deterministic variables and the only source of scatter is the random orientation of the grain boundaries. It is assumed that cavitation occurs in a constrained way, which means that the nucleation of voids on a grain boundary leads to failure of the grain boundary.

Figure 8 shows typical examples of the damage patterns obtained for different stress levels. For low stress levels, failed grain boundary facets are oriented preferentially around an angle of 45° with respect to the loading direction, corresponding to the maximum shear stress, which is responsible for grain boundary sliding. With increasing stress, facets with orientations deviating significantly from 45° begin to cavitate. These simulations were performed without any experimental background. The aim was to show the flexibility of the stochastic model. From Figure 8 it can be seen that, with the same underlying type of grain boundary structure as in Figure 2, very different damage patterns result as a consequence of the different damage mechanisms.

Meanwhile, we have learned that recent experimental observations on creep and grain boundary sliding were performed [8], which might perhaps be used to adapt the damage model used in the simulation procedure.

7 Conclusions

In this paper, some ideas concerning the statistical modelling of damage on a microstructural level and its relation to macroscopic quantities are given.

Two examples using different kinds of approaches for the damage modelling are presented. In the first example, creep damage of an austenitic steel is modelled in a phenomenological way according to experimental observations. This leads to patterns of simulated grain boundary failure, which are in qualitative agreement with experimental findings. A method is indicated how to incorporate micromechanical models in macroscopic relations for creep behaviour. Some qualitative aspects of the experimentally obtained and the simulated damage patterns are shown.

In the second example, a micromechanical model for creep damage of alumina is incorporated into the simulated grain boundary structure.

Characteristic patterns of the different simulation procedures are shown and illustrate the flexibility of the geometrical model which allows to incorporate any kind of grain boundary failure.

As a first step, the results seem to be encouraging. The advantages of the use of stochastic geometry methods in the light of the possible inclusion of more sophisticated models for grain boundary failure and the interaction effects of cavitated grain boundary facets are twofold:

Any kind of grain boundary failure can be handled by the stochastic model which is therefore applicable to various kinds of material. Interaction effects of cavitated grain boundary facets can be handled in a very efficient way, which allows the simulation of realistic configurations without a prohibitively large amount of computing time.

The framework of the stochastic geometry seems therefore to be a very efficient tool for the development of micromechanical damage models and their relation to a macroscopic description of damage.

Acknowledgement: The financial support of the Deutsche Forschungsgemeinschaft under grant No. Mu 466/15-2 is gratefully acknowledged.

8 References

1. M.F. Ashby, B.F. Dyson, Creep Damage Mechanisms and Micromechanisms, *Advances in Fracture Research (Fracture 84),* Vol. 1, 3-30.
2. B.F. Dyson, Continuous Cavity Nucleation and Creep Fracture, *Scripta Metall.* **17** (1983), 31-37.
3. J.W. Hutchinson, Constitutive Behavior and Crack Tip Fields For Materials Undergoing Creep-Constrained Grain Boundary Cavitation, *Acta Metall.* **31** (1983), 1079-1088.
4. H. Riedel, Fracture at High Temperatures, *Springer Verlag, Berlin 1987.*
5. H. Stamm, U. von Estorff, The Use of Creep Damage Measurements in Combination With Material Models for Residual Lifetime Assessment, in: *ECF 8 Fracture Behaviour and Design of Materials and Structures,* Turin, 1990.
6. P.J. Green, R. Sibson, Computing Dirichlet tesselations in the plane, *The Computer Journal* **21** (1978), 168-173.
7. K.S. Chan, R.A. Page, Continuous cavity nucleation by stochastic grain boundary sliding, *J. Mat. Sci.* **25** (1990), 4622-4629.
8. C.R. Blanchard, R.A. Page, Grain Boundary Sliding Microdisplacement Measurements During the Creep of Alumina, *J. Am. Ceram. Soc.* **75** (1992), 1612-1620.

Reliability of Unidirectional Composites Using Genetic Algorithms

David G. Robinson[1], Gregory Toussaint[1]

[1]Department of Aeronautics and Astronautics,
Air Force Institute of Technology, WPAFB OH 45433

Abstract. A number of very successful techniques exist for the estimation of structural reliability under the general headings of first order or second order reliability methods (FORM/SORM). However, these methods rely heavily on gradient search methods for estimation of the safety index. When the failure region can be explicitly defined and is adequately understood, these methods perform extremely well. However, when the limit state function is particularly complex or the design variables are not continuous in support, then gradient search methods are at best difficult and often impossible to apply. This situation occurs frequently when investigating optimal design strategies. Other analysis methods based upon Monte Carlo simulation can be used to estimate probability of exceeding the limit state but lack the sensitivity information and other insights that use of a safety index provide. An alternative method is proposed here based upon genetic algorithms. The method is fundamentally a search method; exploring the interaction between the limit state function and the design variables. Emphasis in this paper is on the application of this method in determining the optimal fiber orientations in a unidirectional composite using the Tsai-Wu failure criteria.

Keywords. genetic algorithms, reliability, composites

1 Introduction

Under the assumption of deterministic loading conditions, a number of classical methods have been developed to identify critical combinations of applied stress. However, even given the reality of the uncertainty of nature and man, investigations into the design implications of uncertainty in material properties and loading conditions have been limited. Due to the character of the manufacturing process associated with composites and the expected extreme operating envelope of the applications, consideration for these uncertainties is critical.

A number of studies have begun to investigate these implications and of particular note is the rather recent work by Miki[1]. In this research Miki discusses the application of advanced first order, second moment methods (FORM) to unidirectional fibrous composites under stochastic stress/strength conditions and investigates the FORM safety index as an indicator of optimum fiber orientation. The results indicated that uncertainty in stress/strength conditions significantly effected the optimum fiber orientation angle; optimum here indicating a minimum probability of failure. The investigation was, however, limited to the search for a single optimum orientation angle, when in fact due to the nature of the failure criteria multiple optimum solutions exist. Miki[2] also recognizes the difficulty in finding the optimum and proposes a stochastic search method to assist in the search. However, again no consideration is given to the existence of multiple solutions.

This paper proposes the use of a biologically based genetic algorithm as a nonlinear search paradigm to assist in identifying the existence of multiple, optimal fiber orientations. While computationally more intensive, multiple solutions are identified and a more accurate estimate of the probability of composite failure results.

2 Reliability Background

2.1 Failure Criteria and Limit State Formulation

While there have been a number of failure criteria proposed for unidirectional composites, probably the most recognized is the Tsai-Wu criteria:

$$1 = F_{xx}S_x^2 + F_{yy}S_y^2 + F_{ss}S_s^2 + 2F_{xy}S_xS_y + F_xS_x + F_yS_y \qquad (1)$$

where:

$$\begin{aligned} F_{xx} &= 1/R_xR_x' & F_x &= 1/R_x - 1/R_x' \\ F_{yy} &= 1/R_yR_y' & F_y &= 1/R_y - 1/R_y' \\ F_{ss} &= 1/R_s^2 & F_{xy} &= F_{xy}^{*}\sqrt{F_{xx}F_{yy}} \end{aligned} \qquad (2)$$

and R_s, R_i, R'_i, and S_i are, respectively, the longitudinal shear strength, tensile and compressive strengths and stress. It is assumed that the factor $F^{*}_{xy} = -1/2$. While this particular failure criteria is the basis for the discussion that follows, any criteria could have been used. For consistency with the work by Miki and the wide spread application, discussion is limited to this particular criteria. Also for consistency, the composite chosen for this investigation is a T300/208 Graphite/Epoxy type composite. The material properties are assumed to be stochastic in nature with the characteristics (μ_i = mean, σ_i = standard deviation) summarized in Table 2.1.1. below[3].

Table 2.1.1. Statistical Characteristics of Strength Properties

units=MPa	R_X	R'_X	R_Y	R'_Y	R_S
μ_R	1500	1500	40	246	68
σ_R	150	150	4	25	7

Figure 2.1.1 is provided to assist in identifying the variables referred to in the discussion. A typical failure envelope defined by the Tsai-Wu failure criteria is presented in Figure 2.1.2 for the case where the applied shear stress S_6=0. The safe operating region is interior to the ellipse, failure exterior to the ellipse, and the line itself defining the limit state or failure surface. It can be shown that for the deterministic loading situation, the optimum fiber angle, θ, is 0 degrees for $S_1>S_2$ and is 90 degrees for $S_1<S_2$.

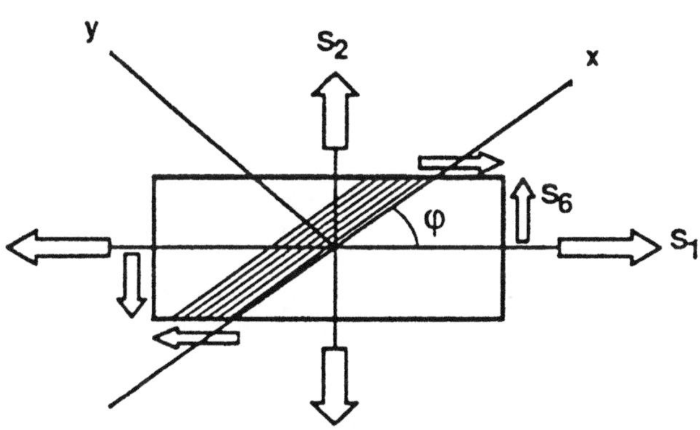

FIGURE 2.1.1 Composite Definition

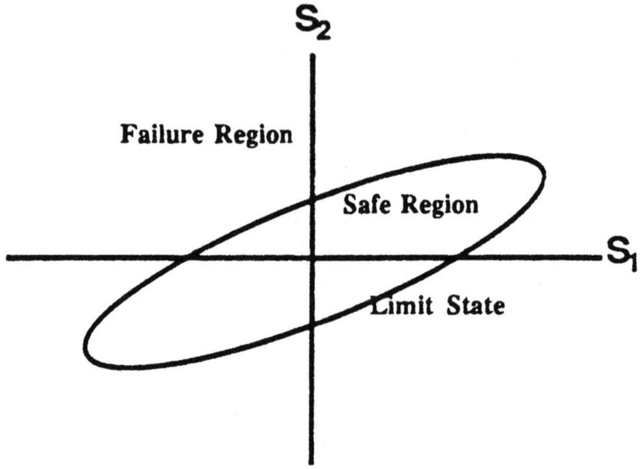

FIGURE 2.1.2 Deterministic Failure Envelope

As demonstrated by Miki[1] these results no longer apply when the loading conditions and material properties are stochastic in nature. Figure 2.1.3. depicts the situation where, again S_6=0, but S_1 and S_2 are random variables. Thus, even though the mean values may fall in the safe region, there exists a finite probability that critical combinations of stress can occur.

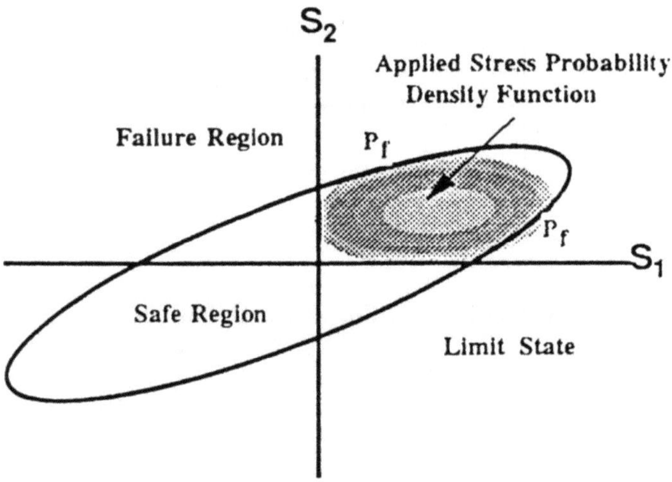

FIGURE 2.1.3 Stochastic Failure Envelope

Note that as the fiber orientation angle changes, the limit state ellipse is rotated through the stress density function. It is desired to have the orientation of the fibers such that the maximum area in this density function is contained by the safe operating region inside the ellipse.

The statistical uncertainties associated with both the strength and stress variables can be considered formally to determine the optimal fiber orientation through the use of FORM reliability design methods. Of particular concern will be the selection of fiber orientation angle θ such that the probability of failure P_f as approximated by the safety index β through the relationship: $P_f \approx \Phi(-\beta)$, is minimized. This relationship is exact for a linear limit state function and Gaussian random variables (neither of which unfortunately exist in the situation considered here). Since there is a one-to-one correspondence between β and P_f, discussion will be limited to determination of β.

2.2 Determination of the Safety Index

Reformulating equation 1 into a formal limit function:

$$g = 1 - F_{xx}S_x^2 + F_{yy}S_y^2 + F_{ss}S_s^2 + 2F_{xy}S_xS_y + F_xS_x + F_yS_y \quad (3)$$

The probability of failure is then given by: $P_f = P[g < 0]$. All random variables are then standardized through the transformation:

$$u_i = \Phi^{-1}[F(Z_i)] \quad (4)$$

where the u_i are referred to as reduced random variables and $\Phi()$ is the standard Gaussian cumulative distribution function. The limit state function can now be expressed in the space of these reduced variables and the safety index is defined as the minimum distance from the origin to the failure surface (in reduced space). The objective is then to find that set of reduced variables on the failure surface such that:

$$\beta = \min_u \sqrt{u^T u} \quad (5)$$

subject to: g=0 and let u^* be that point on the failure surface that satisfies this relationship. Unfortunately due to the nature of the limit state defined by the Tsai-Wu failure criteria, multiple solutions to this minimization exist. For this reason Miki[2] recommends a stochastic search method, coupled with advanced first order second moment methods to increase the likelihood of finding the minimum safety index β from among the multiple possibilities. However, stochastic search methods are plagued with a heavy computational overhead and very limited guarantees of convergence. In addition, the method proposed by Miki is limited to those situations where the limit state function has nicely behaved derivatives. The alternative suggested here involves an approach that provides a robust compromise between power and efficiency and is applicable in those situations where derivatives may not exist. It is proposed that a biologically based search algorithm such as that inherent with genetic algorithms form the basis for such a search paradigm.

3 Genetic Algorithms

3.1 Background

Genetic algorithms are search algorithms based upon the natural selection principle that permeates all naturally occurring ecosystems: that of survival of the fittest. What follows is the very briefest of discussions. For a more thorough discussion, please refer to the excellent book by Goldberg[4]. In addition, genetic algorithms (GA's) have been employed by a number of authors in the search for optimums contained within complex solution spaces. Examples of the application of the method in optimum structural design are contained in Hajela[5].

Genetic algorithms are uniquely different from the more traditional methods of optimization. First, GAs act on a coded set of variables and not on the variables themselves. Second, GA's search the solution space for the best combination of design variables by starting with a sample of the solution space rather than a single point. In addition, GAs do not use derivative information and, further, they explore the solution space based upon stochastically based rules rather than deterministic methods. The foundations for all GA paradigms are based upon three fundamental concepts: natural selection, sexual reproduction and genetic mutation. Each of these will be introduced in turn and then followed by a brief example emphasizing the individual concepts as they apply to the current problem.

Coding A unique characteristic of genetic algorithms is that they do not directly act on elements or variables in the solutions space; rather they act upon a coded set of variables. These coded variables are individually referred to as genes and the vector of variables that present a potential optimum from the solution space, when collectively coded, are referred to as a chromosome. For example, let **x** be a variable that ranges over the interval $[x_{min}, x_{max}]$. This interval can be discretized into 2^n unique values, where **n** is referred to as the resolution of the variable. For example, let $n=4$, $x_{min}=-5$ and $x_{max}=5$. Then the binary string (or gene): {0000} is equivalent to $x=-5$, while {1011}⇔2.33 and {1111}⇔5, etc. If there were two variables x_1 and x_2, ranging over the above interval, then, for example, a solution vector {-4.33,1.67} would be represented by the chromosome: {00011010}.

Natural Selection As noted earlier, a GA is composed of three operators: natural selection, sexual reproduction and genetic mutation. After random construction of an initial chromosome pool, a population from this pool is randomly selected for breeding. While selection is random, the selection is performed in a structured manner. Each chromosome represents a particular point in the solution space and has an associated performance value of f_i. If the total performance of all elements of the population is: $S=\Sigma f_i$, then the probability of selection of a particular element for entrance to the mating pool is given by f_i/S. Thus, those chromosomes most satisfying the performance have a higher likelihood of selection and therefore survival.

Mating and Reproduction Elements from this new gene pool are randomly paired for mating. This mating results in offspring which have genetic information common to both parents. Each parent chromosome consists of a binary string. Segments of each chromosome are chosen, each of random length k, and exchanged between parent chromosomes; these new chromosomes are then considered the offspring or children. The expected length of these genetic segments to be exchanged is proportional to the cross-over rate predetermined by the user. Changes in the cross-over rate can effect the number of generations required to reach a stable population performance. Due to this random exchange of genetic information, the genetic makeup of these 'children' can be significantly different from the original members of the population. This ever changing genetic pattern is what provides biological systems with a propensity towards being robust to their environment and in this analogy leads to an increased likelihood of finding the global optimum.

Mutation The final element in the GA is mutation. While possibly not as significant as selection and reproduction, it does contribute to maintaining variety in the gene pool. Given the new population that results from mating, each chromosome is exposed to the possibility of mutation. Mutation consists of a random change in an element of a chromosome. Each element of every chromosome is presented the likelihood of potentially changing from a 0 to a 1 or vice versa. The likelihood of this event is small, on the order of 0.01, but as pointed out, is realistic in occurrence and does contribute to the variety of the gene pool. Given that an element is identified for mutation, a new element is chosen: either 0 or 1, with equal probability. (This discussion is depicted in Figures 3.1.1 and 3.1.2.)

Initial Seed Population	Pairs Randomly Breed		New Population
1101	0010	0001	0001
1001	1001	1010	1010
1001	1001	1101	1101
0010	1101	1001	1001

FIGURE 3.1.1 Breeding of Solution Chromosomes

New Gene Pool	Performance of gene	Probability ofSurvival
0 0 0 1	5	5/36=.14
1 0 1 0	12	12/36= .33
1 1 0 1	7	7/36= .19
1 0 0 1	12	12/36= .33
	36	

FIGURE 3.1.2 Identification of Chromosomes for Survival

At the end of one generation, a population of potential optimums now exist. A check of the usual convergence criteria is used to determine if, among these potential solutions, suitable optimums exist, or if a new generation needs to be explored.

4 Results and Discussion

In what follows no attempt has been made to explicitly characterize the true probability density function of the loading and material properties. Results are limited to first order, second moment characterization. Extension of the method to include distribution information would be a natural extension and is the subject of current research. However, since the general application of interest is the use of the methods during preliminary design, information regarding the safety index is sufficient at that point in process.

For each of the seven design variables the range (in reduced space) was chosen to be [-5,5]; the cross-over rate was set at 0.6 and the resolution was set at 10. Thus a binary vector of length 70 was used to represent each potential solution. Fine tuning of these parameters is possible for certain classes of problems, but the initial settings above proved to be adequate.

The optimal fiber orientation angle for the load and material characteristics is depicted in Figure 4.1. This plot was generated iteratively investigating orientations angles from 0-90 degrees and observing the resulting safety index. While the final 'best' orientation is important in this simple case, it is informative to observe the progress of the genetic algorithm as the population of potential solutions breed, mutate and otherwise 'struggle' for survival. Figures 4.2 through 4.4 depict a typical series of generations for the fiber orientation angle of 35 degrees.

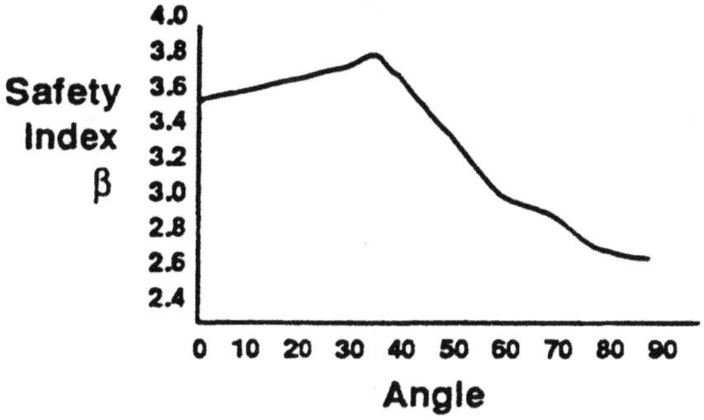

FIGURE 4.1 Optimum Fiber Orientation

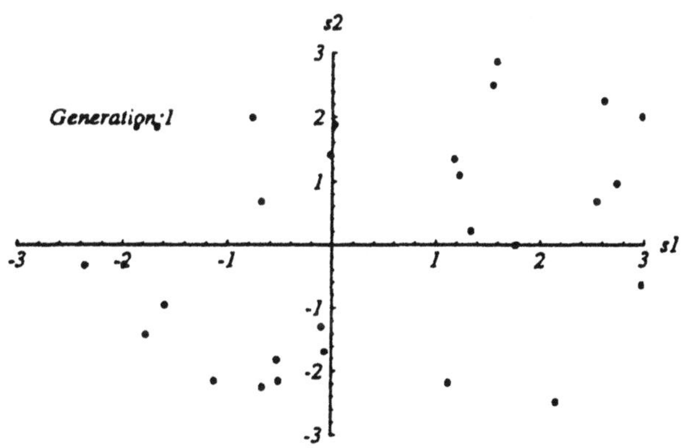

Figure 4.2 Results from Generation 1

Generation 1 consists of a random initial population of solutions (as evidenced by the very obvious scatter throughout the portion of the design space depicted). However, by generation 25, the population characteristics begin to coalesce and it is evident that multiple solutions potentially exist.

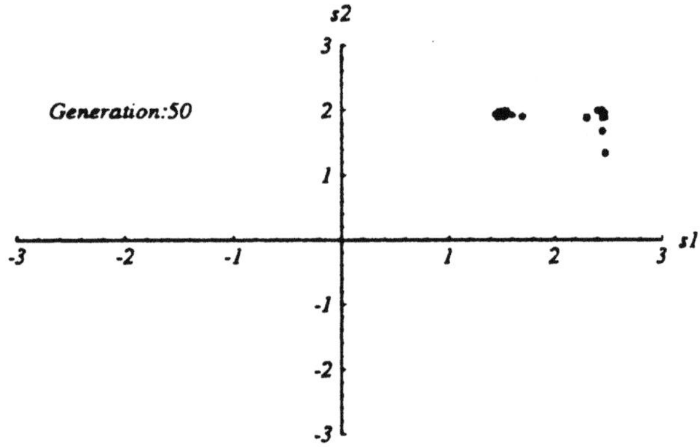

Figure 4.3 Results from Generation 50

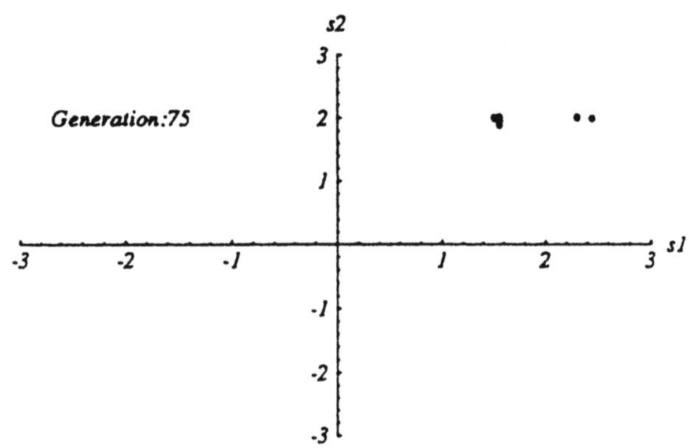

Figure 4.4 Results from Generation 75

Generation 50 provides a very clear picture of multiple solution regions each of which may be the result of more traditional FORM methods. Finally, by generation 75, all sub optimum solutions have been dominated.

It is recognized that, while the single optimum is important (and is the unique goal of the works by Miki), by relying solely on this single value the probability of exceeding the limit state may be grossly underestimated. Through proper consideration of the multiple solutions evidenced in intermediate generations, a more accurate picture of the probability of failure can be provided. An approach currently under investigation is the use of simple Ditlevsen bounds to combine the multiple safety indexes that become evident at some intermediate generation.

5 Conclusions

If a well behaved limit state function is available, then classical FORM methods are extremely efficient in identifying the associated safety index. If, however, the limit state function is extremely complicated (and possibly discontinuous) and/or the support for the design variables is discrete, then genetic algorithms provide a means of exploring the design space in efficient manner. While more computationally intensive then the traditional FORM methods, they do provide a capability where none existed before. Further, while perhaps not providing as accurate a final result as Monte Carlo methods, the computational burden is significantly less. In addition, the possibility of multiple solutions can be investigated. Finally, for continuous performance functions, but where complexity is such that multiple safety indexes may exist, a hybrid method combining GA's and FORM methods (when possible) is the recommended approach for investigating optimum design characteristics where system reliability is the final performance measure.

Research is continuing in this area; initial results from the application of these methods in the preliminary design of aircraft wings indicate that, in the presence of stochastic material and loading conditions, probabilistic based design criteria leads to significant weight savings, and further, results in a design that is more robust to changes in the required operating envelope.

References

1. Miki, M., Murotsu, Y., Tanaka, T., Shao, S. ,"Reliability of Unidirectional Fibrous Composites," *AIAA Journal*, Vol. 28, No. 11, 1990, pp. 1980-1986.
2. Miki, M., Murotsu Y., Tanaka T. , "Optimum Fiber Angle of Unidirectional Composites for Load with Variations", *AIAA Journal*, Vol. 30, No. 1, 1992, pp. 189-196.
3. Tsai, S. W., Hahn, H. T. *Introduction to Composite Materials*, Technomic, Westport, CT, 1980, p. 280.
4. Goldberg, D., *Genetic Algorithms in Search, Optimization and Machine Learning*, Addison-Wesley, Massachusetts, 1989.
5. Hajela, P. "Genetic Search - An Approach to the Nonconvex Optimization Problem," *AIAA Journal*, Vol. 9, No. 7, 1990, pp. 1205-1210.

Random Vibration of Structures Under Non-Gaussian Excitation

P. Śniady[1] and R. Sieniawska[2]

[1,2] Institute of Civil Engineering, Technical University of Wrocław, Wybrzeże Wyspiańskiego 27, 50-370 Wrocław, Poland

Abstract. In this paper the stochastic vibrations of structures excited by non-gaussian loads are considered. Some different types of excitation process are assumed. Among others, some new solutions are presented for dynamics of the structure response under sample discontinuous loads: the train of overlapping pulses (modelled by filtered Poisson process), the train of non-overlapping pulses (modelled by the pure birth stochastic process), random excitation with on-off brakes effect and jump-discontinuous non-interrupted stochastic process.

Keywords. Dynamics, non-gaussian stochastic processes, load models

1 Introduction

In the stochastic dynamics and dynamic reliability of structures it is usually assumed that the loading process - stationary or non-stationary - is sample continuous in time and Gaussian. In many real situations the loading process has a more complicated form. These phenomena occur in earthquakes, dynamics of bridges under a vehicular traffic flow[1,2,3], driving piles, the action of some types of machines, etc. The vibration of many engineering structures is caused by series of short-duration shocks. Few papers are concerned with the influence of sample discontinuous loads. In these, as a load model such discontinuous stochastic processes are usually used: "shot noise", filtered Poisson process or a "telegraph signal" [4,5,6,7,8,9,10].

In this paper the stochastic vibrations of structures excited by non-gaussian loads are considered. Some different types of excitation process are assumed. Among others some new solutions are presented for dynamic response of the structure under sample discontinuous loads:

a) the train of overlapping pulses (modelled by filtered Poisson process),
b) the train of non-overlapping pulses (modelled by the pure birth stochastic process),
c) random excitation with on-off brakes effect,
d) jump-discontinuous non-interrupted stochastic process.

2 Random Train of Pulses

2.1 Linear system

With the normal mode approach, the motion of a linearly elastic structure can be described by the set of uncoupled equations

$$\ddot{y}_n(t) + 2\alpha_n \dot{y}_n(t) + \omega_n^2 y_n(t) = p_n F(t) \tag{1}$$

The excitation is assumed to be a train of arbitrary pulses with random amplitudes and time duration, occurring at random times. Thus, the excitation *F(t)* can be represented as a filtered Poisson process

$$F(t) = \sum_{k=1}^{N(t)} A_k S(t, t_k, T_k) \tag{2}$$

in which the function $S(t, t_k, T_k)$ describes the deterministic shape of the pulses *(k=1,2,...,N(t))* occurring at random times t_k, the quantity T_k is the random time duration of the k-th pulse and A_k is its random amplitude; *N(t)* is the Poisson process with parameter λ. The solution of equation (1) is also a filtered Poisson process and can be obtained in the form of a Stieltjes stochastic integral with respect to the Poisson process *N(t)* as

$$y_n(t) = \int_0^t A(\tau) Z_n(t, \tau, T(\tau)) dN(\tau) \tag{3}$$

where the function $Z_n(t, \tau, T(\tau))$ is the structure normal mode response at time t to the pulse $S(t, \tau, T(\tau))$. The function $Z_n(t, \tau, T)$, where $T=T(\tau)$, can be obtained from the integral equation

$$Z_n(t, \tau, T) = \begin{cases} Z_n^{(1)}(t, \tau, T) = p_n \int_\tau^t h_n(t-\xi) S(\xi, \tau, T) d\xi & \text{for } t - T \le \tau \le t \\ Z_n^{(2)}(t, \tau, T) = p_n \int_\tau^{\tau+T} h_n(t-\xi) S(\xi, \tau, T) d\xi & \text{for } 0 \le \tau \le t - T \end{cases} \tag{4}$$

where $h_n(t-\xi) = \Omega_n^{-1} \exp[-\alpha_n(t-\xi)] \sin \Omega_n (t-\xi)$ is the pulse response function and $\Omega_n^2 = \omega_n^2 - \alpha_n^2$ is the damped natural frequency.

By analogy to equation (3) the velocity $\dot{y}_n(t)$ of the response of the structure has the form

$$\dot{y}_n(t) = \int_0^t A(\tau)\dot{Z}_n(t,\tau,T)dN(\tau) \tag{5}$$

where

$$\dot{Z}_n(t,\tau,T) = \frac{dZ_n(t,\tau,T)}{dt}.$$

The joint characteristic function of the processes $y_n(t)$ and $\dot{y}_n(t)$ has the form [9]

$$\psi_{Q_1Q_2}(Q_1,Q_2,t) = E\{\exp[i(Q_1y_n(t)+Q_2\dot{y}_n(t))]\} =$$
$$= \exp\{\int_0^t \int_{t-\tau}^{\infty} \int_A \exp\left[ia(Q_1Z_n^{(1)}(t,\tau,T)+Q_2\dot{Z}_n^{(1)}(t,\tau,T)\right]g_{AT}(a,T,\tau)\lambda(\tau)dadTd\tau +$$
$$+\int_0^t \int_0^{t-\tau} \int_A \exp\left[ia(Q_1Z_n^{(2)}(t,\tau,T)+Q_2\dot{Z}_n^{(2)}(t,\tau,T)\right]g_{AT}(a,T,\tau)\lambda(\tau)dadTd\tau - \int_0^t \lambda(\tau)d\tau\} \tag{6}$$

where $i=\sqrt{-1}$, E[ ] denotes the expected value and $g_{AT}(a,T,t)$ is the joint probability density function of random variables A and T. The expression (6) can be used to find the joint cumulants (semi-invariants) $\kappa_{r,s}(y_n,\dot{y}_n,T)$ of the random processes $y_n(t)$ and $\dot{y}_n(t)$ from the equation[4]

$$\kappa_{r,s}(y,\dot{y},t) = \frac{1}{i^{r+s}}\left(\frac{\partial^{r+s}}{\partial Q_1^r \partial Q_2^s}\ln\psi_{Q_1Q_2}(Q_1,Q_2,T)\right)\Bigg|_{Q_1=Q_2=0}. \tag{7}$$

The cumulants have the form

$$\kappa_{r,s}(y,\dot{y},t) = \int_0^t \int_{t-\tau}^{\infty} \int_A a^{r+s}[Z_n^{(1)}(t,\tau,T)]^r[\dot{Z}_n^{(1)}(t,\tau,T)]^s g_{AT}(a,T,\tau)\lambda(\tau)dadTd\tau +$$
$$+\int_0^t \int_0^{t-\tau} \int_A a^{r+s}[Z_n^{(2)}(t,\tau,T)]^r[\dot{Z}_n^{(2)}(t,\tau,T)]^s g_{AT}(a,T,\tau)\lambda(\tau)dadTd\tau. \tag{8}$$

The joint probability density function of the processes $y_n(t)$ and $\dot{y}_n(t)$ can be obtained from the expression (6) by the FFT method or can be expressed by the Gram-Charlier series in the case of the one-dimensional probability density function and by the Longuet-Higgins series for the two-dimensional case[6].

In the last case one obtains

$$p_{y_n\dot{y}_n}(y_n,\dot{y}_n,t)=\left\{2\pi(\kappa_{2,0}\kappa_{0,2}-\kappa_{1,1}^2)^{\frac{1}{2}}\right\}^{-1}\exp\left[-\frac{1}{2}(\alpha_1^2-2\beta_{1,1}\alpha_1\alpha_2+\alpha_2^2/(1-\beta_{1,1}^2)\right]$$
$$\cdot\left[1+\frac{1}{6}(\beta_{3,0}H_{3,0}+3\beta_{2,1}H_{2,1}+3\beta_{1,2}H_{1,2}+\beta_{0,3}H_{0,3}+\ldots\right] \tag{9}$$

where

$$\alpha_1=\alpha_1(t)=\left[y_n(t)-\kappa_{1,0}(t)\right]\kappa_{2,0}^{-1/2}(t),$$
$$\alpha_2=\alpha_2(t)=\left[\dot{y}_n(t)-\kappa_{0,1}(t)\right]\kappa_{0,2}^{-1/2}(t),$$
$$\beta_{m,n}=\kappa_{m,n}(\cdot)/\left[\kappa_{2,0}^m\kappa_{0,2}^n\right]^{1/2}$$

and $H_{m,n}$ is a two-dimensional analogue of the Hermit polynomial

$$H_{m,n}\exp\left[-\frac{1}{2}(\alpha_1^2-2\beta_{1,1}\alpha_1\alpha_2+\alpha_2^2)/(1-\beta_{1,1}^2)\right]=$$
$$=(-1^{m+n})\frac{\partial^{m+n}}{\partial\alpha_1^m\partial\alpha_2^n}\exp\left[-\frac{1}{2}(\alpha_1^2-2\beta_{1,1}\alpha_1\alpha_2+\alpha_2^2)/(1-\beta_{1,1}^2)\right]. \tag{10}$$

Knowledge of the joint probability density function allows for finding the crossing rate $\overline{\eta}_+(r,t)$ of the threshold r from Rice's formula

$$\overline{\eta}_+(r,t)=\int_0^\infty \dot{y}_n p_{y_n\dot{y}_n}(r,\dot{y}_n,t)d\dot{y}_n. \tag{11}$$

The cumulants $\kappa_k(y_n,t)$ can be used to investigate the coefficient of skewness and curtosis.

2.2 Non-linear system

As an example of the dynamics of non-linear system we consider vibration of the Duffing oscillator. The equation of motion for the oscillator under random train of pulses has the form

$$\ddot{y}(t)+2\alpha\dot{y}(t)+\omega^2 y(1+\varepsilon y^2)=\sum_{k=1}^{N(t)}A_k S(t,t_k,T). \tag{12}$$

Assuming that the time duration of the pulses T is deterministic for some shape functions of the pulses $S(t,t_k,T)$ the equation (12) can be converted into the set of first-order differential equations (see also[10])

$$\dot{x}_1 = x_2$$
$$\dot{x}_2 = -2\alpha x_2 - \omega^2 x_1(1+\varepsilon x_1^2) + x_3 \qquad (13)$$

where

$$y = x_1, \qquad \dot{y} = x_2$$

and $x_3 = x_3(t)$ is the load process obtained as a result from solving an added linear equation, treated as a linear filter which depend on the impulse shape function.
For rectangular pulses (Fig.2.2.1a)

$$S(t,t_k,T) = H(t-t_k) - H(t-t_k-T) \qquad (14)$$

where $H(t)$ is the unit step function (Heviside's function), the function x_3 satisfies the equation

$$\dot{x}_3(t) = \sum_k A_k[\delta(t-t_k) - \delta(t-t_k-T)] \qquad (15)$$

where $\delta(t)$ is the delta-Dirac function.
For the pulses of exponential decreasing shape (Fig.2.2.1b)

$$S(t,t_k,T) = e^{-\beta(t-t_k)} \qquad (16)$$

we have

$$\dot{x}_3(t) = -\beta x_3(t) + \sum_k A_k\delta(t-t_k)\ . \qquad (17)$$

For the pulses which have a sinusoidal shape (Fig.2.2.1c) (see[3]]

$$S(t,t_k,T) = \sin\frac{2\pi}{T}(t-t_k) - \sin\frac{2\pi}{T}(t-t_k-T) \qquad (18)$$

the function x_3 satisfies the equation

$$\ddot{x}_3(t) + \frac{2\pi}{T}x_3(t) = \sum_k A_k[\delta(t-t_k) - \delta(t-t_k-T)] \qquad (19)$$

or

$$\begin{aligned} \dot{x}_3(t) &= x_4(t) \\ \dot{x}_4(t) &= -\frac{2\pi}{T}x_3(t) + \sum_k A_k[\delta(t-t_k) - \delta(t-t_k-T)]. \end{aligned} \tag{20}$$

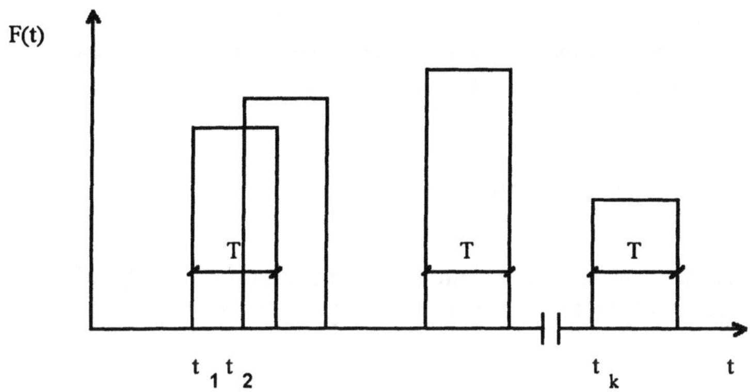

2.2.1a

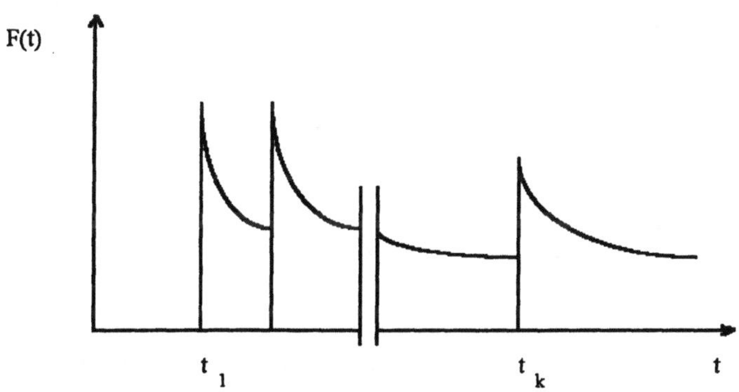

2.2.1b

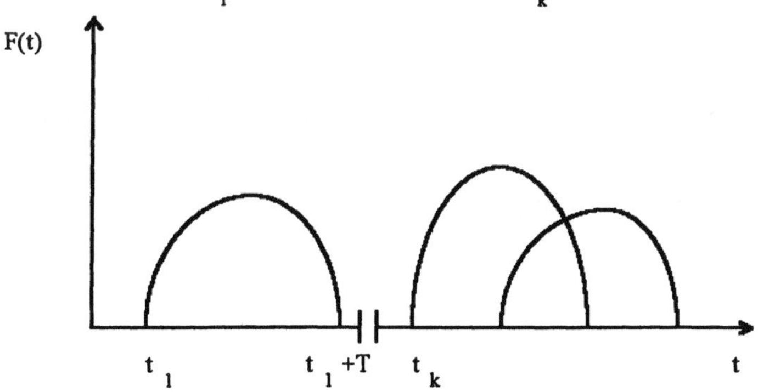

2.2.1c

Fig.2.2.1. Filtered Poisson load process. Different shapes of pulses

Taking into account that

$$\sum_k A_k \delta(t-t_k) = \int_A aN(dt,da) \tag{21}$$

where $N(dt,da)$ is a random Poisson measure[13,14,15], which represents the number of pulse arrivals within the time interval $(t,t+dt)$. The equation (13) with the added equation (15) or (17) or (20) can be shown as the Poisson-driven Markov vector process

$$dX(t) = B(t)dt + \int_A C(a,t)N(dt,da). \tag{22}$$

As an example let us consider a system loaded by random set of pulses of exponential decreasing shape. From eqs. (13) and (17) we obtain

$$\begin{aligned} dx_1 &= x_2 dt \\ dx_2 &= [-2\alpha x_2 - \omega^2 x_1(1+\varepsilon x_1^2) + x_3]dt \\ dx_3 &= -\beta x_3 dt + \int_A aN(dt,da). \end{aligned} \tag{23}$$

Using the extension of Ito's differential rule for the Poisson-driven Markov process[13,14,15]

$$df(t,x) = \left[\frac{\partial f(t,x)}{\partial t} + \sum_i \frac{\partial f(t,x)}{\partial x_i} b_i(t)\right]dt + \int_A [f(t,x+c(a,t)) - f(t,x)]N(dt,da) \tag{24}$$

and after substituting $f(x,t) = x_1^k x_2^l x_3^r$ one can obtain a set of equation for the moments

$$\begin{aligned} \frac{dm_{klr}}{dt} &= km_{k-1,l+1,r} - 2\alpha l m_{k,l,r} - \omega^2 l m_{k+1,l-1,r} - \omega^2 l m_{k+3,l-1,r} + l m_{k,l-1,r+1} - \beta r m_{k,l,r} + \\ &+ \lambda\left[\sum_{j=0}^{r} \binom{r}{j} m_{0,0,j} E[A^{r-j}] - m_{o,o,r}\right] \end{aligned} \tag{25}$$

where

$$m_{klr} = E[x_1^k x_2^l x_3^r].$$

The equations (25) constitute a infinite set of equations, which can be solved for example by using the moment-neglect closure technique, i.e. by neglecting the moments of higher order than k,l,r.

2.3 Piece-wise stationary and non-stationary excitation process

The excitation process *F(t)* in the equation (1) is assumed to be the jump-discontinuous non-interrupted stochastic process (Fig.2.3.1a) (piece-wise process). More precisely, it is assumed that the random times $t_1, t_2, \dots, t_n$ divide the stochastic process into independent parts.

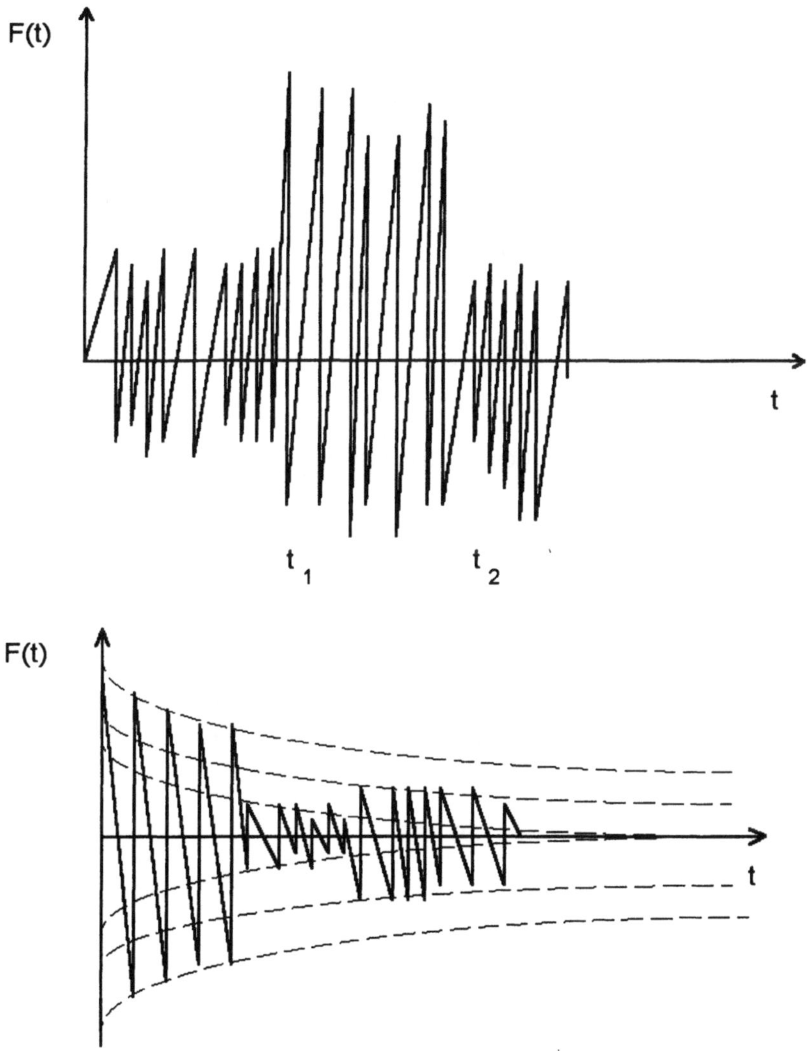

2.3.1a

2.3.1b

Fig. 2.3.1. Piece-wise stationary and non-stationary excitation process

The excitation process is doubly stochastic and can be presented in the form (Fig. 2.3.1)

$$F(t) = g(t)X_{N(t)}(t) \tag{26}$$

where $g(t)$ is a deterministic function. $N(t)$ describes the Poisson stochastic process with parameter λ, which gives the number of discontinuity points $t_1, t_2, \ldots, t_N$ within the time interval $(0,t)$. $X_{N(t)}(t)$ is the continuous stochastic process within the time interval (t_n, t_{n+1}). From equation (26) it follows that the excitation process is represented by a denumerable set of continuous stochastic processes[11]. Let us assume for simplicity that the expected value of such an excitation process is equal to

$$E[F(t)] = g(t)E[X_{N(t)}(t)] = 0 \tag{27}$$

and that its covariance function is known

$$C_{FF}(t_1,t_2) = \begin{cases} 0 & \text{for } N(t) > 0 \\ g(t_1)g(t_2)C_{XX}(t_1,t_2) & \text{for } N(t) = 0 \end{cases} \tag{28}$$

where $t = |t_1 - t_2|$.

In this case the covariance function of the structure response described by equation (1) can be expressed as

$$C_{y_n y_n}(t_1,t_2) = \int_0^{t_1}\int_0^{t_2} h_n(t_1-\tau_1)h_n(t_2-\tau_2)g(\tau_1)g(\tau_2)\cdot \\ \cdot C_{XX}(\tau_1,\tau_2)exp(-\lambda|\tau_1-\tau_2|)d\tau_1 d\tau_2. \tag{29}$$

The spectral density function of the normal modes for the non-stationary random excitation process using the evolutionary spectra[16] can be obtained from the relation

$$\Phi_{y_n y_k}(t,\omega) = A_n(t,\omega)A_k^*(t,\omega)\Phi_F(\omega) \tag{30}$$

where

$$A_n(t,\omega) = \int_0^t h_n(\xi)g(t-\xi)e^{i\omega\xi}d\xi \tag{31}$$

$$\Phi_F(\omega) = \frac{1}{2\pi}\int_{-\infty}^{\infty} C_{XX}(\tau)e^{-|\tau|}e^{-i\omega\tau}d\tau \tag{32}$$

where the asterisk denotes the complex conjugate.

If $g(t)=1$ then the excitation process is stationary. Some particular solutions for a stationary excitation process are presented in [11].

2.4 Non-overlapping train of pulses. Intermitted continuous random process

Let us consider the situation when the loading process is switched on for some time and then is switched off, then again is switched on over and over again, with the moments when it is switched on or off being random. This type of loading can also be treated as a train of the non-overlapping forces or pulses of random time duration separated by periods of random length with no load (Fig.2.4.1). It is assumed that the duration of every force (pulse) and the time moment when the load is off are mutually independent random variables which have the exponential distribution with parameters λ and μ, respectively.

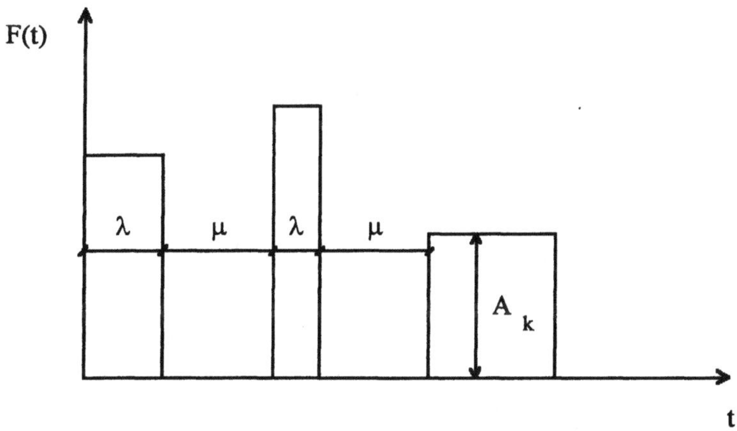

Fig. 2.4.1.. Non-overlapping pulses load process

The loading process $F(t)$ described above can be represented as

$$F(t)=\frac{1}{2}A_{N(t)}[1+(-1)^{N(t)}] \tag{33}$$

where $A_{N(t)}$ are random amplitudes, mutually independent, characterised by their expected values $E[A_{N(t)}]=E[A]=const,\quad E[A_{N(t)}^2]=E[A^2]=const$ and $N(t)$ is the Markov (birth) process which presents a number of changes in the loading process (all the on and off points).

The $N(t)$ process is considered as a Markov process involving two states:

$$S_o(t)=\{N(t)=0,2,4,...\} \text{ and } S_1(t)=\{N(t)=1,3,5,...\}.$$

$P_p(t) = P\{N(t) \in S_0(t)\}$ denotes the probability that *N(t)* has even values and $P_R(t) = P\{N(t) \in S_1(t)\}$ denotes that *N(t)* has odd values. The probability $P_P(t)$ satisfies the equation [12]

$$\dot{P}_P(t) + (\lambda + \mu) P_P(t) = \mu. \tag{34}$$

Solving the equation (34) one obtains

$$P_p(t) = P\{N(t) = 0,2,4,\ldots\} = (\frac{1}{1+\rho})\left[\rho + e^{-\lambda(1+\rho)t}\right], \tag{35}$$

$$P_R(t) = P\{N(t) = 1,3,5,\ldots\} = (\frac{1}{1+\rho})\left[\rho - e^{-\lambda(1+\rho)t}\right], \tag{36}$$

where $\rho = \frac{\mu}{\lambda}$ and $P_R(t) = 1 - P_p(t)$.
Taking into account the expressions (35) and (36) we can find

$$E\left[(-1)^{N(t)}\right] = \frac{1}{1+\rho}\left[2e^{-\lambda(1+\rho)t} - (1-\rho)\right], \tag{37}$$

$$E\left[(-1)^{N(t_i)}(-1)^{N(t_j)}\right] = E\left[(-1)^{N(t_i,t_j)}\right] = \frac{1}{1+\rho}\{(1-\rho)^2 + \\ + 4\rho e^{-\lambda(1+\rho)(t_j - t_i)} + 2(1-\rho)[e^{-\lambda(1+\rho)t_j} - e^{-\lambda(1+\rho)t_i}]\} \tag{38}$$

$$P\{N(t_i) = N(t_j)\} = P\{N(t_i,t_j) = 0\} = \\ = \frac{1}{1+\rho}\left\{\rho e^{-\lambda(t_j - t_i)} + e^{-\mu(t_j - t_i)} + e^{-(\mu t_i + \lambda t_j)} - e^{-(\lambda t_i + \mu t_j)}\right\} \tag{39}$$

and

$$P\{N(t_i) \neq N(t_j)\} = P\{N(t_i,t_j) \neq 0\} = 1 - P\{N(t_i,t_j) = 0\}, \quad \text{where } t_i > t_j. \tag{40}$$

The expected value and the correlation function of the response function (1) can be found from the equations

$$E[y_n(t)] = \frac{1}{2} p_n E[A] \int_0^t h_n(t-\tau)\left\{1 + E\left[(-1)^{N(\tau)}\right]\right\} d\tau \tag{41}$$

and

$$E[y_k(t_1)y_n(t_2)] = K_{y_k y_n}(t_1,t_2) = \frac{1}{4} p_k p_n E^2[A]\{\int_0^{t_1}\int_0^{t_2} H_{kn}(t_1,t_2,\tau_1,\tau_2)P[N(\tau_1) \neq$$

$$\neq N(\tau_2)]\{1+E[(-1)^{N(\tau_1)}]+E[(-1)^{N(\tau_2)}]\}d\tau_1 d\tau_2 +$$

$$+\int_0^{t_1}\int_0^{\tau_1} H_{kn}(t_1,t_2,\tau_1,\tau_2)E[(-1)^{N(\tau_1,\tau_2)}]P[N(\tau_1) \neq N(\tau_2)]d\tau_1 d\tau_2 +$$

$$+\int_0^{t_1}\int_{\tau_1}^{t_2} H_{kn}(t_1,t_2,\tau_1,\tau_2)E[(-1)^{N(\tau_1,\tau_2)}]P[N(\tau_1) \neq N(\tau_2)]d\tau_1 d\tau_2 +$$

$$+\frac{1}{4} p_k p_n E[A^2]\int_0^{t_1}\int_0^{t_2} H_{kn}(t_1,t_2,\tau_1,\tau_2)P[N(\tau_1) =$$

$$= N(\tau_2)]\{2+ E[(-1)^{N(\tau_1)}]+E[(-1)^{N(\tau_2)}]\}d\tau_1 d\tau_2 \tag{42}$$

where $H_{kn}(t_1,t_2,\tau_1,\tau_2) = h_k(t_1-\tau_1)h_n(t_2-\tau_2)$.

From equations (37-40) and (41), (42) follows that the steady state solutions have forms

$$\lim_{t\to\infty} E[y_n(t)] = E[y_n(\infty)] = \frac{p_n E[A]}{\omega_n^2}\frac{\gamma}{\lambda} \tag{43}$$

$$\lim_{t\to\infty} E[y_n^2(t)] = E[y_n^2(\infty)] = \frac{1}{4} p_n^2 E^2[A]\frac{\mu}{\mu+\lambda}\{\frac{\lambda+\mu}{\lambda}\frac{1}{\omega_n^4} - \frac{\mu}{\lambda}J_1(\alpha_n,\lambda) - J_1(\alpha_n,\mu) +$$

$$+4J_1(\alpha_n,\lambda+\mu) - 4\frac{\mu}{\mu+\lambda}J_1(\alpha_n,2\lambda+\mu) - \frac{\lambda}{\mu+\lambda}J_1(\alpha_n,\lambda+2\mu)\} +$$

$$+ p_n^2 E[A^2]\frac{\mu}{\mu+\lambda}\{\frac{\mu}{\lambda}J_1(\alpha_n,\lambda) + J_1(\alpha_n,\mu)\} \tag{44}$$

where

$$J_1(\alpha_n,\beta) = \frac{1}{[(\alpha_n+\beta)^2+\Omega_n^2][(\alpha_n-\beta)^2+\Omega_n^2]}\left[1+\frac{\beta(\omega_n^2+\beta^2-4\alpha_n^2)}{2\alpha_n\omega_n^2}\right].$$

In the case of the set of separated pulses it is possible - what was shown above - to analyse the two first probabilistic moments, but the analysis of the next moments is very complicated both in analytical and numerical cases.

Let us consider the case when the loading process can be described as (Fig.2.4.2)

$$F(t) = \frac{1}{2} X_{N(t)} [1 + (-1)^{N(t)}] \tag{45}$$

where $X_{N(t)}(t)$ is a family of mutually independent stochastic processes.

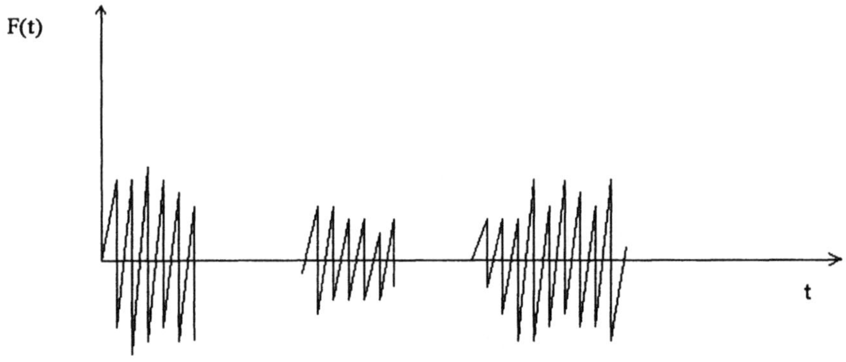

Fig. 2.4.2. The pattern of excitation with on-off breaks effect.

The correlation analysis of the vibratory system (1) under excitation process (45) is presented in the paper [12].

If $\mu << \omega$ and $\lambda << \omega$ then we can assume the crossing rate of the level r in the approximate form

$$\begin{aligned} \nu_+(r,t) &\cong P\{N(t) = 0,2,4,\ldots\} \int_0^\infty \dot{y}_n p_{y\dot{y}}(r, \dot{y}_n, t) d\dot{y}_n = \\ &= \frac{1}{1+\rho}\left[\rho + e^{-\lambda(1+\rho)t}\right] \int_0^\infty \dot{y}_n p_{y\dot{y}}(r, \dot{y}_n, t) d\dot{y}_n . \end{aligned} \tag{46}$$

Let us assume that $X_{N(t)}(t) = X(t)$ where $X(t)$ is the stationary Gaussian stochastic process. In this case from (46) we obtain

$$\nu_+(r,t) = \frac{1}{1+\rho}\left[\rho + e^{-\lambda(1+\rho)t}\right] \frac{\sigma_{\dot{y}_n}}{2\pi\sigma_{y_n}} e^{-\frac{r^2}{2\sigma_y^2}} \tag{47}$$

where σ_{y_n} and $\sigma_{\dot{y}_n}$ are the standard deviations of the responses y_n and $\dot{y}_n$, respectively .

Assuming the covariance function $C_{XX}(t) = \sigma_F^2 \delta(t)$ we obtain

$$\nu_+(r,t) = \frac{1}{1+\rho}\left[\rho + e^{-\lambda(1+\rho)t}\right]\frac{\omega_n}{2\pi} e^{\frac{-2r^2 \alpha_n \omega_n^2}{\sigma_F^2}} . \tag{48}$$

3 Conclusions

In many real situations the loading process has very complicated form. For example the vibration of many engineering structures is caused by series of excitations. In this case the excitation process should be described by using not only continuous stochastic process but also by point stochastic process. In this paper the loading process is assumed as a train of overlapping pulses (modelled by filtered Poisson process), the train of non-overlapping pulses (modelled by the pure birth stochastic process), excitation with on-off brakes effect, piece-wise non-stationary excitation process. When the load is modelled by the filtered Poisson process one can obtain the probabilistic moments (cumulants) both in linear and non-linear problems. The probability density function can also be determined. For the other types of load models only the correlation analysis is possible and the analytical expressions for the probabilistic moments of higher order could be rather unrealizable.

Acknowledgement

This research was supported by the SCIENTIFIC COMMITTEE, Warsaw, under grant 772309203.

References

1. Sieniawska R., Śniady P. (1990), First passage problem of the beam under a random stream of moving forces, *Jour. of Sound and Vibr.*, 136, 2, 177-185.
2. Sieniawska R., Śniady P. (1990), Life expectancy of highway bridges due to traffic load, *Jour. of Sound and Vibr.*, 140, 1, 31-38.
3. Paola M.D., Ricciardi G. (1992), Vibration of a bridge under a random train of moving loads, *Proc. of the 6th Spec. Conf. Probabilistic Mechanics and Structural and Geotechnical Reliability*, Denver, 136-139.
4. Lin Y. K. (1967), *Probabilistic Theory of Structural Dynamics*, New York: McGraw-Hill
5. Roberts J.B. (1965), The response of Linear Vibratory Systems to Random Impulses, *Jour. of Sound and Vibr.*, 1965, 2, pp. 375-390.
6. Roberts J.B. (1972), System Response to Random Impulses, *Jour. of Sound and Vibr.*, 24, pp. 23-34.

7. Parzen E. (1962), *Stochastic Processes,* San Francisco: Holden-Day.
8. Mazur-Śniady K., Śniady P. (1986), Dynamic Response of Linear Structures to Random Streams of Arbitrary Impulses in Time and Space, *Jour. of Sound and Vibr.,* 110, 1, pp. 59-68.
9. Śniady P. (1989), Dynamic Response of Linear Structures to a Random Stream of Pulses, *Jour. of Sound and Vibr.,* 131, 1, 91-102.
10. Iwankiewicz R., Nielsen S.R.K., Thoft-Christensen P. (1990), Dynamic response of non-linear systems to Poisson-distributed pulse trains; Markov approach, *Nonlinear Structural Systems under Random Conditions,* ed. F. Casciati, I. Elishakoff, J.G.,J.B. Roberts, Elsevier, Amsterdam, 223-238.
11. Mironowicz W., Śniady P. (1990), Vibrations of Linear Structures due to Jump-Discontinuous, Non-interrupted, Stochastic Processes, *Earthquake Eng. and Struct. Dyn.,* vol. 19, 577-582.
12. Gładysz M., Śniady P. (1992), Stochastic Vibration of Linear Structures due to Sample Discontinuous Load, *Jour. of Sound and Vibr.,* 158, 1, 1-10.
13. Gikhman I.I., Skorokhod A.V. (1972), *Stochastic Differential Equations,* Springer-Verlag, Berlin.
14. Snyder D.I. (1975), *Random Point Processes,* Wiley, New York.
15. Sobczyk K.(1991), *Stochastic Differential Equations,* Kluwer Academic Publishers.
16. Piestley M.B. (1967), Power Spectral Analysis of Nonstationary Random Processes, *Jour. of Sound and Vibr.,* 6, 86-97.

Design Sensitivity Random Eigenvalue Problem in Dynamics and Buckling of Structures

Paweł Śniady and Stanislaw Żukowski

Institute of Civil Engineering, Technical University of Wrocław,
Wybrzeże Wyspiańskiego 27, 50-370 Wrocław, Poland

Abstract. In the paper, the problem of the sensitivity of the probabilistic characteristics of the eigenvalue problem and buckling is considered. Using the linearization method, relationships defining the sensitivity of the expected values of eigenvalues and eigenvectors and of the covariance and the standard deviation of eigenvalues to changes in structural parameters and in the probabilistic characteristic of these parameters are derived.

Keywords. Design sensitivity, eigenvalue problem, probabilistic method.

1 Introduction

The development of computer technology has made studying of the structure's sensitivity to changes in its design parameters possible. From the theoretical point of view, the problem reduces to the investigation of the structure's response variation (e.g. displacements, efforts, frequencies of freevibration, critical forces etc.) relative to the structural parameters. In recent years, some papers have been devoted to the analysis of the structure's sensitivity (e.g.[1,2]). In most of the papers, the problem of sensitivity is considered in deterministic terms. There are very few papers that treat the problem in probabilistic categories [3-6]. In many mechanical problems, the first step of the analysis is to solve an eigenvalue problem. The different solutions for the above problem in the random terms were presented among others in the papers [7-16]. The problem is of significant practical importance, for instance, in the dynamics of foundations and support structures under machines, in seismic dynamics, and also in the buckling of structures.

In the paper, the problem of the design sensitivity of the probabilistic characteristics of the eigenvalue problem of a random parameters system is considered. The random eigenvalue problem is solved using a first order method. The application the theoretical solution of the design sensitivity of the probabilistic characteristics of the eigenvalue problem in dynamics and buckling structural analysis is presented.

2 The Random Eigenvalue Problem

Let the random eigenvalue problem of a structure be described by the matrix equation

$$(\mathbf{K}-\lambda_i\mathbf{B})\mathbf{w}_i = 0\ , \tag{2.1}$$

which can represent both dynamics [7, 9-12] and buckling structural problems [8]. The quantities **K**, **B**, λ_i and $\mathbf{w}_i$ in the equation (2.1) are assumed to be random.

In dynamics problems **K** is a stiffness matrix, **B** is an inertia matrix, parameter $\lambda_i = \omega_i^2$, where ω_i is the i-th frequency of the free vibration, $\mathbf{w}_i$ is the i-th vector of the main coordinates, and i = 1,2,...,n; n is the rank of matrix **K**.

In the buckling problem of structures, the quantities **K** and $\mathbf{w}_i$ are the same as in the dynamics problem, **B** is a geometrical stiffness matrix, and parameter λ_i is connected with axial forces in the structure.

Using the first order method, relationship (2.1) can be presented in the form

$$\left(\overset{\circ}{\mathbf{K}}-\overset{\circ}{\lambda}_i\overset{\circ}{\mathbf{B}}\right)\overset{\circ}{\mathbf{w}}_i+\left(\overset{\circ}{\mathbf{K}}-\overset{\circ}{\lambda}_i\overset{\circ}{\mathbf{B}}\right)\tilde{\mathbf{w}}_i+\left(\tilde{\mathbf{K}}-\overset{\circ}{\lambda}_i\tilde{\mathbf{B}}-\tilde{\lambda}_i\overset{\circ}{\mathbf{B}}\right)\overset{\circ}{\mathbf{w}}_i = 0\ , \tag{2..2}$$

where signs $\circ$ and $\sim$ denote mean values and random fluctuations.

By performing the expected value operation on relationship (2.2), one obtains a homogenous system of equations

$$\left(\overset{\circ}{\mathbf{K}}-\overset{\circ}{\lambda}_i\overset{\circ}{\mathbf{B}}\right)\overset{\circ}{\mathbf{w}}_i = 0 \tag{2.3}$$

from which mean eigenvalue $\overset{\circ}{\lambda}_i$ and corresponding mean eigenvectors $\overset{\circ}{\mathbf{w}}_i$ can be determined.

By pre-multiplying equation (2.2) by transposed eigenvector $\overset{\circ}{\mathbf{w}}_i$ and taking into account

$$\overset{\circ}{\mathbf{w}}_i^{T}\left(\overset{\circ}{\mathbf{K}}-\overset{\circ}{\lambda}_i\overset{\circ}{\mathbf{B}}\right)=\left(\overset{\circ}{\mathbf{K}}-\overset{\circ}{\lambda}_i\overset{\circ}{\mathbf{B}}\right)\overset{\circ}{\mathbf{w}}_i = 0\ , \tag{2.4}$$

the relationship

$$\overset{\circ}{\mathbf{w}}_i^{T}\left(\tilde{\mathbf{K}}-\overset{\circ}{\lambda}_i\tilde{\mathbf{B}}-\tilde{\lambda}_i\overset{\circ}{\mathbf{B}}\right)\overset{\circ}{\mathbf{w}}_i = 0 \tag{2.5}$$

is obtained from which one may determine the random fluctuation of the eigenvalue

$$\tilde{\lambda}_i = \frac{\tilde{k}_i - \overset{\circ}{\lambda}_i \tilde{b}_i}{\overset{\circ}{b}_i} = \frac{\tilde{d}_i}{\overset{\circ}{b}_i}, \tag{2.6}$$

where $\tilde{k}_i = \overset{\circ}{\mathbf{w}}_i^T \tilde{\mathbf{K}} \overset{\circ}{\mathbf{w}}_i$, $\tilde{b}_i = \overset{\circ}{\mathbf{w}}_i^T \tilde{\mathbf{B}} \overset{\circ}{\mathbf{w}}_i$, $\overset{\circ}{b}_i = \overset{\circ}{\mathbf{w}}_i^T \overset{\circ}{\mathbf{B}} \overset{\circ}{\mathbf{w}}_i$.

On the basis of relationship (2.6), the covariance of two arbitrary eigenvalues has the form

$$\mathrm{Cov}(\lambda_i, \lambda_j) = E\left[\tilde{\lambda}_i \tilde{\lambda}_j\right] = \frac{E\left[\tilde{d}_i \tilde{d}_j\right]}{\overset{\circ}{b}_i \overset{\circ}{b}_j} = \frac{\mathrm{Cov}(d_i, d_j)}{\overset{\circ}{b}_i \overset{\circ}{b}_j}, \tag{2.7}$$

and the variance of an eigenvalue has the form

$$\mathrm{Var}(\lambda_i) = \left(\sigma_{\lambda_i}\right)^2 = E\left[\tilde{\lambda}_i, \tilde{\lambda}_i\right] = \frac{E\left[\tilde{d}_i, \tilde{d}_i\right]}{\left(\overset{\circ}{b}_i\right)^2} = \frac{\mathrm{Var}(d_i)}{\left(\overset{\circ}{b}_i\right)^2}, \tag{2.8}$$

where $k_i = \mathbf{w}_i^T \mathbf{K} \mathbf{w}_i$, $b_i = \mathbf{w}_i^T \mathbf{B} \mathbf{w}_i$, $d_i = k_i - \lambda_i b_i$.

The stiffness matrix K, the matrix B, eigenvalues λ_i and eigenvectors $\mathbf{w}_i$ are functions of the parameters of the structure Z_α ($\alpha = 1, 2, \ldots, m$; m stands for the number of parameters defining the structure).

Using the linearization method, the stiffness matrix and the matrix B will be presented in the form of Taylor series expansions relative to the structural parameters in the neighbourhood of mean values with the linear terms retained

$$\begin{aligned} \mathbf{K} &= \mathbf{K}(Z_1, \ldots, Z_m) + \sum_\alpha \overset{\circ}{\mathbf{K}}_{,\alpha} \tilde{Z}_\alpha , \\ \mathbf{B} &= \mathbf{B}(Z_1, \ldots, Z_m) + \sum_\alpha \overset{\circ}{\mathbf{B}}_{,\alpha} \tilde{Z}_\alpha , \end{aligned} \tag{2.9}$$

where $\overset{\circ}{\mathbf{K}}_{,\alpha} = \left.\frac{\partial \mathbf{K}}{\partial Z_\alpha}\right|_{Z_1 = \overset{\bullet}{Z}_1, \ldots, Z_m = \overset{\bullet}{Z}_m} = \frac{\partial \overset{\circ}{\mathbf{K}}}{\partial \overset{\circ}{Z}_\alpha}$,

$\overset{\circ}{\mathbf{B}}_{,\alpha} = \left.\frac{\partial \mathbf{B}}{\partial Z_\alpha}\right|_{Z_1 = \overset{\bullet}{Z}_1, \ldots, Z_m = \overset{\bullet}{Z}_m} = \frac{\partial \overset{\circ}{\mathbf{B}}}{\partial \overset{\circ}{Z}_\alpha}$,

$\overset{\circ}{Z}_\alpha$ is the expected value of variable Z_α.

Assuming on the basis of relationships (2.9) that

$$\overset{\circ}{\mathbf{K}} = \mathbf{K}\left(\overset{\circ}{Z}_1,\ldots,\overset{\circ}{Z}_m\right), \qquad \tilde{\mathbf{K}} = \sum_\alpha \overset{\circ}{\mathbf{K}}_{,\alpha}\, \tilde{Z}_\alpha\,,$$
$$\overset{\circ}{\mathbf{B}} = \mathbf{B}\left(\overset{\circ}{Z}_1,\ldots,\overset{\circ}{Z}_m\right), \qquad \tilde{\mathbf{B}} = \sum_\alpha \overset{\circ}{\mathbf{B}}_{,\alpha}\, \tilde{Z}_\alpha\,, \tag{2.10}$$

the random fluctuations in relationhip (2.6) can be presented in the form

$$\tilde{k}_i = \sum_\alpha \overset{\circ}{k}_{i\alpha}\, \tilde{Z}_\alpha\,, \qquad \tilde{b}_i = \sum_\alpha \overset{\circ}{b}_{i\alpha}\, \tilde{Z}_\alpha\,, \qquad \tilde{d}_i = \sum_\alpha \overset{\circ}{d}_{i\alpha}\, \tilde{Z}_\alpha\,, \tag{2.11}$$

where $\overset{\circ}{k}_{i\alpha} = \overset{\circ}{\mathbf{w}}{}_i^T\, \overset{\circ}{\mathbf{K}}_{,\alpha}\, \overset{\circ}{\mathbf{w}}_i\,,\quad \overset{\circ}{b}_{i\alpha} = \overset{\circ}{\mathbf{w}}{}_i^T\, \overset{\circ}{\mathbf{B}}_{,\alpha}\, \overset{\circ}{\mathbf{w}}_i\,,\quad \overset{\circ}{d}_{i\alpha} = \overset{\circ}{k}_{i\alpha} - \overset{\circ}{\lambda}_i\, \overset{\circ}{b}_{i\alpha}\,.$

By putting relationships (2.11) into dependences (2.6), (2.7) and (2.8), the random fluctuation of eigenvalue λ_i, the covariance of any two eigenvalues λ_i and λ_j and eigenvalue variance can be defined by the relationships

$$\overset{\circ}{\lambda}_i = \sum_\alpha \overset{\circ}{\lambda}_{i,\alpha}\, \tilde{Z}_\alpha\,,$$
$$\mathrm{Cov}\left(\lambda_i,\lambda_j\right) = \sum_\alpha\sum_\beta \overset{\circ}{\lambda}_{i,\alpha}\, \overset{\circ}{\lambda}_{j,\beta}\, \mathrm{Cov}\left(Z_\alpha, Z_\beta\right), \tag{2.12}$$
$$\mathrm{Var}(\lambda_i) = \sum_\alpha\sum_\beta \overset{\circ}{\lambda}_{i,\alpha}\, \overset{\circ}{\lambda}_{i,\beta}\, \mathrm{Cov}\left(Z_\alpha, Z_\beta\right),$$

where $\overset{\circ}{\lambda}_{i,\alpha}$ is defined by relationship (3.2) as the derivative of the eigenvalue relative to the mean value of structural parameter Z_α; $\mathrm{Cov}\left(Z_\alpha, Z_\beta\right) = r_{\alpha\beta} V_\alpha V_\beta\, \overset{\circ}{Z}_\alpha\, \overset{\circ}{Z}_\beta$; $r_{\alpha\beta}$, V_α, V_β are the correlation coefficient and the variation coefficients of variables Z_α and Z_β.

Formulas (2.12) will constitute the basis for the analysis of the sensitivity of the probabilistic characteristics of the eigenvalue problem.

3 The Sensitivity of the Probabilistic Characteristics of Eigenvalues and Eigenvectors

The investigation of the response sensitivity of a system to a change in a structural parameter according to the deterministic approach reduces to the investigation of the first or second order derivative of the response to this parameter. In the case of random design features, the problem becomes more complex. Here, one should study the derivative of a relevant probabilistic feature of the structure's reaction to the selected probabilistic feature of the structural parameter.

In further considerations we will study the sensitivity of the expected values, the variance, and the standard deviations of eigenvalues and of the values of the expected eigenvectors to a change in the expected value, the variance or the standard deviation of a structural parameter Z_α.

The expected value of eigenvalue λ_i after relationship (2.3) has been pre-multiplied by transposed eigenvector $\overset{\circ}{\mathbf{w}}_i$ can be written as

$$\overset{\circ}{\lambda}_i = \frac{\overset{\circ}{k}_i}{\overset{\circ}{b}_i}, \tag{3.1}$$

where $\overset{\circ}{k}_i = \overset{\circ}{\mathbf{w}}_i^T \mathbf{K} \overset{\circ}{\mathbf{w}}_i$, $\overset{\circ}{b}_i = \overset{\circ}{\mathbf{w}}_i^T \mathbf{B} \overset{\circ}{\mathbf{w}}_i$.

By differentiating relationship (3.1) relative to structural parameter $\overset{\circ}{Z}_\alpha$ and adopting the symbols used in (2.11), one obtains the following form of the sensitivity of expected eigenvalue $\overset{\circ}{\lambda}_i$ to a change in parameter $\overset{\circ}{Z}_\alpha$

$$\overset{\circ}{\lambda}_{i,\alpha} = \frac{\overset{\circ}{d}_{i\alpha}}{\overset{\circ}{b}_i}. \tag{3.2}$$

In order to determine the sensitivity of eigenvector $\overset{\circ}{\mathbf{w}}_i$ to changes in structural parameter $\overset{\circ}{Z}_\alpha$, relationship (2.3) was differentiated relative to that parameter, whereby the following equation was obtained

$$\left(\overset{\circ}{\mathbf{K}} - \overset{\circ}{\lambda}_i \overset{\circ}{\mathbf{B}}\right) \overset{\circ}{\mathbf{w}}_{i,\alpha} = -\left(\overset{\circ}{\mathbf{K}} - \overset{\circ}{\lambda}_i \overset{\circ}{\mathbf{B}}\right)_{,\alpha} \tilde{\mathbf{w}}_i. \tag{3.3}$$

Since $\det\left(\overset{\circ}{\mathbf{K}} - \overset{\circ}{\lambda}_i \overset{\circ}{\mathbf{B}}\right) = 0$, equation (3.3) has a solution (is not contradictory) only when the vector that makes up its right side is orthogonal to the vector that constitutes the solution of a homogenous equation, i.e. to eigenvector $\overset{\circ}{\mathbf{w}}_i$ and the solution has the form

$$\overset{\circ}{\mathbf{w}}_{i,\alpha} = \overset{\circ}{\mathbf{x}}_i + a\,\overset{\circ}{\mathbf{w}}, \tag{3.4}$$

where $\overset{\circ}{\mathbf{x}}_i$ is any particular solution of equation (3.3), and a is any parameter.

After equation (3.3) has been pre-multiplied by transposed eigenvector $\overset{\circ}{\mathbf{w}}_i$, one finds that the left side of the obtained equality is equal to zero. Thus the following relationship holds

$$\overset{\circ}{\mathbf{w}}_i^T\left(\overset{\circ}{\mathbf{K}}-\overset{\circ}{\lambda}_i\overset{\circ}{\mathbf{B}}\right)_{,\alpha}\overset{\circ}{\mathbf{w}}_i=0\ , \tag{3.5}$$

which means that equation (3.3) is not contradictory.

Since one nonzero component, e.g. $\overset{\circ}{w}_{ji}$, of an eigenvector can be assumed arbitrarily (is independent of structural parameters), the ratios of the remaining components to the selected one may depend on the structural parameters. This reduces the study of the eigenvector sensitivity to the determination of the derivative of the vector normalized in such a way that the value of the selected component relative to which the sensitivity is investigated is equal to unity. When the eigenvector is given in the form

$$\overset{\circ}{\mathbf{w}}_i=\left(\frac{\overset{\circ}{\mathbf{w}}_i}{\overset{\circ}{w}_{ji}}\right)\overset{\circ}{w}_{ji}\ , \tag{3.6}$$

the form of its derivative analogous to (3.4) is defined by the relationship

$$\overset{\circ}{\mathbf{w}}_{i,\alpha}=\left(\frac{\overset{\circ}{\mathbf{w}}_i}{\overset{\circ}{w}_{ji}}\right)_{,\alpha}\overset{\circ}{w}_{ji}+\left(\frac{\overset{\circ}{\mathbf{w}}_i}{\overset{\circ}{w}_{ji}}\right)\overset{\circ}{w}_{ji,\alpha}\ . \tag{3.7}$$

Taking into account that $\overset{\circ}{w}_{ji}=1$, we obtain $\overset{\circ}{w}_{ji,\alpha}=a=0$ and $\overset{\circ}{\mathbf{w}}_{i,\alpha}=\overset{\circ}{\mathbf{x}}_i=\left[\overset{\circ}{w}_{1i,\alpha},\dots,0,\dots,\overset{\circ}{w}_{ni,\alpha}\right]^T$, which means that $\overset{\circ}{\mathbf{w}}_{i,\alpha}$ is a particular solution of equation (3.3) such that the j-th selected component equals zero. The eigenvector sensitivity defined in this way eliminates therefore all the degrees of freedom of solution (3.4), and it is specified unequivocally by the choice of a component relative to which the vector is normalized.

The numerical solution of equation (3.3) can be obtained by solving the substitute equation

$$\mathbf{A}\overset{\circ}{\mathbf{w}}_{i,\alpha}=\mathbf{q}_i\ , \tag{3.8}$$

where A is a matrix which was created with matrix $\left(\overset{\circ}{\mathbf{K}}-\overset{\circ}{\lambda}_i\overset{\circ}{\mathbf{B}}\right)$ by zeroing the j-th column and the j-th row and assigning unity to the element at their intersection, and q is a vector which was created with vector $-\left(\overset{\circ}{\mathbf{K}}-\overset{\circ}{\lambda}_i\overset{\circ}{\mathbf{B}}\right)_{,\alpha}\overset{\circ}{\mathbf{w}}_i=-\left(\overset{\circ}{\mathbf{K}}_{,\alpha}-\overset{\circ}{\lambda}_i\overset{\circ}{\mathbf{B}}_{,\alpha}-\overset{\circ}{\lambda}_{i,\alpha}\overset{\circ}{\mathbf{B}}\right)\overset{\circ}{\mathbf{w}}$ by zeroing the j-th element.

The sensitivity of the covariance of eigenvalues λ_i and λ_j and the sensitivity of the variance of eigenvalue λ_i to the change of parameter $\overset{\circ}{Z}_\gamma$ are defined by the relationships obtained as a result of the differentiation of relationships (2.12) relative to that parameter

$$\begin{aligned}\mathrm{Cov}_{,\gamma}\left(\lambda_i,\lambda_j\right)=&\sum_\alpha\sum_\beta \overset{\circ}{\lambda}_{i,\alpha\gamma}\,\overset{\circ}{\lambda}_{j,\beta}\, r_{\alpha\beta} V_\alpha V_\beta\, \overset{\circ}{Z}_\alpha\, \overset{\circ}{Z}_\beta+\\&+\sum_\alpha\sum_\beta \overset{\circ}{\lambda}_{i,\alpha}\,\overset{\circ}{\lambda}_{j,\beta\gamma}\, r_{\alpha\beta} V_\alpha V_\beta\, \overset{\circ}{Z}_\alpha\, \overset{\circ}{Z}_\beta+\\&+\sum_\alpha\sum_\beta \overset{\circ}{\lambda}_{i,\alpha}\,\overset{\circ}{\lambda}_{j,\beta}\, r_{\alpha\beta} V_\alpha V_\beta X,\end{aligned}\tag{3.9}$$

$$\begin{aligned}\mathrm{Var}_{,\gamma}\left(\lambda_i\right)=&\sum_\alpha\sum_\beta \overset{\circ}{\lambda}_{i,\alpha\gamma}\,\overset{\circ}{\lambda}_{i,\beta}\, r_{\alpha\beta} V_\alpha V_\beta\, \overset{\circ}{Z}_\alpha\, \overset{\circ}{Z}_\beta+\\&+\sum_\alpha\sum_\beta \overset{\circ}{\lambda}_{i,\alpha}\,\overset{\circ}{\lambda}_{i,\beta\gamma}\, r_{\alpha\beta} V_\alpha V_\beta\, \overset{\circ}{Z}_\alpha\, \overset{\circ}{Z}_\beta+\\&+\sum_\alpha\sum_\beta \overset{\circ}{\lambda}_{i,\alpha}\,\overset{\circ}{\lambda}_{i,\beta}\, r_{\alpha\beta} V_\alpha V_\beta X,\end{aligned}\tag{3.10}$$

where $X=\begin{cases}\overset{\circ}{Z}_\alpha & \text{for } \gamma=\beta \text{ and } \alpha\neq\beta,\\ \overset{\circ}{Z}_\beta & \text{for } \gamma=\alpha \text{ and } \alpha\neq\beta,\\ 2\overset{\circ}{Z}_\gamma & \text{for } \gamma=\alpha=\beta,\\ 0 & \text{for } \gamma\neq\alpha \text{ and } \gamma\neq\beta.\end{cases}$

The second derivative of eigenvalue $\overset{\circ}{\lambda}_{i,\alpha\gamma}$ relative to the structural parameters still remains undetermined in the above relationships. To determine it, let us differentiate relationship (3.5) relative to parameter $\overset{\circ}{Z}_\gamma$

$$\overset{\circ}{\mathbf{w}}_i^T\left(\overset{\circ}{\mathbf{K}}-\overset{\circ}{\lambda}_i\,\overset{\circ}{\mathbf{B}}\right)_{,\alpha\gamma}\overset{\circ}{\mathbf{w}}_i+2\,\overset{\circ}{\mathbf{w}}_i^T\left(\tilde{\mathbf{K}}-\overset{\circ}{\lambda}_i\,\tilde{\mathbf{B}}\right)_{,\alpha}\overset{\circ}{\mathbf{w}}_{i,\gamma}=0\tag{3.11}$$

and relationship $\overset{\circ}{\mathbf{w}}_i^T\left(\tilde{\mathbf{K}}-\overset{\circ}{\lambda}_i\,\tilde{\mathbf{B}}\right)\overset{\circ}{\mathbf{w}}_{i,\gamma}=0$ relative to parameter $\overset{\circ}{Z}_\alpha$

$$\overset{\circ}{\mathbf{w}}_i^T\left(\overset{\circ}{\mathbf{K}}-\overset{\circ}{\lambda}_i\,\overset{\circ}{\mathbf{B}}\right)_{,\alpha}\overset{\circ}{\mathbf{w}}_{i,\gamma}+\overset{\circ}{\mathbf{w}}_i^T{}_{,\alpha}\left(\tilde{\mathbf{K}}-\overset{\circ}{\lambda}_i\,\tilde{\mathbf{B}}\right)\overset{\circ}{\mathbf{w}}_{i,\gamma}=0\ .\tag{3.12}$$

By incorporating (3.12) into (3.11), the following is obtained

$$\overset{\circ}{\mathbf{w}}_i^T\left(\overset{\circ}{\mathbf{K}}-\overset{\circ}{\lambda}_i\,\overset{\circ}{\mathbf{B}}\right)_{,\alpha\gamma}\overset{\circ}{\mathbf{w}}_i+2\,\overset{\circ}{\mathbf{w}}_i^T{}_{,\alpha}\left(\tilde{\mathbf{K}}-\overset{\circ}{\lambda}_i\,\tilde{\mathbf{B}}\right)\overset{\circ}{\mathbf{w}}_{i,\gamma}=0\ .\tag{3.13}$$

Hence the desired second derivative of an eigenvalue can be presented in the form

$$\overset{\circ}{\lambda}_{i,\alpha\gamma} = \overset{\circ}{\lambda}_{i\alpha,\gamma} = \overset{\circ}{d}_{\alpha\gamma i} - \overset{\circ}{\lambda}_{i,\alpha} \overset{\circ}{b}_{i\gamma} - \overset{\circ}{\lambda}_{i,\gamma} \overset{\circ}{b}_{i\alpha} \,, \tag{3.14}$$

where

$$\overset{\circ}{d}_{i\alpha\gamma} = \overset{\circ}{\mathbf{w}}_i^T \left(\overset{\circ}{\mathbf{K}}_{,\alpha\gamma} - \overset{\circ}{\lambda}_i \overset{\circ}{\mathbf{B}}_{,\alpha\gamma} \right) \overset{\circ}{\mathbf{w}}_i - 2\, \overset{\circ}{\mathbf{w}}_i^T{}_{,\alpha} \left(\overset{\circ}{\mathbf{K}} - \overset{\circ}{\lambda}_i \overset{\circ}{\mathbf{B}} \right) \overset{\circ}{\mathbf{w}}_{i,\gamma} = \overset{\circ}{k}_{i\alpha\gamma} - \overset{\circ}{\lambda}_i \overset{\circ}{b}_{i\alpha\gamma} \,,$$

$$\overset{\circ}{k}_{i\alpha\gamma} = \overset{\circ}{\mathbf{w}}_i^T \overset{\circ}{\mathbf{K}}_{\alpha\gamma} \overset{\circ}{\mathbf{w}}_i - 2\, \overset{\circ}{\mathbf{w}}_i^T{}_{,\alpha} \overset{\circ}{\mathbf{K}}_{i\alpha\gamma} \overset{\circ}{\mathbf{w}}_{i,\gamma} \,,$$

$$\overset{\circ}{b}_{i\alpha\gamma} = \overset{\circ}{\mathbf{w}}_i^T \overset{\circ}{\mathbf{B}}_{\alpha\gamma} \overset{\circ}{\mathbf{w}}_i - 2\, \overset{\circ}{\mathbf{w}}_i^T{}_{,\alpha} \overset{\circ}{\mathbf{B}}_{i\alpha\gamma} \overset{\circ}{\mathbf{w}}_{i,\gamma} \,.$$

The sensitivity of the standard deviation of an eigenvalue to changes of any quantities is

$$\sigma_{\lambda_i},_{\gamma} = 0.5 \frac{\mathrm{Var},_{\gamma}(\lambda_i)}{\sigma_{\lambda_i}} \,. \tag{3.15}$$

The sensitivity of the covariance of eigenvalues λ_i and λ_j and the sensitivity of the variance of eigenvalue λ_i to a change in standard deviation σ_γ of variable Z_γ have the form

$$\mathrm{Cov},_{\sigma_\gamma}(\lambda_i, \lambda_j) = \sum_\alpha \sum_\beta \overset{\circ}{\lambda}_{i,\alpha} \overset{\circ}{\lambda}_{j,\beta}\, r_{\alpha\beta} X \,, \tag{3.16}$$

$$\mathrm{Var},_{\sigma_\gamma}(\lambda_i) = \sum_\alpha \sum_\beta \overset{\circ}{\lambda}_{i,\alpha} \overset{\circ}{\lambda}_{i,\beta}\, r_{\alpha\beta} X \,, \tag{3.17}$$

where $X = \begin{cases} \overset{\circ}{Z}_\alpha & \text{for } \gamma = \beta \text{ and } \alpha \neq \beta, \\ \overset{\circ}{Z}_\beta & \text{for } \gamma = \alpha \text{ and } \alpha \neq \beta, \\ 2\overset{\circ}{Z}_\gamma & \text{for } \gamma = \alpha = \beta, \\ 0 & \text{for } \gamma \neq \alpha \text{ and } \gamma \neq \beta. \end{cases}$

The sensitivity of the covariance of eigenvalues λ_i and λ_j and the sensitivity of the variance of eigenvalue λ_i to a change of variance $\mathrm{Var}(Z_\gamma)$ of variable Z_γ have the form

$$\mathrm{Cov},_{\mathrm{Var}(Z_\gamma)}(\lambda_i, \lambda_j) = \overset{\circ}{\lambda}_{i,\gamma} \overset{\circ}{\lambda}_{j,\gamma} \,, \tag{3.18}$$

$$\mathrm{Var},_{\mathrm{Var}(Z_\gamma)}(\lambda_i) = \left(\overset{\circ}{\lambda}_{i,\gamma} \right)^2 . \tag{3.19}$$

4 An Example

A computer program has been developed on the basis of the derived relationships. Some results of test analyses of the sensitivity of the expected values of eigenvalues and the sensitivity of the variance of eigenvalues for the system with two degrees of freedom shown in Fig.4.1 are presented in the form of diagrams (Figs 4.2-4.5).

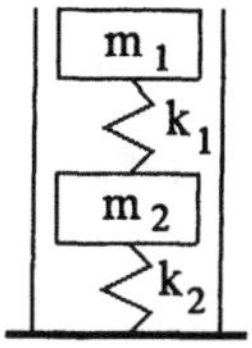

Fig. 4.1. Dynamic system

The analyses were made assuming the folowing basic parameter values: k_1 = 10000-100000 N/m, k_2 = 200000 N/m, m_1 = 3000 kg, m_2 = 9000 kg and the coefficient of variation equals 0.05. The relationship beween stiffness k_1 and the sensitivity of the above mentioned quantities to changes in rigidity k_1 is shown in figs 4.2 , 4.3, 4.4 and 4.5.

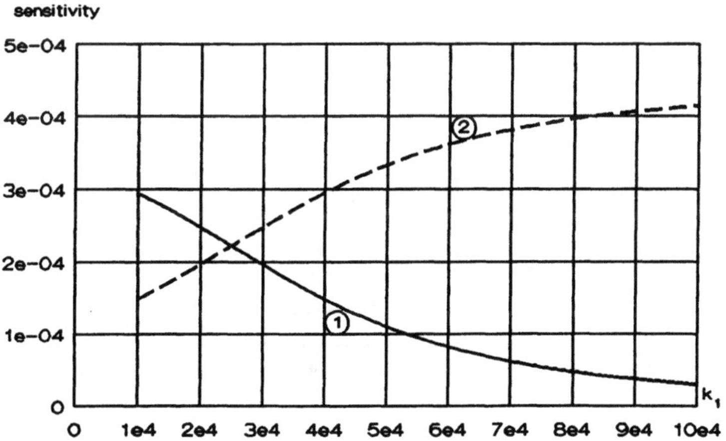

Fig. 4.2. Sensitivity of the expected value of eigenvalue with respect to stiffness k_1 as a function k_1: curve 1 shows $\overset{\circ}{\lambda}_{1,k_1}(k_1)$, curve 2 shows $\overset{\circ}{\lambda}_{2,k_1}(k_1)$

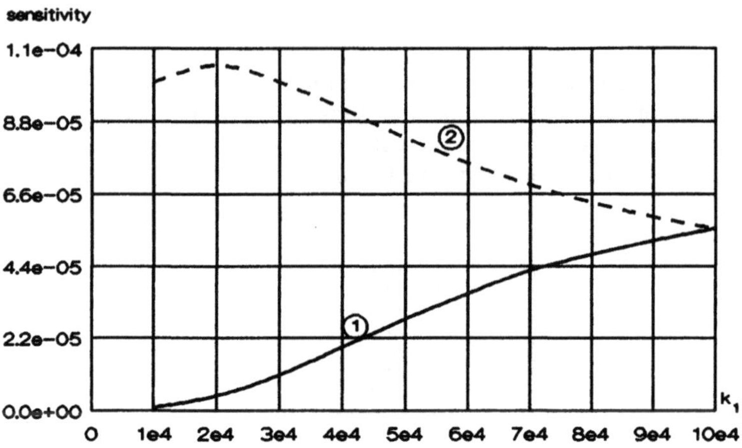

Fig. 4.3. Sensitivity of the expected value of eigenvalue with respect to stiffness k_2 as a function k_1: curve 1 shows $\mathring{\lambda}_{1,k_2}(k_1)$, curve 2 shows $\mathring{\lambda}_{2,k_2}(k_1)$

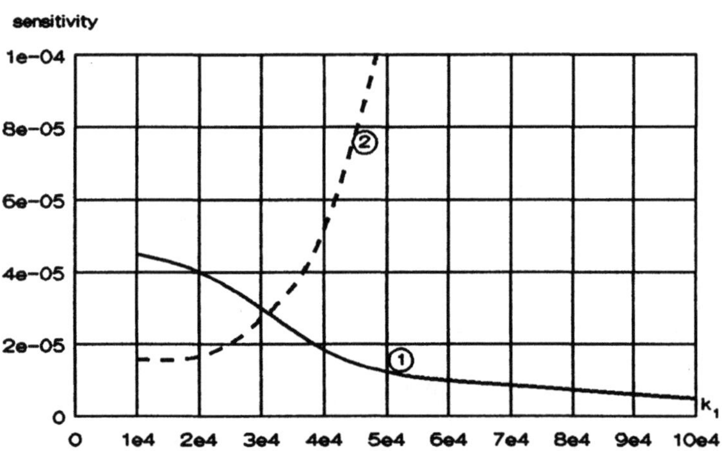

Fig. 4.4. Sensitivity of the variance of eigenvalue with respect to stiffness k_1 as a function k_1: curve 1 shows $Var_{,k_1}(\mathring{\lambda}_1(k_1))$, curve 2 shows $Var_{,k_1}(\mathring{\lambda}_2(k_1))$

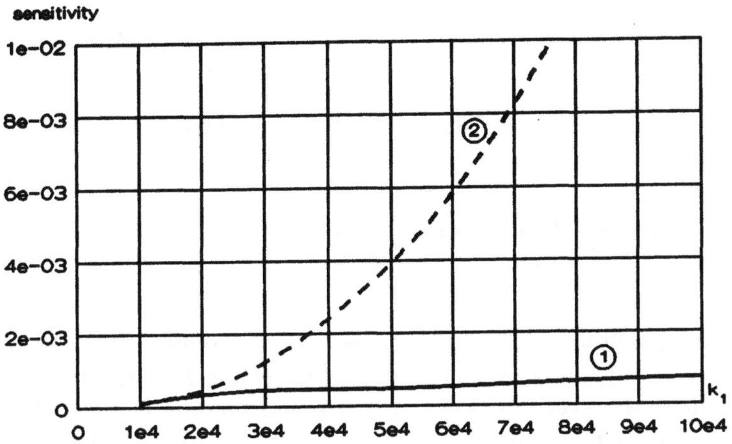

Fig. 4.5. Sensitivity of the variance of eigenvalue with respect to the mass m_1 as a function k_1: curve 1 shows $\mathrm{Var}_{,m_1}(\mathring{\lambda}_1(k_1))$, curve 2 shows $\mathrm{Var}_{,m_1}(\mathring{\lambda}_2(k_1))$

5 Final Remarks

A probabilistic solution being a result of linearization served as the basis for the presented analysis of the sensitivity of the random eigenproblem. It is possible to obtain more accurate solutions by assuming a greater number of terms of the expansions of random variables into the Taylor series. By assuming an expansion with the accuracy to quadratic terms retained, random variable X $(X = \mathbf{K}, \mathbf{B}, \mathbf{w}_i, \lambda_i)$ can be given the form

$$X = X\left(\mathring{Z}\right) = \sum_{\alpha} \mathring{X}_{,\alpha} \tilde{Z}_\alpha + 0.5 \sum_{\alpha} \sum_{\beta} \mathring{X}_{,\alpha\beta} \tilde{Z}_\alpha \tilde{Z}_\beta \ , \tag{5.1}$$

where quantities $X\left(\mathring{Z}\right)$, $\mathring{X}_{,\alpha}$ are defined analogously as relationships (2.9),

whereas $X_{,\alpha\beta} = \left.\dfrac{\partial^2 X}{\partial Z_\alpha \partial Z_\beta}\right|_{Z_1=\dot{Z}_1,\dots,Z_m=\dot{Z}_m}$.

After the expansions of quantities K, B, w and λ_i into series of type (5.1) have been put into equation (2.1) and the terms containing fluctuations to the power higher than the second one have been neglected or after the left side of relationship (2.1) has been expanded into a series of type (5.1), the relationship becomes

$$\left(\mathring{\mathbf{K}}-\mathring{\lambda}_i\mathring{\mathbf{B}}\right)\mathring{\mathbf{w}}_i+\sum_{\alpha}\left(\left(\mathring{\mathbf{K}}-\mathring{\lambda}\mathring{\mathbf{B}}\right)\mathring{\mathbf{w}}_i\right)_{,\alpha}\tilde{Z}_\alpha+0.5\sum_{\alpha}\sum_{\beta}\left(\left(\mathring{\mathbf{K}}-\mathring{\lambda}_i\mathring{\mathbf{B}}\right)\mathring{\mathbf{w}}_i\right)_{,\alpha\beta}\tilde{Z}_\alpha\tilde{Z}_\beta=0\,. \tag{5.2}$$

By performing the expected value operation on equation (5.2), the following relationship is obtained

$$\left(\mathring{\mathbf{K}}-\mathring{\lambda}_i\mathring{\mathbf{B}}\right)\mathring{\mathbf{w}}_i+0.5\sum_{\alpha}\sum_{\beta}\left(\left(\mathring{\mathbf{K}}-\mathring{\lambda}_i\mathring{\mathbf{B}}\right)\mathring{\mathbf{w}}_i\right)_{,\alpha\beta}\operatorname{Cov}\left(Z_\alpha,Z_\beta\right)=0\,. \tag{5.3}$$

The desired quantities $\mathring{\lambda}_i$, $\mathring{\lambda}_{i,\alpha}$, $\mathring{\lambda}_{i,\alpha\beta}$ can be determined from (5.3) in an iterative manner when the solution presented in section 2 is assumed as the first approximation. However, the solution given in section 2 is sufficient for the problem of sensitivity analysis since in this case we want to know if a small change in a structural parameter (here the probabilistic characteristic of this parameter) has a small or great influence on the change in the system response. When an expansion of type (5.1) is used, the expected value of any quantity X is equal to

$$E[X]=X\left(\mathring{Z}_1,\ldots,\mathring{Z}_m\right)+0.5\sum_{\alpha}\sum_{\beta}\mathring{X}_{,\alpha\beta}\operatorname{Cov}\left(Z_\alpha,Z_\beta\right). \tag{5.4}$$

The additional term that occurs in the above formula, in comparison with relationship $E[X]=X\left(\mathring{Z}_1,\ldots,\mathring{Z}_m\right)=\mathring{X}$ assumed in formula (2.12), can be treated as a measure of accuracy of the solution.

References

1. Frank P. M., Introduction to system sensitivity theory, Acad. Press, New York 1978,
2. Haug E. J., Choi, K. K. & Komkov V., Design sensitivity analysis of structural systems, Acad. Press 1985,
3. Socha L., The sensitivity analysis of stochastic non-linear dynamical systems, Jour. of Sound and Vibr., 110(2) (1986) 271-288,
4. Soong T. T. & Chen Y. Q., Stochastic response sensitivity of secondary systems to primary structural uncertainties, Structural Safety, 6 (1989) 311-321,
5. Madsen H.O., Krenk S., & Lind N. C., Methods of structural safety, 1986, Prentice-Hall, Inc. Englewood Cliffs.,
6. Der Kiureghian A. & Ke J. B., The stochastic finite element method in structural reliability, Probabilistic Engin. Mech. 3(2) (1988) 83-91,
7. Collins J.D., & Thompson W.T.,The eigenvalue problem for structural systems with statistical properties, AIAAJ, 7(4) (1969,) 642-648,

8. Shinozuka M. & Astil J., Random eigenvalue problems in structural analysis, AIAAJ, 10(4) (1972) 456-462,
9. Hisada T. & Nakagiri S., Role of the stochastic finite element method in structural safety and reliability, 4th ICOSSAR, 1985, I385-I394,
10. Mironowicz W. & Śniady P., Dynamics of machine foundations with random parameters, Journ. of Sound and Vibr., 112(1) (1987) 23-30,
11. Iyengar R.N. & ManoharC.S., Probability distribution of the eigenvalues of the random string equation, Jour. of Applied Mech. 56(3) (1989) 202-207,
12. Vanmarcke E., Shinozuka M., Nakagiri S., Schueller G. & Grigoriu M., Random fields and stochastic finite elements, Struc. Safety, 3 (1986) 143-166,
13. Nakagiri S., Takopatake H. & Tani S., Uncertain eigenvalue analysis of composite laminated plates by the stochastic finite element method, Jour. of Engineering for Industry, 109 (1987) February, 9-12.
14. Ibrahim R.A., Structural dynamics with parameter uncertainties, Appl. Mech. Rev. 40, 3 (1987) 309-328.
15. Benaroya H., Random eigenvalues and structural dynamic models in "Stochastic structural Dynamics 1", ed. Y.K. Lin, I. Elishakoff, Springer-Verlag 1990, 11-32.
16. Śniady P. & Żukowski S., Safety estimation of the rod structures due to loss of stability (in polish), Arch. of Civ. Engineering, XXXVII, 2 (1991) 165-177.

A Probalistic Model for the Accumulation of Fatigue Damage in Composite Laminates

P. D. Spanos[1] and J. D. Rowatt[2]

[1]L. B. Ryon Chair in Engineering, Rice University, P.O. Box 1892, Houston, TX 77251, USA
[2]Research Assistant, Department of Mechanical Engineering and Materials Science, Rice University, P.O. Box 1892, Houston, TX 77251, USA

Abstract. A stochastic model for the accumulation of fatigue damage in composite laminates in terms of stiffness loss is proposed. The model considers the accumulation of fatigue damage as an evolutionary stochastic process and represents the distribution of stiffness loss in a laminate as a function of the number of load cycles. Stiffness loss is modeled as an embedded discrete time, finite state, and nonstationary stochastic process; it utilizes a discrete time, finite state, stationary Markov model in conjunction with a time transformation-condensation method. The principle damage mechanisms associated with stiffness loss are examined within the framework of the so-called critical element model. The applicability of the proposed model is demonstrated by considering the accumulation of fatigue damage in terms of changes in the dynamic longitudinal secant modulus of graphite/epoxy laminates cycled under constant amplitude, sinusoidal, tension-tension fatigue loads.

Keywords. Composite laminates, fatigue, stiffness loss, Markov chains.

1 Introduction

Under fatigue loading the response of a composite laminate is governed by a complex cumulative damage process. This process is due to the interaction of several distinct damage mechanisms including matrix cracking, delamination, and fiber breakage. These damage mechanisms interact to produce changes in the strength and stiffness of a laminate on both local and global levels. The magnitude of the changes in the material properties of the laminate and their effect on response quantities such as the internal state of stress in a laminate vary from laminate to laminate. Consequently, response quantities dependent on the material properties of a laminate constitute nonstationary stochastic processes.

One option for characterizing such response quantities is to treat them as functions of the material properties of a laminate. This work focuses on the development of nonstationary stochastic models for changes in the stiffness components of a laminate by extending a versatile class of cumulative damage models as proposed by Bogdanoff and Kozin[1]. The models are utilized within the framework of the so-called critical element model proposed by Reifsnider and Stinchcomb[2].

2 Material Considerations

A great deal of effort has been directed by numerous researchers to characterize the complex damage accumulation process in composite laminates subject to fatigue loadings. The typical chronology of damage development as noted by Jamison, et al.[3] among others includes a sequence of matrix related effects and fiber breakage leading ultimately to laminate fracture. More specifically, dispersed intralaminar matrix cracking in the off-axis plies of a laminate creates preferred sites for secondary matrix cracking, delamination, and fiber breakage. These preferred sites induce a progressive localization of damage which acts as a catalyst for laminate fracture.

The critical element model provides a rational means for addressing the complex damage accumulation process by utilizing a representative volume approach wherein the response of the representative volume element is assumed to characterize the response of the laminate as a whole. The representative volume element is divided into critical and subcritical elements based on their contribution to the response of the representative volume element.

The subcritical elements are those material elements that cause a redistribution of stress in a laminate due to the changes in the local geometry of the laminate. The magnitude of the stress redistribution is typically determined by measuring or predicting changes in the stiffness components of a laminate as stiffness changes are proportional to stress redistribution[4]. The critical elements are those material elements which control the failure or fracture of the laminate. The critical elements describe the state of the laminate itself and are characterized by phenomenological information on their response such as S-N diagrams. To this extent the critical element model is a hybrid model in which changes in the internal stress state in a laminate, induced by subcritical element damage, are used to characterize the state of the material in terms of the residual strength of the critical elements. For typical high modulus fiber reinforced laminates, the subcritical elements are matrix cracking and delamination. The critical elements are the load bearing plies of the laminate.

From a practical standpoint, the use of the critical element model requires that stiffness changes due to matrix cracking and delamination are readily attainable. Such changes may be measured directly in a laboratory or modeled analytically. Models such as those proposed by Reifsnider[5], O'Brien[6,7], and Talreja[8] provide a means for determining stiffness changes in a laminate due to matrix cracking and delamination from nondestructive test data often attainable in the field. However, to characterize the variation in stiffness changes from laminate to laminate a stochastic model is needed. Such a model would allow for a more versatile and realistic version of the critical element model to be developed.

3 Probabilistic Modeling of Stiffness Loss

3.1 Preliminary Remarks

Bogdanoff and Kozin[1] have developed a versatile class of phenomenological stochastic cumulative damage models based on the theory of Markov chains. There are several underlying assumptions in the models. Specifically, the time over which the load acts is discretized and is measured in units of duty cycles. Damage is assumed to monotonically increase with the number of duty cycles and takes on discrete values (states). The initial state of damage and the state of damage at failure may be random, and the level of damage is considered only at the end of each duty cycle. Furthermore, it is assumed to depend only on the severity of the duty cycle itself and the level of damage prior to the duty cycle.

To illustrate the structure and the proposed extension of the Markov chain models, consider a laminate subjected to fatigue loading in which damage accumulation is monitored by measuring losses in the longitudinal stiffness of the laminate. Let the levels of stiffness loss be discretized into a finite number of states indexed on the nonnegative set of integers $i = 1, 2, \ldots, b$, with the magnitude of the stiffness loss increasing as the index of the state and state b representing an absorbing or "failure" state. From Markov chain theory the fatigue damage accumulation process can then be modeled as

$$\underline{\mathbf{p}}_x = \underline{\mathbf{p}}_0 \prod_{i=1}^{x} \underline{\mathbf{P}}_i, \tag{1}$$

where $\underline{\mathbf{p}}_0$ and $\underline{\mathbf{p}}_x$ are $(1 \times b)$ row vectors denoting the probability mass functions for the initial distribution of stiffness loss and the distribution of stiffness loss after x duty cycles respectively. The symbols $\underline{\mathbf{P}}_i$ denote $(b \times b)$ probability transition matrices whose elements p_{ij} represent the probability of stiffness loss being in state j after a duty cycle, given that stiffness loss was in state i before the duty cycle. The conservation of probability requires that $\sum_{i=1}^{b} p_0(i) = 1$, $\sum_{i=1}^{b} p_x(i) = 1$, and $\sum_{j=1}^{b} p_{ij} = 1$ for $i = 1, 2, \ldots, b$. Furthermore, the quantities $p_0(i), p_x(i)$, and the p_{ij} must be nonnegative. Recalling the assumption that the damage in a component is monotonically increasing restricts the form of $\underline{\mathbf{P}}$ to upper triangular.

Equation 1 is valid for both stationary and nonstationary cumulative damage processes. However, for stationary cumulative damage process, where the elements of the probability transition matrices are constant for each duty cycle, Equation 1 reduces to

$$\underline{\mathbf{p}}_x = \underline{\mathbf{p}}_0 \underline{\mathbf{P}}^x. \tag{2}$$

For nonstationary cumulative damage processes, Equation 1 requires that a probability transition matrix be specified for each duty cycle. To circumvent this requirement, a time transformation-condensation method may be employed to transform a given stationary Markov chain model into a nonstationary model.

3.2 Time Transformation-Condensation

Consider that a stationary Markov chain model of the form of Equation 2 is known along with the empirical distribution function, $F_Y(y)$, for a nonstationary cumulative damage process. The premise of a time transformation is to transform the cumulative distribution function, $F_X(x)$, of a stationary Markov chain model into the cumulative distribution function, $F_{\hat{Y}}(y)$, of a nonstationary model such that $F_{\hat{Y}}(y)$ is an accurate description of $F_Y(y)$. This is accomplished by considering a nonlinear polynomial transformation of the form

$$y = g(x) = a_1 x + a_2 x^2 + \cdots + a_n x^n, \tag{3}$$

with $g(0) = 0$. For uniqueness purposes, this transformation is valid only for those values of x where $\frac{dg(x)}{dx}$ is monotonically increasing or decreasing.

The unknown coefficients, a_i for $i = 1, 2, \ldots, n$, are determined by finding the n, x and y, abscissa pairs corresponding to a given set of n ordinate values for $F_X(x)$ and $F_Y(y)$. This yields a set of n ordered pairs $\{(x_1, y_1), (x_2, y_2), \ldots, (x_n, y_n)\}$. Placing the respective pairs into Equation 3 yields a set of n linear equations in the unknown coefficients $a_1, a_2, \ldots, a_n$ whose solution defines the time transformation. Once the transformation is determined, $F_X(x)$ is transformed point by point into $F_{\hat{Y}}(y)$ utilizing Equation 3.

Given a nonstationary Markov chain model as described by

$$\underline{\mathbf{P}}_y = \underline{\mathbf{P}}_0 \prod_{i=1}^{y} \underline{\mathbf{Q}}_i, \tag{4}$$

condensation is employed to determine the nonstationary probability transition matrices, $\underline{\mathbf{Q}}_i$, given a time transformation $y = g(x)$ and a stationary Markov chain model as specified by Equation 2 with a probability transition matrix $\underline{\mathbf{P}}$. Condensation is required to preserve the assumption of time being measured in discrete units with respect to both the stationary and nonstationary Markov chain models. That is, the values of x and y used with Equations 2 and 4 must be discrete. To implement the condensation method, let $x_1, x_2, \ldots$ be determined from $i = g(x_i)$ for $i = 1, 2, \ldots$ Given the x_i's the following intervals are formed $(0, x_1], (x_1, x_2], \ldots$ Then, the nonstationary probability transition matrices, $\underline{\mathbf{Q}}_i$, are defined by

$$\underline{\mathbf{Q}}_i = \begin{cases} \underline{\mathbf{I}} & \text{if } (x_{i-1}, x_i] \text{ contains no integer} \\ \underline{\mathbf{P}} & \text{if } (x_{i-1}, x_i] \text{ contains one integer} \\ \underline{\mathbf{P}}^2 & \text{if } (x_{i-1}, x_i] \text{ contains two integers} \\ \vdots & \end{cases} \tag{5}$$

where $\underline{\mathbf{I}}$ is the $(b \times b)$ identity matrix. The matrices $\underline{\mathbf{Q}}_i$ can then be used in conjunction with an initial distribution of damage, $\underline{\mathbf{p}}_0$, to determine the distribution of damage, $\underline{\mathbf{p}}_y$, after y duty cycles utilizing Equation 4.

3.3 Extension of the Markov Chain Models

In the preceding section it was discussed how to utilize the time transformation - condensation method for determining nonstationary Markov chain models given $F_X(x)$, $F_Y(y)$, and a time transformation $y = g(x)$. In their development of the Markov chain models, Bogdanoff and Kozin[1] determine $F_X(x)$ from an arbitrary stationary Markov chain model. This introduces an artificial element into the nonstationary model since damage is indexed on state alone without regard as to what the states physically represent. Furthermore, the nonstationary Markov B-model may not be unique as many nonstationary cumulative damage processes may have the same distribution for the time to failure.

Bogdanoff and Kozin[1] have shown for stationary cumulative damage processes, however, how to construct a stationary Markov chain model based on the second order moment data for the times to reach subcritical levels of damage as determined by monitoring some physically measurable quantity, such as stiffness loss. This procedure allows for the development of a unique stationary Markov chain model in which the states of the model can be directly related to the measured quantity. This method is extended in the present study to allow the procedure to be used along with the time transformation-condensation method. This extension allows for the determination of a nonstationary Markov chain model in which both the states of the stationary and nonstationary models can be directly related to a physically measurable quantity, in this case stiffness loss. Specifically, the second order moment data used to estimate the parameters of the stationary Markov chain model are taken proportional to the second order moment data for stiffness loss due to the nonstationary fatigue damage accumulation process. A nonstationary Markov chain model can then be constructed using the time transformation-condensation method and used to determine the evolutionary distribution of damage in terms of stiffness loss.

To elucidate the procedure assume that the mean and variance on the time to reach j distinct levels of stiffness loss are known or computed from the empirical distribution functions, $F_{Y_i}(y)$ for $i = 1, 2, \ldots, j$, for the nonstationary fatigue damage accumulation process. Let these quantities be denoted by μ_{Y_i} and $\sigma^2_{Y_i}$. An arbitrary value for the mean value of the first level of stiffness loss, μ_{X_1}, of a stationary Markov chain model is then chosen and the ratio between μ_{X_1} and μ_{Y_1} is determined denoting this ratio by $\hat{r}$. Given this ratio, the means for the $(j-1)$ remaining damage levels are found as

$$\mu_{X_i} = \hat{r}\mu_{Y_i} \quad \text{for} \quad i = 2, 3, \ldots, j. \tag{6}$$

The variances $\sigma^2_{X_i}$ are determined by requiring that the coefficient of variation for each stiffness loss level with respect to the stationary Markov chain model,

$\delta_{X_i} = \sigma_{X_i}/\mu_{X_i}$, is equal to the corresponding coefficient of variation for the nonstationary cumulative damage process being modeled, $\delta_{Y_i} = \sigma_{Y_i}/\mu_{Y_i}$. Stated mathematically,

$$\sigma^2_{X_i} = (\delta_{Y_i}\mu_{X_i})^2 \quad \text{for} \quad i = 1, 2, \ldots, j. \tag{7}$$

Imposing this equality establishes a one-to-one relationship between the damage states in the stationary model to those in the nonstationary model. It should be noted that the value of μ_{X_1} is generally chosen to be less than μ_{Y_1} such that the size of the stationary Markov chain model is reduced making the method more computationally efficient.

Given μ_{X_i} and $\sigma^2_{X_i}$ the probability transition matrix for a stationary Markov chain model can be found by assuming that the probability transition matrix is composed of j distinct submatrices and is of the form

$$\underline{\mathbf{P}} = \begin{bmatrix} [\underline{\mathbf{B}}_1] & & & \\ & [\underline{\mathbf{B}}_2] & & \\ & & \ddots & \\ & & & [\underline{\mathbf{B}}_j] \end{bmatrix}. \tag{8}$$

Assuming that the duty cycles are sufficiently small such that damage can increase by only one state during a duty cycle, the individual submatrices, $[\underline{\mathbf{B}}_i]$, within $\underline{\mathbf{P}}$ have the form

$$[\underline{\mathbf{B}}_i] = \begin{bmatrix} p_i & q_i & & \\ & p_i & q_i & \\ & & \ddots & \ddots \\ & & & p_i \quad q_i \end{bmatrix}. \tag{9}$$

Recalling that $\sum_{j=1}^{b} p_{ij} = 1$ for $i = 1, 2, \ldots, b$ requires that $p_i + q_i = 1$, the parameters to be estimated for each submatrix are the $r_i = p_i/q_i$ and the size of the submatrix $b_i - b_{i-1}$. These parameters can be related to the second order moment data on the time to reach the j damage levels of the stationary Markov chain model using the idea of probability generating functions[1]. Specifically,

$$\mu_{x_i} = \begin{cases} (b_1 - 1)(1 + r_1) & \text{for } i = 1 \\ \mu_{x_{i-1}} + (b_i - b_{i-1})(1 + r_i) & \text{for } i > 1 \end{cases} \tag{10}$$

$$\sigma^2_{x_i} = \begin{cases} (b_1 - 1)r_1(1 + r_1) = r_1\mu_{x_1} & \text{for } i = 1 \\ \sigma^2_{x_{i-1}} + r_i(\mu_{x_i} - \mu_{x_{i-1}}) & \text{for } i > 1. \end{cases} \tag{11}$$

The total number of states in both the stationary and nonstationary Markov chain models is $b = b_j$ with each of the b_i for $i = 1, 2, \ldots, j$ corresponding to a specific stiffness loss level. The other states in the model can be associated with a given level of stiffness loss by simply interpolating between the quantities b_i.

The methodology of the model is now complete. To determine the distribution of stiffness loss in a laminate after y duty cycles, a time transformation based on one of the empirical distribution functions, $F_{Y_i}(y)$, and the corresponding cumulative distribution function, $F_{X_i}(x)$ of a stationary Markov chain model is determined as discussed above. A common guideline is to utilize a stiffness loss level with $\mu_{Y_i} > y$ as the empirical distribution function $F_{Y_i}(y)$ will retain some memory of prior damage as it has accumulated. The value of the cumulative distribution function for the stiffness loss level corresponding to damage state b_i after x duty cycles is

$$F_{X_i}(x) = \sum_{j=b_{(i)}+1}^{b} p_x(j). \tag{12}$$

Given this transformation, the condensation method is then employed to determine the corresponding probability transition matrices, $\underline{\mathbf{Q}}_i$, of the nonstationary Markov chain model according to Equation 5. The nonstationary fatigue damage accumulation process can then be modeled using Equation 4.

4 Numerical Example

To demonstrate its usefulness, the proposed model was applied to data from the literature[3] for changes in the longitudinal secant modulus of T300/5208 and T300/914C graphite/epoxy $[0/90/\pm 45]_S$ angle ply laminates under constant amplitude, sinusoidal, tension-tension fatigue loading. The specimens were cyclically loaded under a stress ratio of $R = 0.1$ with a maximum cyclic stress of 62 percent of their ultimate tensile strengths at a loading frequency of 10 Hz. Unfortunately, a complete set of data in the form of second order statistics and empirical distribution functions on the time to reach various stiffness loss levels prior to failure could not be obtained. Therefore, this example makes use of the data that are available to provide a hypothetical, yet illustrative, example of the proposed modeling procedure.

A representative stiffness loss curve for these laminates and loading configurations has been reported in the literature. For the present example, this curve is taken to show the mean of the stiffness loss versus load cycle relationship for the lot of specimens. The data for the time to reach four stiffness loss levels were used to construct the model. Respectively, these losses were 4, 8, 12, and 16 percent. Due to the lack of data on the variance and the empirical distribution functions of damage prior to and including failure, values for the coefficient of variation were assigned assuming that the scatter in the data increased with the number of load cycles. Furthermore, for a given damage level the data were assumed to follow a two parameter Weibull distribution with shape parameter σ_0 and scale parameter ρ_0. A computer program was written to find the parameters of the stationary B-model and the coefficients of the corresponding time transformations. A transformation of the form of Equation 3 with $n = 4$ was used. The coefficient a_1 was taken as unity and

Table 4.1. Model Input and Output

D_{y_i}	$(\frac{E_1(n)}{E_1})_i$	μ_{y_i}	$\frac{1}{\delta_{y_i}}$	σ_{i_0}	ρ_{i_0}	b_i	r_i
1	0.96	1400	4.00	4.54	1533.34	13	3.17
2	0.92	4600	3.75	4.23	5058.78	20	15.33
3	0.88	6550	3.50	3.92	7234.65	22	33.82
4	0.84	7650	3.25	3.61	8488.30	23	38.29

the other coefficients were found for values of $F_X(x)$ and $F_Y(y)$ of 0.25, 0.50, and 0.95. The model inputs and outputs are summarized in Table 4.1. The values of $(\frac{E_1(n)}{E_1})_i$ and b_i can be used to determine the relationship between stiffness loss and damage state for the nonstationary Markov chain model by interpolation.

The resulting nonstationary Markov chain model was used to compute the cumulative distribution functions for the 4, 8, 12, and 16 percent stiffness losses which were compared to the assumed Weibull distributions for the losses and those predicted by a stationary Markov chain model. The results are shown in Figure 1 for the 4 and 8 percent stiffness losses. From these results it can be seen that the nonstationary model provides a much better fit to the Weibull distribution than the stationary model especially in the upper and lower tails of the distribution. The fit in the lower tail regions may be improved by utilizing smaller values of $F_X(x)$ and $F_Y(y)$ for determining the time transformation.

As a further measure of the validity of the model, the mean and standard deviation for the number of duty cycles required to reach the 4 and 8 percent stiffness losses as computed by both the nonsttionary and stationary models are shown in Table 2. Comparing these values with those in or derived from Table 1 indicates that both models perform well in terms of second order moment data with the nonstationary model yielding slightly lower mean values and the stationary model yielding slightly higer mean values than the data suggests. Further, both models slightly underestimated the standard deviation in each case. These results are somewhat misleading, however, since the cumulative distribution function for the nonstationary model provides a much better fit to the experimental data. This illustrates the importance of utilizing more that the first few moments of distribution as an indicator of the merit of a particular model.

Table 4.2. Comparison of Means and Standard Deviations

Model Type	Stiffness Loss	Mean	Standard Deviation
Nonstat.	4 %	1393	324
Nonstat.	8 %	4587	1158
Stat.	4 %	1412	327
Stat.	8 %	4638	1178

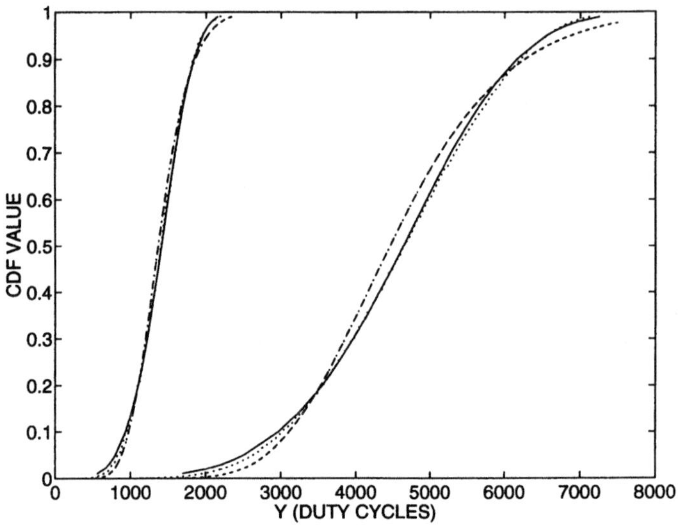

Figure 4.1. Comparison of the experimental (solid), and cumulative distribution functions predicted by a stationary model (dashed) and a nonstationary model (dotted) for 4 and 8 percent losses in secant modulus. 1 duty cycle = 10 load cycles.

5 Concluding Remarks

The model proposed herein provides an effective means for characterizing the accumulation of fatigue damage in composite laminates utilizing changes in the stiffness components as a damage indicator. The results of the model are encouraging in several respects. The most obvious of them is the capacity of the model to provide a much better fit to experimental data over that of a stationary model. This is particularly true for the tails of the experimental distributions where interest is often centered. Furthermore, the model is readily implementable in a numerical code, and given the size and structure of the probability transition matrices is computationally efficient. Results such as these and others warrant further development of the model.

It is hoped that this stiffness based characterization of fatigue damage may be incorporated into the mathematical statement of the critical element model to develop a practical stochastic version of the critical element model in terms of the residual strength of a laminate.

References

1. Bogdanoff, J. L., and Kozin, F., *Probabilistic Models of Cumulative Damage*, John Wiley & Sons, New York, NY, 1985.

2. Reifsnider, K. L., and Stinchcomb, W. W., "A Critical-Element Model of the Residual Strength and Life of Fatigue-Loaded Composite Coupons," *Composite Materials: Fatigue and Fracture*, ASTM STP 907, American Society for Testing and Materials, 1986, pp. 298-313.

3. Jamison, R. D., Schulte, K., Reifsnider, K. L., and Stinchcomb, W. W., "Characterization and Analysis of Damage Mechanisms in Tension-Tension Fatigue of Graphite/Epoxy Laminates," *Effects of Defects in Composite Materials*, ASTM STP 836, American Society for Testing and Materials, 1984, pp. 21-55.

4. Highsmith, A. L., and Reifsnider, K. L., "Stiffness-Reduction Mechanisms in Composite Laminates," *Damage in Composite Materials*, ASTM STP 775, American Society for Testing and Materials, 1982, pp. 103-117.

5. Reifsnider, K. L., "Some Fundamental Aspects of the Fatigue and Fracture Response of Composite Materials," *Proceedings of the* 14th *Annual Meeting of the Society of Engineering Science*, Lehigh University, Bethlehem, PA, November 14-16, 1977, pp. 373-383.

6. O'Brien, T. K., "Characterization of Delamination Onset and Growth in a Composite Laminate," *Damage in Composite Materials*, ASTM STP 775, American Society for Testing and Materials, 1982, pp. 140-167.

7. O'Brien, T. K., "Analysis of Local Delamination and Their Influence on Composite Laminate Behavior," *Delamination and Debonding of Materials*, ASTM STP 876, American Society for Testing and Materials, 1985, pp. 282-297.

8. Talreja, R., *Fatigue of Composite Materials*, Technomic, Lancaster, PA, 1987.

Analysis of Structural Control Robustness: Reliability Methods

B.F. Spencer, Jr.,[a] M.K. Sain,[b] C.-H. Won[b] and L.M. Barroso[c]

[a] Dept. of Civil Engrg. and Geo. Sci., Univ. of Notre Dame, Notre Dame, IN 46556, USA
[b] Dept. of Elec. Engrg., Univ. of Notre Dame, Notre Dame, IN 46556, USA
[c] Dept. of Civil Engrg., Stanford University, Palo Alto, CA 94305, USA

1 Introduction

In recent years, structural control for civil engineering applications has made remarkable progress toward practical implementation (see, for example, Soong,[1] Housner and Masri,[2] ATC-17[3]). The goal of much of this research is to increase the reliability of these structures against dynamic loadings caused by severe earthquakes and strong winds. A framework within which this design goal can be achieved does not seem to be available.

To be effective in the presence of the inherent uncertainty in engineering control situations, a control strategy must have robust performance and stability characteristics. Today's tools for robust control analysis and synthesis are largely based on H_2/H_∞ theories (see, for example, Spencer, *et al.*,[4,5,6] Suhardjo,[7] and Suhardjo, *et al.*[8,9] The basic design problem is to find a stable control which satisfies the performance constraints for all possible cases, including the worst case. For many problems in structural control, the hard bounds on the uncertain parameters required for a worst case design are not meaningful. A probabilistic description of the uncertainty provides a more natural and realistic portrayal; and, in the case of multiple uncertain parameters, a probabilistic description is essential to avoid unreasonable conservatism and associated cost.

Ray and Stengel[10] have addressed certain aspects of this problem using Monte Carlo simulation (MCS) techniques. While effective for determining the central probabilistic tendencies, the major limitation of this approach is that the MCS techniques are usually computationally intractable for the small limit-state probabilities expected for highly reliable controlled structures. Alternatively, Spencer, *et al.*[11,12] have shown first and second order reliability methods (FORM/SORM) to be appropriate for the probability calculations required in these types of problems.

Several consistent, reliability-based, performance/stability measures are formulated in this paper to assess structural control robustness. To illustrate the concepts, we present several examples of a seismically excited three-story building with active tendon control. This work is an initial step toward developing robust control strategies which maximize the overall reliability of a controlled structure.

2 Problem Formulation

We assume the existence of uncertainty that can be modeled by a q-dimensional vector of random parameters $\mathbf{\Delta}$, with a given mean $\boldsymbol{\mu}_{\Delta}$, covariance $\boldsymbol{\Sigma}_{\Delta}$, and joint probability distribution $F_{\Delta}(\boldsymbol{\delta})$. The state space representation of the equations of motion for an n-degree-of-freedom structure is given by

$$\dot{\mathbf{z}} = \mathbf{A}(\mathbf{\Delta})\,\mathbf{z}+\mathbf{B}(\mathbf{\Delta})\,\mathbf{u}+\mathbf{E}(\mathbf{\Delta})\,\mathbf{w}, \tag{1}$$

with measurement equation

$$\mathbf{y} = \mathbf{C}(\mathbf{\Delta})\,\mathbf{z}+\mathbf{D}(\mathbf{\Delta})\,\mathbf{u}+\mathbf{F}(\mathbf{\Delta})\,\mathbf{v}, \tag{2}$$

where the vector $\mathbf{z}$ is a state vector of displacements and velocities, $\mathbf{A}$ is a matrix composed of the parameters of the structure, $\mathbf{u}$ is a vector of control forces, $\mathbf{B}$ is a matrix specifying the points of application of the control forces, $\mathbf{w}$ is an excitation vector, $\mathbf{E}$ is a matrix specifying the manner in which the excitation affects the structure, $\mathbf{y}$ is a measurement vector, $\mathbf{C}$ is a matrix through which combinations of the states are measured, $\mathbf{D}$ is a matrix specifying the feed-through terms in the measurement, $\mathbf{F}$ is a matrix indicating how the measurement noise affects the measurement, and $\mathbf{v}$ is a measurement noise vector. The vector $[\mathbf{w}' \quad \mathbf{v}']'$ is assumed to be a white noise with mean and joint autocorrelation given by

$$E\begin{bmatrix}\mathbf{w}\\ \mathbf{v}\end{bmatrix} = \mathbf{0}, \tag{3}$$

$$E\left[\begin{Bmatrix}\mathbf{w}(t)\\ \mathbf{v}(t)\end{Bmatrix}\{\mathbf{w}'(t+\tau) \quad \mathbf{v}'(t+\tau)\}\right] = 2\pi\mathbf{S}\,\delta(\tau), \tag{4}$$

where $E[\cdot]$ is the expected value operator, $\mathbf{S}$ is a constant two-sided spectral density matrix, and $\delta(\cdot)$ is the Dirac delta function. Nonwhite noises can be incorporated into the problem formulation by augmenting the equations of motion with appropriate disturbance shaping filters.[13]

For a broad class of control strategies, the closed-loop state space description can be written as

$$\dot{\tilde{\mathbf{z}}} = \mathbf{A}_{\mathrm{cl}}(\mathbf{\Delta})\,\tilde{\mathbf{z}}+\tilde{\mathbf{E}}_{\mathrm{cl}}\begin{Bmatrix}\mathbf{w}\\ \mathbf{v}\end{Bmatrix}, \tag{5}$$

where $\tilde{\mathbf{z}}$ is the augmented state vector (*n.b.*, $\tilde{\mathbf{z}}$ may also contain states of the controller or the disturbance shaping filter), $\mathbf{A}_{\mathrm{cl}}(\mathbf{\Delta})$ is the closed-loop state matrix, and $\tilde{\mathbf{E}}_{\mathrm{cl}}$ is the matrix indicating how the disturbance and measurement noise affect the dynamics of the closed loop system. For example, if we assume that the structure state variables are perfectly measurable and that state feedback control is employed, *i.e.*, $\mathbf{u} = -\mathbf{K}\mathbf{z}$, then $\tilde{\mathbf{z}} = \mathbf{z}$ and the closed-loop state space matrix is given by

$$\mathbf{A}_{\mathrm{cl}}(\mathbf{\Delta}) = \mathbf{A}(\mathbf{\Delta}) - \mathbf{B}(\mathbf{\Delta})\,\mathbf{K} \quad , \tag{6}$$

where $\mathbf{K}$ is the control gain matrix.

Stability is essential to any controlled system. The following section details probabilistic measures of stability robustness for a linear controlled system such as Eq. (1). The subsequent sections propose a tractable measure of performance robustness in terms of the structural response and the required control action.

2.1 Probabilistic Stability Measures

For the class of structures described in the previous section, the stability limit-state probability, p_s (*i.e.*, the probability that the controlled structure is unstable), depends on the real parts of the eigenvalues of the closed-loop state space matrix $\mathbf{A}_{cl}$. The system is unstable if the real part of any of the eigenvalues fails to be negative. The limit-state probability p_s is then given by

$$
\begin{aligned}
p_s &= P\{\bigcup_{j=1}^{m} \mathrm{Re}\,[\lambda_j(\mathbf{\Delta})] \geq 0\} \\
&= P\{\max_j \ \mathrm{Re}\,[\lambda_j(\mathbf{\Delta})] \geq 0\},
\end{aligned}
\tag{7}
$$

where the λ_j are roots of $\det[\mathbf{A}_{cl}(\mathbf{\Delta}) - s\mathbf{I}] = 0$. In terms of the probability density function $f_{\mathbf{\Delta}}(\boldsymbol{\delta})$, the stability limit-state probability is

$$
p_s = \int\limits_{\max \ \mathrm{Re}[\lambda_j(\boldsymbol{\delta})] \geq 0} \!\!\cdots\!\! \int f_{\mathbf{\Delta}}(\boldsymbol{\delta})\, d\boldsymbol{\delta} = \int\limits_{\bigcup_{j=1}^{m} \mathrm{Re}[\lambda_j(\boldsymbol{\delta})] \geq 0} \!\!\cdots\!\! \int f_{\mathbf{\Delta}}(\boldsymbol{\delta})\, d\boldsymbol{\delta}\,.
\tag{8}
$$

The system whose limit-state probability is described by Eqs. (7) and (8) can be viewed as a classical reliability problem for a series of components with limit-state functions $g_j^s(\mathbf{\Delta}) = -\mathrm{Re}\,[\lambda_j(\mathbf{\Delta})]$. If any of the components fail (*i.e.*, $g_j^s(\mathbf{\Delta}) \leq 0$), then the entire system is considered failed (*i.e.*, the structure is unstable). First and second order reliability methods (FORM/SORM) are well-suited to calculate the associated limit-state probability, *i.e.*, the probability of system instability, as formulated in this section. Madsen, *et al.*[14] and Melchers[15] provide more detailed discussions of FORM/SORM approaches.

In addition to control system stability, control performance and the associated control costs must be examined in making a robustness assessment of a particular control strategy.

2.2 Objective Performance Measures

For stochastically excited structures, a tractable measure of performance robustness can be given in terms of the stationary structural responses. For example, if one is concerned with the RMS responses σ_{z_i} exceeding specified target values σ_{0_i}, then a series of performance limit-state functions can be defined as

$$
g_i^p(\mathbf{\Delta};\mathbf{u}) = \sigma_{0_i} - \sigma_{z_i}(\mathbf{\Delta};\mathbf{u})\,.
\tag{9}
$$

Here, the notation $\sigma_{z_i}(\mathbf{\Delta};\mathbf{u})$ is chosen to reinforce the fact that the RMS response depends on both the parametric uncertainty $\mathbf{\Delta}$ and the chosen control strategy $\mathbf{u}(\theta)$, $0 \le \theta \le t$. We can define the performance limit-state probability p_{p} as

$$p_{\mathrm{p}} = P\{\bigcup_{i=1}^{\nu} g_i^{\mathrm{p}}(\mathbf{\Delta};\mathbf{u}) \le 0\}, \tag{10}$$

where ν is the number of critical responses. Thus, p_{p} represents the probability of exceedance of one or more of the target RMS structural responses. In terms of the probability density function $f_{\mathbf{\Delta}}(\boldsymbol{\delta})$, the performance limit-state probability can be written as

$$p_{\mathrm{p}} = \int_{\bigcup_{j=1}^{\nu} \sigma_{z_i}[\boldsymbol{\delta};\mathbf{u}] \ge \sigma_{0_i}} \!\!\!\cdots\!\int f_{\mathbf{\Delta}}(\boldsymbol{\delta})\, d\boldsymbol{\delta} \quad , \tag{11}$$

which can be viewed as a reliability problem for a series of components $g_i^{\mathrm{p}}(\mathbf{\Delta};\mathbf{u})$. To evaluate this expression, FORM/SORM can be employed.

For the linear dynamical system in Eq. (5), the RMS responses, or equivalently the response covariance matrix $\mathbf{\Sigma}_{\tilde{z}} = E[\tilde{\mathbf{z}}\tilde{\mathbf{z}}']$, are obtained as the solution of the Lyapunov equation.[16]

2.3 Control Actuation Requirements

The control actuation required to achieve a certain performance must also be factored into control robustness assessment. Similarly to the previous section, the measure of control actuation requirements adopted herein is given in terms of the stationary RMS control action. The control limit-state probability (*i.e.*, the probability of the RMS control actions exceeding specified actuation levels) is defined as

$$p_{\mathrm{c}} = P\{\bigcup_{i=1}^{r} \sigma_{u_i}[\mathbf{\Delta}] \ge \sigma_{0u_i}\}, \tag{12}$$

where $\sigma_{u_i}[\mathbf{\Delta}]$ represents the ith stationary RMS control action, and σ_{0u_i} is the target control action level. Equation (12) can be written in terms of the probability density function $f_{\mathbf{\Delta}}(\boldsymbol{\delta})$ as

$$p_{\mathrm{c}} = \int_{\bigcup_{j=1}^{r} \sigma_{u_i}[\boldsymbol{\delta}] \ge \sigma_{0u_i}} \!\!\!\cdots\!\int f_{\mathbf{\Delta}}(\boldsymbol{\delta})\, d\boldsymbol{\delta} \,. \tag{13}$$

FORM/SORM can be employed to evaluate this expression. The probability integral in Eq. (8) is rewritten as

$$p_{\mathrm{c}} = P\{\bigcup_{i=1}^{r} g_i^{\mathrm{c}}(\mathbf{\Delta}) \le 0\}, \tag{14}$$

where the limit-state function is given by

$$g_i^{\mathrm{c}}(\mathbf{\Delta}) = \sigma_{0u_i} - \sigma_{u_i}[\mathbf{\Delta}], \tag{15}$$

and can be viewed as a reliability problem for a series of components $g_i^{\mathrm{c}}(\mathbf{\Delta})$. The stationary RMS control action required for evaluation of $g_i^{\mathrm{c}}(\mathbf{\Delta})$ is a linear function of the response covariance matrix $\mathbf{\Sigma}_{\tilde{z}}$, which is obtained from the solution of the Lyapunov equation.

3 Numerical Examples

Consider the three-story, single-bay building subjected to a one-dimensional earthquake excitation $\ddot{x}_g$ with tendon controller and depicted in Fig. 6. Assuming a simple shear frame model for the building, the governing equations of motion are

$$\mathbf{M}_s\ddot{\mathbf{x}} + \mathbf{C}_s\dot{\mathbf{x}} + \mathbf{K}_s\mathbf{x} = \mathbf{B}_s u - \mathbf{M}_s\mathbf{\Gamma}_s\ddot{x}_g \tag{16}$$

$$\begin{bmatrix} m_1 & 0 & 0 \\ 0 & m_2 & 0 \\ 0 & 0 & m_3 \end{bmatrix}\begin{Bmatrix} \ddot{x}_1 \\ \ddot{x}_2 \\ \ddot{x}_3 \end{Bmatrix} + \begin{bmatrix} c_1+c_2 & -c_2 & 0 \\ -c_2 & c_2+c_3 & -c_3 \\ 0 & -c_3 & c_3 \end{bmatrix}\begin{Bmatrix} \dot{x}_1 \\ \dot{x}_2 \\ \dot{x}_3 \end{Bmatrix} + \begin{bmatrix} k_1+k_2 & -k_2 & 0 \\ -k_2 & k_2+k_3 & -k_3 \\ 0 & -k_3 & k_3 \end{bmatrix}\begin{Bmatrix} x_1 \\ x_2 \\ x_3 \end{Bmatrix}$$

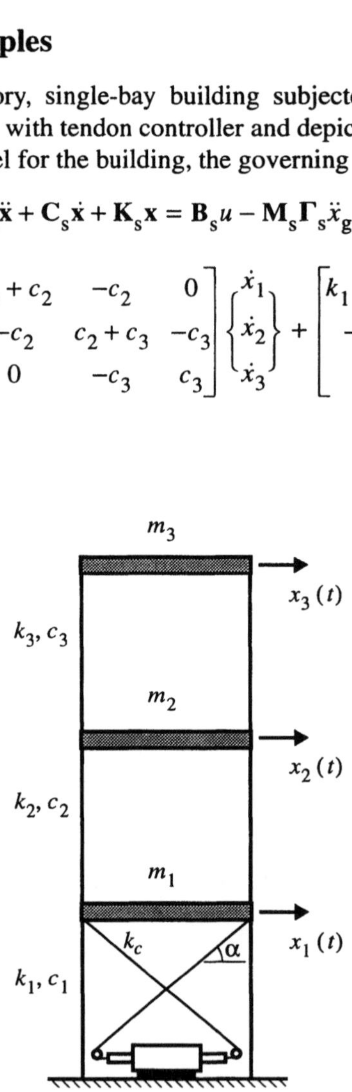

Fig. 1. Three degree-of-freedom structure with active tendon control.

$$= \begin{Bmatrix} 0 \\ 0 \\ -4k_c \cos\alpha \end{Bmatrix} u - \begin{bmatrix} m_1 & 0 & 0 \\ 0 & m_2 & 0 \\ 0 & 0 & m_3 \end{bmatrix} \begin{Bmatrix} 1 \\ 1 \\ 1 \end{Bmatrix} \ddot{x}_g, \qquad (17)$$

where m_i, c_i, k_i are the mass, damping and stiffness, respectively, associated with the ith floor of the building, and u is the position of the hydraulic actuator. Defining the state vector of the system as $\mathbf{z} = [\mathbf{x}' \ \ \dot{\mathbf{x}}']'$, the state space matrices in Eq. (1) are then

$$\mathbf{A} = \begin{bmatrix} \mathbf{0} & \mathbf{I} \\ -\mathbf{M}_s^{-1}\mathbf{K}_s & -\mathbf{M}_s^{-1}\mathbf{C}_s \end{bmatrix}, \quad \mathbf{B} = \begin{bmatrix} \mathbf{0} \\ \mathbf{M}_s^{-1}\mathbf{B}_s \end{bmatrix}, \text{ and } \mathbf{E} = \begin{bmatrix} \mathbf{0} \\ -\mathbf{\Gamma}_s \end{bmatrix}. \qquad (18)$$

For this example, a time delay with a magnitude $\tau = 36$ msec is assumed to be present in the control loop which can be modeled as shown in Fig. 2. Herein we employ a third-order Padé approximation which was shown by Sain, *et al.* [17] to model adequately the time delay in the neighborhood of the crossover frequency for this example structure.

The controller is assumed to be known and deterministic, and the variables modeling the mass, damping, stiffness and time delay are assumed to be Gaussian random variates with a 10% coefficient of variation. Table 1 provides a summary of the model parameters. Note that the nominal parameters were chosen to match closely the characteristics of the experimental structure reported by Chung, *et al.*,[18] in that it has similar modal frequencies and dampings.

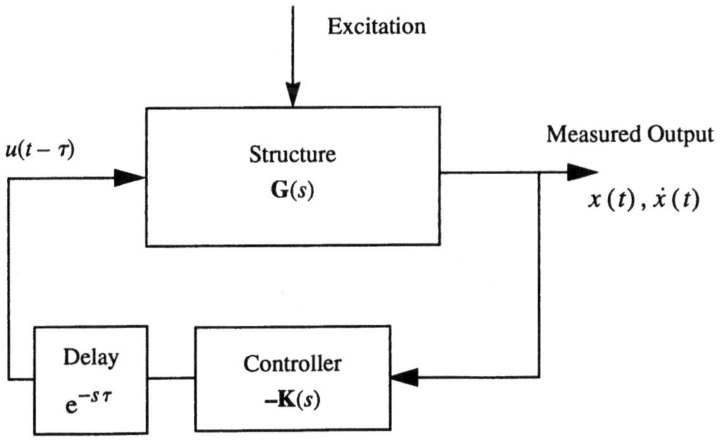

Fig. 2. Block diagram representation for the single degree-of-freedom structure with a delayed controller.

Table 1: Model parameters for the three-story structure.

	Mean, μ	*Std. Dev.* σ		*Mean,* μ	*Std. Dev.* σ
m_1 (lb-sec^2/in)	5.6	0.56	k_1 (lb/in)	5034	503.4
m_2 (lb-sec^2/in)	5.6	0.56	k_2 (lb/in)	10965	1096.5
m_3 (lb-sec^2/in)	5.6	0.56	k_3 (lb/in)	6135	613.5
c_1 (lb-sec/in)	2.6	0.26	τ (msec)	36	3.6
c_2 (lb-sec/in)	6.3	0.63	k_c (lb/in)	2124	0.0
c_3 (lb-sec/in)	0.35	0.035	α (degrees)	36	0.0

Attention will be restricted to the Kalman filter control strategies because of their superior robustness properties.[11,12,17] As in Chung, *et al.*,[18] the controller is obtained by assuming that the ground acceleration can be modeled as a Gaussian white noise and minimizing the quadratic performance index given by

$$J = \lim_{T \to \infty} \frac{1}{T} E\left[\int_0^T (\mathbf{x}'\mathbf{K}_s\mathbf{x} + \gamma k_c u^2)\, dt\right] \quad . \tag{19}$$

Three Kalman filter control strategies will be examined for the three-story building. The first, termed the "structural state feedback control," assumes that the displacement and velocity of the three building floors (*i.e.*, all structural states) are available for control determination. Here, we have chosen $\gamma = 200$. The other two control strategies are based on the measured response being the acceleration of the building floors, *i.e.*, acceleration feedback control strategies.[4,6] Here, the associated case in which $\gamma = 200$ is considered, as well as a lower authority controller in which $\gamma = 5000$. In all cases, the ratio of the intensity of the disturbance to that of the measurement noise was chosen to be 25, and the disturbance and the noise are assumed to be statistically independent.

Figure 3 shows the cumulative probability distribution for the maximum of the real parts of the eigenvalues of the controlled structure. These results were obtained via Monte Carlo simulation (MCS), which is well suited for determining the central tendencies of a random event. Note that the median value for the maximum of the real parts of the eigenvalues is the most negative for the acceleration feedback case in which $\gamma = 200$. If one looks only at the nominal behavior of the system, this result may have led one to believe that the stability robustness of this acceleration feedback control strategy is better than for the other two controllers. When we examine the upper tails of the distribution, however, this conclusion is not so tenable. MCS is generally not a tractable approach for determining the probability of the occurrence of rare events such as the instability of this system, and FORM/

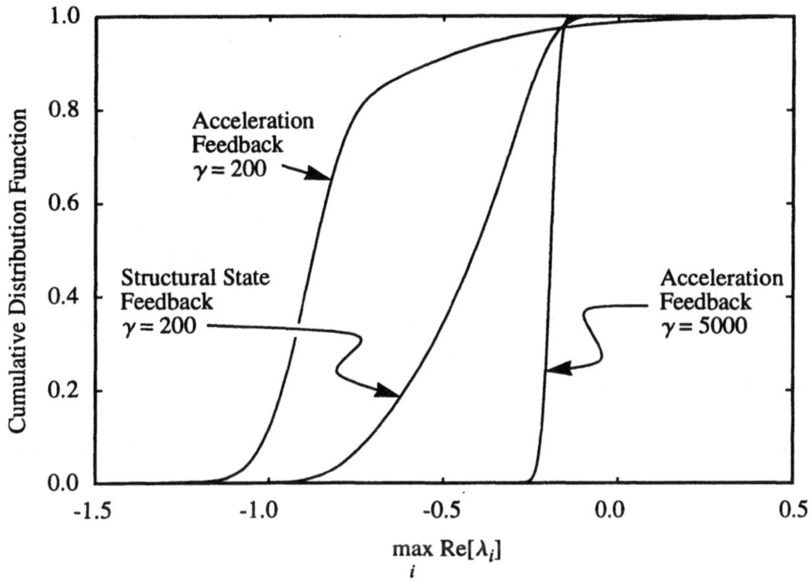

Fig. 3. Cumulative probability distribution of the maximum real part of the eigenvalues of the closed loop dynamical matrix.

SORM[19] is employed to more closely examine the stability limit-state probabilities, *i.e.*, the probability that the system is unstable. Table 2 provides a summary of the SORM results for the stability limit-state. Because multiple design points exist which contribute significantly to the limit-state probability, system reliability methods are employed.[11] For the structural state feedback controller, we see that the stability limit-state probability is only 1.635e–11, indicating that the system possesses great stability robustness. (Note that estimating this probability would be nearly impossible using simple MCS.) The limit-state probability for the corresponding acceleration feedback control strategy is 1.074e–02, which indicates a significant reduction in the stability robustness of the system using this output feedback approach.

There are two natural approaches to increasing the stability robustness of the acceleration feedback strategy to an acceptable level. Because this control strategy is quite authoritative, the first approach to increasing the stability robustness of the acceleration feedback control strategy is to design the controller more conservatively. Table 2 presents results for such a controller with $\gamma = 5000$, where we see that the stability limit-state probability is reduced to 3.080e–05, which is a more acceptable result. The second approach is to decrease the time delay in the system. Spencer, *et al.*[5] have presented experimental results which indicate that time delays of less than 1 ms can be readily obtained. Thus, if the random time delay were to have a mean of 1 ms and a coefficient of variation of 10%, a dramatic increase in the stability robustness can be achieved. For this case, the stability limit state probabilities are 8.599e−34 and 7.769e−24 for $\gamma = 200$ and $\gamma = 5000$, respectively. To understand fully the implications of these results, one must also examine the performance robustness of the controllers.

Figures 4 and 5 present the cumulative probability distributions for the top floor RMS displacement response and the RMS control action, respectively, obtained via MCS. As seen in Fig. 4, all three controllers significantly reduce the uncontrolled response. The controller in which $\gamma = 200$ produces the greatest reduction in the top floor RMS displacement response. However, the more conservative acceleration feedback control strategy still produces a significant reduction in the RMS response (*i.e.,* nominally, a 47% reduction in the RMS response is achieved over the uncontrolled structure). Examining Fig. 5, we see that the required RMS control action is very similar for the two more authoritative controllers, whereas the control action required by the second acceleration feedback controller is nominally 65% smaller.

To look at the extreme responses, FORM/SORM is again employed. We have chosen the critical response to be the top floor displacement. Thus, the performance and the control limit-state probabilities are respectively given by

$$p_{\mathrm{p}} = P\{\sigma_{x_3} \geq \sigma_0\}\text{ , and } p_{\mathrm{c}} = P\{\sigma_u \geq \sigma_{0u}\}\ . \qquad (20)$$

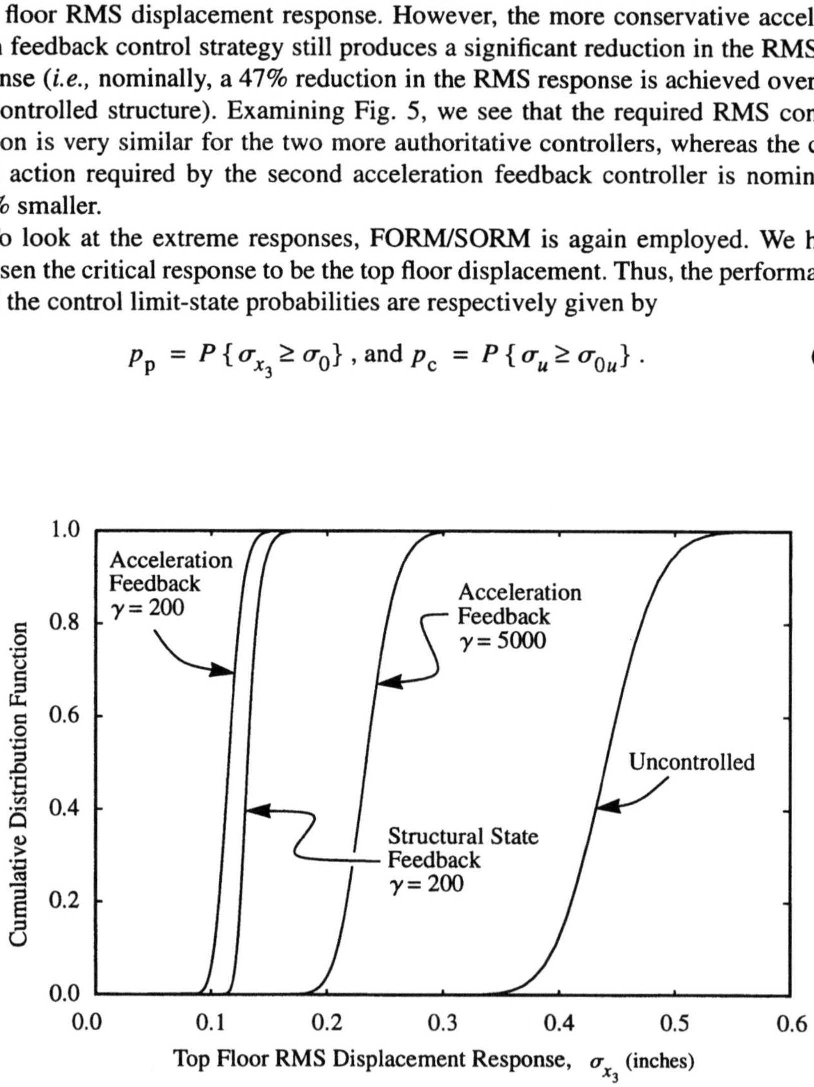

Fig. 4. Cumulative probability distribution function of the top floor RMS displacement response.

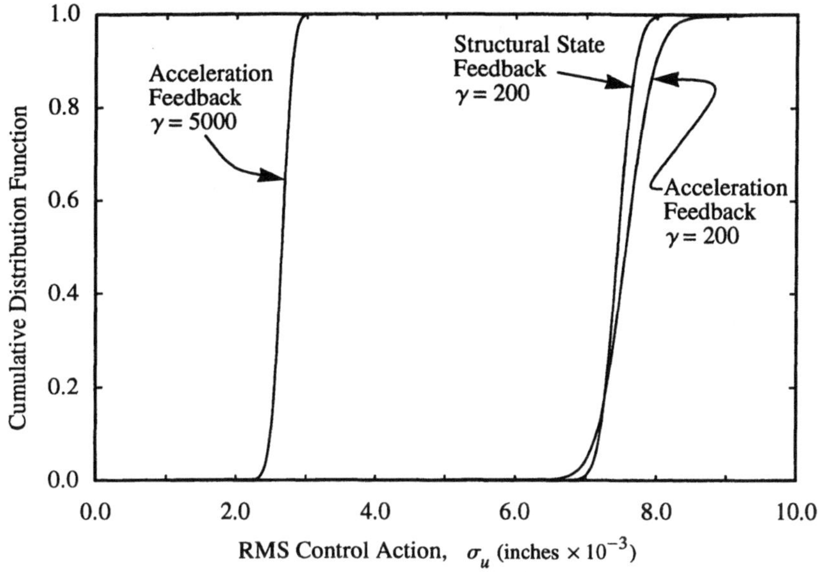

Fig. 5. Cumulative probability distribution function of RMS control action.

Table 3 shows the SORM results for the response performance robustness. For structural state feedback control the limit-state probability of SORM displacement response being greater than 0.16 in. is 1.384e–03, while it is 2.213e–04 for the acceleration feedback control ($\gamma = 200$). For the acceleration feedback case with $\gamma = 5000$ we find $P(\sigma_{x_3} \geq 0.30) = 9.631\text{e}{-02}$. Note that the squared importance factors are similar in all three cases; the mass of the third floor being most critical and k_1 or m_2 being second or third in importance.

To complete the picture of the robust performance, we need to consider control performance as well. Table 4 shows that larger extremes occur in the RMS control action of acceleration feedback control ($\gamma = 200$) than structural state feedback control or the acceleration feedback ($\gamma = 5000$) control. It is interesting to note that the most critical squared importance factors of the structural state feedback case are masses of each floor, while for the more authoritative acceleration feedback case ($\gamma = 200$) the most critical factor is the stiffness of the third floor, k_3, whereas for the more conservative controller ($\gamma = 5000$), the most critical importance factor is associated with the stiffness of the first floor k_1.

4 Conclusions

Analysis methods have been presented which provide objective robustness measures for comparing the capabilities of competing structural control strategies in an inherently uncertain environment. By example, we have shown that certain control strategies are significantly more robust to parametric uncertainties. For the single-

Table 2: Summary of SORM stability robustness results for three-story building.

	Reliability Index, β	p_s	Squared Importance Factors (%)									
			τ	m_1	m_2	m_3	c_1	c_2	c_3	k_1	k_2	k_3
Structural State Feedback Control, $\gamma = 200$	6.634	1.635e–11	23	24	8	11	21	0	3	5	0	5
Acceleration Feedback Control, $\gamma = 200$	2.299	1.074e–02	11	43	6	0	0	2	0	1	36	0
Acceleration Feedback Control, $\gamma = 5000$	4.007	3.080e–05	7	41	5	16	1	0	0	5	1	24

Table 3: Summary of SORM response performance robustness results for three-story building.

	Reliability Index, β	p_p	Squared Importance Factors (%)									
			τ	m_1	m_2	m_3	c_1	c_2	c_3	k_1	k_2	k_3
Structural State Feedback Control, $\gamma = 200$, $\sigma_0 = 0.16$ in.	2.432	1.384e–03	1	6	13	37	0	0	0	29	7	8
Acceleration Feedback Control, $\gamma = 200$, $\sigma_0 = 0.16$ in.	3.513	2.213e–04	1	7	12	45	0	0	0	20	6	8
Acceleration Feedback Control, $\gamma = 5000$, $\sigma_0 = 0.30$ in.	2.980	1.440e–03	0	6	14	44	2	0	0	17	6	10

Table 4: Summary of SORM control performance robustness results for three-story building.

	Reliability Index, β	p_c	*Squared Importance Factors (%)*									
			τ	m_1	m_2	m_3	c_1	c_2	c_3	k_1	k_2	k_3
Structural State Feedback Control, $\gamma = 200$, $\sigma_{0u} = 8.5\text{e–}03$ in.	4.446	4.369e–06	11	10	21	37	1	0	0	7	11	3
Acceleration Feedback Control, $\gamma = 200$, $\sigma_{0u} = 8.5\text{e–}03$ in.	2.530	5.701e–03	11	20	5	16	0	0	0	8	0	40
Acceleration Feedback Control, $\gamma = 5000$, $\sigma_{0u} = 3.0\text{e–}03$ in.	3.118	9.120e–04	2	2	0	1	7	1	0	85	1	0

degree-of-freedom example, the Kalman filter controller has been shown to achieve a greater reliability, in terms of smaller likelihood of extreme RMS displacement response, while maintaining a probability of instability two orders of magnitude lower than the phase-corrected control. Moreover, the probability of the occurrence of extreme control action is the smallest for the Kalman filter controller. For the example employing a three-story building, the Kalman filter controller employing all of the structural states proved to have excellent robustness characteristics. The acceleration feedback control strategies also had good robustness properties. Reduction of the system time delay to levels that are readily achievable via current hardware technology allowed the acceleration feedback control strategies to yield very reliable and authoritative controllers, thus making practical implementation of these controllers very promising.

With the probabilistic robustness measures developed herein, a control designer has the ability to make more realistic trade-off choices among reliability, stability, and uncertainty, and as a result, to use more effectively the available resources in a given structural control situation.

5 Acknowledgment

This research is partially supported by National Science Foundation Grants No. BCS 90-06781 and No. BCS 93-01584.

6 References

1. Soong, T.T., *Active Structural Control: Theory and Practice*, Longman Scientific and Technical, Essex, England, 1990.
2. Housner, G.W. and Masri, S.F., Eds., *Proc. of the U.S. National Workshop on Structural Control Research*, USC Publications No. M9013, University of Southern California, October 25–26, 1990.
3. ATC 17-1, *Proc. on Seismic Isolation, Passive Energy Dissipation, and Active Control*, Applied Technology Council, Redwood City, California, 1993.
4. Spencer Jr., B.F., Suhardjo, J. and Sain, M.K., Frequency Domain Control Algorithms for Civil Engrg. Applications. *Proc. of the Int. Workshop on Tech. for Hong Kong's Infrastructure Development*, Hong Kong, December 19-20, 1991, pp. 169-178.
5. B.F. Spencer Jr., B.F., Dyke, S., Sain, M.K. and Quast, P., Acceleration Feedback Control Strategies for Aseismic Protection. *Proc. of the American Control Conf.*, San Francisco, California, 1993, pp. 1317–1321.
6. Spencer Jr., B.F., Suhardjo, J. and Sain, M.K., Frequency Domain Optimal Control Strategies for Aseismic Protection. *J. of Engrg. Mech., ASCE,* to appear.
7. Suhardjo, J., *Frequency Domain Techniques for Control of Civil Engineering Structures with Some Robustness Considerations*. Ph.D. Dissertation, Univ. of Notre Dame, Dept. of Civil Engrg. 1990.

8. Suhardjo, J., Spencer Jr., B.F. and Kareem, A., Active Control of Wind Excited Buildings: A Frequency Domain Based Design Approach. *J. of Wind Engrg. and Ind. Aerodynamics*, **41–44** (1992) 1985–1996.
9. Suhardjo, J., Spencer Jr., B.F. and Kareem, A., Frequency Domain Optimal Control of Wind Excited Buildings. *J. of Engrg. Mech., ASCE*, **118(12)** (1992) 2463–2481.
10. Ray, L.R. and Stengel, R.F., A Monte Carlo Approach to the Analysis of Control System Robustness. *Automatica*, **29(1)** (1993) 229–236.
11. Spencer Jr., B.F., Sain, M.K., Kantor, J.C. and Montemagno, C., Probabilistic Stability Measures for Controlled Structures Subject to Real Parameter Uncertainties. *Smart Materials and Structures*, **1** (1992) 294–305.
12. Spencer Jr., B.F., Sain, M.K., Kaspari, D.C. and Kantor, J.C., Reliability-Based Design of Active Control Strategies. *Proc., ATC-17-1 Seminar on Isolation, Passive Energy Dissipation and Active Control*, San Francisco, California, March 11–12, 1993, pp. 761–772.
13. Suhardjo, J., Spencer Jr., B.F. and Sain, M.K., Feedback–Feedforward Control of Structures Under Seismic Excitation. *Structural Safety*, **8** (1990) 69–89.
14. Madsen, H.O., Krenk, S. and Lind, N.C., *Methods of Structural Safety*, Prentice-Hall, Inc., Englewood Cliffs, New Jersey, 1986.
15. Melchers, R.E., *Structural Reliability, Analysis and Prediction*, Ellis Horwood Series in Civil Engrg., Halsted Press, England, 1987.
16. Soong, T.T. and M. Grigoriu, *Random Vibration of Mechanical and Structural Systems*, Prentice Hall, Englewood Cliffs, New Jersey, 1992.
17. Sain, P.M., Spencer Jr.,B.F., Sain, M.K. and Suhardjo, J., Structural Control Design in the Presence of Time Delay. *Proc., 9th Engrg. Mech. Conf.*, College Station, Texas, May 25–27 , 1992, pp. 812–815.
18. Chung, L.L., Lin, R.C., Soong, T.T., and Reinhorn, A.M., 1989, Experiments on Active Control for MDOF Seismic Structures. *J. of Engrg. Mech., ASCE*, **115(8)** (1989) 1609–1627.
19. PROBAN, Version 3, Veritas Sesam Systems, Det norske Veritas, Norway, 1993.

Probalistic Structural Analyses of Ceramic Gas Turbine Components

Tetsuo Teramae[1], Seiichi Hamada[1] and Jynichi Hamanaka[2]

[1] Engineering Research Center, Tokyo Electric Power Co., Inc., 2-4-1 Nishi-Tsutsujigaoka, Chofu-Shi, Tokyo 182, Japan
[2] Research Institute, Ishikawajima-Harima Heavy Industries Co., Ltd., 3-1-15 Toyosu, Koto-Ku, Tokyo 135, Japan

Abstract. Tokyo Electric Power Company (TEPCO), with the cooperation of gas turbine manufacturers and ceramic component manufacturers, has conducted research to apply ceramic components to the hot parts of gas turbines, such as the combustors, the stator vanes and the rotor blades.

Although ceramics have a very attractive combination of physical and mechanical properties at high temperatures well over 1300°C, their toughness is very low, and their strength shows wider scatter compared to metal materials. These undesirable properties cause the necessity of "design by analysis" of ceramic components.

As a part of TEPCO's project, therefore, the authors have conducted research to develop a strength evaluation system of structural ceramic components using a probabilistic approach. A computer program, GFICES (Gas turbine - FIne Ceramics Evaluation System), was developed on the basis of statistical strength theory using the two parameter Weibull probability distribution. This program is coupled as a postprocessor to the general purpose structural analysis program MSC/NASTRAN and the dynamic stress analysis program DYNA3D. Main analytical functions of GFICES are the reliability analyses corresponding to fast fracture, static fatigue, dynamic fatigue and the effect of proof testing. The time dependent fracture analyses are based on the assumption of subcritical crack growth (SCG). In addition, several supplemental functions are added in GFICES such as the application of bi-modal Weibull probabilistic theory, the evaluation of aging degradation effects of ceramics on fast fracture, and the reliability evaluation of components subject to non-proportional transient stress history. Bi-modal Weibull probabilistic theory is incorporated to take account of both volume distributed flaws and surface distributed flaws of ceramic components.

As an application of GFICES, the reliability of a ceramic rotor blade subject to thermal stress and centrifugal stress is analyzed. It is shown that GFICES can very effectively evaluate the strength reliability of ceramic components.

1. Introduction

To cope with the increasing demand of electric power, much research and development has been performed in the electric power industry. Specifically, work on applying advanced cooling technologies and advanced heat resistant materials has been extensively conducted in order to raise the thermal efficiency of power generating gas turbines. The application of ceramics is one of the most promising ways to raise the thermal efficiency of gas turbines, and several projects have been undertaken currently in the U.S.A., Europe and Japan.

Since 1984, Tokyo Electric Power Company (TEPCO), with the cooperation of gas turbine manufacturers and ceramic component manufacturers, has conducted research to apply ceramic components to the hot parts of gas turbines, such as the combustors, the stator vanes and the rotor blades [1].

Ceramics, especially the silicon based non-oxide ceramics such as silicon carbide (SiC) and silicon nitride (Si_3N_4), have a very attractive combination of physical and mechanical properties at high temperatures well over 1300°C. The application of these ceramics enables an increase of the turbine inlet temperature and thereby raises the thermal efficiency of gas turbines.

Although ceramics have attractive properties at high temperatures, their toughness is very low, and their strength shows wider scatter compared to metal materials. These undesirable properties cause the necessity of "design by analysis" for the application of ceramic components. For this reason, several programs have been developed to evaluate the structural reliability of ceramic components.

Gyekenyesi et al. have developed the program SCARE (Structural Ceramics Analysis and Reliability Evaluation) [2-3]. SCARE includes the Weibull and Batdorf statistical failure theories for both volume flaws and surface flaws. They have applied the two-parameter Weibull fracture strength distribution model to calculate the probability of fast fracture of ceramic components and have used the principle of independent action and Batdorf's shear sensitive as well as shear-insensitive crack theories for multiaxial stress states. Lamon et al. have developed a reliability analysis program CERAM which can take account of both fast fracture and time dependent fracture [4]. To predict mixed mode fracture strength, it adopts the Weibull approach, the Batdorf approach and the multiaxial elemental strength model, which is based on the criterion of maximum strain energy release rate. Stürmer et al. also have developed a failure probability analysis program CERITS for fast fracture, and CERITS-L for time dependent fracture [5].

Since 1986, as a part of TEPCO's project, the authors also have conducted research to develop a strength evaluation system of structural ceramic components using a probabilistic approach. The computer program GFICES (Gas turbine - FIne Ceramics Evaluation System) has been developed on the basis of statistical strength theory using the two parameter Weibull probabilistic distribution [6-8].

In this paper, the authors would like to report on the theoretical background and the system of GFICES and present numerical results on a ceramic rotor blade to show the usefulness of GFICES.

2. Theoretical background of GFICES

2.1 Probabilistic strength distribution

Experimental fracture strength of ceramic materials is usually measured by 3-point or 4-point bending tests,and the failure probability P_f is approximated by the Weibull probabilistic distribution for volume flaws as

$$P_f = 1 - \exp\left[-\frac{1}{V_o} \int_v \frac{(\sigma - \sigma_u)^m}{\sigma_o} dV \right] , \tag{1}$$

where σ is the applied tensile stress, σ_0 is the scale parameter, m is the Weibull modulus which measures the degree of strength variability, σ_u is the threshold stress, V is the stressed volume and V_0 is the reference volume.

Although this three parameter Weibull distribution has a rigorous basis, the determination of these parameters experimentally is ambiguous. In this study, therefore, the two parameter Weibull probabilistic approach is adopted. Here the threshold stress σ_u is taken as zero.

For surface flaws, a similar equation can be obtained by executing a surface integral instead of a volume integral.

2.2 Multiaxial fracture criterion

In the failure analysis of brittle materials, it is required to establish the fracture criterion for multiaxial stress states, and several models have been proposed. Among them are the principle of independent action (PIA), the criterion of critical coplanar strain energy release rate Gc, and the equivalent stress based on the maximum in the strain energy release rate Gmax.

GFICES fundamentally adopts the Gc criterion for penny shaped cracks,and the equivalent normal stress Z is defined by normal stress σ_r and shear stress τ in the crack plane [9] as

$$Z = \left[\sigma_R{}^2 + \frac{4}{(2 - \nu)^2} \tau^2 \right]^{1/2} , \tag{2}$$

where

$$\sigma_r = \sigma_1 \ell_1^2 + \sigma_2 \ell_2^2 + \sigma_3 \ell_3^2 \quad , \tag{3}$$

$$\tau^2 = \sigma_1^2 \ell_1^2 + \sigma_2^2 \ell_2^2 + \sigma_3^2 \ell_3^2 - (\sigma_1 \ell_1^2 + \sigma_2 \ell_2^2 + \sigma_3 \ell_3^2)^2 \quad , \tag{4}$$

$$\begin{aligned} \ell_1 &= \sin\phi\cos\theta \quad , \\ \ell_2 &= \sin\phi\sin\theta \quad , \\ \ell_3 &= \cos\phi \quad . \end{aligned} \tag{5}$$

Here ν is Poisson's ratio, and σ_1, σ_2 and σ_3 denote principal stresses as shown in Fig. 1. Although this equivalent stress is defined theoretically, each ceramic is considered to have its own fracture characteristics. Oda et al. [10] have proposed the modified Gc criterion in which the modified equivalent stress Z′ is expressed by introducing the parameter β as

$$Z' = [\sigma_R^2 + (\beta\tau)^2]^{1/2} \quad . \tag{6}$$

Mixed mode fracture toughness supported the value of β to be 0.5 [7]. GFICES includes this modified Gc criterion by regarding β as an experimental constant.

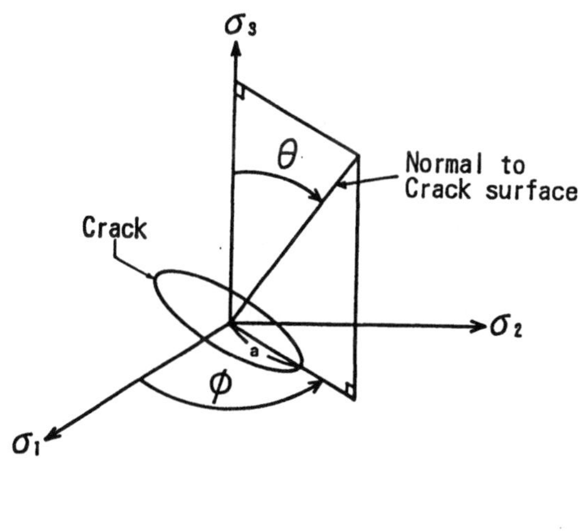

Fig. 1 Penny shaped crack in principal stress space

2.3 Failure Probability of Fast Fracture

According to the Gc criterion, the failure probability of fast fracture for volume flaws is given as

$$P_f\,(\sigma_N) = 1 - \exp\left\{ -\left(\frac{1}{m}\,!\right)^m \left(\frac{\sigma_N}{\sigma_{ref}}\right)^m \frac{1}{V_{ref}} \int_V f\,(\sigma)\,dV \right\} , \tag{7}$$

where σ_N denotes nominal stress, $f(\sigma)$ accounts for the effect of a random orientation of the penny shaped crack and is given as

$$f(\sigma) = \frac{2}{\pi I} \int_0^{\frac{\pi}{2}} \int_0^{\frac{\pi}{2}} \left(\frac{Z}{\sigma_N}\right)^m H\,(\sigma_R)\,\sin\phi\,d\phi\,d\theta \quad , \tag{8}$$

$$I = \int_0^{\frac{\pi}{2}} \left[\cos^4\phi + \frac{4}{(2-\nu)^2} \sin^2\phi\cos^2\phi \right]^{\frac{m}{2}} \sin\phi\,d\phi \quad . \tag{9}$$

Here, H (σ_R) is the step function, which is defined as

$$H(\sigma_R) = \begin{cases} 1 & (\sigma_R > 0) \\ 0 & (\sigma_R \le 0) \end{cases} , \tag{10}$$

and σ_{ref} means the average tensile strength of uniformly tensiled specimen with the volume of V_{ref} and is related to σ_0 and V_0 by

$$\sigma_{ref} = \left(\frac{V_o}{IV_{ref}}\right)^{\frac{1}{m}} \sigma_o \left(\frac{1}{m}\,!\right) . \tag{11}$$

Average strength σ_N, standard deviation of strength S_σ and the effective volume V_{eff} are respectively given as

$$\overline{\sigma}_N = \left(\frac{1}{F}\right)^{\frac{1}{m}} \left(\frac{1}{m}\,!\right) , \tag{12}$$

$$S_\sigma = \left[\left(\frac{1}{F}\right)^{\frac{2}{m}} \left(\frac{2}{m}\,!\right) - \overline{\sigma}_N^{\,2} \right]^{\frac{1}{2}} , \tag{13}$$

$$V_{eff} = \int_V f\,(\sigma)\,dV \quad , \tag{14}$$

where

$$F = \left(\frac{1}{m}\,!\right)^m \left(\frac{1}{\sigma_{ref}}\right)^m \frac{1}{V_{ref}} \int_v f(\sigma)\, dV \quad . \tag{15}$$

Similarly, the following equations can be obtained for surface flaws by replacing volume integrals with surface integrals:

$$P_f(\sigma_N) = 1 - \exp\left[-\left(\frac{1}{m}\,!\right)^m \left(\frac{\sigma_N}{\sigma_{ref}}\right)^m \frac{1}{A_{ref}} \int_A f(\sigma)\, dA \right] , \tag{16}$$

$$A_{eff} = \int_A f(\sigma)\, dA \quad . \tag{17}$$

2.4 Failure Probability of Time Dependent Fracture

Due to the subcritical crack growth (SCG) of ceramics, the failure probability increases gradually when a ceramic component is subjected to long-term loading. Linear fracture mechanics yields the following relation between fracture stress and crack length for the penny shaped crack:

$$K_{IC} = C\sigma_o\sqrt{a_o} = C\sigma\sqrt{a} \quad , \tag{18}$$

where K_{IC} denotes the fracture toughness, and a_0 is the crack length corresponding to fracture stress σ_0. Using this relation, Eq. (1) can be rewritten as the following probabilistic distribution of crack length:

$$P_f = 1 - \exp\left[-\frac{1}{V_o} \int_v \left(\frac{a_o}{a}\right)^{\frac{m}{2}} dV \right] \quad . \tag{19}$$

Subcritical crack growth rate of ceramics can be expressed in terms of the stress intensity factor K as

$$\frac{da}{dt} = BK^n \quad . \tag{20}$$

The stress intensity factor depends on applied stress and crack length as

$$K = CZ\sqrt{a} \quad . \tag{21}$$

From Eqs. (20) and (21), the relation between time t and crack length a is obtained as

$$t = \frac{1}{B\,(CZ)^n}\left(\frac{2}{n-2}\right)\left[a_i^{\frac{2-n}{2}} - a^{\frac{2-n}{2}}\right], \tag{22}$$

where a_i denotes initial crack length. The index n is generally much greater than 1 for structural ceramics, so the fracture time t_f is given approximately by the equation

$$t_f = \left(\frac{2}{n-2}\right)\left[BC^n\,Z^n\,a_i^{\frac{n-2}{2}}\right]^{-1}. \tag{23}$$

Now, from Eqs. (7) and (19), the probabilistic distribution of initial crack length a_i is postulated as

$$P\,(a_r < a_i) = 1 - \exp\left\{-\frac{1}{V_o}\frac{2}{\pi}\int_V\int_0^{\frac{\pi}{2}}\int_0^{\frac{\pi}{2}}\left(\frac{a_o}{a_r}\right)^{\frac{m}{2}}\sin\phi\,d\phi\,d\theta\,dV\right\}. \tag{24}$$

From Eqs. (23) and (24), the probabilistic distribution of failure time t_f is given by the equation

$$\begin{aligned} P_f\,(\sigma_N : t_f) = 1 - \exp\Bigg\{ & -\left(\frac{1}{m}\,!\right)^m\left(\frac{\sigma_N}{\sigma_{ref}}\right)^m \\ & \times\left(\frac{(n-2)}{2}\,BC^2\,\sigma_N{}^2K_{IC}{}^{n-2}\,t_f\right)^{\frac{m}{n-2}}\frac{1}{V_{ref}}\int_V g(\sigma)dV\Bigg\}, \end{aligned} \tag{25}$$

where

$$g\,(\sigma) = \frac{2}{\pi I}\int_0^{\frac{\pi}{2}}\int_0^{\frac{\pi}{2}}\left(\frac{Z}{\sigma_N}\right)^{\frac{nm}{n-2}} H\,(\sigma_R)\sin\phi\,d\phi\,d\theta\;. \tag{26}$$

Average fracture time t_f and standard deviation of fracture time S_{tf} are given as

$$\bar{t}_f = \left(\frac{1}{L}\right)^{\frac{n-2}{m}}\left(\frac{n-2}{m}\,!\right), \tag{27}$$

$$S_{tf} = \left[\left(\frac{1}{L}\right)^{\frac{2(n-2)}{m}}\left(\frac{2\,(n-2)}{m}\,!\right) - \bar{t}_f^{\,2}\right]^{\frac{1}{2}}, \tag{28}$$

where

$$L = \left(\frac{1}{m}\,!\right)^m \left(\frac{\sigma_N}{\sigma_{ref}}\right)^m \left(\frac{(n-2)}{2}\,BC^2\,\sigma_N{}^2 K_{IC}{}^{n-2}\right)^{\frac{m}{n-2}} \frac{1}{V_{ref}} \int_v g(\sigma)dV \quad . \qquad (29)$$

It is easy to obtain similar equations for surface flaws.

2.5 Failure Probability of Dynamic Fatigue

If a ceramic is subjected to repeated loading and is damaged by SCG, the aforementioned time dependent approach can be used to estimate the failure probability of dynamic fatigue. For simplicity, uniaxial stress $\sigma(t)$ is considered to be of the form

$$\sigma(t) = \sigma_m \cdot h(t) \quad . \qquad (30)$$

Then, by introducing the equivalent time $\bar{t}$ defined by the equation

$$\bar{t} = \int_o^t (h(t))^n\, dt \quad , \qquad (31)$$

the failure probability can be easily calculated by considering the case of stress σ_m acting for the equivalent time $\bar{t}$ instead of actual time t. The number of failure cycles is obtained by dividing the real failure time by the period.

2.6 Proof testing

Proof testing is one of the direct methods of removing low strength ceramic components. The proof load is greater than the normal operating load, and the components which survive are judged to be more reliable.

If the modes of the operating load and the proof load are the same, the failure probability can be simply expressed. When proof testing is done at the nominal stress σ_p, the failure probability of fast fracture under the nominal stress σ_N is given as

$$P_f\,(\sigma_N : \sigma_p) = \frac{P_f\,(\sigma_N) - P_f\,(\sigma_p)}{1 - P_f\,(\sigma_p)}\, H\,(\sigma_N - \sigma_p) \quad . \qquad (32)$$

Average strength $\bar{\sigma}_N$ is given as

$$\bar{\sigma}_N\,(\sigma_p) = \frac{1}{1 - P_f\,(\sigma_p)} \{\bar{\sigma}_N\,(O) - \sigma_p{}^*\} \quad , \qquad (33)$$

where

$$\sigma_p^* = Fm\int_0^{\sigma_p} \sigma_N^m \exp\{1 - F\sigma_N^m\}\, d\sigma_N \quad . \tag{34}$$

Hamanaka et al. also have formulated the failure probability of fast fracture when the proof testing mode is different from the operating loading mode [11].

2.7 Bi-modal Weibull probabilistic theory

Ceramic components generally contain both volume flaws caused by manufacturing processes and surface flaws caused by machining to make the required shape. Both types of flaws are theoretically treated at the same time by bi-modal Weibull probabilistic theory which is based on the weakest link theory.

For fast fracture, the failure probability under the nominal stress σ_N is given as

$$P_f = 1 - (1-P_{fV})\,(1-P_{fA}) \quad , \tag{35}$$

where P_{fV} and P_{fA} denote the failure probability of fast fracture corresponding to volume flaws and surface flaws respectively.

For time dependent fracture, the failure probability at time t_f is given by

$$P_f\,(\sigma_N : t_f) = 1 - (\,1-P_{fV}(\sigma_N : t_f)\,)\,(\,1-P_{fA}(\sigma_N : t_f)\,) \quad , \tag{36}$$

where $P_{fV}(\sigma_N : t_f)$ and $P_{fA}(\sigma_N : t_f)$ denote the time dependent failure probabilities corresponding to volume flaws and surface flaws,respectively.

2.8 Fast Fracture considering the material degradation due to SCG

When a ceramic component is subjected to long-term loading, it deteriorates due to subcritical crack growth. Consequently, the failure probabilities for both time dependent fracture and fast fracture increase. For simplicity, it is assumed that a ceramic component is subjected to the uniaxial stress σ_i varying stepwise as shown in Fig. 2. From Eq. (22), the crack length a_M at time t_M is related with initial crack length a_0 as

$$a_o^{\frac{2-n}{2}} - a_M^{\frac{2-n}{2}} = \left(\frac{n-2}{2}\right) BC^n \sum_{i=1}^{M} \sigma_i^n \Delta t_{i-1} \quad . \tag{37}$$

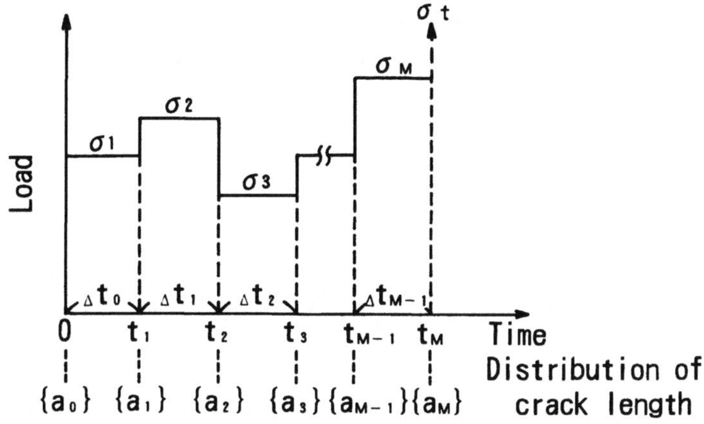

Fig. 2 Stepwised loading pattern

Therefore, the failure probability at time t_M for operating stress σ_t is given by

$$P_f\,(\sigma_t : t_M) = P_{rob}\left\{ a_M > \left(\frac{K_{IC}}{C\sigma_t}\right)^2 \right\}$$

$$= P_{rob}\left\{ a_o > \left[\left(\frac{K_{IC}}{C\sigma_t}\right)^{2-n} + \left(\frac{2-n}{2}\right) BC^n \sum_{i=1}^{M} \sigma_i^{\,n} \Delta t_{i-1} \right] \right\} \; . \tag{38}$$

It can be easily seen that σ_M instead of σ_t should be used if σ_t is less than σ_M. From Eq. (24), the failure probability can be rewritten as

$$P_f\,(\sigma_t : t_M) = 1 - \exp\left\{ -\frac{1}{V_o}\,\frac{2}{\pi} \int_V \int_0^{\frac{\pi}{2}} \int_0^{\frac{\pi}{2}} \left(\frac{\sigma_t^{\,*}}{\sigma_o}\right)^m \sin\phi\, d\phi\, d\theta\, dV \right\} \; , \tag{39}$$

where

$$\sigma_t^{\,*} = \alpha_t\, \sigma_t \; , \tag{40}$$

$$\alpha_t = \left\{ 1 + \left(\frac{n-2}{2}\right)\left(\frac{BC^2}{K_{IC}^{\,2}}\right) \sigma_t^{\,2} K_{IC}^{\,n} \sum_{i=1}^{M} \left(\frac{\sigma_i}{\sigma_t}\right)^n \Delta t_{i-1} \right\}^{\frac{1}{n-2}} \; . \tag{41}$$

The coefficient α_t expresses the degree of material degradation. This result can be extended for multiaxial stress states using the following modified equivalent stress Z_t^* instead of Z in Eq. (8)

$$Z_t^* = \alpha_{tZ} Z_t \quad , \tag{42}$$

where

$$\alpha_{tZ} = \left\{ 1 + \left(\frac{n-2}{2}\right)\left(\frac{BC^2}{K_{IC}^{\,2}}\right) Z_t^{\,2} K_{IC}^{\,n} \sum_{i=1}^{M} \left(\frac{Z_i}{Z_t}\right)^n \Delta t_i \right\}^{\frac{1}{n-2}} . \tag{43}$$

This function can be applied for the reliability analysis of ceramic components subject to a series of non-proportional long-term loadings.

2.9 Failure probability of fast fracture for transient stress history

Transient loadings such as thermal loading and seismic loading cause stresses which are transient and non-proportional. In these cases, it is necessary to take stress histories into account in the failure analysis, and the highest equivalent stress in each part of a ceramic component should be used for a conservative estimate of reliability. The failure probability for fast fracture with a transient stress history can be calculated from Eqs. (7) and (8) in which the equivalent stress is calculated as the maximum in the stress history for each direction of randomly oriented crack.

This function of GFICES is very useful since gas turbines may be subjected to severe transient thermal loads due to the instantaneous temperature drop in combustion gases at trip conditions.

3. System

Fig. 3 shows the flow diagram of the strength evaluation system of ceramic gas turbine components. GFICES plays an important role in this system; it is coupled as a postprocessor to both the general purpose structural analysis program MSC/NASTRAN and the dynamic stress analysis program DYNA3D.

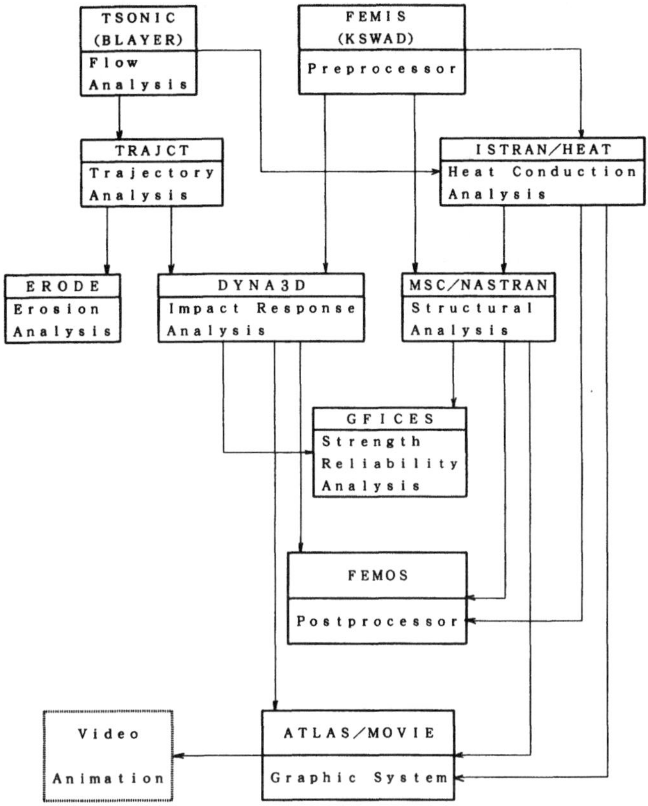

Fig. 3 Strength evaluation system of ceramics gas turbine components

4. Numerical analyses of a ceramic rotor blade

TEPCO has been conducting the cooperative research with Hitachi, Ltd. to develop a first stage ceramic rotor blade for a 20MW class gas turbine with inlet temperature of 1300°C [12].

The candidate material of the blade is a silicon-nitride of which material properties are shown in Table 1. The blade is directly attached to a metal rotor disc as shown in Fig. 4. Metal pads are inserted to convey the centrifugal force of the blade to the metal disc. A small amount of cooling air runs along the gap between the pad and the platform. The configuration of the dovetail was determined by precise stress analyses to lower the peak stress.

Table 1 Material properties of the silicon nitride

		Temperature (°C)							
		R.T	200	400	600	800	1000	1200	1400
Density	[g/cm³]	3.5							
Young's modulus	[Gpa]	300	290	290	290	290	280	280	280
Poisson's ratio		0.26	0.26	0.26	0.26	0.26	0.26	0.26	0.26
Coefficient of thermal expansion	[1/°Cx10⁻⁶]	-	1.9	2.8	3.0	3.3	3.4	3.5	3.5
Thermal conductivity	[W/m · K]	63	50	34	29	25	25	21	21
Specific heat	[J/g · K]	0.67	0.84	1.01	1.13	1.22	1.26	1.34	1.34
JIS 4-point bending strength	[MPa]	790					770	770	760
Weibull modulus		20							

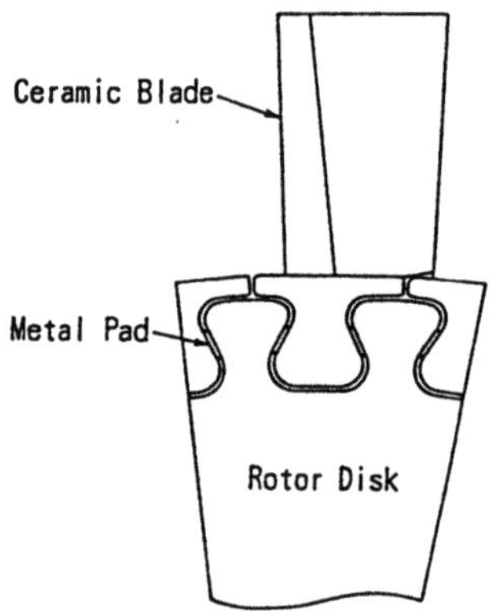

Fig. 4 Structure of the ceramic rotor blade

The strength of this ceramic rotor blade has been confirmed through several kinds of experimental tests such as room temperature spin test, thermal loading test and high temperature spin test. The room temperature spin test was conducted at up to 120% of normal design rotational speed. The thermal loading test was carried out in a static condition using a cascade of five ceramic rotor blades which were subjected to combustion gases at the design condition. The high temperature spin test was conducted using the high temperature developing unit (HTDU) which is a one stage testing gas turbine. Table 2 summarizes the fundamental specifications of the design condition and these experimental tests.

Table 2 Specifications of the design condition and experiments of the ceramic rotor blade

Condition	Gas temperature °C	Pressure ata	Rotation rpm
Design condition	1200	9.5	10800
120% room temperature spin test	R.T	0*1	12960
Thermal loading test	1200	8.3	0*2
High temperature spin test	1100	3.0	10800

Notes *1: Vacuum, *2: Static condition

GFICES was applied to estimate the reliability of this ceramic rotor blade. The analyses were confined to the fast fracture for volume flaws. Preceding the reliability analyses, stress analyses were performed. The normal design condition causes the peak centrifugal stress of 220 MPa at the root of the dovetail, and thermal loading yields a peak thermal stress of 322 MPa near the center of the leading edge on its design trip condition. Similarly, stress analyses were conducted corresponding to experimental conditions. Following these stress analyses, the reliability analyses for volume flaws were performed. Table 3 summarizes the safety factors which are defined as the ratio of the load corresponding to a failure probability of 0.5 to the operating load. As shown in this table, the safety factor of the normal design condition was found to be raised almost twice by conducting the 120% spin test at room temperature. The design trip condition is the severest one for the ceramic rotor blade; its safety factor is 1.57.

The thermal loading test was proven to be effective in raising the reliability on the trip condition. Although it is very difficult to realize the thermal condition of the gas turbine experimentally, it is thought to be quite necessary to establish a proof testing method for the thermal condition of a gas turbine in order to raise the reliability of ceramic gas turbine components.

5. Conclusion

In this paper, the authors presented the theoretical background of the program GFICES which can evaluate the reliability of ceramic gas turbine components. Numerical analyses on a ceramic rotor blade of a 20MW class gas turbine show the usefulness of this program.

Although ceramics are very promising for structural components exposed to elevated temperatures, there still exist some problems for gas turbine applications. The most crucial point is their low toughness, since it lowers the reliability of ceramic structural components. In addition, ceramics fail suddenly without any sign of fracture, and it is very difficult to determine the degree of deterioration

Table 3 Numerical results of the ceramic rotor blade

Condition	Proof test condition	Safety factor
Room temperature spin test		
120% of normal design speed	----------	1.54
Thermal loading test		
Steady state	----------	4.90
Trip condition	----------	1.87
High temperature spin test		
Steady state	----------	2.30
	120% room temperature spin test ----------	4.31
Trip condition	----------	2.42
Design condition		
Normal condition	----------	2.21
	Steady state of thermal loading test ----------	2.21
	120% room temperature spin test and steady state of high temperature spin test -----	4.27
Trip condition	----------	1.57
	Trip condition of thermal loading test ----------	2.11
	120% room temperature spin test and trip condition of high temperature spin test ---	2.16

of ceramic components by inspections. Therefore, "design by analysis" becomes quite important to estimate the strength and lifetime of ceramic components. It is obvious that reliable predictions are only possible if material characteristics are well known. In this respect, it must be noted that a material database of structural ceramics should be accumulated in order to validate the reliability analyses.

Reference

1. Y. Hara, T. Tsuchiya, F. Maeda, I. Tsuji, and K. Wada, "Development of Ceramic Components for a Power Generating Gas Turbine", ASME TURBO EXPO 1991, Orlando.
2. J.P. Gyekenyesi, "SCARE: A Postprocessor Program to MSC/NASTRAN for Reliability Analysis of Structural Ceramic Components", Journal of Engineering for Gas Turbines and Power, Vol. 108, 1986, pp. 540-546.
3. J.P. Gyekenyesi, and N.N. Nemeth, "Surface Flaw Reliability Analysis of Ceramic Components with the Scare Finite Element Postprocessor Program", Gas Turbine Conference and Exhibition, California, 1987, 87-GT-69.
4. J. Lamon, "Statistical Approaches to Failure for Ceramic Reliability Assessment", J. Am. Ceram. Soc., Vol. 71, 1988, pp. 106-112.

5. G. Stürmer, A. Schulz, and S. Wittig, "Lifetime Prediction for Ceramic Gas Turbine Components", Journal of Engineering for Gas Turbines and Power, Vol. 115, 1993, pp. 70-75.
6. J. Hamanaka, Y. Hashimoto, A. Ookoshi, and N. Watanabe, "A Reliability Evaluation System for Ceramic Gas Turbines", JSME International Journal, Vol. 30, 1987, pp. 408-413.
7. J. Hamanaka, T. Teramae, S. Hamada, M. Nakatsuka, H. Ezumi, and Y. Hara, "The Reliability Evaluation System for a Ceramic Gas Turbine (2nd Report - Fast Fracture Evaluation Considering Aging Degradation of Components)", in Japanese, Transactions of JSME, Vol. 55, 1989, pp. 1935-1941.
8. H. Hamada, and T. Teramae, "Reliability Evaluation System for Ceramic Gas Turbine Components", in Japanese, Journal of the Society of Materials Science, Vol. 39, 1990, pp. 76-81.
9. Y. Matsuo, in Japanese, Transactions of JSME, Vol. 49, 1983, pp. 46-52.
10. I. Oda, M. Matsui, T. Soma, M. Masuda, and N. Yamada, "Fracture Behavior of Sintered Silicon Nitride under Multiaxial Stress States", in Japanese, Journal of Ceramic Society of Japan, Vol. 96, 1988, pp. 539-545.
11. J. Hamanaka, Y. Nakamura, S. Hayashi, K. Sakai, and F. Takemasa, "Proof Testing of Ceramic Components", in Japanese, Transactions of JSME, Vol. 53, 1987, pp. 1638-1643.
12. T. Machida, H. Miyata, K. Wada, Y. Furuse, and T. Teramae, "Development of ceramic Rotor for Industrial Gas Turbine", 1991 Yokohama International Gas Turbine Congress.

Robust Structural System Design Using A System Reliability-Based Design Optimization Method

T.Y. Torng[1] and R.J. Yang[2]

[1] Southwest Research Institute (SwRI), 6220 Culebra Road, San Antonio, TX 78228, USA
[2] FORD Motor Company, Village Plaza, Suite 1100, 23400 Michigan Avenue, Dearborn, MI 48124, USA

Abstract. The need for designing a high-reliability robust structural system has been increasing recently due to the demand for greater quality and reliability. This robust structural system design must satisfy the performance/cost/weight objective as well as the system reliability constraint, which is used to account for uncertainty existing in different failure models, loading conditions, and material properties. These uncertainties are modeled as random variables with proper distribution types, means, and standard deviations. For design variables which have uncertain bounds, random means are defined. To expedite the robust design process, approximate reliability constraints are used in the design optimization search process. The proposed system reliability-based design optimization method will be used to solve two demonstration examples,and the results will be compared with conventional design optimization methods.

Keywords. Reliability, probability of failure, robust structural system, optimization, constraint, design, cost

1 Introduction

The ultimate goal in structural design is to produce a robust structural system that satisfies performance/cost/weight objectives as well as system reliability constraints, which are used to account for uncertainty existing in different failure models, loading conditions, and material properties. To deal with these uncertainties, reliability technology provides tools for formal assessment and analysis. To meet these robust design objectives, optimization technology plays an important role. The combination of reliability and optimization technologies is a viable way to design high-reliability robust structural systems.

Recent works using reliability-based structural optimization technologies (Frangopol, 1984; Thoft-Christensen and Murotsu, 1986; Sorensen and Thoft-Christensen, 1986; Lee and Wirsching, 1987; Nicolaidis and Burdisso, 1988; Maglaras and Nikolaidis, 1990) can be summarized as follows:

Maximize/minimize the objective function, OBJ(**X**), subject to constraints:

$$P(g_i(\mathbf{X}) \geq 0) \geq (1 - p_i), \qquad i = 1,\dots,L, \tag{1}$$

where $\mathbf{X}$ is the vector of n random variables, $P(\cdot)$ denotes the probability of the event $(\cdot)$, g_i represents the i-th limit state function, $(1-p_i)$ is the reliability goal for the i-th constraint or failure mode, and L represents the total number of constraints.

In general, there are three major obstacles in solving this problem: (1) How to solve complex problems that require a computation-intensive program; (2) How to solve system reliability problems accurately and efficiently; and (3) How to reduce total computational effort within the design optimization process.

In the aforementioned papers [1-6], a mean value first order method or a first order second moment method was proposed for component reliability calculations; and a bounding approach or a failure mode decomposition method was proposed for system reliability calculations. These methods are designed for solving simple problems with closed-form function. Therefore, for real life problems that require the use of computation-intensive programs such as a finite element program, total required computational efforts become excessive.

To overcome this difficulty, the proposed method uses an advanced mean value method (AMV)[7,8], which has shown its capability of solving reliability for complex problems. Furthermore, a probabilistic fault tree analysis (PFTA)[9,10] method is used to calculate the system reliability and system sensitivity.

In the design optimization process, several design iterations are required in order to search for the optimal design. In each design iteration, both reliability and non-reliability constraints need to be calculated several times, therefore, for complicated problems, total computational effort becomes excessive even with efficient probabilistic methods. To improve the efficiency, an approximate function is used to represent the complex component or system reliability constraint. In other words, for each design iteration, only one reliability calculation is required.

This study proposes a useful tool for designing a high-reliability robust system. In addition, this method can also be used to identify a design with minimal performance variation or be applied to a multidisciplinary design. Two examples will be used to demonstrate the proposed method. The results will be compared with those using other design optimization methods.

2 Robust Structural System Design Definition

A robust structural system design, as shown in Fig. 2.1, must be insensitive to uncertainty incurred from material properties, environmental conditions, manufacturing variations, etc. A smaller system variation must be achieved in order to reduce the possible failure[11,12].

To achieve this robust structural system design, the first initial step is to define appropriate objectives and constraints. In general, each design optimization problem has an objective function (performance/weight/cost) to be minimized or maximized as follows:

$$OBJ(\mathbf{Y},\mathbf{X}), \tag{2}$$

subject to the reliability constraints (component or system)

$$P(g_i(\mathbf{Y},\mathbf{X})\geq 0)\geq(1-p_i), \qquad i=1,\ldots,L, \tag{3}$$

where **Y** is the vector of m design random variables, **X** is the vector of n random variables, $P(\cdot)$ denotes the probability of the event $(\cdot)$, g_i represents the i-th limit state function (component or system), $(1-p_i)$ is the reliability goal for i-th constraint or failure mode, and L represents the total number of constraints.

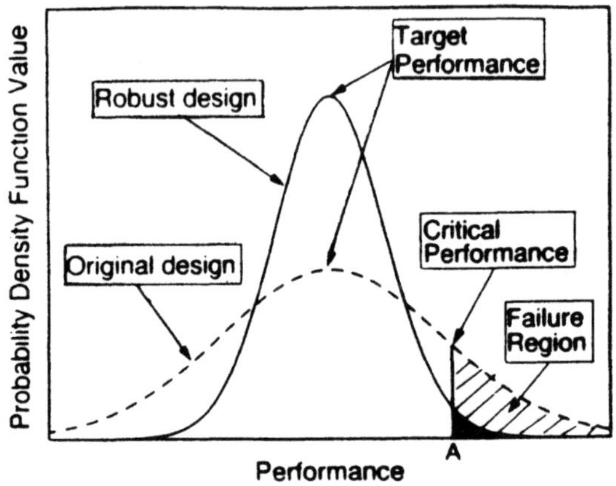

Fig. 2.1. Robust Structural System Definition

3 Uncertainty Modeling

In designing engineering structures, there are often uncertainties associated with material properties as well as geometric properties, external forces, boundary conditions, and optional environment. These uncertainties can be broadly classified as being either reducible or irreducible[13,14]. Irreducible uncertainties are intrinsic to nature and beyond our ability to control such as material properties. Reducible uncertainties are attributed to estimation error (or lack of data), modeling error, and human error, and cannot be completely avoided, however, can be attenuated by additional testing, more accurate modeling, and better quality control measures.

With proper statistic tools, an irreducible type of uncertainty can be modeled as random variable with selected distribution type, mean, and standard deviation. For a reducible type of uncertainty, the same statistical tools may be used to determine their statistics, however, the resulting means or standard deviations may be random themselves because of lack of data or human error. To account for this additional uncertainty, a noisy random variable, which is often called the modeling error factor, is assumed.

The effect of an irreducible type of uncertainty is reflected in the calculated reliability, whereas the effect of a reducible type of uncertainty is reflected in a confidence bound on the calculated reliability result[14]. In other words, with the existence of a reducible type of uncertainty such as uncertain mean and standard deviation, the reliability itself becomes a random variable.

In addition to both reducible and irreducible uncertainties, uncertain design parameters must be properly modeled. As shown in Fig. 3.1, these design parameters usually have bounded ranges and inherent uncertainties. According to Fig. 3.1, these design parameters can be modeled as the reducible type of uncertainties. In other words, each uncertain design parameter will be modeled as a random variable with a random mean value (for bounded range) and a fixed/random standard deviation (for inherent uncertainty). With this setup, the system reliability can be adjusted to reach the design goal by adjusting the means or standard deviations of the design variables.

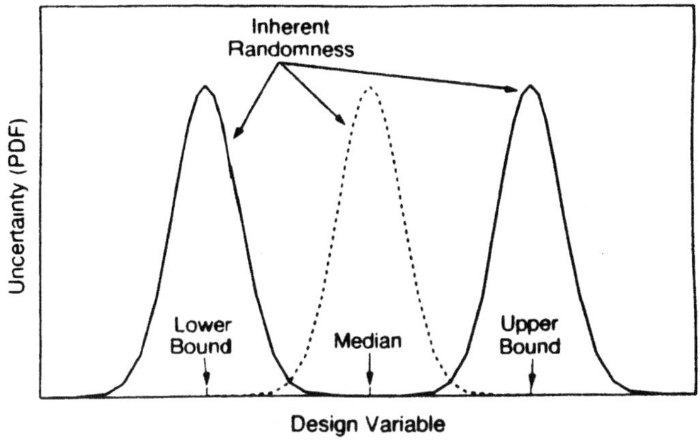

Fig. 3.1. Uncertain Design Parameter

4 Reliability Constraint Function Definition

Reliability constraint functions or limit state functions are used to represent failure mechanisms such as yield failure, fracture failure, buckling failure, or others. These functions can be constructed through the response (or performance) function, Z,

$$Z = Z(\mathbf{X}, \mathbf{Y}) \quad , \tag{4}$$

where $\mathbf{X}$ represents the random variables, and $\mathbf{Y}$ represents the design random variables. This Z-function can be a simple closed-form function or a complicated model which requires the use of computer intensive programs to solve.

To compute the reliability or probability of failure, it is required to define a critical failure event. This failure event is defined when Z-function value being less than or greater than a critical response value, z_0. In other words, the reliability (equal chance) constraint function or limit state function, g, becomes

$$g(\mathbf{X}, \mathbf{Y}) = Z(\mathbf{X}, \mathbf{Y}) - z_0 = 0 \quad . \tag{5}$$

The limit-state, $g(\mathbf{X}) = 0$, separates the variable space into "failure" and "safe" regions. When the equal chance constraint function becomes unequal, i.e., $g < 0$ or $Z \le z_0$, the reliability or the probability of failure, p_f can be calculated as

$$p_f = Prob(g(\mathbf{X}, \mathbf{Y}) \le 0) = Prob(Z(\mathbf{X}, \mathbf{Y}) - z_0 \le 0) \quad . \tag{6}$$

For simple closed-form g-function, the reliability computation is straightforward. To calculate an implicitly defined g-function, because each g-function calculation becomes time-consuming, the selected probabilistic method must be efficient, robust, and reasonably accurate. Furthermore, when considering a system reliability problem, p_f calculation becomes more complex than a component problem because it is constructed by multiple g-functions, i.e.,

$$g_{system} = F[g_1, g_2, g_3, \ldots\ldots g_i] \quad , \tag{7}$$

where **F** represents a logic function between all limit state functions, g_i. To illustrate this logic function, a fault tree is used. The probability of system failure, p_{sf}, can be denoted as

$$P_{sf} = P[g_{system} < 0] = P[F(g_1, g_2, \ldots\ldots, g_i) < 0] \quad . \tag{8}$$

5 Reliability Constraint Function Calculation

In general, the structural reliability analysis method is developed to solve a limit-state (constraint) function, $g(\mathbf{X})$. Given the joint probability density function, $f_{\mathbf{X}}(\mathbf{x})$, the probability of failure can be formulated as

$$p_f = P(g \le 0) = \int_{\Omega} \ldots \int f_{\mathbf{X}}(\mathbf{x}) d\mathbf{x} \quad , \tag{9}$$

where Ω is the failure region. This multiple integral is in general very difficult to evaluate. Alternatively, a Monte Carlo solution provides a convenient, but usually time-consuming solution.

The first step in the current reliability analysis methods requires the transformation of a generally dependent, random vector **X** into an independent, standardized normal vector **u**. The Rosenblatt transformation has been suggested when the joint distribution is available[14,15]. If only the marginal distributions and the covariances are known, a transformation can be made to generate a joint normal distribution that satisfies the given correlation structure.

By transforming $g(\mathbf{X})$ to $g(\mathbf{u})$, the most probable point (MPP) in the u-space, $\mathbf{u}^*$, is located. $\mathbf{u}^*$ is the point that defines the minimum distance, β, from the origin ($\mathbf{u} = 0$ point) to the limit-state surface. This point is most probable because it has maximum joint probability density on the limit-state surface, as shown in Fig. 5.1. The MPP may be found by using optimization, advanced mean value (AMV), or other iteration schemes.

Next, the $g(\mathbf{u})$ or $g(\mathbf{X})$ function is approximated by a polynomial that approximates the true function in the vicinity of the MPP. Once the approximate function is obtained, the associated failure probability can be computed. If the $g(\mathbf{u})$ formulation is used, several analytical solutions are available for linear and quadratic functions.[15] For example, the first-order reliability method (FORM) estimate is

$$P(g \le 0) \approx \Phi(-\beta), \tag{10}$$

and the asymptotic second-order reliability method (SORM) estimate, derived by Breitung[16], is

$$P(g \le 0) \approx \Phi(-\beta) \prod_{j=1}^{n-1} (1 + \beta\kappa_j)^{-1/2}, \quad \beta \to \infty \quad , \tag{11}$$

where $\Phi(\cdot)$ is the cdf of a standard normal distribution and κ_j, $j = 1, \ldots n-1$ are the main curvatures of the limit-state surface at $\mathbf{u}^*$.

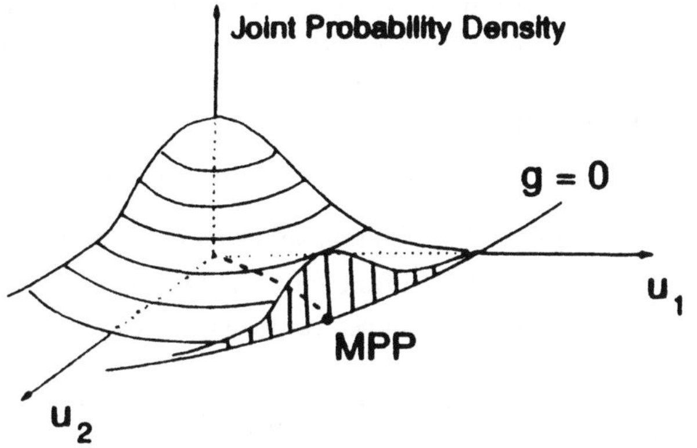

Fig. 5.1. Illustration of a Most Probable Point

5.1 Component Reliability Calculation

AMV or AMV plus iteration (AMV+) methodologies, discussed in greater detail in Refs. [7,8], is suitable for solving complex problems. Its methodology is summarized as follows.

Assume that the Taylor's series expansion of performance function, Z exists at the mean values. The Z function can be expressed as

$$\begin{aligned} Z(\mathbf{X}) &= Z(\mu) + \sum_{i=1}^{n} \frac{\partial Z}{\partial X_i} \cdot (X_i - \mu_i) + H(\mathbf{X}) \\ &= a_0 + \sum_{i=1}^{n} a_i X_i + H(\mathbf{X}) \\ &= Z_1(\mathbf{X}) + H(\mathbf{X}) \quad , \end{aligned} \tag{12}$$

where the derivatives, a_i, are evaluated at the mean values, μ, Z_1 is a random variable representing the sum of the first-order terms, and $H(\mathbf{X})$ represents the higher-order terms. In general, the coefficients a_i can be computed by numerical differentiation, and the minimum required number of Z-function calculations is $(n+1)$ for n random variables. Since Z_1 is explicit and linear, its cdf can be computed efficiently.

The AMV method reduces the truncation error by replacing the higher-order terms $H(\mathbf{X})$ by a simplified function $H(Z_1)$ dependent on Z_1. Ideally, the $H(Z_1)$ function should be based on the exact most probable point (MPP) locus of the Z function to minimize the truncation error. The AMV procedure simplifies this approach by using the MPP of Z_1.

5.2 System Reliability Calculation

A probabilistic fault tree analysis (PFTA) method[9,10] was proposed to solve the system reliability problem. This PFTA method first uses a fault tree to model all failure mechanisms based on their logic relationship. These failure mechanisms may be dependent when they share the same common random variables. As an example, consider a simple fault tree[10] as shown in Fig. 5.2. Its system failure, P_{sf}, can be denoted as follows:

$$P_{sf} = P[\{(g_1 < 0) \cap (g_2 < 0)\} \cup (g_3 < 0) \cup (g_4 < 0)] \quad , \qquad (13)$$

where g_1 and g_2 represent the vibration failure model, g_3 represents the stress failure model, and g_4 represents the fracture failure model. Each bottom event represents a closed-form function or a complex computer model.

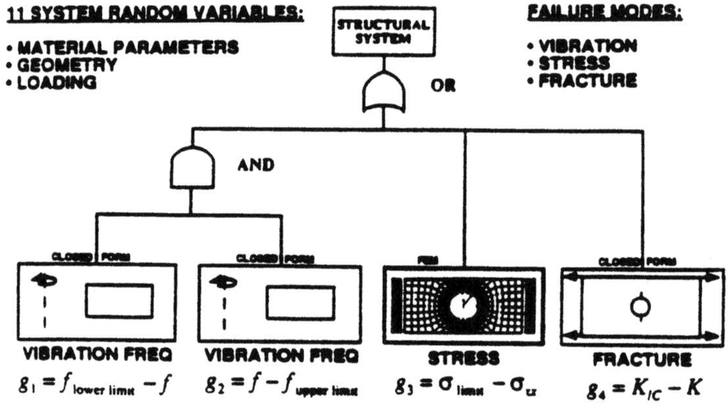

Fig. 5.2. A Simple Fault Tree

To solve this problem, the most probable point of each bottom event can be identified using the AMV or AMV+ method first. Next, approximate functions are constructed based on the calculated MPPs. These approximate functions will replace the original implicitly defined functions and be used as initial sampling functions.

With all the approximation functions determined, the fault tree becomes an explicitly defined system. Before calculating the system reliability, this system reliability problem can be further simplified if important events can be identified. To eliminate those relatively insignificant events, a dominant sampling sequence for all bottom events is developed based on their failure probabilities and the logic of the fault tree. Next, an adaptive importance sampling method (ASM), discussed in greater detail in Refs. [9,10], is proposed to solve the system reliability.

6 Approximate Reliability Constraint Function

An important by-product of each reliability analysis is the probability sensitivity factors, α, which are defined in the transformed standard normal space (u-space), i.e.,

$$\alpha = \frac{\partial \beta}{\partial u} = \frac{\partial \beta}{\partial \Phi^{-1}(F_X(x))} \quad , \tag{14}$$

where β represents the safety index value, Φ^{-1} represents the inverse standard normal cdf, and $F_X(x)$ represents the cdf for the original random variable, X. By comparing the absolute values of these sensitivity factors, the most important random variables can be identified.

As discussed in the uncertainty modeling section, all design random variables will be modeled with uncertain means (or standard deviations). With these uncertain means, the reliability itself becomes a random function of these uncertain parameters. To measure the effect caused by these uncertain parameters, the probabilistic sensitivity factors, with respect to these uncertain design parameters and reliability (safety index, β), can be derived as follows:

$$\frac{\partial \beta}{\partial \mu_X} = \frac{\partial \beta}{\partial u_X} \cdot \frac{\partial u_X}{\partial \mu_X} = \alpha \cdot \frac{\partial u_X}{\partial \mu_X} \tag{15}$$

$$\frac{\partial \beta}{\partial \sigma_X} = \frac{\partial \beta}{\partial u_X} \cdot \frac{\partial u_X}{\partial \sigma_X} = \alpha \cdot \frac{\partial u_X}{\partial \sigma_X} \quad , \tag{16}$$

where μ_X and σ_X represent mean and standard deviation of random variable X, respectively. Since u_X is a function of μ_X and σ_X, $\partial u_X/\partial \mu_X$ and $\partial u_X/\partial \sigma_X$ can be derived.

With both $\partial u_X/\partial \mu_X$ and $\partial u_X/\partial \sigma_X$ values evaluated, it is possible to construct an approximate reliability constraint function, $g_A(\mu_X, \sigma_X)$, as

$$\begin{aligned} g_A(\mu_X, \sigma_X) &= \beta_0 + \sum_{j=1}^{n} \frac{\partial \beta}{\partial \mu_{X_j}} \cdot \left(\mu_{X_j} - \mu_{X_{j0}}\right) \\ &\quad + \sum_{k=1}^{n} \frac{\partial \beta}{\partial \sigma_{X_k}} \cdot \left(\sigma_{X_k} - \sigma_{X_{k0}}\right) - \Phi^{-1}(1 - p^*) \\ &= C_0 + \sum_{j=1}^{n} C_j \cdot \mu_{X_j} + \sum_{k=1}^{n} C_k \cdot \sigma_{X_k} \quad , \end{aligned} \tag{17}$$

where β_0 is the safety index result, $\mu_{X_{j0}}$ is the j-th initial mean value, $\sigma_{X_{k0}}$ is the k-th initial standard deviation value, $\Phi^{-1}(\cdot)$ represents the inverse normal cumulative distribution function (cdf), $(1 - p^*)$ is the selected reliability goal , and C_0, C_j, and C_k are constant.

For system reliability sensitivity, $\partial u_X/\partial \mu_X$ and $\partial u_X/\partial \sigma_X$ values can not be derived directly from probabilistic sensitivity factors. In other words, there is no closed-form relationship for creating the approximate system reliability constraint.

To efficiently compute a reasonably accurate system sensitivity, the simplest strategy is to construct an approximate function at the MPP of each bottom event. By using all these approximate functions instead of the original complex failure functions, the probability sensitivity, with respect to mean value and standard deviation, $\partial u_X/\partial \mu_X$ and $\partial u_X/\partial \sigma_X$, can be derived. Total computational effort is reduced greatly because these values are calculated based on analytical (or closed-form) approximate functions.

7 System Reliability Based Design Optimization Method

A step-by-step analysis procedure for the proposed system reliability-based design optimization method is summarized as follows:

a. Construct an optimization design problem as listed in Eqs. 2 and 3 based on the robust structural system requirements.
b. Use the old design point as the initial design point, $\mathbf{X}_0$, if available. Otherwise, use the mean value of design random variables as the initial design point.
c. Construct an approximation function for either component or system constraint function g_i at the initial design point by using the following steps:
   c1. Evaluate the safety index, β_0, for the ith reliability constraints at $\mathbf{X}_0$ by using advanced mean value (AMV) method.
   c2. Construct an approximate constraint function for either component or system reliability constraint based on the safety index, β with respect to the mean values and standard deviations of those design random variables. By using the strategy discussed earlier, the probabilistic sensitivity factors between the design variables, and the safety index, $\partial\beta/\partial\mu_X$ and $\partial\beta/\partial\sigma_X$, can be calculated by PFTA and be used to construct a safety index approximation function, $\beta_{Ai}(\mathbf{X})$.

$$\beta_{Ai}(\mathbf{X}) = \beta_0 + \sum_{j=1}^{n} \frac{\partial\beta}{\partial\mu_{X_j}} \cdot \left(\mu_{X_j} - \mu_{X_{j0}}\right) + \sum_{k=1}^{n} \frac{\partial\beta}{\partial\sigma_{X_k}} \cdot \left(\sigma_{X_k} - \sigma_{X_{k0}}\right)$$

$$= C_0 + \sum_{j=1}^{n} C_j \cdot \mu_{X_j} + \sum_{k=1}^{n} C_k \cdot \sigma_{X_k} \quad , \qquad (18)$$

   where $\mu_{X_{j0}}$ is the j-th initial mean value, $\sigma_{X_{k0}}$ is the k-th initial standard deviation value, and C_0, C_j, and C_k are constant.
   c3. Construct an approximate reliability constraint (or a limit state) function, $g_{Ai}(\mathbf{X})$, i.e.,

$$g_{Ai}(\mathbf{X}) = C_0 + \sum_{j=1}^{n} C_j \cdot \mu_{X_j} + \sum_{k=1}^{n} C_k \cdot \sigma_{X_k} - \Phi^{-1}(1-p_i) \quad , \qquad (19)$$

   where Φ^{-1} is the inverse normal cdf, and p_i is the limit of probability of failure.
   c4. Construct other approximation functions, $g_{Ai}(\mathbf{X})$s, for other constraint functions. Notice that if the same finite element calculation can be applied to other g functions, those approximation functions should be constructed simultaneously.
d. Based on the objective function and all approximation functions, an appropriate optimizer[18] is used to find the optimal design, $\mathbf{X}_1$.
e. Repeat steps 2-4 until the convergence criteria is met or the number of iteration limit is reached.

8 Example One

A simple cantilever beam problem as shown in Fig. 8.1 is used to illustrate the proposed procedure. This design optimization problem is defined to minimize the weight of the beam, therefore, the objective function is

$$F(L,b,h)=L\cdot b\cdot h \quad , \tag{20}$$

subject to yield and deflection reliability constraints

$$P(g_1\geq 0)\geq 0.999 \tag{21}$$

$$P(g_2\geq 0)\geq 0.999 \quad , \tag{22}$$

with

$$g_1=R_L-\frac{6QL}{bh^2} \tag{23}$$

$$g_2=\Delta_L-\frac{6QL^3}{Ebh^3} \quad , \tag{24}$$

where R_L is the material yield strength, Δ_L is the deflection limit, Q is the static loading, E is the Young's Modulus, L is the length of the beam, b is the width of the beam, and h is the height of the beam.

To illustrate the importance of the dependency problem between failure models, this problem will also be solved by assuming a system reliability constraint approach. The system reliability constraint is defined as

$$1-P[(g_1\leq 0)\cup(g_2\leq 0)]\geq 0.999 \tag{25}$$

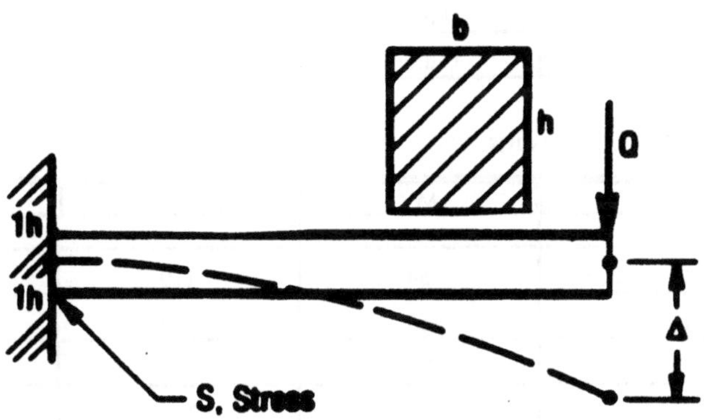

Fig. 8.1. Example One

A total of seven random variables are defined in this optimization problem as shown in Table 8.1. Among these random variables, only mean values of length, width, and height of the beam are considered as design random variables. Reducible type of uncertainty (e.g., modeling error) will not be considered here. The bounds of these uncertain means are defined in Table 8.2. The standard deviations of these means are assumed fixed. Finally, the required reliability goal is 0.999.

This problem is first solved by using the system reliability approach concept without considering dependency between failure models. The most probable point (MPP) is solved by using several methods which include the mean value method, the advanced mean value method, and the first-order reliability method. The final objective function values, as shown in Table 8.3, are very close for all three methods. However, the total required number of g-function calculations is changed greatly. The mean value based method uses the least number of g-function calculations, however, the probability of failure solution may not be very accurate. Both FORM and AMV methods produce accurate probability of failure calculation; but FORM uses a lot more g-function calculations.

Table 8.1. Random Variables

Var Name	Dist	Mean	Std Dev
R_L (ksi)	Normal	120.	8.46
Δ_L (in)	Normal	0.75	0.042
Q (kips)	Normal	1.888	0.283
E (ksi)	Normal	3.E4	3.E3
L (in)	Normal	18.0	0.9
b (in)	Normal	1.0	0.05
h (in)	Normal	1.5	0.075

Table 8.2. Design Random Variables

Var Name	Dist	Upper Bound	Lower Bound
μ_L (in)	Uniform	20.0	16.0
μ_b (in)	Uniform	1.25	0.8
μ_h (in)	Uniform	1.8	1.25

Table 8.3. COV = 5% Results

Method	MV	AMV	FORM
Iteration no.	9	11	9
g_1-function calculation no.	54	110	324
g_2-function calculation no.	63	279	378
Final p_{f1}	8.98E-4	9.79E-4	1.E-3
Final p_{f2}	3.66E-5	3.22E-6	2.78E-6
OBJ value	24.3976	24.0782	24.2729
L value	16	16	16
b value	.84714	.83605	.84281
h value	1.8	1.8	1.8

In traditional reliability-based optimization methods, the design parameters are usually assumed to be a deterministically bounded value without considering the randomness of these design parameters. To understand the impact of uncertain design parameters, the same problem is solved by changing the standard deviations of the length, the width, and the height to a negligible value (0.005% COV) and a higher number (10%). All three AMV results are listed in Table 8.4. As expected, the final objective (OBJ) function value increases as the coefficient of variation value increases.

Table 8.4. Results Comparison

COV	0.005%	5%	10%
Final p_{f1}	1.E-3	9.79E-4	7.31E-4
Final p_{f2}	4.29E-9	3.22E-6	3.95E-4
OBJ value	22.0717	24.0782	31.5933
L value	16	16	16
b value	0.8	0.83605	1.097
h value	1.72435	1.8	1.8

To validate the accuracy of these reliability results, a Monte Carlo simulation approach is used to solve the same problem. The total g-function calculations required using the Monte Carlo method, listed in Table 8.5, is much larger than the one using the AMV method.

Table 8.5. Monte Carlo Results For 5% COV case

Method	MONTE	AMV
Total samples	93370	11
Accepted samples	10000	
g_1-function calculation no.	3.36E6	110
g_2-function calculation no.	3.92E6	279
OBJ value	24.88727	24.07823
L value	16.092	16.0
b value	0.86983	0.83605
h value	1.778	1.80

To illustrate the idea of dependency, the system reliability of the beam is calculated at the mean value of all design variables. Both independent (Eqs. 21 and 22) and dependent (Eq. 25) are calculated and listed in Table 8.6. To calculate independent case, the AMV method is used first to compute the reliability for each failure model. Next, both reliability results are used to calculate the system reliability. For the dependent case, the PFTA method is used to calculate the system reliability based on both approximation functions created at the MPP identified by AMV. Notice that the dependency between yield and deflection failure models comes from the common random variables. As expected, with consideration of dependency, the system reliability result is less conservative.

To check if dependency will effect the final design, the same design problem is solved by using the system reliability constraint function (Eq. 25). The final design based on the system approach and previous result (without dependency) are listed in Table 8.7. Both results are closed for this problem because yield failure becomes the only dominate failure model. In other words, the dependency problem becomes less important in this example. However, when both failure models are equally important, the final system reliability result will be effected.

Table 8.6. System Reliability Approach

Type	Independent	Dependent (PFTA)
System Reliability	1.06E-1	8.7E-2

Table 8.7. Design with/without Dependency

Constraint	Eqs. 21 & 22	Eq. 25
no. of iteration	11	8
g_1-function calculation no.	110	80
g_2-function calculation no.	279	175
system failure, p_{f2}	9.82E-4	9.87E-4
OBJ value	24.0782	24.2735
L value	16	16
b value	0.83605	0.84283
h value	1.8	1.8

9 Example Two

A ten member cantilever truss shown in Fig. 9.1 is used. The problem here is to find the cross-sectional area of each member of the truss to minimize its weight, subject to stress, displacement, and member size constraints. The optimal results for this problem have been solved by several optimization techniques as discussed in Ref. [17]. The main goal of using this example here is to show the importance of including reliability constraints when the design parameters themselves become random.

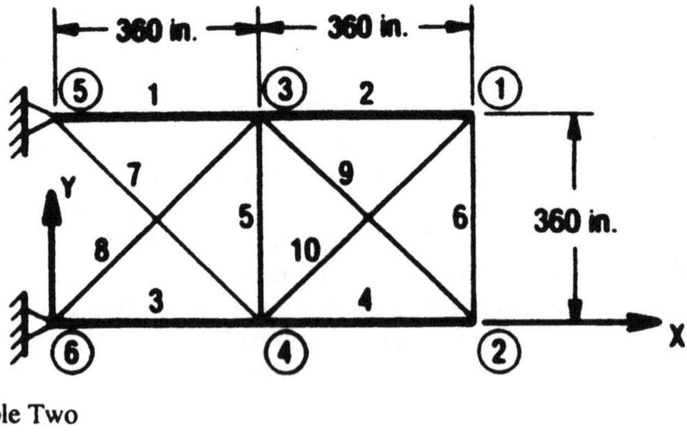

Fig. 9.1. Example Two

In this example, there are ten stress constraints, eight displacement constraints, and lower-bound constraints on each of the design variables, which makes a total of twenty-eight constraints. For the stress constraints, the limit is 25 ksi in tension or compression; for the displacement constraints, the limit is two inches along each degree of freedom; for the lower-bound constraints the cross-sectional area is 0.10 in.^{2} The rest of the important parameters include material density equal to 0.10 lb/in.^{3} and Young's modulus of elasticity equal to 10^4 ksi. The design parameters for this example are the cross-sectional area in each member.

To show the importance of including reliability constraints, the stress constraints and displacement constraints are used to evaluate reliability when the cross-sectional area becomes random. This assumption is actually necessary for these cross-sectional areas because uncertainty does exist during the manufacturing process. Therefore, for the stress constraints, the reliability is defined when the calculated stress falls within the 25 ksi tension/compression limit. For displacement limit, the reliability is defined when the calculated displacement less than the displacement limit 2 in. Here, the system failure is defined when any one of the constraints fails.

Without considering the reliability constraints, this problem is first solved by using CONMIN program[18]. The final objective function value after ten iterations is 5644. This results will be used as the initial design point for the proposed reliability analysis. To calculate the reliability of this design, the cross-sectional area results from previous deterministic run are used as the mean values of normally distributed random variables. These mean values will have the lower-bound constraints at 0.1 in.^{2} As for the standard deviations of these areas, they are considered to be 3%, 5% or 7.5% of the mean values. In other words, the coefficient of variations (COV) for these normal random variables are equal to 3%, 5%, or 7.5%, respectively. The random variables data (for COV = 0.05 case) are shown in Table 9.1.

Table 9.1. Random Variables For COV = 0.05 Case

Var Name	Dist	Mean	Std Dev
A_1 (in.^{2})	Normal	29.536	1.4768
A_2 (in.^{2})	Normal	2.7327	0.1366
A_3 (in.^{2})	Normal	29.456	1.4728
A_4 (in.^{2})	Normal	13.589	0.6795
A_5 (in.^{2})	Normal	0.1000	0.0050
A_6 (in.^{2})	Normal	2.1663	0.1083
A_7 (in.^{2})	Normal	14.045	0.7023
A_8 (in.^{2})	Normal	19.184	0.9592
A_9 (in.^{2})	Normal	19.236	0.9618
A_{10} (in.^{2})	Normal	3.0002	0.1500

With consideration of these defined design random variables, the reliabilities of these eighteen constraints need to be evaluated. To calculate the system failure probability, it is necessary to consider the dependency between failure conditions. However, to simplify the analysis, a full dependent assumption is used for this problem so that only the worst component reliability constraint limit needs to be concerned. With this assumption, the system failure probability for the initial design is calculated for different COVs and is shown in Table 9.2. The results indicate the original design's reliability equals 94.7, 83.4, and 74.1 %, for COV equals 3%, 5%, and 7.5%, respectively. In other words, if the reliability goal of this design is .99 or .999, the initial design needs to be redesigned.

Table 9.2. System Failure Probability

COV	0.030	0.050	0.075
P_f	5.30E-2	1.66E-1	2.59E-1

To study the impact of different reliability goal and COV of design random variables, several design problems have been solved. The objective function (cost) results for these problems are shown in Table 9.3. As expected, the objective function (cost) value becomes larger when higher reliability goal or COV value are used and vice versa. Using higher reliability goal, the cost of design will be increased, however, a greater customer satisfaction can be achieved. Using higher COV value, the cost of design will be increased, however, the cost of manufacturing will be reduced. Therefore, the balance between these factors needs to be studied carefully.

Table 9.3. Objective Results For Different Reliability Goals

Reliability Goals	COV = 0.030	COV = 0.050	COV = 0.075
0.999	5716*	5867	6075
0.9999	5762	5950	6212
0.99999	5804	6032	6330

* Each OBJ result requires 5-10 CONMIN iterations

10 Summary

An advanced system reliability-based design optimization method was presented. The method was designed primarily for the complex structures that often require the use of finite element codes. From both demonstration examples, the proposed method not only reduces the number of finite element analyses, but also gives very accurate results. In example one, the proposed method was compared with the conventional Monte Carlo method to show its efficiency. In addition, this example was solved by considering the dependency between the failure mechanisms. In example two, the original design from the optimization program was used to compute its reliability corresponding to different levels of uncertainties. As a result, the original design may have very poor quality or reliability. Therefore, the proposed method should be applied in order to design a robust system that satisfies design's performance/cost/weight objectives as well as system reliability constraints of uncertain design parameters.

11 Acknowledgement

This research was supported by FORD Motor company. The support is gratefully acknowledged.

12 References

1. Frangopol, D.M., "Interactive Reliability-Based Structural Optimization," Computer & Structures, Vol. 19, No. 4., pp. 559-563, 1984.
2. Thoft-Christensen, P., and Murotsu, Y., Application of Structural Systems Reliability Theory, Springer-Verlag, Berlin, 1986.
3. Sorensen, J.D., and Thoft-Christensen, P., "Recent Advances in Optimal Design of Structures form a Reliability Point of View," 9th Advances in Reliability Technology, 1986.
4. Lee, S.J., Reliability-Based Optimal Structural and Mechanical Design, Dissertation, University of Arizona.
5. Nikolaidis E., and Burdisso, R., "Reliability Based Optimization: A Safety Index Approach," Computers & Structures, Vol. 28, No. 6, pp. 781-788, 1988.
6. Maglaras, G.K., and Nikolaidis E., "Integrated analysis and design in Stochastic optimization," Structural Optimization, Vol. 2, pp. 163-172, 1990.
7. Wu, Y.-T., Millwater, H.R., and Cruse, T.A., "An Advanced Probabilistic Structural Analysis Method for Implicit Performance Functions,'' AIAA Journal, Vol. 28, No. 9, pp. 1663-1669, 1990.
8. Wu, Y.-T., Burnside, O.H., and Cruse, T.A., "Probabilistic Methods for Structural Response Analysis," Computational Probabilistic Methods, AMD-Vol. 93, ASME; Computational Mechanics of Reliability Analysis, W.K. Liu and T. Belytschko (eds.), Elmepress International, Ch. 7, 1989.
9. Wu, Y.-T., Torng, T.Y., Millwater, H.R., Fossum, A.F., and Rheinfurth, M.H., "Probabilistic Methods for Rotordynamics Analysis," presented at the 1991 SAE Conference, April, 1991.
10. Torng, T.Y., Wu, Y.-T., and Millwater, H.R., "Structural System Reliability Calculation Using A Probabilistic Fault Tree Analysis Method," Proceedings of the 33rd AIAA/ASME/ASCE/AHS/ASC Structures, Structural Dynamics and Materials Conference, Dallas, April 13-15, 1992.
11. Hsieh, C.C., and Oh, K.P., "MARS: A Computer-Based Method for Achieving Robust System," FISITA 92 Conference, London, UK, June 7-11,1992.
12. Oh, H.L., "Variation Tolerance Design," Proc. of ASME Winter Annual Meeting, Production Engineering Division, Vol. 115, pp. 137-146, 1987.
13. Wirsching, P.H., "Reliability Methods in Mechanical and Structural Design," Class notes at the University of Arizona.
14. Ang, A.H.-S., and Tang, W.H., Probability Concept in Engineering Planning and Design, Volume II: Decision, Risk, and Reliability, New York, John Wiley & Sons, Inc., 1984.
15. Madsen, H.O., Krenk, S., and Lind, N.C., Methods of Structural Safety, Englewood Cliffs, New Jersey, Prentice Hall, Inc., 1986.
16. Breitung, K., "Asymptotic Approximations for Probability Integrals," Probabilistic Engineering Mechanics, Vol. 4, No. 4, 1989.
17. Haug, E.J., and Arora, J.S., Applied Optimal Design, New York, John Wiley & Sons, Inc., 1979.
18. Vanderplaats, G.N., "CONMIN - A fortran program for constrained function minimization user's manual", TM X 62 282, NASA, 1973.

Adaptive Importance Sampling (AIS)-Based System Reliability Sensitivity Analysis Method

Y.-T. Wu

Southwest Research Institute, 6220 Culebra Road
San Antonio, Texas 78238, USA

Abstract. This paper presents an adaptive importance sampling (AIS) method that can be used to efficiently compute reliability and reliability sensitivities for complicated structures that require computer-intensive calculations. The AIS approach starts from an initial approximate failure domain and incrementally changes the sampling limit state surface to reach a sampling domain that is slightly greater than the failure domain to minimize over-sampling in the safe region. Several reliability sensitivity measures are proposed that can be computed directly and easily from the AIS-based failure points and limit-state surfaces. These probability sensitivities can be used to identify key random variables and to adjust design parameters to achieve reliability-based objectives. The proposed methodology is demonstrated using a turbine blade reliability analysis example problem.

Keywords. Adaptive importance sampling, probabilistic mechanics, reliability, reliability sensitivity, structural reliability method

1 Introduction

The analysis and design of advanced, high-performance structures requires accurate, computer-intensive calculations using numerical algorithms such as the finite element (FE) method. To ensure high reliability, reliability testing of structures is desirable. However, it is often impractical to perform sufficient numbers of tests. To overcome the problem, numerical modeling can be used to replace or supplement reliability testing. A straightforward Monte Carlo method, useful for reliability analysis of simple analytical models, becomes too time-consuming to be practical for large FE or other numerical models. Thus, efficient reliability and reliability sensitivity calculation algorithms are needed for analyzing complicated structural models.

Efficient and accurate computational reliability methods for a single limit state (or performance function) have been developed and successfully applied to aerospace propulsion structures.[1-3] Computational methods have also been developed recently for system reliability problems.[4] Reliability sensitivity calculation methods are well developed for single limit state problems. For multiple limit states, available reliability sensitivity calculation methods are primarily based on bounds or asymptotic results of series and parallel systems.

Reliability sensitivity measures are typically based on certain approximation points, e.g., design points, and are valid only in the first-order approximation sense.[5] In this paper, a method is proposed to directly and accurately calculate system reliability sensitivities based on the AIS-generated sampling points and limit-state surfaces.

2 Overview of Limit-State Approach

A limit-state or failure function $g(\mathbf{X})$ is a function of a vector of basic random variables, $\mathbf{X} = (X_1, X_2, ..., X_n)$, with $g(\mathbf{X}) = 0$ being the limit-state surface that separates the variable space into two regions: failure ($g \leq 0$) and safe ($g > 0$). Given the joint probability density function, PDF, $f_X(\mathbf{x})$, the probability of failure is the probability in the failure domain Ω and is given by

$$p_f = \int_{\Omega} \cdots \int f_X(\mathbf{x}) d\mathbf{x} . \tag{1}$$

The above multiple integral can be computed using a straightforward standard Monte Carlo procedure. However, for complicated g-functions that require intensive finite element analyses, more efficient methods are required.

A widely used procedure performs reliability analysis in a coordinate system of an independent, standardized normal vector $\mathbf{u}$.[6] By transforming $g(\mathbf{X})$ to $g(\mathbf{u})$ using a proper distribution transformation, the most probable point (MPP) in the u-space, $\mathbf{u}^*$, is a point that defines the minimum distance from the origin ($\mathbf{u} = 0$) to the limit-state surface $g(\mathbf{u}) = 0$. This point, also called design point in the structural reliability literature, is most probable (in the u-space, not necessarily in the X-space) in the sense that it has maximum joint PDF on $g(\mathbf{u}) = 0$.

Once the MPP is identified and an approximate, linear or quadratic, function is developed around the MPP, the probability estimate can be calculated easily. The sensitivity factors, α_i, are the directional cosines of the unit vector from the origin to the MPP. The sensitivity factors provide first-order information on the importance of the individual random variable. Other sensitivity measures with respect to a distribution parameter or a limit-state function parameter can be estimated based on the sensitivity factors and the distribution transformation.[5] Using the limit-state approach, the primary computational effort is typically spent on locating the MPP and developing the polynomial function. This is particularly true for complicated g-functions.

2.1 Approximate Linear Limit-State Function

There are many algorithms available for searching for the MPP.[4,5] This paper uses the advanced mean-value (AMV)-based method[2] for finding the MPP and defining the approximate linear g-function in the X-space. Given a performance function $Z(\mathbf{X})$, the AMV model is

$$Z_{AMV} = Z_{MV} + H(Z_{MV}) , \tag{2}$$

where

$$Z_{MV} = Z(\mu_X) + \sum_{i=1}^{n} \left(\frac{\partial Z}{\partial X_i} \right) \cdot \left(X_i - \mu_{X_i} \right) \quad = a_o + \sum_{i=1}^{n} a_i X_i , \tag{3}$$

and $H(Z_{MV})$ is defined as the difference between the values of Z_{MV} and Z calculated on the Most Probable Point Locus (MPPL) of Z_{MV}. Note that the MPPL of Z_{MV} is generally a nonlinear curve. For a complicated structural analysis, the construction of Equation 2 may be time consuming, but its MPPL can be computed easily.

The derivatives in Equation 3 are evaluated at the mean values by numerical differentiation or other methods. Based initially on Equation 2 and by keeping track of the MPPL,[8] the exact MPP for a limit state $Z = z$ (i.e., $g = Z - z$) can be computed to establish the X-based, first-order model using the equation

$$Z_{AMV+} = Z(\mathbf{x}^*) + \sum_{i=1}^{n} \left(\frac{\partial Z}{\partial X_i} \right) \cdot (X_i - x_i^*) \quad = b_o + \sum_{i=1}^{n} b_i X_i = z \ . \tag{4}$$

where $\mathbf{x}$ is the corresponding value of $\mathbf{u}^*$.

It is important to note that the variables X_i are generally non-normal and dependent, therefore, Equation 4 is different from the first-order reliability method (FORM) model, which is linear in the u-space. The advantage of Equation 4 is that it can be used to provide an initial estimate of the curvature (about the MPP) in the u-space. It is particularly useful when the curvature is primarily caused by the non-normal to normal transformation, which has been observed to be true for many application problems.

2.2 Multiple Limit States

For multiple limit-state problems, the u-space approach can be applied to all the g-functions. Reliability analysis requires additional calculation of the joint probabilities. Because of the complexities involved, approximate and simplified methods such as the probability bounding approaches and asymptotic results are commonly used. From the literature, the accuracies of these approximation methods seem to be problem dependent, and must be used with caution.

Reliability sensitivity analyses are also much more difficult than a single limit state because there are multiple MPP's. The sensitivity factors derived from the individual g-function cannot be used to derive the system reliability sensitivities because the importance of the individual g-function cannot be quantified easily. This paper proposes using the approximate limit states to develop importance sampling regions. Both the system reliability and its sensitivity are computed based on the AIS samples and the AIS-based limit-state surfaces. The AIS approach eliminates the computational difficulties presented in the approximate analytical approach.

3 Adaptive Importance Sampling (AIS)

First consider a single limit state. If the true limit-state surface cannot be well represented by Equation 4, the probability estimate may be inaccurate. Therefore, it is useful to improve/check the approximate limit-state surface by importance sampling. Several such importance sampling schemes have been proposed that focus sampling in the probability-critical regions to increase sampling efficiency.[9-14] An adaptive importance sampling concept originally developed for a rotor instability analysis problem has been extended recently for general system reliability problems.[15-16,4]

The adaptive surface approach has the following features: (1) the sampling density is proportional to the joint PDF of the random variables, (2) the initial sampling domain is the failure side of the system limit-state surface constructed from an approximate, parabolic surface, (3) additional samples are added after incrementally changing the curvatures of the parabolic surface at the most probable point, and (4) the final sampling domain contains, but is only slightly greater than, the failure domain.

The goal of the above adaptive surface approach is to minimize over-sampling in the safe region. Ideally, we do not want to perform any sampling in the safe region. However, the only way to detect the limit-state surface is by passing the limit-state surface. To minimize unnecessary sampling, the chosen adaptive surface for AIS should be flexible enough to represent closely the exact limit-state surface. Significant efficiency gain relative to the conventional Monte Carlo method can be achieved if the sampling is performed in regions very close to the failure regions.

There are several possibilities in choosing and varying the sampling limit-state surface, as illustrated in Figure 3.1, where three kinds of sampling boundary surfaces are shown. The sphere-based method is the simplest to implement and is the most robust in the sense that it can account for multiple MPP's by decreasing the radius. However, the sphere-based method is inefficient for large n. Figure 3.2 shows the allowable maximum number of variables to maintain a high sampling efficiency relative to the standard Monte Carlo method. In the figure, p is the probability outside of the sphere. The efficiency factor is 1/p.

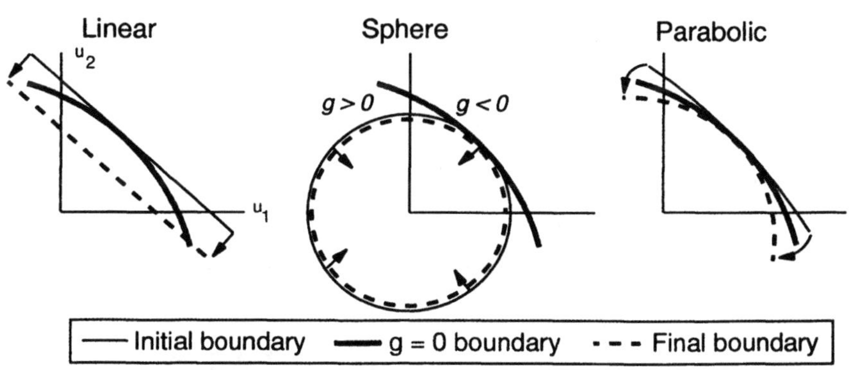

Fig. 3.1. Illustration of the AIS Concept

The linear-surface-based method is easy to implement, and the sampling efficiency is not a function of n. However, it is not as robust as the sphere-based method and is not in general as efficient as the curvature-based method.

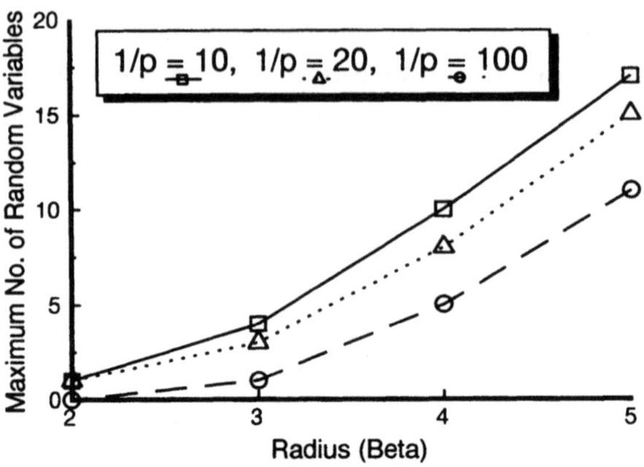

Fig. 3.2. Effect of the Number of Random Variables to the Sampling Efficiency Using the Sphere-Based AIS Method

The curvature-based method changes the curvature of the parabolic surface at the MPP. While unable to fully account for multiple MPP's, this method is very efficient relative to the other two methods and, in general, has the ability to closely approximate a nonlinear limit state. The curvature-based AIS concept is further illustrated in Figure 3.3, in which the parabolic surface is rotationally symmetric about the vector OP that passes through the origin and the MPP. In general, the sampling domain can be adjusted by increasing or decreasing the curvatures of the parabolic surface. By decreasing the curvatures to the extreme (-∞), the sampling region will cover the entire space. However, if β is correctly calculated, the region inside the sphere need not be sampled. Therefore, the AIS bound is a sphere, as indicated in Figure 3.3. For additional checking, the distance to the MPP can also be reduced in case the calculated β is in error.

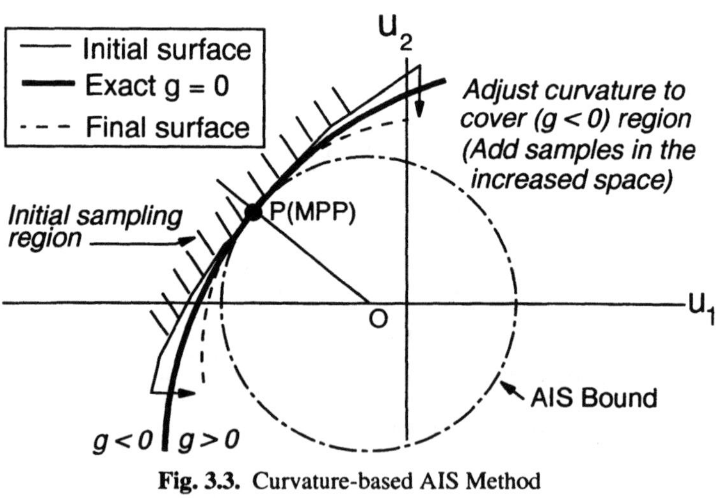

Fig. 3.3. Curvature-based AIS Method

4 AIS for a Single Limit-State

Assume that an initial second-order limit state is available in the form

$$g(\mathbf{u}) = \nabla g(\mathbf{u}^*)^{\mathrm{T}}(\mathbf{u}-\mathbf{u}^*) + \frac{1}{2}(\mathbf{u}-\mathbf{u}^*)^{\mathrm{T}}\mathbf{H}(\mathbf{u}^*)(\mathbf{u}-\mathbf{u}^*), \tag{5}$$

where $\nabla g(\mathbf{u}^*)$ is the gradient at the MPP, and $\mathbf{H}$ is the Hessian matrix containing second-order derivatives. Because Equation 5 will be used to define an initial sampling domain for the AIS, only a rough estimate is needed. The linear g(**X**)-function in Equation 4 can be transformed to a second-order (without cross-product terms) g(**u**)-function. Other methods such as the curve-fitting approach may be used.[17]

The transformation procedure to develop a parabolic surface is well documented in the literature[5,18] and is given by the equation

$$g = \beta - v_n + \sum_{i=1}^{n-1} \lambda_i v_i^2, \tag{6}$$

in which v_i are independent standardized normal variables, and λ_i are related to the main curvatures by the relation $\kappa_i = 2\lambda_i$. The curvature-based AIS procedure uses Equation 6 to develop the initial adaptive sampling boundary and generate the samples in the $g < 0$ domain. The corresponding probability in the sampling region, p_S, can be numerically computed using a second-order reliability method[18] or a convolution method.[4]

To change the sampling space, the curvatures λ_i are changed to λ_i'. The changes in λ_i can be made individually or simultaneously based on a selected probability increment Δp_S. Define the changed or perturbed-limit state surface as

$$g' = \beta + \sum_{i=1}^{n-1} \lambda_i' v_i^2, \tag{7}$$

where the superscript (prime) indicates a perturbed condition.

Let N_1 be the initial number of samples determined based on a confidence interval and an error bound. During the sampling process, this number will be adjusted based on the calculated probability of failure in the initial sampling region. The number of failure points in S is denoted as N_1^+. Given a set of λ_i', the sampling is applied to an increased sampling ΔS region. A predetermined number of samples, ΔN, in the ΔS region, is calculated using

$$\Delta N = (p_S' - p_S)N_1 = \Delta p_S N_1 \,. \tag{8}$$

where p_S' is the probability in the total sampling region S'. By computing the original g-function, the number of failure points, ΔN^+, within ΔS, can be determined and the updated probability of failure estimate becomes

$$\hat{p}_f = p_S' \cdot \hat{p} = p_S' \cdot \frac{N_1^+ + \Delta N^+}{N_1 + \Delta N}. \tag{9}$$

Where $\hat{p}$ is the conditional probability of failure in S'. The perturbation procedure is repeated until no more failure points are observed after at least one perturbation.

Assuming that $\hat{p}$ has a normal distribution, for a $(1-\alpha)$ confidence interval, the error bound, γ (in percentage), is given by

$$\gamma = \Phi^{-1}\left(1 - \frac{\alpha}{2}\right)\sqrt{\frac{1-\hat{p}}{\hat{p}N}}, \tag{10}$$

where $\Phi(\bullet)$ is the standard normal distribution function. Equation 10 can be used to calculate the required samples, N, given an estimated $\hat{p}$. Ideally, we wish to apply AIS to a region very close to the exact failure region, resulting in $\hat{p} \approx 1$. This goal, however, cannot be achieved because the sampling region must be greater than the failure region to indicate that the failure region has been covered sufficiently. In addition, the parabolic surface may not be a sufficiently flexible surface to represent the true surface.

The convergence procedure used in the demonstration example is as follows:

(1) Generate samples in the initial sampling domain S. Begin with a small sample (e.g., 10 points) to estimate $\hat{p}$. Then gradually increase the samples until the error bound γ is acceptable for a selected confidence level.

(2) Reduce the curvatures (average curvatures are used in this paper) such that the increased probability Δp_s is a fraction (e.g., 10%) of the initial probability p_s. Generate samples only in the increment ΔS.

(3) Update the probability estimate. Stop the analysis if the probability estimate converges. Otherwise, decrease the curvature again.

(4) As an option, (in case the MPP is is in error), slightly reduce the minimum distance (i.e., shift the parabolic surface towards the origin) and generate additional samples. If the MPP is correct, ΔS will be in the safe region. Stop the analysis if no failure points are observed in ΔS.

5 AIS for System Reliability

In this paper, a system reliability problem will be defined using a fault tree, which provides a convenient and systematic way to manage multiple failure modes. The limit-state functions are defined in the bottom events. Sequential failures can be modeled using the PRIORITY AND gate. A sequence of g-functions, corresponding to a sequence of updated structural configurations with load redistribution, can be explicitly or implicitly defined in the bottom events.

A failure mode can involve one or more limit states. By adding all the failure modes, and therefore all the limit states, the system limit-state surface can be constructed piece by piece. The AIS procedure for system reliability analysis requires the construction of multiple parabolic surfaces. In principle, it is a straightforward extension of the concept for one limit state. The difficult part is to develop a procedure to add failure regions.

One approach is to add samples progressively starting from the most important limit state, based on the approximate functions. This approach is ineffective for cases where system failure is governed by the joint effects from several limit states. In such cases, none of the limit states can be considered as a dominate one because the MPP of the individual limit state is not a likely event for a system failure. A more effective computational procedure adds samples progressively based on failure modes. The initial sampling regions are based on the approximate (Equation 4) limit states and the increased sampling regions are based on the perturbed limit states.

Figure 5.1 illustrates the above curvature-based AIS procedure for a system with two failure modes that involve three limit states. The probability of failure statement is

$$p_f = P\{(g_1 < 0) \cup [(g_2 < 0) \cap (g_3 < 0)]\}\,. \tag{11}$$

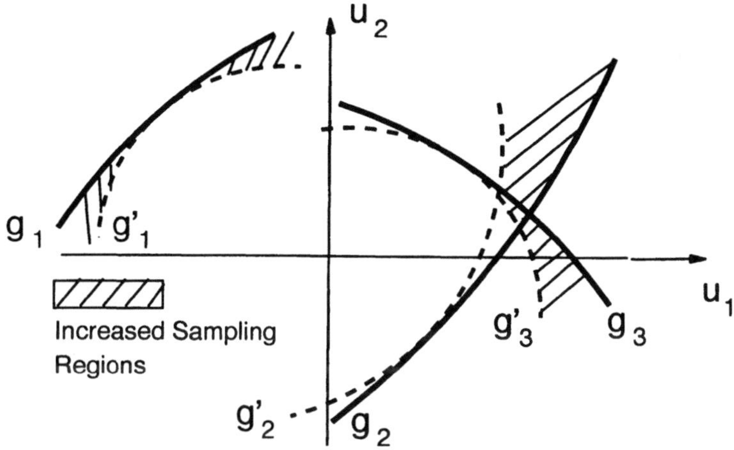

Fig. 5.1. AIS Method for System Reliability Analysis

The above adaptive approach for system reliability can be easily modified for use with the sphere-based and the linear surface-based methods. However, the curvature-based AIS is recommended for its overall performance in efficiency and robustness and is used in the demonstration example.

6 Reliability Sensitivity with Respect to a Distribution Parameter

When a distribution parameter is changed, the sensitivity of p_f, abbreviated as p, with respect to a distribution parameter, θ, can be evaluated using

$$\frac{dp}{d\theta} = \int_{\Omega} \cdots \int \frac{df_X}{d\theta}\, dx\,. \tag{12}$$

Therefore,

$$\frac{dp/p}{d\theta/\theta} = \int_{\Omega} \cdots \int \frac{\theta}{p} \frac{df_X}{f_X d\theta} f_X dx\,, \tag{13}$$

or

$$\frac{d\ln p}{d\ln\theta} = E\left[\frac{d\ln f_X}{d\ln\theta}\right]_{\Omega}, \tag{14}$$

where the expected value is evaluated using the PDF in the failure region. Because the AIS approach uses the same PDF, the probability sensitivity can be computed using those AIS points in the failure region. No additional g-function calculations are required.

Two useful dimensionless probability sensitivity coefficients, S, are

$$S_\sigma = \frac{d\ln p}{d\ln\sigma}\ , \text{ and} \tag{15}$$

$$S_\mu = \frac{d\ln p}{d\mu/\sigma}\ . \tag{16}$$

For independent normal variables, these coefficients can be analytically derived. Let μ_i and σ_i be the mean and the standard deviation, respectively, of the normal random variable i. It can be shown that

$$\frac{d\ln p}{d\ln\sigma_i} = E[u_i^2 - 1]_\Omega\ , \tag{17}$$

and

$$\frac{d\ln p}{d\mu_i/\sigma_i} = E[u_i]_\Omega\ . \tag{18}$$

Assuming a small change in θ, denoted as $\Delta\theta$, results in a small change in p, Δp. The change in probability can be approximated by

$$\Delta p \doteq p \cdot S \cdot \frac{\Delta\theta}{\sigma}\ . \tag{19}$$

A random variable approaches a deterministic value if σ approaches 0, i.e., $\Delta\sigma \doteq -\sigma$. In this case,

$$\Delta p \doteq -pS_\sigma\ , \tag{20}$$

which suggests that S_σ can be used to rank the importance of the random variable uncertainty (σ_i).

Similarly, if we reduce μ by one σ for a random variable,

$$\Delta p \doteq -pS_\mu\ . \tag{21}$$

which suggests that S_μ can be used to rank the importance of the uncertainty in the mean values.

The above AIS-based sensitivity analysis approach is particularly useful for system reliability problems in which there are multiple MPP's. With the AIS method, the sensitivity calculations are straightforward.

7 Numerical Example

A numerical example is presented to demonstrate the AIS-based sensitivity analysis. The structure is a turbine blade subject to high temperatures and pressures. Three failure modes are considered. The fault tree representation of the structural failure is given in Figure 7.1.

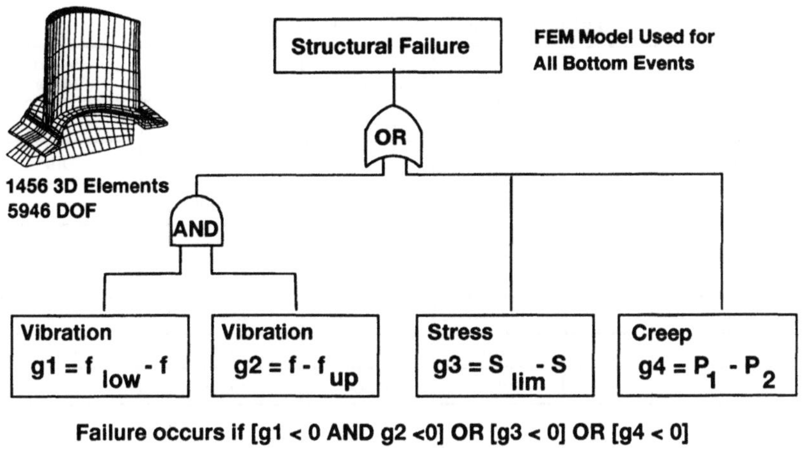

Fig. 7.1. Turbine Blade Reliability Analysis Model

The failure modes are:

1. The first natural frequency falls within an operating range, causing an intolerably large response. The two limit-state functions are

$$g_1 = f_{low} - f \quad \text{and} \quad g_2 = f - f_{upper}. \tag{22}$$

2. The von Mises stress exceeds a limit. The limit-state function is

$$g_3 = \sigma_{\max} - \sigma\,, \tag{23}$$

where σ is the von Mises stress at a highly stressed location. In this analysis $\sigma_{\max} = 110$ ksi.

3. Creep at a specified time-to-rupture exceeds the Larson-Miller parameter. The limit state function is given by the equation

$$g_4 = B_0 + B_1 \log\sigma + B_2 \log\sigma^2 + B_3 \log\sigma^3 - T(C + \log t)\,, \tag{24}$$

where B_0, B_1, B_2, B_3, and C are material coefficients, T is the absolute temperature, and t is the service time. In this example, $t = 75$ hours.

The blade has directional material properties. The random variables include three material orientation angles, elastic modulus, Poisson's ratio, shear modulus, a creep material coefficient (B_0), and the density. The random variable definitions and the failure modes to which they contribute are given in Table 7.1.

The finite element model consists of 1456 three-dimensional elements. The blade is subjected to centrifugal, pressure, and temperature loadings. The temperature-dependent material effect is included in the analysis.

The analysis was carried out using a probabilistic analysis computer program.[19,20] The AMV iteration algorithm (denoted as AMV+) was used for each bottom event. The reliability results of the AIS analysis, based on a 90% confidence interval and a 10% error bound, are shown in Table 7.2.

Table 7.1. Random Variable Definitions

	Mean	Std. Dev.	Distribution	Failure Mode
Material Orientation θ_z	.05236°	0.067544°	Normal	All
Material Orientation θ_y	-.03491°	0.067544°	Normal	All
Material θ_x	.08727°	0.067544°	Normal	All
Elastic Modulus E^a	18.36E6	0.4595E6	Normal	All
Poisson's Ratio ν^a	.386	0.00965	Normal	All
Shear Modulus G^a	18.63E6	0.4657E6	Normal	All
Creep Coef. B_0	86.0	0.086	Normal	Creep
Density ρ	.805E-3	.493E-5	Normal	Freq.

[a] At room temperature

Table 7.2. System Reliability Results (< 10% error, 90% confidence)
$P_f = Prob.\{[(g_1 < 0) \cap (g_2 < 0)] \cup (g_3 < 0) \cup (g_4 < 0)\}$

Reliability Analysis Method	Structural Models	Number of Random System Samples	Probability of Failure	Total CPU[a]
AIS	Approx. (AMV+)	85 (73 failures)	0.018	3.7 hrs
AIS	Finite Element	117 (85 failures)	0.022	25 hrs

[a] Includes AMV+ solution time

Table 7.3, taken from Reference 20, lists the coefficients of the final MPP-based linear approximations (see Equation 4). These coefficients were used in generating the AIS samples and computing the reliability sensitivities. The AIS-based sensitivity analysis was performed based on the AMV+ result, which is reasonably close to the AIS result using the finite element model.

Table 7.3. Limit State Definitions at the MPP's

Random Variable		θ_Z	θ_Y	θ_X	E	ν	G	B_0	ρ
Coef.	b_0	b_1	b_2	b_3	b_4	b_5	b_6	b_7	b_8
Freq.	.140E4	-.287E3	-.304E3	.525E2	-.103E-3	-.233E4	-.358E-4	0.0	.276E7
Stress	.151E6	-.368E5	-.437E5	.152E3	-.124E-2	-.207E6	-.187E-2	0.0	0.0
Creep	-.781E5	.997E3	.389E4	-.591E3	-.245E-3	-.119E4	-.113E-3	.100E4	0.0

8 Numerical Example Results

Previous analysis indicated that the g_2 limit state has an insignificant effect on the reliability result. To simplify the analysis, g_2 was removed, leaving only three limit states in the problem.

Based on the equations defined by the data in Table 7.3, 85 AIS samples were generated in which 73 samples were in the failure region. These failure points were used to compute the reliability sensitivity coefficients; results are shown in Table 8.1.

The 73 failure points were split into three failure modes. There were three common points between the stress failure mode and the frequency failure mode, indicating a small correlation with approximately a joint probability of 0.00074.

Based on the sampling points for the individual failure modes, the reliability sensitivity coefficients were computed; these results are also included in Table 8.1. The first three significant random variables are identified in the table.

Table 8.1. AIS-based Reliability Sensitivity Coefficients

Failure Mode	p	β	N	Coef.	θ_z	θ_y	θ_x	E	ν	G	B_o	ρ
Freq.	.00528	2.55	22[b]	u^*	0.78	**0.83**	-0.144	**1.91**	**0.90**	0.67		-0.55
				S_μ	0.64	**1.08**	-0.16	**1.96**	**1.45**	0.71		-0.83
				S_σ	0.18	**0.59**	0.48	**3.14**	**2.09**	0.45		0.53
Stress	.00233	2.83	12[b]	u^*	**1.57**	**1.87**	-0.01	0.36	**1.27**	0.55		
				S_μ	**1.58**	**2.09**	0.24	0.36	**1.35**	0.99		
				S_σ	**1.85**	**4.16**	-0.52	0.23	**1.76**	1.07		
Creep	.00978	2.33	42	u^*	-0.50	**-1.95**	0.29	**0.84**	0.08	-0.39	**-0.64**	
				S_μ	-0.57	**-2.33**	0.41	**0.86**	0.19	0.41	**-0.65**	
				S_σ	0.43	**4.63**	0.01	**0.57**	0.34	0.19	**0.46**	
System	.0181	2.09[a]	73[c]	S_σ	0.49	**3.48**[d]	0.07	**1.24**	**0.91**	0.44	0.19	0.16
				S_μ	0.44	**-0.72**	0.23	**1.08**	**0.65**	0.58	-0.35	-0.37

a: Calculated from $\beta = \Phi^{-1}(P_f)$.
b: Three common points in the frequency and stress failure modes.
c: Total AIS samples is 85.
d: Numbers in bold indicate the top three coefficients in magnitude.

The sensitivity coefficients can be used to predict probability changes due to small changes in μ or σ. The results of several test cases based on the most significant variables, θ_y and E, are listed in Table 9.1. Fairly good agreement between the predicted and the exact solutions was observed for up to a 20 percent change in the distribution parameters.

9 Other Reliability Sensitivities

9.1 Sensitivity with Respect to a g-function Parameter

In reliability-based design, some design parameters are to be optimized. These parameters might be deterministic, such as tightly controlled geometries. To compute reliability sensitivity with respect to such a parameter, assume the parameter to be random, with a small standard deviation. The assumed σ should be small enough not to produce a dominant S_σ. On the other hand, the assumed σ should be large enough so that S_μ can be estimated using a reasonable number of AIS samples.

Table 9.1. Probability Prediction Using Sensitivity Coefficients

% Changes (in σ)		Probability of failure, p	
		Predicted based on S	Exact
$\Delta\sigma/\sigma$	-5%	0.0149	0.0152
	-10%	0.0118	0.0127
(θ_y)	-20%	0.0055	0.0090
	-5%	0.0171	0.0161
$\Delta\mu/\sigma$	-10%	0.0161	0.0153
(E)	-20%	0.0142	0.0138

Based on the assumed σ, the correct probability can be estimated by reducing σ to 0, assuming that Δp is reasonably small with respect to p. Work is in progress to determine criteria for selecting a proper standard deviation.

9.2 Sensitivity with Respect to Distribution Types

In many design and analysis situations, available data are insufficient to define probability distributions confidently. In some cases, the means and the standard deviations can be reasonably assumed, but the distribution types are uncertain. For example, it is uncertain whether a Weibull or a log-normal is a better distribution. In such cases, it is useful to assume a (reference) distribution and, without having to redo the analysis, estimate the reliability assuming other distributions.

The difficulty in applying the u-space limit-state approach to this problem is that a different distribution may change the MPP's significantly. Therefore the approximate limit-state functions developed from the reference distribution must be insensitive to the MPP's.

To solve the problem, an approach is proposed based on finding the best quadratic $g(\mathbf{u})$-functions in the u-space, but transform them back to $g(\mathbf{X})$ before changing a distribution. The goal is to determine sufficiently accurate $g(\mathbf{X})$ functions insensitive to distribution transformations. The proposed procedure is as follows:

(1) Obtain a best set of quadratic $g(\mathbf{u})$-functions. This is achieved by starting from the final AIS parabolic surfaces, $g'(\mathbf{u})$, and re-adjusting the curvatures based on the AIS sampling information (i.e., both safe and failure points) to closely approximate the true limit-state surfaces. The generated approximate system limit-state function is denoted as $g''(\mathbf{u})$.

(2) The approximate system limit-state function in the original X-space is defined by transforming $g''(\mathbf{u})$ to $g''(\mathbf{X})$.

(3) The $g''(\mathbf{X})$-function will be treated as the "exact" function from which to perform reliability analyses for various distribution assumptions including the reference assumption.

(4) Re-apply the AIS procedure to all the different assumptions. To minimize numerical error caused by insufficient sample size, apply the same sampling points in the u-space to all the different assumptions.

The above-proposed approach appears to be conceptually sound. Work is in progress to test its performance.

10 Summary

The adaptive importance sampling (AIS) method is a computational tool that allows the sampling to be focused on critical failure regions. Using a turbine blade reliability analysis example, it has been demonstrated that the AIS method can be used to compute reliability and reliability sensitivities efficiently and accurately. Several reliability sensitivity measures and calculation procedures were proposed based on the AIS-generated sampling points and quadratic limit-state surfaces. The AIS-based reliability sensitivity analysis method can be used to support reliability/risk-based designs and decisions and is particularly suitable for analyzing computer-intensive models.

11 Acknowledgement

This work is partially supported by the PSAM (Probabilistic Structural Analysis Methods) program,[21] sponsored by the NASA Lewis Research Center. The support of C. C. Chamis at LeRC is gratefully acknowledged. The original finite element model and random variable definitions were developed by the Rocketdyne division of the Rockwell Corporation. Harry Millwater and Tony Torng at SwRI provided computational support in solving the demonstration example.

12 References

1. Wu, Y.-T., Millwater, H.R., and Cruse, T.A., "An Advanced Probabilistic Structural Analysis Method for Implicit Performance Functions," *AIAA Journal*, Vol. 28, No. 9, pp. 1663-1669, 1990.
2. Rajagopal, K.R., Keremes, J., Ho, H., and Orient, G., "A Probabilistic Approach to the Evaluation of Fatigue Damage in a Space Propulsion System Injector Element," *Proceedings of the 33rd Structures, Structural Dynamics, and Materials Conference*, pp. 625-639, 1992.
3. Torng, T.Y., Wu, Y.-T., and Millwater, H.R., 1992, "Structural System Reliability Calculation Using a Probabilistic Fault Tree Analysis Method," *Proceedings of the 33rd Structures, Structural Dynamics, and Materials Conference*, pp. 603-613.
4. Wu, Y.-T., "An Adaptive Importance Sampling Method for Structural System Reliability Analysis," *Reliability Technology*, Proceedings of the Symposium on Reliability Technology, ASME 1992 Winter Meeting, edited by T.A. Cruse, ASME AD-Vol. 28, pp. 217-231, 1992.
5. Madsen, H.O., Krenk, S., and Lind, N.C., *Methods of Structural Safety*, Englewood Cliffs, New Jersey, Prentice Hall, Inc., 1986.
6. Ang, A.H.-S., and Tang, W.H., *Probability Concepts in Engineering Planning and Design, Volume II: Decision, Risk, and Reliability*, New York: John Wiley & Sons, Inc., 1984.
7. Wu, Y.-T., Burnside, O.H., and Cruse, T.A., "Probabilistic Methods for Structural Response Analysis," *Computational Mechanics of Reliability Analysis*, W.K. Liu and T. Belytschko (eds.), Elmepress International, Ch. 7., 1989.

8. Wu, Y.-T., Gureghian, A.B., Codell, R.B., and Sagar, B., "Sensitivity and Uncertainty Analysis Applied to One-Dimensional Transport in a Layered Fractured Rock. Part II: Probabilistic Methods Based on the Limit-State Approach," *Nuclear Technology*, Vol. 104, No. 2, pp. 247-308, November 1993.
9. Harbitz, A., "An Efficient Sampling Method for Probability of Failure Calculation," *Structural Safety*, Vol. 3, pp. 109-115, 1986.
10. Bjerager, P., "Probability Integration by Directional Simulation," *Journal of the Engineering Mechanics*, 114(8), ASCE, 1987.
11. Ditlevsen, O., Bjerager, P., Olesen, R., and Hasofer, A., "Directional Simulation in Gaussian Processes," *Probabilistic Engineering Mechanics*, Vol. 3, No. 4, pp. 207-217, 1988.
12. Melchers, R.E., "Importance Sampling in Structural Systems," *Structural Safety*, Vol. 6, pp. 3-10, 1989.
13. Melchers, R.E., "Radial Importance Sampling for Structural Reliability," *Journal of Engineering Mechanics*, Vol. 116, No. 1, Paper No. 24288, pp. 189-203, 1990.
14. Engelund, S., and Rackwitz, R., "Comparison of Some Importance Sampling Techniques in Structural Reliability," *Proceedings of the 1992 ASCE Specialty Conference on Probabilistic Mechanics*, pp. 108-111, 1992.
15. Wu, Y.-T., and Burnside, O.H., "Computational Methods for Probability of Instability Calculations," *Proceedings of the 31st Structures, Structural Dynamics, and Materials Conference*, pp. 1081-1091, 1990.
16. Wu, Y.-T., Torng, T.Y., Millwater, H.R., Fossum, A.F., and M.H. Rheinfurth, "Probabilistic Methods for Rotordynamics Analysis," SAE Aerospace Atlantic 1991 Conference, *SAE Transactions*, 1991.
17. Der Kiureghian, A., Lin, H.-Z., and Hwang, S.J., "Second-Order Reliability Approximations," *Journal of Engineering Mechanics*, Vol. 113, No. 8, Paper No. 21727, pp. 1208-1225, 1987.
18. Tvedt, L., 1990, "Distribution of Quadratic Forms in Normal Space — Application to Structural Reliability," *Journal of Engineering Mechanics, ASCE*, Vol. 116, No. 6, pp. 1183-1197, 1990.
19. Millwater, H.R., Wu, Y.-T., Torng, Y., Thacker, B., Riha, D., and Leung, C., "Recent Development of the NESSUS Probabilistic Structural Analysis Computer Program," *Proceedings of the 33rd Structures, Structural Dynamics, and Materials Conference*, pp. 614-624, 1992.
20. Millwater, H.R., and Wu, Y.-T., "Computational Structural Reliability Analysis of a Turbine Blade," presented at the 1993 International Gas Turbine Conference, May 1993.
21. Chamis, C.C., "Probabilistic Structural Analysis Methods for Space Propulsion System Components," *Probabilistic Engineering Mechanics*, Vol. 2, No. 2, pp. 100-110, 1987.

Time Dependent System Reliability Under Cumulative Damage

Qiang Xiao[1] and Sankaran Mahadevan[1]

[1]Department of Civil and Environmental Engineering, Vanderbilt University, Nashville, TN 37235, USA

ABSTRACT

This paper reviews methods for time-variant system reliability estimation of brittle and ductile structures, and proposes a computational approach for brittle structures. In order to model the cumulative damage of the structural system over time, the proposed method groups the damage states into damage levels ranging from intact structure to system collapse. Such grouping is based on the number of the failed members in the structure. For the sake of application of the proposed method to relatively large structures, the significant damage states are identified from the significant failure sequences obtained by branch and bound method. The expressions for transition probabilities from one damage level to another are derived. The Markov chain model is used to calculate the probabilities of each damage level at any time, including the system failure probability. Two numerical examples are used to illustrate the proposed method.

KEYWORDS

random process, probability, failure, structures, reliability, safety, systems engineering, damage states, failure paths

1 Introduction

The reliability of a structure varies over its lifetime, due to repetitive load applications causing accumulated damage, degradation of material resistance over time, corrosion, wear etc. Time dependent reliability has received some attention in research, and estimation methods have been developed for two types of structural failure: ductile and brittle. Time dependent system reliability for ductile structures can be evaluated by identifying the significant failure modes and applying outcrossing rate analysis. For brittle structures, based on the Markov Chain model, a currently available method evaluates time dependent system reliability by identifying the stable configurations of the structure (damaged or intact) that can carry the loads. This method has limited application because a large number of stable configurations are involved in the computation even for simple structures and there is no systematic search procedure yet to identify the significant stable configurations and to reduce the computational effort. Therefore, an alternative method based on the failure path approach is proposed in this paper. The branch and bound method is used to identify significant failure sequences, and these are combined to describe damage levels ranging from intact structure to total collapse. Formulas for calculating the time-dependent probabilities of different damage levels are presented. The proposed method is examined through two numerical examples.

2 Assumptions

The following assumptions are made:

- The loadings on the structure are Poisson pulse processes as shown in Fig. 2.1. The load occurrence time follows a Poisson process and the duration and load intensity are random variables. The product of a mean occurrence rate ν and a mean duration μ_d varies from zero to one. Independence is assumed between the load intensity and duration in each occurrence as well as from occurrence to occurrence, and among different load processes.

- Dynamic effect is neglected.
- The material resistances are idealized as rigid-plastic and time-invariant for ductile structures, and elastic-brittle and randomly change with load application for brittle structures. This assumption is not realistic, and is used only for the sake of simplicity in developing the proposed method; computational methods are available to relax this assumption for realistic applications[6].
- No repair is made during the lifetime.

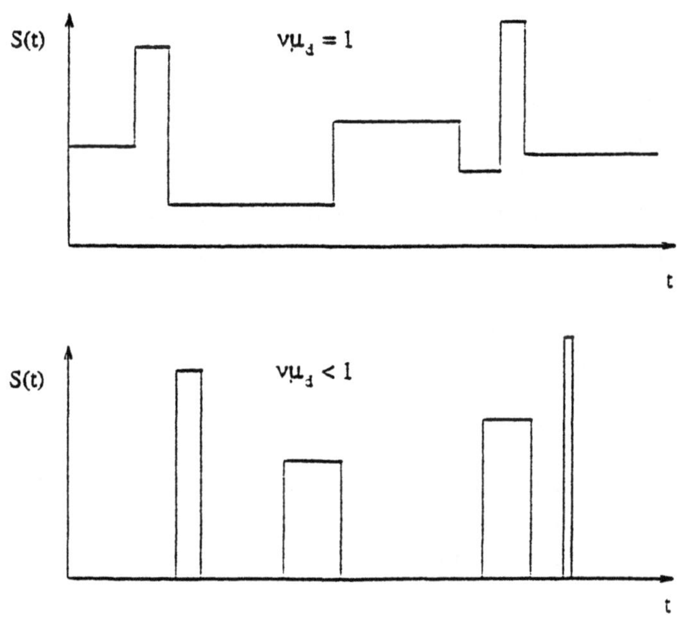

Figure 2.1: Poisson pulse process

3 Ductile System Reliability

System failure probability of ductile structures is evaluated through the union of significant failure modes. Wen and Chen proposed the following method for outcrossing rate analysis of ductile systems[8].

Based on a Poisson outcrossing violation of performance criterion assumption, a good estimate of the reliability of the structure is

$$L(t) \approx exp[-\nu(\boldsymbol{r}_o)t] \tag{1}$$

where $\nu(\boldsymbol{r}_o)$ is the mean outcrossing rate given the resistance $\boldsymbol{R}=\boldsymbol{r}_o$.

The approximate method to evaluate $\nu(\boldsymbol{r}_o)$ for loads of the square wave type has been proposed[4]

$$\nu(\mathbf{r_0}) = \sum_{i=1}^{n} \nu_i\{P(\mathbf{s}^i \subset F) - P[(\mathbf{s}^i \subset F) \cap (\mathbf{s} \subset F)]\} \tag{2}$$

where $\mathbf{s}$ is the vector of load intensity random variables; $\mathbf{s}^i$ is the vector of load intensity random variables given there is a change in the ith load component; F is the failure domain which is the union of the significant segments of the limit state of a desired member for given values of component resistances $\boldsymbol{R}=\boldsymbol{r}_o$. For the Poisson pulse load process, $\nu(\boldsymbol{r}_o)$ is obtained by load coincidence method[8].

Eq. 1 gives the conditional reliability for $\boldsymbol{R} = \boldsymbol{r}_o$. The unconditional reliability is therefore,

$$L(t) = \int_R exp[-\nu(\boldsymbol{r})t] f_R(\boldsymbol{r}) d\boldsymbol{r} \tag{3}$$

where f_R = joint density function of the resistance variables. Analytical solution is generally difficult and numerical integration becomes impractical for $n \geq 4$. A first order approximation of Eq. 3 has been obtained[6].

4 Brittle System Reliability

The existing method to estimate the time-dependent reliability of brittle structures is based on identifying all the stable configurations of the structure[9]. Since the number of stable configurations for a large structure is very large, and since there is no systematic procedure to search for significant stable configurations, such a method is difficult to apply. On the other hand, there is an alternate approach for system reliability, namely the failure path approach, within which a systematic failure sequence generation strategy such as the branch and bound method is available. This paper takes advantage of the branch and bound method, and extends it to time-variant reliability analysis.

4.1 Embedded Markov Chain Model

The Markov chain model assumes that the probability of transition from state I to state J is related only to the present state I of the system, not how this state was reached[3]. The structure may collapse due to damage accumulation from load application to load application. The state of the system may change only at the occurrence (application) of loads.

The Markov chain model works very well for single load application. When multiple loads are applied, overlapping between load durations becomes important. Since the load intensities are assumed to be independent from occurrence to occurrence within each load and among different loads, the state of the structure at any load application can be assumed to depend only on that at the previous load application and not the ones before. This is especially true when $\nu\mu_d \ll 1$, because in that case the probability of simultaneous occurrence of two or more loads is small. Even if there is simultaneous occurrence, the probability that one of the loads continues with the same intensity during two consecutive occurrences of another load is very small. Under these conditions, one can approximately model the changes in the state of the structure as a Markov chain from application to application if such dependence is ignored. In the following, the damage levels will be defined as states of the system in a Markov chain.

4.2 Division of Damage Levels of the Structure

A statically indeterminate structure usually goes through many states of damage before it collapses under the load processes. Similarly, the significant failure sequences identified by the failure path approach in time-invariant system reliability analysis also go through many partial failure sequences (each partial failure sequence corresponds to a damage state) before reaching complete failure or collapse of the structure[1, 2, 5]. This similarity facilitates the use of the failure path approach to identify the significant damage states that occur over time in the structure.

The proposed method divides the progressive failure into different damage levels. Each damage level includes many damage states of the system. Division is made as follows:

1. Choose the significant failure sequences by the failure path approach.
2. The progressive failure of the structure is partitioned into n damage levels according to the stages of the selected significant complete failure sequences. In other words, one damage level involves a group of damage states of the system. During the process of damage accumulation, the structure is considered to be in one of those damage levels. Such grouping could be in terms of the number of failed members, accumulated failure probability etc.

The damage levels based on the division of significant failure sequences instead of stable configurations is expected to be more efficient than Wen and Chen's method and may be applied to relatively large brittle structures[9]. However, the proposed method may not be as efficient as Wen and Chen's method for ductile structures because the number of the significant failure sequences is much more than that of the significant failure modes; each significant failure mode usually corresponds to many significant failure sequences in a ductile structure[8]. Moreover, the probability of different damage levels at time t are by-products of the proposed method, which are useful for the purpose of maintenance of the structure.

4.3 Transition Probability

Considering m load processes, the transition probability a_{JI} from level I to level J can be obtained as follows[9]

$$a_{JI} = \sum_{i=1}^{m} \bar{a}^{i}_{JI} + \sum_{i \neq j}^{m} \sum^{m} \bar{a}^{ij}_{JI} + \ldots \tag{4}$$

in which

$$\bar{a}^{i}_{JI} = \frac{\nu_i}{\nu} [\prod_{j \neq i} (1 - \nu_j \mu_{d_j})] a^{i}_{JI} \tag{5}$$

$$\bar{a}^{ij}_{JI} = \frac{\nu_i \nu_j}{\nu} [\prod_{k \neq j \neq i} (1 - \nu_k \mu_{d_k})] \mu_{d_i} a^{ij}_{JI} \tag{6}$$

$\nu = \sum_{i=1}^{m} \nu_i$. a^{i}_{JI}: is the probability of transition from level I to J given that only load $S_i(t)$ is "on"; a^{ij}_{JI} is the probability of transition from level I to J

given that only loads $S_i(t)$ and $S_j(t)$ are "on" and with the corresponding sequence [i.e.,$S_i(t)$ followed by $S_j(t)$].

In the following, "node" means a member in failure sequences. Based on the identified failure sequences, term a^i_{JI}, a^{ij}_{JI} can be calculated as follows:
(i). For the case of $I = J$, which means the probability of remaining at the Ith damage level, a_{II} is expressed as:

$$a_{II} = P[(\cap_{s_1=1}^{N^I} \bar{E}_{s_1}^{I+1}) \cap (\cap_{s_2=1}^{N^{If}} \bar{E}_{s_2}^{If})] \tag{7}$$

where $\bar{E}_{s_1}^{I+1}$ is the safety domain of node s_1 at the $(I+1)$th damage level which has N^I nodes. $\bar{E}_{s_2}^{If}$: is the safety domain of node s_2 which is partitioned from the node at the Ith damage level and whose failure results in the failure of the structure; N^{If} is the number of such nodes and is set to be zero if there are no such nodes.
(ii). For the case of $I < J$, a_{JI} is expressed as:

$$a_{JI} = P[\cup_{m=1}^{N_{path}^{JI}}((\cap_{k_1=1}^{N_{node}^{JI}} E_m^{I+k_1}) \cap (\cap_{k_2=1}^{N_m^{J+1}} \bar{E}_{k_2}^{J+1}) \cap (\cap_{k_3=1}^{N_m^{Jf}} \bar{E}_{k_3}^{Jf}))] \tag{8}$$

where N_{path}^{JI} is the number of the failure paths from the Ith damage level to the Jth damage level. N_{node}^{JI}: is equal to $(J - I)$, which is the number of the nodes in failure path m from the Ith damage level to the Jth damage level and $E_m^{I+k_1}$ is the failure domain of the node in failure path m at the $(I+k_1)$th damage level. N_m^{J+1}: is the number of the nodes branched forward from the node of level J in failure path m. $\bar{E}_{k_2}^{J+1}$: is the safety domain of such nodes. $\bar{E}_{k_3}^{Jf}$: is the safety domain of node k_3 which branches from the node at the Jth damage level and whose failure results in the failure of the entire system; N_m^{Jf} is the number of such nodes.

Eqs. 7 and 8 are explained with assistance of Fig. 4.3.1. The computation of the union of intersections in Eq.8 is not easy and its value can be approximated by the second or third order bounds[7]. For the sake of exploration of the new method, however, such computation is done by simulation in this paper.

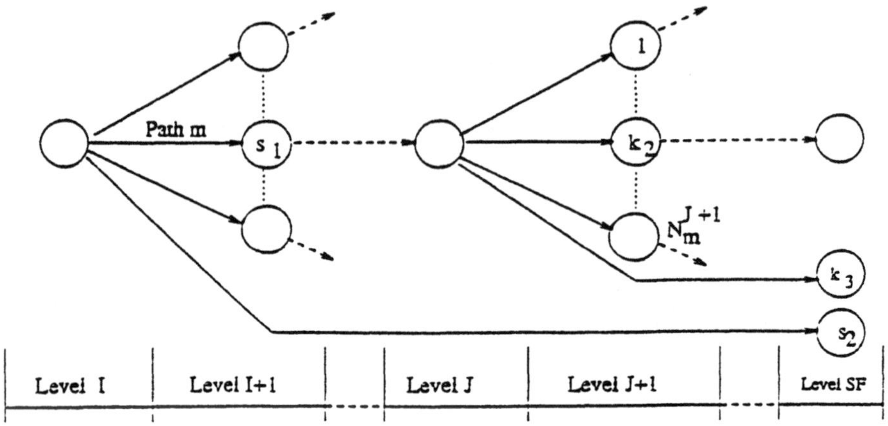

Figure 4.3.1: Explanation for a_{IJ} and a_{JI} of time invariant analysis

A probability transition matrix (ptm) can model the transition from damage level to damage level. The ptm has the following form:

$$[A] = \begin{bmatrix} a_{00} & 0 & 0 & \dots & 0 \\ a_{10} & a_{11} & 0 & \dots & 0 \\ a_{20} & a_{21} & a_{22} & \dots & 0 \\ . & . & . & \dots & . \\ a_{n0} & a_{n1} & a_{n2} & \dots & a_{nn} \end{bmatrix} \tag{9}$$

in which the upper triangle of zeros is due to no repair during cumulative damage. The nth damage level stands for system collapse. Thus the term a_{nn} equals to one.

4.4 Probability of Damage Levels

Since the loading occurrences are modeled by independent Poisson processes, the occurrence time for the changes of the damage levels is also a Poisson process in which the Markov Chain is embedded. The probabilities of different levels of structural damage at time t are therefore (i). For a Poisson square wave process ($\nu_i \mu_{d_i} = 1$).

$$\{P(t)\} = \sum_{k=0}^{\infty} [A]^{k+1} \{P(0)\} \frac{(\nu t)^k}{k!} e^{-\nu t} = [A][Q][e^{-\nu t(1-u_i)}][Q]^{-1}\{P(0)\} \quad (10)$$

(ii). For Poisson pulse processes with $\nu_i \mu_{d_i} < 1$.

$$\{P(t)\} = \sum_{k=0}^{\infty} [A]^{k} \{P(0)\} \frac{(\nu t)^k}{k!} e^{-\nu t} = [Q][e^{-\nu t(1-u_i)}][Q]^{-1}\{P(0)\} \quad (11)$$

where $[Q]$ is the eigenvector of $[A]$; $[e^{-\nu t(1-u_i)}]$ is the diagonal matrix; u_i is the ith eigenvalue of $[A]$; $\{P(0)\}$ is the column vector of the initial probability of the damage levels of the system, for example

$$\{P(0)\} = \{1.0, 0.0, ..., 0.0\}^t. \quad (12)$$

This means that no damage exists at t=0. The first element in $\{P(0)\}$ corresponds to the initial probability of the intact structure.

Therefore, the system failure probability is the nth element in $\{P(t)\}$.

5 Numerical Examples

5.1 Daniels System

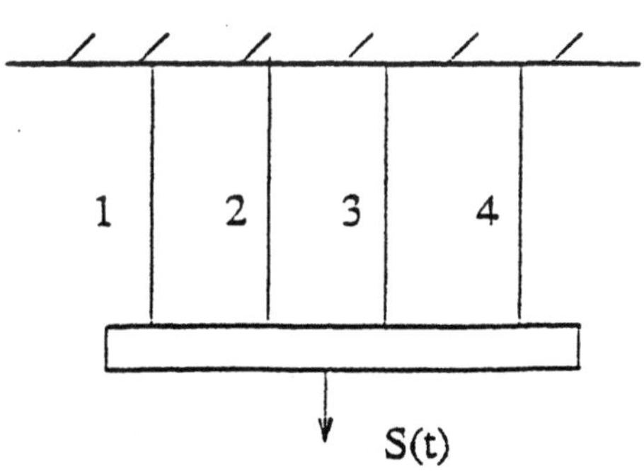

Figure 5.1.1: Four-bar truss

Table 5.1.1: Statistics of variables for a 4- bar daniel system

	R_1 *kip*	R_2 *kip*	R_3 *kip*	R_4 *kip*	S *kip*
Mean	0.5	0.7	2.0	1.0	1.2
C.O.V.	0.2	0.3	0.2	0.1	0.4

The four-bar system shown in Fig. 5.1.1 is now evaluated for time variant system reliability. The statistics of the problem are shown in Table 5.1.1. The loading is assumed to be a Poisson square wave process with occurrence rate ν of 1.0 per year. The significant failure sequences and the grouping of the damage states are shown in Fig. 5.1.2. The values in this figure are the path probabilities. Its transition probability matrix is obtained by Eqs. 7 and 8:

$$[A] = \begin{bmatrix} 0.86619 & 0. & 0. & 0. & 0. \\ 0.08134 & 0.65599 & 0. & 0. & 0. \\ 0.03008 & 0.28528 & 0.93803 & 0. & 0. \\ 0.00751 & 0.02546 & 0.02728 & 0.89979 & 0. \\ 0.01488 & 0.03327 & 0.03429 & 0.10021 & 1. \end{bmatrix} \tag{13}$$

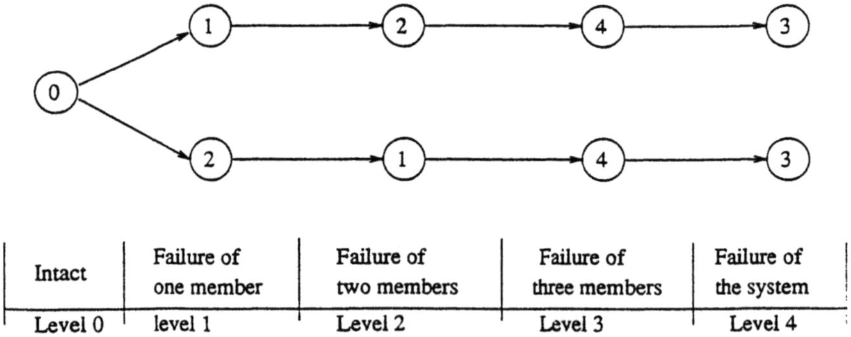

Figure 5.1.2: Division of damage levels for a four-bar parallel system

Grouping of the different damage states into a few damage levels (based on number of failed members) results in system failure probability estimates shown in Fig. 5.1.3. It is observed that the proposed method shows excellent agreement with Monte Carlo simulation estimates (100,000 samples). The probabilities of individual damage states and system collapse are shown in Fig. 5.1.4.

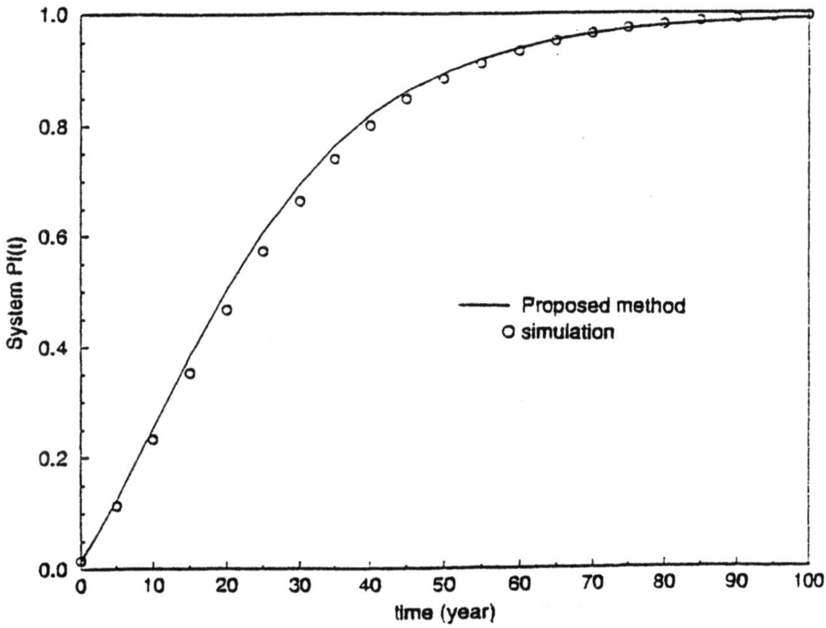

Figure 5.1.3: Four-bar truss: system failure probability

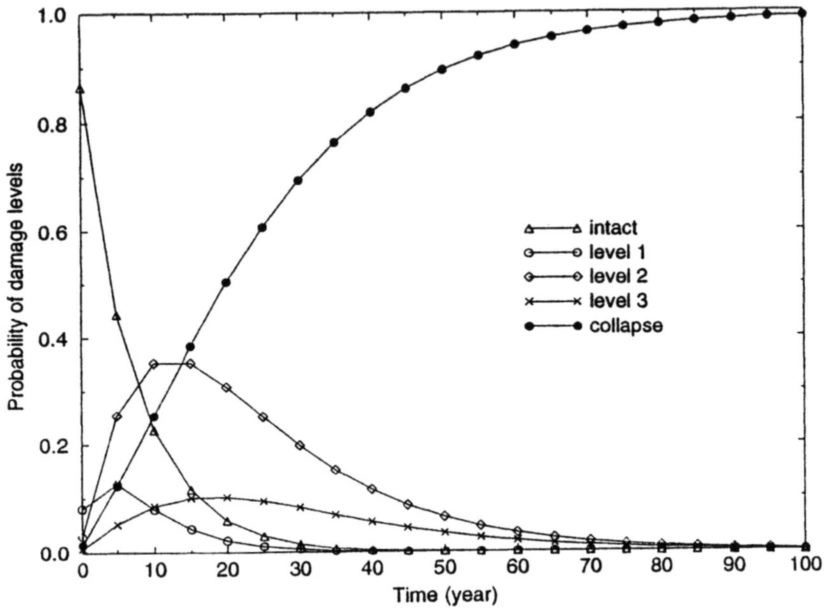

Figure 5.1.4: Four-bar truss: probabilities of damage states

5.2 Ten-bar Truss

A ten-bar truss is shown in Fig. 5.2.1. The statistics of the problem are shown in Table 5.2.1. The significant failure sequences identified by the branch and bound method are shown in Fig. 5.2.2 when two loads are "on". For sequence $4 \rightarrow 6$, the actual complete failure sequence is $4 \rightarrow 6 \rightarrow 9$. Since the conditional failure probability of member 9 given the failure of member 4 and 6 is almost equal to one, member 6 is taken as the last member of that complete failure sequence. Two loading cases are considered:

Horizontal load only

The load is assumed to be a Poisson square wave process with occurrence rate ν of 0.5 per year. The system failure probability is obtained by Eq.10. Again, the results in Fig. 5.2.3 show excellent agreement between the proposed method and Monte Carlo simulation for the case of a single load only.

Both horizontal and vertical loads

This numerical example illustrates the system failure probability under the interaction of two load processes. Both loads are assumed to be Poisson pulse processes. When both the loads are applied, usually two factors are important: (i) load sequence, and (ii) overlapping between the load intensities during simultaneous occurrences. For this particular example, the limit state equations for the various limit states show that the load sequence is not important. Two cases are studied: (i) $\nu_i \mu_{d_i}$ is very small ($\nu_1 = 0.5, \mu_{d_1} = 0.005; \nu_2 = 1.0, \mu_{d_2} = 0.002$) and (ii) $\nu_i \mu_{d_i}$ is very large (μ_{d_1} and μ_{d_2} are increased to 0.8 and 0.2 respectively); this implies frequent simultaneous occurrences of the two loads. The ptm is obtained by Eq. 4 and the system failure probability by Eq. 11. The results in Fig. 5.2.4 show that the proposed method gives very good results for the first case, but conservative results for the second case. This is largely because the overlapping between the load intensities is not included in Eq.11. Therefore further work is necessary to include the effects of load sequence and overlapping in the proposed method.

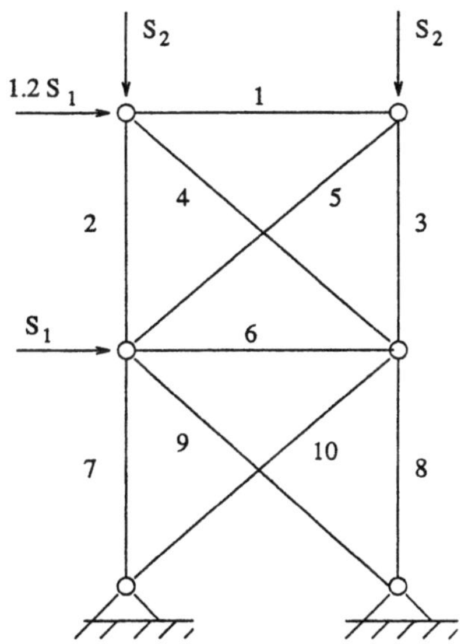

Figure 5.2.1: Ten-bar truss

Table 5.2.1: Statistics of variables for a ten-bar truss

	R_1 *kip*	R_2 *kip*	R_3 *kip*	R_4 *kip*	R_5 *kip*	R_6 *kip*	R_7 *kip*	R_8 *kip*	S_1 *kip*	S_2 *kip*
Mean	1116	864	97.2	360	324	180	432	972	125	20
C.O.V.	0.15	0.15	0.15	0.15	0.15	0.15	0.15	0.15	0.3	0.2
Element	7, 8	6	9	2, 3	1	4	5	10		

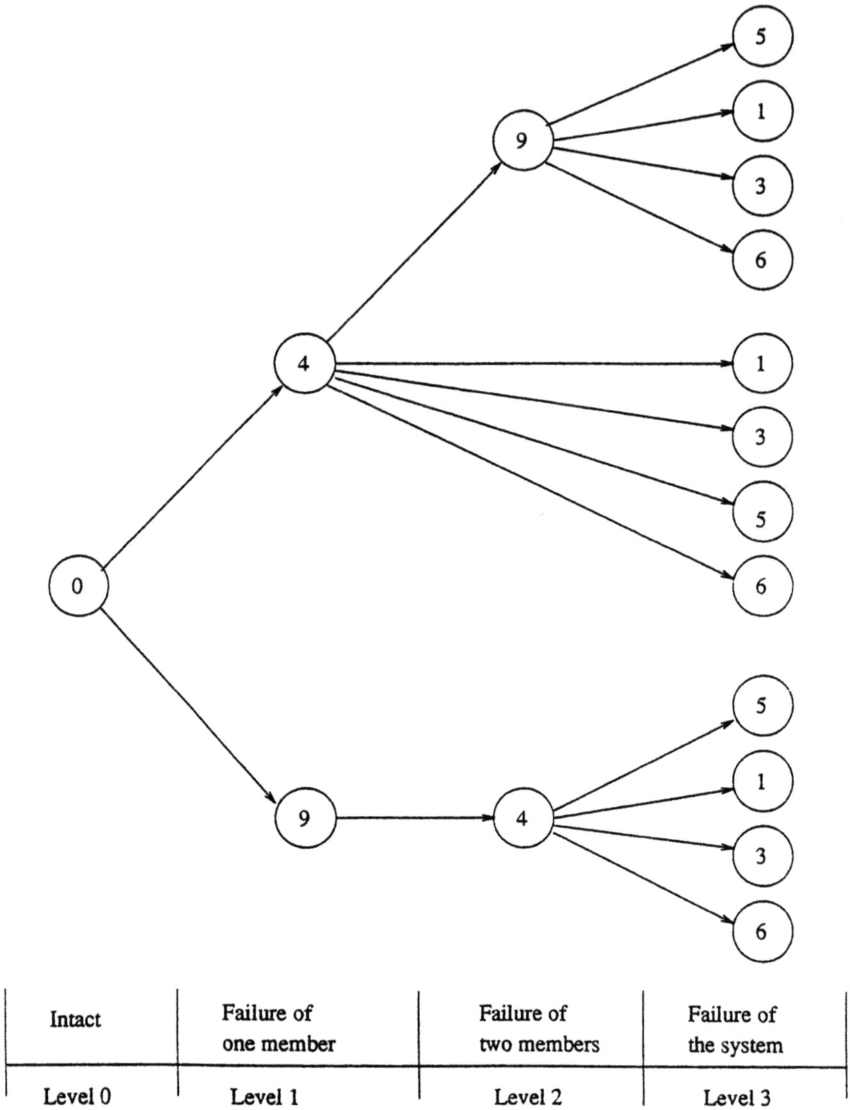

Figure 5.2.2: Ten-bar truss: failure sequences and damage levels

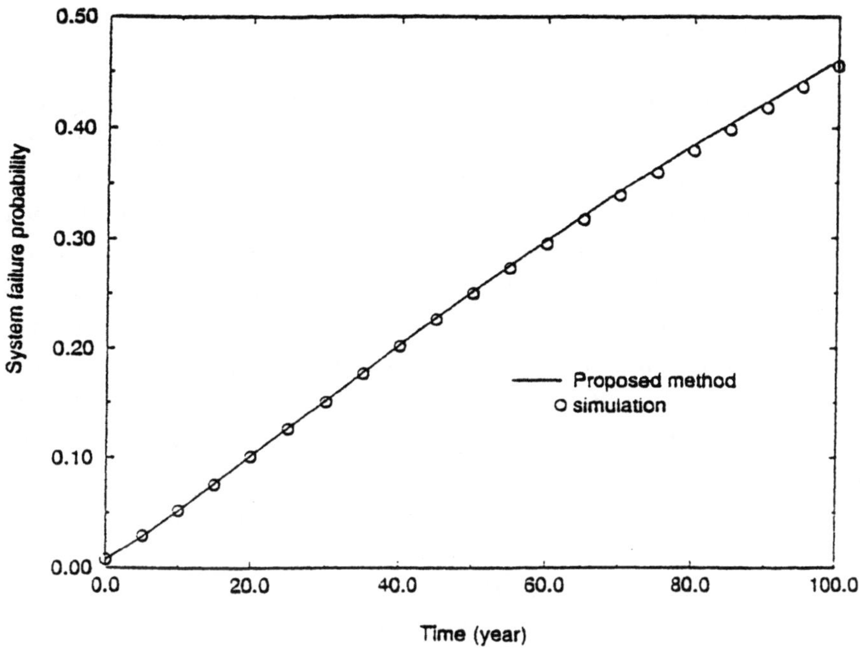

Figure 5.2.3: Time-variant reliability: horizontal load only

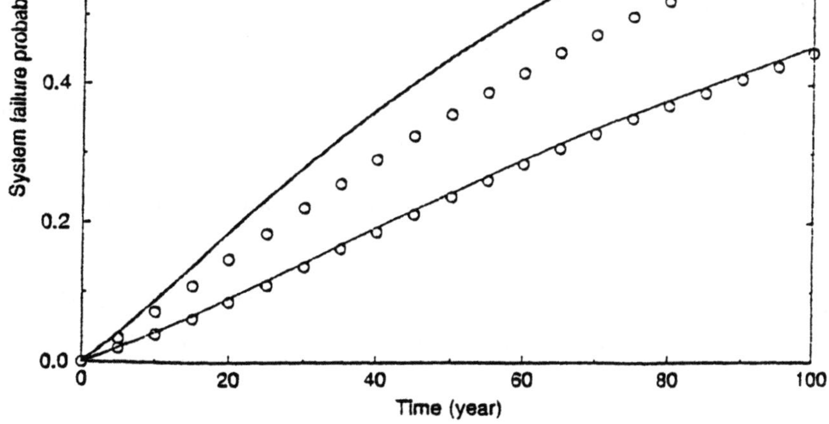

Figure 5.2.4: Time-variant reliability: load sequence not important

6 Conclusion

A computational method to estimate the time-variant system reliability of structures has been proposed in this paper. The method is based on the use of failure sequences and the grouping of damage states into damage levels. The significant failure sequences are identified using the branch and bound method. The damage states along all these sequences are grouped into damage levels. Using the Markov chain assumption, the proposed method is able to compute the probability of different damage levels and system failure at time t. Such partial as well as total failure probability information is useful for the purpose of maintenance of the structure. The method is applicable to relatively large structures due to the systematic failure sequence identification strategy employed. The resultant probability transition matrix is found to be more compact and to represent the accumulation of damage better than existing methods. The method appears to perform well under single load and also under multiple loads when the load sequence or overlapping among their intensities are not important. Further work is required to incorporate these effects in the proposed method such that it may comprehensively address such effects.

References

[1] Moses, F., System reliability developments in structural engineering *Struc. Safety*, **1** (1982) 3-13.

[2] Melchers, R.E. & Tang, L.K., Dominant failure modes in stochastic structural systems *Struc. Safety*, **2** (1984) 127-143.

[3] Bogdanoff, J.L. & Kozin, F., *Probabilistic Models of Cumulative Damage*, John Wiley & Sons, Inc., New York, 1985.

[4] Rackwitz, R., Reliability of system under renewal pulse loading, *J. Struc. Eng.*, ASCE, **110** (1985) 1175-1184.

[5] Thoft-Christensen, P. & Murotsu, Y., *Application of Structural Systems Reliability Theory*, Springer-Verlag Berlin, Heidelberg, 1986.

[6] Wen, Y.K. & Chen, H-C., On fast integration for time variant structural reliability, *Prob. Eng. Mech.*, **2** (1986) 156-839.

[7] Hohenbichler, M., Gollwitzer, S., Kruse, W. & Rackwitz, R., New light on first- and second- order reliability methods, *Struc. Safety*, **4** (1987) 267-284.

[8] Wen, Y.K. & Chen, H-C., System reliability under time varying loads: I, *J. Engrg. Mech.*, ASCE, **115** (1989) 808-823.

[9] Wen, Y.K. & Chen, H-C., System reliability under time varying loads: II, *J. Engrg. Mech.*, ASCE, **115** (1989) 824-839.

Reliability Against Fracture with Uncertain Crack Geometry

Y. Zhang and A. Der Kiureghian

Department of Civil Engineering, University of California,
Berkeley, CA 94720

Abstract. A first-order finite element reliability method is used to predict the probability of brittle fracture in a continuum with uncertain crack geometry. The limit-state function is expressed in terms of the J-integral, and a finite element formulation is developed and implemented to calculate the J-integral and its gradient with respect to variables defining the crack geometry. In the neighborhood of the crack, an active finite element mesh is used as a function of the crack geometry. Plastic deformation at the crack's front is accounted for by an incremental elasto-plastic finite element procedure. The proposed method is demonstrated through an example.

Keywords: crack propagation, finite element method, fracture mechanics, J-integral, mesh, probability theory, reliability theory, uncertainty analysis

1. Introduction

Fracture is a structural failure mechanism caused by unstable propagation of a crack. Due to the random nature of material properties and loads, fracture may occur unexpectedly during the service life of a structure. Therefore, in assessing structural safety, consideration must be given to probabilistic prediction of failure due to fracture.

The unstable propagation of a crack is related to the stress conditions in the vicinity of the crack tip. In fracture mechanics, these conditions are represented by variables such as stress intensity factors or the J-integral. When these variables reach their critical values, fracture is assumed to have occurred. In a given structure, these variables are functions of the loading, material properties, and crack geometry, which are often uncertain and vary unpredictably during the service life of the structure. The critical values are determined through experiments, where specified stress conditions are imposed on specimens with an existing crack. Applied to the crack in a structure where the stress condition is much more complicated, the critical value may also be seen as an uncertain variable.

The advent of finite element reliability methods[1] now makes it possible to investigate the fracture in a structure within a probabilistic framework, fully accounting for all the uncertainties involved. A conventional reliability problem is formulated in terms of a limit-state function separating the outcome space of the random variables representing uncertain quantities into safe and failure domains. The probability of failure is expressed as the integral of the joint probability density over the failure domain. Since the limit-state function is usually irregular, a direct evaluation of the integral is not possible. The Monte Carlo simulation method can be used, but it often requires prohibitively large amounts of computation to accurately predict the small failure probabilities that are involved. To overcome these difficulties, various approximate methods have been proposed. In the First-Order Reliability Method (FORM), an approximation of the probability integral is obtained by replacing the integration boundary by a first-order approximation in a transformed space of standard normal variates. The approximation is carried out at points of the boundary that have minimum distance to the origin in the standard normal space. These points, known as design points, have maximal probability densities among all points in the failure domain and, hence, their neighborhoods make dominant contributions to the probability integral, this being the reason for performing the surface approximations there. By the definition of the design points, a nonlinear programming problem is postulated, which can be solved numerically by employing a constrained optimization algorithm. The finite element method is called for to provide the values of the limit-state function and its gradient, which are required in the solution algorithm. Use of the finite element method has opened new frontiers for application of FORM to structures with increasing complexity in their geometry and behavior.

In this paper, the J-integral is used to characterize the conditions at the crack tip. Brittle fracture is assumed to occur when the J-integral reaches its critical value, thus providing a natural formulation of the limit-state function for reliability analysis. A finite element formulation is developed to calculate the J-integral and its gradient with respect to random variables representing all uncertain quantities. Special attention is given to the uncertainty in the crack geometry. To handle this problem, an adaptive finite element mesh that is a function of variables defining the crack geometry is constructed in the crack zone. Plastic deformation at the crack's front is accounted for by employing an incremental elasto-plastic finite element procedure. A new finite element formulation for the response gradients with respect to geometric variables is developed and incorporated into an existing general-purpose finite element code. The validity and efficiency of the proposed method are demonstrated through an example.

2. Limit-State Function and Reliability Analysis

By convention, the limit-state function $g(\boldsymbol{x})$, where $\boldsymbol{x}$ denotes the set of random variables, is formulated in such a manner that $g(\boldsymbol{x}) \leq 0$ describes the failure event. The failure event here is unstable crack propagation, namely brittle fracture, which is mathematically described by the exceedance of the J-integral above a critical threshold, J_c. Hence, the limit-state function may be written as

$$g = J_c(\boldsymbol{x}) - J(\boldsymbol{x}), \tag{1}$$

where J denotes the value of the J-integral and both J and J_c are shown as implicit functions of the relevant random variables. Since large scale yielding in the front of the crack is expected, an elasto-plastic formulation (without unloading) is used to describe the J-integral in the form

$$J = \int_C (W n_1 - \sigma_{ij} n_j \frac{\partial u_i}{\partial y_1}) ds \tag{2}$$

with the strain energy density

$$W = \int_0^{\varepsilon_{ij}} \sigma_{ij} d\varepsilon_{ij}, \tag{3}$$

where u_i are the displacements, ε_{ij} and σ_{ij} are the strains and stresses, respectively, C is a closed contour around the crack tip, n_i are the components of the contour's normal, and y_i are the rectangular Cartesian coordinates. We assume C is invariant, but all other variables may be functions of the random variables $\boldsymbol{x}$. The derivative of g with respect to a variable x can be derived by taking the derivative of the above equations with respect x, yielding

$$g' = J_c' - J' \tag{4}$$

and

$$J' = \int_C (W' n_1 - \sigma_{ij}' n_j \frac{\partial u_i}{\partial y_1} - \sigma_{ij} n_j \frac{\partial u_i'}{\partial y_1}) ds, \tag{5}$$

where $(.)'$ denotes the partial derivative with respect to x. Provided with g and g', FORM analysis is carried out as follows.

Employing the above limit-state function, the probability of fracture is expressed as

$$P = \int_{g(\boldsymbol{x}) \leq 0} p(\boldsymbol{x}) d\boldsymbol{x}, \tag{6}$$

where $p(\boldsymbol{x})$ is the probability density function of $\boldsymbol{x}$. In FORM, the reliability index, β, is computed as $\beta = \|\boldsymbol{u}^*\|$ if $G(0) > 0$ and $\beta = -\|\boldsymbol{u}^*\|$ if $G(0) \leq 0$, where $\boldsymbol{u}^*$ is the solution to the constrained optimization problem

$$\min \{ \frac{1}{2} \|\boldsymbol{u}\|^2 \mid G(\boldsymbol{u}) = 0 \}, \tag{7}$$

where $\boldsymbol{u}$ is the vector of standard normal variates and $G(\boldsymbol{u})$ is the limit-state function expressed in terms of $\boldsymbol{u}$. The standard normal variates are obtained from a one-to-one mapping $\boldsymbol{u} = \boldsymbol{u}(\boldsymbol{x})$ of the basic variables $\boldsymbol{x}$, which in general could be non-normal and dependent.[2] The function $G(\boldsymbol{u})$ represents the transformation of the limit-state function $g(\boldsymbol{x})$ from the original space, i.e., $G(\boldsymbol{u}) = g(\boldsymbol{x}(\boldsymbol{u}))$. The first-order estimate of the probability of failure is $P_1 = \Phi(-\beta)$, where $\Phi(.)$ denotes the standard normal cumulative probability. In the case of multiple design points, well known bounds on P are obtained in terms of the β values and unit normals at each design point.[3]

A number of optimization algorithms are available for solving Eq. 7.[4] The most efficient of these algorithms employ the gradient of the limit-state function. Hence, for the limit-state function in Eq. 1, the success of the FORM analysis lies in accurate and efficient computation of g and g'.

3. Finite Element Method for Computation of J and J'

In the finite element method for analysis of the fracture mechanics problem, the crack is often approximated by a traction-free boundary, and the J-integral is calculated along a contour C defined by a set of finite element nodes, as exemplified in Fig. 1. The nodal displacements, strains and stresses are solved, and the values between any two nodes are approximated by a linear interpolation. The J-integral is then numerically evaluated along the contour C. Since the J-integral is path independent, the contour C should not be taken too close to the crack tip where the finite element solution may not be sufficiently accurate. Due to high stress concentration in the vicinity of the crack tip, plastic deformation is in presence. This is accounted for by utilizing elasto-plastic elements in this vicinity and by employing an appropriate incremental solution procedure.

Employing a conventional finite element matrix notation, let $\boldsymbol{u}$ denote the vector of nodal displacements, and $\boldsymbol{\varepsilon}$ and $\boldsymbol{\sigma}$ denote the element stresses and strains, respectively. These quantities are related through the compatibility and constitutive equations

$$\boldsymbol{\varepsilon} = \boldsymbol{B}\boldsymbol{u} \tag{8}$$

and

$$\boldsymbol{\sigma} = \boldsymbol{D}(\boldsymbol{\varepsilon} - \boldsymbol{\varepsilon}^p), \tag{9}$$

respectively, where $\boldsymbol{B}$ is the strain operator, $\boldsymbol{D}$ is the constitutive matrix of elastic properties, and $\boldsymbol{\varepsilon}^p$ is the vector of plastic strains. The solution is obtained by solving the equilibrium equation

$$\boldsymbol{R} - \boldsymbol{P} = 0, \tag{10}$$

where $\boldsymbol{P}$ is the vector of external forces, and

$$\boldsymbol{R} = \sum_e \boldsymbol{R}^e \tag{11}$$

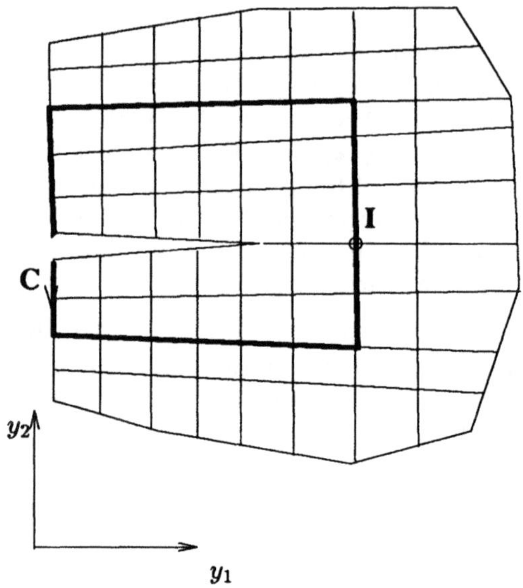

Fig. 1. A crack and the contour C in a finite element mesh

is the global vector of internal resisting forces, where $\boldsymbol{R}^e$ are elemental resisting forces and $\sum_e$ denotes the assembly over all the elements. $\boldsymbol{R}^e$ are further expressed as

$$\boldsymbol{R}^e = \int_{\Omega^e} \boldsymbol{B}\boldsymbol{\sigma} d\Omega = \int_{\square} \boldsymbol{B}\boldsymbol{\sigma} j d\square, \tag{12}$$

where Ω^e is the element domain, $\square$ is the domain in natural coordinates into which the first integral is transformed, and j is the determinant of the Jacobian of the coordinate transformation. Solving Eq. 10 for $\boldsymbol{u}$, $\boldsymbol{\varepsilon}$ and $\boldsymbol{\sigma}$ are obtained from Eqs. 8 and 9, which are then projected to the nodes on the contour C by a nodal projection approach. The nodal strains and stresses, say at node I, are expressed as

$$\boldsymbol{\varepsilon}_I = \sum_e \int_{\square} N_I^e \boldsymbol{\varepsilon} j d\square \tag{13}$$

and

$$\boldsymbol{\sigma}_I = \sum_e \int_{\square} N_I^e \boldsymbol{\sigma} j d\square, \tag{14}$$

where N_I^e is the shape function of element e associated with node I. Once the nodal solutions are available, the J-integral is obtained by performing approximate line integration.

To calculate the derivative of J with respect to a random variable x, the derivatives $\boldsymbol{u}'$, $\boldsymbol{\varepsilon}'$, $\boldsymbol{\varepsilon}^{p\prime}$, and $\boldsymbol{\sigma}'$ need to be solved first. Taking the derivative of Eqs. 8 and 9, the following formulas are obtained

$$\boldsymbol{\varepsilon}' = \boldsymbol{B}'\boldsymbol{u} + \boldsymbol{B}\boldsymbol{u}' \tag{15}$$

and

$$\boldsymbol{\sigma}' = \boldsymbol{D}'(\boldsymbol{\varepsilon} - \boldsymbol{\varepsilon}^p) + \boldsymbol{D}(\boldsymbol{\varepsilon}' - \boldsymbol{\varepsilon}^{p\prime}). \tag{16}$$

Detailed formulas for computing $\boldsymbol{\varepsilon}^{p\prime}$ with a direct differentiation approach have been developed by Zhang and Der Kiureghian,[5] which will not be repeated here. Instead, we will concentrate on deriving formulas for computing the derivatives of stresses and strains, accounting for possible variation in the boundary of the element.

Taking derivative of Eq. 10, we obtain

$$\boldsymbol{K}\boldsymbol{u}' = \boldsymbol{P}' - \left.\frac{\partial \boldsymbol{R}}{\partial x}\right|_{\boldsymbol{u}}, \tag{17}$$

where $\boldsymbol{K}$ is the tangent stiffness matrix and $\left.\frac{\partial \boldsymbol{R}}{\partial x}\right|_{\boldsymbol{u}}$ is the partial derivative with respect to x with $\boldsymbol{u}$ fixed. In a paper by Liu and Der Kiureghian dealing with a nonlinear elastic problem,[6] for x denoting a geometric parameter (e.g., the coordinate of a node), the partial derivative $\left.\frac{\partial \boldsymbol{R}}{\partial x}\right|_{\boldsymbol{u}}$ was calculated using a finite difference scheme. Here, since elasto-plastic deformation is in presence and a continuous derivative is not available at all points, an analytical solution should be employed. Proceeding with direct differentiation, we have

$$\left.\frac{\partial \boldsymbol{R}}{\partial x}\right|_{\boldsymbol{u}} = \sum_e \left.\frac{\partial \boldsymbol{R}^e}{\partial x}\right|_{\boldsymbol{u}} \tag{18}$$

and

$$\left.\frac{\partial \boldsymbol{R}^e}{\partial x}\right|_{\boldsymbol{u}} = \int_{\square}\left(\left.\frac{\partial \boldsymbol{B}}{\partial x}\right|_{\boldsymbol{u}} \boldsymbol{\sigma} j + \boldsymbol{B}\left.\frac{\partial \boldsymbol{\sigma}}{\partial x}\right|_{\boldsymbol{u}} j + \boldsymbol{B}\boldsymbol{\sigma}\left.\frac{\partial j}{\partial x}\right|_{\boldsymbol{u}}\right)d\square. \tag{19}$$

Note that the above derivatives are all computed with respect to the natural coordinate system, which remains fixed. Eq. 17 can now be solved for $\boldsymbol{u}'$, after which $\boldsymbol{\varepsilon}'$, $\boldsymbol{\varepsilon}^{p\prime}$ and $\boldsymbol{\sigma}'$ can be solved at all nodes. For the projected nodal strains and stresses, the formulas take the form

$$\boldsymbol{\varepsilon}_I' = \sum_e \int_{\square}(N_I^e \boldsymbol{\varepsilon}' j + N_I^e \boldsymbol{\varepsilon} j')d\square \tag{20}$$

and

$$\boldsymbol{\sigma}_I' = \sum_e \int_{\square}(N_I^e \boldsymbol{\sigma}' j + N_I^e \boldsymbol{\sigma} j')d\square. \tag{21}$$

All the above derivatives are computed by directly differentiating the relevant matrices of each element. With the derivatives of nodal stresses and strains computed, J' in Eq. 5 is computed by line integration along C.

The above formulation is implemented in the general-purpose finite element code FEAP by Taylor,[7] and all the subroutines for a quadrilateral four node element are developed. This code, in conjunction with the general-purpose reliability code CALREL,[8] is used to solve the example problem described in the following section.

4. Reliability of a Cracked Plate

Consider a cracked plate subjected to a uniformly distributed edge load as shown in Fig. 2. The material properties are assumed to be deterministic: the Young's modulus is $E = 20,000kN/cm^2$, the Poisson's ratio is $\nu = 0.3$, the yield stress is $\sigma_y = 25kN/cm^2$, the linear kinematic hardening parameter is $H_{kin} = 600kN/cm^2$, and the critical value of the J-integral is $J_c = 0.1 \times 10^{-3}kN/cm$. The intensity of the edge load is also deterministic: $p = 25kN/cm$. The plate is assumed to be in a state of plane strain.

A crack in a structure is formed through a complex mechanical process involving many uncertain factors, such as random inhomogeneity of the medium and random loading. As a result, the geometry of the crack is often uncertain. For the present illustrative example, a probabilistic model of the crack is assumed. We assume the crack is composed of 10 linear segments, and each segment is geometrically defined by its length a_i and its angle θ_i with respect to the preceding segment, as shown in Fig. 2. The stochastic model of the crack is fully defined by assigning a joint distribution to the random variables $a_i, \theta_i, i = 1, ..., 10$. The relative angles θ_i are assumed to be statistically independent and identically distributed, each having a normal distribution with mean $0°$ and standard deviation $4°$. The segment lengths a_i are assumed to be jointly lognormally distributed with identical means $0.02cm$, identical standard deviations $0.0022cm$, and correlation matrix

$$\rho = \begin{pmatrix} 1.0 & 0.4 & 0.0 & 0.0 & \cdots & 0.0 & 0.0 \\ 0.4 & 1.0 & 0.4 & 0.0 & \cdots & 0.0 & 0.0 \\ \vdots & \vdots & \vdots & \vdots & \ddots & \vdots & \vdots \\ 0.0 & 0.0 & 0.0 & 0.0 & \cdots & 0.4 & 1.0 \end{pmatrix}. \tag{22}$$

As can be seen, the model assumes that only the lengths of neighboring segments are correlated.

Obviously the adopted model for the crack geometry is hypothetical and does not relate to the physical processes that have caused the crack in the first place. However, this model is adopted here for the purpose of illustration, since the main objective of the paper is to demonstrate the applicability of the

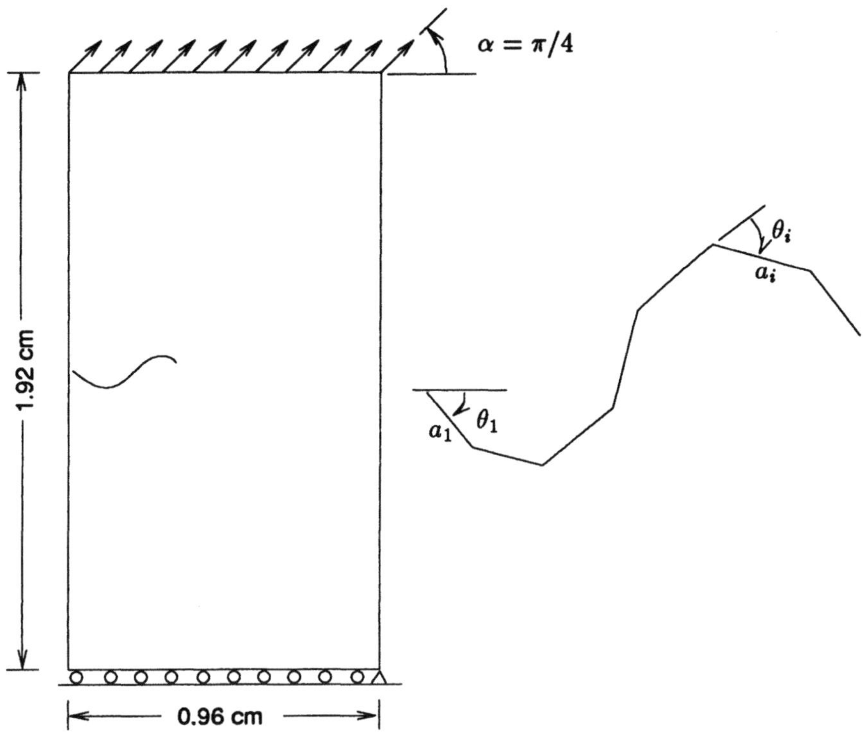

Fig. 2. Idealized crack geometry within the example plate

finite-element FORM method to fracture mechanics problems with random crack geometry. Presently, attempts are being made to describe the random crack geometry through the use of continuum damage models in conjunction with stochastic finite element analysis.

The finite element mesh must account for the uncertain nature of the crack geometry. We assume the nodal points enclosed by the contour C, denoted $\boldsymbol{y}^c$, are random variables, which are dependent on quantities a_i and θ_i, i.e., $\boldsymbol{y}^a = \boldsymbol{y}^a(\boldsymbol{a}, \boldsymbol{\theta})$. A simple linear interpolation between the fixed nodes on C and the nodes defining the crack is used to construct this relationship. This formulation assures that the nodal coordinates are continuous functions of the random variables. It also maintains proper shapes for the elements for all likely values of the variables a_i and θ_i.

The finite element mesh for the entire plate at the mean values of the random variables (mean geometry of the crack) is shown in Fig. 3. The detail of the mesh near the crack and contour C at mean values is shown in Fig. 4(a). During the course of finding the design point, various outcomes of

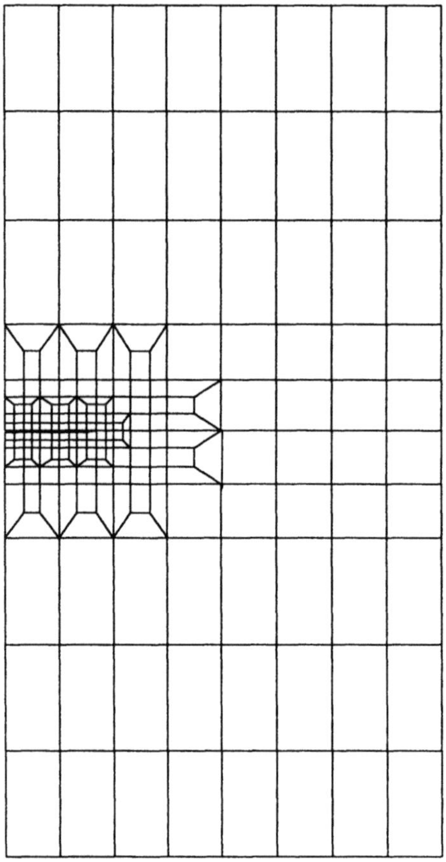

Fig. 3. Finite element mesh

the random variables a_i and θ_i are considered, causing systematic changes in the coordinates of the nodes enclosed by C. In Fig. 4(b), the active mesh at a selected point during the course of the solution is illustrated. It can be seen that the elements maintain proper shapes.

Rewriting the J-integral as $J = J(\boldsymbol{y}^a(\boldsymbol{a}, \boldsymbol{\theta}))$, the derivatives of J are expressed as

$$\frac{\partial J}{\partial a_i} = \sum_k \frac{\partial J}{\partial y_k^a} \frac{\partial y_k^a}{\partial a_i} \quad and \quad \frac{\partial J}{\partial \theta_i} = \sum_k \frac{\partial J}{\partial y_k^a} \frac{\partial y_k^a}{\partial \theta_i}. \tag{23}$$

The terms $\frac{\partial J}{\partial y_k^a}$ are responsible for most of the required computational effort. Hence, the active mesh should be restricted to as small a region as is possible.

For FORM analysis, the modified HLRF algorithm as described in [3] is utilized. However, at each iteration a line search is performed with small

Table 1 Results of FORM analysis

Var.	Design Point		Sensitivity Vectors	
	$\mathbf{x}^*$	$\mathbf{u}^*$	δ	η
θ_1	-2.3005	-0.57514	0.2459	-0.1414
θ_2	-1.9651	-0.49127	0.2092	-0.1028
θ_3	-1.6716	-0.41791	0.1772	-0.0741
θ_4	-1.4071	-0.35178	0.1486	-0.0523
θ_5	-1.1736	-0.29339	0.1236	-0.0363
θ_6	-9.4991	-0.23748	0.0998	-0.0237
θ_7	-7.4482	-0.18621	0.0782	-0.0146
θ_8	-5.3420	-0.13355	0.0560	-0.0075
θ_9	-3.3556	-0.08389	0.0352	-0.0030
θ_{10}	-1.3438	-0.03359	0.0141	-0.0005
a_1	0.0215	0.71391	-0.2078	-0.1332
a_2	0.0220	0.70674	-0.2110	-0.1894
a_3	0.0220	0.69502	-0.2074	-0.1863
a_4	0.0220	0.69122	-0.2083	-0.1857
a_5	0.0220	0.68256	-0.2046	-0.1803
a_6	0.0220	0.67577	-0.2051	-0.1783
a_7	0.0219	0.66107	-0.1989	-0.1691
a_8	0.0219	0.64787	-0.1994	-0.1652
a_9	0.0217	0.56335	-0.1893	-0.1388
a_{10}	0.0210	0.29773	-0.1409	-0.0598

steps to make sure that the finite element mesh remains feasible. The solution to the design point converges in 19 iterations. The reliability index is $\beta = 2.29$ and the first-order probability of fracture is $P_f = 1.1\%$. To verify this result, Monte Carlo simulation is carried out resulting in the reliability index estimate $\beta = 2.40$ and fracture probability $P_f = 0.82\%$, both with the coefficient of variation 0.06. These results indicate that the FORM solution is reasonably accurate. Other interesting results from the FORM analysis are listed in Tab. 1. The sensitivity measures $\delta_i = \sigma_i \frac{\partial\beta}{\partial\mu_i}$ and $\eta_i = \sigma_i \frac{\partial\beta}{\partial\sigma_i}$ are the scaled sensitivities of β with respect to the mean μ_i and standard σ_i of each random variable, respectively. It is clear that the length and relative angle of the initial segments of the crack have stronger influences on the reliability index than those near the tip of the crack. In Figs. 5a and 5b, the geometry of the crack and the zone of plastic deformation are shown for values of the random variables at the mean point and at the design point, respectively.

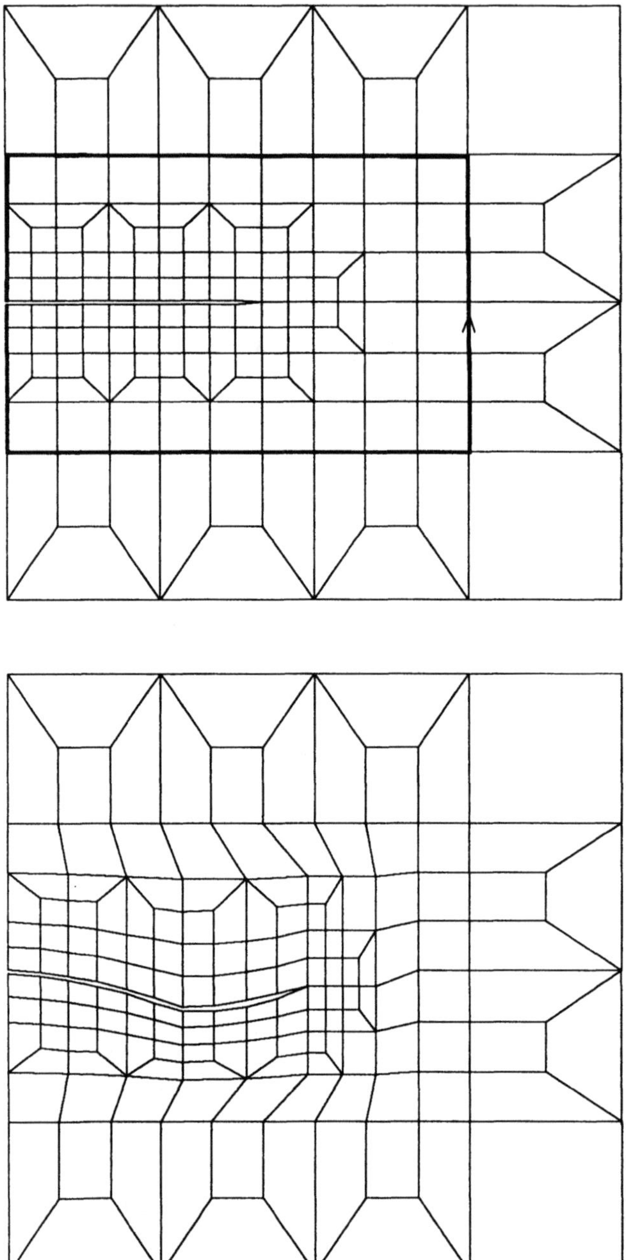

(a)

(b)

Figure 4. Finite element mesh near the crack: (a) straight crack and contour C, (b) arbitrary crack geometry

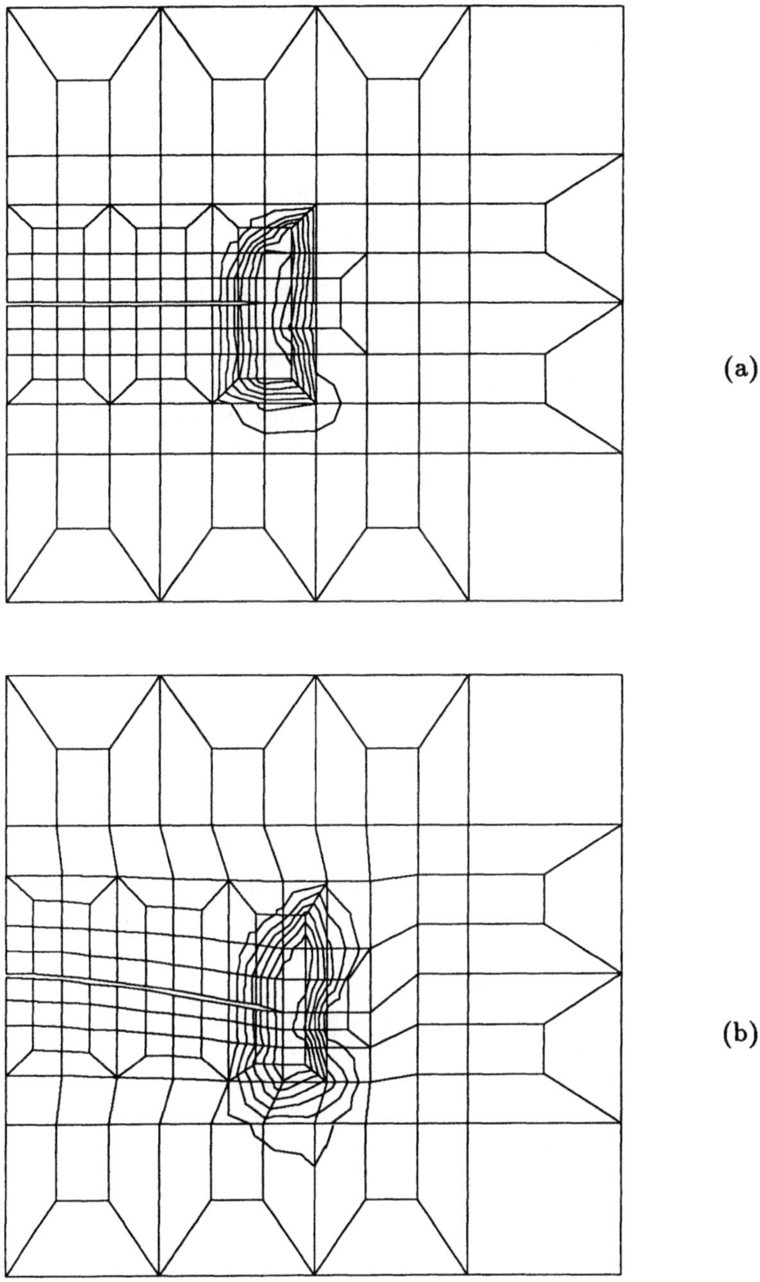

Figure 5. Mesh and plastic zone at (a) mean point, (b) design point

At the design point, the crack geometry must be regarded as the most likely geometry that could lead to fracture of the plate.

5. Conclusion

A finite element reliability method is investigated for a fracture mechanics problem with uncertain crack geometry. An active mesh is introduced to model the variable crack geometry such that the J-integral becomes a continuous function of variables defining the crack geometry. Finite element formulations for the J-integral and associated gradients are fully developed and implemented in the finite element code FEAP. The probabilistic analysis is carried out by FORM, using the general-purpose reliability code CALREL. The validity of the FORM result is checked by comparison with Monte Carlo simulation results. In addition to the probability of fracture, FORM provides the sensitivities of the reliability index with respect to variables defining the crack geometry, and the most likely configuration of the crack. It is concluded that FORM in conjunction with the finite element method provides an effective tool for probabilistic fracture mechanics analysis, accounting for the uncertain nature of crack geometry.

References

1. Der Kiureghian, A. and Ke, B.-J., The stochastic finite element method in structural reliability, *Probabilistic Engineering Mechanics*, 3(2) (1988) 83–91.
2. Hohenbichler, M., and Rackwitz, R., Non-normal dependent vectors in structural safety, *J. Eng. Mech.*, ASCE, 107(6) (1981) 1227–1238.
3. Ditlevsen, O., Narrow Reliability Bounds for Structural Systems, *J. Struct. Mech.*, 7(4) (1979) 453–472.
4. Liu, P-L. and Der Kiureghian, A., Optimization algorithms for structural reliability, *Structural Safety*, 9(3) (1991) 161–177.
5. Zhang, Y. and Der Kiureghian, A., Dynamic response sensitivity of inelastic structures, *Comp. Meth. Appl. Mech. Eng.*, 108(1) (1993) 23–36.
6. Liu, P-L. and Der Kiureghian, A., Finite element reliability of geometrically nonlinear uncertain structures, *J. Eng. Mech.*, ASCE, 117(8) (1991) 1806–1825.
7. Zienkiewicz, O.C. and Taylor, R.L., *The Finite Element Method*, Vol. 1, McGraw-Hill, London, 1989.
8. Liu, P.-L., Lin, H.-Z. and Der Kiureghian, A., CALREL user manual, *Report No. UCB/SEMM-89/18*, Structural Engineering, Mechanics and Materials, Department of Civil Engineering, University of California, Berkeley, CA, April, 1989.

Dynamic Snap-Through and Snap-Buckling of Shear-Deformable Panels in a Random Environment

F. Ziegler [1], R. Heuer [1] and H. Irschik [2]

[1] Civil Engineering Department, Technical University of Vienna, Wiedner Hauptstraße 8-10/E201, Austria, A-1040

[2] Institute of Technical Mechanics and Foundations of Mechanical Engineering, Johannes-Kepler-University Linz, Linz-Auhof, Austria, A-4040

Abstract. Random vibrations are considered for the cases of shear-deformable plates with initial curvature and of elastically buckled plates. The nontrivial generalization of the flat plate vibrations is expressed by the fact that "small amplitude" vibrations exist about the curved equilibrium position together with the snap-through and snap-buckling type large amplitude vibrations about the flat position. The problem of snap-buckling in the large deflection random response, i.e., the loss of stability accompanied in simple and regular structures by the change from a symmetric to an antimetric configuration, is studied in detail. The geometrically nonlinear panel vibrations are treated by applying Berger´s approximation to the generalized von Karman-type plate equations considering hard hinged supports of the straight boundary segments of skew or even more generally shaped polygonal plates. Shear deformation is considered by means of Mindlin´s kinematic hypothesis. A distributed lateral force loading is applied, and additionally, the influence of thermal prestress is taken into account as well as a randomly fluctuating temperature. A multi-mode approach and the Galerkin procedure are applied to the boundary value problem. The result of the projection and of a transformation is a set of nonlinearly coupled ordinary differential equations (ODEs) driven effectively by random forces and with potential restoring forces. For reasons of convergence, a light viscous modal damping is added at this stage. By means of a nondimensional formulation and introducing the eigen-time of the basic mode of the associated linearized problem renders a unifying result with respect to the planform of the panel. With the simplifying approximation of the action of the random environment by effective forces modeled by uncorrelated, zero-mean wide-band noise processes, and by considering the set of modal equations to be finite, the Fokker-Planck-Kolmogorov (F.P.K.–) equation for the transition probability density of the generalized coordinates and velocities is derived. The probability of first dynamic snap-through is derived for a single mode approximation with the influence of higher modes taken into account. Using the two-mode expansion, the probability distribution of the asymmetric snap-buckling is also evaluated. All probabilistic results obtained for the complex panel structures to be considered hold independently (in the sense of similitude) of their special planform. They are cast in the form of graphs with a structural parameter varying in a wide range.

Keywords. Nonlinear vibrations of plates and shallow shells, snap-through, snap-buckling, Fokker-Planck-equation

1 Introduction

Crandall and Zhu[1] sketch the historical development of the application of stochastic processes in dynamics and provide references to back the proper approximation of the response by a Markov process if the excitation is approximated by (physical) white noise. Drift and diffusion coefficients of the Fokker-Planck-Kolmogorov (F.P.K.-) equation are derived by limit operations that are directly applied to the equations of motion of the dynamic system. The latter equations are possibly extended to include linear filters towards the excitation (the differential equations hold only for sufficiently smooth response due to broad-band excitation). For nonlinear vibrations see e.g., Crandall[2], Ariaratnam[3], Caughey[4,5], Bolotin[6] and Lin[7]. Equivalent linearization, Spanos[8], and the closure method of the moment equations are the most broadly applicable approximate analytical procedures. Otherwise numerical simulation and the Monte Carlo method remain generally applicable at the price of high computing costs. See also[9]. Up to now, the F.P.K.–equation was used to determine the transition probability density function of the response of rather simple structures like strings[3], beams[10], rectangular plates[11] and shallow arches, Ariaratnam et al.[12–14] and[15]. The Authors[16–18] recently extended that classical approach to more complex plane and shallow structures with polygonal boundaries. Their strategy is applied in this paper to shallow shells and elastic plates in the post-buckling configuration when forced by random pressure loads, taking into account the shear deformations. That means composite structures are also included in the analyses. Nonlinear vibrations of shallow arches and buckled beams in the elastic range have been considered by Seide[19], Penketkorn et al.[20], Locke and Mei[21] and Abou-Rayan and Nayfeh[22] by means of either simulation techniques or with the closure scheme and also by discretization with FEM.

2 Multimodal approach for nonlinear vibrations of shallow shells

Geometrical nonlinearity is considered by means of the kinematic assumptions for the midsurface strains of the shallow shell according to the von Karman and Tsien[23]. Taking into account the effect of shear, the distribution of strain through the thickness of the shell is assumed according to the Reissner-Mindlin first order shear deformation theory, see[24,25]. Using these kinematic assumptions when setting up the strain energy the contributions due to the in-plane, bending, and shear deformation become, respectively,

$$U_m = \frac{1}{2} D \int_A I_e^2 \, dA \quad , \tag{2.1}$$

$$U_b = \frac{1}{2} K \int_A \left\{ \left[\psi_{x,x}^2 + \psi_{y,y}^2 + \frac{1}{2} (\psi_{x,y} + \psi_{y,x})^2 \right] \right.$$

$$\left. + \nu \left[2\psi_{x,x} \psi_{y,y} - \frac{1}{2} (\psi_{x,y} + \psi_{y,x})^2 \right] \right\} dA \ , \tag{2.2}$$

$$U_s = \frac{1}{2s} \int_A \left[(w_{,x} + \psi_x)^2 + (w_{,y} + \psi_y)^2 \right] dA \ , \tag{2.3}$$

where w denotes the deflection, and ψ_x, ψ_y are the cross-sectional rotations. I_e stands for the first invariant of the midsurface strain tensor. The contribution of the second invariant is neglected in Eq.(2.1) following Berger[26]: Berger´s assumption is a reasonable well-behaving approximation for structures with immovable in-plane boundary conditions. It renders exact results in the case of a shallow shell strip and has been confirmed for circular plates by Schmidt[27] using a perturbation technique and the von Karman equations. Immovable in-plane boundary conditions are considered throughout the paper. In Eqs. (2.1)–(2.3), A denotes the shell area projected onto the (x, y) – plane. D, K, 1/s denote the effective stiffness in the case of symmetrically laminated shells composed of transversely isotropic layers, as given by the equations

$$(D, K) = \sum_{k=1}^{N} \int_{z_{k-1}}^{z_k} \frac{E_k}{(1-\nu_k^2)} (1, z^2)\, dz \ , \quad \nu = \frac{1}{K} \sum_{k=1}^{N} K_k \nu_k \ , \tag{2.4}$$

$$\frac{1}{s} = \kappa^2 \sum_{k=1}^{N} \int_{z_{k-1}}^{z_k} G_{ck}\, dz \ , \tag{2.5}$$

and ν is an effective Poisson´s ratio. The shear factor κ^2 is commonly set to 1.2.

The portion of the strain energy due to thermal heating is given by

$$U_\theta = -(1+\nu) \left[K \int_A \kappa_\theta (\psi_{x,x} + \psi_{y,y})\, dA + D \int_A (n_\theta^* I_e - C_\theta^*)\, dA \right], \tag{2.6}$$

where n_θ^* and κ_θ denote effective mean temperature and effective thermal curvature, respectively:

$$\left(n_\theta^*, C_\theta^*\right) = \frac{1}{D(1+\nu)} \sum_{k=1}^{N} \int_{z_{k-1}}^{z_k} \frac{E_k}{(1-\nu_k)} \left(\alpha_k \theta, \alpha_k^2 \theta^2\right) dz \ , \tag{2.7}$$

$$\kappa_\theta = \frac{1}{K(1+\nu)} \sum_{k=1}^{N} \int_{z_{k-1}}^{z_k} \frac{E_k}{(1-\nu_k)} \alpha_k \, \theta \, z \, dz \ . \tag{2.8}$$

The imposed temperature distribution $\theta(x, y, z; t)$ corresponds to a proper solution of the equation of heat conduction assuming the deformation rate dependent source term to be negligible in a thermally decoupled dynamic theory.

Considering the lateral pressure loading p perpendicular to the shallow shell surface and neglecting rotatory as well as in-plane inertia, the field equations of this shear-deformable Berger-type shallow shell theory have been derived by Heuer[28] using Hamilton's principle. For thermally stressed and buckled plates see[16, 17, 29]. As a result, the deflection turns out to be governed by the single fourth-order differential equation, μ is the averaged mass per unit of area

$$K\,(1+s\,n)\,\Delta\Delta w - n\left[\Delta w - 2(H - Ks\,\Delta H)\right] - Ks\,\mu\,\Delta\ddot{w} + \mu\ddot{w} = p - Ks\,\Delta p - (1+\nu)\,K\,\Delta\kappa_\theta \ , \tag{2.9}$$

where n characterizes a time variant isotropic in-plane force that is constant throughout the plate domain and is related to the deflection by the averaging integral

$$n = -D\left[\frac{1}{2A}\int_A w\,(\Delta w - 4H)\,dA + \bar{n}_\theta^*\right] , \tag{2.10}$$

$$\bar{n}_\theta^* = \frac{(1+\nu)}{A}\int_A n_\theta^*\,dA \ . \tag{2.11}$$

H is the initial Gaussian curvature of the shallow shell. Plates and shells with straight edges and with boundary conditions of the hard hinged type are considered in refs.[16-18]. In a multimodal approach, the deflection, the lateral force loading, and the thermal curvature of the shallow shell are expanded into the orthogonal set of eigenfunctions w_j^* of the corresponding linearized flat plate problem where $n = 0$. Thus,

$$w(\mathbf{x},t) = \sum_{j=1}^{N} c_j\, q_j^*(t)\, w_j^*(\mathbf{x}) \ , \tag{2.12}$$

$$p(\mathbf{x},t) = \sum_{j=1}^{N} f_{pj}^*(t)\, \bar{p}_j(\mathbf{x}) = \sum_{j=1}^{N} f_{pj}^*(t)\, c_{pj}\, w_j^*(\mathbf{x}) \ , \tag{2.13}$$

$$\kappa_\theta(\mathbf{x},t) = \sum_{j=1}^{N} f_{\theta j}^*(t)\, \bar{\kappa}_{\theta j}(\mathbf{x}) = \sum_{j=1}^{N} f_{\theta j}^*(t)\, c_{\theta j}\, w_j^*(\mathbf{x}) \tag{2.14}$$

The coefficients c carry the appropriate dimensions and may be chosen freely. The superscript (*) stands for nondimensional quantities. It has been shown by Irschik[30] that the eigenfunctions of those shear deformable simply supported plates with polygonal planform are governed by a set of second-order Helmholtz-differential equations with homogeneous Dirichlet's boundary conditions,

$$\Delta w_j^* + \alpha_j w_j^* = 0\,,\;\; j = 1, 2 \ldots N\,, \tag{2.15}$$

$$\Gamma:\;\; w_j^* = 0\;. \tag{2.16}$$

For homogeneous plates that are rigid in shear, the eigenvalues are directly proportional to the linear natural frequencies at reference temperature by the common factor $\sqrt{\rho h/K}$, otherwise see Eq. (2.18) for j = 1. Using Eq. (2.12) as a Ritz-approximation for the solution of Eq. (2.9) and running through the Galerkin procedure give the set of coupled non-linear ODEs; a light linear viscous modal damping is added at this stage of discretization, λ_j^*:

$$q_j^{*''} + 2\lambda_j^* q_j^{*'} + a_j^* q_j^* + \delta_{1j} \sum_{k=1}^{N} b_{jk}^* q_k^{*2} + 2\, b_{1j}^* q_1^* q_j^*$$

$$+ q_j^* \sum_{k=1}^{N} e_{jk}^* q_k^{*2} = f_{pj}^* + f_{\theta j}^* \frac{(1 + s^* \alpha_j^*)}{\alpha_j^*} + f_{n_\theta}^* \bar{n}_\theta^* \delta_{1j}\;. \tag{2.17}$$

The non-dimensional form of Eq. (2.17) has been derived by scaling the time with the fundamental natural frequency of the linearized plate problem,

$$t^* = \omega_{1L0}^p\, t\,, \qquad \omega_{1L0}^p = \left[K\alpha_1^2 / \mu(1{+}Ks\alpha_1)\right]^{1/2}\,, \tag{2.18}$$

where $(\;)' = \dfrac{\partial}{\partial t^*}$. Furthermore, in the derivation of Eq. (2.17) it has been assumed that the initial curvature of the shallow shell is proportional to the basic eigenmode of the linearized plate problem,

$$H = \tfrac{1}{2}\, C\, \alpha_1 w_1^*\,, \tag{2.19}$$

where $C\alpha_1$ measures the intensity of initial curvature. In that case the coefficients in Eq. (2.17) turn out to be given by

$$a_j^* = \alpha_j^{*2}\frac{(1+s^*)}{(1+s^*\alpha_j^*)} - \bar{n}_\theta^* D^* \alpha_j^* \frac{(1+s^*)}{R_0^2\alpha_1} + C^{*2}D^*\beta_1^*(1+s^*)\,\delta_{1j}\ , \quad (2.20)$$

$$b_{1j}^* = \frac{1}{2}C^*D^*\sqrt{\beta_1^*}\,\alpha_j^*(1+s^*)\ , \quad (2.21)$$

$$e_{jk}^* = \frac{1}{2}D^*\alpha_j^*\,\alpha_k^*(1+s^*)\ , \quad (2.22)$$

$$f_{n_\theta}^* = \frac{C^*D^*\sqrt{\beta_1^*}\,(1+s^*)}{R_0^2\,\alpha_1}\ , \quad (2.23)$$

with the following similarity numbers:

$$\alpha_j^* = \frac{\alpha_j}{\alpha_1}\ ,\quad s^* = Ks\alpha_1\ ,\quad D^* = \frac{D R_0^2}{K}\ ,$$

$$C^* = \frac{C}{R_0}\ ,\quad \beta_1^* = \frac{1}{A}\int_A w_1^{*2}dA\ . \quad (2.24)$$

R_0 denotes a characteristic length. Structures having the same similarity numbers result in an identical nondimensional response being sure or random. Any possibly irregular polygonal planform of the shell enters only via the parameters α_j^* and β_1^* of the simple linear boundary value problems as determined by Eqs. (2.15) and (2.16). Thus, only these linear eigenvalue problems have to be solved. To finally derive the simple form of Eqs. (2.17), the following choice must be made but without loss of generality:

$$c_j^2 = \frac{R_0^2}{\beta_j^*}\ ,\quad c_{\theta j} = \frac{c_j\alpha_1}{(1+s^*)\,(1+\nu)}\ ,\quad c_{pj} = \frac{c_j\alpha_1^2 K}{(1+s^*)}\ . \quad (2.25)$$

The mean thermal loading $\bar{n}_\theta^*$ affects the coefficient a_1^*, Eq. (2.20), as well as the right hand side of Eqs. (2.17).

Shells without $\bar{n}_\theta^*$ or plates with static thermal prestress given by a non-zero value of $\bar{n}_\theta^*$, including the thermally prebuckled plates, are subsequently analyzed. Furthermore, in the right hand side of Eqs. (2.17) note the mode-number dependent weighting of the effective thermal curvature type force $f_{\theta j}^*$ when compared to the classical lateral effective force-type f_{pj}^*. Either one, a thermal curvature or a lateral force load is considered subsequently, and in both cases the modal forces are commonly denoted by f_j^*, and split into the static and the dynamic part,

$$f_j^* = f_{js}^* + f_{jd}^* \ . \tag{2.26}$$

The mean thermal load $\bar{n}_\theta^*$ and f_j^* are assumed to be given and deterministic, while $f_{jd}^*(t)$ is characterized by a stochastic process. By definition, a generalized potential function

$$V^* = -\left(f_{n_\theta}^* \bar{n}_\theta^* q_1^* + f_{js}^* q_j^*\right) + \frac{1}{2}\sum_{k=1}^{N}\left(a_j^* + 2\, b_{j1}^* q_1^*\right) q_j^{*2} + \frac{1}{4}\sum_{j=1}^{N}\sum_{k=1}^{N} e_{jk}^* \, q_j^{*2} \, q_k^{*2} \ , \tag{2.27}$$

transforms the coupled modal Eqs. (2.17) to the dynamically driven set of ODEs

$$q_j^{*''} + 2\lambda_j^* \, q_j^{*'} + \frac{\partial V^*}{\partial q_j^*} = f_{jd}^* \ . \tag{2.28}$$

It is this standard form of Eqs. (2.28) that makes the application of the classical F.P.K.-formulation of the theory of nonlinear random vibrations possible. That approach, however, until recently was restricted in applications to simple structures, like one-span beams or arches and rectangular plates or cylindrical panels. The nondimensional form of Eqs. (2.28), when properly interpreted has no restrictions of validity with respect to geometry of the planform or the "order of elasticity" of rather complex structures, like shallow composite shells above a irregular polygonal planform, even prestressed and thermally or laterally forced.

3 Nonlinear random vibrations

The stochastic modal dynamic loads $f_{jd}^*(t)$ are approximated by uncorrelated zero-mean Gaussian white noise processes with the autocorrelation, we omit subsequently the superscript (*),

$$\left\langle f_{jd}(t)\, f_{kd}(t+\tau)\right\rangle = 2D_j \, \delta(\tau)\, \delta_{jk} \ , \tag{3.1}$$

while a given deterministic static force load f_{1s} is considered to be affine to the first mode with the static amplitude q_{1s}.

In that classical case the transition probability density $p\left(q_1 \ldots q_N, q_1' \ldots q_N'\right)$ of the system with potential restoring force and light linear modal damping is a solution of the F.P.K.-equation with the initial conditions

$$\sum_j \frac{\partial}{\partial q_j'}\left[\left(\frac{\partial V}{\partial q_j} + 2\lambda_j q_j'\right) p\right] - \sum_j \frac{\partial}{\partial q_j}\left(q_j' p\right) + \sum_j D_j \frac{\partial^2 p}{\partial q_j'^2} = \frac{\partial p}{\partial t} ,$$

$$\lim_{t \to 0_+} p\left(q_1 .. q_N, q_1' .. q_N', t\right) = \delta\left(q_1 - q_{1s}\right) \delta q_2 \ldots \delta q_N \, \delta q_1' \ldots \delta q_N' . \qquad (3.2)$$

The stationary solution of Eq. (3.2) in a mechanical setting has been derived by Caughey[4],

$$p\left(q_j .. q_N, q_j' .. q_N'\right) = \bar{B} \exp\left\{-\frac{1}{\bar{\bar{K}}}\left(\frac{1}{2}\sum_j q_j'^2 + V\right)\right\} , \qquad (3.3)$$

with the additional assumption of constant ratios of intensities and light damping,

$$D_j / \left(2\lambda_j\right) = \bar{K} = \text{const. for all } j . \qquad (3.4)$$

Multiple integration over the velocity space gives the n-dimensional probability density function of the generalized coordinates

$$p\left(q_j \ldots q_N\right) = B \exp\left\{-V/\bar{K}\right\} , \qquad (3.5)$$

where B is determined through normalization,

$$\int_{-\infty\ \text{N-fold}}^{+\infty} \cdots \int_{-\infty}^{+\infty} p\left(q_j \ldots q_n\right) \prod_j dq_j = 1 \Rightarrow B . \qquad (3.6)$$

The mean-square response of the j-th modal deflection equals the second moment that results from the cross moments by putting j = k,

$$\left\langle q_j q_k \right\rangle = \int_{-\infty\ \text{N-fold}}^{+\infty} \cdots \int_{-\infty}^{+\infty} q_j q_k \, p\left(q_j \ldots q_n\right) \prod_j dq_j . \qquad (3.7)$$

Considering the nonlinear vibrations of flat plates, the influence of the higher modes upon $\langle q_1^2 \rangle$ has been studied in detail in ref.[16]. As an extension of these results, Fig. 1 illustrates the influence of the initial curvature and shape, measured by C, Eq. (2.19), of shallow shells with $\alpha_j/\alpha_1 = j^2, j = 1, 2, \ldots$, upon $\langle q_1^2 \rangle$. The influence of the higher modes is taken into account. These results refer to a shell with polygonal boundary, without any deterministic lateral load and with mean temperature kept at zero, when vibrating about the undeformed configuration.

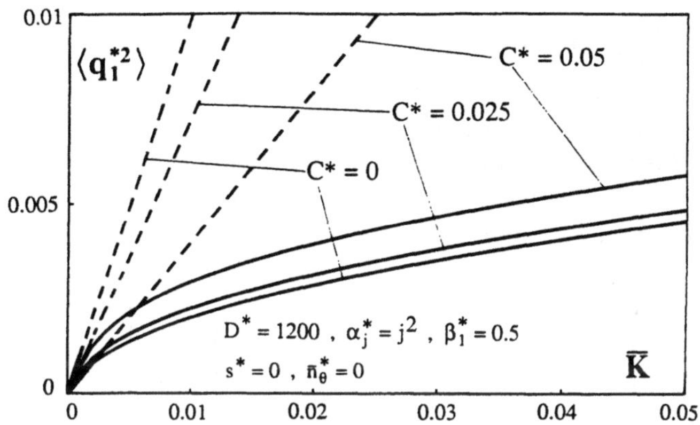

Fig. 1. Linear (– – –) and nonlinear (——) nondimensional mean-square response $\langle q_1^{*2} \rangle$ as a function of $\bar{K}$ according to Eqs.(3.5) and (3.7) . The graph is valid for shallow shells with $a_j^* = j^2, j = 1, 2, \ldots N$, regardless of their special polygonal planform. $C^* = C / R_0$.

It is well-known, however, that the vibrations about the static equilibrium position of shallow shells or those of thermally pre-buckled plates may show snap-through behavior. The probability of first occurrence of snap-through in a time interval T has been approximately determined by Ariaratnam and Sankar[12] for SDOF-systems of the form

$$q_1'' + 2\,\lambda_1\, q_1' + \frac{\partial V}{\partial q_1} = f_{1d} \ . \tag{3.8}$$

Using Eqs. (2.27) with N = 1 and (3.8), V can be identified in the present case for shallow shells

$$V(q_1) = \frac{e_{11}}{4}\, q_1^4 + b_{11}\, q_1^3 + \frac{a_1}{2}\, q_1^2 - \left(f_{1s} + f_{n_\theta}\, \bar{n}_\theta\right) q_1 \ . \tag{3.9}$$

The case of thermally buckled plates can be treated along the same lines by setting, see Heuer et al.[17],

$$a_1 = 12q_{1\theta}^2 \ , \ b_{11} = -6q_{1\theta} \ , \ e_{11} = 6 \ ,$$
$$q_{1\theta}^2 = \frac{1}{6}(12\,\bar{n}_\theta/\alpha_1 h^2 - 1) \ , \ f_{n_\theta} = 0 \ . \tag{3.10}$$

Assume the intensities in Eq. (3.4) to be sufficiently small,

$$\bar{K} << h \ , \ h = V\left(q_{1B}\right) - V\left(q_{1A}\right) , \tag{3.11}$$

where B refers to the intermediate barrier of unstable equilibrium position where the potential takes on its maximum, and A indicates the initial (stable) equilibrium configuration q_{1s}, Fig. 2. In that case the following approximation has been derived in ref.[12], Eq. (31):

$$P_T = \frac{T}{2\pi}\left(-\frac{V_A''}{V_B''}\right)^{1/2}\left[\left(\lambda_1^2 - V_B''\right)^{1/2} - \lambda_1\right]\exp\left(-\frac{h}{\bar{K}}\right) , \tag{3.12}$$

$$V'' \equiv \frac{\partial^2 V}{\partial q_1^2} = \left(3\,e_{11}\,q_1^2 + 6\,b_{11}\,q_1 + a_1\right)\Bigg|_{\substack{q_1 = q_{1A} \\ q_1 = q_{1B}}} , \tag{3.13}$$

and $\frac{\partial V}{\partial q_1} = 0$ renders the equilibrium positions $q_{1A} < q_{1B} < q_{1C}$ that exist under the action of f_{1s}. Snapping occurs with probability P_T from q_{1A} to q_{1C}, Fig. 2.

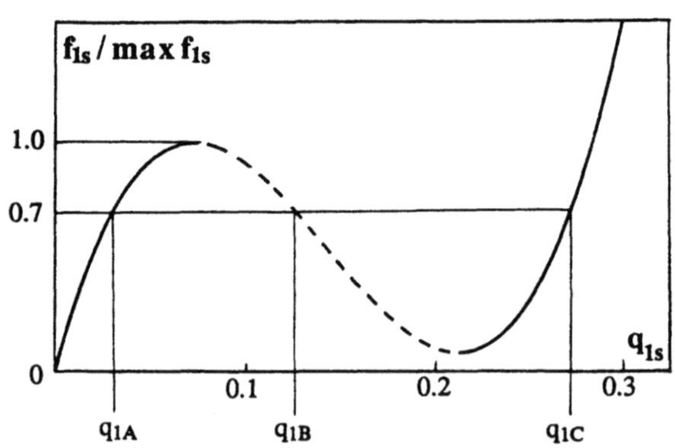

Fig. 2. Static force deflection curve. Snap-through problem.
$D^* = 300, \beta_1^* = 0.5, s^* = 0.2, \bar{n}_\theta^* = 0.$

In the present case Eq. (3.12) is evaluated, and the result is plotted in Fig. 3 for shear deformable shallow shells. The probability P_T of snap-through is presented as a function of the static force load ratio $f_s = f_{1s} / \max f_{1s}$, where $\max f_{1s}$ is the static snap-through load. No mean thermal pre-load is taken into account. For thermally pre-buckled plates that are rigid in shear, P_T has been evaluated as a function of f_s in ref.[17]. Note, that the SDOF-approximation of Eqs. (3.8)–(3.13) gives a unifying formulation for simply supported plates and shallow shells of arbitrary polygonal planform. The influence of moderate shear deformations on these nondimensional results is rather small. By inspection of Eq. (2.18), however, it is seen that shear deformation increases the actual observation time and therefore increases the probability of snap-through.

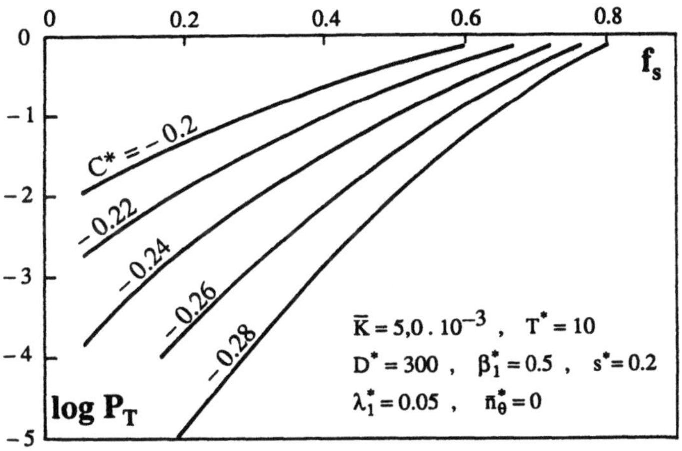

Fig. 3. Probability of first occurrence of snap-through P_T as a function of $f_s = f_{1s} / \max f_{1s}$ for various values of C*. The graph is independent of the special skew planform of the shallow shell. $C^* = C/R_0$.

As was pointed out by Ashwell[31], see also Irschik[32] and Rammerstorfer and Ziegler[33], the loss of stability of the static equilibrium position may be accompanied by a bifurcation to a higher mode. For nonlinear random vibrations, this so-called snap-buckling has been studied in detail by Sankar and Ariaratnam[13,14] starting with a two-mode approximation of the form

$$q_1'' + 2\lambda_1 q_1' + \frac{\partial V}{\partial q_1} = f_{1d} \ , \ q_2'' + 2\lambda_2 q_2' + \frac{\partial V}{\partial q_2} = 0 \ . \tag{3.14}$$

This set of equations can be matched to the current more general formulation of Eqs. (2.27) with N = 2, by putting

$$V(q_1, q_2) = \frac{1}{4} (e_{11} q_1^4 + e_{22} q_2^4) + b_{11} q_1^3 + \frac{1}{2} (a_1 q_1^2 + a_2 q_2^2) + q_1 q_2^2 (b_{12} + \frac{e_{12}}{2} q_1) - (f_{1s} + f_{n_\theta} \bar{n}_\theta) q_1 \ , \tag{3.15}$$

to render the class of systems of shallow shells. Again sufficiently small levels of excitation are assumed

$$\bar{K} << h_1 \ , \ h_1 = V\left(q_{1B_1}, q_{2B_1}\right) - V\left(q_{1A}, 0\right) \ . \tag{3.16}$$

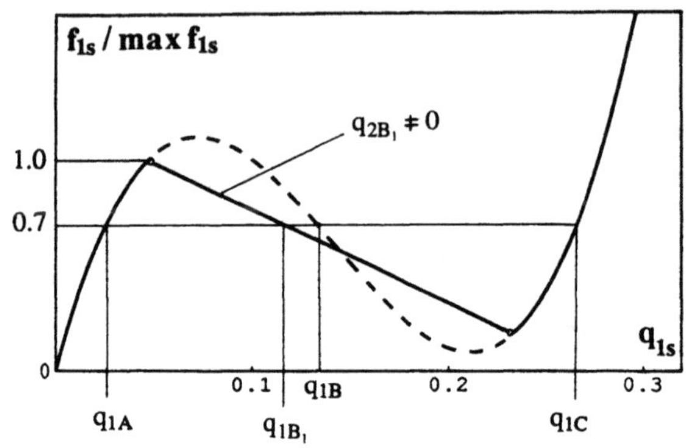

Fig. 4. Static force deflection curve. Snap-buckling problem.
$D^* = 300, \beta_1^* = 0.5, s^* = 0.2, \bar{n}_\theta^* = 0.$

The value of the generalized coordinate at the equilibrium point A is consequently taken from the single-mode analysis. The coordinates of the saddle-point B_1 are the roots of the static equations resulting from Eqs. (3.14), see Fig. 4,

$$\frac{\partial V}{\partial q_1} = 0 \ , \ \frac{\partial V}{\partial q_2} = 0 \ , \tag{3.17}$$

and are given by the following expressions, where $q_{2B_1}^2 \geq 0$:

$$q_{1B_1} = e_{22} \left(f_{1s} + f_{n_\theta} \bar{n}_\theta + a_2 b_{12} / e_{22} \right) \Big/ \left(a_1 e_{22} - 2b_{12}^2 - a_2 e_{12} \right) , \tag{3.18}$$

$$q_{2B_1} = \left[-\left(a_2 + 2\, b_{12} q_{1B_1} + e_{12} q_{1B_1}^2 \right) / e_{22} \right]^{\frac{1}{2}} . \tag{3.19}$$

An evaluation of the equations that result for the case of thermally pre-buckled plates is given in ref.[17]. In that case, the following relations hold

$$a_1 = 12 q_{1\theta}^2\,,\ a_2 = (\alpha_2/\alpha_1 - 1)(\alpha_2/\alpha_1)\,,\ b_{11} = -6 q_{1\theta}\,,$$
$$b_{12} = -6\alpha_2/\alpha_1\, q_{1\theta}\,,\ e_{11} = 6\,,\ e_{12} = 6\alpha_2/\alpha_1\,,$$
$$e_{22} = 6(\alpha_2/\alpha_1)^2\,,\ q_{1\theta}^2 = \frac{1}{6}(12\,\bar{n}_\theta/\alpha_1 h^2 - 1)\,,\ f_{n_\theta} = 0\,. \qquad (3.20)$$

Sankar and Ariaratnam[13, Eq. (24)], in a detailed analysis, derived the following approximation for the probability of first snap-buckling within the time interval T:

$$\bar{P}_T = \frac{T}{2\pi}\cos^2\phi\left[\frac{(V_{11})_A}{-(\bar{V}_{11})_{B_1}}\right]^{1/2} \times\left\{\left[\lambda^2 - (\bar{V}_{11})_{B_1}\right]^{1/2} - \lambda\right\}\exp\left(-\frac{h_1}{\bar{\bar{K}}}\right), \qquad (3.21)$$

where

$$(V_{11})_A = \frac{\partial^2 V}{\partial q_1^2}\bigg|_{q_1 = q_{1A}}\,,\quad (\bar{V}_{11})_{B_1} = \frac{\partial^2 \bar{V}}{\partial s_1^2}\bigg|_{\substack{s_1 = s_{1B_1}\\ s_2 = s_{2B_1}}}\,, \qquad (3.22)$$

$$(\bar{V}_{11})_{B_1} = \frac{1}{2}\left[(V_{11})_{B_1} + (V_{22})_{B_1}\right] - \frac{1}{2}\left\{\left[(V_{11})_{B_1} - (V_{22})_{B_1}\right]^2 + 4(V_{12})_{B_1}^2\right\}^{1/2}, \qquad (3.23)$$

$$(V_{ij})_{B_1} = \frac{\partial^2 V}{\partial q_i\,\partial q_j}\bigg|_{\substack{q_1 = q_{1B_1}\\ q_2 = q_{2B_1}}}\,,\quad i, j = 1, 2\,, \qquad (3.24)$$

and a transformation to the principal coordinates s_1, s_2 has been performed by means of Mohr's circle:

$$\begin{Bmatrix} q_1 \\ q_2 \end{Bmatrix} = \mathbf{D}(\phi)\begin{Bmatrix} s_1 \\ s_2 \end{Bmatrix}, \qquad (3.25)$$

with the angle of rotation ϕ given by the equation

$$\tan 2\phi = 2\left(V_{12}\right)_{B_1} / \left[\left(V_{11}\right)_{B_1} - \left(V_{22}\right)_{B_1}\right] \, . \tag{3.26}$$

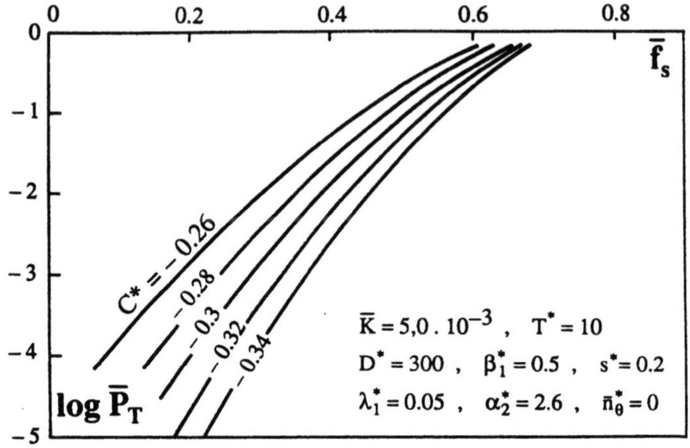

Fig. 5. Probability of first occurrence of snap-buckling $\bar{P}_T$, Eq. (3.17) as a function of $\bar{f}_s = f_{1s} / \max f_{1s}$, for various values of C*. The graph is valid for shear deformable shallow shells with $\alpha_2^* = 13/5$, e.g. right-angled isosceles triangular shell. $C^* = C / R_0$.

Fig. 5 shows the probability of first occurrence of snap-buckling of a shallow shell with $\alpha_2^* = 13/5$ (that holds e.g. for a shell above a right-angled isosceles triangular base with hard hinged supports) as a function of the load ratio $\bar{f}_s = f_{1s} / \max f_{1s}$. The maximum load according to the static Eqs. (3.18) and (3.19), $\max f_{1s}$, results from the condition $q_{2B_1}^2 = 0$.

4 Conclusion

It has been shown that even complex structures like irregularly shaped shallow shells or prebuckled plates designed with composite materials, prestressed, and forced by lateral loads or thermally by a fluctuating temperature field, can be projected onto the space of the linear modes of the associated flat plates. The resulting set of coupled nonlinear ODEs is transformed and finally brought in a nondimensional form that is of the classical type of dynamic equations with potential restoring force. Directly or indirectly through filters approximating the random environment by physical white noise processes makes the classical analysis of the resulting Markov process by means of the F.P.K.-equation possible. Probabilities of snap-through and snap-buckling vibrations in the approximations derived in refs.[13, 14] are calculated in the parameter space. In the case of single mode approximations the results are general. Taking into account the influence of higher modes restricts the results to a class of structures having a definite sequence of the linear natural frequencies divided through the basic frequency.

If a multimode approximation is applied to the dynamic response, the effect of internal resonance should also be considered that may lead to strong nonlinear interaction of the modes. An investigation has been made by Ibrahim and Li[34].

References

1. Crandall, S.H. and Zhu, W.Q., Random Vibration: A Survey of Recent Developments *J. Appl. Mech.*, **50** (1983) 953-62.
2. Crandall, S.H., Random Vibrations of Systems with Nonlinear Restoring Forces. IUTAM-*Symposium on Non-linear Vibrations*. Publish. House of the Ukrainian Academy of Sciences. Kiev, 1961.
3. Ariaratnam, S.T., Response of a Loaded Nonlinear String to Random Excitation *J. Appl. Mech.*, **29** (1962) 483-85.
4. Caughey, T.K., Derivation and Application of the Fokker-Planck Equation to Discrete Nonlinear Dynamic Systems Subjected to White Random Excitation *J. Acoust. Soc. Am.*, **35** (1963) 1683-92.
5. Caughey, T.K., Nonlinear Theory of Random Vibrations. *Advances in Applied Mechanics* **11** (1971) 209-53.
6. Bolotin, V.V., *Statistical Methods in Structural Mechanics*. (Transl. by S. Aroni), Holden Day, San Francisco, 1969 (1st Russian ed. 1961).
7. Lin, Y.K., *Probabilistic Theory of Structural Dynamics*. McGraw-Hill, New York, 1967.
8. Spanos, P.D., Stochastic Linearization in Structural Dynamics *AMR.*, **34** (1981) 1-8.
9. Roberts, J.B., Response of Nonlinear Mechanical Systems to Random Excitation. Part 1: Markov Method. Part 2: Equivalent Linearization and Other Methods *Shock and Vibration Digest*, **13** (1981).
10. Herbert, R.E., Random Vibrations of a Nonlinear Elastic Beam *J. Acoust. Soc. Am.*, **36** (1964) 2090-94.
11. Herbert, R.E., Random Vibrations of Plates with Large Amplitudes *J. Appl. Mech.*, **32** (1965) 547-52.
12. Ariaratnam, S.T., Sankar, T.S., Dynamic snap-through of shallow arches under stochastic loads *AIAA Journal*, **6** (1968) 798-802.
13. Sankar, T.S. and Ariaratnam, S.T., Snap-buckling of shell-type structures under stochastic loading *Int. J. Solids & Structures*, **7** (1971) 655-66.
14. Sankar, T.S. and Ariaratnam, S.T., Instability of stochastically loaded shallow arches in nonsymmetric modes *Int. J. Solids & Structures*, **7** (1971) 1305-16.
15. Sun, J.-Q. and Hsu, C.S., Random vibration of hinged elastic shallow arch *J. Sound and Vibration*, **132** (1989) 299-315.
16. Heuer, R., Irschik, H. and Ziegler, F., Large Amplitude Random Vibrations of Polygonal Plates. IUTAM-*Symposium on Nonlinear Stochastic Mechanics*, eds. N. Bellomo, F. Casciati. Springer-Verlag, Berlin, 1992, 285-96.

17. Heuer, R., Irschik, H. and Ziegler, F., Nonlinear Random Vibrations of Thermally Buckled Skew Plates *Probabilistic Engineering Mechanics* (1993), in press.
18. Heuer, R., Irschik, H. and Ziegler, F., Randomly Forced Nonlinear Vibrations of Layered Skew Plates and Shallow Shells. 34th AIAA/ASME/ASCE/AHS/ASC–Conference, Paper 93-1625-CP, La Jolla, CA, 1993.
19. Seide, R., Snap-Through of Initially Buckled Beams Under Uniform Random Pressure. *Random Vibration-Status and Recent Developments,* eds. I. Elishakoff and R.H. Lyon. Elsevier, Amsterdam, 1986, 403-14.
20. Penketgorn, T., Rojiani, K.B. and Plaut, R.H., Snap-Through of a Shallow Arch Under Random Loading *Probabilistic Engineering Mechanics,* **4** (1989) 114-19.
21. Locke, J. and Mei C.C., Finite Element, Large-Deflection Random Response of Thermally Buckled Beams *AIAA Journal,* **28** (1990) 2125-31.
22. Abou-Rayan, A.M. and Nayfeh, A.H., Stochastic response of a buckled beam to external and parametric random excitations. 34th AIAA/ASME/ASCE/AHS/ASC–Conference, Paper 93-1425-CP, La Jolla, CA, 1993.
23. von Karman, Th. and Tsien, H.S., The buckling of thin cylindrical shells under axial compression *J. Aeronautical Sciences,* **8** (1941) 303-12.
24. Reissner, E., Reflections on the theory of elastic plates *AMR* **38** (1985) 1453-64.
25. Mindlin, R.D., Influence of rotatory inertia and shear on flexural motions of isotropic, elastic plates *J. Appl. Mech.,* **18** (1951) 31-38.
26. Berger, H.M., A new approach to the analysis of large deflection of plates *J. Appl. Mech.,* **22** (1955) 465-72.
27. Schmidt, R., On Berger´s Method in the Nonlinear Theory of Plates *J. Appl. Mech.,* **41** (1974) 521-23
28. Heuer, R., Large Flexural Vibrations of Thermally Stressed Layered Shallow Shells *Nonlinear Dynamics,* **5** (1994) 1-14.
29. Heuer, R., Irschik, H. and Ziegler, F., Multi-modal approach for large natural flexural vibrations of thermally stressed plates *Nonlinear Dynamics,* **1** (1990) 449-58.
30. Irschik, H., Membrane-type Eigenmotions of Mindlin Plates *Acta Mechanica,* **55** (1985) 1-20.
31. Ashwell, D.G., Nonlinear Problems. *Handbook of Engineering Mechanics,* ed. W. Flügge. McGraw Hill, New York 1962. pp. 45-7 – 45-9.
32. Irschik, H., Large Thermoelastic Deflections and Stability of Simply Supported Polygonal Panels *Acta Mechanica,* **59** (1986) 31-46.
33. Ziegler, F. and Rammerstorfer, F.G., Thermoelastic Stability. *Thermal Stresses* III, ed. R.B. Hetnarski. North-Holland, Amsterdam, 1989. Ch. 2, pp. 108-189.
34. Ibrahim, R.A. and Li, W., Structural modal interaction with combination internal resonance under wide-band random excitation *J. Sound and Vibration,* **123** (1988) 473-95.

Springer-Verlag and the Environment

We at Springer-Verlag firmly believe that an international science publisher has a special obligation to the environment, and our corporate policies consistently reflect this conviction.

We also expect our business partners – paper mills, printers, packaging manufacturers, etc. – to commit themselves to using environmentally friendly materials and production processes.

The paper in this book is made from low- or no-chlorine pulp and is acid free, in conformance with international standards for paper permanency.

MIX
Papier aus verantwortungsvollen Quellen
Paper from responsible sources
FSC® C105338

If you have any concerns about our products,
you can contact us on
ProductSafety@springernature.com

In case Publisher is established outside the EU,
the EU authorized representative is:
Springer Nature Customer Service Center GmbH
Europaplatz 3, 69115 Heidelberg, Germany

Printed by Libri Plureos GmbH
in Hamburg, Germany